HUMAN INFECTION WITH FUNGI, ACTINOMYCETES AND ALGAE

BY

ROGER DENIO BAKER

AND

A. ANGULO O. · C. BARROSO-TOBILA · L. M. CARBONELL
R. CÉSPEDES F. · E. W. CHICK · B. M. CLARK · O. DUQUE
G. M. EDINGTON · B. F. FETTER · J. H. GRAHAM · D. J. GUIDRY
R. W. HUNTINGTON, JR. · H. ICHINOSE · G. K. KLINTWORTH
H. I. LURIE · L. N. MOHAPATRA · J. MORENZ · H. S. NIELSEN, JR.
J. C. PARKER, JR. · C. E. PEÑA · P. PIZZOLATO · L. POLLAK
K. SALFELDER · J. SCHWARZ · J. P. WIERSEMA
H. I. WINNER · D. J. WINSLOW

SPECIAL EDITION OF
HANDBUCH DER SPEZIELLEN PATHOLOGISCHEN
ANATOMIE UND HISTOLOGIE III/5

WITH 796 FIGURES
2 IN COLOR

Springer-Verlag Berlin Heidelberg GmbH 1971

Special edition of Handbuch der speziellen pathologischen Anatomie und Histologie, Band III/5

ISBN 978-3-662-12029-3 ISBN 978-3-662-12027-9 (eBook)
DOI 10.1007/978-3-662-12027-9

Buchbinderei: Brühlsche Universitätsdruckerei, Gießen

Preface

Half a century ago our knowledge of mycoses, especially pulmonary mycoses, was rather fragmentary. It was limited to rare case reports as oddities. Accordingly, in the "Handbuch der speziellen pathologischen Anatomie und Histologie" the chapter on lung diseases caused by budding and spore-forming fungi by J. WÄTJEN (Halle) took up as little as 27 pages. Only ARNDT (Göttingen) could report on several cases from which he made his observations on actinomycotic changes of the lungs and pleura.

Since then our knowledge of mycoses has deepened and expanded in an unpredictable manner. This progress was mainly due to research and publications in the USA and South America. In Central Europe the number of cases of mycoses has increased during the last two decades, being reported especially as a second disease in patients with spontaneous or iatrogenic destruction of the bone marrow after treatment of cancer with cytostatic agents.

The number of known types of pathogenic fungi has increased. The knowledge of their types and conditions of growth have given rise to a subspecialty. Therefore, a great need has arisen for a new edition of the chapter on mycoses in the Henke-Lubarsch-Roessle Handbook of Special Pathological Anatomy and Histology.

The publishers as well as the editors had the great fortune to find as author Prof. ROGER D. BAKER, who himself has contributed substantially to the knowledge of human mycoses. It is thanks to Prof. Baker that a large number of eminent scientists could be won as co-authors. Today, under his guidance, a complete handbook on human infection with Fungi, Actinomycetes and Algae has been written. At the present time no other book exists which describes so extensively the fundamentals of human mycology, including taxonomy, clinical aspects, epidemiology, and pathological anatomy.

For all the great progress of the past decades we should not forget that 50 years ago MAX ASKANAZY wrote, for the textbook of General Pathology edited by LUDWIG ASCHOFF, a substantial and thoughtful chapter on fungi as germs. Among his illustrations there is a marvelous colorprint of an aspergilloma of the lungs. This, an early milestone, should not get lost among the abundance of genuine and supposed newer discoveries.

Zürich/Heidelberg 1971 Prof. Dr. E. UEHLINGER

Contents

Unusual and Rare Mycoses (Chapter XXII—XXIII)

List of Authors

ALBERTO ANGULO O., M.D.
Profesor Titular, Director del Instituto de Anatomía Patológica Facultad de Medicina, Universidad Central e Instituto National de Tuberculosis
Correos de Sabana Grande
Apartado 50647
Caracas, Venezuela

ROGER DENIO BAKER, M.D.
Professor of Pathology
Rutgers Medical School
New Brunswick, New Jersey 08903, USA
Formerly Professor of Pathology, Louisiana State University, School of Medicine
New Orleans, Louisiana, USA

CESAR BARROSO-TOBILA, M.D., Jefe
Dermatología y Dermapathología
Policlinica Maracaibo
Av. 8 (Santa Rita) Esq. Calle 71
Maracaibo, Venezuela

LUIS M. CARBONELL, M.D.
Jefe del Departamento de Microbiología, Instituto Venezolano de Investigaciones Científicas (I.V.I.C), Apartado 1827
Caracas, Venezuela

RODOLFO CÉSPEDES F., M.D.
Profesor de Anatomía Patológica en la Facultad de Medicina, Universidad de Costa Rica, Jefe del Servicio de Anatomía Patológica del Hospital San Juan de Díos y Hospital Central del Seguro Social
Apartado 3275, San José
Costa Rica

ERNEST W. CHICK, M.D., Professor
Department of Community Medicine
Mycology Program, University of Kentucky, College of Medicine
Lexington, Kentucky 40506, USA
Formerly Professor and Chairman
Division of Preventive Medicine
West Virginia, Medical Center of Morgantown, West Virginia, USA

BETTY M. CLARK, M.B., M.R.C. Path.
Associate Professor
Department of Medical Microbiology
University College Hospital
Ibadan, Nigeria

OSCAR DUQUE, M.D.
Profesor y Jefe del Departamento de Patología, Facultad de Medicina
Universidad de Antioquia
Medellín, Colombia

GEORGE M. EDINGTON, C.B.E., M.D. M.R.C. Path.
Professor and Head of the Department of Pathology
University College Hospital
Ibadan, Nigeria

BERNARD F. FETTER, M.D.
Professor of Pathology
Duke University, Medical Center
Durham, North Carolina 27706, USA

JAMES H. GRAHAM, M.D.
Professor of Medicine (Dermatology)
Chairman, Division of Dermatology
Professor of Pathology, Department of Pathology, California College of Medicine, University of California
Irvine, California, USA
Address at Orange County Medical Center
101 South Manchester Avenue
Orange, California 92668, USA

D.J. GUIDRY, Ph.D.
Associate Professor of Microbiology
Louisiana State University
Medical Center
1542 Tulane Avenue
New Orleans, Louisiana 70112, USA

ROBERT W. HUNTINGTON, JR., M.D.
Clinical Professor of Pathology
University of Southern California
School of Medicine
Los Angeles, California. USA
Address: Pathologist, Kern County General Hospital
Bakersfield, California 93305, USA

HERBERT ICHINOSE, M.D.
Associate Professor of Pathology
Tulane University, School of Medicine
1430 Tulane Avenue
New Orleans, Louisiana 70112, USA

GORDON K. KLINTWORTH, M.D., Ph. D.
Associate Professor, Department of Pathology, Duke University
Medical Center
Durham, North Carolina 27706, USA

Harry I. Lurie, B.Sc., M.B., Ch.B.
M.R.C. Path., Professor of Pathology
Medical College of Virginia
Richmond, Virginia 23219, USA

L.N. Mohapatra, M.D., Dip. Bact.
(London), Professor of Microbiology
All-India Institute of Medical Sciences
New Delhi 16, India

J. Morenz, M.D.
Professor of Medical Microbiology
and Epidemiology
Medizinische Akademie Magdeburg
301 Magdeburg, Leipziger Straße 44, GDR

Harry S. Nielsen, Jr., Ph.D.
Formerly Assistant Professor of
Microbiology, Duke University
Medical Center. Presently Director
California Allergenics Laboratories
2602 First Avenue, Suite 104
San Diego, California 92103, USA

Joseph C. Parker, Jr., M.D., M.S.
Formerly Fellow in Neuropathology
Duke University, Medical Center
Presently at Laboratory of Pathology
New England Deaconess Hospital
185 Pilgrim Road
Boston, Massachusetts 02215, USA

Carlos E. Peña, M.D.
Formerly Assistant Professor of
Pathology, University of
Pittsburg, School of Medicine
Presently Assistant Professor
of Pathology, Georgetown University
Medical School
Washington, D.C. 20007, USA

Philip Pizzolato, M.D.
Professor of Pathology, Louisiana
State University, Medical Center
Pathologist, Veterans Administration
Hospital. Address at Veterans
Administration Hospital
Perdido Street
New Orleans, Louisiana 70140, USA

Ladislao Pollak, M.D.
Profesor Asociado, Jefe de la
Cátedra de Microbiología
Escuela Vargas, Facultad de Medicina
Universidad Central e Instituto
National de Tuberculosis
Correos de Sabana Grande
Apartado 50647
Caracas, Venezuela

K. Salfelder, M.D. (Prof. Titular)
Jefe del Departamento de Patología
Instituto de Anatomía Patológica
Universidad de los Andes
Hospital "Los Andes"
Apartado No. 75
Mérida, Venezuela

Jan Schwarz, M.D.
Associate Professor of Pathology
The University of Cincinnati
School of Medicine
The Jewish Hospital
Cincinnati, Ohio 45229, USA

Jan P. Wiersema, M.D.
Formerly Research Pathologist
Leonard Wood Memorial
Special Mycobacterial Diseases
Branch, Geographic Pathology
Division, Armed Forces Institute
of Pathology
Washington, D.C., USA
Presently Pathologist
Good Samaritan Hospital
1425 West Fairview Avenue
Dayton, Ohio 45406, USA

H.I. Winner, M.D., M.R.C.P.
F.C. Path., Professor of Bacteriology
Charing Cross Hospital
Medical School, University of London
13 William IV Street
London WC2N 4DW, England

Donald J. Winslow, M.D.
Formerly Chief, Infectious
Disease Branch, Geographic
Pathology Division, Armed Forces
Institute of Pathology
Washington, D.C., USA
Presently Chief
Laboratory Service
Veterans Administration Center
Bay Pines, Florida 33504, USA

Introduction

At the First Symposium on Mycotic Diseases of The International Academy of Pathology, meeting in London, Professor Dr. E. A. UEHLINGER of the Institute of Pathology of the University of Zürich, Switzerland, invited me to prepare a volume on the pathologic anatomy of the mycoses. This treatise would be a part of the section on the lung of the "Handbuch der speziellen pathologischen Anatomie und Histologie", now edited by Professor UEHLINGER, and would be published in English. It was my good fortune to persuade pathologists, microbiologists, and epidemiologists the World over to contribute chapters on subjects in which they had long experience and special competence.

The result is this authoritative volume of monographs, a number of which presently represent the most comprehensive exposition of "Infection with Fungi, Actinomycetes, and Algae". The investigation of the literature necessary to write the chapters has lead to new insights and correlations concerning the nature, epidemiology, diagnosis, prevention, and treatment of these increasingly significant diseases. This material is indispensable not only to pathologists, but to microbiologists, research workers, dermatologists and other practising physicians.

I owe a debt of gratitude to the co-authors of this volume, to my wife, ELEANOR USSHER BAKER, who helped in the preparation and editing of the volume, and to Springer-Verlag.

Professor ROGER DENIO BAKER, M.D.
Senior Author
New Brunswick, N.J., 1971

Chromoblastomycosis[1]

Chromomycosis

Rodolfo Céspedes F., San José, Costa Rica

With 4 Figures

Definition: Chromoblastomycosis is a chronic cutaneous mycosis, usually without tendency to generalization or deep invasion. It occurs most frequently on a lower extremity of an adult male and forms verrucous nodules.

The disease is called chromoblastomycosis because of the resemblance of the lesions to those of blastomycosis. Since the fungus does not multiply in the tissues by budding but by the formation of internal septa, some students of the disease prefer the term chromomycosis, but this has not been widely accepted. Unfortunately, chromomycosis implies that the disease is pigmented, whereas only the causative organism is pigmented.

Geographical Distribution and Incidence

The disease has been reported from all the continents, but most cases have been seen in the tropical regions of America and Africa. Cases reported from North America have been few.

In 1954 there were 21 cases reported from the United States, 9 of which were from Louisiana (Gardner et al., 1964; Howles et al., 1954; Moore et al., 1943; Romero and Trejos, 1953). In Europe, cases have been reported only in the eastern regions (U.S.S.R. and Finland). A few cases have been seen in Australia, Asia and Japan (Sasano et al., 1961). Until 1964, only 3 cases of chromoblastomycosis had been reported in India (Klokke, 1964). In Costa Rica, a tropical country of 1,400,000 inhabitants, 140 cases had been diagnosed as of May, 1966, and 14 new cases per year were diagnosed from 1961—1966. In Madagascar, 129 cases have been studied in 4 years (Brygoo and Segretain, 1960). In Colombia, Duque has collected 59 cases and observes about 8 new cases every year.

A survey of pathology laboratories in Latin America gives an idea of the number of cases diagnosed by histological examination each year: Buenos Aires, Argentina, 0; La Plata, Argentina, 0; Santos, Brazil, 1; Manaos (Amazonas), Brazil, 1; Recife, Pernambuco, Brazil, 1; Medellín, Colombia, 8; San José, Costa Rica, 14; Santiago, Chile, 0; Concepción, Chile, 0; Quito, Ecuador, 0; San Salvador, El Salvador, 2; Tegucigalpa, Honduras, 1; Chihuahua, México, 0; Monterrey, México, 1; Yucatán, México, 5; Durango, México, 2; México, D.F., México, 1; León, Nicaragua, 1; Panamá, Panamá, 6; Asunción, Paraguay, 3; Ica, Perú, 0; San Juan, Puerto Rico, 1; Aquadilla, Puerto Rico, 2; Valera, Venezuela, 0.

Since chromoblastomycosis is also diagnosed clinically without histological examination, the number of cases must be larger than that indicated in the above inquiry. On the other hand, in a few cases the diagnosis is not suspected clinically and the etiology of the lesion is a surprising finding on histological examination (Kempson and Sternberg, 1963). The reported incidence of the disease will probably increase when the agricultural, developing countries have more and better laboratories for diagnosis, but it is apparent that countries of temperate climates have few cases.

1 Doctors Jorge Salas, Jorge Piza, Saeed Mekbel and Pedro Morera, and Mrs. Cecilia Chacón, assisted in the preparation of the manuscript. The members of the Sociedad Latino Americana de Anatomía Patológica contributed by answering the questionnaire of the survey.

Epidemiology: The condition is usually seen in *adults* and only exceptionally in children. There is no racial predilection, since the disease has been seen in caucasians, orientals, Negroes and even American Indians. All authors agree that chromoblastomycosis is much more frequent in males than in females. In our case material the rate is 10 to 1. Predilection for the male sex is explained by the fact that men do more agricultural work than women and are more exposed to trauma and contamination. Most patients are field and forest workers. There is usually no history of a definite incident that initiated the disease. A few patients tell of antecedent injury with wood or thorns at farm work. Later they notice the appearance of an erythematous papule which then begins to enlarge and fungate. Fungi morphologically and serologically identical with *Phialophora verrucosa* have been isolated from wood pulp (CONANT et al., 1954).

Chromoblastomycosis as a spontaneous disease has not been reported in animals. Experimentally, it can be induced in rabbits, mice and cold-blooded animals (BINFORD et al., 1952; TREJOS, 1963).

Several different species of fungi have been implicated in this disease. It is probable that climatic conditions have something to do with the species of fungus that predominates in a given area. In Madagascar (BRYGOO and SEGRETAIN, 1960) it was established that cases of chromoblastomycosis from an area with little rain (500—600 mm per year) are caused by *Cladosporium carrionii*, while cases from rainy areas (2,200—3,000 mm per year) are caused by *Fonsecaea pedrosoi*. The fungi live in wood or soil, and on passing to the human host probably undergo a process of adaptation for which some species must have a better potential. It is possible that the climatic characteristics of certain geographical regions foster the emergence of certain pathogenic species.

It has been shown (BINFORD et al., 1952) that *Cladosporium trichoides* inoculated into mice and rabbits by the intravenous route, causes fatal cerebral lesions. Intracutaneous inoculation of rats and mice, which are the most susceptible animals, does not produce chronic verrucous or papillomatous lesions, but abscesses that later rupture and heal spontaneously (LASKOWSKI, 1960). With *H. pedrosoi* it has been possible to cause systemic infections. So, in chromoblastomycosis, as in other diseases, the species of fungus, the host route of inoculation, occupation, age and sex, are factors that determine the disease process.

Mycology: Chromoblastomycosis is caused by several pigmented fungi, all of which produce similar lesions in the human skin, so that it is not possible to identify the species by their appearance in tissues. There are five species that are pathogenic for man: 1. *Phialophora verrucosa*; 2. *Fonsecaea (Hormodendrum) pedrosoi*; 3. *Fonsecaea (Hormodendrum) compactum*; 4. *Fonsecaea (Hormodendrum) dermatitidis* and 5. *Cladosporium carrionii*. The etiological agents of chromoblastomycosis grow at room temperature on Sabouraud's medium. Colonies grow slowly and have a dark color, varying from gray to olive to black, with a velvety appearance. There are variants according to the species (Laboratory Procedures in Clinical Mycology 1964).

Carrion stated that the species which most frequently produces lesions of chromoblastomycosis is *Fonsecaea (Hormodendrum) pedrosoi*, but in the light of recent investigation (BRYGOO and SEGRETAIN, 1960), especially in geographical pathology, this may have to be changed. In the opinion of Duque the classification of the species of fungi that cause chromoblastomycosis should be submitted to revision.

Mycological Diagnosis: A simple preparation of scrapings from the lesion in 10% KOH with warming on a glass slide, without the addition of any stain,

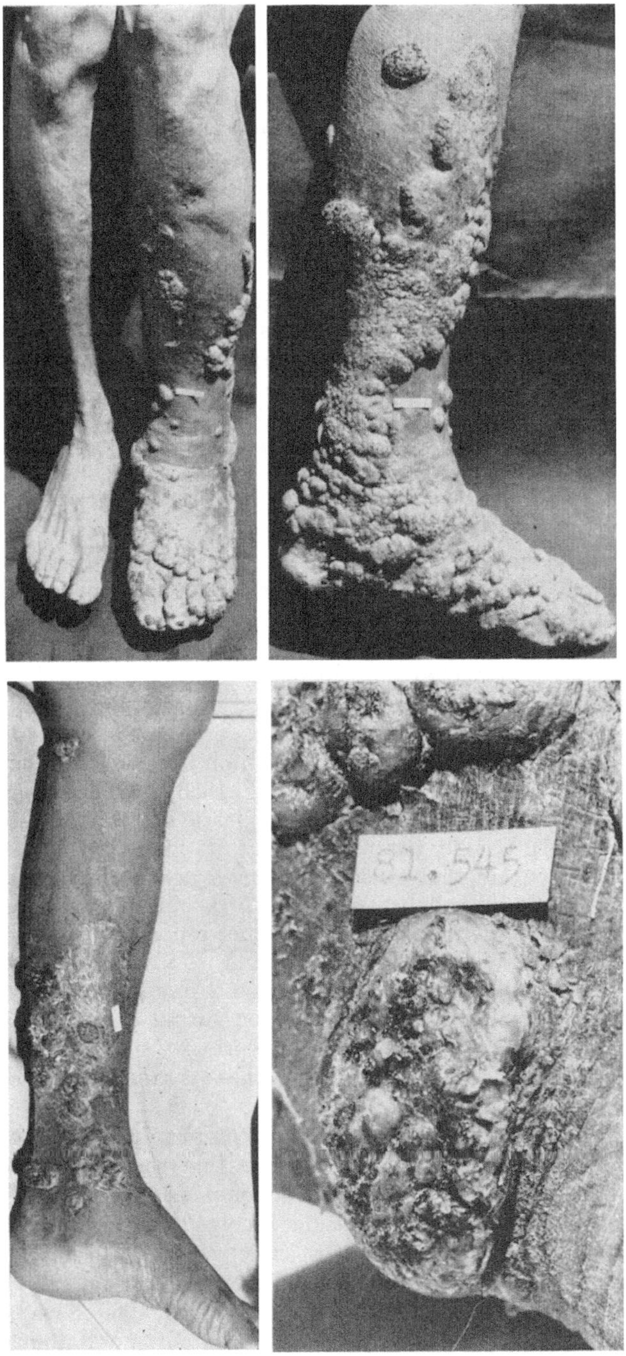

A B C D

Fig. 1. Chromoblastomycosis. A. The lesions affect a single lower limb and consist of confluent, verrucous nodules. Note the edema of the skin between the nodules. B. Detail of A to show "mossy skin". C. A lesion of the lower half of the leg, with a metastasis to the inner aspect of the knee. D. A well circumscribed cauliflower lesion of the patient in A and B suffices for diagnosis. Dark brown, spherical bodies with thick walls (Fig. 2B) and often with septa are easily seen. Sometimes the fungus cell gives origin to germinative tubules of thick and septated mycelia (Fig. 2A). The dark brown, spherical bodies, measuring from 7 to 10 microns approximately, are also easily recognized in histological sections with any stain, and even without staining. The organisms are either in microabscesses or in giant cells (Fig. 4D and E). It is helpful to add penicillin and streptomycin to the media to suppress bacterial growth when culturing material from lesions.

For species identification, it is necessary to use meticulous cultural procedures with special media at room temperature, and also slide cultures that allow observation of the type of sporulation. There are four types of sporulation: phialophora, hormodendrum, cladosporium and acrotheca (Fig. 2C, D, E, F, G). All four types can be found in a single species! The work of typing takes weeks. Physiological characteristics are used to complete species identification.

Serological methods have no important practical application, but it is possible to demonstrate the existence of antibodies (Emmons et al., 1963; French and Russell, 1953; Gamet and Brottes, 1963; Gardner et al., 1964; Howles et al., 1954; Kempson and Sternberg, 1963; Kokke, 1964; Laboratory Procedures in Clinical Mycology, 1964; Laskowski, 1960). Experimental inoculation of rats, rabbits and cold-blooded animals is not used in clinical practice for mycological diagnosis.

Pathogenesis: A history of local trauma is present in only a small number of cases of chromoblastomycosis. Most patients do not seek medical care until the lesion has been present for many years and they do not recall a history of trauma. The disease spreads locally over a period of several years. Sometimes satellite lesions appear, with unaffected skin surrounding the original lesion. These satellites are probably the result of lymphatic spread. No regional lymph node involvement has been reported except in cases with superimposed bacterial infection. On the other hand, there have been reports of lymph node involvement by chromoblastomycosis, without skin lesions (Tsai et al., 1966). The process rarely invades the subcutaneous tissue, a fact which allows plastic surgeons to resect the lesion entirely (Fig. 4A, B). Usually extensive lesions are found in only one lower limb (Fig. 1A). It is infrequent that other cutaneous regions are affected. The incubation period in man is difficult to establish, for a history of antecedent trauma seldom exists. Trejos (personal communication) reported a case of autoinoculation. The lesions appeared a few days later, and, at the time of resection several weeks later, measured 11 mm in diameter and 2 mm in depth. Microscopically there were already typical microabscesses and fungal cells. We have also seen metastatic lesions appear 15—20 cm from the primary site, 90 days after surgical resection of chromoblastomycosis (Fig. 1C). These attained a diameter of 35 mm over a period of 5 months.

The disease sometimes takes 15—20 years completely to involve a limb. That it eventually involves an entire limb suggests that the microorganisms do not provoke antibody formation in sufficient amount as to limit the spread of the infection. The fact that the disease does not involve the viscera is in favor of a low virulence of the causative organism. Most cases reported as cerebral abscesses, or hepatic granulomas of chromoblastomycotic origin, are probably examples of cladosporiosis (Binford et al., 1952; Riley and Mann, 1960; Sasano et al., 1961; Sugawara et al., 1964; Watson, 1962). Rarely, chromoblastomycosis is associated with a deep mycosis; Saenz and Morera reported a case of South American blastomycosis with pulmonary involvement in a patient who had chromoblastomycosis of a lower extremity for over 30 years. The blastomycosis was cured by medical treatment, whereas the chromoblastomycosis progressed to the point of

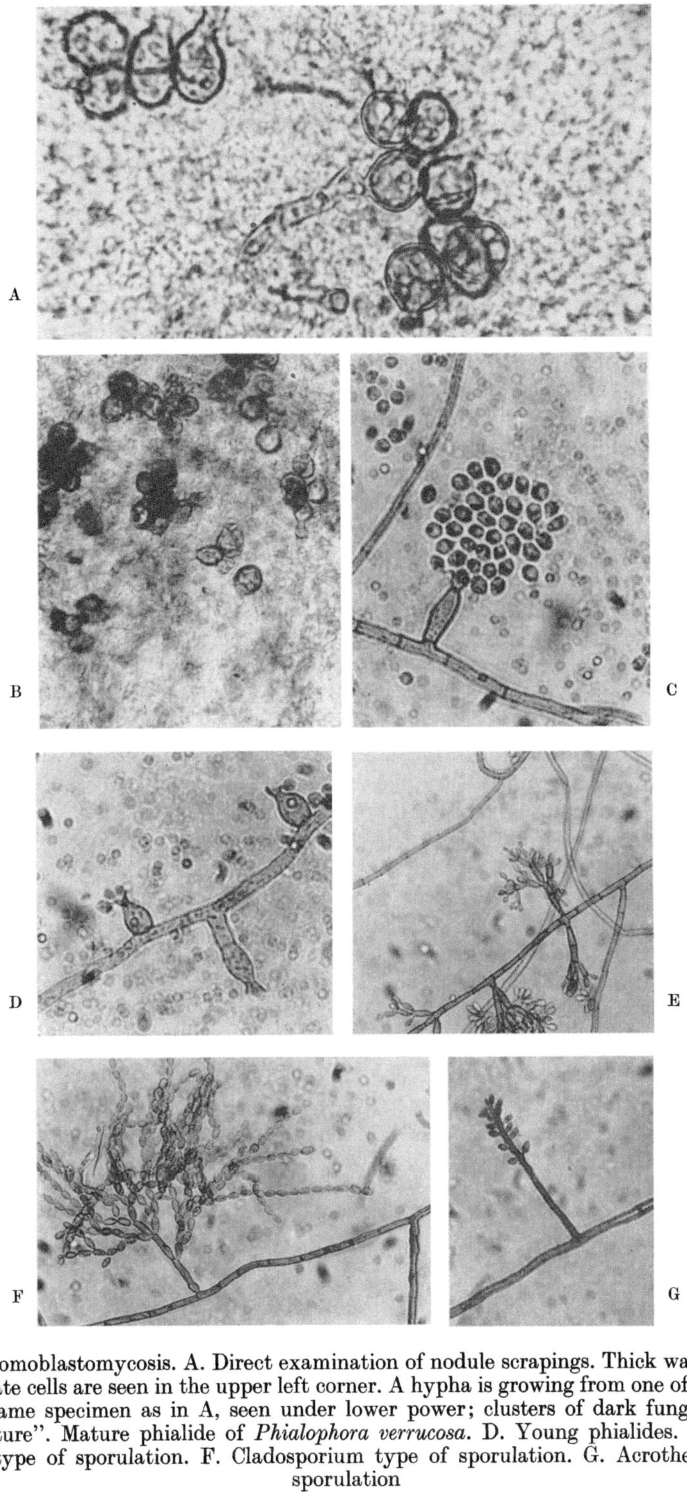

Fig. 2. Chromoblastomycosis. A. Direct examination of nodule scrapings. Thick walled fungus cells. Septate cells are seen in the upper left corner. A hypha is growing from one of the fungus cells. B. Same specimen as in A, seen under lower power; clusters of dark fungus cells. C. "Slide culture". Mature phialide of *Phialophora verrucosa*. D. Young phialides. E. Hormodendrum type of sporulation. F. Cladosporium type of sporulation. G. Acrotheca type of sporulation

making amputation necessary. Recurrence of the lesion appeared in the stump several months later.

No cases of *transmission* of the disease from man to man are known, nor are there reported cases of spontaneous infection in animals.

As stated before, antibodies may be detected in the serum of chromoblastomycosis patients, by means of a complement fixation test (LASKOWSKI, 1960). In clinical practice, no skin reaction is used for the diagnosis of the disease.

Gross and Microscopic Appearances: The basic lesion is a warty, discrete, firm nodule (Fig. 1D). It may be single or multiple, sometimes coalescing to give the appearance of "a mossy skin" (Fig. 1B). The skin surface may or may not be ulcerated and may be keratinized. The intervening skin may be edematous (Fig. 1A) or atrophic; usually it is hard and cyanotic. The early single lesion looks much like that of cutaneous leishmaniasis or chronic sporotrichosis (Fig. 3A). If there is superimposed bacterial infection, the surface of the lesion is wet with serosanguineous or purulent, foul-smelling exudate.

If resection is performed and the defect covered with a skin graft, the lesion often recurs in the grafted skin in the following months. This recurrence does not preclude a new resection, followed by another skin graft.

The low power microscopic appearance shows marked papillomatosis, hyperkeratosis and acanthosis (Fig. 4C). There is a granulomatous infiltrate- both in the upper and lower dermis, but the hypodermis is spared (Fig. 4B). This allows for radical surgery. The edges of the lesion are sharply defined (Fig. 4A, B). The dermal granulomas may resemble tubercles, with clusters of epithelioid cells surrounding brown fungus cells, single or multiple (Fig. 4E). Septa within the fungus spherules are not always seen in histologic sections; sometimes only peripheral portions of the fungus cells are visible; but the brown color facilitates their identification. Microabscesses are also seen, and consist of masses of polymorphonuclear leukocytes surrounded by histiocytes or epidermal epithelial cells. Within these abscesses one may see the chromoblastomycotic spherules (Fig. 4D). Among the tubercles and microabscesses there is a pleomorphic infiltrate composed of lymphocytes and plasma cells, less than in leishmaniasis, some eosinophils, multinucleated giant cells, either of LANGHANS or foreign-body type, and a slight fibroblastic proliferation. Fungus cells are often seen within the giant cells. Chromoblastomycosis does not usually tend to spontaneous healing, and may show slight fibrosis and a scant blood vessel proliferation. Although the usual picture is that of epidermal hyperplasia, atrophic areas are seen. One may observe hyphae and germinal tubes within the debris of the horny layer.

Differential Diagnosis: Grossly, chromoblastomycosis appears as a verrucal dermatitis which may be confused with leishmaniasis, chronic forms of sporotrichosis, North American and South American blastomycosis; less frequently, it may have to be distinguished from yaws, syphilis, leprosy and tuberculosis. The differential diagnosis cannot be made grossly. It requires biopsy and culture. When fungus cells are present, the microscopic diagnosis is easily made although chromoblastomycosis and leishmaniasis sometimes occur together (MEDINA and BORELLI, 1958).

Keloidal blastomycosis (Lobo's disease) may be easily ruled out because of the absence of epithelial hyperplasia and the presence of marked fibrous proliferation, as well as abundant non-pigmented organisms. In leishmaniasis, epithelial hyperplasia is less severe than in chromoblastomycosis; the granulomas show histiocytes and plasma cells, but there are no abscesses. If no parasites are seen and the microscopic picture is not diagnostic, the formalin-picrofuchsin stain of GALLEGO may be helpful in the search for organisms of leishmaniasis.

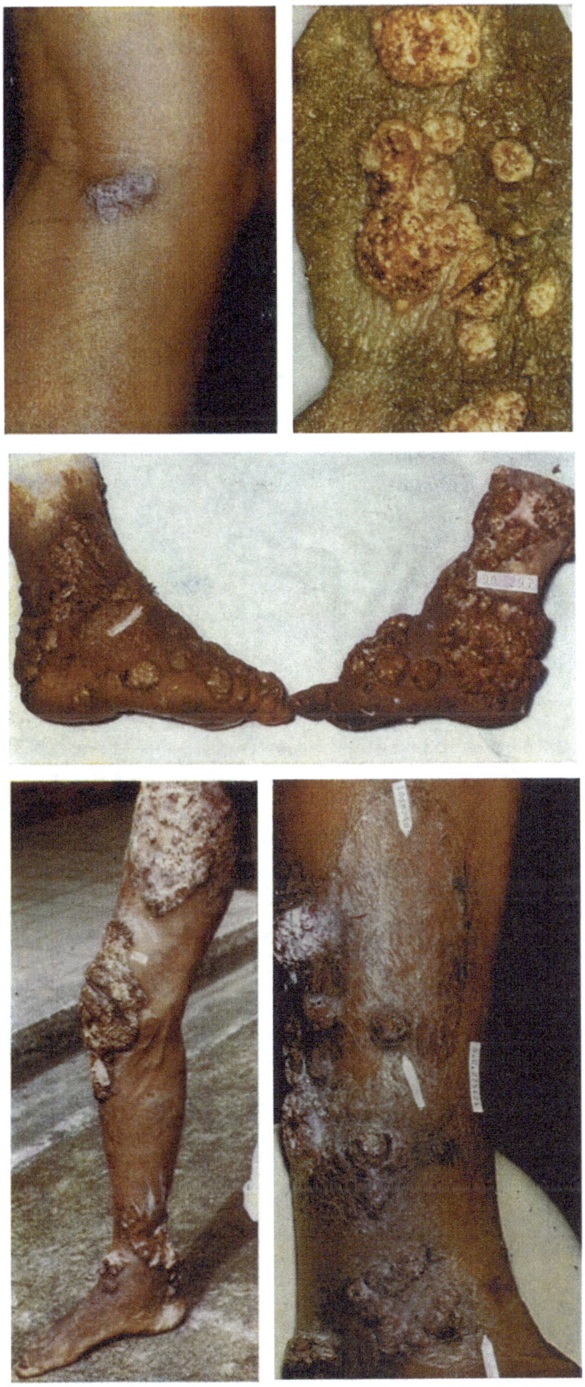

Fig. 3. Chromoblastomycosis. A. Young lesion (90 days), quite similar to those of leishmaniasis and sporotrichosis. B. Resected verrucous nodules. C. Two amputated feet. D. Widespread lesions involving the entire lower extremity. E. Recurrence after resection; the arrows point to the skin grafts and the new lesions

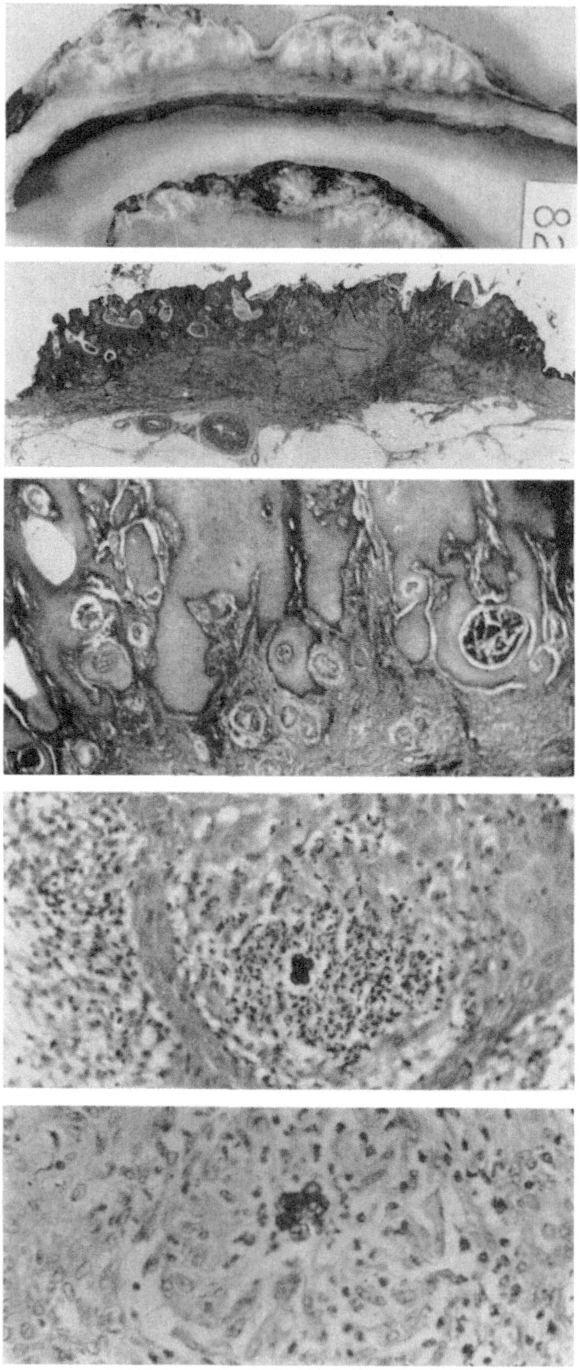

A

B

C

D

E

Fig. 4. Chromoblastomycosis. A. Section through verrucous nodules. Note the well-defined demarcation. B. Same specimen as in A; section stained with H & E (2×). The inflammatory infiltrate does not extend into the subcutaneous tissue. The nodule is sharply limited. C. Low-power microscopic field showing hyperkeratosis, ancanthosis, papillomatosis, and abscesses, both "intraepidermal" and dermal. D. Microabscess surrounded by epidermis and containing spherules of chromoblastomycosis. E. Tubercle made up of epitheloid cells showing central fungus cells, one of which is septate

The most difficult differential diagnosis is between chromoblastomycosis and sporotrichosis when no microorganisms are seen in the sections. It must be borne in mind that the predominant cell in *sporotrichosis* is the *neutrophil*, and eosinophils are absent. In *chromoblastomycosis*, on the other hand, microabscesses are surrounded by *histiocytes*, and there are eosinophils. These criteria are not enough to establish a diagnosis with certainty. Better results are obtained by multiple sectioning in search of the causative agent, and also the periodic acid-Schiff stain, which may bring out the microorganism in cases of sporotrichosis. Direct examination and culture may also be of great help. South American and North American blastomycosis show characteristic blastospores which are easily seen with special stains.

The tuberculoid form of leprosy may be confused with chromoblastomycosis; however, there is neither epithelial hyperplasia nor papillomatosis, nor microabscesses, but there are tubercles, or granulomas (Ash and Spitz, 1945; Baker, 1966; French and Russel, 1953; Gamet and Brottes, 1963; Thys, 1964). Syphilis has a characteristic microscopic picture and no tendency to abscess formation.

Lastly, there remains the problem of cutaneous granulomas without demonstrable etiologic agents. This cannot be solved by the pathologist. Sometimes immunologic methods are of value.

Clinical Picture: In the tropics, the patient usually seeks medical care only after many years of evolution of the disease that began as a papule, then became a wart and spread slowly, in spite of treatment with ointments and electrofulguration. Finally, the patient requests an amputation, not because of the pain or itching, but because of the repugnant appearance of the lesion, which he considers depressing and incurable (Fig. 3C). When the disease is widespread, and there is involvement of a whole limb, nothing can be done. If it affects a foot, no shoe can be worn.

Clinical Case: A farmer, aged 74 years, had had chromoblastomycosis of his left foot and the lower third of his left leg for 5 years. The lesions were completely resected in December, 1965. Skin grafts measuring 15 by 17 cm were applied to the foot and medial aspect of the leg (Fig. 3E). Three months later there was recurrence of the mycosis in the lower third of the leg, with progressive invasion of the grafts; a month later a metastatic lesion appeared on the medial aspect of the knee (Fig. 1C). Photographs 1C and 3E were taken in August, 1966, eight months after the operation. The lesions were again completely excised, including subcutaneous tissues (Fig. 3B and Fig. 4A, B).

A new skin graft was performed. No recurrence had occured after 4 years.

Prognosis: The disease does not heal spontaneously, and evolves slowly over a period of 15—20 years. It does not kill or spread to internal organs. If diagnosed in time, radical resection down to hypodermis may be curative. Resections not sufficiently deep are commonly followed by recurrences. No satisfactory drug therapy is available.

References

Akün, R.: Eine chromoblastomykoseähnliche Pilzkrankheit beim Pferde. Zbl. Path. **90**, 294 (1953). — Ash, Spitz: Pathology of Tropical Diseases, an Atlas. Philadelphia: W.B. Saunders Co. 1945. — Audebau, G., Escudié, A., Courmes, E.: Les Mycoses Humaines en Guadeloupe. Premiers cas de chromoblastomycose et de sporotrichose. Résultats d'une enquete histoplasminique. Bull. Soc. Path. exot. **57**, 1012 (1964). — Baker, R.D.: Fungus Infections. In: Anderson's Pathology, Fifth ed. St. Louis: C.V. Mosby Co. 1966. — Binford, C.H., Thompson, R.K., Gorham, M.E., Emmons, C.W.: Mycotic brain abscess due to *Cladosporium trichoides*, a new species. Amer. J. clin. Path. **22**, 535 (1952). — Brygoo, E.R., Segretain, G.: Étude clinique epidemiologique et mycologique de la chromoblastomycose a Madagascar. Bull. Soc. Path. exot. **53, 443** (1960). — Carrion, A.L.: The specific fungi of chromoblastomycosis. Puerto Rico J. publ. Hlth **15**, 340 (1940). — Conant, N.F., Smith, D.T., Baker,

R.D., Callaway, J.L., Martin, D.S.: Manual of Clinical Mycology, Second ed. Philadelphia and London: W.B. Saunders Co. 1954. — Duque, O.: Cromoblastomicosis. Revisión general y estudio de la enfermedad en Colombia. Antioquía méd. **11**, 499 (1961). — Emmons, C.W., Binford, C.H., Utz, J.P.: Medical Mycology. Philadelphia: Lea & Febiger 1963. — French, A.J., Russell, S.R.: Chromoblastomycosis. Report of a case recognized in Michigan, apparently contracted in South Carolina. A.M.A. Arch. Derm. Syph. **67**, 129 (1953). — Gamet, A., Brottes, M.: Les chromoblastomycoses au Cameron. Bull. Soc. Path. exot. **56**, 117 (1963). — Gardner, J.T., Pace, B.F., Freeman, R.G., et al.: Chromoblastomycosis in Texas. Report of four cases. Tex. St. J. Med. **60**, 913 (1964). — Howles, J.K., Kennedy, C.B., Garvin, E., Brueck, J.W., Buddingh, G.J.: Chromoblastomycosis: Report of nine cases from a single area in Louisiana. A.M.A. Arch. Derm. Syph. **69**, 83 (1954). — Kempson, R.L., Sternberg, W.H.: Chronic subcutaneous abscesses caused by pigmented fungi, a lesion distinguishable from cutaneous chromoblastomycosis. Amer. J. clin. Path. 39, 598 (1963). — Klokke, A.H.: Chromomycosis. J. Indian med. Ass. **43**, 340 (1964). — Laboratory Procedures in Clinical Mycology. Dept. of the U.S. Army Technical Manual, Feb., 1964. — Laskowski, L.F., Jr.: Chromomycosis (chromoblastomycosis, dermatitis verrucosa). Basic techniques and newer concepts in clinical diagnostic medical mycology. Catholic Hospital Association, St. Louis, Mo., 153—157, Oct., 1960. — Martin, D.S., Baker, R.D., Conant, N.F.: Case of verrucous dermatitis caused by *Hormodendrum pedrosoi* (chromoblastomycosis) in North Carolina. Amer. J. trop. Med. **16**, 593 (1936). — Medina, Rafael, Borelli, Dante: Leishmaniasis tegumentaria asociada a cromomicosis. Acta méd. Venez. **6**, 202 (1958). — Moore, M., Cooper, Z.K., Weiss, R.S.: Chromomycosis (chromoblastomycosis), report of two cases. J. Amer. med. Ass. **122**, 1237 (1943). — Moss, E.S., McQuown, A.L.: Atlas of Medical Mycology. Baltimore: Williams & Wilkins Co. 1960. — Niño, F.L.: Lecciones de Micología y Micopatología Médica. Cajica, Buenos Aires, 1959. — Romero, A., Trejos, A.: La cromoblastomicosis en Costa Rica. Rev. Biol. trop. (S. José) **1**, 95 (1953). — Riley, O., Mann, S.H.: Brain abscess caused by *Cladosporium trichoides*. Amer. J. clin. Path. **33**, 525 (1960). — Sáenz, L., Morera, P.: Sobre un caso de blastomicosis suramericana asociada a cromomicosis. Acta méd. costarric. **6**, 55 (1963). — Salomon, L.M., Berman, H.: Amphotericin B and electrodesiccation for chromoblastomycosis. Arch. Derm. **87**, 492 (1963). — Sasano. N., Okamoto, T., Takahashi, T., Suzuki, S.: An autopsy case of primary chromoblastomycosis arising from the internal organs. Dark-brown granulomas in the liver and the brain without skin symptoms, observed in a smoking child 3 years old. Tohoku J. exp. Med. **73**, 180 (1961). — Simpson, F.W., Harington, C., Barnetson, E.J.: Chromoblastomycosis: A report of six cases. J. Path. Bact. **55**, 191 (1943). — Solano, E.: Cromomicosis. Acta méd. costarric. **9**, 77 (1966). — Sugawara, M., Sobajima, Y., Tamura, H.: A case of generalized chromoblastomycosis. Acta path. jap. **14**, 239 (1964). — Thys, A.: Personal observation on the histopathology of Congolese mycoses. Ann. Soc. belge Méd. trop. **44**, 887 (1964). — Trejos, A.: *Cladosporium carrionii* n. sp. and the problem of Cladosporia isolated from chromoblastomycosis. Rev. Biol. trop. **2**, 75 (1954). ~ Cromoblastomicosis experimental en Bufo marinus. Rev. Biol. trop. **1**, 39 (1963). ~ Communicación personal. — Tsai, C.Y., Lu, C., Wang, L.T., Hsu, T.L., Sung, J.L.: Systemic Chromoblastomycosis due to *Hormodendrum dermatitidis* (Kano) Conant. Report of the first case in Taiwan. Amer. J. clin. Path. **46**, 99 (1966). — Watson, K.C.: Cerebral chromoblastomycosis. J. Path. Bact. **84**, 233 (1962). — Zapater, R.C.: Atlas de diagnóstico micológico (aplicación del Laboratorio) VI + 251. Buenos Aires: El Ateneo 1962.

Cladosporiosis of the Central Nervous Tissue

Oscar Duque, Medellín, Colombia

With 17 Figures

Definition: Cladosporiosis may be defined as infection of the central nervous system and meninges by *Cladosporia*, and the fungi causing chromoblastomycosis (*Cladosporium, Hormodendrum, Phialophora, Fonsecaea*).

Historical: In 1911 Guido Banti presented to the Academia Medico Fisico Fiorentina the case of a woman who died with symptoms of brain tumor. The autopsy disclosed multiple nodular dark brown areas in the cerebrum and cerebellum formed by granulomatous inflammation with necrosis. In these lesions brown septate hyphae and spherical forms of a fungus were found. Cultures of this microorganism were studied by Saccardo who published its description and biological characteristics in 1912 with the name of *Torula bantiana*. It produced brain lesions when inoculated into rabbits. The observations of Banti and Saccardo remained largely ignored until Borelli (1960), and Stigliani called attention to it. In 1952 Binford and associates described the case of a man who suffered from headache and drowsiness. An abscess was found in the left frontal lobe of his brain which again consisted of a granulomatous inflammation with necrosis and purulent material from which a fungus was grown. This was studied by Emmons (in Binford et al.) who noted that by its morphology and brown pigment it should be classified with *Cladosporia*. He named it *Cladosporium trichoides* "signifying its hyphal growth in human brain tissue". There can be little doubt today that both the cases of Banti and Binford were caused by the same fungus. The figures of *T. bantiana*, published by Saccardo, are very similar to those of *Cl. trichoides*; both are highly pathogenic for laboratory animals, and both grew well at 37 °C, unlike other species of saprophtic *Cladosporia* which are not thermophilic. Emmons has pointed out that the spore size of *Torula bantiana*, as registered by Saccardo, is larger than that of *Cladosporium trichoides*; however, one wonders if the difference could not be within the limits of strain variation. Borelli (1960) proposed a new combination to designate the species: *Cladosporium bantianum* (Saccardo, 1912; Borelli, 1960). Both names will be used here to avoid the question of priority which will have to be solved by the proper authorities.

A total of 31 acceptable cases of cladosporiosis has been reported (see Table 1). Fourteen of the cases were caused by *C. trichoides* making this species the most important agent of cladosporiosis (Banti, Binford, King, McGill, Dereymäker, Watson, Watson and Lines, Garcin (1957), Riley, Barnola, Duque, Coudert, Chevrel, Dastur). Two cases were caused by *Hormodendrum pedrosoi* (Fukushiro, Lucasse); the fungus isolated from the case of Lucasse was studied by Vanbreuseghem who described it as *C. trichoides* but commented that it produced the acrotheca type of sporulation. Borelli (1956) re-examined this fungus and concluded that it was *H. pedrosoi*. The cases of Shimazono, and Tsai were due to *Hormodendrum dermatitidis*. *Cladosporium gougeroti* was the causative agent in the case of Aufdermaur. In the other 12 cases the fungus was not isolated but diagnosis was established by the morphology and characteristic brown color of the fungi in tissue sections (Bonne, two Indonesian cases; Garcin, 1949; Franca-Netto, Bobra, Manson, Symmers, Horn, Sasano, Duque, Warot, Bagchi). The case of Wybel, cited by some authors, is doubtful as no mention is made of the fungus being pigmented.

Table 1. *Reported cases of cladosporiosis*

Case No. Age, Sex	Chief Manifestations	Associated Conditions	Outcome and Duration	Post Mortem Findings: Central Nervous System	Post Mortem Findings: Meningitis	Post Mortem Findings: Other Organs	Organism Responsible
Case 1 43, F	Symptoms of brain tumor. Mental deterioration. Vomiting, headache, coma	Patient was deaf	Death 6 months after onset	Multiple, melanoma-like granulomas; brain and cerebellum		Nothing of note	*Torula bantiana*
Case 2 23, F	Fever, paralysis of left arm and leg, and R. sixth and seventh cranial nerves		Death a few weeks after onset of symptoms	Meningeal and pontine granulomas	Nodular	Nothing of note	Not isolated
Case 3 20, M	Cachexia; right hemiplegia; papilloedema	Scabies; decubitus ulcer	Death 1 month after admission	Hemorrhagic left frontal granuloma	?	Nothing of note	Not isolated
Case 4 37, M	Fever, headache; R. hemiplegia, papilloedema, leucocytosis, cachexia		Death 8.5 mos. after onset of symptoms	Ependymitis. Discrete lesions in left frontal lobe. Hydrocephalus	Nodular thickening of basal meninges	Not examined	Not isolated
Case 5 22, M	Headache, drowsiness, fever, leucocytosis		Free from symptoms 2 years after operation	3 cm left frontal abscess shelled out	No		*Cl. trichoides*
Case 6 47, M	One month history of headache, fever, dysarthria; R. hemiplegia, convulsions	Arteriosclerosis, hypertension. Left otorrhea	Death about 3 months after first symptoms	Large left frontal lobe abscess and multiple secondary abscesses	?	Arteriosclerosis	*Cl. trichoides*
Case 7 34, M	Fever, headache, convulsions, diplopia, meningism, paresis		Death 1 year after onset of symptoms		Chronic sclerosing	Old rheumatic mitral valve disease	Not isolated
Case 8 42, M	3 mos. history of headache, tiredness, disorientation; ataxia, meningism		Death 4 months after onset of symptoms	Small abscesses in brain	Basal meningitis	Hyphae in lungs (contaminant?)	*Cl. gougeroti*
Case 9 10, M	4 mos. history of fever, anemia, convulsions, papilloedema	Filariasis	Death 9 months after first symptoms	Ependymitis and small brain abscesses	Basal meningitis	Nothing of note	*H. pedrosoi* (Borelli, 1956)
Case 10 27, F	4 mos. history of vomiting, headache, meningism, fever, R. hemiplegia		Death about 10 mos. after onset of symptoms	2 chronic left lobe abscesses. Two secondary abscesses		No autopsy	*Cl. trichoides*
Case 11 32, M	5 mos. history of headache, vomiting, papilloedema, ataxia, failing sight		Death 7 months after onset of symptoms	Fibrino purulent basal meningitis (operation)		No autopsy	*Cl. trichoides*

Case 12 37, F	One mos. history of R. hemiplegia; Jacksonian epilepsy	Old laceration of scalpp syphilis, skin abscess	Death 7.5 months after first symptoms	Left parietal abscess (removed). Large secondary abscess	Post surgical infection of meninges	Lung infarct	*Cl. trichoides*
Case 13 8, F	6 mos. history of chromoblastomycosis of the face; then headache, fever, paralysis of increasing severity	chromoblastomycosis of skin of face	Death 12 mos. after appearance of skin lesions	Multiple dark green granulomas in brain, cerebellum, pons	Nodular meningitis. Arteritis	Autopsy confined to CNS	*H. pedrosoi*
Case 14 30, M	Fever, headache, meningism, mental deterioration		Death about 24 days	2 cm encysted abscess in right frontal lobe	No	Horseshoe kidney	"Hormodendr." probably *Cl. trichoides*
Case 15 3, M	Fever; R. hemiplegia; swollen abdomena, leucocytosis	Neglected, smoking, allotriophagic child	Death 1 month after onset of symptoms	Multiple disseminated dark brown nodules	Nodular basal meningitis	Dark brown nodules in liver, kidney and lymph nodes	Not isolated
Case 16 57, M	10 mos. history of mental deterioration. Hemiplegic strokes; papilloedema	Hypertension	Sudden death 10 mos. after onset of symptoms	3 cm left parieto-occipital abscess with extension to chiasm	?	Arteriosclerosis and hypertension	Not isolated
Case 17 44, M	2 weeks history of weakness of l. limbs, then l. hemiplegia, meningism		Death 6 weeks after first symptoms	Large right occipital, left frontal and temporal abscesses	No	Pulmonary emphysema. Bantu siderosis	Not isolated
Case 18 22, F	1 month history of fever, loss of weight associated with polyarteritis. Left hemiplegia following cortisone therapy	Polyarteritis nodosa, cortisone treatment	Death about 5 mos. after first symptoms	Right fronto-parietal abscess (infected infarct?)	No	Healed and healing lesions of arteritis nodosa	Not isolated
Case 19 68, F	Fever, opisthotonus, coma	Mild diabetes. Neurogenic hypernatremia	Death within 3 mos. after onset of symptoms	Multiple nodular granulomatous lesions	No	Left fibrinous exudate with hyphae	Not isolated
Case 20 24, M	3 episodes of seizures; cough, fever, headache, vomiting	Dental caries. Scarred right ear drum	Well 3 mos. after operation	Small purulent mass in right frontal lobe	No		*Cl. trichoides*
Case 21 23, F	3 months history of headache. Left hemiplegia		Progressive worsening sudden death	Gray black area surrounding abscess cavity of 5 cm r. fronto parietal	No	Small mycotic abscess in right lung	*Cl. trichoides* (brain and lung)

Reported cases by:

1 = BANTI (1911); 2 = BONNE et al. (1948); 3 = BONNE et al. (1948); 4 = GARCIN et al. (1949); 5 = BINFORD et al. (1952); 6 = KING and COLLETE (1952); 7 = FRANCA NETTO et al. (1953); 8 = AUFDERMAUR et al. (1954); 9 = LUCASSE et al. (1954); 10 = SEGRETAIN (1955) and GARCIN (1957); 11 = DEREYMAKER and DE SOMER (1955); 12 = MCGILL and BRUECK (1956); 13 = NISHI (1956), OKUDAIRA (1956) and FUKUSHIRO (1957); 14 = WATSON and LINES (1957); 15 = SASANO et al. (1958, 1961); 16 = BOBRA (1958); 17 = MANSON (1958); 18 = SYMMERS (1960); 19 = HORN et al. (1960); 20 = RILEY and MANN (1960); 21 = BARNOLA and ANGULO (1961)

Table 1 (continued)

Case No. Age, Sex	Chief Manifestations	Associated Conditions	Outcome and Duration	Post Mortem Findings: Central Nervous System	Post Mortem Findings: Meningitis	Post Mortem Findings: Other Organs	Organism Responsible
Case 22 22, M	Headache, fever, vomiting 20 day, then vertigo, staggering gait, anisocoria	"sore throat"	Discharged 1 month after operation	Abscess in left parietal lobe	No		*Cl. trichoides*
Case 23 29, M	Laceration of face and scalp 9 mos. prior to admission. Headache, vomiting, ataxia, dysarthria, meningism, papilloedema		Death 9 mos. after accidental laceration of scalp	Granular ependymitis Hydrocephalus	Basal granular meningitis	Autopsy confined to CNS	Not isolated
Case 24 40, M	2 episodes of epileptic seizures. Paresis of left face and limbs; vomiting, meningism	Sarcoidosis	Death 3 mos. after first symptoms	3 black nodules in right frontal lobe resected surgically		No autopsy	Not isolated
Case 25 25, M	15 days history of facial paralysis, dysarthria; then aphasia, R. hemiplegia		Death 3 weeks after onset; 5 days after operation	3 cm abscess in left frontal lobe		No autopsy	*Cl. trichoides*
Case 26 54, M	2 months history of fever, headache, paresis, vomiting, convulsions, papilloedema		Death 10 mos. after first symptoms	Mass in right parietal area with multiple abscesses; resected			Not isolated
Case 27 18, M	18 months history of headache, impaired vision, dizziness, meningism, papilloedema	Right middle ear infection, 3 years prior to admission	Discharged in good condition after operation. No follow up	Large abscess in right temporo-occipital area	No		*Cl. trichoides*
Case 28 12, M	Syndrome of intracraneal hypertension		Well 10 mos. after operation	Multiple abscesses in left temporal lobe	No		*Cl. trichoides*
Case 29 30, F	Headache, vomiting, facial paralysis, confusion, nystagmus, left hemiplegia		Death about 3 mos. after operation	Multiple disseminated abscesses	Nodular basal meningitis	Slight fibrosis of pancreas	*H. dermatitidis*
Case 30 19, F	Weakness of right leg. Epileptic seizures; pyramidal signs; involvement of cranial nerves	Jaundice and lymph node enlargement 3 years prior to admission	Death 7 mos. after onset of neurological symptoms	Large greenish nodule in left frontal lobe. Multiple disseminated nodules	Basal and spinal	Mycotic granulomas in liver, kidney pancreas, lymph n.	*H. dermatitidis*
Case 31 24, M	Paresthesia, headache, vomiting, hemiparesis, convulsions, leucocytosis	Scaly skin eruption from which "Cladosporium sp." was recovered	Death about 8 mos. after first symptoms	3 encapsulated purulent nodules in left fronto-parietal region		No autopsy	*Cl. trichoides* (brain), Cladosporium sp. (skin)

Reported cases by:

22 = Duque (1961); 23 = Duque (1961); 24 = Warot et al. (1961); 25 = Chevrel et al. (1962); 26 = Bagchi et al. (1962); 27 = Watson (1962); 28 = Coudert et al. (1962); 29 = Shimazono et al. (1963); 30 = Tsai et al. (1966); 31 = Dastur et al. (1966).

Epidemiology: Cases of cladosporiosis have been reported from Europe, North and South America, Africa, India, Indonesia, Japan and Taiwan. Both sexes are about equally affected. The disease occurred in a Japanese smoking child of 3 years (SASANO), and in an 8 year old Japanese girl (FUKUSHIRO), but most of the cases were in adults. White, negro, oriental and mixed races are susceptible.

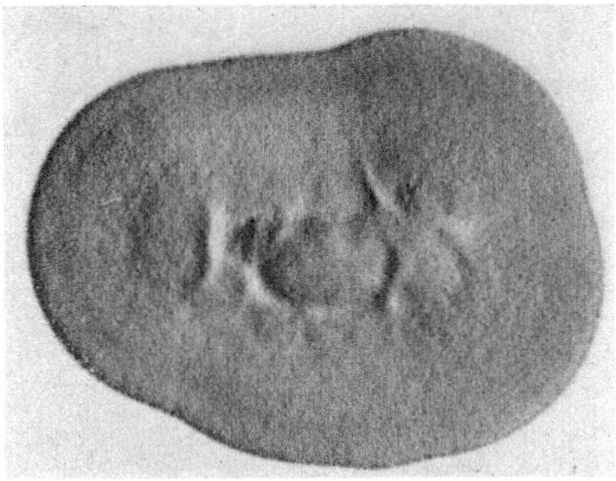

Fig. 1. *Cladosporium trichoides* (bantianum). Three weeks old culture in Sabouraud's glucose agar at room temperature

Mycology: *Cladosporium trichoides* grows well in Sabouraud's glucose agar at room temperature producing black-grey to olive-brown velvety colonies that may reach 4 cm in diameter after a month in a plate culture; folds may be present in older colonies (Fig. 1). Unlike other *Cladosporia*, *C. bantianum* grows well at 37 °C and has no proteolytic action on Loeffler's medium. In culture, hyphae are brown, septate, 1—2 μ in diameter, smooth and doubly walled. Conidiophores

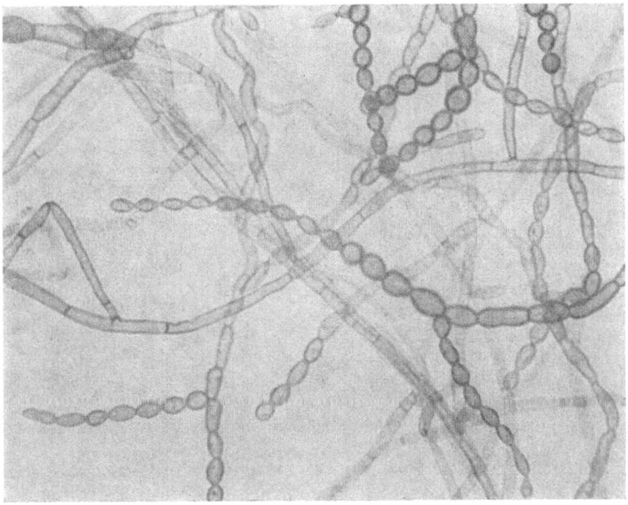

Fig. 2. *Cladosporium trichoides*, slide culture

vary in length, straight or shield shaped, bearing one or several long, branched chains of elliptical or fusiform conidia 2—4 μ in diameter, 5—20 μ in height; younger conidia are produced at the ends of chains, basal conidia may be septate (Figs. 2 and 3). In pus from the lesions, the fungus appears as olive green to brown septate branched hyphae with bulbous extremities (Fig. 4). In tissues the hyphal phase is abundant (Fig. 5), but vesicular and large spherical forms, like the "sclerotic cells" of chromoblastomycosis of the skin may be also observed (Fig. 6).

C.trichoides is highly pathogenic for mice, rats, guinea pigs, and rabbits when inoculated into these animals from cultures by intracerebral or intravenous routes. It kills mice within 5—25 days, growing so abundantly in the brain that the

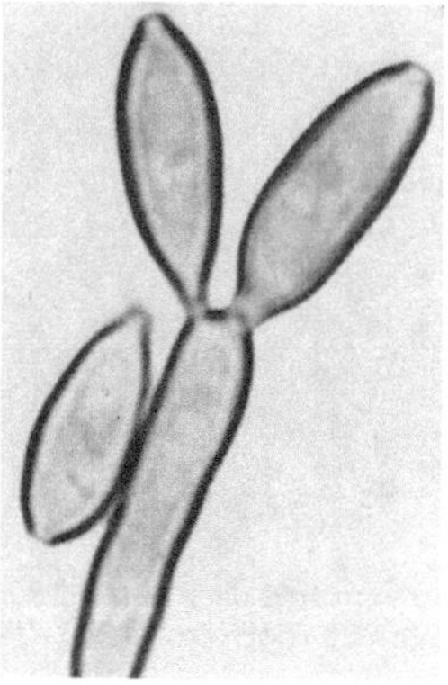

Fig. 3. *C.trichoides*, slide culture. Conidiophore with two terminal conidia. ×2500

4

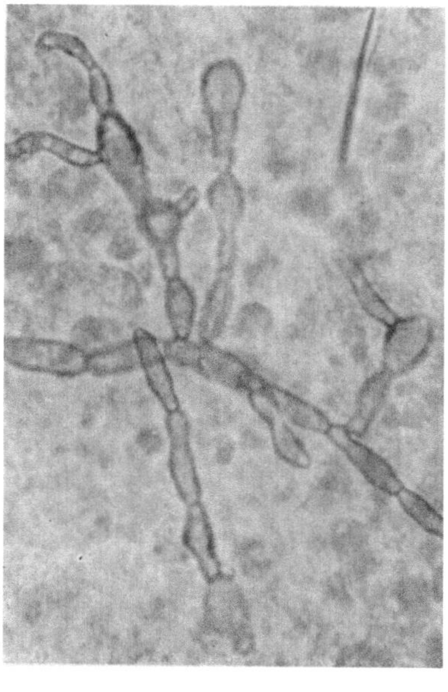

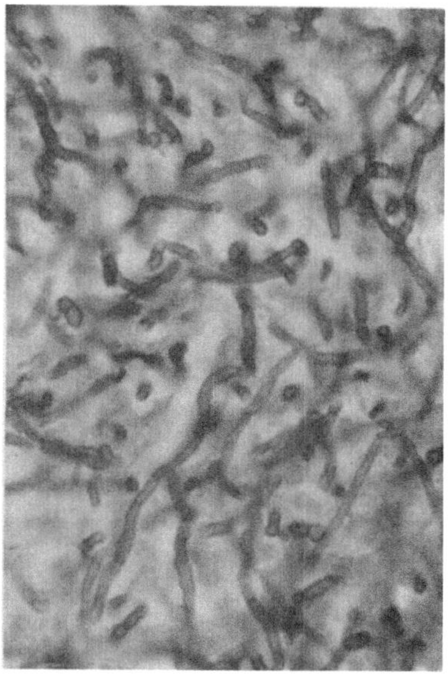

5

Fig. 4. Septated, branched hyphae with bulbous extremities of *C.trichoides* as observed in pus from a brain abscess. Unstained preparation. ×600

Fig. 5. Abundant hyphal growth of *C.trichoides* in brain lesion. H & E, ×600

lesions present a characteristic black greenish color (Fig. 7). Direct inoculations into animals of material from the human brain lesions is less successful. The fungus is distinctly neurotropic; when inoculated intravenously the main lesions are always in the brain although limited growth may be observed in other organs. Intraperitoneal inoculations in new born mice (DUQUE, 1961, 1963) produce small

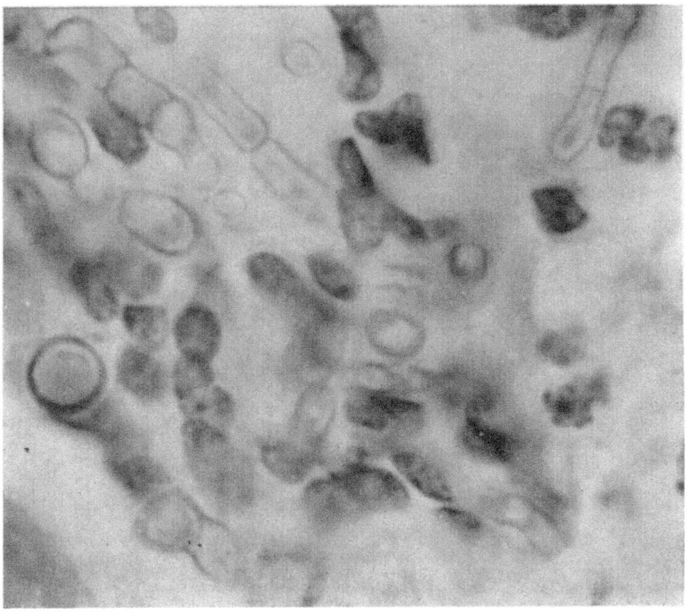

Fig. 6. Hyphae and spheric forms ("sclerotic cells") of *C.trichoides* in brain lesion. H & E, × 1250

granulomatous nodules in the serosa and severe lesions in the brain. Because of the close morphological similarity between *C.trichoides* and one of the etiological agents of chromoblastomycosis, *Cladosporium carrionii*, the question arose as to whether or not other dematiaceous fungi could cause brain lesions also. In 1958 IWATA reported that the fungus isolated from the case of FUKUSHIRO, an 8 year old Japanese girl with chromoblastomycotic lesions of the skin of the face and metastatic lesions in the brain, was *Hormodendrum pedrosoi*. Then, DUQUE, and FELGER demonstrated experimentally that *H.pedrosoi*, *H.compactum*, *Phialophora verrucosa*, and *H.carrionii* produce brain lesions when inoculated intracerebrally or intravenously in new born or cortisone treated mice. The neurotropism of these fungi and the severity of the brain lesions that they produce varied from very intense with *H.pedrosoi* to mild and limited with *C.carrionii*, but, unlike *C. bantianum*, these microorganisms did not reach the brain when inoculated intraperitoneally. The case of AUFDERMAUR caused by *Cladosporium gougeroti*, and those of SHIMAZONO, and TSAI due to *Hormodendrum dermatitidis* enlarged the number of dematiaceous fungi causing chromoblastomycosis that may also infect the central nervous system.

Cladosporium trichoides cannot be differentiated in tissue sections from *H.pedrosoi* or the other fungi of chromoblastomycosis. All of these microorganisms produce pigmented hyphae and spherical forms in the lesions. They are Schiff positive and stain black with Grocott's methenamine silver stain, but these special stains offer little advantage as the fungi are readily demonstrable in unstained or in ordinary H & E preparations because of their brown color. Cultures are necessary to establish the identity of the species in this group of fungi. Studies of *C.trichoides* and *C.carrionii* with immuno-fluorescence methods by AL DOORY showed that both fungi react with a number of dematiaceous species in addition to the homologous organism, but selective dilution and absorption resulted in conjugates specific for each of the two species.

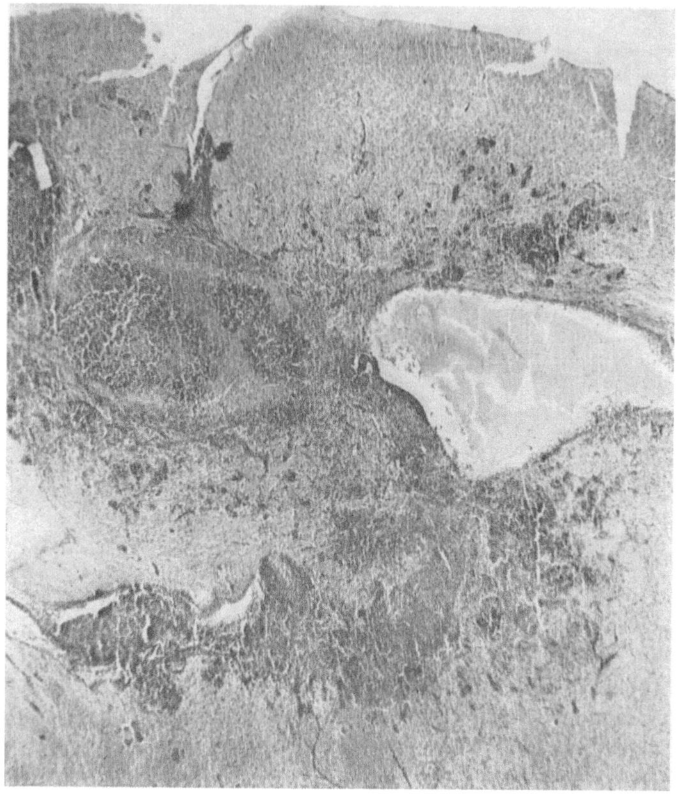

Fig. 7. Brain lesions produced by intracerebral inoculation of *C. trichoides* in a new born mouse. H & E, ×20

Pathogenesis: The portal of entry of *Cladosporium trichoides* is unknown. Direct inoculation was suspected in the cases of MCGILL and DUQUE, which had sustained trauma to the face and scalp. In three cases (KING, RILEY, WATSON) there was inflammation or healed lesions of one of the ears, and BANTI's patient was deaf. KING isolated *C. bantianum* from the inner ear of his case, but whether the infection of the ear was primary or secondary to the brain lesion in these cases cannot be determined. There is some evidence that, as in other systemic mycoses, the portal of entry may be respiratory; *Cladosporium trichoides* was recovered from lung lesions in the case of JAFFE and BARNOLA (reported by BARNOLA). Hyphae were seen in lung lesions by HORN, and in the lungs and other organs by SASANO, but no cultures were made in these last two cases. Hyphae were also seen in the lung of the case of AUFDERMAUR due to *C. gougeroti*. The work of FERNANDEZ-BAQUERO has thrown new light on the role of the respiratory tract in chromoblastomycosis. He isolated *Hormodendrum pedrosoi* from the bronchial washings of 4 patients with cutaneous chromoblastomycosis; the identity of the fungus was established by its cultural characteristics and by inoculation into 6 men who developed typical lesions at the sites of injection. Furthermore, as more cases of disseminated chromoblastomycosis are reported (SUGAWARA, AZULAY) it is becoming increasingly evident that hematogenous dissemination of its etiological agents may be a feature of the disease. The close biological similarity between

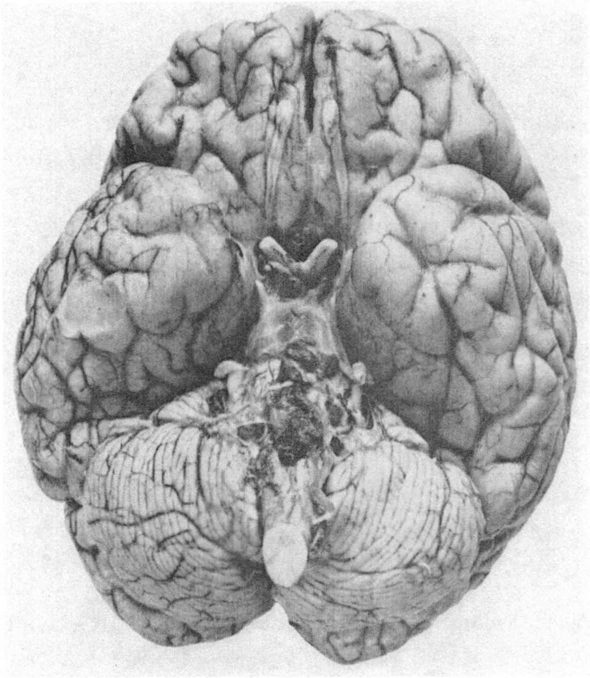

Fig. 8. Meningitis due to *Hormodendrum dermatitidis* (case of Shimazono et al.). Black nodules scattered on the basal surface of the brain and brain stem

C.bantianum and the species causing chromoblastomycosis is emphasized by the fact that in the case of Dastur (studied also by Desai), *C.trichoides* infection of the brain was associated with a scaly, olivaceous dermatitis and a subcutaneous abscess; in these skin lesions brown septate hyphae were found. Emmons (1963) mentions that this fungus was identified in a lesion of the skin of the abdomen in a patient that had no signs of cerebral mycosis. These facts lead one to to suppose that *C.trichoides*, and the fungi of chromoblastomycosis, may reach the brain by hematogenous dissemination from an infected wound, or from the respiratory tract. The presence of arteritis in some of the reported cases of cladosporiosis lends weight to this hypothesis.

C.trichoides (bantianum) does not seem to be a merely opportunistic agent of disease like certain fungi that invade only the debilitated individual. In most of the cases of cladosporiosis the patients were in good health until the onset of cerebral symptoms. Exceptions are the case of Horn which occured in a diabetic, and the patient of Symmers who had been receiving high doses of corticoesteroids for polyarteritis nodosa. Hypertension was present in the cases of King, and of Bobra. No relation with a particular occupation has been reported.

Pathologic Anatomy: The gross and microscopic appearances of the *central* nervous system lesions caused by *C.trichoides*, *C.gougeroti*, *H.pedrosoi*, and *H.dermatitidis* are entirely similar. Two main forms of cladosporiosis have been described, meningeal and cerebral, but lesions in both meninges and brain may be frequently seen. Meningitis has always been basal (Dereymaker, Lucasse, Tsai, Aufdermaur, Bonne, Garcin, 1949; Duque, 1961; Shimazono), consisting of fibrous thickening with or without granules or small nodules; when

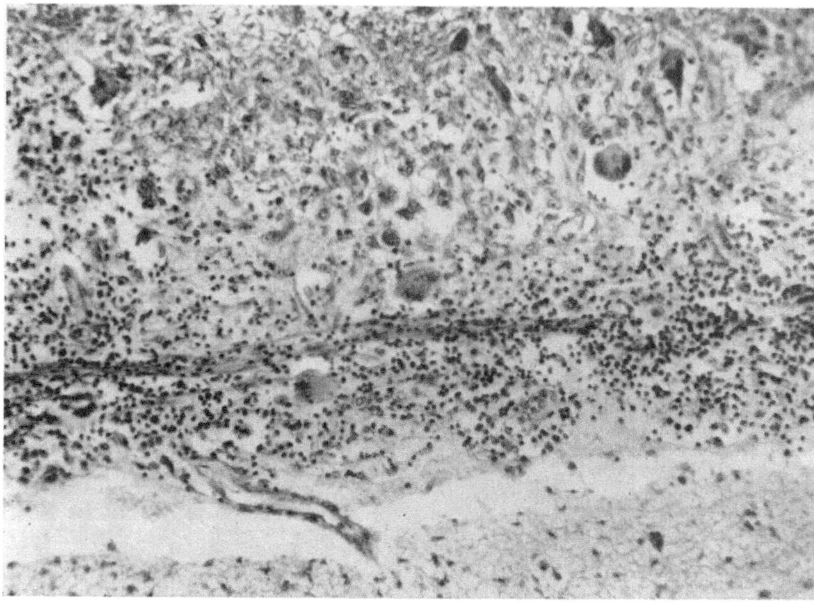

Fig. 9. Cladosporiosis. Granulomatous meningitis. H & E, $\times 120$

these were present they were described as grey, green, or black-red in color (Fig. 8). The lesions may extend around the brain stem and even down to the cervical spinal cord, and it may be accompanied by fibrinopurulent exudate. Involvement of the ependyma is frequent in these cases, presenting as small disseminated granules on the walls of the ventricles. Choroid plexus may also be affected, and hydrocephalus was described in the cases of GARCIN, and of DUQUE. Microscopically the lesion consists of a granulomatous inflammation with abundant giant cells of the foreign body and LANGHANS types, and an admixture of polymorphonuclear leucocytes and lymphocytes (Fig. 9). Microabscesses are frequently present; the degree of fibrosis is variable, probably in accord with the virulence and duration of the process. Superficial involvement of the nervous tissue is always present either as small granulomas and abscesses, or in the form of non specific inflammatory reaction. Coexistence of meningitis and deep cerebral lesions were found by TSAI, AUFDERMAUR, GARCIN, and SHIMAZONO. Severe arteritis and partial destruction of the roots of cranial nerves by the inflammatory process may be found. In these lesions the fungus grows mainly in the form of hyphae but spherical forms are also present.

In the cerebral form of the disease the brain is mainly affected, but cerebellum and brain stem may also be involved. The lesion occurs slightly more frequently on the left than on the right side, and it may be single or multiple, varying in size from a few milimeters to 5 cm in diameter. It appears as a firm pigmented nodule, sometimes with a necrotic or purulent center, or as a poorly limited necrotic or abscessed area surrounded by an irregular area of pigmented granulation tissue (Figs. 10 and 11); a true capsule is rarely found. Pigmentation is due in part to the pigment of the fungus itself; it has been described as brown, greenish, grey, or black-red but it may be so intense as to be taken for metastasis of a melanoma (BANTI, WAROT). Microscopic structure of the lesion is that of a granuloma with abundant giant cells, polynuclear and eosinophil leucocytes, and lymphocytes.

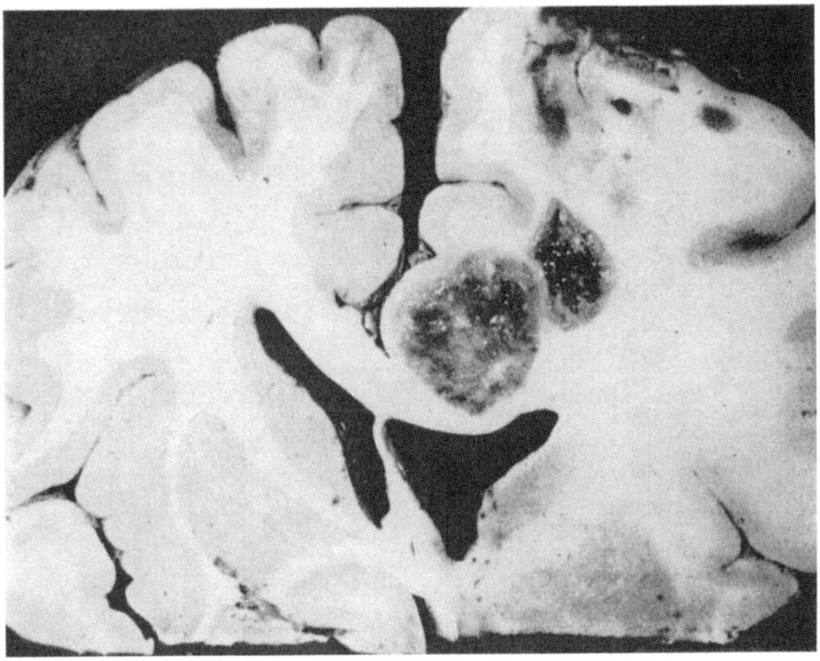

Fig. 10. Cerebral lesions produced by *C. bantianum (trichoides)*. Case of King and Collette. Multiple abscesses and distortion of ventricular system

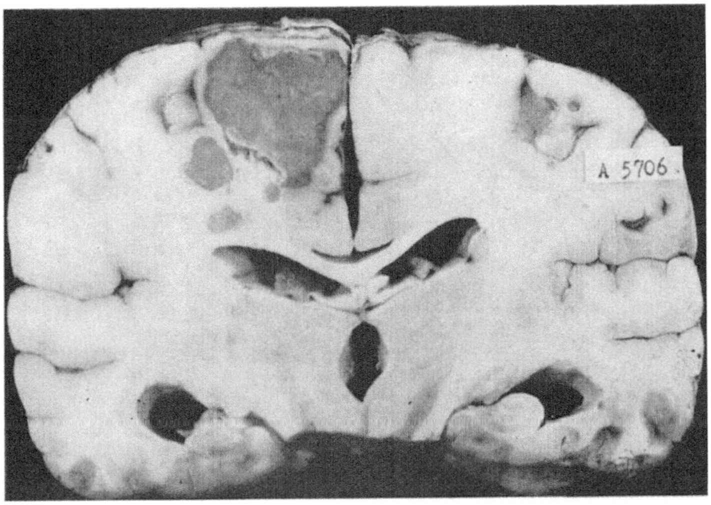

Fig. 11. Lesions caused in the brain by *Hormodendrum dermatitidis*. Case of Tsai et al.

Tubercle formation is uncommon but microabscesses and areas of necrosis are found scattered in the granulomatous tissue (Fig. 12). Proliferation of glia cells and areas of scar tissue are found in older lesions (Fig. 13). Condensation of glia and collagen towards the periphery may give the appearance of a true capsule but this usually fails to confine the growth of the fungus. Occasionally the core of the

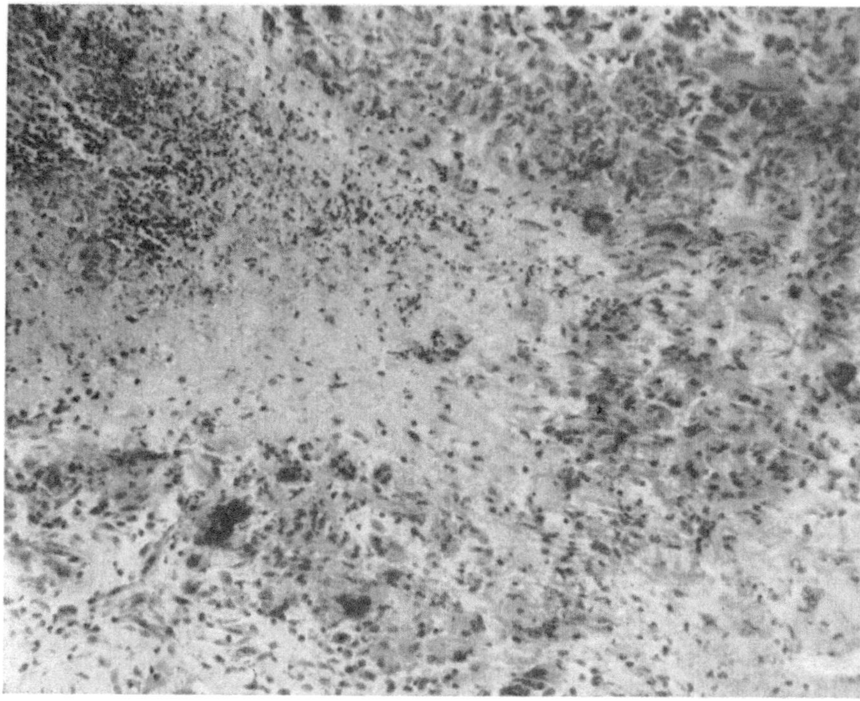

Fig. 12. Cerebral lesions of *C. trichoides*. Purulent exudate (upper left) is confined by necrotic tissue (center left); this area is surrounded by granulomatous inflammation with many giant cells. H & E, ×150

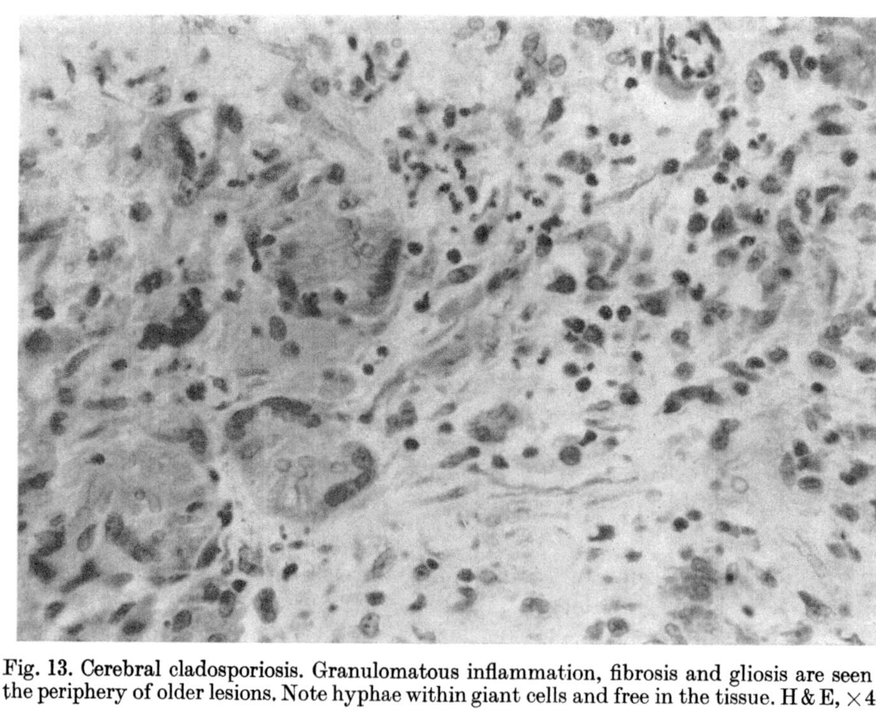

Fig. 13. Cerebral cladosporiosis. Granulomatous inflammation, fibrosis and gliosis are seen in the periphery of older lesions. Note hyphae within giant cells and free in the tissue. H & E, ×450

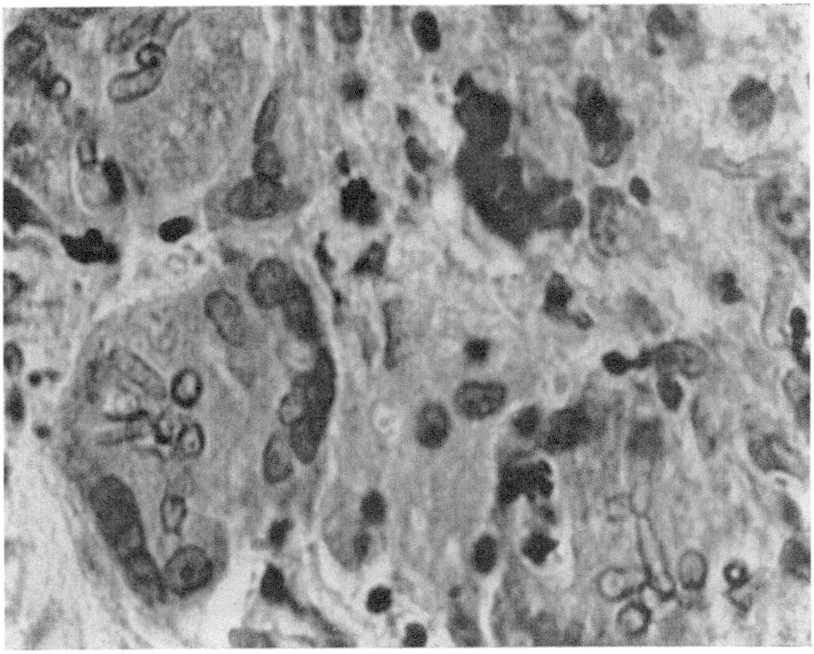

Fig. 14. Detail of Fig. 13 showing the fungus within giant cells. H & E, ×1250

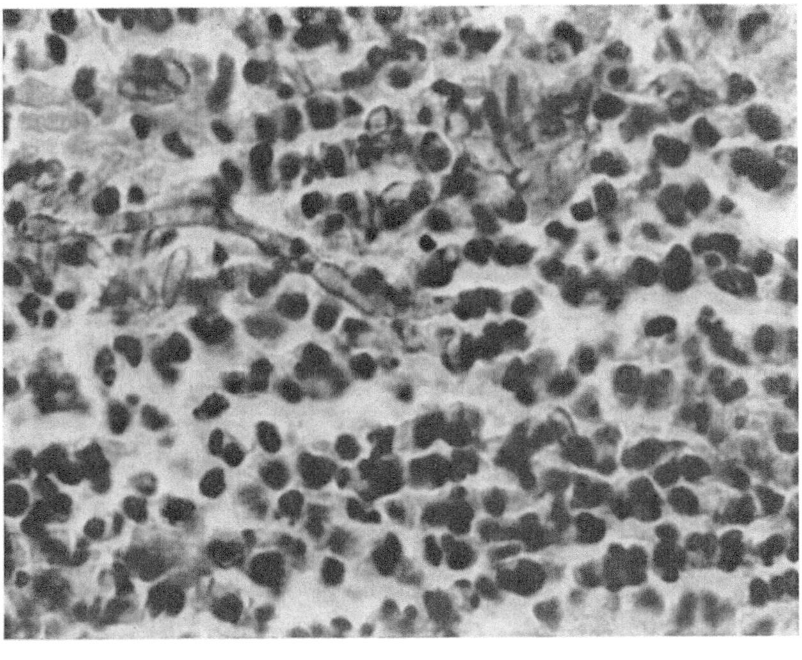

Fig. 15. Septate hyphae of *C. trichoides* in a microabscess. Brain lesion. H & E, ×1250

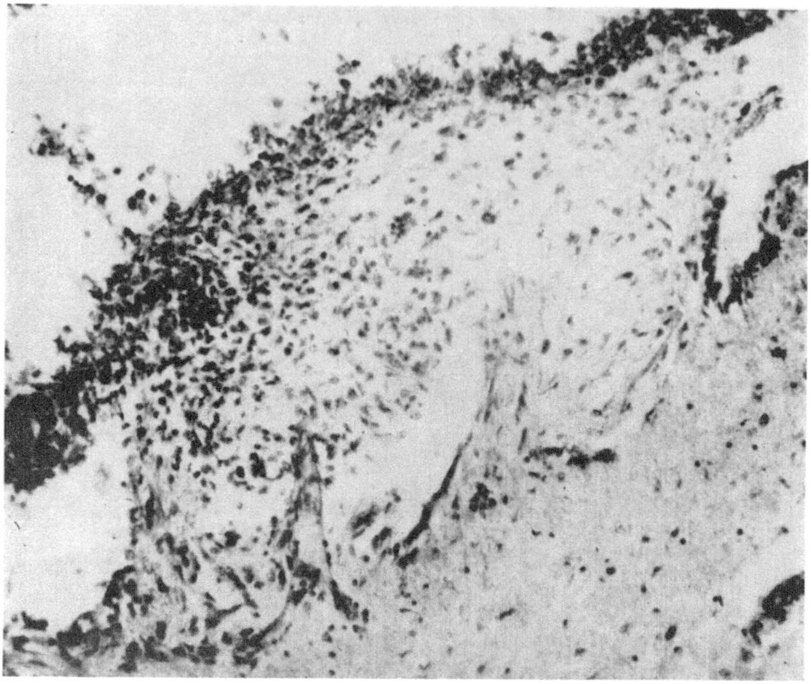

Fig. 16. Cerebral cladosporiosis. Proliferative ependimitis. H & E. ×200

lesion is surrounded by necrotic brain tissue permitting the surgeon to "shell-out" the granuloma as in the cases of BINFORD, and KING and COLLETTE. Hyphae are abundant in the lesions; they are seen within the giant cells (Fig. 14), or free in the tissue (Fig. 15), clearly septate, pigmented, measuring about 4 μ in diameter. Branching is rarely seen in tissue sections but it is easily demonstrable in fresh preparations. The large spheric elements ("sclerotic cells", "chlamydospores") are usually less numerous, although they were predominant in the case of FRANCA-NETTO. Granulomatous and proliferative inflammation of the ependyma and choroid plexus is found frequently, accompanying the cerebral lesion (Figs. 16 and 17).

Clinical Picture: The disease may have a sudden onset and a rapid course of less than a month. More commonly, symptoms appear gradually and the process extends for a period of from several months to two years. Symptoms are those of intracranial hypertension, space occupying lesion, and meningismus. Irregular fever is common; headache is mentioned in the majority of the cases, sometimes accompanied by nausea, vomiting and vertigo. Paralysis of the face and extremities develops in most cases, preceded by a period of sudden or gradual paresis. Papillary edema has been mentioned in 9 patients, and loss of weight in 12. In one third of the cases there have been convulsions. The symptoms of meningitis are frequently obscured by signs of cranial nerve involvement. In 14 cases the cerebrospinal fluid was examined; it was normal in 5. In the rest it was described as turbid, pressure variable, proteins generally increased, but data on sugar and chloride was inconsistent. Cells were generally increased, up to 4,450 per cmm, mostly polymorphonuclear leucocytes. The fungus has never been seen in the spinal fluid and its inoculation has been negative. In only one instance (case of

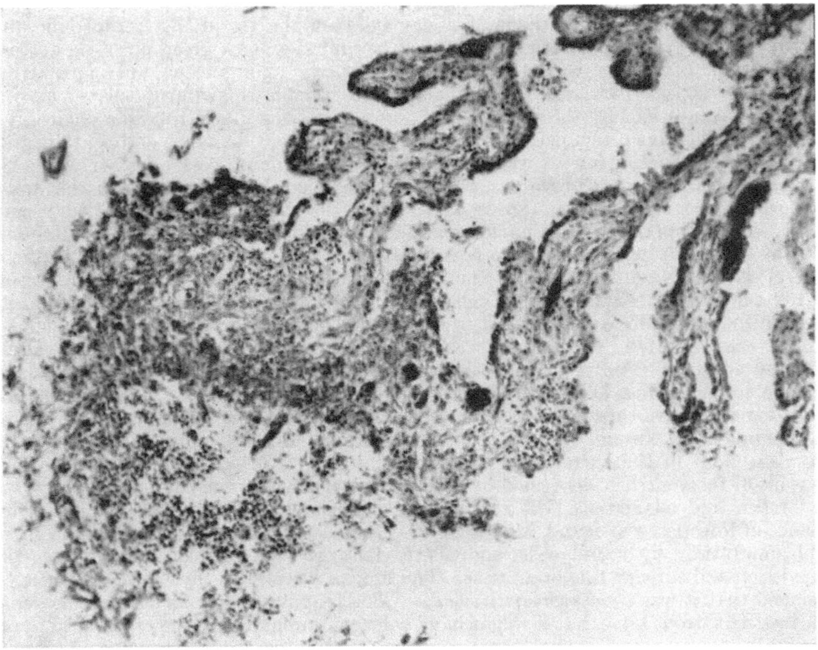

Fig. 17. Section of choroid plexus revealing involvement with granulomatous inflammation. Case of meningo-encephalitis caused by *C.trichoides*. H & E, ×200

DASTUR, studied by DESAI) has the fungus been isolated from the cerebrospinal fluid. Leucocytosis of the peripheral blood is frequently observed, with counts as high as 21400 per cmm, with an increased number of neutrophils. Ventriculograms and carotid angiography appear to be most succesful diagnostic procedures.

It is clear from this brief description that cladosporiosis has no specific diagnostic features that permit it to be separated from other cases of brain abscess or meningitis of different etiology. Cladosporiosis is a serious disease, 25 of the cases having ended in death. The patient of BINFORD was seen again two years after operation, free from symptoms, and this is the only case that could be considered as cured. COUDERT'S patient was well 10 months after operation, and in three other cases the patients were discharged in good condition but were not seen again. No effective therapy is available; penicillin, wide spectrum antibiotics, mycostatin, amphotericin B, izoniazide, chloramphenicol, cortisone, and potassium iodide have been tried without succes.

Summary of Illustrative Cases

Case of KING and COLLETTE (*Cladosporium trichoides*)

A 47-year-old white laborer was admitted with severe headaches. A short time after the onset of purulent drainage from the left ear the patient experienced headaches in the occipital and parietal regions. It was then noted that his speech was becoming inarticulate. A month after the onset of symptoms, weakness of the right extremities developed. Physical examination disclosed arteriosclerosis and a blood pressure of 156/100; dulled mental state and a motor speech aphasia were noted. There was also moderately severe right hemiparesis with Babinsky response. Laboratory studies were within normal limits. A left carotid angiogram showed an avascular area in the left frontal lobe. Puncture of the right ventricle yielded 25 ccm of clear fluid; only a few drops of fluid could be obtained from the left ventricle. Roentgenograms disclosed a mass in the frontal lobe near the midline.

A left frontal flap was reflected. The brain was not under tension. A firm mass was palpated subcortically near the midline about halfway between the tip of the frontal lobe and the Rolandic fissure. Needle aspiration of the mass produced a thick green pus. The abscess was shelled out in toto. After the operation speech became spontaneous and more articulate. There was no actual paralysis but the patient was unable to perform isolated movements of the right arm or leg. By the time of discharge the patient was sitting in a chair but could not walk. The specimen removed at operation was 3.5 cm in diameter and contained a light green pus from which a fungus was grown. The wall of the abscess was 3 cm in thickness. Microscopic examination showed a marked glial reaction with a rich round cell and some polymorphonuclear infiltration. Three weeks later the patient was readmitted because of convulsions; he was drowsy and could not talk. The right hemiplegia was now profound, the neck was stiff; early papilledema was present. Leucocyte count was 13,800 with a differential count of 82% polymorphonuclears. Spinal fluid contained 6 cells per cmm, 60 mg protein and 102 mg of sugar. Bacteriological cultures remained sterile. Ventriculogram demonstrated a mass in the posterior part of the left frontal lobe. About 20 ccm of a thick gelatinous green material was aspirated. The patient's course was one of steady decline. The temperature fluctuated from 98°—102°F. The patient expired two weeks after his second admission. Penicillin and potassium iodide were given without any noticeable effect. Postmortem findings included marked arteriosclerotic cardiovascular disease. In the brain a large hemorrhagic area was present at the site of enucleation of the original abscess; multiple daughter abscesses were also found in the surrounding white matter (Fig. 10), with little capsule formation. The walls of the ventricle were lined by a thin layer of pus and the choroid plexus on the left was swollen and edematous. The middle ears and paranasal sinuses were opened but no evidence of infection was found. Microscopically, besides a moderate glial reaction, there was a mild infiltration by round cells, and many giant cells were noted. Hyphae were present in the lesion and surrounding brain tissue. The fungus was studied by Dr. C. W. EMMONS who concluded that it was *Cladosporium trichoides*. This fungus was grown also in cultures taken from the left petrous bone, but it might have been contaminated by the removal of the brain.

Case of FUKUSHIRO et al. (*Hormodendrum pedrosoi*)

An 8-year-old Japanese girl consulted for verrucous lesions of the skin of the nose and right cheek of a year's duration. Previously she had suffered from tinea corporis caused by *Trichophyton ferrugineum*. She was in good physical condition. Tuberculin, trichophytin, and sporotrichin tests were negative, but she reacted with chromoblastomycetin antigen. Biopsy of the skin lesions revealed typical changes of chromoblastomycosis. *Hormodendrum pedrosoi* grew in cultures taken from these lesions. Three months after her first visit the girl developed neurological symptoms consisting of headaches, fever, nausea, and facial paresis that progressed to left facial paralysis and, two months later, to paralysis of the four extremities, accompanied by dysphasia. Leucocyte count was 12,000, sedimentation rate 65 mm, spinal fluid clear with pressure of 20 mm, Pandy test positive, and 10 cells per cmm; cultures of the fluid were negative. The patient's condition worsened steadily and she died of respiratory paralysis 9 months after the onset of neurological symptoms. Chlorophenol and sulphanilamide were administered without improvement. At autopsy multiple "black-green granulomatous tumors" were found in the basal meninges, cerebral hemispheres, cerebellum and pons. The larger one of the nodules, about 4 cm in diameter, was found in the left internal capsule. *Hormodendrum pedrosoi* pathogenic for mice was recovered from these lesions. Microscopically the lesions consisted of a granulomatous inflammation with many giant cells and polymorphonuclear leucocytes, and necrosis. Brown hyphae and spheric forms of the fungus were abundant in the lesions.

Case of TSAI et al. (*Hormodendrum dermatitidis*)

A 19-year-old female student from Taiwan became icteric in September 1961; jaundice fluctuated thereafter, and in March 1963 several swellings in her neck were observed. Streptomycin was given without benefit. Five months later, soreness and numbness developed in her right leg. On admission she exhibited mild jaundice; lymph nodes were palpated in the right side of her neck, and there was hepato-splenomegaly. White cell count 8,450 with 10% eosinophils. Liver tests revealed moderately altered hepatic function. Tuberculin was weakly positive; sedimentation rate was 120 mm. Lumbar puncture yielded clear fluid with a pressure of 120 mm, 23 cells predominantly lymphocytes, total protein 20 mg, sugar 60 mg, chloride 123 mEq. Cholecystography failed to visualize the gall bladder or bile ducts. Weakness of her right leg became progressive and Jacksonian seizures of the right hand developed. A lymph node biopsy revealed granulomatous inflammation with brown hyphae in the giant cells. *Hormodendrum dermatitidis* was obtained from a lymph node culture. Amphotericin B therapy was begun but because of severe side effects prednisolone was administered. In spite of treatment her general condition deteriorated progressively. Anemia increased, serum bilirubin rose to 11 mg and a high fever developed. In November she complained of abdominal pains, and this was followed by signs of acute peritonitis which became localized with abscess formation. Fungus was not seen in this abscess. A needle biopsy of the liver disclosed a few granulomatous lesions with multiple pigmented hyphae. In addition to the Jacksonian seizures and bilateral pyramidal signs, symptoms of involvement of cranial nerves appeared. An E.E.G. revealed generalized cortical abnormalities and severe involvement of the left central and parietal portions. She became comatose and died in April 1964. Iodide and Mycostatin therapy did not alter the course of the disease. Autopsy: no lesions were found in the lungs; fibrinopurulent exudate and adhesions were found between the peritoneum and the abdominal viscera, without perforation of hollow organs. The liver was cirrhotic, with white scarred patches, and intrahepatic pigment stones were found. Marked fibrosis and miliary brown spots were seen in the intrapancreatic portion of the common bile duct. Almost all the lymph nodes of the body presented yellow brown nodules. Over the leptomeninges of the spinal cord, brain stem and hypothalamic portions there were numerous raised black or red-black patches measuring up to 1 cm in diameter. Multiple green yellow nodular lesions were found in the cerebrum (Fig. 11), and cerebellum. The largest of these lesions, 7 cm in diameter, was found near the superior frontal pre- and post central gyri of the left hemisphere. Microscopically, the lymph node lesions consisted of granulomas with yellow brown hyphae and round structures of the fungus; this was found also in the lesions around the bile duct. Histologic changes in the brain consisted of conglomerate granulomas with or without fibrous capsule, and necrotic areas. Real abscesses were seen in a few daughter nodules adjacent to the large lesion in the mid brain; pigmented hyphae and some round bodies were found in these lesions. The ependyma of the fourth ventricle was covered by a thick layer of granulomatous tissue. The exuberant growth of the fungus in the meningeal lesions resembled a pure culture of the organism. Granulomatous angiitis in vessels invaded by the fungus was observed. *Hormodendrum dermatitidis* was grown from the various lesions. It produced abscesses in the brain, liver and spleen of mice inoculated intravenously.

Acknowledgements: The author wishes to express his appreciation to the Williams and Wilkins Company, of Baltimore, publishers of the American Journal of Clinical Pathology for their kind permission to use the pictures in the papers by Dr. Tsai, and by the author, for illustrations N° 2, 3, 4, 9, 11, 13, 16, and 17. The Editors of the Johns Hopkins Medical Journal graciously granted their permission to the author to use one of the brain photographs of Dr. King's paper for illustration N° 10. — Professor R. Fukushiro and Dr. Y. Shimazono, of Kanazawa University, provided the author with the beautiful photograph of the brain for illustration N° 8. Dr. A.B. King, of Sayre, Pennsylvania, kindly allowed the author to use the picture for illustration N° 10. Dr. C.Y. Tsai, of the National Taiwan University graciously furnished the author with the potograph for illustration N° 11. To all these distinguished colleagues the author is deeply indebted.

References

Al Doory, Y., Gordon, M.A.: Application of fluorescent antibody procedures to the study of pathogenic dematiaceous fungi. J. Bact. **86**, 332 (1963). — Aufdermaur, M., Piller, M., Fischer, E.: Sporotrichose des Hirns. Schweiz. med. schr. **84**, 167 (1954). — Azulay, R.D., Serruya, J.: Hematogenous dissemination in chromoblastomycosis. Arch. Derm. **95**, 57 (1967). — Bagchi, A., Aikat, B.K., Barua, D.: Granulomatous lesion of the brain produced by *Cladosporium trichoides*. J. Indian med. Ass. **38**, 602 (1962). — Barnola, J., Angulo, A.: Cladosporiosis profunda. Mycopathologia (Den Haag) **15**, 422 (1961). — Binford, C.H., Thompson, R.K., Gorham, M.E., Emmons, C.W.: Mycotic brain abscess due to *Cladosporium trichoides*, a new species. With mycologic report by C.W. Emmons. Amer. J. clin. Path. **22**, 535 (1952). — Bobra, S.T.: Mycotic abscess of the brain probably due to *Cladosporium trichoides*. Report of the fifth case. Canad. med. Ass. J. **79**, 657 (1958). — Bonne, C., Bros, G., Ver Haart, W.J.C.: Localized lesions of the meninges and the brain caused by a brown fungus with septate hyphae of unknown nature. Med. Maandbl. (Batavia) **1**, 465 (1948). — Borelli, D.: Cladosporiosis profunda. Revisión de una cepa. Bol. venez. Lab. Clin. **1**, 29 (1948). ~ *Torula bantiana*, agente di un granuloma cerebrale. Riv. Anat. pat. **17**, 615 (1960). — Chevrel, M.L., Javalet, A., Doby, J.M., Doby-Dubois, M., Louvet, M.: A propos d'une observation de cladosporiosis cérébrale. Ann. Anat. path. **7**, 607 (1962). [This same case was also studied by Doby-Dubois, M. et al.: Ann. Parasit. hum. comp. **37**, 644 (1962)]. — Coudert, J., Allegre, G., Tomassi, M., Battesti, M.R.: Un cas lyonnais de mycose cérébrale due á *Cladosporium trichoides* (Emmons, 1952). Etude anatomo-clinique et mycologique. Rev. lyon. Méd. **11**, 847 (1962). [This case was also studied by Tommasi, M. et al.: Arch. Anat. path. **12**, 48 (1964)].— Dastur, H.M., Chaukar, A.P., Rebello, M.D.: Cerebral chromoblastomycosis due to *Cladosporium trichoides* (bantianum). I. A review and case report. Neurology (Bombay) **14**, 1 (1966) (See DESAI). — DESAI, S.C., BHATIKAR, M.L., MEHTA, R.S.: Cerebral chromoblastomycosis due to *Cladosporium trichoides* (bantianum). II. Mycopathologic investigation of brain abscess and skin involvement. Neurology (Bombay) **14**, 6 (1966). — DEREYMAEKER, A., DE SOMER, P.: Arachnoidite fibrinopurulente cérebelocervical due á une moisissure (*Cladosporium*). Acta neurol. belg. **55**, 629 (1955). — DOBY-DUBOIS, M., CHEVREL, M.L., DOBY, J.M., ROBERT, Y.: Abcès mycosique cérébral humain par *Cladosporium trichoides*. Ann. Parasit. hum. comp. **37**, 644 (1962). — DUQUE, O.: Cladosporiosis cerebral experimental. Rev. lat.-amer. Anat. pat. **7**, 101 (1963). ~ Meningoencephalitis and brain abscess caused by *Cladosporium* and *Fonsecaea*. Review of the literature, report of two cases, and experimental studies. Amer. J. clin. Path. **36**, 505 (1961). — EMMONS, C.W., BINFORD, C.H., UTZ, J.P.: Medical Mycology. Chapter 27, Cladosporiosis, pp. 330 to 338. Philadelphia: LEA & FEBIGER 1963. — FELGER, C.E., FRIEDMAN, L.: Experimental cerebral chromoblastomycosis. J. infect. Dis. **111**, 1 (1962). — FERNÁNDEZ-BAQUERO, G., REAUD, B.: Cultivo del *Hormodendrum pedrosoi* del material obtenido por lavado bronquial en cuatro enfermos de cromoblastomicosis. Bol. Soc. Cuba. Derm. Sif. **17**, 99 (1960). — FERNÁNDEZ-BAQUERO, G., BARQUIN-LOPEZ, P., REAUD-LESCAY, B.: Cromoblastomicosis experimental. Bol. Soc. cuba. Derm. Sif. **18**, 19 (1961). — FRANCA-NETTO, A.S., DE BRITTO, T., DE ALMEIDA, F.P.: Cromomicose do sistema nervoso. Etudo anatomoclinico de um caso. Arch. Neuropsiquiat. (S.Paulo) **11**, 265 (1953). — FUKUSHIRO, R., KAGAWA, S., NISHIYAMA, S., TAKAHASHI, H.: Un cas de chromoblastomycose cutanée avec métastase cérebrale mortelle. Presse méd. **65**, 2142 (1957). [This same case was reported by several Japanese authors: NAGAHATA, M.: Pediatr. Univ. (Tokyo) **2**, 75 (1958). NISHI, M.: Trans. Soc. path. jap. **45**, 566 (1956). OKUDAIRA, M.: Acta path. jap. **6**, 207 (1956)]. — GARCIN, R., MARTIN, R., BERTRAND, I., GRUNER, J., TOURNEUR, R.: Mycose méningo-épendymaire. Etude anatomo-clinique. Presse méd. **57**, 1201 (1949). — GARCIN, R., MARTIN, R., MANINGAND, G., BERTRAND, I., GODLEWSKI, S., SUREAU, B.: Mycose cérébrale á *Cladosporium trichoides*. Sem. Hôp. Paris **33**, 2282 (1957). [The fungus from this case was studied also by SEGRETAIN, G. et al.: Ann. Inst. Pasteur **89**, 465 (1955)]. — HORN, I.H., WILANSKY, D.L., HARLAND, W.A., BLANK, F.: Neurogenic hypernatremia with mycotic brain granuloma due to *Cladosporium trichoides*. Canad. med. Ass. J. **83**, 1314 (1960). — IWATA, K., WADA, T.: Mycological studies on the strains isolated from a case of chromoblastomycosis with metastasis in central nervous system. Jap. J. Microbiol. **1**, 355 (1957). — KING, A.B., COLLETTE, F.S.: Brain abscess due to *C.trichoides*. Report of the second case due to this organism. Bull. Johns Hopk. Hosp. **91**, 298 (1952). — LUCASSE, C., CHARDOME, J., MAGIS, P.: Mycose cerebrale par *Cladosporium trichoides* chez un indigène de Conge Belge, et note mycologique sur *Cladosporium trichoides* EMMONS (1952), par R. Vanbreuseghem. Ann. Soc. belge Méd. trop. **34**, 475 (1954). — MANSON, M.D.E.: Chromoblastomycotic brain abscess in a South African Bantu. Report of a case. S. Afr. J. Lab. clin. Med. **4**, 283 (1958). — MCGILL, H.C., BRUECK, J.W.: Brain abscess due to *Hormodendrum* species. Arch. Path. **62**, 303 (1956). — RILEY, O., MANN, S.H.: Brain abscess caused by *C.trichoides*. Review of 3 cases and report of fourth case. Amer. J. clin. Path. **33**, 525 (1960). — SACCARDO, P.A.: Torula fungine bantiana. Ann. Mycol. **10**, 320 (1912) (reproduced in STIGLIANI, q.v.). — SASANO, N., OKAMOTO, T., TAKAHASHI, T., SUZUKI, S.: An autopsy case of primary chromoblastomycosis arising from the internal organs. Dark brown granulomas in the liver and the brain without skin symptoms, observed in a smoking child 3 years old. Tohoku J. exp. Med. **73**, 180 (1961). [Published originally in Japanese: Trans. Soc. path. jap. **47**, 400 (1958)]. — SEGRETAIN, G., MARIAT, F., DROUHET, E.: Sur *Cladosporium trichoides* isolé d'une mycose cérébrale. Ann. Inst. Pasteur **89**, 465 (1955). — SHIMAZONO, Y., ISAKI, K., TORII, H., OTSUKA, R., FUKUSHIRO, R.: Brain abscess due to *Hormodendrum dermatitidis* (Kano) Conant 1953. Report of a case and review of the literature. Folia psychiat. neurol. jap. **17**, 80 (1963). [This case was reported also by FUKUSHIRO, R.: Saishin Igaku **16**, 580 (1961)]. — STIGLIANI, R.: Particolaritá istologiche della torulose encefalica nei preparati originali Bantiana della prima osservazione conosciuta. Arch. De Vecchi Anat. pat. **36**, 329 (1961). — SUGAWARA, M., SOBAJIMA, Y., TAMURA, H.: A case of generalized chromoblastomycosis. Acta Path. jap. **14**, 239 (1964). — SYMMERS, W.ST.C.: A case of cerebral chromoblastomycosis (Cladosporiosis) occurring in Britain as a complication of polyarteritis treated with cortisone. Brain **83**, 37 (1960). — TSAI, C.Y., LÜ, Y,C., WANG, L.T., HSU, T.L., SUNG, J.L.: Systemic chromoblastomycosis due to *Hormodendrum dermatitidis* (Kano) Conant. Amer. J. clin. Path. **46**, 103 (1966). — WAROT, P., GALIBERT, P., MEIGNIE, S., DELANDTSHEER, J.M., PETIT, H.: Mycose cérébrale á symptomatologie tumorale (*Cladosporium trichoides* probable). Rev. neurol. **105**, 489 (1961). — WATSON, K.C., LINES, G.M.: Brain abscess due to the fungus *Hormodendrum*. S. Afr. med. J. **31**, 1081 (1957). — WATSON, K.C.: Cerebral chromoblastomycosis. J. Path. Bact. **84**, 233 (1962). — WYBEL, R.E.: Mycosis of cervical cord following intrathecal penicillin therapy. Report of a case simulating cord tumor. Arch. Path. **53**, 167 (1953).

Subcutaneous Abscesses due to Brown Fungi

HERBERT ICHINOSE, New Orleans/Louisiana USA

With 19 Figures

Introduction

The subcutaneous abscesses described in this chapter are apparently caused by members of the same general group of organisms that also give rise to chromoblastomycosis. However, the two diseases show distinct differences in clinical manifestations and biologic course. In addition, the organisms in the lesions of chromoblastomycosis are almost exclusively yeast forms in contrast to the organisms in the subcutaneous abscess which characteristically exhibit dimorphism.

Definition of the Disease

The chronic subcutaneous abscesses caused by pigmented fungi are solitary lesions covered by normal epidermis. The lesions are often misdiagnosed clinically and the causative organisms have rarely been cultured. The infection remains localized and the lesions are readily cured in most instances by surgical excision. The patients frequently have unrelated serious diseases.

Cases of Subcutaneous Abscesses

The following is a listing of reported cases of such subcutaneous abscesses from 1908 to the present in literature available to the author:

YOUNG, J.M. and ULRICH, E., Memphis, Tenn./USA (1953)	1 case
MEAD, M. and RIDLEY, M.F., Brisbane, Australia (1957)	1 case
NITYANANDA, K., Colombo, Ceylon (1962)	1 case
MARIAT, F., et al., Paris, France (1967)	1 case
KEMPSON, R.L. and STERNBERG, W.H., New Orleans, La./USA (1963)	7 cases

Two other cases (DUKE nos. 2492 and 3326) included in an antigenic study by NIELSEN and CONANT have not been reported in detail. Five additional cases have been made available to the author by several pathologists including Dr. R.J. REED, Dr. F. HARRIS, and Dr. W.H. STERNBERG. The pathologic and clinical data to be presented have been derived largely from these 5 cases (ICHINOSE, unpublished data) and a restudy of the 7 cases reported by KEMPSON and STERNBERG.

Mycology

Organisms from the lesions have rarely been cultured. In two instances, the organisms were identified as *Hormiscium dermatitidis* (NIELSEN and CONANT; SCHNEIDAU). The preliminary report on organisms from another lesion was *Cladosporium sp.* (SAMUELS, personal communication.) In two other cases, the organisms were classified as *Sporotrichum gougerotii* by YOUNG and ULRICH and *Phialophora gougerotii* by MARIAT et al. The latter authors consider their isolate identical to that reported by YOUNG and ULRICH. They also refer to BORRELLI's opinion that this organism is synonymous with *H.dermatitidis*. It should also be noted that CARRION and SILVA studied the organisms from the lesion reported by YOUNG and ULRICH as well as organisms from two other sources and reclassified *Sporotrichum gougerotii* as *Cladosporium gougerotti*. These opinions indicate that the lesions are caused by closely related fungi. Because of the small number of lesions from which organisms have been cultured, it is premature to assume that similar lesions might not be caused by other organisms.

Clinical

The patients were adults usually in their fifth or sixth decade. About half of the patients had some serious systemic disease unrelated to the fungal infection. Three patients had mild diabetes mellitus and two of these had severe hypertension and congestive heart failure. The third patient had severe cerebral atherosclerosis, three cerebral vascular accidents, and atherosclerosis obliterans of the legs. A fourth patient had minor infections of axillary sweat glands and a strong family history of diabetes mellitus. Another patient had dysglobulinemia, long standing lymphadenopathy, and serous cavity effusions. The patient reported by YOUNG and ULRICH was being treated for pulmonary tuberculosis. Similarly, the patient described by MARIAT et al. suffered from bilharziasis and the fungus disease was thought to be an example of an "opportunistic" infection. In this regard, the statement by KEMPSON and STERNBERG is appropriate: "Although the series of cases complaine is too small for definite conclusions, it seems that subcutaneous chromoblastomycosis infections may tend to occur in patients with metabolic and debilitating diseases."

No important symptoms were attributed to these infections. The patients frequently regarded the lesions as small calluses and brought them to the attention of the physician for cosmetic reasons.

The subcutaneous abscesses probably developed from inconspicuous puncture wounds made by contaminated plant material. Wood splinters were present in three of the lesions and a fourth contained what appeared to be a portion of a thorn. The rapid healing which is characteristic of small skin punctures, in all likelihood, accounts for the normal epidermis overlying the lesions. Several of the patients lived in rural areas and engaged in agricultural work. Another patient raised flowers including roses as an avocation. The onset and duration of the lesions could rarely be accurately stated by the patients. The shortest reported duration was four days and the longest was two years.

On physical examination the lesions were situated in the distal parts of the upper extremities including the volar surfaces of the fingers and palm, the dorsum of the hand, the wrist and the forearm. Less commonly, the lesions were on the lower extremity, e.g., dorsum of the foot (NITYANANDA), the ankle, the leg near the lateral malleolus and the knee. The lesions were discrete solitary masses clinically situated in the subcutis. The overlying epidermis was normal in all instances, and no ulceration or sinus tracts were observed. An exception to this generalization was a sinus tract that developed between the lesion and the skin surface after the lesion had been aspirated for the second time (YOUNG and ULRICH). The lesions were less than 2 cm in diameter. They frequently exhibited a soft fluctuant consistency and, therefore, were commonly misdiagnosed as sebaceous cysts or lipomas. When located near joints, the possibility of Baker's cyst or ganglion was occasionally considered. Sometimes, the lesions were firm giving some basis for a tentative diagnosis of giant cell tumor of tendon sheath. When a history of previous trauma to the area of the lesion could be elicited, the possibility of foreign body granuloma was also considered.

Gross Pathology

The typical specimen was a solitary cystic lesion surrounded by a fibrous capsule to which subcutaneous adipose tissue was adherent (Fig. 1). The cyst

1 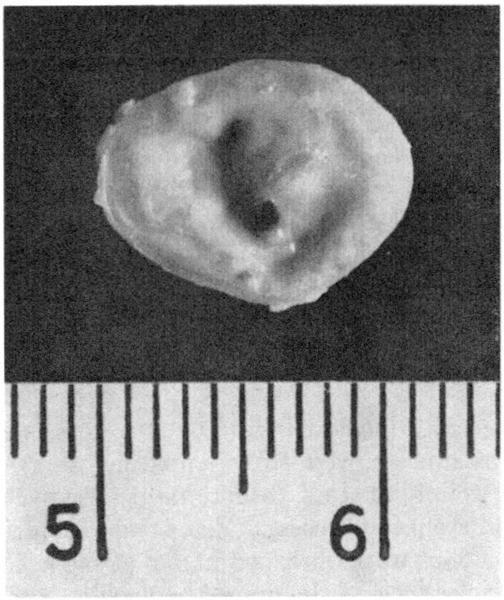

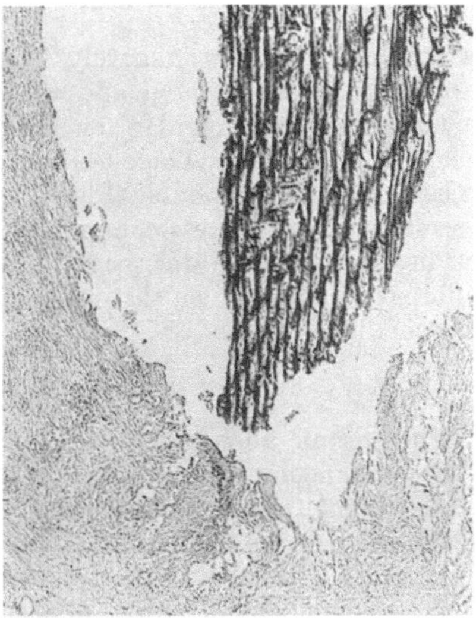 2

3 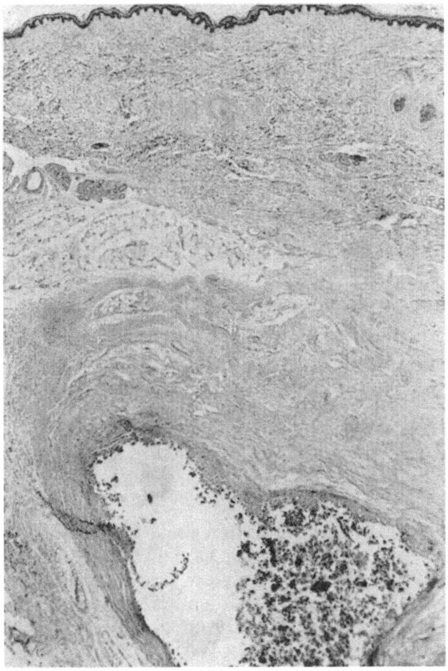

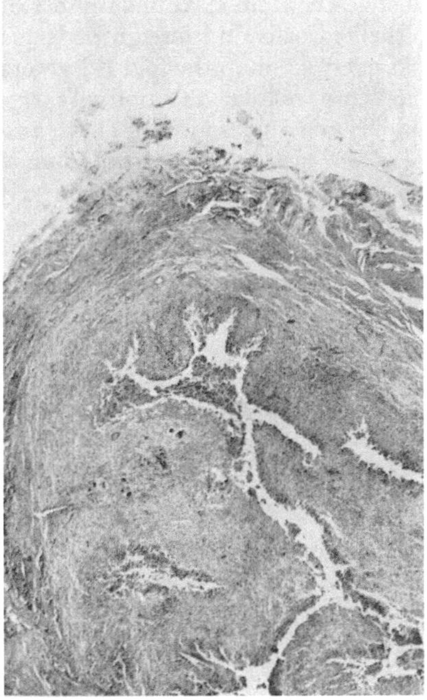 4

Fig. 1. Cross specimen, cut section surface: The specimen shows an abscess cavity containing whitish purulent exudate and showing a darker appearing granulation tissue lining. The lining is indistinct on the right side of the lesion. An intact fibrous capsule is present. The black circular body in the interior of the abscess cavity is a cut section of a sliver of plant material. Formalin fixed, ×4 approximately

Fig. 2. Foreign body: A foreign body is present in the abscess cavity and shows histologic features of plant tissue. Gridley stain, ×105

Fig. 3. Subcutaneous abscess, fluctuant stage: The lesion is situated in the subcutaneous layer of the skin and the overlying epidermis and dermis are normal. A fibrous capsule is present around the abscess. (Refer to Acknowledgements.) H & E, ×8

Fig. 4. Subcutaneous abscess, stellate abscess stage: The lesion shows a stellate outline of the abscess cavity with a festooned margin. Purulent exudate is present in the lumen. An intact fibrous capsule is present. H & E, ×15 cavity measured approximately 1 cm in diameter, was filled with viscid yellow exudate or cheesy material, and was lined by firm yellowish tan tissue. Rarely, the lining also showed granular areas of blackish discoloration probably representing colonies of organisms. Three lesions contained a sliver of foreign material believed to be a small wood splinter (Figs. 1 and 2) and one contained a foreign body that resembled a thorn. Less commonly, the lesion failed to show a cavity. In these instances, the nodule was made up of firm yellowish tan tissue contained within a fibrous capsule. When the overlying epidermis was also received, it showed no abnormalities.

Histopathology

The chronic abscess was characteristically located in the deeper dermis and adjacent subcutis. *Histologically*, the lesions showed three apparent stages of development: (1) tuberculoid phase, (2) stellate abscess, (3) fluctuant abscess. The combination of the fluctuant (Fig. 3) and stellate abscesses (Fig. 4) were typically found in most lesions although all three stages were observed in one of the lesions. In the tuberculoid stage, clusters of tubercles formed the main nodule (Fig. 8). The tubercles contained abundant organisms and were composed predominantly of epithelioid cells and foreign body giant cells. Small foci of coagulation and fibrinoid necrosis of epithelioid cells were present and were associated with a chronic and acute cellular reaction. As these foci of necrosis apparently increased in number and coalesced, a stellate abscess developed. The festooned outline of the lesion was still composed predominantly of epithelioid cells. The giant cells and occasional microabscesses were nearly always seen in that portion of the granulation tissue bordering the necrotic debris. This border showed sporadic fibrinoid and coagulation necrosis (Fig. 5). In this granulation tissue, capillaries showing prominent endothelial cells coursed toward the necrotic center. Occasional xanthoma cells were present and some lesions showed hyalinization of the granulation tissue (Fig. 6). With apparent progressive necrosis of granulation tissue, there was a loss of the stellate outline resulting in a more nearly spherical fluctuant abscess. Epithelioid cells continued to form a complete inner lining of the cavity which was considerably attenuated. The exudate filling the abscess cavity was composed predominantly of degenerating neutrophils. Occasional giant cells

---

Fig. 5. Abscess wall: The lining of the abscess is formed by granulation tissue rich in epithelioid cells and giant cells. A small microabscess is present near the upper part of the left margin of the picture. Fibrinoid necrosis of portions of the granulation tissue are present in the center and the left upper corner of the picture. Inflammatory cells, predominantly neutrophils, fill the abscess cavity and show varying stages of degeneration. The fibrous capsule is intact and exhibits mild chronic inflammation. H & E, $\times 105$

Fig. 6. Abscess wall: The granulation tissue lining shows hyalinization of the luminal edge in occasional lesions. A group of xanthoma cells are present in the lower right corner of the picture. A paucity of neutrophils and predominance of lymphocytes and other chronic inflammatory cells are noted. This particular lesion contained only a small number of organisms. H & E, $\times 360$

Fig. 7. Unstained section: The organism has an olive-brown color and is readily detected without staining. The aberrant hypha shows a "rounding out" of the second segment from the left. The fourth segment exhibits a similar change. These alterations are interpreted as stages in transformation to chlamydospores. Unstained, $\times 875$

Fig. 8. Subcutaneous abscess, tuberculoid stage: The nodular lesion is formed by a conglomeration of tubercules best demonstrated in the lower portion of the picture. It is contained within a fibrous capsule. Large numbers of organisms are present and appear as black dots of various sizes. The arrow is directed at one of the larger groups of organisms. PAS, $\times 83$

contained phagocytized foreign matter that resembled non-viable plant material. The chronic cellular infiltrate of leucocytes was predominantly composed of mature lymphocytes. Occasional aggregates of eosinophils were also present when the lymphocytic infiltrate was heavy. Only a few plasma cells were present. A dense collagenous capsule surrounded the lesion in all three developmental stages. No organisms were present in this fibrous capsule. In the tissue situated just exterior to the fibrous capsule, chronic inflammatory cellular reaction was present around blood vessels and was most prominent in the region deep to the lesion.

5
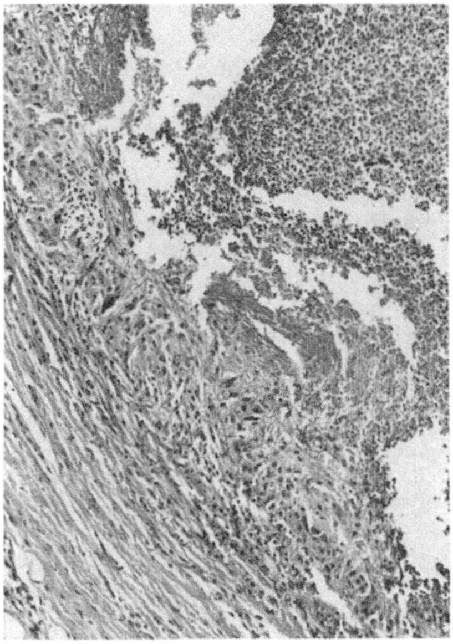

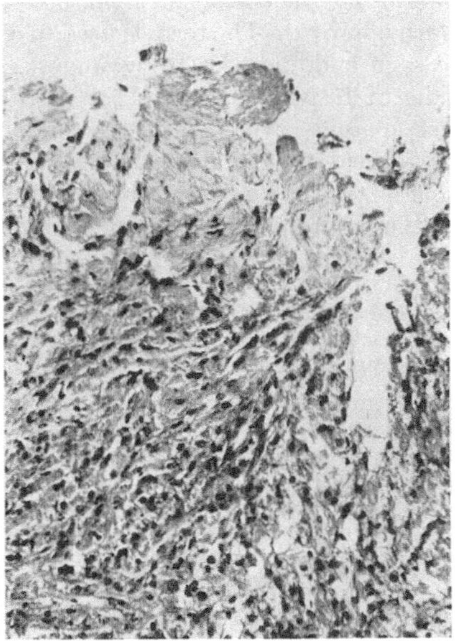
6

7
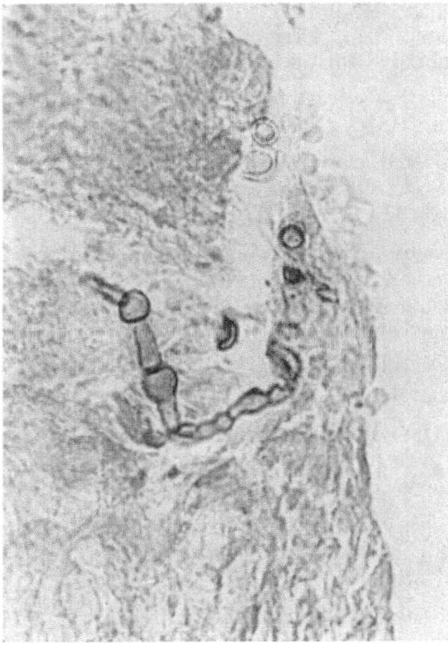

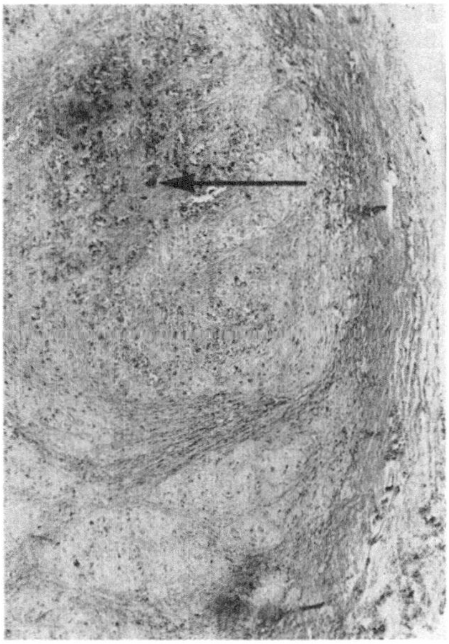
8

Figs. 5—8

Organisms in Tissue

Terminology: Since the tissue phase of these organisms has not been previously described in detail, appropriate terminology has not as yet been firmly established. Therefore, in the following description the application of certain terms is arbitrary and, likewise, somewhat conjectural. The terminology employed by others in describing the tissue phase of the pigmented fungi is used whenever it appears appropriate.

By definition, a "blastospore" is a "spore that arises by budding" (Snell and Dick). Since terminal bud-fission was frequently observed at the ends of chains of organisms, the use of the term, blastospore chains, was adapted to describe this pattern of growth. The term is used in cognizance of the nearly identical sequential process of budding with the youngest cell at the apex in the Homodendrum type conidia of *H. pedrosoi* and *P. gougerotii* in culture. However, in these chains the oldest spore is attached to a conidiophore and, therefore, can be classically distinguished from the blastospore chain. In a larger sense, even the conidial chains of these two organisms might be considered a form of blastosporulation since they share phenomenon of terminal spore formation with blastospore chains.

Fig. 9. A cluster of chlamydospores is present within a degenerating giant cell. Two spores in the upper left portion of the cluster are budding. These organisms and those in Fig. 10 were present in the exudate of a fluctuant portion of the abscess. PAS, ×835

Fig. 10. A group of organisms has formed branching chains. Most of the organisms have thick dark staining walls and are interpreted as chlamydospores having developed from blastospores. The chain forming the left upper extremity of this group of organisms shows branching near its apex and terminal bud formation. The cells from which these buds have arisen show thin walls and are smaller in size than the chlamydospore to which the cells are attached. Septation is present in the right extremity of this group of organisms. PAS, ×835

Fig. 11. A short chain of organisms is present and includes a chlamydospore (sclerotic cell). The three cells attached to the inferior aspect of the chlamydospore are smaller in size and are believed to be blastospores. PAS, ×835

Fig. 12. Three spores forming a chain are separated by thick transverse disjunctors. The largest cell (sclerotic cell) is probably the oldest and exhibits a thickened wall. It has a disjunctor in its upper extremity and probably broke off from a larger chain. The lower two cells are blastospores and the most inferior cell has probably been formed most recently. PAS, ×835

Fig. 13. A tubular type disjunctor separates the lowest cell from the cell situated in the middle of the chain. PAS, ×835

Fig. 14. An inverted L shaped structure is interpreted as a group of non-viable, degenerating organisms. Apparent shrinkage and distortion of overall shape are exhibited. The overall size is comparable to the organisms in Figs. 11, 12, and 13. Gridley, ×840

Fig. 15. A branching filament is partly enveloped by a coating with proteinaceous protuberances forming an asteroid body. The left upper extremity consists of three spores, the largest being a chlamydospore. The latter appears to have originated from a hypha and subsequently given rise to a short blastospore chain. Gridley, ×840

Fig. 16. A growth tube has developed from a large chlamydospore. The latter is attached by a tubular structure to two other spores. PAS, ×1410

Fig. 17. A chain of spores shows the member on the left extremity having formed a filament. Septation is present in the third and fourth cells from the right. The left portion of the picture was photographed at a different level of focus from the right part. PAS, ×1410

Fig. 18. Two hyphae have germinated from a fairly large chlamydospore. This particular germinating pattern has been frequently observed in epidermal scales and exudate from verrucous lesions of chromoblastomycosis. PAS, ×835

Fig. 19. Abundant hyphae are present. Several of these are extracellular in location. Aberrant filaments are more plentiful than "classic" hyphae. Some of the hyphae are believed to have formed after the specimen had been removed from the patient. PAS, ×692

Tubaki employs the term, blastospore, for the conidia of Homodendrum type in a recent treatise. He is of the opinion that the use of the term in this sense has been accepted by many workers.

Description: The pigmented fungal organisms in our material showed considerable morphological variation. In unstained sections (Fig. 7) as well as in

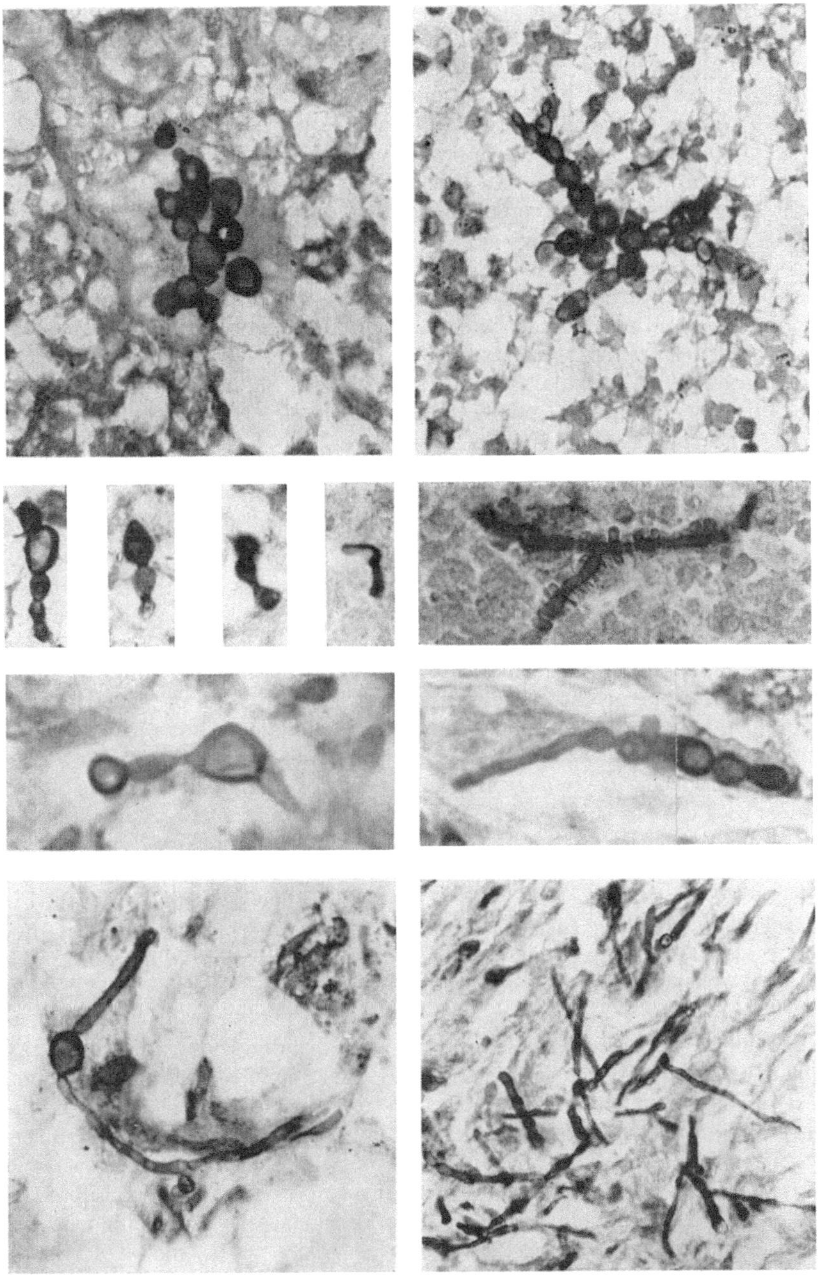

9 10 11 15 16 17 18 19

Figs. 9—19

sections stained with hematoxylin and eosin, the organisms exhibited an olive-brown color. With the PAS stain, the organisms were stained purple pink, and with Gridley stain, the organisms appeared dark purple. The Gomori methanamine silver stain was also employed to some advantage. Since organisms were plentiful (Fig. 8), more than 100 organisms and groups of organisms were available for study in all except two lesions. In the latter, only a few sparse organisms could be found even on careful study of all the available sections. Though the organisms were generally plentiful, the massive growths forming dense "mulberry clusters" in histologic sections of chorioallantoic membrane studied by Moore and the "black granules" found in a mycetoma-like lesion of the hand by Lewis et al. were not present in the subcutaneous abscess.

Chlamydospores, synonymous with sclerotic cells[1], were present in all instances and showed characteristic thickened walls and occasional septation. They were frequently grouped in conglomerate aggregations in giant cells. Some of the cells of a sclerotic cell aggregate occasionally gave rise to small buds (Fig. 9). More frequently, the sclerotic cells were attached to a chain of fairly uniform sized spherical fungal cells believed to be chains of blastospores. The chains occasionally were branched and propagated by sequential budding occurring at the ends of the chain (Fig. 10). Organisms propagating by budding have been previously observed in tissue sections of the subcutaneous abscess (Young and Ulrich). The branching chains of blastospores are identical to those described in culture (Carrion, 1950b). The spores which appeared to have been formed more recently were smaller in size in contrast to the older cells which were of the same size or larger than the parent cell (Figs. 10, 11, 12). Thick disjunctors were present, and structures interpreted as tubular disjunctors were also observed on rare occasion (Figs. 12, 13). Sclerotic cells and blastospore chains were the prevalent forms and were present in all lesions. Some of the blastospores in a chain had thickened cell walls and probably represent conversion to chlamydospores. Growth tubes and hyphae arose from chains as well as from isolated chlamydospores (Figs. 16, 17, 18). Such examples of germination are identical to those noted in epidermal scales and exudate of chromoblastomycosis lesions (Carrion, 1950a; Merrin; Da Fonseca). In the rare lesions containing abundant hyphae, the hyphae were commonly found in extracellular sites in the granulation tissue of the abscess wall (Fig. 19). This observation, as well as the finding of significant numbers of chlamydospores with growth tubes, suggests that at least some of the hyphae might have been formed after the lesions had been surgically excised. The hyphae were segmented, had fairly uniform width, and showed occasional branching, usually at acute angles. There were some fungal cells which appeared to be aberrant hyphae with segmants showing variable dimensions and tapering at points of junction (Figs. 7, 19). These aberrant hyphae were usually more abundant than the "classic" hyphae and were found predominantly in giant cells as were most of the sclerotic bodies and blastospore chains. Organisms were also present in the exudate filling the abscess cavities and were usually widely distributed. They frequently exhibited degenerative changes consisting of distortion, shrinkage, and loss of some of their chromatic characteristics (Fig. 14). In the hematoxylin and eosin preparations, the organisms stained dark purple, and in unstained sections they showed a loss of their natural olive-brown color. Other organisms in the exudate retained normal morphologic features and occasionally exhibited acidophilic encrustations enveloping their exterior surfaces representing one type of asteroid formation (Fig. 15).

1 These two terms are employed in the sense used by Carrion (1950a) with the realization that both are not consistent with classic terminology.

Summary and Comment

Chlamydospores (sclerotic cells) divided by septation as well as gave rise to chains of blastospores. Of these two methods of multiplication, blastosporulation was more frequently observed. Some of these spores showed conversion back to chlamydospores. Some of the aberrant hyphal forms may represent a similar adaptive phenomenon. In this regard, SILVA has demonstrated the development of chlamydospores from "rounded out" and thick-walled hyphal segments as well as from occasional conidia in organisms growing *in vitro* and *in vivo*. The chlamydospores and blastospore chains were the prevalent fungal cells and probably represent the most favorable adaptation to the adverse environment created by the host tissue reaction. Degenerating organisms were present in the exudate filling the abscess cavities. Single chlamydospores as well as those situated in blastospore chains gave rise to hyphae. At least some of the hyphae may have formed after the lesions had been surgically removed. This opinion is partly supported by the observation that exposure to light favored abundant hyphal growth of *C. mansonii* in contrast to the near absence of these structures when cultures were incubated in the dark (SUSSMAN et al.). In addition, the continued growth of the organisms in tissue in spite of the specimens having been fixed in formalin might be considered on the basis of WEIDMAN and ROSENTHAL's findings. These workers added 10% formaldehyde to a dextrose bouillon culture of *F. pedrosoi* and observed continued growth of the colonies after several months. Under these conditions it is interesting that they found that the organisms were converted from yeast forms to straight hyphae, and conidia were absent.

Differential Diagnosis: Clinically, the fungus lesion most closely mimicking the subcutaneous abscess caused by the brown fungi is the rare case of subcutaneous nocardiosis (THORLAKSON and RUSNAK). The lesions are not attached to the normal overlying epidermis, may be solitary, and are associated with a history of trauma. The clinical features of the lesion may remain unchanged for long periods of time. Similarly, one of the rare instances of localized cutaneous histoplasmosis appeared as a painless nodule after having been previously incised. The lesion became ulcerated a little over 18 months later (SYMMERS, 1956). In disseminated chromoblastomycosis, rare skin lesions are subcutaneous abscesses. Some of these lesions are believed to be metastatic in origin. However, the possibility that these lesions may also have arisen from primary innoculations cannot be excluded.

In the early stages of their development, several different fungal diseases may show a close clinical resemblance to the subcutaneous abscesses caused by the brown fungi. The subcutaneous phycomycosis lesions of Indonesia (JOE and ENG) and parts of Africa as well as the early lesions of primary cutaneous coccidioidomycosis (LEVAN and HUNTINGTON), sporotrichosis, chromoblastomycosis, mycetoma pedia and manus fail to show ulceration and may present as solitary asymptomatic nodules initially. Also, the history of skin trauma may be obtained with some regularity in these other infections. Atypical acid-fast organism infections may occasionally present in an identical manner (BROCK et al.).

Treatment and Prognosis: The lesions were readily dissected and removed surgically under local anesthesia. However, in two instances the cavity ruptured during the procedure resulting in the spillage of the abscess contents into the surgical wound. It is noteworthy that, in these two instances as well as in all the remaining cases, no recurrences were noted during follow-up periods ranging up to nine years (KEMPSON and STERNBERG; ICHINOSE). The patients also failed to show clinical indications of systemic spread of the infection. This lack of local or systemic spread was further confirmed at autopsy in the patient who died with dysglobulinemia. Another patient died seven years after removal of the lesion. This patient died after a series of cerebral vascular accidents documented by carotid angiograms. No autopsy was performed but clinical evidence of disseminated fungus infection was absent.

The patients have a good prognosis in regard to the mycotic infection since it is localized and the contents of the abscess cavity apparently have a poor potential for infectivity. The latter observation suggests that the luminal contents containing degenerating organisms might be less likely to yield a positive culture than the tannish granulation tissue lining of the cavity. This may be especially applicable to lesions containing few organisms.

General Comment: The clinical manifestations of chromoblastomycosis include a variety of cutaneous lesions classified by PARDO-CASTELLO et al. into five different groups: Verrucous, tuberculoid, syphiloid, psoriasiform, cicatricial and elephantiasic lesions. This listing does not include the lesion presenting as a solitary subcutaneous abscess covered by normal epidermis. No lesion of this character is described in several more recent studies including a report of 129 cases of chromoblastomycosis by BRYGOO and SEGRETAIN. The indolent progressive course of chromoblastomycotic lesions differs also from the nearly static and quiescent behavior of the subcutaneous abscesses. For these reasons, the separation of the subcutaneous abscess caused by pigmented fungi from the general category of chromoblastomycosis made by KEMPSON and STERNBERG as well as by BAKER appears to be warranted.

In rare instances subcutaneous abscesses are seen in patients with disseminated chromoblastomycosis. These lesions are believed to be metastatic, and although they exhibit the same histopathologic changes as the subcutaneous abscess we have seen, they are associated with widespread lesions typical of ordinary chromoblastomycosis (AZULAY and SERRUYA). Two of these lesions described by CARRION and KOPPISCH were present in subcutaneous tissue and a third lesion was found in the deep fascia of the right thigh. The possibility that at least some of these lesions are due to primary innoculations has not been excluded. However, further support for some of these lesions being metastatic is found in a report by KAKOTI and DEY. Their patient developed extensive lesions of the labia majora and lymph nodes of both inguinal regions. The nodes were necrotic and apparently resembled tuberculous lesions grossly. The same organisms were isolated from the vulvar lesions as well as from the nodes. Tissue sections and smears of the nodes also showed these organisms. The massive deep abscesses also found in rare patients with chromoblastomycosis should likewise be set apart from the subcutaneous abscesses. PEDROSO and GOMES found bacteria as well as fungi in such a lesion from the neck which measured 6×2 cm.

The medical significance of the subcutaneous abscess has not been determined with certainty because of the small number of cases studied up to the present time. As far as is known, the infection remains localized and is innocuous. This is somewhat surprising in view of the presence of unrelated serious diseases in many of these patients. One explanation for this limited infection is that the organisms may have such low virulence that they are unable to take further advantage of the lowered host resistance after having established the initial lesion. The observation that the abscess contents may be spilled into the surgical wound during attempts at excision without initiating a recurrence is somewhat in favor of this explanation. Indeed it is possible that some of these lesions sterilize themselves as is suggested by the extremely small number of fungi in two lesions and the presence of degenerating organisms in many lesions. On the other hand, pigmented fungus infections may be quite aggressive in certain instances. FUKU-

SHIRO et al. presented the case of an 8 year old child who developed lesions of the right cheek and ala nasi due to *H. pedrosoi*. Several months later the child died of brain abscesses caused by the same organism. At least 13 brain lesions were present and were believed to have been metastatic from the sores on the face. Although this is an isolated case, its validity and significance are difficult to ignore. A second observation of some pertinence is found in the experiences with experimental induction of subcutaneous abscesses in mice and rats using chromoblastomycosis organisms. The abscesses usually rupture and heal spontaneously (CONANT et al.). By contrast, the subcutaneous abscesses produced by IWATA and WADA penetrated the thoracic wall and extended into the lungs of mice. The organisms had been obtained from FUKUSHIRO et al. 's case. These workers had added a small amount of carboxy-methyl cellulose to the usually employed saline suspension of cultured organisms. The cellulose polymer may have been the critical factor responsible for the aggressive behavior of the lesions. On rare occasion, an intracutaneous innoculation in the mouse has apparently given rise to a brain lesion (WATSON).

The wide variety of organisms found in the subcutaneous abscess provides a bountiful opportunity to study the tissue phase of these fungi. Organisms showing authentic budding are found with regularity, a phenomenon not observed in the lesions of chromoblastomycosis. The blastospore chains studied by RAJAM et al. appear to be an exception. The conversion of blastospores to chlamydospores, likewise, seems to occur frequently. In addition, the germination of the spores as documented by the formation of growth tubes can be observed. Although both yeast and mycelia are characteristically present in the subcutaneous abscess, this dimorphism is not limited to that particular lesion. Dimorphism is also characteristic for organisms in lesions of cladosporiosis, the infection of brain and meninges by chromoblastomycosis organisms (SYMMERS, 1960). It is not known what factors are directly responsible for the development of two growth forms of the same organism in these diseases.

Acknowledgements: The author acknowledges the interest and inspiration of Drs. R.J. REED, CHARLES E. DUNLAP, and W.H. STERNBERG. Dr. STERNBERG also provided one of the figures and pathologic material on seven of the cases, two of which were originally contributed by Dr. FRIEDRICHS HARRIS. Dr. J.T. MCQUITTY provided clinical follow-up on one case. Drs. LORRAINE FRIEDMAN and JOHN D. SCHNEIDAU, JR. (Tulane Department of Microbiology, Mycology Section) as well as Drs. EMMA MOSS and MARION HOOD (Charity Hospital, New Orleans, Louisiana) gave assistance in evaluation of fungi in tissue; each has some reservations about the terminology employed and, since they do not all agree, the responsibility becomes solely that of the author. Dr. RENÉE CORNILLE translated two of the French articles. Mr. RAYMOND JOHNSON made the photographs. Figure C is used with the permission of the Williams and Wilkins Co., Baltimore, Maryland and Drs. R.L. KEMPSON and W.H. STERNBERG. It is a reproduction of Fig. 1, page 599, American Journal of Clinical Pathology, Vol. 39, No. 6, 1963.

References

AZULAY, R.D., SERRUYA, J.: Hematogenous dissemination in chromoblastomycosis; report of a generalized case. Arch. Derm. **95**, 57 (1967). — BAKER, R.D., in ANDERSON, W.A.D., ed.: Pathology, ed. 5, Saint Louis, C.V. Mosby Co., 1966, Chapter 12, Fungus Infections, p. 310. — BROCK, J.M., KENNEDY, C.B., CLARK, W.G.: Cutaneous infection with atypical mycobacterium; report of a case. Arch. Derm. **82**, 918 (1960). — BRYGOO, E.R., SEGRETAIN, G.: Etude clinique épidémiologique et mycologique de la chromoblastomycose à Madagascar. Bull. Soc. Path. exot. **53**, **443** (1960). — CARRION, A.L.: Chromoblastomycosis. Ann. N.Y. Acad. Sci **50**, 1255 (1950a). ~ Yeastlike dematiaceous fungi infecting the human skin; special reference to so-called *Hormiscium dermatitidis*. Arch. Derm. Syph. (Chic.) **61**, 996 (1950b). — CARRION, A.L., KOPPISCH, E.: Observations on dermatomycosis in Puerto Rico; report on a case of chromoblastomycosis. Puerto Rico J. publ. Hlth. **9**, 169 (1933). — CARRION, A., SILVA, M.: Sporotrichosis; special reference: a revision of so-called *Sporotrichum gougerotti*. Arch. Derm. **72**, 523 (1955). — CONANT, N.F., SMITH, D.T., BAKER, R.D., CALLAWAY, J.L.,

Martin, D.C.: Manuel of Clinical Mycology, ed. 2, p. 278. Philadelphia: W.B. Saunders Co. 1954. — DaFonseca, O.: Sur l'état actuel de la question des chromoblastomycoses. Presse méd. **48**, 133 (1940). — Fukushiro, R., Kagawa, S., Nishiyama, S., Takahashi, H.: Un cas de chromoblastomycose cutanée avec métastase cérébrale mortele. Presse méd. **65**, 2142 (1957). — Ichinose, H.: Unpublished data. — Iwata, K., Wada, T.: Mycological studies on the strains isolated from a case of chromoblastomycosis with a metastasis in the central nervous system. Jap. J. Microbiol. **1**, 355 (1957). — Joe, L.K., Eng, N.T.: Subcutaneous phycomycosis: a new disease found in Indonesia. Ann. N.Y. Acad. Sci. **89**, 4 (1960). — Kakoti, L.M., Dey, N.C.: Chromoblastomycosis in India. J. Indian med. Ass. **28**, 351 (1957). — Kempson, R.L., Sternberg, W.H.: Chronic subcutaneous abscesses caused by pigmented fungi, a lesion distinguishable from cutaneous chromoblastomycosis. Amer. J. clin. Path. **39**, 598 (1963). — Levan, N.E., Huntington, R.W., Jr: Primary cutaneous coccidioido-mycosis in agricultural workers. Arch Derm. **92**, 215 (1965). — Lewis, G.M., Hopper, M.E., Sachs, W., Cormia, F.E., Potelunas, C.B.: Mycetoma-like chromoblastomycosis affecting the hand; further findings and comparative mycologic studies. J. invest. Derm. **10**, 155 (1948). — Mariat, F., Segretain, G., Destombes, P., Darrasse, H.: Kyste sous-cutane mycosique (Phaeo-sporotrichose) à *Phialophora gougerotii* (Matruchot, 1910) Borelli, 1955, obsereyé au Sénégal. Sabouraudia **5**, 209 (1967). — Mead, M., Ridley, M.F.: Sporotrichosis and chromoblastomycosis in Queensland. Med. J. Aust. **1**, 192 (1957). — Merrin, J.A.: Weitere Beobachtungen über den Erreger der europäischen Chromoblastomycosis. Arch. Derm. Syph. (Chic.) **166**, 722 (1932). — Moore, M.: The virulence of strains of *Phialophora verrucosa* determined by innoculating chorio-allantoic membranes of chick embryos. J. invest. Derm. **5**, 411 (1942). — Nielson, H.S., Jr., Conant, N.F.: Practical evaluation of antigenic relationships of yeastlike dematiaceous fungi. Sabouraudia **5**, 283 (1967). — Nityananda, K.: Chromoblastomycosis in Ceylon. Derm. Trop. **1**, 184 (1962). — Pardo-Castello, V., Leon, E.R., Trespalacios, F.: Chromoblastomycosis in Cuba. Arch. Derm. Syph. (Chic.) **45**, 19 (1942). — Pedroso, A., Gomes, J.M.: Four cases of dermatitis verrucosa produced by *Phialophora verrucosa*. Ann. paulist. de med. e. cir. **9**, 53 (March), 1920, as cited by F.D. Weidman and L.H. Rosenthal. Reference 21. — Rajam, R.V., Kandhari, K.C., Thirumalachar, M.J.: Chromoblastomycosis caused by a rare yeast like dematiaceous fungus. Mycopathologia (Den Haag) **9**, 5 (1958). — Samuels, M.S.: Personal communication, Feb. 1968. — Schneidau, J.D., Jr.: Personal communication, Feb. 1968. — Silva, M.: The parasitic phase of the fungi of chromoblastomycosis: development of sclerotic cells *in vitro* and *in vivo*. Mycologia **49**, 318 (1957). — Snell, W.H., Dick, E.A.: A Glossary of Mycology, p. 20. Cambridge: Harvard University Press 1957. — Sussman, A.S., Lingappa, Y., Bernstein, I.A.: Effect of light and media upon growth and melanin formation in *Cladosporium mansoni*. Mycopathologia (Den Haag) **20**, 307 (1963). — Symmers, W.St.C.: Localized cutaneous histoplasmosis. Brit. med. J. **2**, 790 (1956). ~ A case of cerebral chromoblastomycosis (cladosporiosis) occurring in Britain as a complication of polyarteritis treated with cortisone. Brain **83**, 37 (1960). — Throlakson, R.H., Rusnak, C.H.: Subcutaneous nocardiosis. Canad. med. Ass. J. **95**, 224 (1966). — Tubaki, K.: Sporulating Structures in Fungi Imperfecti. In: G.C. Ainsworth, and A.S. Sussman, eds.: The Fungi; An Advanced Treatise. New York and London: Academic Press 1966. Vol. II, The Fungal Organism, Chapter 4, p. 116. — Watson, K.C.: Cerebral chromoblastomycosis. J. Path. Bact. **84**, 233 (1962). — Weidman, F.D., Rosenthal, L.H.: Chromoblastomycosis: a new and important blastomycosis in North America; report of a case in Philadelphia. Arch. Derm. Syph. (Chic.) **43**, 62 (1941). — Young, J.M., Ulrich, E.: Sporotrichosis produced by *Sporotrichum gougerotii*; report of a case and review of the literature. Arch. Derm. **67**, 44 (1953).

Candidosis

H. I. Winner, London, Great Britain

With 33 Figures

1. General

The intense interest in this subject in recent years is reflected in the great volume of published literature on all its aspects, especially the clinical and experimental. (Winner and Hurley, 1964).

a) Definition

Candidosis (candidiasis, moniliasis) is disease due to infection by *Candida albicans* and other species of candida.

The commonest variety of the disease is "thrush" or infection of the mucous membrane of the oral cavity or the vagina. Less commonly, however, the skin and any visceral organ may be affected.

b) Epidemiology

Systematic studies on the epidemiology of the disease are scanty. Superficial infection is contracted as a result of direct contact with infected persons. There is some evidence (Clayton and Noble, 1966) that airborne contamination may also occur. Infection of the gut frequently follows the ingestion of infected saliva. Systemic disease, due to bloodstream contamination, may arise by invasion of the tissues from superficial infection, or by direct implantation of the organism into the bloodstream during parenteral therapy or surgical operation.

c) Incidence and Predisposing Factors

Oral thrush is common in infants and children; it also frequently occurs in adults.

Vaginal thrush occurs in women of all ages, but particularly in the child bearing age and in pregnancy.

All types of the disease occur more commonly in the following conditions:

1. Physiological: Pregnancy, Early infancy
2. Traumatic: Maceration of skin, Allergic conditions of the skin, Chemical damage to mucous membranes
3. Endocrine disorders: Diabetes mellitus, Hypothyroidism, Hypoparathyroidism, Hypoadrenalism
4. Malnutrition and malabsorption syndrome.
5. Malignant disease, especially leukaemias.
6. Agranulocytosis and aplastic anaemia.
7. Post-operative states.
8. Antibacterial antibiotic therapy.
9. Therapy by immunosuppressive drugs.

2. Mycology

Candida albicans is a yeast-like fungus which exists in three morphological forms: yeast cells or blastospores, mycelium or pseudomycelium, and chlamydospores. For the purpose of this article, mycelium and pseudomycelium are regarded as identical.

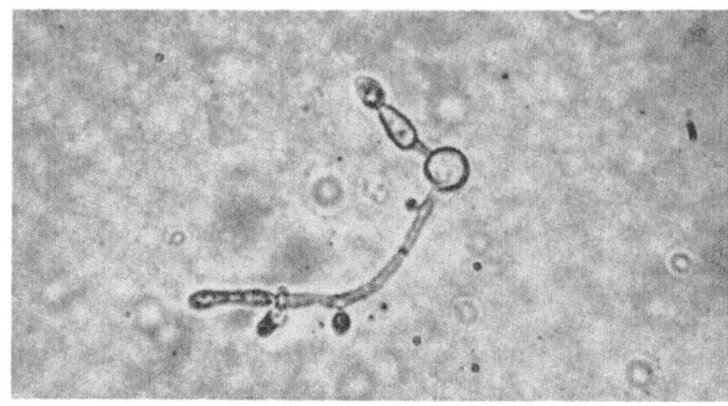

Fig. 1. *C. albicans* mycelium, chlamydospore, and blastospores. Unstained preparation after 24 h incubation on corn-meal agar at room temperature (×360). (From WINNER and HURLEY, 1964; Fig. 5, p. 24)

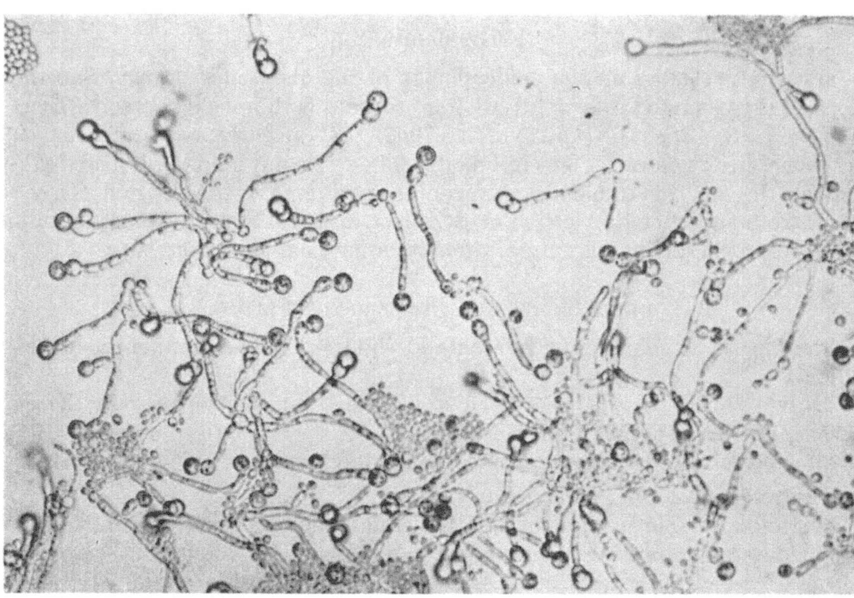

Fig. 2. *C. albicans* mycelium, blastospores, and chlamydospores. Unstained preparation after 24 h incubation on "Oxoid" corn-meal agar at 26°C (×480). (From WINNER and HURLEY, 1964; Fig. 6, p. 25)

Yeast cells and mycelium are the forms found in lesions in human cases. There is controversy about whether the yeast or the mycelium is the invasive form, or whether both may be invasive; the evidence is inconclusive. Chlamydospores are thick-walled, spherical, refractile structures found on artificial culture in poor media; their significance and function are unknown (Figs. 1, 2, 3).

Approximately 25% of cases of candidosis are due to species other than *C. albicans*. The most important are *C. stellatoidea*, *C. tropicalis*, *C. parapsilosis*,

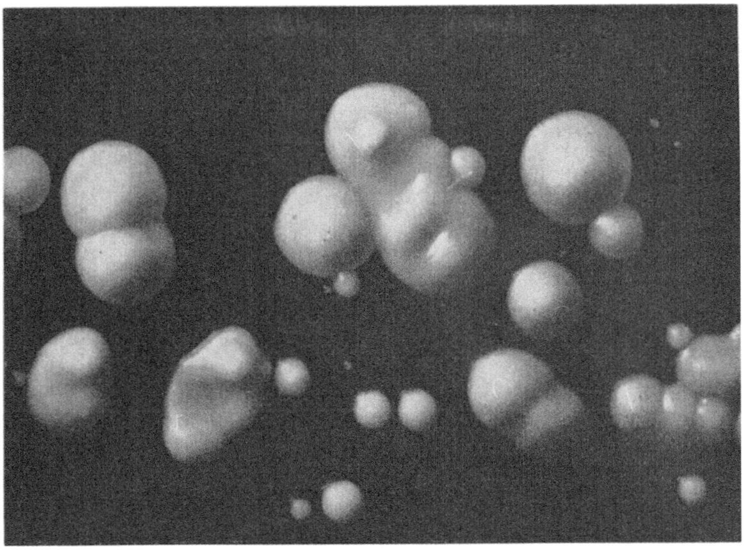

Fig. 3. Colonial appearance of *C.albicans* after 24 h incubation on peptone maltose agar at 37°C (×6). Note creamy consistency of colonies. (From WINNER and HURLEY, 1964; Fig. 10, p. 41)

C.pseudotropicalis, *C.guilliermondii*, and *C.krusei*. The species are all about the same size; they differ somewhat in morphology, in biochemical characteristics and in antigenic structure. The clinical pictures and pathology of the diseases they cause are broadly similar and for convenience may be considered together.

3. Diagnosis of Candida Infections

a) Identification from Clinical Specimens

From superficial lesions smears from swabs or loops of exudate should be examined microscopically both fresh and after gram staining. Direct cultures should be made on to Sabouraud's or other suitable media. On Sabouraud's agar all types of candida show a growth within 24 h after incubation at 37°C. The species may be identified by further cultural, biochemical and serological tests.

Portions of tissues suspected of infection by candidas should whenever possible be cultured in the fresh state before fixation. It is impossible to identify a candida with certainty, or even to assign an evident fungus to the candida genus, unless cultures are made from the material. In the absence of a culture, the morphological appearances of the genus are not sufficiently characteristic to enable more than a presumptive identification to be made from the examination of stained sections.

On the other hand, experienced histologists can identify certain features of the candida genus as seen in stained sections of histological tissues which enable them to name the genus, if not the species, with a fair degree of confidence.

C.albicans may be seen in tissue sections either in the mycelial or blastospore form; usually, but not always, both are present. When this is so, identification of the genus candida may be inferred since the dimorphic appearance is characteristic. However, the possibility of a dual infection must be borne in mind. Further difficulties arise when either the blastospore or the mycelial form is present by

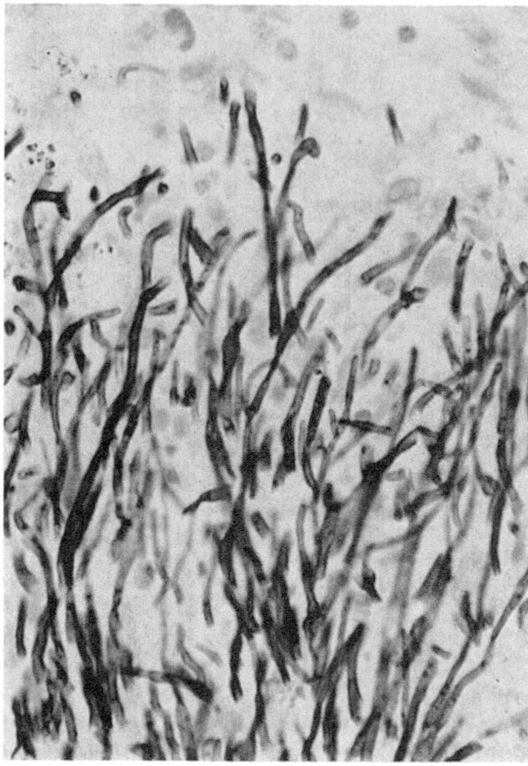

Fig. 4. *C.albicans* in the margin of a liver abscess in a case of septicaemic candidosis. Close-set, septate, rather narrow and angular hyphae; branching at various angles (Hexamine silver, ×930)

Table 1

Septa	Candida +	Aspergillus +	Mucor —
Section	Circular varying calibre	Circular, uniform calibre	Flat
Thickness	Varying; numerous constrictions	Uniform	Great variation mainly very broad. Numerous constrictions
Folds	Few	Few	Many, may give appearance of septa
Angle of branching	Various	Dichotomous	Mainly perpendicular

itself, and in such circumstances anything more than a tentative diagnosis is usually impossible in the absence of cultural confirmation.

Blastospores of *C.albicans* are spherical or slightly ovoid; all other candida species are more oval or elongated than *C.albicans*. The amount of tissue reaction is variable, and there may be none.

The candidas do not stain sharply with haemotoyxlin and eosin, but they can easily be stained by the modified periodic-acid-Schiff reaction and by silver impregnation methods such as Grocott's methenamine-silver technique. They are strongly gram positive.

Differentiation from *Cryptococcus neoformans* may be made by the absence of a capsule and the presence of filaments.

The mycelium of candida is usually uniform in thickness, with little branching; straight filaments are characteristic. Filaments of candida may be differentiated from those of mucor by the features shown in Table 1 (SYMMERS, 1964) (Fig. 4).

b) Serological Diagnosis

In cases of suspected deep-seated candidosis, in which it has not proved practicable to culture the organism from the lesion, attempts have been made to assist the diagnosis by the detection of antibodies in the patient's serum. Care is needed in the interpretation of serological tests, since the occurrence of candida antibodies is widespread throughout the human population even in the absence of clinical infection (WINNER, 1955).

On the other hand, a persistently high titre is undoubtedly supporting evidence of a candida infection. Various techniques have been tried, of which the most commonly used has been the agglutination reaction using killed whole candida cells as the antigen. Complement fixation and precipitin reactions have also been used, but are technically more exacting and the formulation of standard antigens is a difficulty. Moreover the results are less easy to control by comparison with "control" subjects who show no manifestations of infection.

At present much research is in progress on the elucidation of the antigenic structure of the candidas, the preparation of standardised antigens, and the significance of different types of antibody in infected subjects. This should lead to an increase in the value of serological methods of diagnosis.

c) Other Methods of Diagnosis

Skin tests: Dermal hypersensitivity to *C. albicans* was reported by RAMEL (1925) and by RAVAUT and RABEAU (1929), who showed that normal people would give positive skin reactions to extracts of candida. Since then, there have been many attempts to assess the significance of such reactions, and particularly to use them as a means of detecting latent infection. However, too many normal subjects have shown positive reactions (SCLAFER and FLAVIN, 1954; SCLAFER, 1957). LEWIS et al. (1937), showed that 43% of patients with candidosis gave negative skin reactions, and 46% without the disease gave positive reactions.

These results indicated that the test was useless as a means of diagnosis of infections. The possibility that they might be due to unsuitable antigenic preparations was propounded by AKIBA et al. (1957). They recommended the use of purified antigens. On the other hand, in the absence of exact chemical definition of the antigens, it is difficult to see how purified antigens can be obtained. MAIBACH and KLIGMAN (1962) concluded that the development of skin allergy in dermal candidosis was too inconstant to be of value in diagnosis. Moreover, skin reactions are not specific; patients infected with other fungi may cross-react to extracts of candida.

To summarise, it is doubtful whether tests for hypersensitivity have any place today in the diagnosis of deep-seated infection.

Radiology: Radiology has been tried as an ancillary method of diagnosis in suspected disease of the chest and the gastrointestinal tract. So far as chest radiology is concerned, there is no means of differentiating radiologically between infections with candida and those by other micro-organisms. Some authors have claimed (MARSH, 1959; KAUFMAN et al. ,1960) that the radiological appearances in candidosis of the oesophagus are sufficiently characteristic to warrant a confident diagnosis. It seems, however, that radiology, no matter how well performed, must still be confirmed by the isolation of the organisms from the lesions.

4. Pathogenesis of Candida Infections

It has repeatedly been observed that candida infections, both superficial and deep-seated, are particularly liable to occur in patients whose immune mechanisms are disturbed for any reason, or whose microbial flora has been assaulted by antimicrobial therapy. A list of the predisposing clinical conditions associated with candidosis is given earlier in this chapter. Such a list, however, cannot be exhaustive. Indeed, the first manifestation of one of the disorders listed may be the clinical appearance of a candida infection. The diagnosis of such an infection is therefore an indication for a searching clinical examination, including all ancillary tests. As WILSON (1962) points out, *C. albicans* is a better clinician and more able to detect patients with immunological abnormalities than are human physicians. In adults hitherto regarded as previously normal, the appearance of even a superficial candida infection is a manifestation of either some local damage at the site of the infection or of some systemic disorder. The only exceptions are the appearance of vaginal thrush in non-pregnant women, or candidosis anywhere in the body in pregnant women. Early infancy, that is to say, the first few weeks of extrauterine life, is another state in which immunity patterns are too immature to withstand tissue invasion by this organism, and there is no need to look for any predisposing cause of candidosis in a small infant.

In all other cases, the normal immune mechanisms of the individual are sufficient to withstand tissue invasion by candida. Such invasion only occurs in the breakdown of immune mechanisms, either locally or generally, and the appearance of a candida lesion is a clear indication for a search for such a breakdown.

5. Clinical Features and Pathology of Candida Infections

The commonest form of candidosis is thrush of the oral or vaginal mucous membranes. A more generalised infection, affecting surface areas in many parts of the body, is occasionally seen. Thus the skin and nails may be affected as well as the mucous membranes. Generalised candidosis, as this superficial disease is called, is usually associated with one of the predisposing conditions to which we have referred, and more specifically, may be part of a generalised endocrine deficiency syndrome sometimes called the *endocrine candidosis syndrome* (see later). Candida granuloma ("monilial granuloma") is a variety of superficial generalised candidosis.

Generalised candidosis is usually superficial. However, it may also be associated with the presence of the organisms in the bloodstream, and may therefore be a manifestation of septicaemic candidosis. The latter, as would be expected, is a much more serious and frequently fatal condition. The organism may set up metastatic lesions in almost any organ of the body, most commonly the kidneys. The heart valves may be affected in patients with pre-existing valvular lesions.

a) Infections of the Mouth

The commonest clinical manifestation of *C. albicans* infection is oral thrush, which occurs frequently in infants and in debilitated adults.

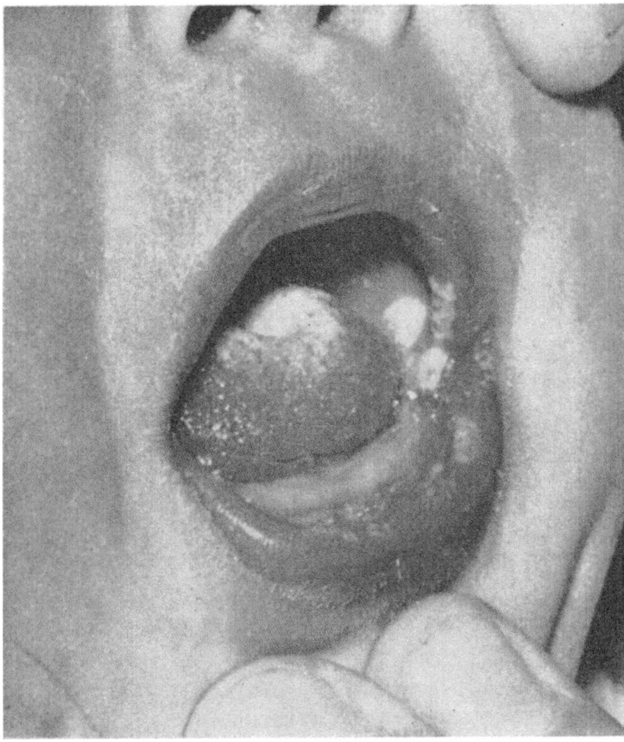

Fig. 5. Oral thrush in a child of eight days (From WINNER and HURLEY, 1964; Fig. 18, p. 95)

Clinical features: The disease first appears in the form of small white patches on the lining of the gums, the side of the tongue, the buccal mucosa, and the mucosa of the throat. The lesions coalesce and form pearly-white elevated patches which look like curdled milk, but, unlike these, can be removed only with difficulty. They are painless with no swelling, and they may heal within a few days of starting treatment. Occasionally erosion and ulceration may occur with necrosis and granuloma formation (Fig. 5).

Pathology: In oral thrush, the essential lesion is oedema of the infected mucous membrane. An oedematous plaque forms from the desquamating epithelial cells. This forms a pseudomembrane which consists of shed desquamating epithelium mixed with keratin, fibrin, necrotic tissue, food debris, leucocytes and bacteria, all matted together and anchored down to the living epithelium by the fungal hyphae. There may be little or no tissue response (LEHNER, 1966) (Fig. 6).

Candida leukoplakia. Some cases of oral leukoplakia are associated with, if not due to, candida infection (CAWSON and LEHNER, 1968). Clinically this may be indistinguishable from other forms of leukoplakia, though candida may readily be detected in smears of the exudate and in biopsy sections. Cases infected with candida also show a raised antibody titre to this organism. While some cases have responded to local antimycotic therapy, however, it is also possible that the changes initiated in the mucosa by candida infection may in some cases be progressive and irreversible, even after the infection has been removed (Fig. 7 and 8).

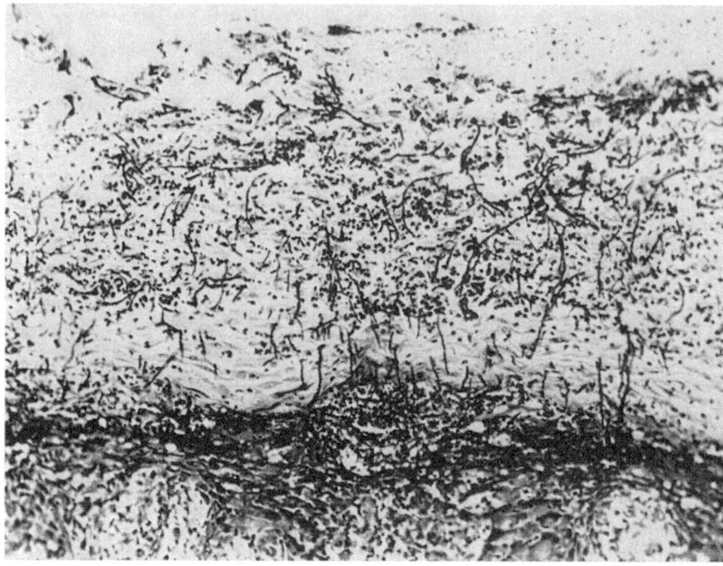

Fig. 6. Oral thrush. Loosely knit oedematous plaque of desquamating epithelial cells invaded by hyphae and infiltrated by inflammatory cells (PAS, ×120)

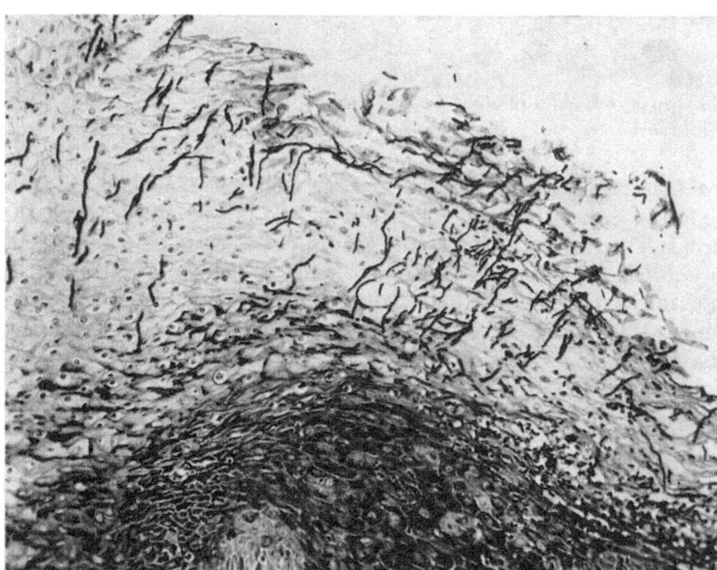

Fig. 7. Candida leukoplakia. Hyphae invading the parakeratotic zone. Oedema and inflammatory cells at the junction of the parakeratotic zone and the stratum spinosum (PAS, ×120)

The histological picture shows a marked parakeratosis, with hyphal invasion. There is an inflammatory reaction along the junction of the parakeratotic zone with the stratum spinosum. Some degree of acanthosis is present. Chronic inflammatory changes are also found in the corium.

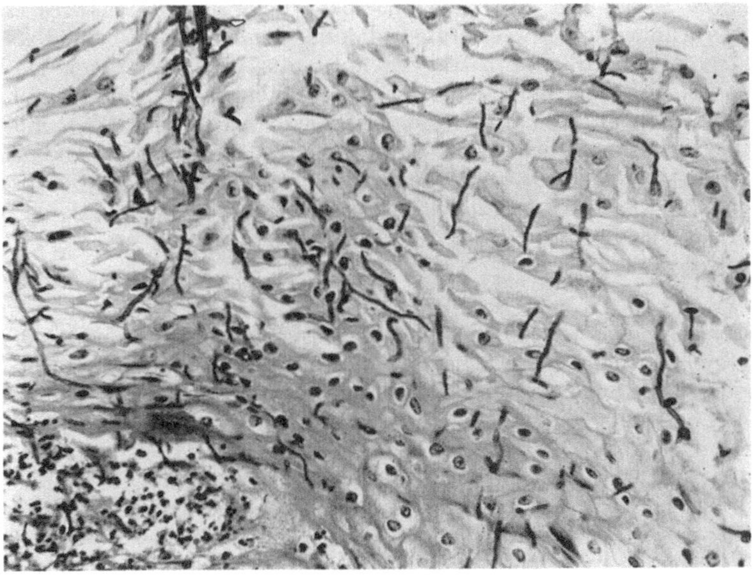

Fig. 8. Candida leukoplakia. Higher power from same lesions as Fig. 7, to show details of plaque (PAS, ×120)

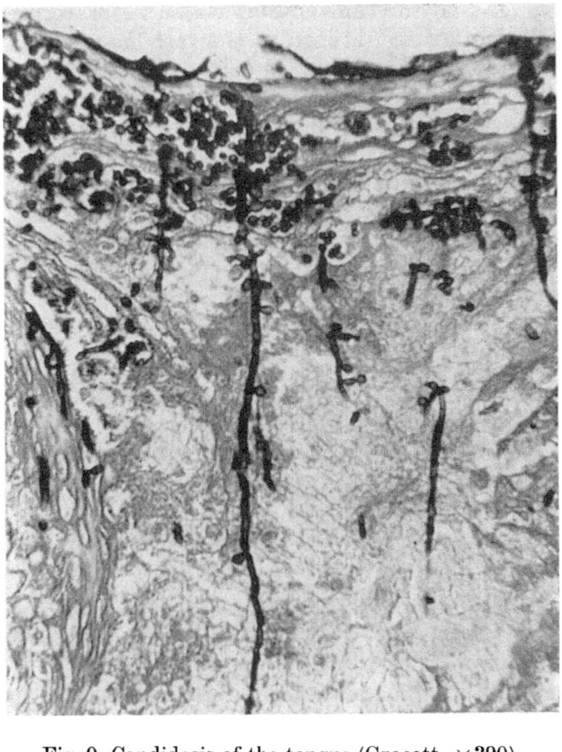

Fig. 9. Candidosis of the tongue (Grocott, ×390)

Lehner (1966) and Cawson (1966) have described the changes associated with other clinical varieties of oral candidosis. Acute atrophic candidosis is often associated with antibiotic therapy ("antibiotic sensitive tongue"). In this condition the pseudomembrane may be shed, leaving an intensely painful atrophic red mucosa. Denture stomatitis is frequently associated with candida infection; in fact, Cawson asserts that this is the commonest form of oral candidosis. It is characterised by erythema and oedema of the upper-denture bearing part of the mucosa, which becomes bright red and may be sharply differentiated from the surrounding mucosa at the margins of the denture (Fig. 9).

b) Skin

Maceration: The most important predisposing factor in candidosis of the skin is *maceration*, or damage by moisture. Those whose occupation involves frequent immersion of the hands in water are particularly liable to candidosis of the hands. Such occupational groups include housewives in whom candidosis of the fingers and chronic paronychia are common. Fishmongers, shoemakers, chefs and barmen are other groups subject to this occupational disease.

Another source of the damaging moisture is perspiration which is unable to evaporate freely, as in intertriginous lesions anywhere in the body. These are liable to infection by candida and other microbes. Soldiers are subject to candidosis of the feet.

In *infants*, the diaper region may be kept wet by the infants' excreta. Diaper rashes due to candida infection are common. They have been described by Bound (1956). The buttocks and inner thighs are most commonly involved, but the genitalia and lower abdomen may also be affected. At first the lesions consist of dull red areas which tend to coalesce. Flat vesicles soon form, after which desquamation occurs. Bound considers that the rash may easily be distinguished from two other common napkin rashes, namely, excoriated buttocks and ammonia dermatitis. Another type of illness was described by Robinson (1957). This is a sudden flare up of acute dermatitis in cases of infantile eczema.

Perlèche is a form of maceration of the corners of the mouth first described by Lemaistre (1886) among inhabitants of the Limousin district who moistened their lips with their tongue after sipping wine.

Candida invasion of intertriginous skin may occur in the axillae, the submammary regions, the umbilicus, the webs of the fingers and toes, the limb flexures, and the vulval area.

Candidosis may also occur in the *external ear* (La Touche, 1966).

Candida Granuloma: A well defined and serious skin infection in older children is *candida granuloma* (monilial granuloma). This takes the form of superficial inflammation with hyperkeratosis which may be very marked, giving rise to crust formation and sometimes to the formation of horns.

In the case described by Hauser and Rothman (1950) the patient was a well-developed and well-nourished boy of seven. Apparently there were no predisposing factors. Multiple horns developed on the skin, biopsy of which showed hyperkeratosis and acanthosis. The corium was filled with a dense infiltration of lymphocytes, plasma cells, polymorphs, and foreign body giant cells. The infiltrate extended downwards to the subcutis and was perifollicular and periglandular in some areas. Filaments thought to be hyphae were seen in the stratum corneum. Some years after this case report, Rothman (1959) further reported that the patient had suddenly developed diabetic coma and that he later died of diabetes.

In their comments on this and other reported cases, Hauser and Rothman considered that there might be a constitutional defect in defence mechanisms. They showed that the organisms which cause candida granuloma were not more pathogenic than other strains of candida and they noted, as additional evidence of subnormal resistance, the early age of death in patients with candida granuloma. NITYANANDA (1959) pointed out the early age of onset of the disease. On the other hand, several cases have been described in adults (ROCKWOOD and GREENWOOD, 1934; DANBOLT, 1940; CARSLAW et al., 1966).

Skin Allergy: Next to maceration, the most important predisposing cause of candidosis of the skin is allergy. OWEN et al. (1939) showed that spontaneous skin infections by *C. albicans* might lead to the development of "id" reactions, which could be acute or chronic. Invasion by candida has also often been noted in skin affected by allergy due to other causes.

c) Lungs

The frequency with which *C. albicans* can be cultured from the sputum, in apparent cases of acute and chronic chest infection, has led to much discussion on its possible pathogenic role in the lungs.

The mere fact of isolating *C. albicans* from the sputum is no indication of its possible pathogenic role in the chest. *C. albicans* is often present in the mouths of healthy people and it can appear in their sputum by contamination from the mouth. Moreover, unless the sputum specimen is fresh when cultured, and unless it is collected into a sterile container, the possibility of aerial contamination exists.

As it occurs so often in the mouths of healthy people, *C. albicans* will obviously be found no less frequently in the sputum of those with lung disease, and the numerous sanatorium and hospital surveys confirm this. Moreover, its occurrence, here as elsewhere in the body, is increased in patients on antibiotic therapy.

For there to be a reasonable assurance that organisms recovered from the *sputum* have come from the lungs, the following criteria should be fulfilled.

1. The specimen should be collected into a sterile container.

2. The specimen must either be refrigerated until it is cultured, or it must be cultured soon after collection — within an hour or two. ORIE (1946) recommended that it be washed.

3. Either the organisms should be grown frequently or invariably from the sputum and not from the mouth or throat or postnasal swabs, or they should be grown from a bronchial cast, plug or membrane from a bronchiectatic cavity or pulmonary abscess.

Any patients may harbour the organism in their mouths or throats whether or not they are suspected cases of pulmonary candidosis. Unless the above criteria are fulfilled, therefore, the presence of the organism in the sputum of such patients should be discounted, so far as the diagnosis of pulmonary candidosis is concerned. It is from such patients that a bronchial aspiration specimen may reasonably be demanded in order to confirm that the *C. albicans* has truly come from their lungs. It is not sufficient to cleanse the mouth.

There are, of course, many cases in which *C. albicans* is one of several types of fungus which are apparently playing a saprophytic role in the lung. Such cases are those of cavitation or bronchiectasis, where there is likely to be much stagnant organic material on which yeasts and other fungi may thrive. The bizarre name of "pulmonary intercavitary fungus ball" was given to this condition, in which *C. albicans* is rarely, if ever, the only fungus present (WIESE and BIXBY, 1941; GELLI, 1954).

No aspect of the pathogenicity of *C. albicans* has given rise to more controversy than the part it plays in lung disease. Apparent cases of pulmonary candidosis have been reported from most parts of the world. However, many experienced workers in this field have been inclined to decry the importance of *C. albicans* as a pulmonary pathogen, and indeed to deny it can ever cause primary lung disease. SHREWSBURY (1936) considered that the organism could not cause pneumonia, and that the extent of lung pathology produced by this yeast was secondary thrush of the bronchi. His view was endorsed by RIDDELL and CLAYTON (1958). Similar views have also been expressed by ROBERTSON (1945), FRISK (1952, and ORIE (1952).

This standpoint does not, of course, dispose of the importance of *C. albicans* as a lung pathogen. It is often difficult, and it may be impossible, to determine in any particular case whether the role of *C. albicans* in the lungs is that of a primary or a secondary invader. Moreover, this point, though important, does not seem to be crucial (OBLATH et al., 1951). The important question is not whether the role of *C. albicans* is primary or secondary, but whether it causes disease in the lungs at all. If it only does so as a secondary invader, it resembles several pathogenic bacteria in this respect, for Friedlander's bacillus and *Haemophilus influenzae* are often secondary invaders. This fact does not diminish their potential danger as pathogens.

If *C. albicans* can cause lung disease, this role is by no means incompatible with Shrewsbury's view, endorsed by RIDDELL and his colleagues, that the organism causes secondary thrush of the bronchi. It may, in such circumstances give rise to chronic bronchitis and hence pave the way for bacterial invasion of the lungs. It may give rise to allergic symptoms resembling asthma. It may cause bronchial obstruction leading to pulmonary collapse and its sequelae. It may cause mycomas in lungs affected by pulmonary tuberculosis (SCHULTE, 1957). If the organism fulfils any of these roles, whether as a primary or as a secondary invader, it is a lung pathogen. Moreover, in a small number of cases, of which that reported by SMITH and ARMEN (1955) is the most convincing, the organism has been shown to cause pneumonia, contrary to the view of SHREWSBURY and others.

The controversy surrounding this subject appears to be due to the chronic nature of the disease, and to the facts that histopathological studies are made late in the course of the illness, or post-mortem, if they are made at all, and that there is no clear-cut association of a particular clinical and a particular pathological picture at all stages of the disease. If pneumonia occurs, it occurs at a stage at which the patient does not die, rarely has a lung resected, and hence is unable to present a specimen for histological study. Inferences as to the possible role of *C. albicans* in causing pulmonary consolidation, the chief matter in dispute, have, therefore, had to be made retrospectively. None the less, they have been made, and with justifiable confidence.

Diagnostic Criteria: Neither the clinical features, or serological or radiological evidence are more than adjuvants in the diagnosis of pulmonary candidosis. For the reasons already given, cultural evidence too, is suspect, though clearly negative culture rules out the possibility of pulmonary candidosis. In the opinion of the present writer, the following are the necessary criteria for the diagnosis of primary pulmonary candidosis:

1. *C. albicans* is obtained consistently and in large amounts from the sputum or bronchoscopic swabbings.
2. *M. tuberculosis* is not found.
3. No other cause of disease is found.
4. Candida is demonstrated in histological sections of the lungs obtained from operation specimens, biopsy, or necropsy.

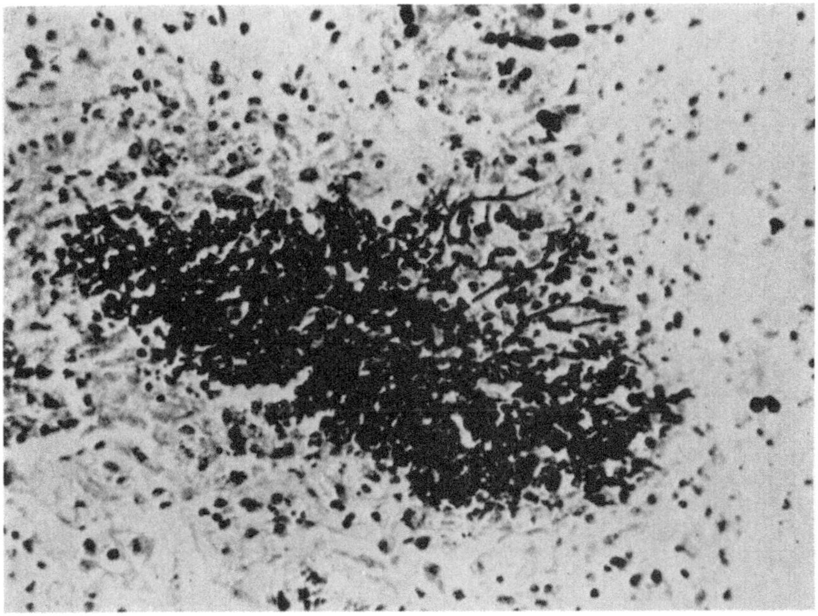

Fig. 10. Candidosis of the lung in a fatal case of septicaemic candidosis (PASH, ×107). (From the same case as Figs. 17, 23 and 33; from WINNER and HURLEY, 1964; Fig. 19, p. 110)

Using these admittedly exacting criteria, a number of convincing cases have been reported. (HURWITZ and YESNER, 1948; SMITH and ARMEN, 1955; OBLATH et al., 1951; FALKMER and WISING, 1955).

In addition to the above cases where candida has caused disease of the lung by direct invasion, candida may also be a cause of asthma. Moreover, it is a common cause of "secondary thrush of the bronchi" (SHREWSBURY, 1936).

In cases of septicaemic candidosis, *C. albicans* may invade the lungs as it does other organs (Fig. 10).

d) Gut

Candida infection of the oesophagus, stomach, and small or large intestines may occur as part of systemic candidosis, or in association with the predisposing conditions already mentioned, especially agranulocytosis, aplastic anaemia, and leukaemia.

Macroscopically the changes associated with candidosis of the alimentary tract are of two types. The mucous membrane may be covered by a thick, shaggy, cream-coloured confluent membrane for much of its length; for example, the whole of the oesophagus may be so covered. This membrane strips easily, revealing greyish-brown plaques of confluent ulceration. The second type of change consists of more localised, discrete and scattered multiple lesions, which may be few or many and may occur in any part of the gut. These are greyish-brown raised plaques or areas of shallow ulceration, and may be covered by a rough friable brown exudate. Mycotic lesions of the second type may be small and may be overlooked at necropsy unless the gut is carefully examined.

The histology of the mycotic lesions of the bowel was described by BEEMER et al. (1954) and by LANNIGAN and MEYNELL (1959). Both mycelium and yeast forms of candida were present, the latter predominating in the case of BEEMER

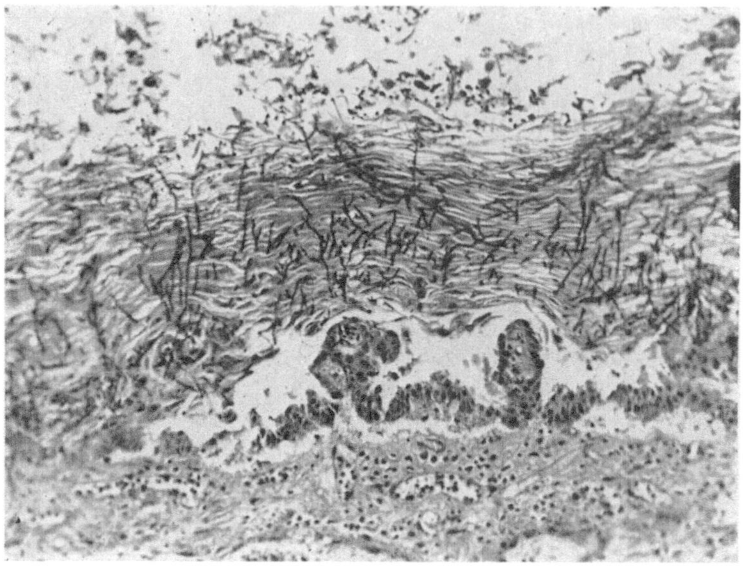

Fig. 11. Candidosis of the oesophagus (PAS, ×120)

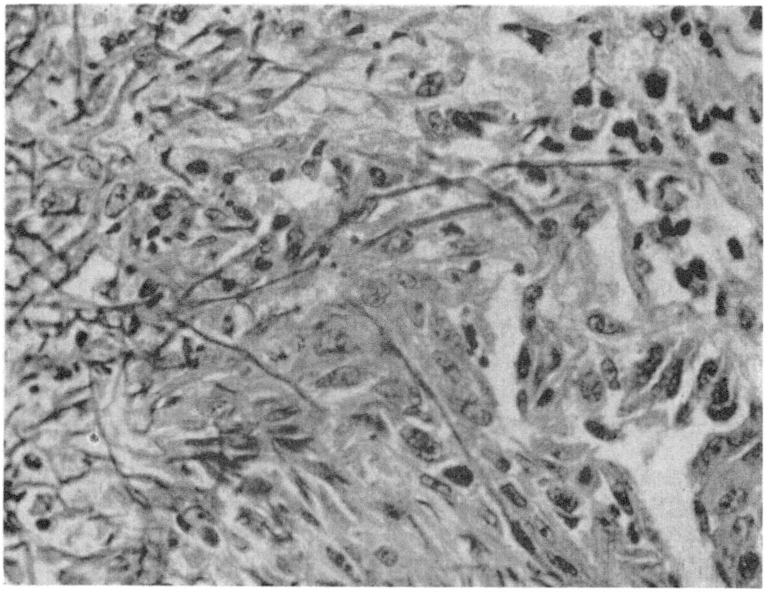

Fig. 12. Proliferating mycelium in the wall of the oesophagus. From a fatal case of disseminated candidosis in a child of 6 treated for acute leukaemia (PAS, ×450). (From WINNER and HURLEY, 1964; Fig. 21, p. 129)

et al. These authors noted a cellular reaction consisting of round cells and neutrophils, with some fibroblastic proliferation. LANNIGAN and MEYNELL reported a reaction which was predominantly lymphocytic and mononuclear, with variable numbers of plasma cells (Figs. 11, 12, 13).

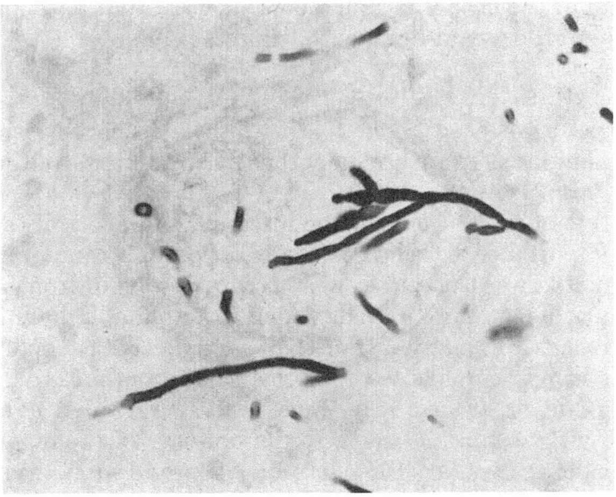

Fig. 13. Candidosis of the stomach. *C. albicans* invading a thrombosed venule in the submucosa in the vicinity of an ulcerating carcinoma, the surface of which was covered with the fungus (Hexamine silver, ×600)

e) Vagina

Candida albicans is now the commonest cause of vaginal discharge. However, in vaginal thrush there may be vulvitis with little or no discharge. The vulva is slightly swollen and tender, reddish blue or intensely red, sometimes with a grey surface. The epithelium may be injured from scratching. The area feels warm both subjectively and to the touch. Small whitish areas ("thrush spots") of caseous material may be observed, giving a granular appearance. The discharge may be thin and yellow, often with white flakes.

The diagnosis is readily confirmed by the microscopic examination and culture of smears of the discharge.

f) Candidosis and Endocrine Disorders

There is a well marked association between candidosis of the skin or mucous membranes and various endocrine disorders. This is most marked in the case of diabetes mellitus, and in many patients the first manifestation of this disease has been the appearance of a superficial candida infection. Other cases of candida infection have been reported in association with hypoadrenalism, hypoparathyroidism, and hypothyroidism. Sometimes two or more of these disorders have been combined in the same patient who is usually a child. The association has been so well marked as to be given the title "*Candida endocrine syndrome*" (WHITAKER et al., 1956; LEHNER, 1966).

It is probable that the precipitating cause for the attack by candida on the host is different in different conditions. In the case of diabetes mellitus, an obvious factor might appear to be the abnormally high glucose content of the blood and tissues, which would be expected to favour the growth of candida. Probably this explanation is too facile; certainly it is now well recognised that changes in carbohydrate metabolism may precede by several years the development of clinical diabetes, and in this pre-diabetic period a candida infection may give warning of the impending disease (WILSON and PLUNKETT, 1965).

In other endocrine disorders it is no easier to find an explanation for the candida infection. Presumably there is some widespread disruption of the immunity mechanisms of the host. LEHNER (1966) suggests that auto-immunity may be a factor, but the rather scanty evidence is indirect.

g) Systemic (Septicaemic) Candidosis

Various terms have been given to this disease, such as acute disseminated (septicaemic) candidosis. There are numerous recent reviews and reports (WINNER and HURLEY, 1964; SYMMERS, 1966).

Diagnosis: One of the reasons for the increasing numbers of cases diagnosed is the increasing importance attached to the isolation of a candida. All too often in the past, if a candida was isolated from a necropsy lesion or from the bloodstream, it was regarded as a contaminant. This belief was undoubtedly mistaken in many cases, and has resulted in many instances of candida septicaemia being missed. It cannot be too strongly emphasized that the isolation of a candida from a lesion, whether during life or at necropsy, must not be regarded as an insignificant finding and full weight must be given to the presence of the organism.

Clinical varieties: SYMMERS (1964) has made a useful subdivision of the cases as follows:

1. Septicaemic candidosis occurring without recognised predisposing factors.
2. Septicaemic candidosis resulting from direct inoculation of the fungus into the body.
3. Septicaemic candidosis occurring as a complication of other diseases or of their treatment.

So far as the first category is concerned, the history of many of the cases described suggests that the underlying predisposing cause was missed. There seems no reason to suppose that candida septicaemia is an exception to what has already been propounded, that the disease does not occur in people hitherto well except in conditions of "physiological unorthodoxy", such as early infancy and pregnancy.

The direct inoculation of the fungus into the bloodstream is a curiously frequent precipitating cause of the disease. This has been known to follow injections of various kinds, including therapeutic substances and narcotic drugs; several cases have followed open-heart surgery (see later).

A large number of cases have been described of septicaemic candidosis arising as a complication of other diseases or of their treatment.

The disease has been described in both adults and children. It is associated with the predisposing causes already listed on page 731. Since such conditions are likely to be present in almost any group of hospital patients, the possibility of such patients developing septicaemic candidosis must always be borne in mind, especially during intense antibacterial or immunosuppressive therapy.

Portal of Entry: As has been stated, the portal of entry of the fungus may well be intravenous, when therapy has been given by this route. In other cases, it seems likely that the organism invades from the alimentary tract, where it is often present as a commensal. Mycotic ulcers of the gut may be small and may be overlooked at necropsy unless carefully sought (LUDLAM and HENDERSON, 1942). This may account for the apparently low incidence of alimentary tract lesions in the cases reviewed above.

Other cases are apparently secondary to superficial candidosis. In some patients who have developed systemic candidosis the portal of entry is obscure. Possibly the organism has entered the bloodstream by direct invasion through the wall of a blood vessel, a hypothesis which may be supported by the appearance of histological sections of the tissues of infected patients and experimentally infected animals.

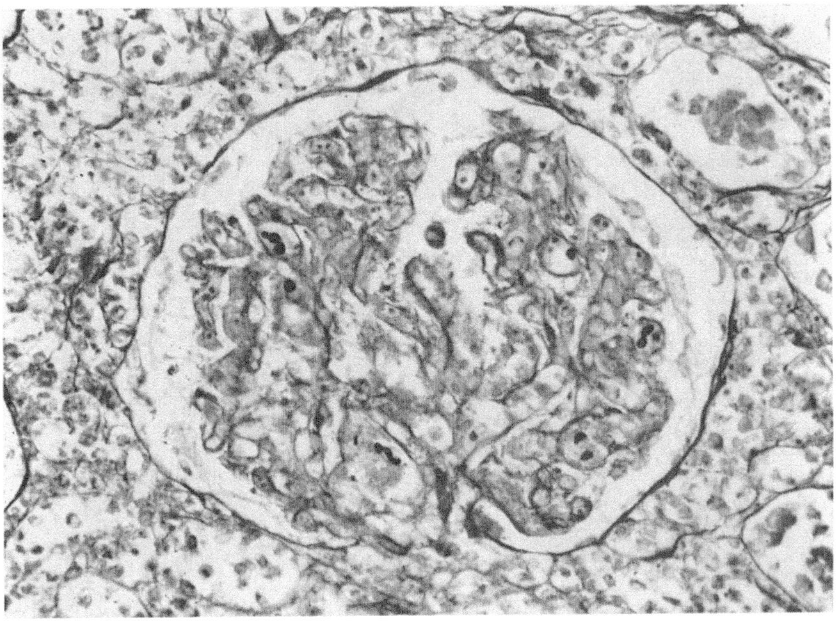

Fig. 14. Renal candidosis, showing yeasts in glomerulus (Hexamine silver, ×310)

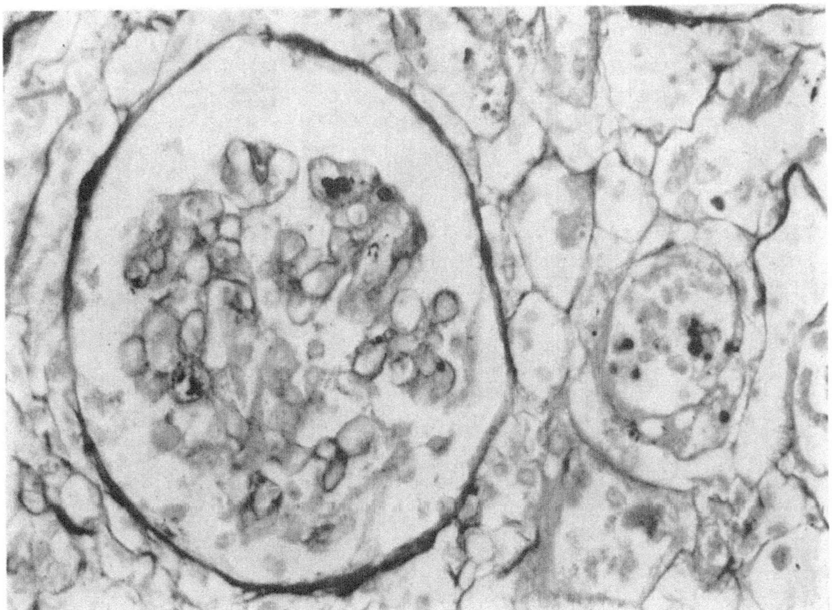

Fig. 15. Renal candidosis. Several fungal cells in a glomerulus and in a nearby dilated capillary (Hexamine silver, ×400)

Pathology: The general picture is that of an acute severe widespread mycotic infection. The organ most frequently involved is the kidney, which shows numerous areas consisting of fungus in both the mycelial and the yeast-like phase, and

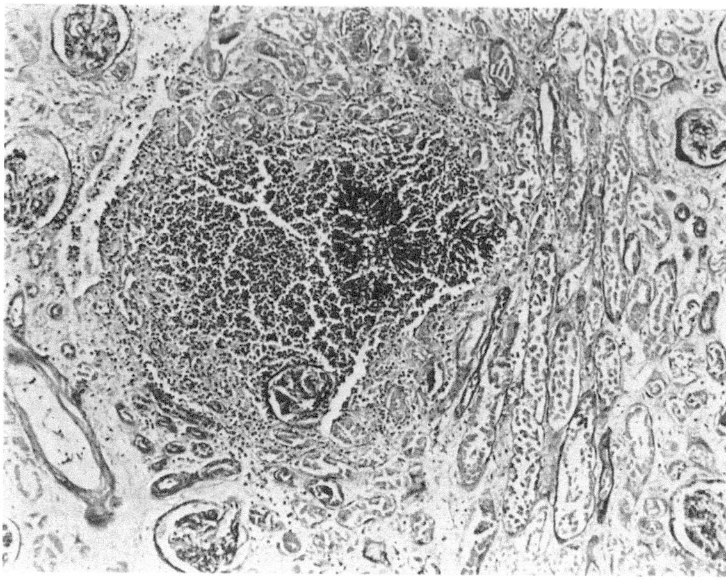

Fig. 16. Acute nephritic abscess. Candida is present in the centre of the lesion (PAS, ×65)

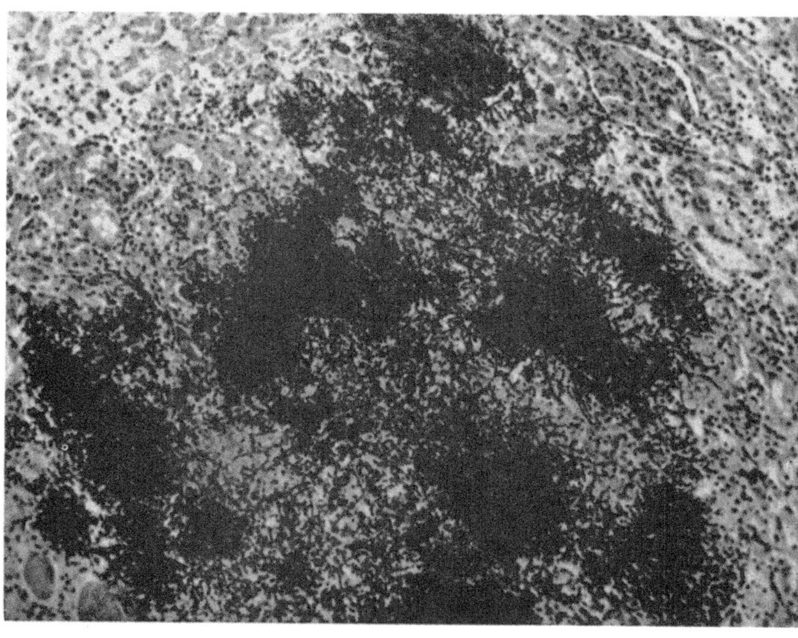

Fig. 17. Areas of fungal proliferation in the kidney. Note absence of cellular reaction. From the same case as Figs. 10, 23 and 33 (PAS, ×86). (From WINNER and HURLEY, 1964; Fig. 27, p. 148)

inflammatory cells. Some of the lesions are large enough to be seen with the naked eye. The lesions are usually confined to the cortex, but proliferating fungi may also occasionally be seen in the medulla. In some cases there is a marked poly-

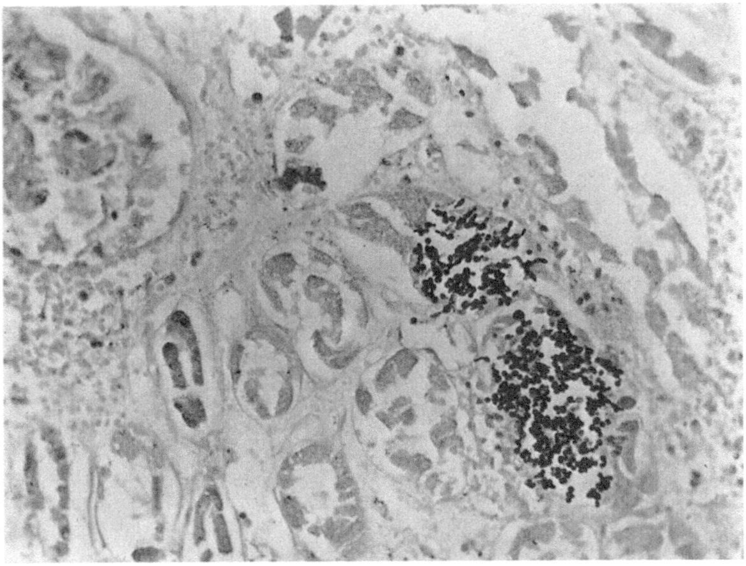

Fig. 18. Candida in kidney lesion (Gram stain, ×220)

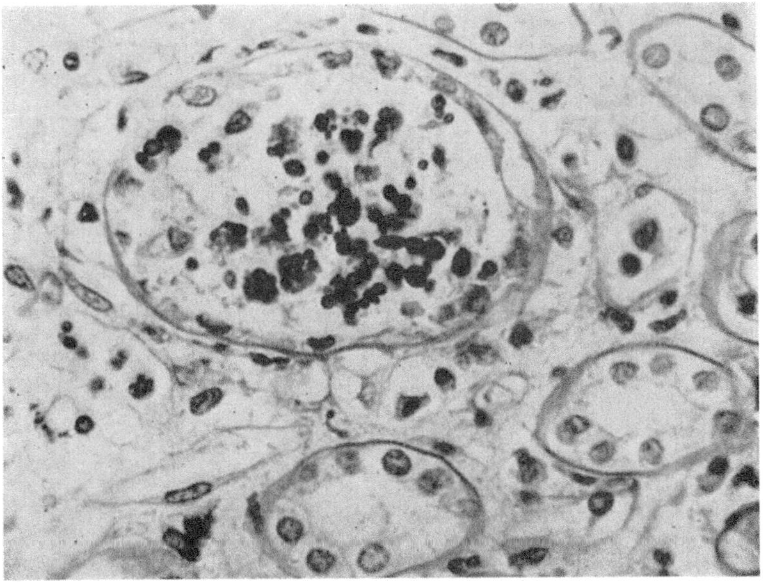

Fig. 19. Budding yeasts and cellular debris in a renal tubule. From a fatal case of disseminated candidosis in a woman of 50 (PASH, × 550). (From WINNER and HURLEY, 1964; Fig. 25, p. 146)

morphonuclear reaction to the fungus and the lesions are mycotic abscesses; in others, cellular reaction is minimal or absent (LEHNER, 1964) (Figs. 14—22).

Sometimes the fungus provokes a basal meningitis, in which small nodules, often less than 1 mm in diameter, are to be found close to blood vessels. The inflammatory exudate consists largely of lymphocytes and plasma cells, though a

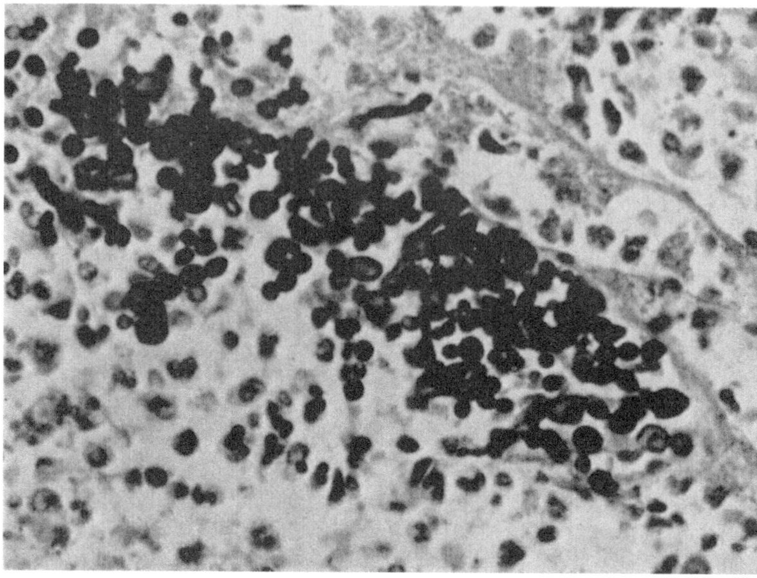

Fig. 20. Budding yeasts in renal tubule. From the same case as Fig. 19 (PASH, ×650). (From WINNER and HURLEY, 1964; Fig. 26, p. 147)

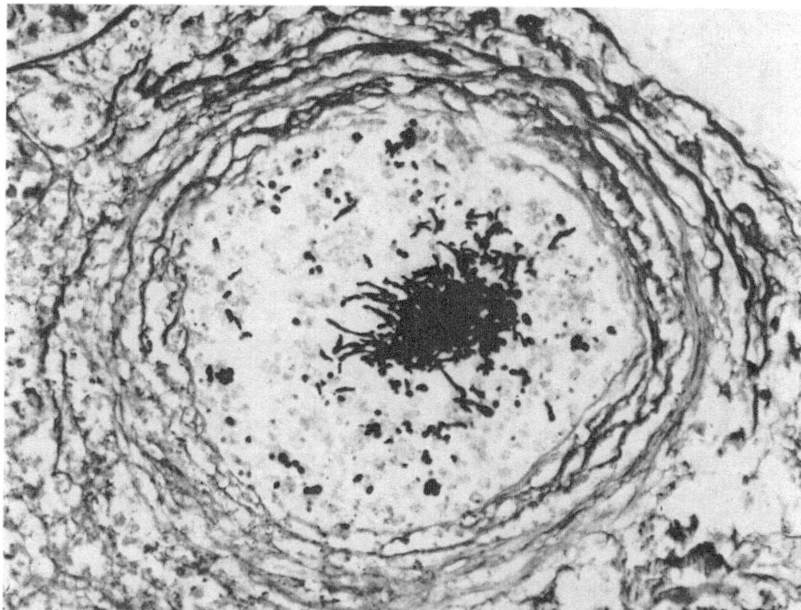

Fig. 21. Candida in fresh cellular thrombus in a small artery in a kidney (Hexamine silver, ×250)

few polymorphonuclear leucocytes may be present. The brain occasionally shows minute abscesses and granulomata, which may contain multinucleate giant cells (FETTER et al., 1967).

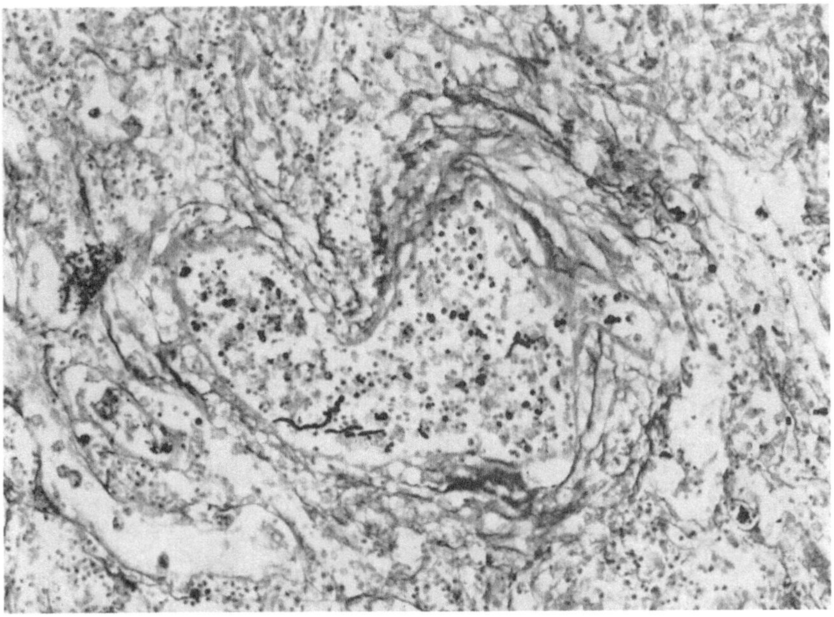

Fig. 22. Candida in a kidney venule (Hexamine silver, ×210)

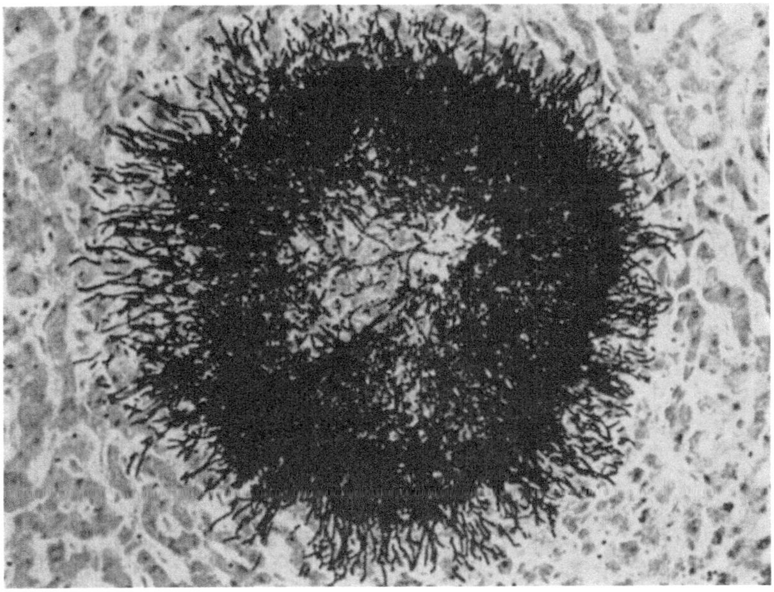

Fig. 23. Colony of Candida in the liver. There is no tissue reaction. From the same case as Figs. 10, 17 and 33 (PASH, ×130). (From WINNER and HURLEY, 1964; Fig. 28, p. 149)

Other organs which may be involved are the liver, lungs, blood vessels, spleen, thyroid, bone marrow, retina, skeletal muscles, and urinary bladder. These lesions do not show any particular features other than those which have already been described. Some of them are illustrated here (Figs. 23—28).

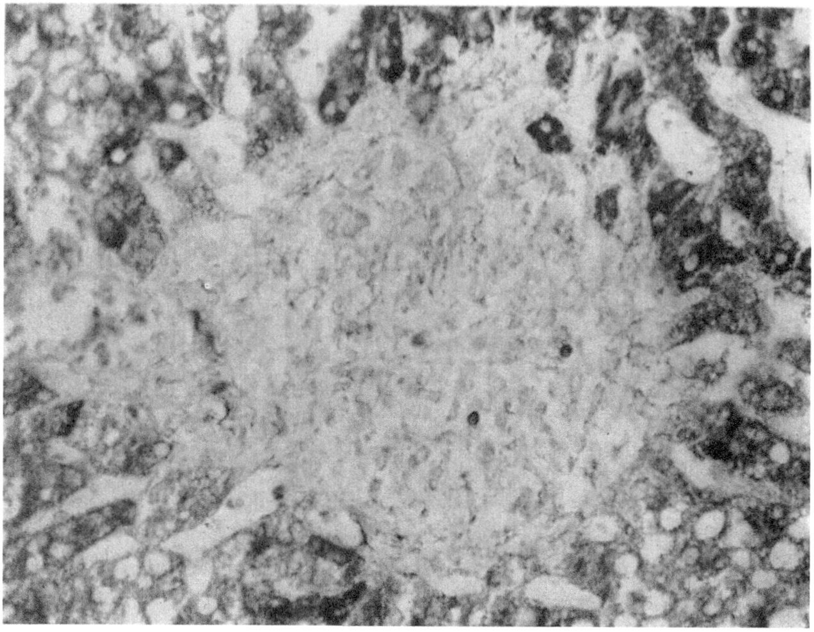

Fig. 24. Candidosis of the liver: A small focus of necrosis containing two candida cells in the liver of a patient with septicaemic candidosis (Hexamine silver, ×300)

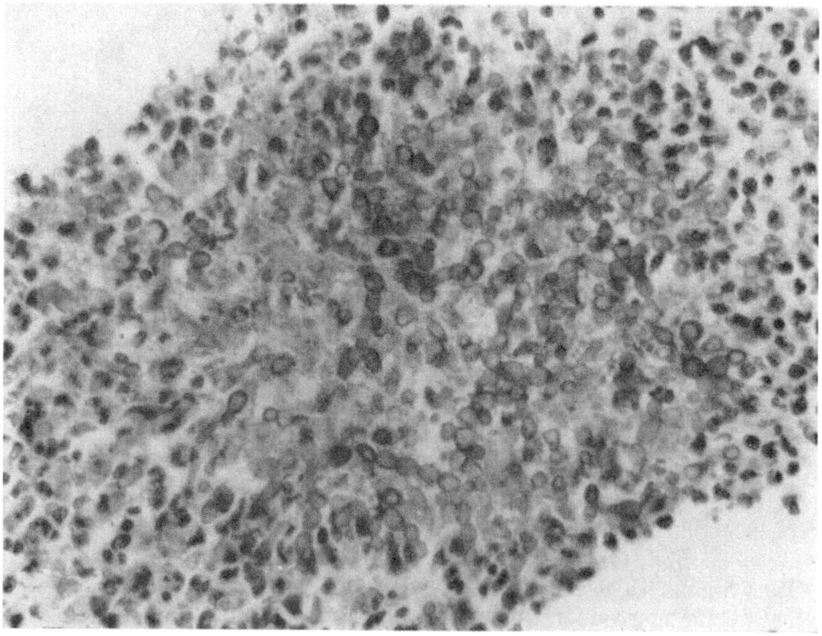

Fig. 25. Acute abscess of liver, showing candida (PASH, ×480)

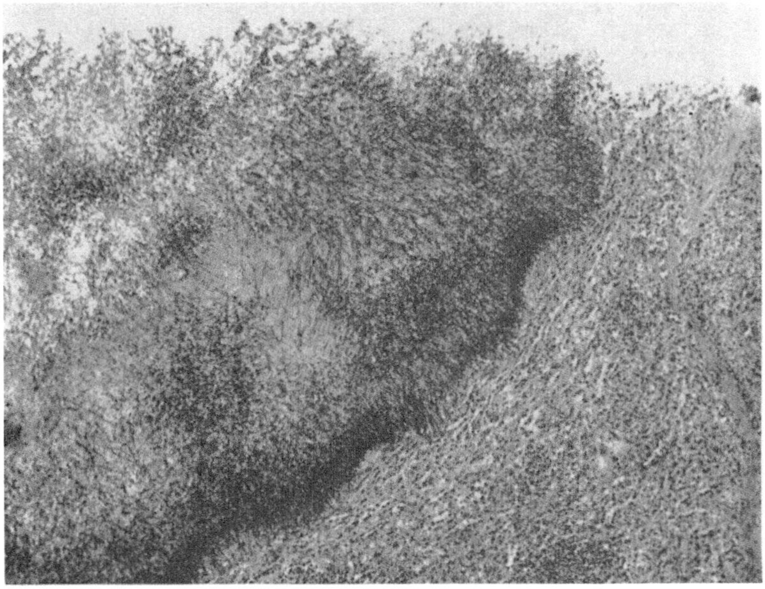

Fig. 26. Candida abscess of the spleen. From a fatal case of disseminated candidosis in a child of 6 treated for acute leukaemia (PASH, ×50). (From WINNER and HURLEY, 1964; Fig. 29, p. 150)

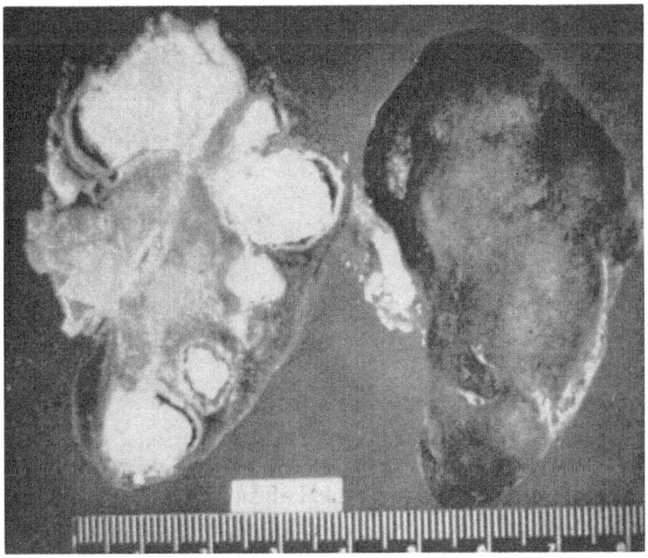

Fig. 27. Multiple infarcts in the kidney by emboli from candida vulvitis (From UTZ, WINNER and HURLEY, 1966; Fig. 5, p. 227)

h) Candida Endocarditis

This is a variant of acute disseminated (septicaemic) candidosis, with involvement of one or more heart valves as the predominant pathological feature. Predisposing factors are pre-existing disease of the heart valves, and more than

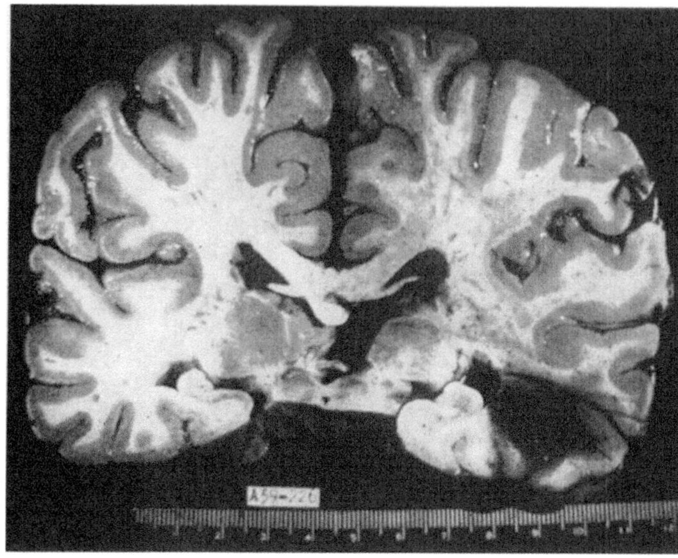

Fig. 28. Infarct of left cerebellum by embolus from candida valvulitis. (From UTZ, ed. WINNER and HURLEY, 1966; Fig. 6, p. 227)

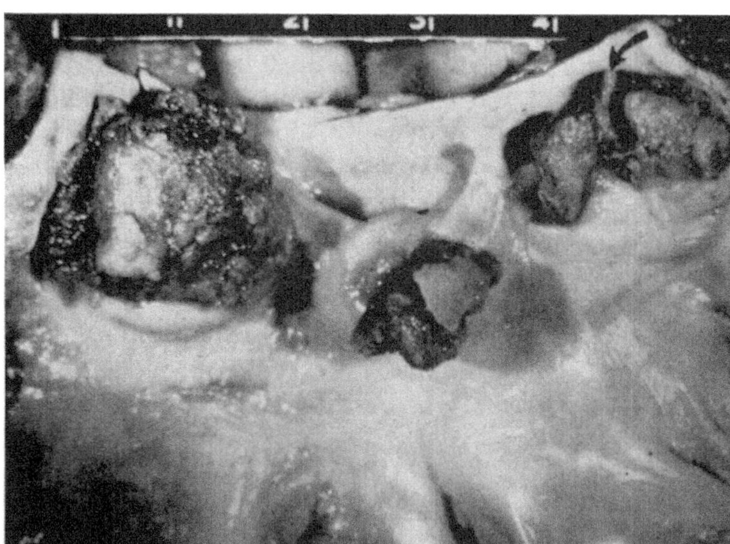

Fig. 29. Candida endocarditis of the aortic valve, viewed with heart opened. The arrow indicates the coronary ostium. (From UTZ, ed. WINNER and HURLEY, 1966; Fig. 3, p. 226)

half the cases have a previous history of rheumatic heart disease; other cases have congenital lesions (PEARL and SIDRANSKY, 1960; ANDRIOLE et al., 1962; KAPLAN and LEVIN, 1967). The disease is always blood-born. In some cases it is iatrogenic, or self-inflicted, in that the infection may get into the bloodstream from intravenous therapy. Some of the self-inflicted cases have occured in "main-line" drug addicts, who may give themselves "shots" of drugs which are infected with candida (WILSON, 1961).

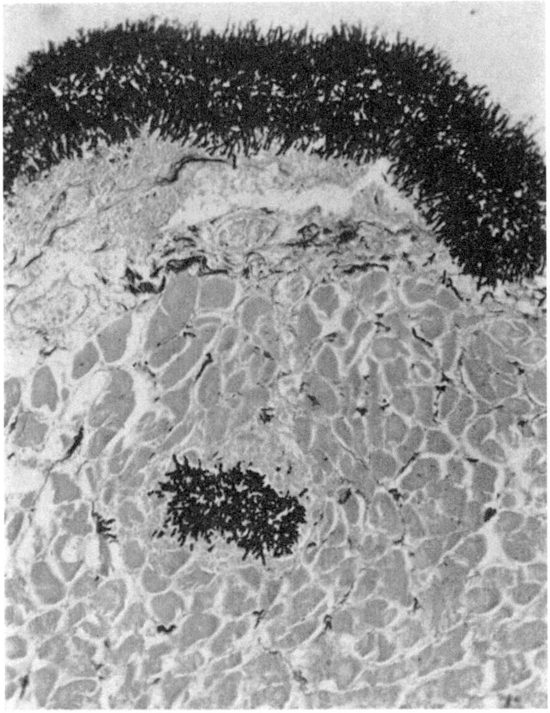

Fig. 30. Gross candida invasion of the pericardium and endocardium (Hexamine silver, ×120)

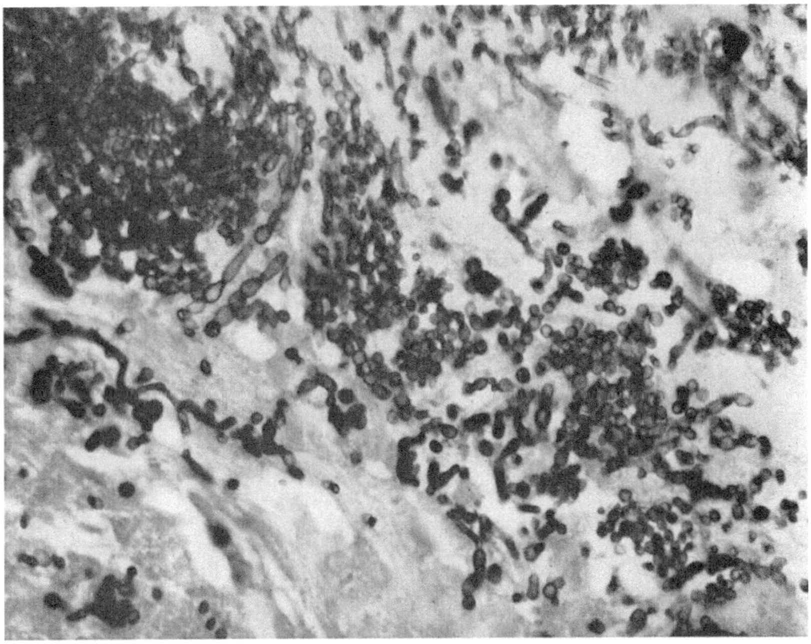

Fig. 31. *C.albicans* vegetation on mitral valve and invading the fibrotic tissue of the cusp (PASH, ×400)

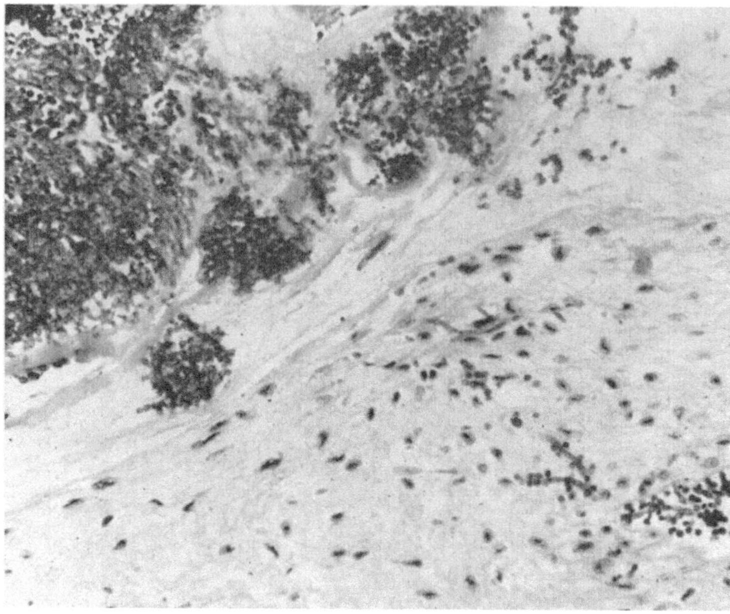

Fig. 32. Candida endocarditis (PASH, ×200)

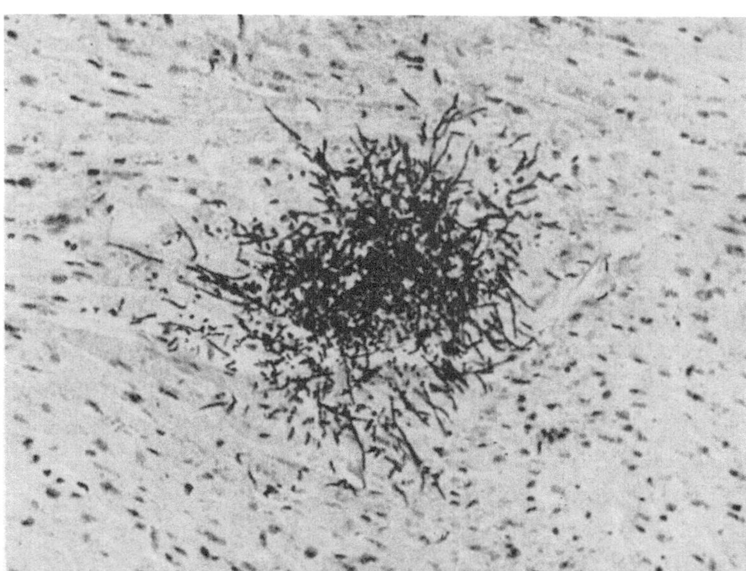

Fig. 33. Colony of *C. albicans* in the myocardium. From the same case as Figs. 10, 17 and 23 (PASH, ×160). (From WINNER and HURLEY, 1964; Fig. 32, p. 166)

Other cases have followed open-heart surgery for congenital or acquired valvular disease (ANDRIOLE et al., 1962; JAMSHIDI et al., 1963; CLIMIE and RACHMANINOFF, 1965) (Figs. 29—33).

Clinical Manifestations: The clinical manifestations of candida endocarditis are those of any infective endocarditis, together with those of a deep mycotic infection. Usually there has been some preceding systemic disorder, most commonly rheumatic heart disease. Special predisposing factors are those of other deep mycoses, that is to say, debilitating disease such as diabetes mellitus, antibiotic therapy or steroid therapy. The commonest clinical features are fever; abdominal pain; nausea and vomiting; heart failure, occurring in patients with recent or old cardiac murmurs; splenic or hepatic enlargement; and embolic manifestations. Embolism to the large arteries has been reported in one-third of the cases reviewed by ANDRIOLE et al. (1962).

Pathology: Necropsy or operation specimens of the heart valves show the presence of large friable vegetations. They are usually larger than those seen in bacterial endocarditis, which may, of course, precede mycotic infection. The incidence of infection on the various heart valves does not appear to differ from that in bacterial endocarditis.

Sections of the valves show filamentous and yeast forms of candida, as illustrated. Candida lesions may also be seen on the myocardium and pericardium.

i) Candidosis of the Urinary Tract

Candidosis of the urinary tract is a variety of the septicaemic disease (WINNER and HURLEY, 1964). It is clear both from experimental studies and from examination of cases of systemic infection that candida has a predilection for the kidney. It can apparently multiply in the parenchyma of that organ, probably more rapidly than in any other part of the body. The carriage of viable candida into the kidney by the bloodstream may therefore be sufficient in itself to initiate a candida lesion. It is not known whether pre-existing kidney disease is necessary for the successful establishment of candida; the large size of the organism may well produce blockage of the renal arterioles and hence bring about an embolic lesion which could favour the further multiplication of candida. Experimental studies give some colour to this idea (MACKINNON, 1936; EVANS and WINNER, 1954).

Unlike cases of candida septicaemia, which always occur in patients who are known to be suffering from some other constitutional disorder, candida infection of the urinary tract has been reported as a primary disease in those hitherto thought to be healthy (ALBERS, 1953; GUZE and HALEY, 1958; WINNER and HURLEY, 1964). This observation, of course, does not rule out the possibility, or, in the present author's view, the probability that the affected patients have been subject to some undetected disorder which has enabled the candida to change from a commensal to an invasive prasite.

Diagnosis: It appears to be a melancholy fact that many cases of candidosis of the urinary tract have been missed because the finding of *C. albicans* in urine specimens has been disregarded as insignificant. When searching for the cause of an apparent clinical urinary infection, clinicians (and pathologists) have perhaps been excessively obsessed by the overwhelming importance of coliform bacilli, faecal streptococci, and tubercle bacilli. Another factor is the unhygienic and septic conditions under which urine specimens are collected, which obviously may favour aerial contamination. A combination of these two factors has led to a situation in which any yeasts found in urine specimens are throught to be of no account.

This is far from the case, as the recent emergence of urinary candidosis as a clinical entity has shown. In all cases of suspected urinary infection, where no other microbiological cause has been demonstrated, a candida or other yeast isolated from the urine must be regarded as being the cause until this is proved otherwise.

Experimental studies have confirmed that candida in the bloodstream of animals can produce a chronic pyelonephritis (HURLEY and WINNER, 1963).

Thrush of the urinary bladder: This has been reported as a cause of acute cystitis (MOULDER, 1946; SNYDER, 1948; SAUER and METZNER, 1948). A distinct variety of this disease has been the appearance of candida in "fungus ball of the bladder" (CHISHOLM and HUTCH, 1961).

6. Experimental Aspects of Candida Infections

Experimental aspects of candida infections have attracted many workers. The reasons for this are not hard to seek; the organism is accessible and easily handled, and it is not difficult to manipulate the host in such a way to as produce variations in immunity which might be expected to favour the onset of the disease.

In these days of antibiotics and immunosuppressive therapy it is obviously important to achieve a better understanding of the mechanisms whereby an "opportunistic" pathogen such as candida may cause disease; for this reason, experimental studies of candidosis have an importance far beyond the boundaries of the subject itself.

The earliest such studies, dating back well over a hundred years, were attempts to produce disease of the mucous membranes in animals and humans. These experiments led to attempts to reproduce systemic candidosis. Workers as early as the end of the 19th century reported that the organ most affected in systemic candidosis was the kidney. The experimental disease produced in laboratory animals by the intravenous injection of *C.albicans* has been described in great detail by several groups of workers. A comprehensive bibliography is to be found in the monograph by WINNER and HURLEY (1964).

At the time of writing, the main lines of research on candida infection in various centres may be summarised as follows:

1. Studies of variations in the parasite — differences in the pathogenicity of different species of candida, and of different strains of single species. (HURLEY, 1967).

The different species of candida may now be placed in a descending order of pathogenicity, as follows: *C.albicans*, *C.stellatoidea*, *C.tropicalis*, *C.parapsilosis*, *C.pseudotropicalis*, *C.guilliermondii*, and *C.krusei*. An elegant demonstration of the differential pathogenicity of different species of candida is that of STANLEY and HURLEY (1967).

At the time of writing it is still doubtful whether the different strains of a particular candida species show different pathogenicity .Current studies of *Candida albicans* Types A and B, however, suggest that Type A is more pathogenic than Type B.

2. Studies of variations in the host:

(a) *Modification of host resistance* by external factors such as irradiation, toxic poisoning, surgical operations, damage to skin and mucous membranes, immunosuppressive and antibacterial therapy. The object of this work is to reproduce in the host conditions similar to those in which clinical candidosis may occur. A recent review with additional information is that of HURLEY (1966).

(b) Study of *individual host factors* such as the effect of phagocytes and of antibodies on the candida cell. (LOURIA and BRAYTON, 1964) (1 and 2).

3. Studies of possible *pathogenic mechanisms*, such as the production of toxins by candida. (WINNER and HURLEY, 1964).

4. Detailed study of the *candida cell*, both intact and altered by disruption or exposure to antifungal antibiotics.

(a) Metabolism.

(b) Antigenic analysis (STALLYBRASS, 1964, 1965).

(c) Ultrastructure studies.

5. Experimental studies of mechanisms of immunity and allergy.

Work on some of these latter aspects is in progress at the time of writing, and it is premature to give references to these.

Acknowledgments for Illustrations: The author acknowledges with thanks the provision of Figs. 4, 13, 14, 15, 21, 22, 24, 25, 31 and 32 by Professor W. ST. C. SYMMERS; Figs. 6, 7, 8, 9, and 11 by Professor R. A. CAWSON; Figs. 16, 18, and 30 by Dr. T. LEHNER; Figs. 27, 28 and 29 by Dr. J. P. UTZ; Fig. 2 by Dr. D. W. R. MACKENZIE. Figs. 1, 2, 3, 5, 10, 12, 17, 19, 20, 23, 26, and 33 are reproduced from "Candida albicans" by H. I. WINNER and R. HURLEY (Churchill, London, 1964), by permission of the publishers. Figs. 28 and 29 are reproduced from "Symposium on Candida Infections", edited by H. I. WINNER and R. HURLEY (Livingstone, Edinburgh, 1966), by permission of the publishers. Fig. 27 is reproduced from the American Journal of Medicine, by permission of the publishers.

References

AKIBA, T., IWATA, K., INOUYE, S.: Studies on the serologic diagnosis of deep-seated Candidiasis. Jap. J. Microbiol. **1**, 11 (1957). — ALBERS, D. D.: Monilial infection of (the) kidney (s): case reports. J. Urol. (Baltimore) **69**, 32 (1953). — AMON, H., SCHREYER, W.: Sepsis mycotica (Candidiasis) mit disseminierter Thrombarteriitis. Frankf. Z. Path. **71**, 370 (1961). — ANDRIOLE, V. T., KRAVETZ, H. M., ROBERTS, W. C., UTZ, J. P.: Candida endocarditis: Clinical and pathologic studies. Amer. J. Med. **32**, 251 (1962). — BEEMER, A. M., PRYCE, D. M., RIDDELL, R. W.: *Candida albicans* infection of the gut. J. Path. Bact. **68**, 359 (1954). — BEUTHE, D.: Candida-Sepsis. Wiss. Z. Univ. Jena, Math.-nat. Reihe 4, S. 199 (1955). ~ Candidasepsis. Zbl. allg. Path. path. Anat. **93**, 241 (1955). — BICHEL, J. STENDERUP, A.: Investigaciones experimentales acerca del efecto de la monilia (Candida albicans) sobre la linfopoyesis del raton. Folia clin. int. (Barcelona) **5**, 273 (1955). — BOUND, J. P.: Thrush napkin rashes. Brit. med. J. **1**, 782 (1956). — CAPLAN, H.: Monilial (Candida) endocarditis following treatment with antibiotics. Lancet **II**, 957 (1955). — CARSLAW, R. W., RICHARDSON, Anne, CAMERON, J. A.: Monilial granuloma in an adult, successfully treated with amphotericin B. Brit. J. Derm. **78**, 242 (1966). — CAWSON, R. A.: Chronic Oral Candidosis, denture stomatitis, and chronic hyperplastic candidosis. In: "*Symposium on Candida infections*", edited by H. I. WINNER and R. HURLEY, Livingstone, Edinburgh, pp. 138—153 (1966). — CAWSON, R. A., LEHNER, T.: Chronic hyperplastic candidiasis — Candidal leukoplakia. Brit. J. Derm. **80**, 9 (1968). — CHISHOLM, E. R., HUTCH, J. A.: Fungus ball (*Candida albicans*) formation in the bladder. J. Urol. (Baltimore) **86**, 559 (1961). — CLAYTON, YVONNE M., NOBLE, W. C.: Observations on the epidemiology of *Candida albicans*. J. clin. Path. **19**, 76 (1966). — CLIMIE, A. R. W., RACHMANINOFF, N.: Fungal (Candida) Endocarditis following open-heart surgery. J. thorac. cardiovasc. Surg. **50**, 431 (1965). — DANBOLT, N.: Deep cutaneous moniliasis, a fatal case of a peculiar type. Acta derm.-venereol. (Stockh.) **21**, 98 (1940). — DEBRÉ, R., MOZZICONACCI, P., DROUHET, E., DROUHET, V., HOPPELER, A., GRUMBACH, R., HABIB, R.: Les infections a Candida chez le nourrisson. Ann. paediat. (Basel) **184**, 129 (1955). — DUBOIS-FERRIÈRE, H., FEUARDENT, R., GOLDSCHLAG, H., BURSTEIN, L., BOUVIER, C., BÖHNI, H.: Septicémies à muguet (candida albicans) favorisées par les antibiotiques et la cortisone au cours d'hémopathies malignes. Helv. med. Acta **22**, 477 (1955). — EVANS, W. E. D., WINNER, H. I.: The histogenesis of the lesions in experimental moniliasis in rabbits. J. Path. Bact. **67**, 531 (1954). — FALKMER, S., WISING, P. J.: Fatal bronchopulmonary moniliasis. Acta med. scand. **151**, 117 (1955). — FETTER, B. F., KLINTWORTH, G. K., HENDRY, W. S.: *Mycoses of the Central Nervous System*. Baltimore: Williams & Wilkins Co. 1967. — FRISK, A.: Lungcandidos (monilos) och antibioticabehandlung. Svenska Läk.-Tidn. **49**, 1493 (1952). — GAVALLÉR, BELA DE: Contribucion a la patogenicidad de las monilias en recién nacidos a base de los hallazgos anátomo-pathológicos. Rev. sudamer. Morf. **10**, 117 (1952). — GELLI, G.: Micosi (candidosi) polmonare in forma cavitaria; (particolare riferimento al trattamento chirurgico). Riv. Clin. pediat. **54**, 155 (1954). — GERLÓCZY, F., SCHMIDT, K., SCHOLZ, M.: Beiträge zur Frage der Moniliasis im Säuglingsalter. Ann. paediat. (Basel) **187**, 119 (1956). — GIESSLER, G., GULOTTA, F.: Moniliasis des Zentralnervensystems. Zbl. allg. Path. path. Anat. **105**, 433 (1964). — GUZE, L. B., HALEY, L. D.: Fungus infections of the urinary tract. Yale J. Biol. Med. **30**, 292 (1958). — HAUSER, F. V., ROTHMAN, S.: Monilial granuloma: report of a case and review of the literature. Arch. Derm. Syph. (Chic.) **61**, 297 (1950). — Hoffmann, D.H.: Die experimentelle endogene Entzündung des Augeninneren durch Candida albicans. Ophthalmoskopische, histologische und mikrobiologische Studien zum Ablauf der Infektion beim Kaninchen. Ophthalmologica (Basel) Vol. 151 Suppl. Basel: S. Karger 1966. — Hoffmeister, W., Dickgiesser, F., Götting, H.: Tierexperimentelle und serologische Untersuchungen zur Diagnostik und Therapie der Infektion mit Candida albicans. Dtsch. Arch. klin. Med. **198**, 499 (1951). — Hurley, R.: Experimental infection with *Candida albicans* in modified hosts. J. Path. Bact. **92**, 57 (1966). — Hurley, R.: The Pathogenic Candida species: A Review. Rev. med. vet. Mycot. **6**, 159 (1967). — Hurley, R., Winner, H.I.: The pathogenicity of Candida tropicalis. J. Path. Bact. **84**, 33 (1962). — Hurley, R., Winner, H.I.: Experimental renal moniliasis in the mouse. J. Path. Bact. **86**, 75 (1963). ~ Pathogenicity in the genus Candida. Mycopathologia (Den Haag) **24**, 337 (1964). — Hurwitz, A., Yesner, R.: A report of a case of localized bronchopulmonary moniliasis successfully treated by surgery. J. thorac. Surg. **17**, 26 (1948). — Jamshidi, A., Pope, R.H., Friedman, N.H.: Fungal Endocarditis Complicating Cardiac Surgery. Arch. intern. Med. **112**, 370 (1963). — Kärcher, K.H.: Experimentelle Untersuchungen zur Pathogenität und biologischen Wirkung der Candida albicans an Mensch und Tier. Arch. klin. exp. Derm. **202**, 424 (1956). ~ Die Candidamykose (Candidosis, Candidiasis Moniliasis, Oidiomykosis), aus Pilzkrankheiten der Haut durch Hefen, Schimmel, Actinomyceten, Handbuch der Haut- und Geschlechtskrankheiten von J. Jadassohn, Ergänzungswerk Bd. IV/4, S. 1—74. Berlin-Göttingen-Heidelberg: Springer 1963. — Kaplan, B., Levin, S.E.: Candida Endocarditis Complicating Congenital Heart Disease: A case report. Med. Proc. **13**, 614 (1967). — Kaufman, S.A., Scheff, S., Levene, G.: Esophageal moniliasis. Radiology **75**, 726 (1960). — Khmelnitsky, O.K.: Pathologico-anatomical changes in visceral candidiasis. Arkh. Pat. (Mosk.) **24**, 20 (1962). — Lannigan, R., Meynell, M.J.: Moniliasis in acute leukaemia. J. clin. Path. **12**, 157 (1959). — La Touche, C.J.: Candida in Chronic Ear Infections. In: "*Symposium on Candida infections*", edited by H.I. Winner and R. Hurley, Livingstone, Edinburgh, pp. 154—160 (1966).— Lehner,T.: Systemic Candidiasis and Renal Involvement. Lancet **I**, 1414 (1964). ~ Immunofluorescent investigation of Candida albicans antibodies in human saliva. Arch. oral Biol. **10**, 975 (1965). ~ Classification and Clinico-Pathological Features of Candida Infections in the Mouth. In: "*Symposium on Candida Infections*", edited by H.I. Winner and R. Hurley, Livingstone, Edinburgh, pp. 119—137 (1966). — Lemaistre, J.: Etude sur l'air de la ville de Limoges: de la perlèche: du Streptococcus plicatilis. J. Soc. Méd. Pharm. Haute-Vienne **10**, 41 (1886). — Lewis, G.M., Hopper, M.E., Montgomery, R.M.: Infections of the skin due to Monilial albicans. I. Diagnostic value of intradermal testing with a commercial extract of Monilial albicans. N.Y. St. J. Med. **37**, 878 (1937). — Louria, D.B., Brayton, R.G.: A substance in blood lethal for *Candida albicans*. Nature (Lond.) **201**, 309 (1964). ~ Behaviour of Candida cells within leukocytes. Rev. med. vet. Mycol. **66**, 956 (1964). — Ludlam, G.B., Henderson, J.L.: Neonatal thrush in a maternity hospital. Lancet **1942**, **I**, 64. — Mackinnon, J.E.: Caracteres y grado de la virulencia experimental de las Torulopsidáceas de la subfamilia micotorulaceas (monilias). An. Fac. Med. Montevideo **21**, 320 (1936).— Maibach, H.I., Kligman, A.M.: The biology of experimental human cutaneous moniliasis (*Candida albicans*), Arch. Derm. **85**, 233 (1962). — Mankowski, Z.T.: Occurrence of malignancy, collagen diseases and Myasthenia gravis in the course of experimental infections with Candida albicans. Mycopathologia (Den Haag) **19**, 1 (1962). — Marsh, A.P.: Oesophageal moniliasis. Amer. J. Roentgenol. **82**, 1063 (1959). — Masshoff, W., Adam, W.: Histomorphologie der experimentellen Candida-Infektion. Arch. klin. exp. Derm. **204**, 416 (1957).— Miyake, M., Okudaira, Akiba, T., Inoue, S.: An immunopathologic study on deep seated candidiasis, especially on the precipitin reaction using the infected tissue extract as antigen. Acta path. jap. **10**, 25 (1960). — Moulder, M.K.: Thrush of the urinary bladder: case report. J. Urol. (Baltimore) **56**, 420 (1946). — Nezelof, C., Sarrut, S.: Infections mortelles à candida chez le nouveau-né et le nourrisson. Six observations anatomohistologiques. Ann. Pédiat. **1957**, 531. — Nilsby, I., Norden, A.: Studies of the occurrence of Candida albicans. Acta med. scand. **133**, 347 (1949). — Nityananda, K.: Monilial Granuloma. Brit. med. J. **1**, 690 (1959). — Oblath, R.W., Donath, D.H., Johnstone, N.G., Kerr, W.J.: Pulmonary moniliasis. Ann. intern. Med. **35**, 97 (1951). — Orie, N.G.M.: De aanwezigheid en de betekenis van gisten in de luchtwegen. Thesis. Utrecht 1946. ~ Candida (Monilia) infection of the respiratory tract. Dis. Chest. **22**, 107 (1952). — Owen, C.R., Anderson, M.B., Henrici, A.T.: Allergy in Monilia and yeast infections. Mycopathologia (Den Haag) **1**, 10 (1939). — Pearl, M.A., Sidransky, H.: Candida endocarditis: two new cases with a review of 12 previously reported cases. Amer. Heart J. **60**, 345 (1960). — Ramel, E.: Beitrag zur Kenntnis der Hautblastomykose mit besonderer Berücksichtigung der Allergieerscheinungen. Arch. Derm. Syph. (Berl.) **148**, 218 (1925). — Ravaut, P., Rabeau, H.: Parakeratoses eczematiformes provoquees par des injections intradermiques de levurine. Presse méd. **37**, 372 (1929). — Rey, J.C., Natin, I.: Moniliasis pulmonar. Pren. méd. argent. **1953**, 2306. — RIDDELL, R.W., CLAYTON, Y.M.: Pulmonary mycoses occurring in Britain. Brit. J. Tuberc. **52**, 34 (1958). — ROBERTSON, R.F.: Pulmonary moniliasis. Edinb. med. J. **55**, 274 (1945). — ROBINSON, R.C.V.: Cutaneous moniliasis in infants. J. Pediat. **50**, 721 (1957). — ROCKWOOD, E.M., GREENWOOD, A.M.: Monilial infection of the skin: report of a fatal case. Arch. Derm. Syph. (*Chic.*) **29**, 574 (1934). — ROTHMAN, S.: Some unusual forms of cutaneous moniliasis. Arch. Derm. **79**, 598 (1959). — RUHRMANN, G., ADAM, W.: Tödliche pulmonale Moniliasis eines Säuglings. Virchows Arch. path. Anat. **327**, 273 (1955). — SAUER, H.R., METZNER, W.R.T.: Thrush infection of the urinary bladder. J. Urol. (Baltimore) **59**, 38 (1948). — SCHABERG, A., HILDES, J.A., WILT, J.C.: Disseminated candidiasis. Arch. intern. Med. **95**, 112 (1955). — SCHULTE, H.W.: Die lokalisierte Candidamykose in Kombination mit der Lungentuberkulose. Tuberk.-Arzt **11**, 751 (1957). — SCLAFER, J.: L'allergie à Candida albicans (clinique; diagnostique; traitement). Sem. Hôp. (Paris) **33**, 1330 (1957). — SCLAFER, J., FLAVIAN, N.: Étude de l'allergie cutanée a Candida albicans chez les enfants. Sem. Hôp. (Paris) **30**, 1557 (1954). — SHREWSBURY, J.F.D.: Secondary thrush of the bronchi. Quart. J. Med. **5**, 375 (1936). — SKOBEL, P., JORKE, D., SCHABINSKI, G.: Akute, generalisierte Mykose. Sepsis durch Candida pseudotropicalis. Münch. med. Wschr. 194 (1955). — SMITH, G.M., ARMEN, R.M.: Pulmonary moniliasis treated by brilliant green aerosol: report of a case. Ann. intern. Med. **43**, 1302 (1955). — SNYDER, O.C.: Thrush of urinary bladder; case report. Urol. cutan. Rev. **52**, 80 (1948). — SOLOV, K.A.: Candida pneumopathies in neonates (Candidaerkrankungen der Lungen bei Neugeborenen). Arkh. Pat. (Mosk.) **26**, Nr. 1, S. 60—63 mit engl. Zusammenfassung (1964). — STALLYBRASS, F.C.: Candida precipitins. J. Path. Bact. **87**, 89 (1964). ~ Serotypes of Candida albicans. Nature (Lond.) **207**, 220 (1965). — STANLEY, VALERIE C., HURLEY, ROSALINDE: Growth of Candida species in culture of mouse epithelial cells. J. Path. Bact. **94**, 301 (1967). — STIEFEL, E., ANDRÉU URRA, J., LAZÁRO, J.: A propósito de la candidasis pulmonar, antigua moniliasis. Rev. esp. Tuberc. **22**, 225 (1953). — SYMMERS, W.ST.C.: The tissue reactions in deepseated fungal infection. The role of histological examination in mycological diagnosis. Ann. Soc. belge Méd. trop. **44**, 869 (1964). ~ Septicaemic candidosis. In: "Symposium on Candida infections", edited by H.I. WINNER and R. HURLEY, Livingstone, Edinburgh, pp. 196—212 (1966). — THJÖTTA, TH., AMLIE, R.: Three cases of infection with Monilia (Castellania) tropicalis Castellani. Acta path. microbiol. **24**, 161 (1947). — TOBLER, R., COTTIER, H.: Familiäre Lymphopenie mit Agammaglobulinämie und schwerer Moniliasis. Die "essentielle Lymphocytophthise" als besondere Form der frühkindlichen Agammaglobulinämie. Helv. paediat. Acta **13**, 313 (1958). — VIALATTE, J., SATGE, P., DROUHET, E., WOLF: La forme granulomateuse de la moniliase cutanée. Arch. franç. Pédiat. **19**, 37 (1962). — WHITAKER, J., LANDING, B.H., ESSELBORN, V.M., WILLIAMS, R.R.: The syndrome of familial juvenile hypoadrenocorticism, hypoparathyroidism and superficial moniliasis. J. clin. Endocr. **16**, 1374 (1956). — WIESE, E.R., BIXBY, E.W.: Bronchiectasis associated with monilia simulating pulmonary tuberculosis. Clinical Pathologic study. J. Lab. clin. Med. **26**, 624 (1941).—WILSON, J.W.: Abstract of discussion of "The biology of experimental human cutaneous moniliasis (*Candida albicans*)" (MAIBACH and KLIGMAN). Arch. Derm. **85**, 254 (1962). — WILSON, J.W., PLUNKETT, O.A.: The Fungous Diseases of Man, p. 175. University of California Press Berkeley: 1965. — WILSON, R.M.: Candidal endocarditis. J. Amer. med. Ass. **177**, 332 (1961). — WINNER, H.I.: A study of *Candida albicans* agglutinins in human sera. J. Hyg. (Lond.) **53**, 509 (1955). ~ Experimental moniliasis in the guinea-pig. J. Path. Bact. **79**, 420 (1960). — WINNER, H.I., HURLEY, ROSALINDE: *Candida albicans*. London: J. & A. Churchill (1964). — WINNER, H.I., HURLEY, R.: Symposium on Candida infections, pp. 226—227. Edinburgh: E & S Livingstone, Ltd. (1966).

Aspergillosis

CARLOS E. PEÑA, Washington/D.C., USA

With 35 Figures

Introduction

Aspergillus, a ubiquitous fungus with thin branching hyphae and stout sporulating fruiting heads, may sometimes mount the horse of opportunism to become a dangerous pathogen.

The genus includes 150 recognized species and varieties, of which only a relatively small number are pathogenic. The fungus has the ability to colonize living and dead matter, enjoys a worldwide distribution and possesses a wide spectrum of pathogenicity for man and for animals. In man the infections most often involve the lungs, but can also affect almost any organ of the body. In the veterinary field the fungus' importance stems from its ability to produce abortion in mares, cattle and sheep, to colonize the respiratory tract of birds, and to attack insects. In industry, where its fermentative properties can be utilized, workers are oftentimes exposed to heavy concentrations of air-borne spores.

Definition

Aspergillosis can be broadly defined as any infection or colonization of tissues or cavities by fungi of the genus *Aspergillus*. This includes proliferation in normal or dilated bronchi or caverns, a necrotizing or granulomatous pneumonitis sometimes with hematogenous dissemination, endocarditis, sinusitis, infections of the tegument and certain cases of mycetoma pedis. In addition the spores of the fungus may act in man as an allergen that when inhaled produces a hypersensitivity reaction similar to bronchial asthma.

History

Aspergillus, like many other mycological terms, was coined by MICHELI, a botanist from Florence, in the descriptivist atmosphere of the 18th century. According to him the fungus was thus called because of the similarity of its fruiting heads and the "aspergillum" (from *aspergere*, to sprinkle), a perforated globe used to sprinkle holy water in religious ceremonies. In his treatise "Nova Plantarum Genera", published in 1729, MICHELI listed nine different species, separated mainly on the basis of the characteristics of the fruiting heads, which he proceeded to identify with Latin sentences. This use of sentences to characterize the species is hardly surprising, since the system of binomial nomenclature, devised by LINNAEUS, did not appear until 1753.

Subsequently LINK (1809) described several species of the fungus from decaying vegetation, including *A. flavus*, *A. candidus* and *A. glaucus*. Recognition of the role of *Aspergillus* as an agent of disease waited more than 100 years after its first description, when Pasteur, Koch and others showed that many diseases were caused by microorganisms. In the case of *Aspergillus* this knowledge first came from observations in animals.

The earlier communications of the fungus infections which were probably aspergillosis, were those of MAYER (1815), who found a mold in the lungs of a jay (*Corvus glandarius*), and JÄGER (1816), who observed a similar condition in the lungs of two swans. However, they did not classify the fungus any further. DESLONGCHAMP (1841) also reported on a "mold developed during life" in the air pockets of a duck (*Anas mollissima*), but like his predecessors, did not give a precise classification of the fungus. The first definite account of an *Aspergillus* infection in an animal was that of RAYER and MONTAGNE (1842), who identified *A. candidus* in the air pockets of a tuberculous bullfinch (*Pyrrhula vulgaris*).

All too often HUGHES BENNETT of Edinburgh (1844) has been credited with the first description of a human case of the disease. However, according to the pictures and the description given, the fungus was probably not an *Aspergillus*.

The first human case in which *Aspergillus* seems to have been identified was that of SLUYTER (1847). The patient was a woman who died of a pulmonary disease. In a cavern of one of the lungs, BAUM, LITZMANN and EICHSTEDT found a black mycelial mass. The authors thought that the fungus was a MUCOR, but VIRCHOW believed that it was actually an *Aspergillus*. ROBIN (1853) reviewed the literature and reported *Aspergillus nigrescens* in the air pockets of a pheasant that died of tuberculosis.

VIRCHOW (1856) described four cases of bronchial and pulmonary aspergillosis in patients who died respectively of dysentery, pulmonary inflammation with emphysema, gastric carcinoma and pneumonia. The same year Friedreich in Würzburg recorded a case of "pneumomycosis aspergillina" characterized by the presence of the fungus in a tuberculous cavity. CRAMER in 1859 isolated *A. niger* from ear infections that he reported under the name of "*Sterigmatocystis antacustica*". FRESENIUS described in 1863 *A. fumigatus*, encountered in the respiratory tract of a bustard (*Otis tarda*) from the Frankfurt Zoological Park. WREDEN reported in 1867 two new types, *A. flavescens* and *A. nigricans*, isolated from human ear infections, which he considered to be varieties of *A. glaucus*. Other cases of human pulmonary infections were described by COHNHEIM (1865), FÜRBRINGER (1876) and LICHTHEIM (1882).

Towards the end of the 19th century came the discovery that many microbiological diseases occurring in man could be reproduced in experimental animals. *Aspergillus* was then used as a pioneer tool by GRAWITZ (1880) who produced lesions in the brains of rabbits and dogs by the injection of the fungus into the carotid artery.

The first cases of *Aspergillus* infections observed on this side of the Atlantic were reported by CHARLES BURNETT of Philadelphia, who studied a number of patients with otomycosis (1874, 1879, 1889). OSLER (1887) also recorded the case of a woman who expectorated bodies consisting almost completely of aspergillar hyphae.

During the same period SIEBENMANN (1889) published in Germany his comprehensive monograph on human ear infections, and WHEATON (1890) observed the first case ever published in Britain. SCHUBERT (1885) and MACKENZIE (1893) recorded cases of nasal and maxillary sinus infections.

At the same time French physicians started to show interest in the study of aspergillosis. In 1890 DIEULAFOY et al. first observed the disease in pigeon-crammers ("gaveurs de pigeons"). The duty of a "gaveur" consisted of chewing wet grain which they then fed to the animals, mouth to beak. The authors called the condition "pseudotuberculose mycosique" and described quite adequately its clinical picture, as well as the histopathology of the lesions in birds. Furthermore, they reproduced it experimentally and identified *A. fumigatus* as the etiologic agent. GAUCHER and SERGENT (1894) also observed a case of this condition in a pigeon-crammer. RÉNON's series of studies published in 1893, 1895 and 1897 examined the subject on a wider perspective. In his last study the author discussed four previously published instances of pulmonary aspergillosis in pigeon-crammers (DIEULAFOY et al., 1890; POTAIN and RÉNON, RÉNON, 1893; GAUCHER and SERGENT, 1894). In addition he commented on two cases, husband and wife, that he had observed beforehand (1895), in "peigneurs de cheveux" (hair-combers), who used rye flour for cleaning hair which was used in the manufacturing of wigs. The author also distinguished between primary and secondary disease according to the absence or presence of a pre-existing condition, and was able to compile 68 references on human aspergillosis.

English physicians were also active in this field as exemplified by the 2 cases reported by BOYCE (1893), and by ARKLE and HINDS (1895—1896).

OPPE (1897) recorded in Germany the first case of cerebral involvement, due to intracranial extension of a sphenoid sinusitis. LANGERON (1922) isolated a fungus, apparently *A. terreus*, from a case of otomycosis, which he called *Sterigmatocystis hortai*. Two Portuguese physicians, EGAS MONIZ and LOFF (1931) observed a case of cerebral aspergillosis in a patient from Lisbon.

The most recently described form of aspergillosis, and perhaps the most characteristic, was observed by DÉVÉ (1938). This French physician described the clinical and radiologic, as well as the anatomic features of the entity, which he called intrabronchiectatic megamycetoma.

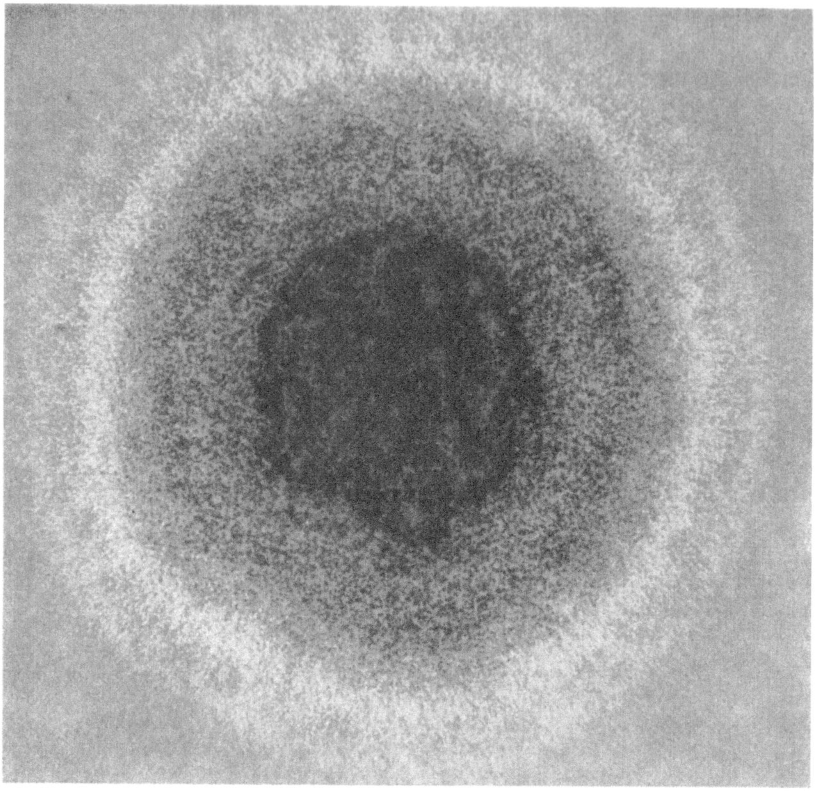

Fig. 1. Culture of *A. flavus* on Sabouraud's medium. Pigmentation proceeds centrifugally in the wake of the mycelial growth

Aspergillus restrictus has been the latest species found to be pathogenic for man and animals (Marsalek et al., 1960).

The study of toxic metabolites of the *Aspergilli* has become quite important in recent years. Early studies in this field were carried out by Ceni and Besta in Italy (1902), by Bodin and Gautier in France (1906), and by Henrici in the United States (1939). These investigations recently culminated in Britain with the discovery of aflatoxins, following outbreaks in 1960 and 1961 of an unknown "X" disease in poults, swine and cattle.

Mycology

Raper and Fennell (1965) recognize 150 species and varieties of *Aspergilli*, that are classified into 18 groups. In human infections of the lung and other viscera the species most frequently encountered is by far *A. fumigatus*, followed by *A. flavus*, *A. niger*, *A. terreus* and *A. versicolor*. In ear and skin infections the most frequent species is *A. niger*.

Direct examination and cultures are the techniques most frequently used in the diagnosis of the disease. Animal inoculations are generally not helpful.

Direct Examination: The suspected fluid (sputum) is placed on a slide and pressed under a coverslip. The use of sodium hydroxide improves the quality of

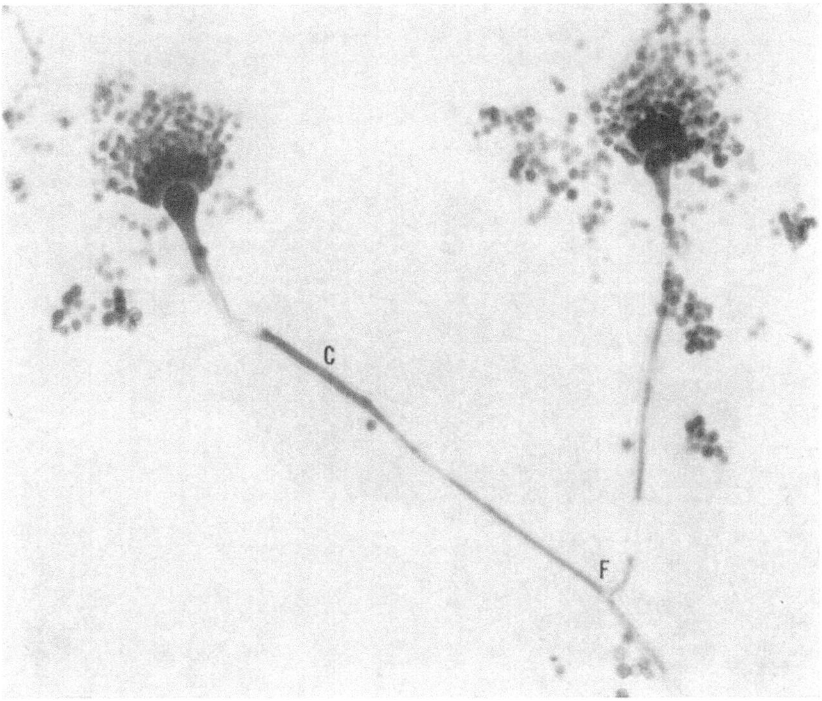

Fig. 2. Foot cell (F), conidiophores (C), and fruiting heads of *A. fumigatus*. Slide culture, lactophenol cotton blue stain. × 450

the preparation. Branching hyphae or fragments, greenish spores (3—5 μ in diameter) and fruiting heads may be seen.

Culture: The material to be examined may be inoculated on Sabouraud's glucose — agar slants with chloramphenicol and kept at temperatures ranging from normal room temperature to 45° C. The fungus grows in 2—5 days, appearing first as white filamentous colonies, that become bluish green or black, according to the species, due to the appearance of fruiting heads and spores (Fig. 1).

Small fragments of mycelium or slide cultures may be stained for microscopic examination with lactophenol cotton blue. The hyphae are seen as very long, gently curving structures, measuring approximately 10 μ in diameter, stained blue and covered by an unstained, somewhat refractile membrane. The septa are sharp and regularly spaced. Successive dichotomous branching is apparent, the branches originating at an angle of 45°. Conidiophores branch off at an angle close to 90°, from hyphae called "foot" cells. The identification of foot-cells is strong evidence that the fungus is an *Aspergillus*.

The fruiting heads are supported by a hypha (conidiophore) that widens progressively towards its distal end, where there is a spherical swelling (the vesicle) (Figs. 2, 3). No line of demarcation separates the vesicle from the conidiophore.

According to the species, the vesicle is surrounded by one or two rows of flask-like projections, the primary and secondary sterigmata to which chains of spores are attached. Brownish-green pigmentation is present in the elements of the fruiting head.

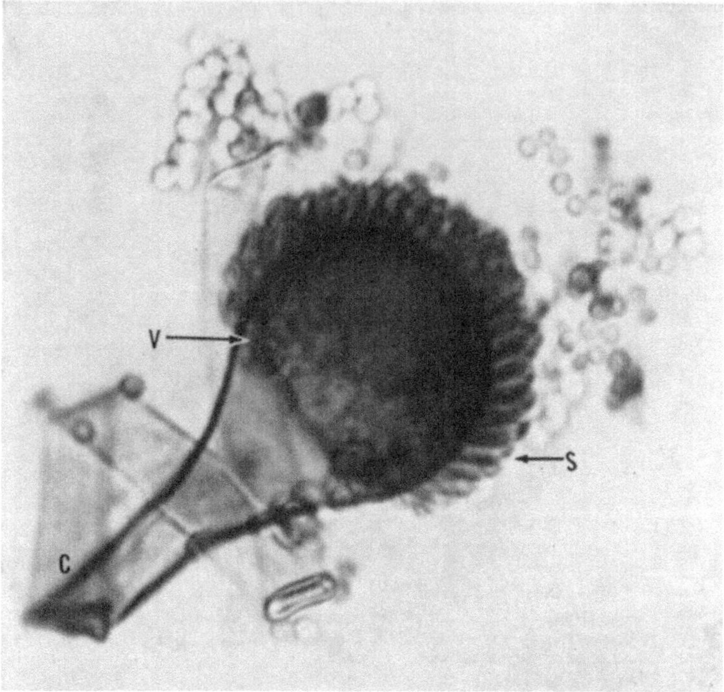

Fig. 3. Fruiting head of *A. fumigatus*. The conidiophore (C), the vesicle (V), a row of sterigmata (S), and a number of spores are seen. Unstained preparation. Approx. × 1200. (Courtesy of Dr. J. C. Sieracki, Pittsburgh, Pa.)

Species differences are related to the manner of implantation of the conidia (radial, columnar), to the shape of the vesicle, to the number of rows of sterigmata, to the shape of the conidia (globose, elliptical) and to the pigmentation of the head.

Further details may be found in manuals and textbooks (RAPER and FENNELL, 1965, or its preceding editions, THOM and RAPER, 1945; THOM and CHURCH, 1926).

Incidence and Epidemiology

Aspergillosis is worldwide in distribution and has been reported from all continents. It is stated that warm, humid climates are favorable for the growth of the fungus, but perhaps with the exception of otomycosis, there does not seem to be any evidence that the disease is more frequent in such regions. Its overall incidence is difficult to ascertain. HEFFERNAN et al. (1966) found aspergillosis in 0.7% of all the autopsies performed in the Johns Hopkins Hospital in the year 1963. CARBONE et al. (1964) reported an incidence of 1.7% in patients dying with *cancer* and of 5.8% in those dying with acute *leukemia*.

No important difference can be detected as to sex and race. However, some localizations are more frequent in men (bronchial aspergilloma) and others in women (infections of the paranasal sinuses). The condition is less frequently reported in children than in adults, although in the young it disseminates more frequently and runs a faster course.

An ever growing body of evidence indicates that in recent years there has been a progressive increase in the incidence of aspergillosis (KEYE and MAGEE, 1956;

TORACK, 1957; SIDRANSKY and PEARL, 1961; BAKER, 1962a; WAHNER et al., 1963; GRUHN and SANSON, 1963; CARBONE et al., 1964; HUTTER et al., 1962; 1964; HEFFERNAN et al., 1966). This higher frequency has been ascribed to the use of new drugs that may prolong the life of patients with low organic resistance, such as antibiotics, or that may themselves affect the individual's organic defenses such as steroids and cytotoxic drugs. Although dissenting opinions still persist (MACARTNEY, 1964) it is evident that the data gathered through the analysis of cases in humans, as well as through experimental observation show that certain chemotherapeutic agents have played an important role in such a rise.

Sources of Infection: *Aspergillus* is found widely distributed in nature and grows in a large number of natural substrates. RÉNON (1895, 1897) reported its isolation from the grain and flour used respectively by pigeon-crammers and hair cleaners. EMMONS (1960, 1962) has reported the occurrence of *Aspergillus fumigatus* in piles of vegetable compost during early stages of decomposition. When this material is exposed to rain, the subsequent microbial fermentation produces a rise in the temperature that supports the growth of the fungus. Contaminated hay and straw are a very important source of pathogenic *Aspergilli*. By means of quantitative methods, GREGORY and BUNCE (1960) were able to show that *A. fumigatus* was one of the main fungi in hay baled with a moisture content of 42%, and that *A. glaucus* was predominant at about 28% of moisture content. Furthermore GREGORY and LACEY (1963) demonstrated that, at a constant wind speed, an obviously moldy hay released 1000 times as many spores as a well preserved hay, and that the number of spores released at a given time increased at higher wind speeds. Other materials that may harbor the fungus are soil (SCHMIDT et al., 1962), household dust (SWAEBLY et al., 1952), bagasse used for litter (HUTSON, 1966), rye flour (WEINER, 1960), oilcake stored in a barn (STRELLING et al., 1966), feed pellets (FORGACS et al., 1954), groundnuts (SARGEANT et al., 1961), hop manure (HINSON et al., 1952), corn, and bark of various tropical fruits (LARA, 1928).

The air-borne spores of *Aspergillus* can readily be isolated by exposing culture plates to the air.

In addition the use of samplers or spore traps allows a quantitative determination that can be expressed as number of colonies obtained per unit volume of air. Isolation of the fungus by air sampling has been carried out in various countries which include France (VALLERY-RADOT et al., 1950), United States (BRUSKIN, 1953), Puerto Rico (PONS et al., 1961), Colombia (ALVAREZ et al., 1965), Wales (HYDE et al., 1949, 1956), and England (BARUAH, 1961; NOBLE and CLAYTON, 1963). VALLERY-RADOT et al., (1950) observed an increase in the number of spores during the winter months. NOBLE and CLAYTON (1963) isolated *A. fumigatus* and *A. niger* from the air in a London hospital and also demonstrated a hundred-fold increase of the number of spores during the autumn and winter months. The fungus was likewise shown to be present in the hospital blankets.

Occupational exposure has been stressed ever since DIEULAFOY et al. (1890) and RÉNON (1895) reported the disease in *pigeon-crammers and wig cleaners*. Fortunately, in this age of the "affluent society" these occupations and their inherent dangers have largely disappeared, although occasional cases are still reported (YESNER and HURWITZ, 1950). More danger seems to exist for farmers who very often come into contact with decaying vegetation or with the materials previously mentioned.

Industrial exposure may be significant in factories where molasses is fermented by *Aspergilli* for the production of citric or other acids. Such process is carried out in large open pans and, when finished, the layer of *Aspergillus* culture is removed by workers who at this moment are exposed to the dust containing great quantities of spores (HOREJSI et al., 1960).

Pathogenesis

In some cases the *Aspergillus* infection is primary but most often is secondary or opportunistic.[1]

Portals of Entry: The air-borne spores may gain entrance into the body in a variety of manners. The usual routes of infection are the ear canal, nose and sinuses, eye, skin and adnexa and respiratory tract, the latter being the most important. Open heart and major vascular surgery provide another portal of entry for a very ominous type of the disease (PORTA, 1959; NEWMAN et al., 1964; DARRELL, 1967; MAHVI et al., 1968). LUKE et al. (1963) have suggested the possibility of inoculation through exchange-transfusions and MAHVI et al. (1968) have indicated as an alternative the penetration of the fungus through a splinter wound. Other possible sites of entrance are a rectal fistula (WELSH and MCCLINTON, 1954), the uterus after a septic abortion (FINEGOLD et al., 1959), and traumatic injuries (ZIMMERMAN, 1950).

The development of an infection depends on external as well as internal factors (Fig. 9). See Table 3.

External Factors: Among the external factors, the number of spores that enter the body, i.e. the/size of the inoculum, seems to be the most important (AUSTWICK, 1965). This in turn is determined by the ability of the fungus to proliferate in the environment (AUSTWICK, 1962a). BARUAH (1961) calculated that the atmospheric concentration of Aspergillus-Penicillium spores in a cow-shed ranged from 1,300 to 12,390,000 spores/m^3. If a constant concentration of 100,000 spores/m^3 is assumed, a cow would inhale 600,000 spores/hour, or 14,400,000 in 24 hours. Under similar conditions a man would inhale 60,000 spores/hour, but under maximum concentration this inhalation would be in the order of 7,000,000 spores/hour.

AUSTWICK and APPLEBY (1965) have estimated that at least 16,000,000 spores are needed to cause an infection in young chickens. Other observations (SIDRANSKY and FRIEDMAN, 1959) suggest that extremely low concentrations of air-borne spores may be sufficient to induce pulmonary infections in cortisone-treated animals. The number of spores needed to produce an infection in man has not been precisely determined.

Internal Factors: The individual factors that may facilitate the onset of the fungal infection are:

1. Acute or chronic diseases that debilitate the individual's mechanisms of defense, such as leukemias, lymphomas, disseminated malignant tumors, leukopenic states and infectious diseases.
2. Therapeutic agents which may alter the patient's reactivity, including adrenal corticosteroids, cytotoxic drugs and radiation therapy, often given to treat the conditions listed above or as immunosuppressors.
3. Drugs that may produce changes in the individual's flora, i.e. wide-spectrum antibiotics.
4. Local disorders which facilitate the entry of the fungus, as tuberculosis, bronchiectasis, bronchial cysts, chronic bronchitis, asthma, silicosis and other pneumoconioses, histoplasmosis and bronchogenic carcinoma.
5. Traumatic injuries that facilitate the entry of the fungus, particularly in

[1] According to the Committee of the International Symposium on Opportunistic Fungus Infections, the term *opportunistic* should be used to denote fungi that invade the tissues of man or animals with predisposing conditions (BAKER, 1962b).

cases of corneal infections and mycetoma pedis. In the following paragraphs some of the more important iatrogenic factors are discussed.

Antibiotics: These drugs may affect the host in a number of ways. There is some evidence that tetracyclines, dihydrostreptomycin and penicillin depress the synthesis of antibodies *in vivo* (STEVENS, 1953), and that chloramphenicol has a similar effect *in vitro* (AMBROSE and COONS, 1963). The possible effects of antibiotics on phagocytosis are less clearly established, although some studies suggest that they may inhibit phagocytosis of bacteria by human leukocytes (MUNOZ and GEISTER, 1950).

Whether these factors, inhibition of antibody synthesis and inhibition of phagocytosis, are operative in human *Aspergillus* infections is not clear at the present time. In clinical practice the main effect of antibiotics is to alter the individual's microflora by allowing the proliferation of fungi in the absence of competing organisms. In addition many patients are already victims of bacterial infections that may also act as predisposing factors by producing local tissue damage.

Adrenal Corticosteroids: Increased levels of adrenal corticosteroids, whether spontaneously occurring (SAYLE et al., 1965) or therapeutically induced, favor the development of *Aspergillus* infections in humans. In the experimental field, SIDRANSKY and FRIEDMAN (1959) observed that cortisone rendered mice inhaling spores of *A. flavus* highly susceptible to fatal pulmonary aspergillosis. In the search for an explanation of this enhanced susceptibility, EPSTEIN et al. (1967) demonstrated, in a similar experiment, that there was a decreased peribronchial leukocytic infiltrate in the cortisone-treated mice as compared to control animals. FRENKEL (1962) has moreover demonstrated that cortisone inhibits inflammatory cellular responses, as well as proliferation of epithelioid and giant cells. At the ultrastructural level, it has been further observed that in cortisone treated animals the lysosomes of alveolar macrophages, unlike those of control animals, fail to attract and to fuse with the phagocytic vacuoles containing spores (MERKOW et al., 1968). This diminished lysosomal activity is considered to be an important factor in the increased susceptibility of cortisone treated animals.

Organ Transplantation and other Factors: The iatrogenic setting favoring mycotic infections reaches its ideal state (from the fungus' point of view) in patients undergoing immunosuppresive therapy for organ transplantation. Although the number of systematic studies is still small it is now clear that patients on such therapy are highly vulnerable to mycotic infections, just as they are to bacterial or viral diseases. Among RIFKIND et al's. 23 cases (1967) with systemic fungal infections complicating renal transplantation, 7 were found to have aspergillosis. The condition was associated in various cases with candidiasis, cytomegalic inclusion disease and *Pneumocystis* infection. In SAUNDERS and BIEBER'S case (1968) of cardiac transplantation the lungs were the site of a diffuse *Aspergillus* infection that involved alveoli, bronchi, arteries and veins. The transplanted heart disclosed fungal invasion of the endocardium and a coronary artery branch, accompanied by focal myocardial necrosis. Fungus lesions were also present in the brain.

Neutropenia, spontaneously occurring or induced by cytotoxic drugs in the treatment of leukemias, lymphomas and other malignancies, has been shown to be frequently associated with fungal invasion (REED, 1935; BAKER, 1962a; FRANCOMBE et al., 1965).

Histopathogenesis: In humans, the incursion of spores, in the majority of cases, does not produce a pathologic process. If the person is debilitated or predisposed, the spores proliferate, and the mycelium invades the tissue to produce an acute necrotizing disease. In persons that repeatedly inhale spores hypersensitivity may result. In other instances, the fungus may find a pre-existing bronchopulmonary cavity, or perhaps create one, and proliferate to form a large mycelial mass constituting the condition known as aspergilloma.

The mechanisms by which the fungus produces tissue injury are less well known. Some strains of *A. fumigatus*, under suitable nutritional conditions, produce several substances known to be toxic to animals, namely fumigacin, fumigatin and gliotoxin (FORGACS and CARLL, 1962). Whether these toxic metabolites are responsible for *in vivo* tissue injury is not clear. Microscopic study of newly developed pulmonary lesions reveals focal necrosis of individual alveolar septa. This *histolytic property* seems to be an important factor and certainly facilitates the invasion of the tissues. However, it seems somewhat incongruous that the most potent toxin-producing species is *A. flavus*, whereas the most commonly infecting one is *A. fumigatus*. It appears likely, therefore, that other factors must be operative.

Dissemination: *Aspergillus* disseminates along blood vessels and seems also to be able to spread along lymphatics, as well as intrabronchially. The fungus has a marked tendency to invade arteries and veins, producing a necrotizing angiitis, thus setting the stage for its hematogenous spread. The organs most frequently involved by metastatic propagation are the heart, brain, kidneys, gastrointestinal tract, liver, thyroid and spleen.

The involvement of lymph nodes is considered evidence for lymphatic spread.

Bronchogenic dissemination from a primary pulmonary focus does not seem to be rare, but little on this subject can be found in the literature. This mechanism seems to be operative in patients with active open lesions as recent abscesses. GOWING and HAMLIN (1960) considered that the fungus can reach the alveoli by aspiration into the respiratory bronchioles with subsequent spread into the associated alveolar sacs.

General Morphology

The response of the human organism to the presence of *Aspergillus* varies from a very mild non-specific inflammatory reaction to a severe acute necrotizing lesion or to a granulomatous inflammation. Minimal responses usually occur when the fungus colonizes a cavity such as the external ear or a bronchus and are characterized by a mild polymorphonuclear or mononuclear leukocytic infiltration.

From the *histopathologic point of view*, the lesions observed in the tissues are of 3 types: 1. necrotizing; 2. suppurative; and 3. granulomatous. The *necrotizing lesions* exhibit a central zone of coagulative necrosis surrounded by a variable amount of polymorphonuclear leukocytes, often pyknotic (Fig. 11), and an outer zone of hemorrhage. The necrotic area often contains vessels displaying invasion by mycelial elements, necrosis of their walls and thrombosis. The suppurative lesions usually have the microscopic appearance of small abscesses with an exudate rich in neutrophils (Fig. 12).

Both necrotizing and *suppurative lesions* are seen most frequently in the lungs, but may be found in the brain, heart, liver, spleen, kidney and other organs. An absolute distinction between these two types of lesions is not always possible since some of them exhibit both necrotizing and suppurative components.

The *granulomatous lesions* are composed of epithelioid and giant cells accompanied by polymorphonuclear and mononuclear leukocytic infiltration. This form

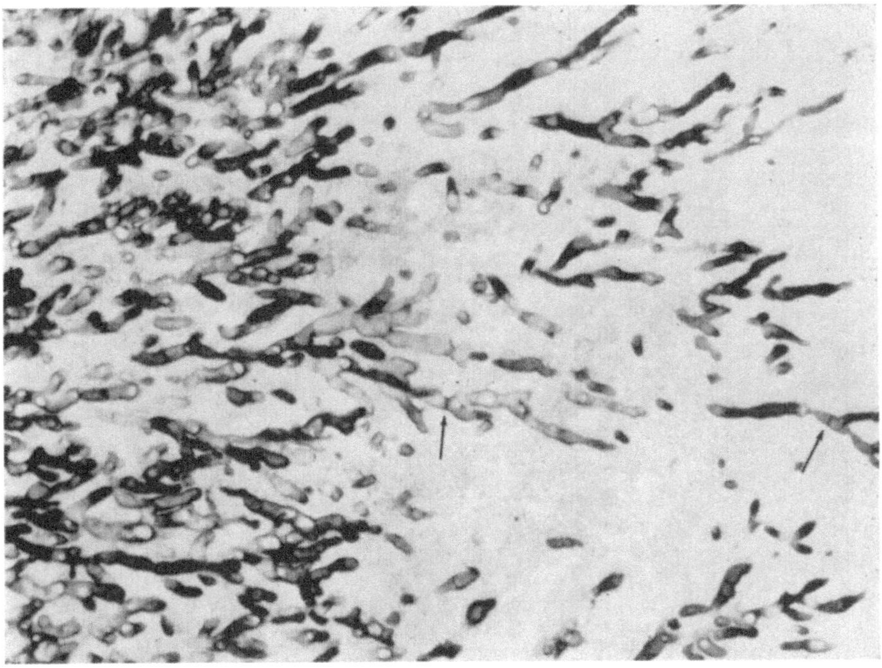

Fig. 4. Morphology of the fungus in tissues. The hyphae are septate (arrows) and exhibit dichotomous branching. Grocott-Gomori, × 450

of reaction is not necessarily chronic and has been observed in cases of acute primary aspergillosis of the lungs (HERTZOG et al., 1949; STRELLING et al., 1966). In other instances, the granulomatous lesions have a marked fibroblastic component. This type of inflammation occurs also in the paranasal sinuses, orbit and cranial cavity.

AUSTWICK (1962b) described still another type of lesion, the "*asteroid*" body. This formation, 50—80 μ in diameter, is composed of a central core of hyphae and a peripheral rim of eosinophilic radiating club-like structures similar to those observed in actinomycosis. The asteroids have been interpreted as the result of the host's immunologic response to the fungus. They are frequently seen in the lungs of cattle but apparently have not been observed in humans. SYMMERS (1962) has observed eosinophilic amorphous deposits of fibrin which form club-like expansions of the surface of human intracavitary aspergillomas. These structures may, however, be of a different nature inasmuch as they are seen only in intraluminal masses.

Morphology of the Fungus in Tissues: The hyphae of the *Aspergilli* are usually well stained with hematoxylin and eosin (H & E), but their morphology is better appreciated with the use of special stains. The most frequently employed are the periodic acid Schiff (PAS), the Gridley, and the Grocott-Gomori methenamine-silver nitrate stains.

The hyphae are septate, have uniform caliber, measure about 7—10 μ in diameter, branch dichotomously at an angle of 45° and are basophilic with H & E stains (Fig. 4). They often display a radial disposition, especially in intracavitary lesions (Fig. 5). This type of hypha is sometimes referred to as "actinomycetoid". In

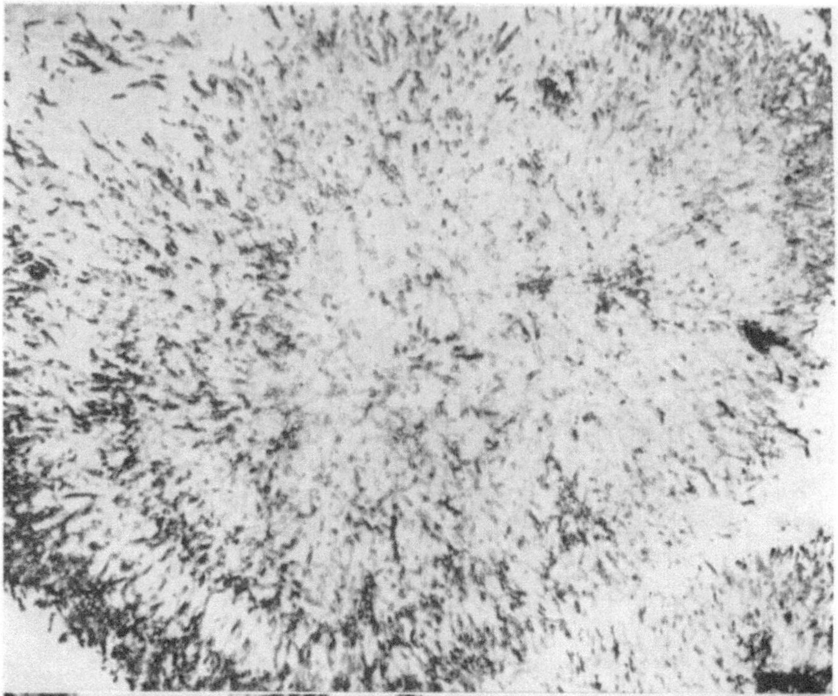

5

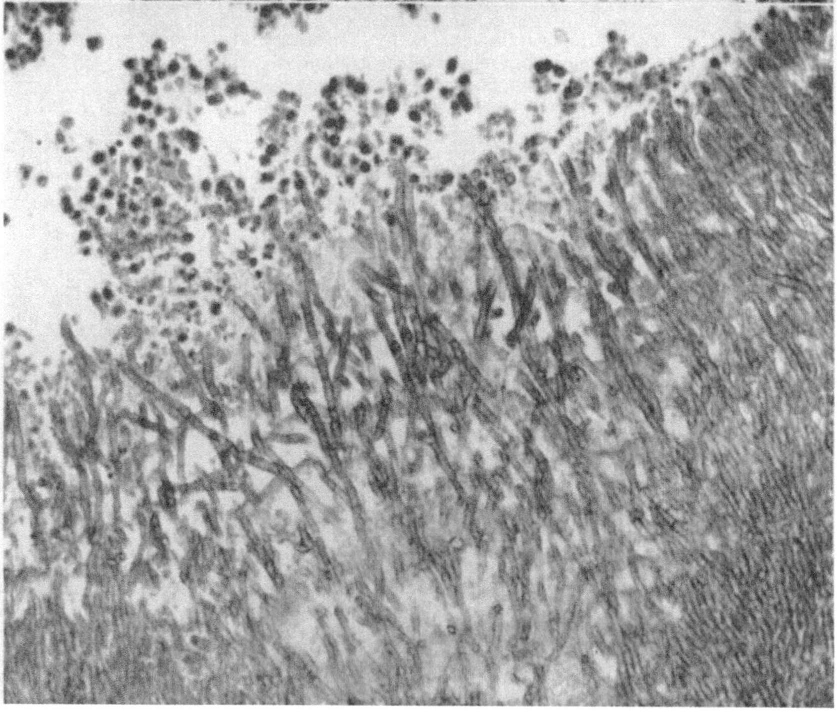

6

Fig. 5. Morphology of the fungus in tissues. Intrabronchial aspergillosis. The hyphae display a radiating arrangement. Grocott-Gomori. Approx. × 100

Fig. 6. Morphology of the fungus in tissues. Invasive hyphae exhibiting parallel growth. H & E, × 200

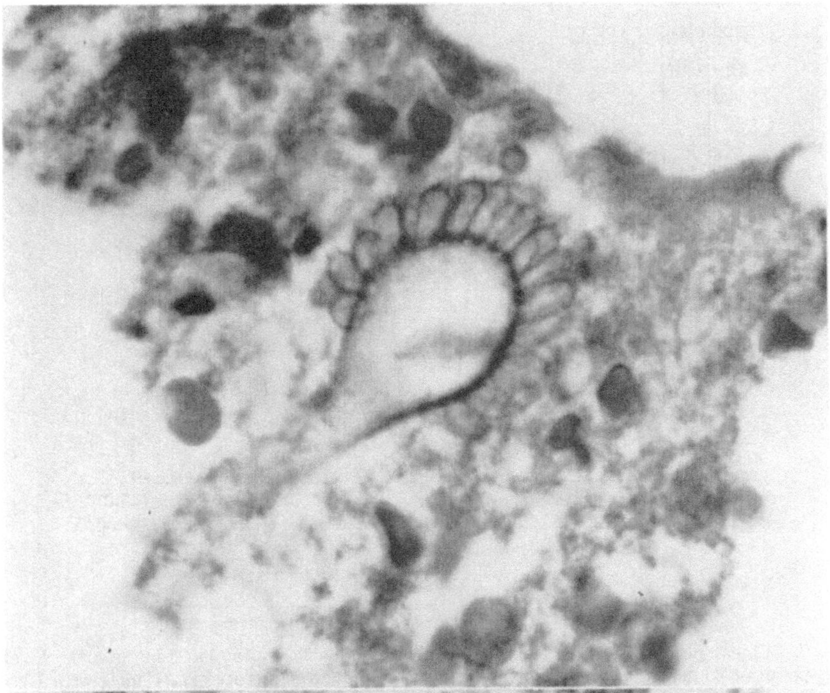

7

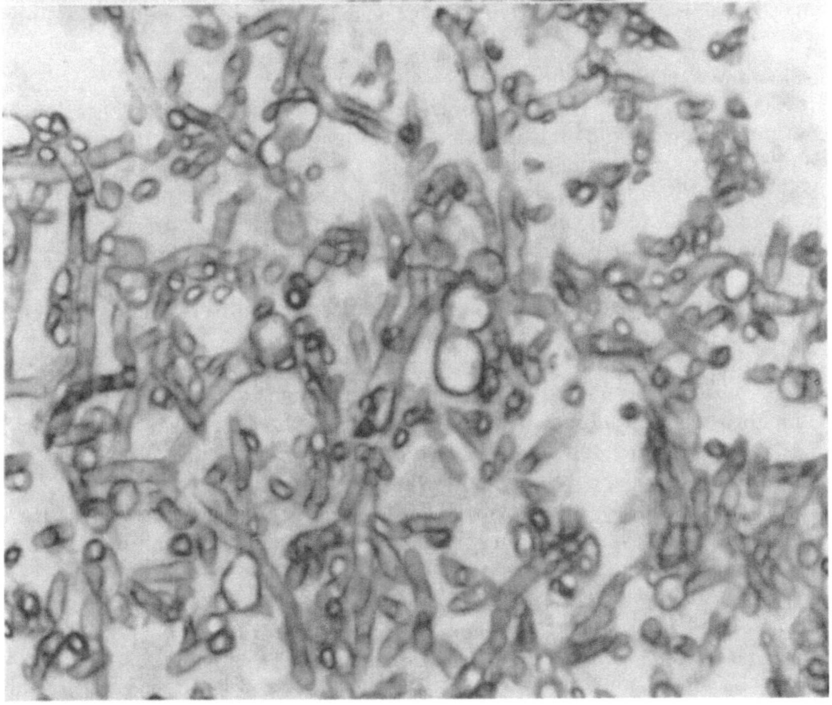

8

Fig. 7. Morphology of the fungus in tissues. This fruiting head from a pulmonary abscess displays a row of sterigmata. H & E, × 1000

Fig. 8. Morphology of the fungus in tissues. Some hyphae show degenerative changes consisting of irregularity and swelling. This appearance is reminiscent of that of a Phycomycete. Grocott-Gomori, × 450

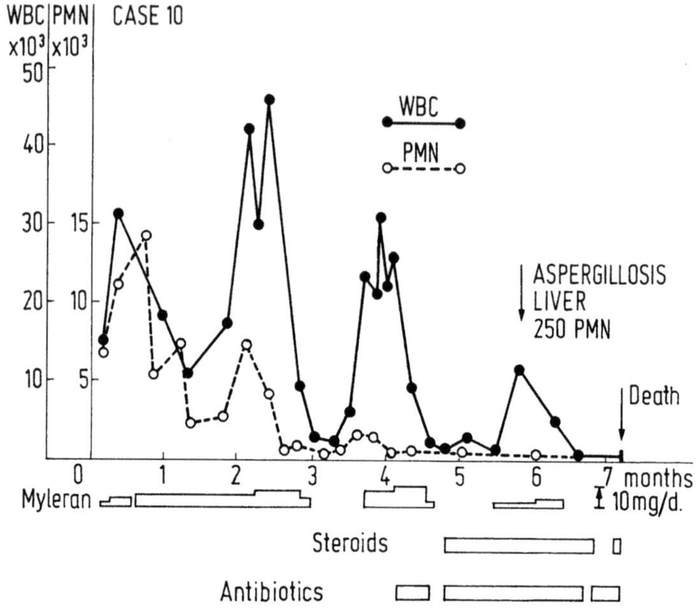

Fig. 9. Clinical plot in a patient with chronic myelocytic leukemia. Aspergillosis developed 6 weeks prior to death, at the time the patient had severe neutropenia. (Courtesy of Dr. R.D. BAKER. Amer. J. clin. Path. **37**, 358 (1962), and The Williams & Wilkins Company)

other instances, especially in invasive lesions, the hyphae are straight, parallel and have very few branches (Fig. 6). Another type of hypha that may occasionally be encountered is an ovoid or somewhat irregular structure, 5—10 μ in diameter, which represents a germinating spore with its primary branch.

The *fruiting heads* of the *Aspergilli* are pathognomonic and, when observed, allow a definite histologic classification of the fungus (Fig. 7). In sections only the conidiophore, the vesicle and a row of sterigmata with variable degrees of brownish-yellow pigmentation can be seen. Spores are difficult to find. These fruiting heads are very rarely seen in tissues, but are often encountered in aerated cavities such as an aspergilloma or in the external ear canal.

The fungus may undergo degenerative changes. In some cases, the hyphae, although present in the tissues, are not stained with H & E. This is especially true in fibrosing granulomatous inflammation of the paranasal sinuses, orbit or cranial cavity. Stains for fungi usually reveal the presence of bizarre hyphal elements with irregular swellings and abortive branching.

In other instances, especially in the center of fungus balls hyphae are wider, somewhat irregular, less basophilic and lose their septa, resembling the hyphae of a *Phycomycete* (Fig. 8).

Differential Diagnosis. Aspergillus in tissues displays similarities with *Candida, Penicillium* and *Phycomycetes.*

The hyphae (pseudohyphae) of *Candida* are thinner than those of *Aspergillus* and exhibit septa at which level the hyphae are constricted. The presence of blastospores is also a distinguishing feature.

Penicillium organisms have uniformly cylindrical septate hyphae 2—5 μ in width that branch at an angle of 25°—45°. Characteristics helpful in distinguishing them from *Aspergilli* are the presence in their hyphae of clear vacuoles alternating with basophilic zones, and the occurence of chlamydospores. *Penicillium,* like *Aspergillus,* may have a radiating growth and invade blood vessels.

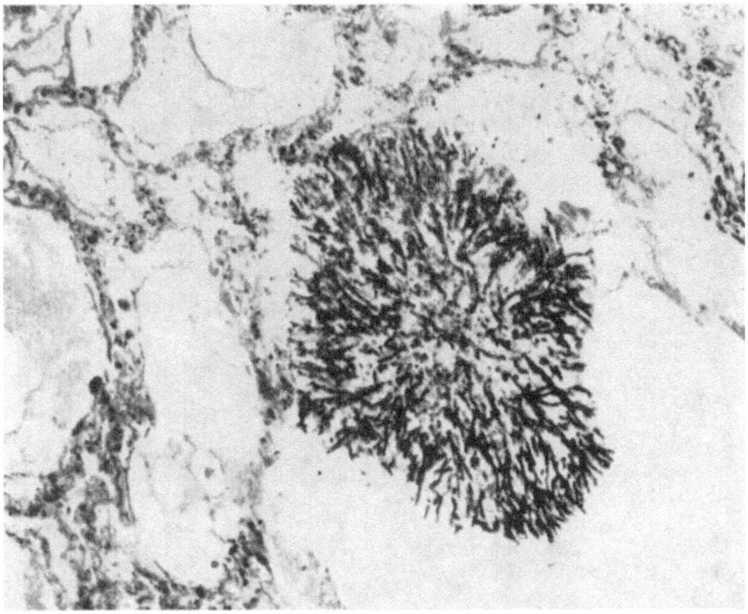

Fig. 10. Tissue changes. Adjacent to the fungus there is minimal necrosis and granulocytic infiltration of individual septa. The mycellium is in a respiratory bronchiole, probably as a result of intrabronchial spread from a pulmonary abscess

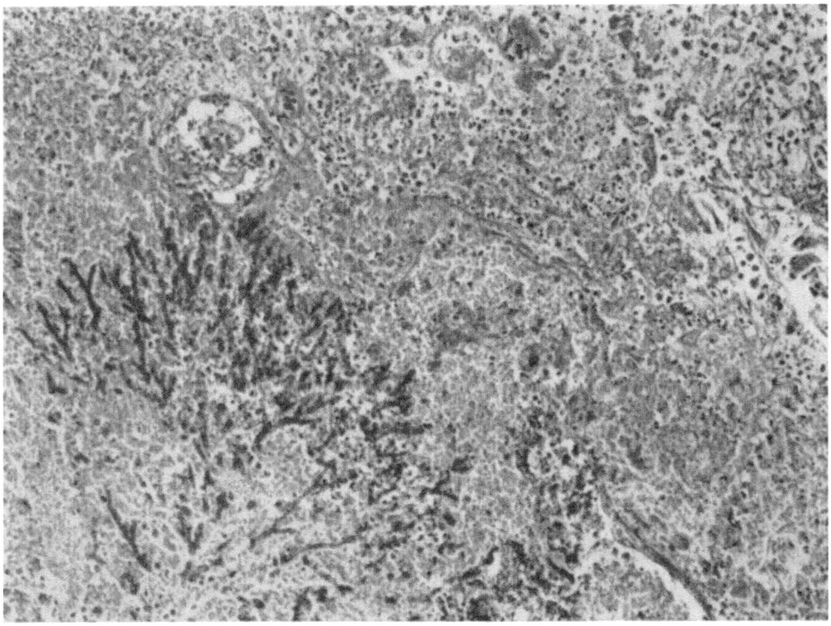

Fig. 11. Tissue changes. This lesion exhibits coagulative necrosis and abundant fibrinous exudate as well as numerous pyknotic neutrophils. H & E, × 100

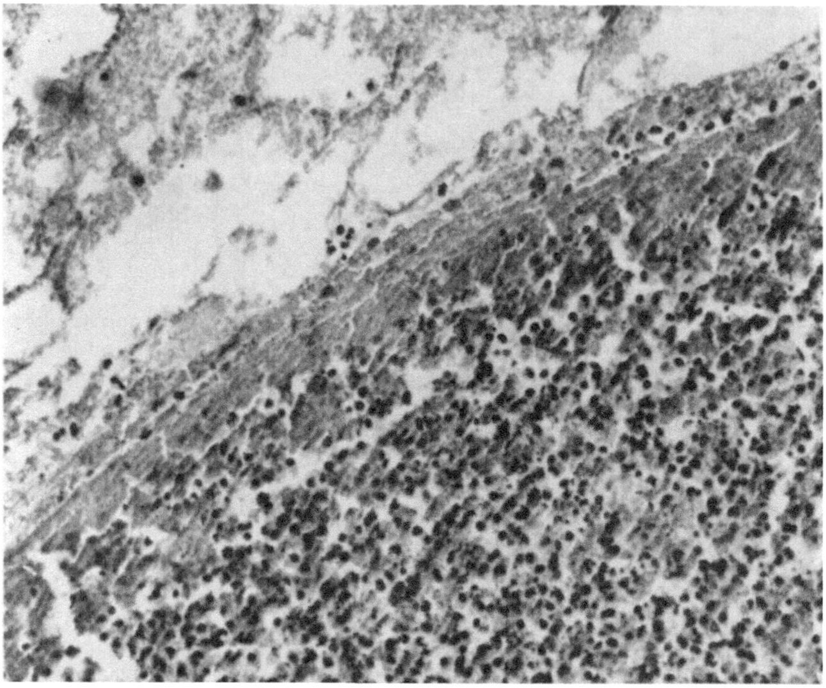

Fig. 12. Tissue changes. Purulent focus in the lung, displaying a sharp demarcation from the surrounding parenchyma. *Aspergillus* hyphae, not visible in the picture, were present in the center of the abscess

The *Phycomycetes* can be distinguished by their irregular wide hyphae that are usually non septate and branch at variable angles.

In spite of the characteristics mentioned the differential diagnosis of these fungi may be quite difficult or impossible unless the organism is identified by culture.

Biopsy. The diagnostic value of biopsies in aspergillosis (and other mycoses) has been discussed by CONANT et al. (1954) and by BAKER (1957). Most cases of aspergillosis in which a clinical diagnosis proves difficult are likely to be deeply seated in the thoracic, abdominal or cranial cavities and are difficult to reach. In infections localized in the nasal fossae, paranasal sinuses and orbit, the biopsy may be the single most important diagnostic procedure.

In these cases, the identification of a fibrosing granulomatous process, often with no visible hyphal elements in the routine stains, is a peremptory indication for the request of special stains. Other locations in which a biopsy may prove helpful are the ear, skin and bronchi. Cultures of the fungus must be done simultaneously. "How many regrets this procedure would save!" (BAKER, 1957).

Clinical Types and Morphologic Findings

Aspergillosis, usually a disease of the respiratory tract, may spread via the blood stream to the heart, central nervous system, liver, spleen and other organs. It may also be a localized infection of the nose and paranasal sinuses, eye, ear and skin.

This section will deal with aspergillosis of the different systems both from the clinical and from the morphologic points of view. Table 1 lists the main features and Table 2 the frequency of organ involvement in 81 cases of disseminated aspergillosis found in the literature.

General reviews on the subject include those of THOM and CHURCH (1926), CONANT et al. (1954), LANDAU et al. (1963), EMMONS et al. (1963), HILDICK-SMITH et al. (1964), BADER (1965), and AUSTWICK (1965).

Respiratory System

The lungs are the organs most frequently affected by the disease. Reviews have been published by WÄTJEN (1931), VADALA (1940), HINSON et al. (1952), FINEGOLD et al. (1959), NAJI (1959), GREER (1962), and WAHNER et al. (1963).

Pulmonary aspergillosis may be acute or chronic, primary or secondary, although this latter differentiation is not always possible (DES AUTELS et al., 1962).

In a number of instances reported the diagnosis has been made solely on the basis of the isolation of the fungus from the sputum and the data available are not sufficient to classify the case (MIERES, 1925; STEELE, 1926; SMITH, W.R., 1934; GARCIA-TRIVIÑO et al., 1935; NAVARRO-BLASCO, 1936; WAHL, 1936; TEIXEIRA, 1941; DONALDSON et al., 1941—1942), although sometimes the cases are claimed to be of primary nature (WAHL, 1928; SCHNEIDER, 1930; SÁNCHEZ-FREIJO, 1935; MOOLTEN, 1938—1939; STOLOW, 1939).

The predisposing disease may be systemic as with leukemias, or localized to the respiratory tract as in tuberculosis (KELMENSON, 1959; RODIN et al., 1963), bronchiectasis and primary or metastatic tumors (YOUNG, 1926; KAMPMEIER et al., 1934; UTZ et al., 1959). In some cases the fungal infection is superimposed on a terminal bacterial bronchopneumonia or on an organizing pneumonitis (FINEGOLD et al., 1959). Other associated conditions reported are leukemic reticuloendotheliosis (VAITHIANATAN et al., 1962), amebic abscess of the liver (GONZÁLEZ-MENDOZA et al., 1962), coal workers pneumoconiosis (HEPPLESTON et al., 1949), kala-azar (MACKIE, 1922), a pneumonectomy space (SOCHOCKY, 1959) and dermatomyositis treated with steroids (STANLEY, 1965).

Clinical Manifestations: The symptomatology of pulmonary aspergillosis, especially the secondary form, is difficult to define. HEFFERNAN et al. (1966) and SIDRANSKY and PEARL (1961) have pointedly observed that no consistent set of symptoms or signs appear which may be helpful to the diagnosis. Most cases are secondary and the mycotic infection is usually overshadowed by the underlying disease.

Secondary Aspergillosis of the lungs is most often an acute process. The main symptoms are cough, purulent expectoration, fever, dyspnea, chest pain and blood-streaked sputum of recent onset, appearing in a patient with an underlying disease. Hemoptysis or blood-streaked sputum is an important symptom and has been related to fungal invasion of blood vessels (SEABURY and SAMUELS, 1963). Physical findings are variable and depend to a large extent on the characteristics of the mycosis as well as on the nature of the underlying local disease. They include wet rales, rhonchi and signs of consolidation which mimic a lobar pneumonia. Local tenderness of the thoracic wall can sometimes be elicited. Chest X-ray examination may reveal numerous foci of infiltration, usually bilateral and discrete, less frequently confluent and of lobar distribution. VAN ORDSTRAND (1940) observed a shadow radiating from the hilum which he compared with a "sun burst". Any or all of the above symptoms may be absent, and surprises in the autopsy room, or at the examination of surgical specimens are indeed common.

Table 1. *Disseminated. Aspergillosis Clinical and pathological findings of 81 cases gleaned from the literature*

Author and Year	Country	Author Case No.	Age Sex	Predisp. Factors	Organs Affected	Culture
Linck (1939)	Germany	—	19, M	Not apparent	Lungs ? brain, leptomeninges	—
Cawley (1947)	U.S.	—	8, M	?	Lungs, brain, dura, mediastinum, heart, liver, spleen, ankle	*A. fumigatus*
Gerstl et al. (1948)	U.S.	2	62, M	Pneumonia, anemia, leukopenia	Lungs, skin	*A. fumigatus*
Zimmerman (1950)	U.S.	2	25, M	Trauma with leg amputation	Heart, kidney	—
—		3	32, F	Old rheumatic heart disease	Heart, adrenal	—
Welsh and McClinton (1954)	U.S.	—	64, M	Rectal fistula, necrosis of ischiorectal tissues, antibiotics	Lungs, duodenum	*A. fumigatus*
Levy et al. (1955)	U.S.	—	24, F	Idiop, anemia, ACTH	Lungs, heart, brain	*Asperg. sp.*
Zimmerman et al. (1955) . .	U.S.	1	Newborn M	Staph. sepsis, antibiotics, steroids	Lungs, brain, liver, pericardium	*A. sydowi*
Welsh et al. (1955)	U.S.	—	18, M	Idiop. neutropenia, S. choleraesuis sepsis, ACTH, steroids, antibiotics	Lungs, heart	*A. flavus*, *A. fumigatus*, *A. nidulans*
Keye et al. (1956)	U.S.	1	23, ?	Systemic lupus	Trachea, bronchi, lung, brain	*A. fumigatus*
—		3	4 mos., Sex ?	Neuroblastoma with metastasis	Lungs, meninges, brain	—
Sherman and Boshes (1956)	P. Rico	5	23, F	Atypical leukemia ? ACTH	Lungs, trachea, heart, brain, dura	*Asperg. sp.*
Kirschtein and Sindransky (1956)	U.S.	—	50, M	Anemia, ACTH, antibiotics, leukopenia	Lungs, heart	*A. flavus*
Akkoyunlu et al. (1957) . .	Turkey	—	20 days M	Not apparent	Lungs, bronchi, brain, leptomeninges	—
Sathmary (1958)	U.S.	—	37, M	Ac. myeloid leukemia, B. subtilis pneumonia, antibiotics, steroids	Lungs, brain, heart, kidney, thyroid	*A. fumigatus*
Grcevic et al. (1959)	U.S.	1	36, M	Antibiotics, steroids	Lungs, bronchi, brain, heart	*Asperg. sp.*
—		2	13, M	Hepatitis, antibiotics, steroids	Lungs, brain, heart, kidneys, pancreas, thyroid, spleen, mediast. nodes	*Asperg. sp.*
—		3	22, M	Antibiotics, febrile disease	Lungs, heart, brain, thyroid, kidney	*Asperg. sp.*
Finegold et al. (1959) . . .	U.S.	9	33, F	Septic abortion, antibiotics	Lungs, heart, spleen, uterus, brain	*A. fumigatus*
Porta (1959)	U.S.	—	14, M	Open-heart surgery (subaortic stenosis)	Heart (aortic valve), aorta, kidneys, periadrenal tissues, aortic embolus	—
Gowing and Hamlin (1960)	U.K.	2	44, M	Ac. myeloid leukemia, steroids, cytotoxics	Lungs, liver, spleen, kidneys, heart, thyroid	*A. fumigatus*
—		4	14, F	Ac. myeloid leukemia, steroids, cytotoxics	Lungs, brain, meninges	—

Allan et al. (1960)	U.S.	—	18 days M	Not apparent	Lungs, liver, spleen, kidneys, heart, thyroid, jejunum, tongue, skin	*A. fumigatus*
Hadorn (1960)	Switzerland	—	25, M	Open-heart surgery (subvalvular aortic stenosis)	Suture line in aorta above valve, myocardium, embolus to femoral artery, brain	*A. fumigatus*
Marsalek et al. (1960)	CSSR	—	60, M	Tuberculosis, amyloidosis	Lungs, kidneys, heart, mediastinal lymph nodes, adrenals	*A. restrictus*
Utz et al. (1961)	U.S.	43	27, M	Not stated	Lungs, skin, prostate, probably kidneys	*A. fumigatus*
Baker (1962)	U.S.	9	70, M	Ac. myeloblastic leukemia, steroids, antibiotics	Brain	—
—		10	10, M	Chr. myeloid leukemia, neutropenia, cytotoxics, antibiotics	Lungs, liver	*Asperg. sp.*
Fraumeni and Fear (1962)	U.S.	—	41, M	Lymphoma and leukemia, steroids, antibiotics, X-ray, cytotoxics	Lungs, heart, spleen, kidney, stomach, psoas, thyroid, brain	*A. fumigatus*
Conen et al. (1962) (Hendrick et al., 1963)	Canada	—	7, F	?	Lungs, brain	*A. fumigatus*
Landau et al. (1963)	U.S.	2	3, M	Ac. lymphobl. leukemia, steroids, X-ray, cytotoxics, antibiotics	Lungs, brain, heart, pancreas, kidneys, thyroid	—
Paradis (1963)	U.S.	—	2 mos. F	Prematurity, cytomegalic incl. dis.	Lungs, brain, heart, eyes, mediast. nodes, small bowel, kidney	*A. fumigatus*
Gruhn (1963)	U.S.	20	68, F	Chr. granulocytic leukemia, neutropenia, steroids, cytotoxics, antibiotics	Lungs, intestine, heart, liver, kidney, brain	—
Luke et al. (1963)	U.S.	—	51, F	Exchange transfusion, antibiotics, steroids	Heart, kidney, thyroid, brain	*A. fumigatus*
Wahner et al. (1963)	U.S.	1	4, M	Staph. pneumonia, antibiotics	Lungs, heart, mediastinal nodes, kidney	*A. fumigatus*
Seabury and Samuels (1963)	U.S.	4	24, M	Sarcoidosis, anemia, steroids, antibiotics	Lungs, kidneys, liver, spleen	*A. fumigatus*
Carbone et al. (1964)	U.S.	7 cases of generalized aspergillosis are included. Underlying diseases were: ac. leukemia 4, lymphoma 2, myeloid metaplasia 1. Organs involved were: Lung 7, central nervous system 5, kidney 3, intestine 3, heart 2, liver 2, thyroid 2, spleen 1. Neutropenia, steroids, antibiotic and cytotoxic drug therapy were considered as possible pathogenic factors.				
Hutter et al. (1964)	U.S.	2	3, F	Ac. leukemia, steroids, ACTH, cytotoxics, antibiotics	Lungs, brain	—
—		6	8, M	Ac. leukemia, X-ray, cytotoxics, antibiotics	Lungs, spleen, kidneys, jejunum	—
—		7	10, M	Ac. leukemia, steroids, cytotoxics, antibiotics	Lungs, heart, kidney, thyroid	*A. fumigatus*

Table 1. Continuation

Author and Year	Country	Author Case No.	Age Sex	Predisp. Factors	Organs Affected	Culture
Hutter et al. (1964)	U.S.	8	10, M	Ac. myelog. leukemia, steroids, cytotoxics, antibiotics	Lungs, kidneys	—
—		10	12, M	Ac. lymph. leukemia, steroids, cytotoxics, antibiotics	Lungs, small bowel	—
		16	43, M	Steroids, cytotoxics, antibiotics, X-ray, lymphosarcoma	Lungs, heart, kidneys, thyroid, muscle, brain	*A. fumigatus*
—		17	56, F	Ac. viral hepatitis, steroids, antibiotics	Lungs, heart	—
—		19	57, M	Ret.-cell sarcoma, steroids, antibiotics, X-ray, cytotoxics	Lungs, kidneys	—
—		28	68, M	Hypoplastic anemia, steroids, antibiotics	Lungs, stomach	—
—		30	69, M	Ac. myel. leukemia, steroids, antibiotics	Lungs, stomach	—
Soergel et al. (1964) . . .	U.S.	—	55, M	Rheumatoid arthritis, steroids, antibiotics, neutropenia	Lungs, heart, kidney	—
Schumacher et al. (1964) . .	U.S.	24	4, M	Ac. lymphocytic leukemia, total body irradiation	Brain, oral cavity	—
—		27	26, M	Ac. monobl. leukemia, antibiotics, cytotoxics, total body irradiation	Lungs, liver	—
Louria et al. (1964)	U.S.	2	10, M	Ac. blastic leukemia, X-ray cytotoxics	Lungs, brain, heart, liver, kidney, thyroid	*A. fumigatus*
Sayle et al. (1965)	U.S.	1	62, F	Ca. of pancreas, Cushing's syndrome, antibiotics	Lungs, heart	—
Francombe et al. (1965) . .	Canada	—	80, M	Pancytopenia (idiopathic), steroids, antibiotics	Lungs, liver, spleen, kidney, bone	*Asperg. sp.*
Redmond et al. (1965) . . .	U.K.	—	6, M	Not apparent	Lungs, ribs, vertebrae	*A. nidulans*
Buttrick et al. (1965) . . .	U.S.	1	15 days F	Prematurity, cytomegalic incl. disease, antibiotics, steroids	Heart, brain, eyes, peritoneum, pleura, small bowel, lymph nodes, kidneys, thyroid	*A. fumigatus*
Heffernan (1966)	U.S.	7	60, F	Ac. myeloid leukemia	Lungs, pleura, mediast. nodes, stomach, liver, spleen, pancreas, kidney, heart, brain	—

—		11	78, M	Pernicious anemia, ac. myel. leukemia	Lungs, stomach (polyps)	—
—		19	17, F	Ac. lymph. leukemia	Lungs, esophagus	—
—		20	71, M	Ac. myel. leukemia	Lungs, myocardium	—
FISHER et al. (1966)	U.S.	—	54, M	T.B.C., pancytopenia, antibiotics, steroids	Lungs, coronary arteries, myocardium, liver, spleen, GI tract, peritoneum	*A. fumigatus*
HUGHES (1966)	U.S.	—	8, F	Staph. aureus and S. typhimurium sepsis	Lungs, brain	*A. fumigatus*
STRELLING et al. (1966) . . .	U.K.	1	4, F	Not apparent	Lungs, mediastinal nodes, spleen, liver	*A. fumigatus*
—		2	1, F	Not apparent	Lungs, mediastinal nodes, spleen	*A. fumigatus*
KHOO (1966).	Singapore	—	2 mos. F	Not apparent	Lungs, brain, heart, kidney, pulmonary artery, small intest., diaphragm, thyroid	*A. fumigatus*
LEDERMAN et al. (1966) . . .	U.S.	—	7, M	Recurrent pneumonitis	Lungs, brain, eye, heart, skin, vertebrae	*A. fumigatus*
DARRELL (1967)	U.S.	—	40, M	Prosthetic replacement mitral valve	Prosthetic mitral valve (thrombus), brain, eye	*A. fumigatus*
RIFKIND et al. (1967) . . .	U.S.	7	25, M	Immunosuppresive therapy (renal transplantation)	Lung, kidney, brain, heart	—
—		8	37, M	Immunosuppresive therapy (renal transplantation)	Lung, liver	—
		15	16, M	Immunosuppresive therapy (renal transplantation)	Lung, brain, heart	—
—		19	28, M	Immunosuppressive therapy (renal transplantation)	Lung, kidney, brain, thyroid, heart	—
—		21	17, M	Immunosuppresive therapy (renal transplantation)	Lung, brain, heart	—
		23	35, M	Immunosuppresive therapy (renal transplantation)	Lung, brain, heart	—
PEÑA et al. (1968)	Colombia	10	19, M	Intoxication by yellow P, antibiotics, steroids	Lungs, kidneys, eye, meninges	—
MAHVI et al. (1968)	U.S.	—	9, M	Open-heart surgery (congenital aortic stenosis), foot wound	Brain stem, aortic embolus	*A. niger*
SAUNDERS and BIEBER (1968)	U.S.	—	55, M	Immunosuppresive therapy (heart transplant)	Lungs, new heart, brain	*Asperg. sp.*

Table 2. *Frequency of Organ Involvement in 81 cases of Disseminated Aspergillosis*

Organ and Nos. of Cases		Organ and Nos. of Cases		Organ and Nos. of Cases	
Lungs	71	Stomach	5	Adrenal	1
Heart	46	Leptomeninges	5	Ankle	1
Brain	44	Dura	2	Chest Wall	1
Kidney	33	Eyes	4	Esophagus	1
Liver	16	Skin	4	Tongue	1
Thyroid	16	Bone	3	Uterus	1
Spleen	14	Skeletal Muscle	3	Oral Cavity	1
Intestine	12	Pancreas	3	Periadrenal Tissues	1
Mediastinal Nodes	8	Spinal Canal	2	Prostate	1

Table 3. *Pre-existing conditions in 81 cases of disseminated Aspergillosis*

Leukemia	25 cases
Lymphoma	5 cases
Immunosuppressive therapy	7 cases
Idiopathic anemia, neutropenia or pancytopenia	6 cases
Sepsis (*Staph.*, *choleraesuis*), pneumonitis	5 cases
Open heart surgery	4 cases
Viral hepatitis	2 cases
Malignant tumors	2 cases
Tuberculosis	2 cases
Prematurity with cytomegalic inclusion disease	2 cases
Systemic lupus erythematosus	1 case
Septic abortion	1 case
Sarcoidosis	1 case
Myeloid metaplasia	1 case
Rheumatoid arthritis	1 case
Yellow P intoxication	1 case
Traumatic amputation of leg	1 case
Rectal fistula	1 case
Exchange transfusion	1 case
Old rheumatic heart disease	1 case
Not apparent	11 cases
Total	*81 cases*

The isolation of *Aspergillus* sp. from the sputum is frequently reported (EDWARDS, 1935; FINEGOLD et al., 1959; FRAUMENI and FEAR, 1962; WAHNER et al., 1963) and, although not necessarily diagnostic, this finding is helpful. However, it must be kept in mind that *Aspergilli* in sputum may be non pathogenic saprophytes.

The infection may lead to death in a period of time ranging from 2 or 3 days to one or two months, or it may enter a chronic phase that exhibits a clinical picture similar to that of tuberculosis. The chronic form of the disease is characterized by a tendency to abscess formation (COOPER, 1946), fibrosis (SILVER et al., 1962), and pleural empyema (WAHNER et al., 1963). Under adequate treatment the patients may recover (CANNON, 1935; PEER, 1960; RUBINSTEIN, 1963).

Primary aspergillosis of the lungs evolves as an acute pneumonitis characterized by a sudden onset of chills, fever, cough, hemoptoic sputum, cyanosis and chest pain. Physical signs of consolidation my be present. The fungus may sometimes be identified in the sputum (DELIKAT and DYKE, 1945). Chest X-rays reveal a discrete or confluent nodular infiltrate of the lung fields. The condition may have a favorable outcome (DELIKAT and DYKE, 1945), may lead rapidly to a suppurative stage (TOIGO, 1960), or prove fatal (HERTZOG, 1949; TOIGO, 1960; STRELLING et al., 1966).

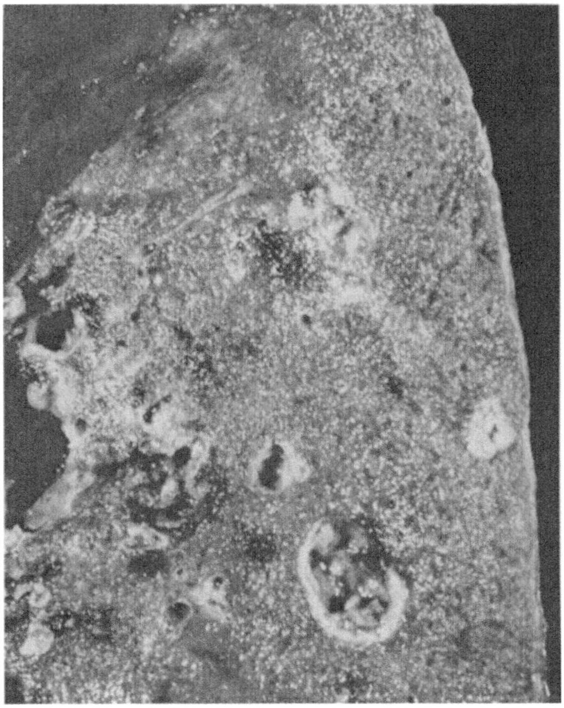

Fig. 13. Pulmonary aspergillosis. The lung parenchyma is consolidated and exhibits numerous well demarcated necrotic foci. (Presbyterian-University Hospital, Pittsburgh, Pa.)

Primary aspergillosis may also have a chronic evolution, with a clinical picture similar to tuberculosis (HETHERINGTON, 1943; HUNT et al., 1961). The main symptoms are cough, weakness, hemoptysis and weight loss. Abscess-like cavities (WILSON et al., 1967) or nodular infiltrations of the lungs are seen on X-rays and the fungus may be found in the sputum. Some patients are asymptomatic and X-ray alterations precede the symptoms by a number of years. In some instances there is a long term history of occupational exposure to grain or dust (COE, 1945).

Morphologic Findings: *Secondary aspergillosis* of the acute type most commonly presents as a necrotizing pneumonitis. The lungs are heavy, weighing 1200—1500 grams, and disclose multiple, poorly demarcated foci of consolidation that measure 2 or 3 cm in diameter. As a rule these foci do not have any specific distribution, and may be more abundant in any pulmonary lobe. On sectioning they typically have a bright white or yellow-colored necrotic center with a hemorrhagic halo (GOWING and HAMLIN, 1960; BAKER, 1962a). Sometimes the lesions have the gross appearance of pulmonary infarcts (CASTLEMAN et al., 1966). On occasion the process is more diffuse and the lungs exhibit consolidation of one or more lobes. There may also be numerous abscesses a few mililmeters in diameter. Large abscesses are not frequent, although they have been reported in a number of cases (ATKINSON et al., 1958; BECH, 1961; FRAUMENI and FEAR, 1962; REDMOND, 1965).

Microscopically the foci of consolidation show coagulative necrosis surrounded by an area of intra-alveolar fibrinous exudate and an outer zone of recent hem-

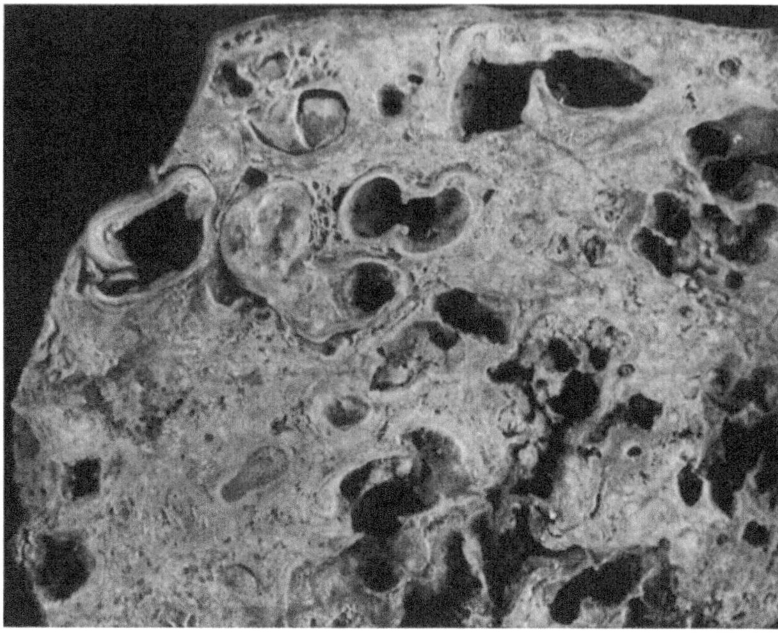

Fig. 14. Double pulmonary infection by *Aspergillus* and *Candida*. There are numerous ectatic bronchi, several intraparenchymatous abscesses and consolidation of the intervening parenchyma

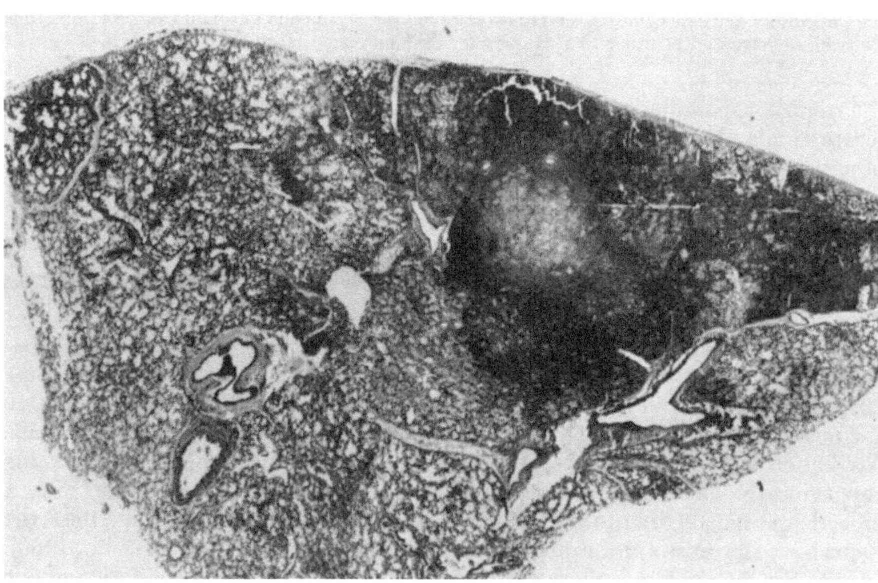

Fig. 15. Pulmonary aspergillosis. This focus reveals a necrotic center with remnants of alveolar septa surrounded by an area of recent hemorrhage. H & E, Approx. × 30

orrhage (Fig. 15, 16). Leukocytes, especially pyknotic neutrophils are found at the edge of the necrotic zone. Large areas of consolidation disclose only an intra-alveolar exudate composed mainly of neutrophils and macrophages.

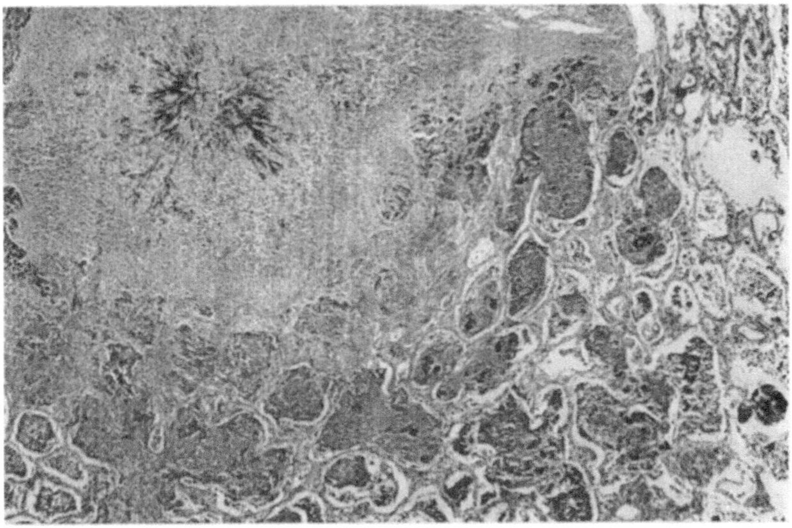

Fig. 16. Pulmonary aspergillosis. A sharply demarcated necrotic focus shows radiating hyphal elements. The non necrotic alveoli contain a proteinaceous material characteristic of pulmonary alveolar proteinosis. The patient also suffered from a chronic granulocytic leukemia. PAS, Approx. × 100. (Courtesy of Dr. R. Hellstrom, Pittsburgh, Pa.)

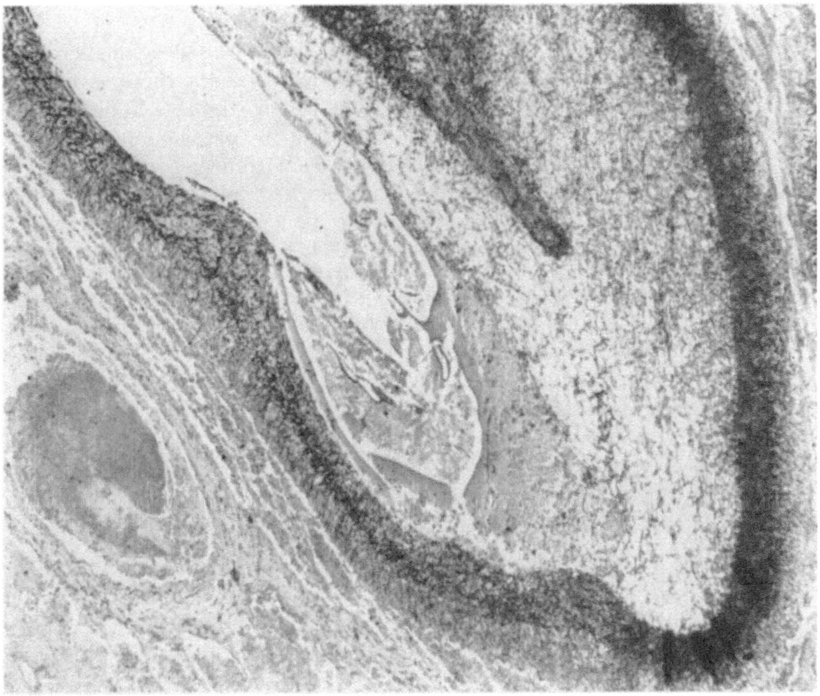

Fig. 17. Double infection by *Aspergillus* and *Candida*. The wall of this pulmonary cavity is lined by a massive fungal growth in which both *Aspergillus* and *Candida* hyphae may be identified at higher magnification. Typical fruiting heads were present in an area and *Candida* sp. was repeatedly isolated from the sputum. (Same case as Fig. 14)

Occasionally the inflammatory reaction is characterized by granulomas of giant and epithelioid cells (KIRSCHTEIN and SIDRANSKY, 1956; TORACK, 1957; REDMOND et al., 1965). Concurrent staphylococcal abscesses may be present (GOWING and HAMLIN, 1960). The inflammatory reaction may be modified by conditions that alter the reactivity of the host. In patients with leukemia and steroid therapy, the neutrophils around foci of coagulation necrosis may be absent, instead of which scattered mononuclear cells are found (GOWING and HAMLIN, 1960).

The chronic form of secondary aspergillosis has similar pathologic characteristics as the acute form, although a more marked tendency for fibrosis and abscess formation is manifest (Fig. 14, 17). In its terminal stages the lesion may be reduced to well demarcated isolated nodular residual lesions as observed by TAKANO et al. (1962).

Primary aspergillosis manifests itself as a granulomatous pneumonitis (HERTZOG, 1949; STRELLING et al., 1966). In these cases there is a diffuse consolidation of both lungs. On sectioning both discrete and confluent grayish-white nodules measuring from 0.1—3 cm in diameter are seen in all lobes. Microscopically they are composed of a center of neutrophils and a peripheral zone of epithelioid and giant cells. Numerous hyphae are present.

In cases having a chronic evolution there is a tendency to increased degrees of fibrosis around granulomas or abscesses (ROSS, 1951). Occasionally the lungs are studded with numerous ragged cavities (LIBRACH, 1957). The case reported by HUNT et al. (1961) was characterized by extensive fibrosis of the lungs that extended into the mediastinum and neck.

Pleura: The alterations of the pleura in all types of aspergillosis reflect the type of changes observed in the lung. In some instances there is a fibrinous pleuritis with variable degrees of organization which contains hyphae (KIRSCHTEIN and SIDRANSKY, 1956). Pleural empyema is occasionally encountered (WAHNER et al., 1963; SCHECHTER et al., 1960). In chronic cases fibrous adhesions may obliterate the pleural cavity to a variable degree. Occasionally, bronchopleural and pericardial fistulas are reported as a postoperative complication (WAHNER et al., 1963; HIDDLESTONE et al., 1954).

Larynx, Treachea, Bronchi: The fungus is sometimes found within normal or ectatic bronchi as 1—2 mm white or yellow granules. The mycelium may or may not invade the bronchial wall which may show little or no inflammatory reaction. In the latter instance it is assumed that the fungus is growing as a saprophyte. In invasive cases, *Aspergillus* often produces an acute necrotizing laryngo-tracheobronchitis with a marked mucosal inflammatory reaction and destruction of the cartilage plates. OKUDAIRA and SCHWARZ (1962) have described in detail a pseudomembranous tracheo-bronchitis previously mentioned in the literature only *en passant* (LARA, 1928). The pseudomembrane consisted of a compact mass of hyphae, mucus, nuclear debris, and white and red blood cells. The adjacent tissues were infiltrated by polymorphonuclear and mononuclear leukocytes. SMITH et al. (1965) described a case in which hemorrhagic ulcers were present in the larynx, trachea and bronchi.

Pulmonary Aspergilloma: *Aspergilloma* is the only form of aspergillosis that manifests itself as a well defined clinical syndrome. SCHWARZ et al. (1961), VILLAR et al. (1962), VILLAR (1963) and a Lancet Editorial (1968) have reviewed the subject. The discussion that follows is based upon the analysis of 126 cases gleaned from the literature.

This form of the disease has the highest incidence in the 4th, 5th and 6th decades with similar frequencies in each decade. The age limits reported are 10 (PIMENTEL, 1966) and 82 years (LEVIN, 1956). The condition is found more frequently in men than in women in the proportion of 2.8:1, although in a recent British Tuberculosis Association survey (B.T.A., 1968), females outnumbered males. Most cases of aspergilloma appear in patients suffering from an underlying disorder that has damaged the bronchopulmonary tree, especially by producing cavitation or necrosis of the parenchyma. The associated conditions are the following: proven tuberculosis (28 cases), bronchiectasis and bronchiectatic cysts (11), histoplasmosis (5), bacterial pneumonia (4), asthma (3), silicosis (2), pulmonary sarcoidosis (2), removed echinococcus cyst of lung (2), bronchiolar carcinoma (1), asbestosis (1), emphysematous bleb (1). Nineteen cases were claimed to be primary. In 47 cases the presence or absence of a previous disease could not be definitely established: in general terms, the authors tended to regard them as secondary, with tuberculosis suspected in a number of them. The predisposing causes may be multiple as in CORPE and COPE's case (1956) in which bronchogenic cystic disease was complicated by a *Salmonella choleraesuis* infection.

Table 4. *Species identified in 34 cases of disseminated Aspergillosis in which cultures were reported*

A. fumigatus	29 cases
A. flavus	2 cases
A. nidulans	2 cases
A. niger	1 case
A. sydowi	1 case
A. restrictus	1 case
Total	36 cases*

* The 2 extra cases are accounted for by a patient that disclosed a triple infection.

How often aspergillomas develop in pre-existing pulmonary cavities is not well known. In a prospective survey by the British Tuberculosis Assocation (B.T.A., 1968), positive precipitin serum reactions to *Aspergillus* antigens were obtained in 25% of patients with quiescent tuberculous cavities larger than 2.5 cm. The radiological appearance was compatible with aspergilloma in 44% of the positive reacting individuals (11% of the total).

Clinical Manifestations: The symptomatology is variable and the manifestations directly ascribable to aspergilloma are often difficult to separate from those of the underlying condition.

Hemoptysis, the most important symptom, and very often the presenting one, is encountered very frequently (76 cases, or 60%). It is generally recurrent and extends for periods of months or even years. Usually it is mild, but occasionally it may be severe (LEVIN, 1956; VANTRAPPEN et al., 1959) and even fatal (HUSEBYE, 1967). SALFELDER et al. (1961) have described the curious occurrence of postural hemoptysis in a young woman. Other common symptoms are, in order of frequency, cough, expectoration, fever, weakness and weight loss. Expectoration of small fragments of mycelium is infrequently observed (IKEMOTO, 1963—1964; WEENS and THOMPSON, 1950). Other symptoms encountered are chest pain (RUEDA et al., 1962), wheezing, night sweats and clubbing of the fingers. In many cases the general state of health is quite well preserved and the patient may even be asymptomatic when the aspergilloma is disclosed by a chest roentgenogram (VELLIOS et al., 1957). In those in which the general condition is poor this can usually be attributed to the underlying disease.

The physical examination is in most patients unrewarding. Signs can be elicited pointing to the presence of a pulmonary cavity or a focus of consolidation.

The prognosis is, in general terms, good, and only a relatively small number of deaths have occurred (17 cases, or 13%). The aspergilloma was the immediate cause of death in one patient (HUSEBYE 1967) and a contributor in 3 others through post-operative complications (DÉVÉ, 1938; FINEGOLD et al., 1959), and pneumonia (MONTES, 1963). In 6 cases the demise was related to the pre-existing diseases, which were cor pulmonale (MACARTNEY, 1964; FERNÁNDEZ-LUNA, 1958), acute respiratory infection (HINSON et al., 1952), cachexia (SCHWARZ et al., 1961), brain abscess (PIMENTEL, 1966), and generalized carcinomatosis (MAYS et al., 1967). In 2 other cases, death was unrelated to the underlying disease or to the fungal infection (PECORA et al., 1960; IKEMOTO, 1963—1964). In the remaining 5 cases the cause of death was not determined (PESLE et al., 1954; LEVIN, 1956; IKEMOTO, 1963—1964; AGUILAR-CELI, 1960).

It is quite difficult to assess the total *duration* of the process. Estimates vary from 3 months (ROBINSON, 1962) to 31 years (MONTES, 1963), but most fall between 1 and 3 years, although in BARIÉTY et al's. case (1957) the symptoms were present only 18 days. It is likely that, if left undisturbed, the process will remain stable for a long time.

The *clinical diagnosis* of pulmonary aspergilloma rests heavily upon the demonstration of the characteristic radiological picture. This is described as a round to oval thin walled cavity in an upper lobe containing a moderately dense mass. A thin air crescent separates the mass from the wall of the cavity. Usually the mass occupies about 80—90% of the cavity, but sometimes it fills only $^1/_3$ or $^1/_4$ of it. The intracavitary mass is typically movable and can be seen to change its position with the movements of the patient, an important diagnostic sign. ARGEN et al. (1962) observed mobility and plasticity of the ball in a case studied by cineradiography. In some cases the mass is not mobile (RILEY et al., 1962). The *X-ray picture*, although highly characteristic, is not absolutely pathognomonic of aspergilloma. Other conditions such as certain phases of a hydatid cyst, inspissated necrotic or purulent material in an abscess or bronchiectatic cavity, an intracavitary blood clot or a Rasmussen aneurysm can produce a similar appearance (WEENS and THOMPSON, 1950). In one case pulmonary angiography showed displacement of the segmental and lobar arteries near the lesion, reduction of their caliber, and retention of the contrast medium (SALFELDER, 1961). In rare occasions a fluid level may be observed (BRAATELIEN et al., 1961; FOUSHEE et al., 1958; GUILFOIL, 1964).

Most often the roentgenographic changes are evident on standard chest films although sometimes they are apparent (and certainly better seen) only on tomograms (COLLIE et al., 1965) (Figs. 18, 19). Occasionally there is calcification of the intracavitary mass, extensive enough to be seen in X-rays as focal nodules or as coral-like structures (PIMENTEL, 1966). Bronchography is not very helpful. In most cases the contrast medium does not enter the cavity (LEVIN, 1956). Occasionally it does, producing displacement of the intraluminal mass (VILLAR et al., 1962) or delineating the shape of a "glass of champagne" (GALUSSIO and MOSCA, 1963). SALFELDER et al. (1961) observed "amputation" of bronchi in one case. Associated bronchiectases can also be demonstrated by this technique (SCHWARZ et al., 1961; BARNOLA et al., 1961).

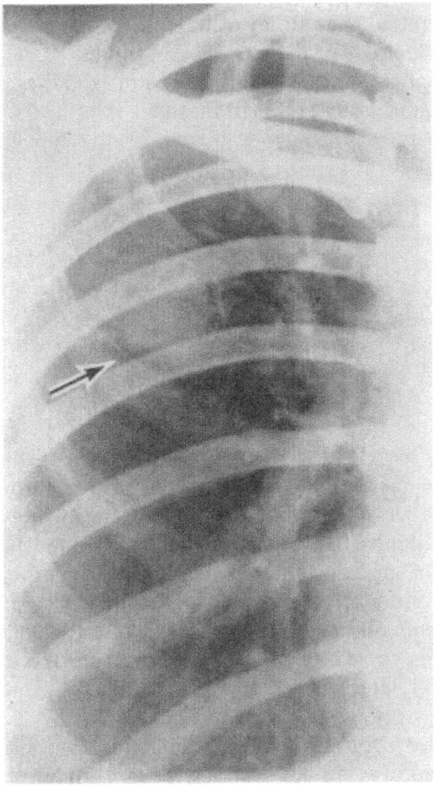

Fig. 19. Intrabronchial aspergilloma. Tomogram. The intracavitary mass, the air meniscus and the cavity wall are clearly visible. Same patient as Fig. 18. (Courtesy of Dr. M. Sánchez-Bertieri, Cali, Colombia). (Originally published in Mycopathologia 1968)

Fig. 18. Intrabronchial aspergilloma. Plain chest film. The upper right lung field displays a thin-walled cavity containing a dense mass, from which it is separated by an air meniscus (arrow). (Courtesy of Dr. M. Sánchez-Bertieri, Cali, Colombia). (Originally published in Mycopathologia 1968)

Bronchoscopy is most of the time non contributory (KESHISHIAN, 1961; ROBINSON et al., 1962; HAUSMANN, 1958). Minor degrees of non-specific inflammation may be observed in the lobar bronchus adjacent to the lesion (ROBINSON et al., 1962; HEMPHILL, 1946) or in the main bronchi (ROBINSON et al., 1962; WEENS and THOMPSON, 1950).

Laboratory Studies: Examination of the sputum directly or by culture, an important diagnostic step, is positive in most instances. Thus in 19 of the 23 cases studied by CAMPBELL and CLAYTON (1964), *A. fumigatus* was obtained.

Blood cell counts are usually not helpful. Among CAMPBELL and CLAYTON'S (1964) 23 cases of pulmonary aspergilloma, only one had more than 1000 eosinophils per cu. mm. This finding agrees with the generally held notion that eosinophilia is not a typical feature in patients with aspergilloma. In isolated cases there is moderate leukocytosis with neutrophilia (VILLAR et al., 1962).

Serum precipitins are an almost invariable finding in patients with aspergilloma (NICHOLSON, 1964; CAMPBELL and CLAYTON, 1964).

Etiologic Agent: *Aspergillus fumigatus* is the fungus most frequently found (34 cases), followed by *A. flavus* (2 cases, ARGEN et al., 1962; RAMIREZ, 1964), *A. niger* (3 cases, VILLAR et al., 1962; GALUSSIO et al., 1963; MONTES, 1963), *A. terreus* (2 cases, PLIHAL et al., 1964; WAHNER et al., 1963), and *A. versicolor* (1 case, CUEVA, 1965) (Table 5). Cultures can be made from the sputum, from bronchial secretions obtained at bronchoscopy, and from the surgical specimen. In many cases it is difficult to obtain any growth from the intraluminal mass after surgery, perhaps due to loss of viability of the fungus. With the possible exception of *Nocardia* (WILHITE, 1966) it is very doubtful that fungi other than *Aspergillus* are capable of producing the clinical-pathological picture of a "fungus ball", although in some early reports an *Actinomyces* (WEENS and THOMPSON, 1950) and a "*Monilia*-like" organism (GRAVES et al., 1951) were incriminated.

Table 5. *Species identified in 43 cases of Aspergilloma in which cultures were reported*

A. fumigatus	34 cases
A. niger	3 cases
A. flavus	2 cases
A. terreus	2 cases
A. versicolor	2 cases
Total	43 cases

Pathogenesis: The exact mechanism by which the fungus implants itself is not fully understood. It is supposed, however, that an airborne fragment of mycelium or spores can be brought into contact with the previously diseased tissue, whereupon it will proliferate and eventually acquire its full dimensions. Sequential chest X-rays have shown the appearance of mycelial masses in previously empty tuberculous cavities or abscesses (FINEGOLD et al., 1959; RILEY et al., 1962). In other cases a known aspergilloma has been observed to enlarge (LEVIN, 1956; RAMSAY, 1960; HUSEBYE, 1967). The lesion grows slowly as evidenced in one of Levin's cases in which the cavity enlarged from 2 to 3 cm and the mass from 1 to 2 cm in 3 years.

One of the conditions that seems necessary for the development and growth of aspergilloma is the presence of a free communication between the site of implantation of the fungus and the airway. The finding of bronchi communicating with the cavity in a great number of cases supports this view. The relationship of *aspergilloma and tuberculosis* is of particular interest. In some cases the fungus colonizes a pre-existing tuberculous cavern. However, this is not always the case, and perhaps because of the presence of bronchiectasis and poor drainage, the fungus may instead colonize an adjacent bronchus (VILLAR et al., 1962). The difficulties encountered in ascertaining this relationship is exemplified by a case that we had the opportunity to study (PEÑA et al., 1968). In this patient the existence of an aspergilloma and/or tuberculosis was suspected on the basis of the radiological picture. No other lesions were visible on chest X-rays. The finding of acid-fast bacilli in the sputum "confirmed" the existence of tuberculosis. Lobectomy showed, however, that the cavity wall was formed of bronchial elements (smooth muscle, respiratory epithelium and mucous glands) with only a moderate inflammatory infiltrate. No evidence of active or healed tuberculosis was present. The possibility was then entertained that atypical *Mycobacteria* could have been growing in an aspergilloma cavity, but definite cultural proof was unfortunately lacking. The case reported by GUILFOIL (1964), in which acid-fast bacilli were isolated twice, is remarkably similar.

MARTIN-LALANDE'S (1961) review of 20 cases of aspergillosis and tuberculosis culled from the literature suggests that *Aspergillus* and *Mycobacteria* are unable to grow simultaneously in the human lung. This investigator points out that when tuberculosis and aspergillosis appear in the same individual, active tuberculous lesions and *Aspergillus* lesions do not occur in the same area. Indeed, MARTIN-LALANDE'S (1961) experimental work indicates that the tubercle bacillus is initially destroyed *in vitro* by the *Aspergillus*. However, this is not an absolute rule, and SÉGRETAIN (1962) has illustrated the seemingly peaceful coexistence of *Mycobacterium* and *Aspergillus* in the same cavity.

Eighteen of the 126 cases reviewed were claimed to be of primary nature (DÉVÉ, 1938; MONOD et al., 1951; YESNER et al., 1950; ROBINSON et al., 1962; BRAATELIEN et al., 1961; ORIE et al., 1960; HOCHBERG et al., 1950; BARIÉTY et al., 1957; FOUSHEE et al., 1958; GALUSSIO et al., 1963; and MONTES, 1963). However, the possibility of occurrence of primary disease is still controversial and perhaps is more widely accepted in Europe than in the United States. MONOD et al. (1951), and PESLE and MONOD (1954) used the term *bronchiectasis — producing aspergilloma ("aspergillome bronchectasiant")* to denote a definite pathogenetic mechanism. A critical evaluation of some of the cases believed to be primary reveals that the evidence on which such an impression was built is often insufficient. However, several reports should be kept in mind. YESNER and HURWITZ's case (1950), a chicken farmer who had ample opportunity for contracting an infection by feeding his flock mouth to beak, did not exhibit evidence of any other disease, clinically or pathologically. BRAATELIEN et al's. 2 cases (1961) had a history of heavy exposure to grain dust; no other significant underlying condition could be detected either clinically or on examination of the resected pulmonary lobe, and the aspergilloma was considered to be primary. It seems, therefore, that aspergilloma occurs as a primary entity even though these cases are rare. It is thought that in the primary disease the repeated exposure to air-borne spores, perhaps of an occupational nature, is important. This would lead to the development of an intrabronchial mycelial mass that after attaining a certain size, would obstruct the airway and produce dilatation of the bronchus through a valve action mechanism.

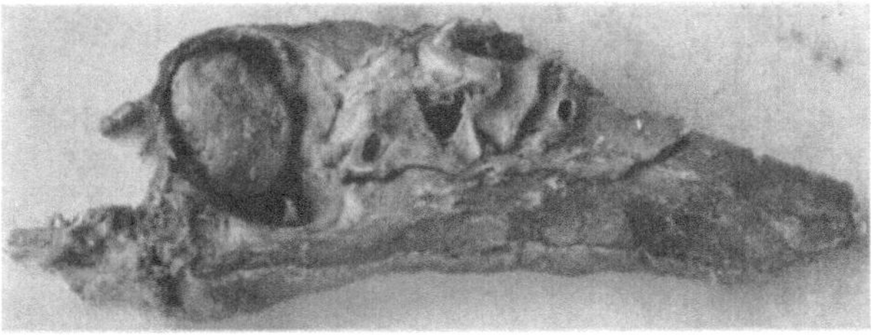

Fig. 20. Intrabronchiectatic aspergilloma. Resected right upper lobe from patient of Fig. 18. A smooth walled cavity contains an unattached ovoid mass. (Originally published in Mycopathologia 1968)

Morphologic Findings: On gross inspection *the classical aspergilloma* appears as a medium sized, round or oval cavity measuring around 3 cm in diameter, located in the apical segment of the right or left upper lobes. Such a lesion has a very regular contour; its walls are white, firm, well defined and usually measure from 1—5 mm in thickness. The inner lining is usually smooth and grayish-white, The material found in the cavity is commonly a yellow-brown, friable, granular, ovoid mass that has a smooth or lobulated surface and does not adhere to the cavity wall (Fig. 20). In the majority of cases, one or several bronchi, measuring 2—3 mm in diameter open into the cavity. However, a great deal of variation exists. The lesion may affect the lower lobes. The size of the cavity varies from 1 (HINSON et al., 1952) to 19 cm (IKEMOTO, 1963—1964) in diameter. The inner surface is sometimes irregular, ragged and rugose and sometimes shaggy. Occasionally there is more than one cavity (SCHWARZ et al., 1961). The area most fre-

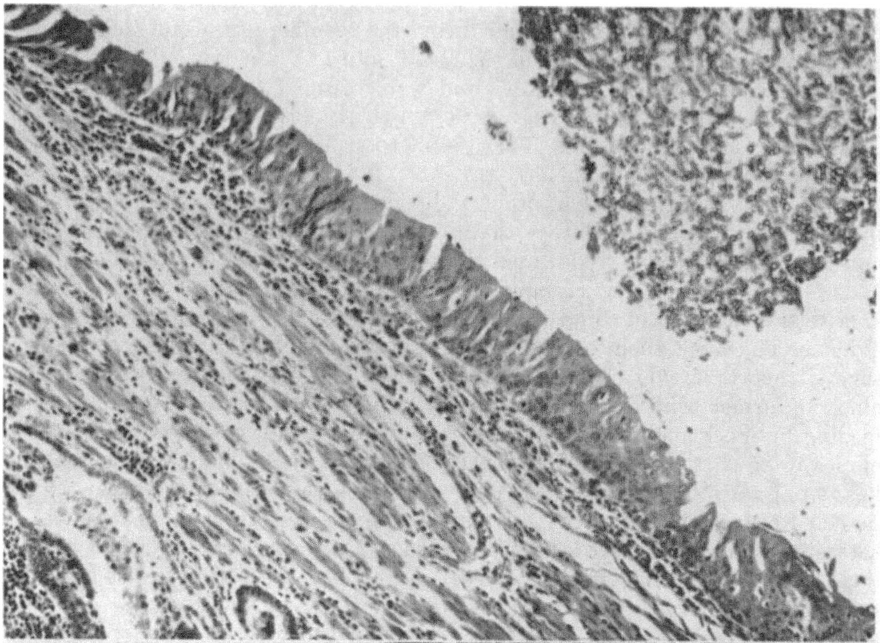

21

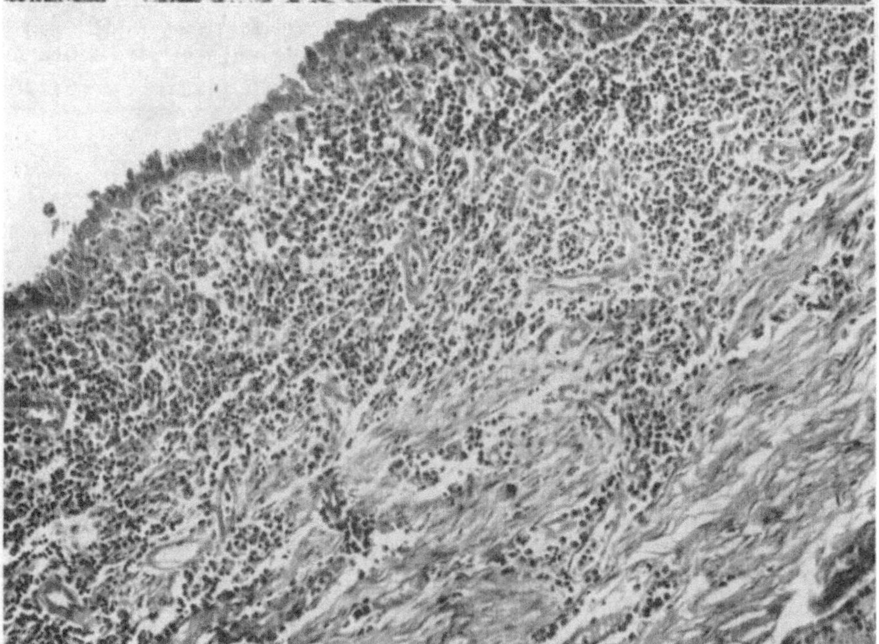

22

Fig. 21. Intrabronchial aspergilloma. From same case as Fig. 18. The wall of the cavity is composed of an epithelial lining, a smooth muscle layer and bronchial glands. In the right upper corner is a fragment of the mycelial mass. The inflammatory reaction is minimal. H & E, × 100

Fig. 22. Wall of pulmonary aspergilloma. In this case the cavity wall is composed of fibrous connective tissue with an abundant mononuclear leukocytic infiltration. This type of lesion probably develops from a healed tuberculous cavity. H & E, × 100 quently involved is the right upper lobe (52 cases), followed by the left upper lobe (40 cases). The condition may be bilateral, involving both upper lobes (6 patients). The lower lobes are less frequently involved (left, 5 cases; right, 4 cases). In 1 case it was located in an interlobar space. In the remaining 18 cases the exact location was not specified. Whether in the upper or in the lower lobes the lesion has a great predilection for an apical segment. The intracavitary contents may vary. There can be several masses of a dark-brown or grayish and pasty or crumbly material. Focal calcification is not unusual (GERSTL et al., 1948; PIMENTEL, 1966), but extensive calcification to the point of the mass becoming a cavernolith or a coral-like structure is rare (VILLAR et al., 1962). In some cases it is not possible to find a communication with a bronchus perhaps due to closure of the drainage by the inflammatory process. Communication of the aspergilloma cavity with the pleural space producing a hydro-pneumothorax has been reported (FERNÁNDEZ-LUNA, 1958).

Microscopically the wall of the cavity has a variable appearance according to its origin (Figs. 21—22). In some cases it exhibits bronchial elements as smooth muscle, cartilage plates and mucous glands. A respiratory epithelium with foci of squamous metaplasia forms the inner lining. This type of cavity certainly originates in saccular or cylindrical bronchiectases, or bronchial cysts. Another kind of cavity is that in which only a fibrotic wall can be identified, often lined by patches of respiratory epithelium. This kind originates probably from tuberculous caverns or other healed conditions. Both types of cavity may display a degree of inflammatory infiltrate varying from mild to marked, in which lymphocytes, plasma cells and monocytes are abundant. Sometimes eosinophils (HEMPHILL, 1946) and polymorphonuclear leukocytes are found. Infrequently, granulomas composed of giant and epithelioid cells are observed, which betray the origin of the cavity: tuberculosis (PECORA and TOLL, 1960; PIMENTEL, 1966), histoplasmosis (PROCKNOW et al., 1960), and sarcoidosis (ADELBERG et al., 1961). The intracavitary mass reveals upon microscopic examination an almost pure intertwining meshwork of hyphae with very little or no inflammatory exudate (Fig. 23). In many cases the hyphae have a radial disposition around several centers, as if growth was proceeding independently in separated colonies. The mycelium often exhibits broad concentric bands, each band having different staining characteristics. This disposition has been referred to as "zonal" growth and probably represents different periods of growth. In such cases the better preserved and more basophilic hyphae are usually located at the periphery, whereas the degenerating ones are located toward the center of the colony and tend to be eosinophilic. In some cases it is possible to identify the fruiting heads of the fungus. Only in rare instances are hyphae observed penetrating the cavity wall (PECORA and TOLL, 1960; ROBINSON et al., 1962; ENJALBERT et al., 1957; FOUSHEE et al., 1958; SNELLING et al., 1963). In 3 such cases (reported in the 3 last references) the presence of the hyphae was associated with a granulomatous reaction. In one case (MONTES, 1963), there was extensive invasion of the pulmonary parenchyma by numerous hyphae.

The lobe where the cavity lies is usually retracted, firm, covered by a thickened pleura, and may show some degree of peri-cavitary fibrosis and collapse. Other portions of the lobe, as well as the rest of the pulmonary parenchyma, may reflect the pre-existing condition, but very often are remarkably free of any pathological process.

Treatment: The treatment of choice for pulmonary aspergilloma seems to be a *lobectomy* (FRIEDMAN et al., 1956; VILLAR et al., 1962) or a segmental resection (KRASNITZ, 1957; LEVENE et al., 1965). IKEMOTO (1963—1964) and RAMIREZ (1964) have administered endobronchial instillations of amphotericin B and

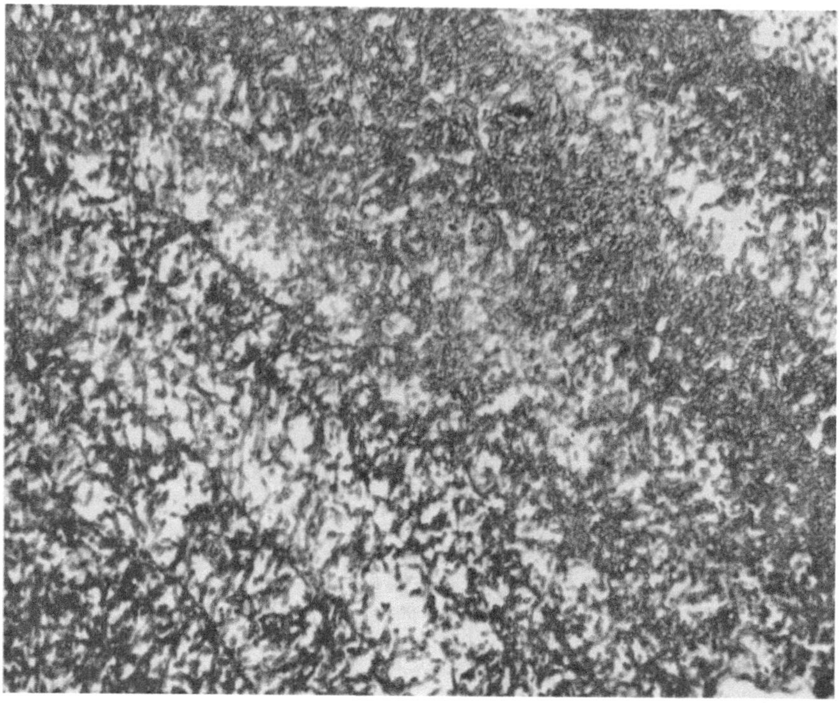

Fig. 23. Intracavitary aspergilloma. The intracavitary mass is composed of an almost pure growth of intertwining hyphae. Zonal growth of the mycelium, probably representing different periods of fungal growth is apparent. H & E, × 100

iodides with some success. More recently, HENDERSON and PEARSON (1968) have advocated local surgical evacuation of the cavity followed by irrigation with natamycin.

Allergic Bronchopulmonary Aspergillosis: *Aspergilli* posses complex antigenic factors, and repeated inhalation of its spores may lead to sensitization (BIGUET et al., 1962), usually manifested as bronchial asthma.

In 1930 BERNTON described the first case of an asthmatic syndrome associated with hypersensitivity to *A. fumigatus*.

Since then numerous reports have appeared illustrating various aspects of the condition (WEINER, 1960; ARGRABITE et al., 1963; ELLIS, 1965; SANERKIN et al., 1966; STARK, 1967; SPOTNITZ et al., 1967; AGBAYANI, 1967; PATTERSON, 1968). It is estimated that about 11% of patients with asthma have allergic pulmonary aspergillosis (HENDERSON et al., 1968). Studies of British workers, especially those of the Brompton group, have done a great deal to clarify the relationship between hyperreactive states, actual *Aspergillus* infection and the various types of antibody response encountered in affected individuals (PEPYS et al., 1964; LONGBOTTOM and PEPYS, 1964; CAMPBELL and CLAYTON, 1964). The subject has been reviewed by PEPYS (1966a, b, c), and by HENDERSON et al. (1968).

Clinical Manifestations: An asthmatic syndrome due to *Aspergillus* sensitization may be either simple or complicated by transient pulmonary infiltrations and blood eosinophilia, a syndrome that has been called "pulmonary eosinophilia".

Clinically, the former is similar to other types of asthma and manifests itself by paroxysms of dyspnea and wheezing followed by cough and expectoration of mucoid material. Symptoms of coryza may be present. The manifestations appear very often during the day, soon after the patient has come into contact with the material harboring the fungus (HINSON et al., 1952). The condition may be present for a long time before the patient seeks medical attention. Instances of occupational disease have been reported (WEINER, 1960).

PEPYS states that in Britain pulmonary eosinophilia associated with asthma is due, in the great majority of cases, to pulmonary aspergillosis (1966a). Clinically there are febrile episodes, wheezing, cyanosis, cough, and, not infrequently, hemoptysis. The sputum may contain characteristic plugs consisting of mucus, hyphae, eosinophils, CHARCOT-LEYDEN crystals and CURSCHMANN spirals (HINSON et al., 1952). Bronchoscopy reveals the presence of a thick mucous secretion blocking bronchi of various calibers. The resulting obstruction may lead to collapse of a pulmonary segment, a lobe, or the entire lung (ELLIS, 1965; SANERKIN, 1966). X-ray examination may disclose areas of collapse and/or consolidation, more frequently in the upper lobes (HENDERSON, 1968). The superimposed pneumonitis as well as the repeated exposure often lead to a chronic phase lasting for months or years, often accompanied by upper lobe contraction, as seen on X-rays (HENDERSON, 1968). A number of patients recover, although the disease is occasionally fatal, as in one of HINSON et al's. patients (1952) who died in status asthmaticus. The condition may be treated with apparent good results by desensitization with emulsified extracts of the fungus (ARGRABITE, 1963), by inhalations of fungistatics (STARK, 1967) or by the use of steroids (PATTERSON, 1968; HENDERSON, 1968).

Etiopathogenesis: Two types of antibodies seem to mediate these clinical manifestations:

1. a *reagin*, or *skin-sensitizing antibody* that is associated with an immediate hypersensitivity response. This antibody can be demonstrated by the production of immediate wheal reactions on skin testing and is responsible for the asthmatic symptoms (PEPYS, 1966a).

2. a *precipitin type* of antibody that can be demonstrated in the patient's serum with the appropriate test antigens.

In addition to the immediate skin hypersensitivity response there may exist a slower type of skin reaction that develops 3—4 hours after testing. This form of response is considered to be a precipitin-mediated reaction of the Arthus type. However, it probably also requires the presence of a reagin.

Although both types of antibodies may occur in simple asthma as well as in asthma with pulmonary eosinophilia, their incidence in the two syndromes is different. In non-selected, uncomplicated asthmatics the prick skin test (immediate) with *A. fumigatus* is positive in about 10% of cases, whereas in cases of asthma with pulmonary eosinophilia the frequency of positivity is about 90% (PEPYS, 1966a). Serum precipitins against *Aspergillus* antigens are present in about 9% of uncomplicated asthmatics (LONGBOTTOM and PEPYS, 1964; PEPYS, 1964), whereas in asthma with pulmonary eosinophilia they appear in 66% or more of the cases (PEPYS, 1966a). It seems that both reagins and precipitins are necessary for the appearance of pulmonary infiltrations (LONGBOTTOM and PEPYS, 1964). PATTERSON et al. (1968) have reported a very illustrative case in whom both reaginic and precipitating antibodies were present while the patient was harboring *A. fumigatus* in his respiratory tract. With treatment, the organism disappeared followed by the decrease of both types of antibodies.

Morphologic Findings: Grossly, the lungs are distended and show marked dilatation of the bronchi (HINSON et al., 1952). These are full of a tenacious mucus and have a thickened mucosa. Microscopically the material occluding the bronchi is composed of mucus, fibrin, Curschmann's spirals, Charcot-Leyden crystals, eosinophils or other inflammatory cells and hyphal elements. A striking change is the very abundant production of mucus by individual cells and by the bronchial glands. No alterations are described in the pulmonary parenchyma.

Farmer's Lung: According to DICKIE and RANKIN (1958) farmer's lung may be defined as an interstitial granulomatous pneumonitis occurring in agricultural workers. Although 25% of patients with this condition have serum precipitins against *A. fumigatus* (PEPYS et al., 1962) no definite relationship of this syndrome with *Aspergillus* has been conclusively demonstrated. On the other hand, it has been shown, on the basis of immunoelectrophoretic techniques, that the serum of approximately 90% of patients reacts with thermophylic *Actinomycetes (Thermopolyspora polyspora* and *Micromonospora vulgaris)* (PEPYS and JENKINS, 1965). This has led to the identification of these fungi as the most important cause of farmer's lung (PEPYS, 1966a, c).

Cardiovascular System

Heart: Involvement of the heart in aspergillosis may result from hematogenous spread, usually originating in the lungs, or from an initial endocardial colonization. The subject has been reviewed by MERCHANT et al. (1958), LANDAU et al. (1963), and NEWMAN et al. (1964). Among the 81 cases of generalized aspergillosis listed in Table 1, 46 had involvement of the heart. The symptomatology referable to the cardiovascular system varies according to the extent of the involvement of endocardium, myocardium and pericardium, and to whether the disease is metastatic or primary. Initial endocardial involvement is encountered in similar circumstances as those in which bacterial endocarditis occurs. They include old rheumatic endocarditis (ZIMMERMAN, 1950) and open-heart surgery either for rheumatic valvulitis with prosthetic replacement of the mitral valve (NEWMAN et al., 1964; DARRELL, 1967) or for congenital aortic or subaortic stenosis (PORTA, 1959; HADORN, 1960; MAHVI et al., 1968). Much more frequently the process is the result of a generalized hematogenous dissemination originating in the lungs. In some instances the portal of entry is a traumatic injury (ZIMMERMAN, 1950) and the infection involves a seemingly healthy endocardium. In rare cases the fungus may reach the pericardium through direct spread from an adjacent pulmonary abscess (FRAUMENI and FEAR, 1962).

Clinical Manifestations: *Endocarditis* tends to dominate the clinical picture in those cases in which the endocardium is initially involved. The clinical picture is similar to that observed in subacute bacterial endocarditis. The most important manifestations are embolization, fever, murmurs and heart failure. Less conspicuous are hematuria and splenomegaly. Embolisms involve quite often large arteries, which is perhaps related to the bulkiness of the mycotic vegetations. The most frequent sites are the aortic bifurcation (with production of a Lériche syndrome), the arteries of the lower extremities (femoral, popliteal), the kidneys, the spleen and the central nervous system (HADORN, 1960; NEWMAN et al., 1964; MAHVI, 1968; ZIMMERMAN, 1950). Embolism to the spleen produces flank pain and splenomegaly. Kidney involvement is manifested by hematuria. Embolism to the central nervous system is attended by the corresponding focal signs. Multiple petechiae of the skin, of the oral mucosa and of the conjunctiva are frequent, appear sometimes in crops, and may be extremely numerous. Osler's nodes of the toes and fingers may coexist. Ocular involvement with rapidly progressing amblyopia is not infrequent (PARADIS et al., 1963; LEDERMAN et al., 1966; DARRELL, 1967). Heart failure may be due to valvular or myocardial lesions and

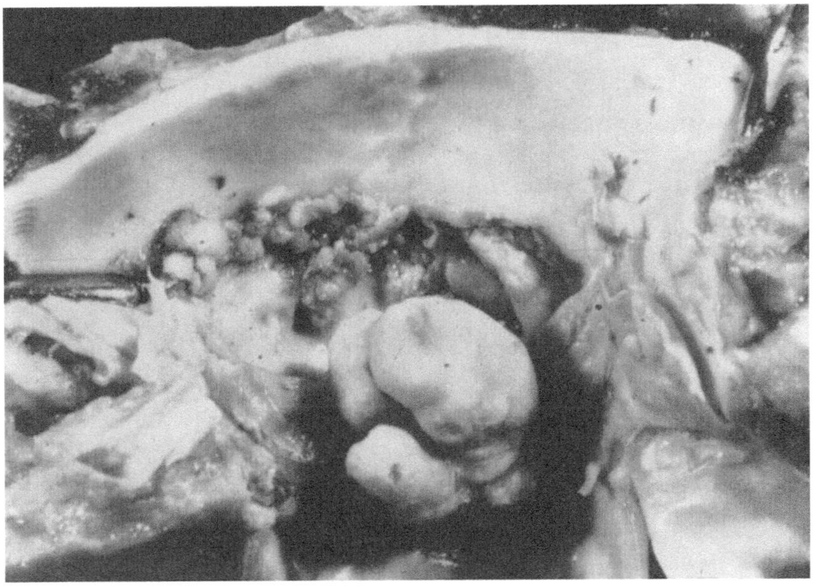

Fig. 24. *Aspergillus* endocarditis. This aortic valve homograft, apparently the primary site of infection, exhibits large polypoid vegetations. Premortem blood cultures were positive for *Aspergillus* sp. (Presbyterian-University Hospital, Pittsburgh, Pa.)

occasionally to pericardial involvement. The presence of murmurs is related to the valvular lesions produced by the *Aspergillus* infection or by the underlying disease. A loud diastolic murmur was present in NEWMAN et al's. case (1964) in which a tongue of thrombus protruded through the aortic valve making it insufficient. A widening of the pulse pressure reflects an aortic insufficiency. The disease is characterized by a rather fast evolution which can be estimated in weeks rather than in months. Terminal shock or sudden death are frequent.

Pericarditis is in some instances an important part of the clinical picture (FRAUMENI and FEAR, 1962), and is manifested by rapid enlargement of the cardiopericardial shadow, electrocardiographic changes, increased venous pressure and pulsus paradoxus.

Myocarditis is usually focal. On occasions it is manifested by heart failure.

With few exceptions (case observed at the Presbyterian-University Hospital of Pittsburgh, see Fig. 24), *Aspergillus* has not been cultured from the blood in life, although it has been recovered from heart blood after death (DARRELL, 1967; LEDERMAN et al., 1963).

Morphologic Findings: The disease affects myocardium and endocardium with approximately the same frequency. The pericardium is perhaps less frequently involved. *Myocardial* foci of necrosis measuring 1 or 2 mm are very frequently encountered. They are visible as white or yellow foci, often subendocardial or subepicardial, that involve the ventricles and sometimes the atria. Microscopically they display coagulative necrosis and variable numbers of hyphae (Fig. 26). Invasion of vessels, with subsequent thrombosis and myocardial necrosis is quite frequent.

Involvement of the *mural endocardium* is characterized by small abscesses and sometimes polypoid vegetations or thrombi in the atria or in the ventricles (LEVY et al., 1955; BUTTRICK et al., 1965; KHOO et al., 1966).

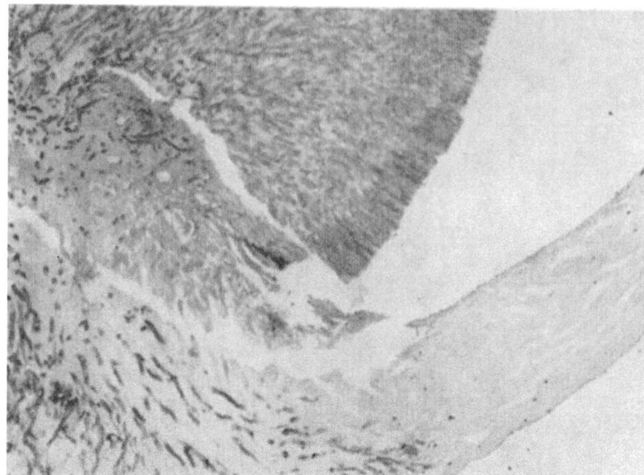

25

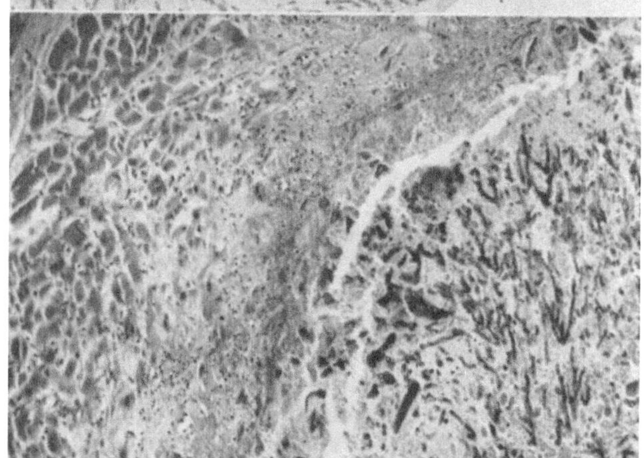

26

Fig. 25. Aortic valve. From same case as Fig. 24. The vegetation is composed of an almost pure growth of hyphae, some of which penetrate deeply into the valve tissue. Grocott-Gomori, × 100. (Presbyterian-University Hospital, Pittsburgh, Pa.)

Fig. 26. Myocardial aspergillosis. The mycelium is surrounded by necrotic myocardium containing pyknotic neutrophils. H & E, × 100

Valvular involvement is the most spectacular lesion of the disorder (Figs. 24, 25). Most frequently the aortic and the tricuspid valves are affected. The region of the mitral valve is less frequently affected, but cases have been reported following prosthetic replacement of the mitral valve (Newmann et al., 1964; Darrell, 1967) and disseminated pulmonary infection (Marsalek et al., 1960). Involvement of the pulmonary valve has apparently not been recorded. The process is characterized by bulky, friable or firm polypoid reddish-brown to gray vegetations which are much larger than those observed in bacterial endocarditis. Generally the vegetations measure 2 or 3 cm but sometimes they reach extremely large sizes (Kirchstein and Sidransky, 1956). There may also be destruction and perforation of the valves or of the interventricular septum (Zimmerman, 1950).

Microscopically the vegetations are composed of a massive mycelial growth (Zimmerman, 1950), sometimes mixed with large numbers of polymorphonuclear leukocytes (Luke et al., 1963), and covered by a fibrinous layer. The points of

27

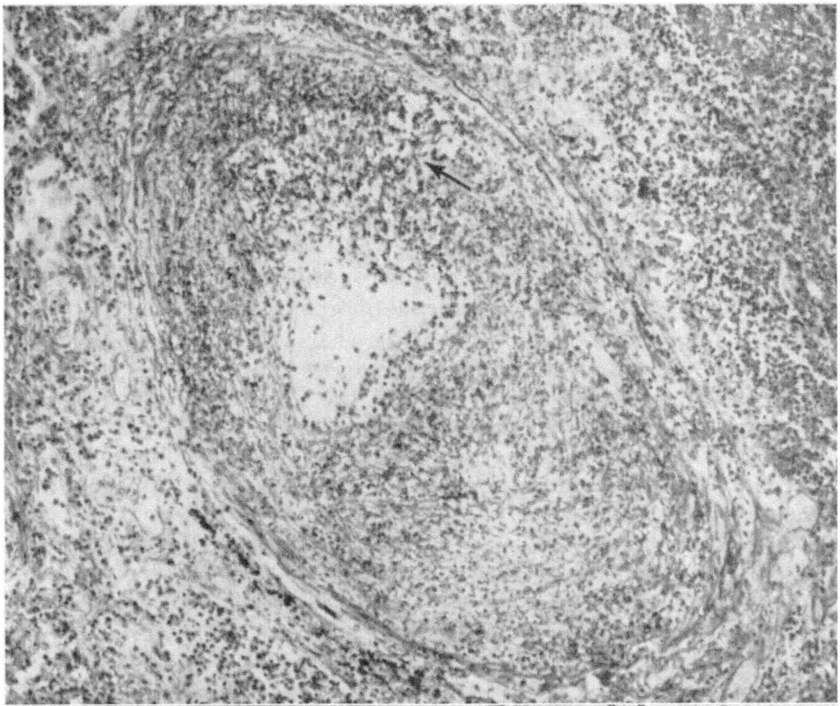

28

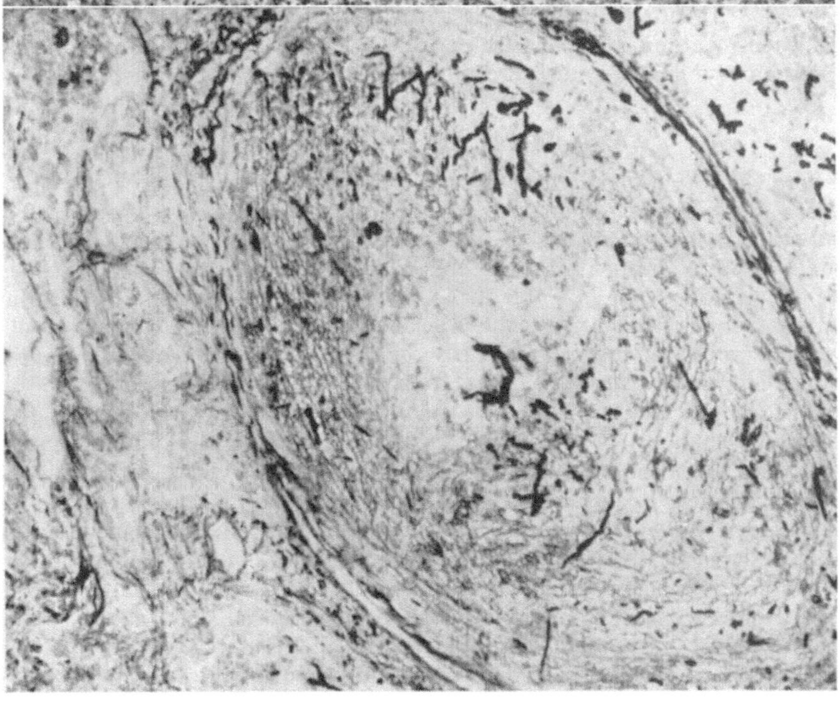

Fig. 27. Thrombosed necrotic vessel. Although hyphae are abundant, they are difficult to see (arrow). PAS, × 100

Fig. 28. Same vessel as in Fig. 27. The hyphae are clearly seen with the Grocott-Gomori stain. × 100 attachment of the vegetations to the valves exhibit necrosis of the collagen and granulocytic infiltration. The valves themselves are the site of edema, inflammatory infiltrate and sometimes show the stigmata of an old rheumatic process. In mural lesions the endocardium is covered by a fibrinous thrombus containing hyphae and neutrophils. The hyphae may extend into the underlying myocardium which shows inflammatory changes.

Pericarditis is frequently characterized by small abscesses that involve the epicardial surface. Occasionally there is purulent effusion and the pericardial membranes are covered by a shaggy fibrinous exudate (FRAUMENI and FEAR, 1962). Microscopically the epicardial abcesses or exudate contain variable numbers of hyphae surrounded by granulocytes and fibrin.

Vessels: As previously noted, one of the characteristics of *Aspergillus* is its tendency to invade small to medium sized arteries and veins, producing necrosis of their walls as well as thrombosis (Figs. 27, 28). This is a very frequent occurrence and GOWING and HAMLIN (1960) observed involvement of small vessels in all of their cases. The coagulative necrosis usually involves the entire circumference of the vessel, but sometimes is segmental. There is also reactive infiltration by polymorphonuclear leukocytes.

Subsequent *thrombosis* is an important factor in the character of the lesions and some of the more necrotizing ones are due to ischemia. In rare instances invasion and thrombosis of coronary branches leads to myocardial infarction (FISHER et al., 1966).

Involvement of large vessels such as the aorta is much more rare but has been reported following open-heart surgery for aortic stenosis (PORTA, 1959; MAHVI et al., 1968; HADORN, 1960). The lesion is characterized by thrombi or polypoid vegetations containing hyphae which adhere to the intima of the vessel and are located a few millimeters distal to the aortic valve.

Central Nervous System

Involvement of the central nervous system in aspergillosis is, in most instances, the result of generalized dissemination of the disease, with the lungs as the initial focus. Reviews on this subject have been published by SCHEIDEGGER (1958) and by FETTER et al. (1967). Among the 81 cases of generalized aspergillosis listed in Table 1, 44 (53%) have involvement of the central nervous system. In addition Table 6 lists 15 cases in which involvement of the central nervous system was apparently the result of direct extension from lesions in the orbit and paranasal sinuses, from self-administered narcotics, and from intrathecal penicillin therapy. In a small number of cases the original site of infection is unknown (DAVID et al., 1951). The routes of dissemination used by *Aspergillus* to reach the central nervous system from distant organs have not been conclusively demonstrated. However, there is little doubt that the blood stream is the pathway followed in cases of systemic spread originating in lung and heart, or in the cases attributed to self-administration of drugs (BURSTON et al., 1963), or to exchange transfusions (LUKE et al., 1963).

"Direct" extension from the orbit, nose or paranasal sinuses may be due to actual invasion and erosion of the bone plate of the cranial base, to extension through the anatomic orifices like the cribriform plate (SCHNYDER, 1948) and optic foramen, or to spread along anastomotic vessels that communicate between the cranial cavity and the orbit or sinuses via the ophthalmic vein. Definite proof of these various possibilities is lacking. However, in WÄTJEN'S (1928) and in MCKEE'S cases (1950), the intracranial involvement was attributed to ethmoid

Table 6. *Aspergillosis of the central nervous system without systemic involvement**

Author and year	Age	Sex	Portal of Entry
OPPE (1897)	37	M	Sphenoid sinus
WATJEN (1928)	66	F	Ethmoid sinuses
MONIZ et al. (1931)	44	F	Eye
GUILLAIN et al. (1935) . . .	33	F	Orbit (?)
MCKEE (1950)	60	M	Sphenoid sinus
DAVID et al. (1951)	42	F	Not apparent
WYBEL (1952)	40	F	Spinal canal (Intrathecal penicillin therapy)
IYER et al. (1952)	23	M	Unknown
IYER et al. (1952)	10 mos.	M	Unknown
JACKSON et al. (1955) . . .	36	M	Nasal sinuses or orbit
JACKSON et al. (1955) . . .	22	F	Orbit
ZISKIND et al. (1958) . . .	35	M	Malar region (?)
BURSTON et al. (1963) . . .	37	M	Self administration of drugs (?)
TVETEN et al. (1965) . . .	57	M	Orbit (?)
SCHNYDER (1948)	71	M	Nasal fossae

* Features of 15 cases culled from the literature.

and sphenoid sinusitis, and the bones of the base of the skull were involved by the mycotic process. Furthermore experimental evidence suggests that *Aspergillus* spores may spread from the nose into the cranial cavity and brain by way of emissary veins (EPSTEIN et al., 1968).

Clinical Manifestations: The clinical picture of aspergillosis of the central nervous system depends in great part on whether the disease is primary or metastatic, as well as on the location, extent and character of the lesions.

In some cases the clinical picture is that of a space-occupying lesion or of an abscess (SCHNYDER, 1948; IYER et al., 1952; JACKSON et al., 1955; BURSTON et al., 1963; TVETEN, 1965; SATHMARY, 1958; HENDRICK et al., 1963; DAVID et al., 1951).

The symptoms or signs more frequently encountered are headache, hemiparesis and convulsions, but there is often fever, paralysis of cranial nerves, and abnormal plantar reflexes. Sometimes there is blurring of the optic discs, vomiting, coma and other signs of increased intracranial pressure. Occasionally loss of memory and personality changes (BOSHES et al., 1956) are present. In other instances the clinical picture is that of a meningitic process (LINCK, 1939; AKKOYUNLU et al., 1957), manifested by fever, convulsions, nuchal rigidity, and in children, tense fontanelle. The neurologic signs are often non-specific (KHOO, 1966; MCKEE, 1950).

In cases in which the disease is due to a direct extension from nose or eye lesions, the symptomatology related to these organs is prominent and precedes the intracranial involvement for months or even years (MONIZ et al., 1931; JACKSON et al., 1955; TVETEN, 1965).

In some instances *Aspergillus* infection of the central nervous system presents itself as rather bizarre syndromes, as a "neuroma" of the trigeminal nerve (ZISKIND et al., 1958), arteritis, thrombosis and rupture of the internal carotid artery with subarachnoid hamorrhage (MCKEE, 1950), spinal block with quadriparesis (WYBEL, 1952), unilateral painless exophthalmos (JACKSON et al., 1955), and chronic encephalitis with progressive dementia and spastic paralysis (IYER et al., 1952).

Examination of the cerebrospinal fluid may show a moderate increase of protein (around 140—190 mg per 100 ml), and a moderate pleocytosis varying from a few cells to 600 or 700 per mm^3, most of which are polymorphonuclear leukocytes. Rare cases (HUGHES, 1966) disclose counts over 1000 cells. In cases of spinal block the protein reaches high levels (WYBEL, 1952). The fungus has not been isolated from the cerebrospinal fluid during life, but post-mortem cultures of the cerebrospinal fluid are occasionally positive for *A. fumigatus* (LEDERMAN et al., 1966).

Carotid angiograms, ventriculograms and electro-encephalograms are useful adjuncts.

In the great majority of cases the prognosis is poor, although fluctuations and temporal improvement may appear for some time. Only two cases have been reported with apparent recovery following the removal of a cerebral *Aspergillus* abscess (DAVID et al., 1951; HENDRICK and CONEN, 1963, — same case reported by CONEN et al., 1962).

Morphologic Findings: The morphologic changes observed in aspergillosis of the central nervous system vary according to whether the infection is the result of hematogenous dissemination or of local spread. In the first instance, the lesions tend to be acute and necrotizing or purulent (Fig. 29). In the second case the process is usually chronic and has a tendency to fibrosis and granuloma formation.

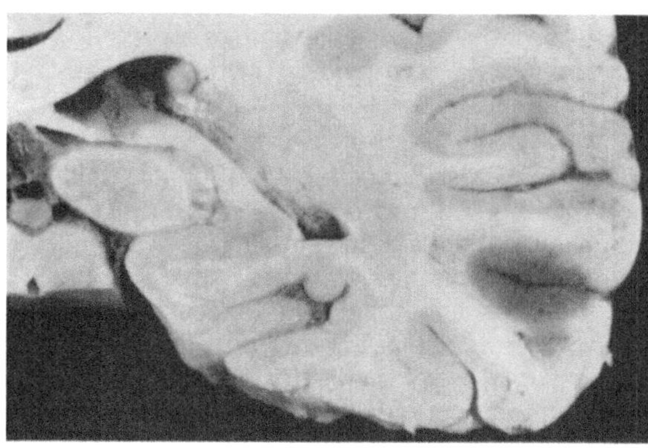

Fig. 29. Embolic *Aspergillus* lesion of the cerebral cortex. This lesion, quite similar to an infarct, is characterized by hemorrhage, necrosis and microscopically, by the presence of hyphae in and outside vessels. (Presbyterian-University Hospital, Pittsburgh, Pa.)

Hematogenous lesions most frequently take the form of multiple abscesses that measure between 0.1 and 2.5 cm (ZIMMERMAN, 1955; SATHMARY, 1958; FRAUMENI et al., 1962; LUKE et al., 1963; HUGHES, 1966), although sometimes they are quite large, measuring 4 or 5 cm (HENDRICK et al., 1963), and producing massive destruction of a hemisphere (LANDAU et al., 1963). The lesions may involve the cerebral cortex, white matter and basal ganglia, and do not have preference for any of the cerebral lobes. The cerebellar hemispheres are likewise frequently affected. A thick fibrous capsule is seen only in rare instances (HENDRICK et al., 1963). Microscopic examination discloses frank pus in the center of the lesion and abundant neutrophilic infiltration at the edges, sometimes accompanied by giant and epithelioid cell granulomas (Fig. 32). Lesions of necrotizing non suppurative type are also frequently seen. They are typically described as white or yellow multiple foci of necrosis with a hemorrhagic component, measuring from a few millimeters to several centimeters. They have a similar distribution to that observed in abscesses. Microscopically they consist of an area of coagulative necrosis with variable amounts of neutrophilic reaction and hemorrhage. Hyphae are commonly seen in the tissues. Vasculitis with thrombosis is quite common both in abscesses and in coagulative necrotic foci.

Brain lesions originating by *local spread* from the orbit, nose or sinuses are, in the majority of cases, of the fibrosing granulomatous type. The more typical examples have been described by IYER et al. (1952), JACKSON et al. (1955), and ZISKIND et al. (1958). These lesions are essentially masses of newly formed inflammatory tissue of rather large size that compress or infiltrate the brain parenchyma and often surround and compress groups of cranial nerves. On inspection they are somewhat irregular, lobulated, poorly demarcated, firm and are often mistaken for infiltrating neoplasms. They are found at the base of the brain, in the orbit, and in the right frontal lobe. On sectioning, these lesions reveal a yellow or gray-white to pink surface and are occasionally soft, cavitated or necrotic. Microscopic examination reveals the masses to be composed of more or less dense fibrous connective tissue infiltrated with collections of lymphocytes, mononuclear cells and eosinophils. The most important characteristic is the presence of foreign body giant cells and epithelioid cells, often arranged in well defined granulomas. Hyphal elements may not be easily found on routine stains. Sometimes the cytoplasm of giant cells contains ovoid or elongated spaces which upon staining with special techniques for fungi, will evidence fragments of hyphae (JACKSON et al., 1955).

The two types of inflammation discussed (purulent-necrotizing and granulomatous) are not unrelated, and may coexist. It seems, however, that the portal of entry of the infection is an important factor in determining the type of inflammatory reaction, which will, in turn, influence the clinical course of the disease.

Examination of the outer surface of the brain reflects the changes present in the cerebral tissues. In some instances there is swelling of the organ accompanied by the usual signs of increased intracranial pressure, as flattening of the convolutions, narrowing of the sulci, herniation of the cerebellar tonsils and uncal grooving. Hemorrhage of the pons and midbrain of the type associated with increased intracranial pressure is occasionally found (BURSTON et al., 1963).

The midbrain (BOSHES et al., 1956), the medulla (GRCEVIC et al., 1959; BUTTRICK et al., 1965) and the pons (IYER et al., 1952) as well as the cranial nerves are at times affected by direct invasion or by compression from an adjacent process.

The *leptomeninges* are frequently affected, although the extent and the importance of the process are quite variable (Fig. 30). In cases of metastatic cerebral infection the meningeal inflammation is often limited to those areas overlying parenchymatous foci. The meningeal inflammation tends to have the same histologic characteristics as the brain lesions. In some cases the meningitis is either generalized or mostly basilar and its symptoms become a prominent part of the clinical picture (LINCK, 1939). In such cases the leptomeninges are cloudy and contain gray or greenish exudate which upon microscopic examination is composed mostly of neutrophils (KHOO et al., 1966) or shows an admixture of inflammatory cells (AKKOYUNLU et al., 1957).

The *dura* may also be the site of focal inflammatory infiltrations in cases of generalized aspergillosis (CAWLEY, 1947; BOSHES et al., 1956). In instances of brain involvement by extension from a neighboring eye or sinus infection, the pachymeninges as well as the leptomeninges are, of necessity, at least focally involved.

The *ventricular system* is seldom the site of any changes. In rare cases, a necrotizing hemorrhagic lesion may rupture into the lateral ventricle (ZIMMERMAN, 1955).

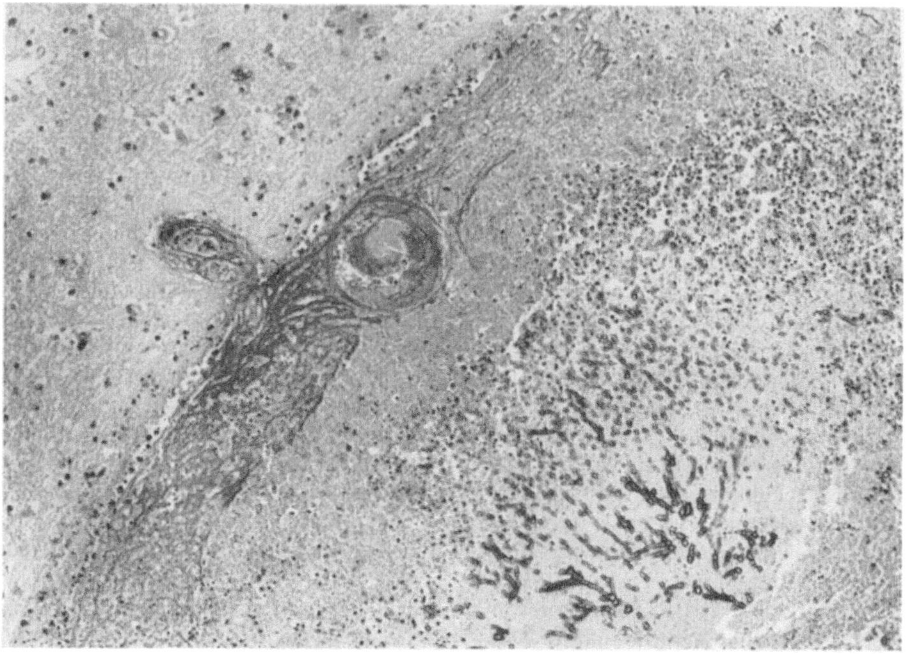

Fig. 30. *Aspergillus* meningitis. In the depth of a cerebral sulcus the leptomeninges are the site of necrosis, neutrophilic and fibrinous exudate and proliferation of the fungus. The molecular cell layer of the cerebral cortex also displays necrosis and leukocytic infiltrate. H & E, × 100 (Presbyterian-University Hospital, Pittsburgh, Pa.)

In cases of basal meningitis with obstruction there can be a moderate and uniform dilatation of the ventricles.

Involvement of the *spinal canal* is rare but has been occasionally reported (WYBEL, 1952; HUGHES, 1966; REDMOND et al., 1965; LEDERMAN et al., 1966) (Fig. 31). The blood-borne infection may affect the leptomeninges and form a well defined subdural mass. In other cases there is invasion of vertebrae, blockage of the spinal canal, and compression of the cord by an epidural mass.

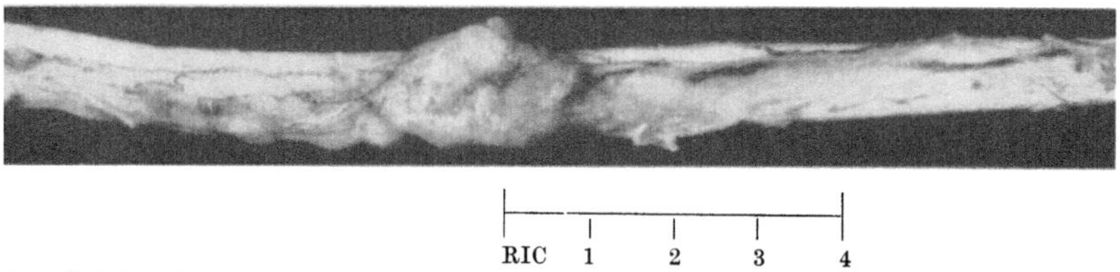

Fig. 31. Subdural extra-medullary aspergillar granuloma compressing the spinal cord. Microscopically the mass is composed mostly of a pure fungal growth and polymorphonuclear leukocytes. (Presbyterian-University Hospital, Pittsburgh, Pa.)

In cases in which the infection has followed intrathecal administration of penicillin there is marked thickening of the tissues around the cervical cord, attended by fibrosis and granulomatous inflammation (WYBEL, 1952). In no instances, perhaps with the exception of LEDERMAN et al's. patient on whom a precise description is not available, has actual mycotic involvement of the spinal cord been described.

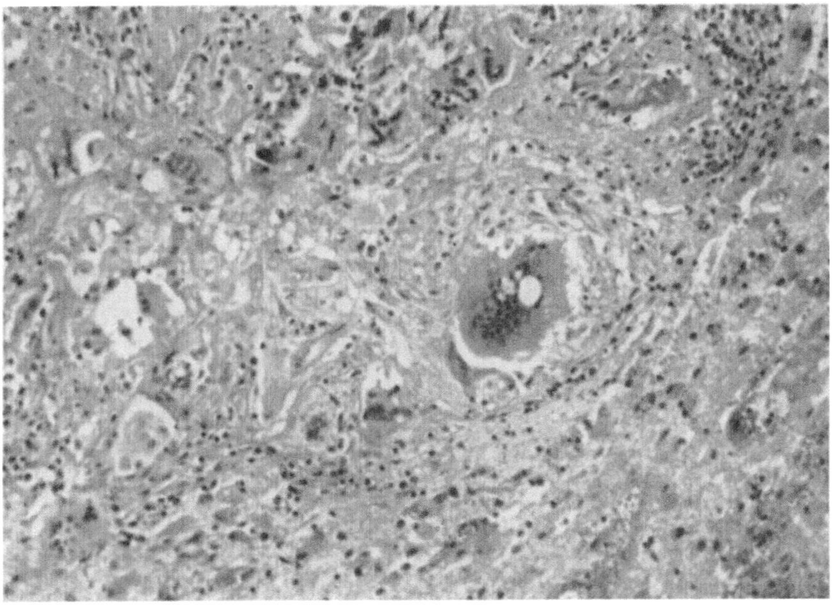

Fig. 32. Fibrogranulomatous reaction in the cerebrum. Foreign-body giant cells, histiocytes, lymphocytes and a moderate fibrous connective tissue proliferation are the main components. H & E, × 100

Involvement of small vessels in the central nervous system is as frequent as in other organs. Major cerebral vessels are rarely affected. In these cases there may be invasion, thrombosis and rupture of the internal carotid or middle cerebral arteries producing subarachnoid hemorrhage or brain infarction (McKee, 1950; Tveten, 1965). Another possibility of vascular injury to the brain is the compression of small or medium-sized arteries produced by tumor-like granulomas, leading to necrosis of the cerebral tissues (Jackson et al., 1955).

Digestive System

Clinical Manifestations: Involvement of the liver (20% of the 81 cases of disseminated aspergillosis reviewed) is likely to produce few clinical manifestations and the majority of these lesions pass undetected. It may however, be manifested by local pain and moderate enlargement of the organ. In rare cases there is jaundice and a tender palpable liver mass due to a fungal abscess (Baker, 1962a).

In a substantial number of cases the *intestine* (especially the *small bowel*), *stomach*, *esophagus* and *tongue*, in that order of frequency are affected. In most instances involvement of the gastrointestinal tract is not clinically apparent. However, the fungus may be present in ulcerations of the stomach (Fraumeni and Fear, 1962), or duodenum (Welsh and McClinton, 1954) and produce erosion of vessels followed by fatal hemorrhage. Rare cases of ulcerative *stomatitis*, with no apparent involvement of other organs, are reported (Whitaker et al., 1933). The *pancreas* is rarely involved.

Morphologic Findings: Infections of the *liver* and *pancreas* are likely to be of necrotizing or purulent character (Fig. 33). Grossly there are multiple white or yellow foci measuring from 0.1—2 or 3 cm sometimes accompanied by a zone of hemorrhage. Microscopically these foci show coagulative necrosis, a variable number of hyphae and polymorphonuclear leukocytic infiltration with abscess formation. Thrombotic vasculitis is common.

In the *stomach*, *duodenum*, *small bowel* and *esophagus* the lesions tend to be necrotizing, ulcerating and hemorrhagic. Microscopically, necrosis, neutrophilic infiltration, hemorrhage and infiltrating hyphae are the main features.

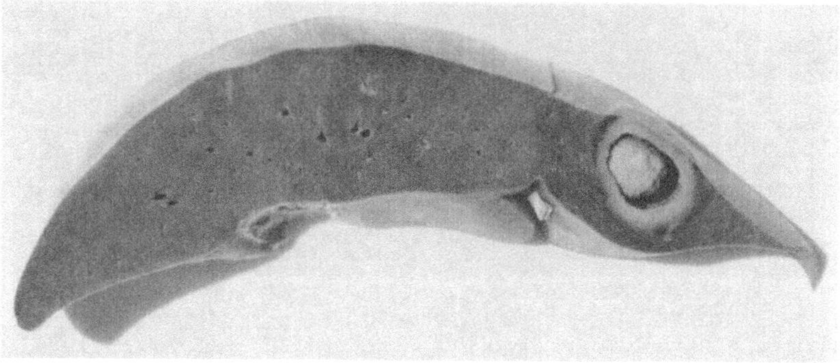

Fig. 33. *Aspergillus* abscess of the liver. The ball in the cavity consists of necrotic liver tissue covered with hyphae. The cavity wall is infiltrated with hyphae. From same patient as Fig. 9. (Courtesy of Dr. R. D. BAKER. Amer. J. clin. Path. **37**, 358 (1962), and The Williams & Wilkins Company)

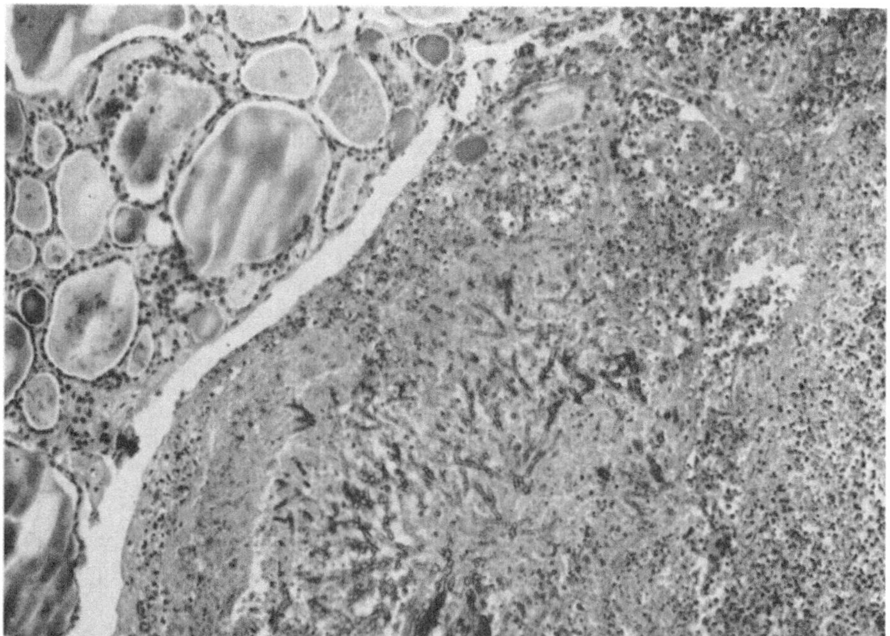

Fig. 34. Aspergillosis of the thyroid gland. The lesion is characterized by abundant neutrophilic infiltrate. H & E, × 100

Endocrine Glands

The *thyroid* is affected in a surprisingly large number of cases (16 out of 81 cases, or 20%), which is similar to that of the liver (Fig. 34). The *adrenal glands* are only rarely affected. No specific symptoms are described in these patients. From the histopathologic point of view the lesions are similar to those of other organs.

Genito-Urinary System

The kidney is frequently involved in generalized aspergillosis (33 out of 81 cases, or 40%). Symptoms are usually absent but there may be lumbar pain and hematuria.

Involvement of the *prostate* is manifested by enlargement and tenderness of the organ (UTZ et al., 1961).

Isolated intraluminal aspergillosis producing obstruction of the urinary tract occurs occasionally. The fungus may form a cast in the renal pelvis (COMMINGS et al., 1962) or lower ureter (FREIDELL et al., 1962). Growth of *Aspergillus* around a calculus of the urinary bladder has also been reported (BROUNST et al., 1963).

Aspergillus vaginitis is rarely observed. Clinically there is discharge and pruritus vulvae (GOLDSTINE, 1933). The vaginal wall exhibits gray-white nodules or a sooty-black pigmentation (WEINSTEIN et al., 1938—1939). Microscopic examination reveals hyphae and necrosis.

In rare instances a post-abortal *uterus* may be the portal of entry for a generalized aspergillosis (FINEGOLD et al., 1959).

Spleen and Lymph Nodes

Involvement of the *spleen* is relatively frequent (14 cases out of 81, or 17%) but does not as a rule produce significant clinical changes.

Involvement of *lymph nodes*, especially of the mediastinal groups is not rare (8 cases out of 81, or 10%) (GRCEVIC et al., 1959; MARSALEK et al., 1960; PARADIS, 1963; WAHNER et al., 1963; HEFFERNAN et al., 1966; STRELLING et al., 1966). No specific symptoms are described.

Grossly and microscopically the lesions are similar to those of other organs.

Musculo-Skeletal System

Clinical Manifestations: Bony structures are rarely affected in systemic aspergillosis. Bones that may be involved are vertebrae and ribs (REDMOND et al., 1965; HUGHES, 1966; SHAW et al., 1936; LEDERMAN et al., 1966), the maxilla (MEHROTRA, 1965), and the base of the skull (HUGHES, 1966).

The clinical manifestations of spinal involvement are kyphosis and destruction of vertebrae, as seen on X-rays. There is infraorbital swelling with invasion of the maxilla.

Skeletal muscles are rarely involved, although lesions of the psoas and diaphragm have been reported. Such processes are not easy to detect clinically and are usually autopsy findings.

Morphologic Findings: Bone lesions are of a destructive and purulent nature. Involved vertebrae frequently collapse. On microscopic examination there are abscesses containing hyphae, sometimes surrounded by groups of giant cells (REDMOND et al., 1965) and a variable degree of fibroblastic reaction. Necrosis is prominent in muscle lesions.

Treatment of Disseminated Aspergillosis: EVANS and BAKER (1959), UTZ et al. (1961), UTZ (1962, 1963a, b) and PROCKNOW (1962) have discussed the treatment of disseminated aspergillosis. It is the consensus that amphotericin B continues to be the drug of choice in cases of serious infection in spite of its toxic potential. Saramycetin (X-5079C) is also capable of significant anti-mycotic activity.

Eye, Adnexa and Orbit

Aspergillosis of the eye may be endogenous, usually the result of hematogenous spread from a distant organ, or exogenous, due to contamination following trauma or surgery.

Endogenous Intraocular Aspergillosis: This form of ocular aspergillosis occurs as a complication of a systemic infection. Well documented examples are rare. Cases based on autopsy studies have been observed by PARADIS et al. (1963), LEDERMAN et al. (1966), and DARRELL (1967). COGAN (1949) and HARLEY et al. (1959), have also reported 2 cases, based on clinical studies. The primary site of infection is usually the lung, but the heart has been implicated in instances following open-heart surgery (DARRELL, 1967).

Clinical Manifestations: Early symptoms include irritating sensation in the eye, tearing, pain, and partial loss of vision. Examination reveals injection of the conjunctiva, chemosis, normal or elevated intraocular tension, hypopion, and gray masses in the vitreous chamber. Usually the infection is unilateral or asymmetrical. Symptoms of involvement of other systems are often present, especially of the central nervous system. The condition has a rapid course and may lead to retinal detachment and extensive destruction of the uvea and retina with total loss of vision.

Morphologic Findings: Grossly there may be cloudiness of the anterior chamber, formation of synechiae and swelling, dislocation or destruction of the lens. The vitreous is frequently opaque and contains exudates. The retina is often detached. Microscopically almost every structure of the eye may be affected. The cornea is edematous or normal. The iris and ciliary body may show leukocytic infiltration. The lens is the site of variable degrees of destruction. Vitreo-retinal abscesses are an important finding (LEDERMAN et al., 1966; DARRELL, 1967). In addition to the necrosis and polymorphonuclear leukocytic infiltration characteristic of such lesion, an outstanding feature is the presence of numerous foreign-body giant cells around the purulent focus. Hyphae are often present in the abscess or in other locations as the anterior chamber or iris. They frequently display an atypical morphology. In some cases the abscesses are initially located between the sclera and the ciliary body, but subsequently extend into the retina and vitreous. The inflammatory process is sometimes diffuse and takes the form of multiple small necrotic foci that involve the iris, ciliary body, retina and vitreous.

Exogenous Aspergillosis: This form of the disease may be localized to the cornea or may be invasive, affecting intraocular structures. The latter usually follows penetrating trauma or intraocular surgery, but may also be due to extension of a corneal infection.

Keratitis: An extensive review of the literature dealing with *Aspergillus* keratitis was published by SCHOLER and SAUBERMAN (1959). The more general subject of mycotic keratitis has been surveyed by ZIMMERMAN (1962), and CHICK et al. (1962). The latter found that cases have been reported in substantial numbers in two periods: from 1879 to 1916 (42 cases), and from 1951 to the time of their writing (84 cases).

The cornea is the most frequent site of ocular aspergillosis, and *Aspergillus* is the most frequent cause of mycotic keratitis. In GINGRICH's compilation (1962) of 109 cases of keratomycosis studied by culture, *Aspergillus* sp. was the most common fungus (38 cases) followed by *Penicillium* (11 cases). *A. fumigatus* is the most frequent species, but *A. flavus, glaucus, niveus* and *versicolor* have also been found.

Aspergillosis of the cornea is almost always preceded by traumatic injury or corneal disease. The injury is usually minor, and often produced by fragments of vegetable matter that may themselves carry the contaminant fungus. The most frequent pre-existing ocular conditions are glaucoma with bullous keratitis, as well as various other ulcers (ZIMMERMAN, 1962). Occasionally, no trauma or predisposing factors are apparent (STERN and KULVIN, 1950). It is the consensus that the incidence of corneal aspergillosis has increased due to the topical use of steroids and perhaps also of antibiotics (MITSUI and HANABUSA, 1955; SCHOLER and SAUBERMAN, 1959; ZIMMERMAN, 1962; SUIE and HAVENER, 1963; FINE, 1965; ALLEN, 1966).

Clinical Manifestations: Corneal aspergillosis is characterized by the development of a corneal ulcer measuring 0.3—0.5 cm in diameter, accompanied by redness of the eye, pain and blurred vision. On examination, the ulcer has the appearance of a slightly raised ovoid, gray plaque with well-demarcated, slightly irregular, rolled or fragmented edges. Hypopion is frequently present. The fungus may be found by examination of smears from the ulcer edge and by culture. The latter method is positive more frequently.

Morphologic Findings: Histopathologic examination of an ulcer reveals necrosis and loss of substance of the cornea accompanied by polymorphonuclear leukocytic infiltration of the tissues. The hyphae may be present in the center of the defect, but are most often seen at the edges. They may penetrate deeply into the corneal stroma producing little or no necrosis. Descemet's membrane may for some time limit the fungal invasion, but is ultimately invaded. Rupture of the ulcer into the anterior chamber produces involvement of the iris and lens, although the inflammatory process occasionally penetrates more deeply.

Corneal ulcerative aspergillosis may be treated with nystatin, but the final result is all too often a corneal scar (MANGIARACINE and LIEBMAN, 1957; SAUBERMAN and SCHOLER, 1959) or sometimes a staphyloma (GINGRICH, 1962).

Exogenous Intraocular Aspergillosis: This form of the disease may be secondary to a mycotic keratitis (CASTROVIEJO and MUÑOZ-URRA, 1921) or to an intraocular surgical procedure. The latter type has become quite troublesome in recent years. Of particular interest is the infection following an intracapsular cataract extraction (FINE, 1962). Postsurgical intraocular fungal infections are an entity with distinctive clinical and histopathological features (FINE and ZIMMERMAN, 1959a, b). Clinically the condition is characterized by a long latent period of days, weeks or even months between operation and onset of symptoms. Pain, redness and blurred vision may appear suddenly. Examination reveals gray white masses in the anterior part of the vitreous, as well as good preservation of light perception. Histopathologic study discloses numerous well demarcated microabscesses located generally in the areas surgically manipulated, i.e. the anterior part of the vitreous and ciliary body. Branching hyphae can be demonstrated. DIAMOND and KIRK (1962) have reported a case following an intracapsular cataract extraction, characterized by massive granulomatous sclerochoroiditis.

Experimental intraocular *Aspergillus* infection has been treated by FINE and ZIMMERMAN (1960) by intraocular injections of nystatin, which they feel might occasionally be indicated in humans as a heroic procedure.

Orbit: The orbit may sometimes be involved by *Aspergillus* infections. Most of the cases are due to extension from processes involving the paranasal sinuses (ADAMS, 1933; WELLER et al., 1960; BAILEY and FULMER, 1961).

Clinical Manifestations: The lesions behave as intraorbital masses producing unilateral exophthalmos (JACKSON et al., 1955), diplopia and compression of the orbital structures. In some instances the process is extremely destructive (WRIGHT, 1927). Propagation into the cranial cavity is a dangerous possibility (BAILEY et al., 1961; JACKSON et al., 1955). The condition may last for years and is quite difficult to treat.

Morphologic Findings: On gross inspection the process is characterized by firm, poorly demarcated, inflammatory masses that compress the neighboring tissues and even produce erosion and destruction of the orbital floor or other bony structures. Microscopically there is a fibrosing granulomatous reaction with foreign body giant cells, neutrophils and mononuclear leukocytes and marked fibroblastic proliferation. The fungus may be difficult to demontrate and special stains should always be used in suspicious cases. *A. fumigatus* and *A. flavus* are cultured from such cases.

Eye Adnexa: Isolated granulomas presenting as an ulceration of the upper eyelid may be due to *A. fumigatus* (HARRELL et al., 1966). Microscopic examination reveals a granulomatous inflammatory reaction and hyphae. Cases of non-specific dacryocystitis, blepharitis, conjunctivitis (ROSENVOLD, 1942; MOSTAFA, 1966), chronic granulomatous blepharitis (BENNETT et al., 1962) and mycotic obstruction of the lacrimal canaliculi (DONAHUE, 1949), due to *A. niger* have also been recorded.

Nose and Paranasal Sinuses

Aspergillosis of the paranasal sinuses and nose is usually the result of an exogenous infection. The earliest reports are those of SCHUBERT (1885), MACKENZIE (1893), HARMER (1913), and TILLEY (1914—1915).

The condition may affect the nose and the maxillary antrum as well as other paranasal sinuses. It usually begins in the maxillary antrum, from which it may extend into the ethmoid, sphenoid or frontal sinuses and into the orbit. Isolated involvement of the nose is apparently very rare.

Table 7 lists the areas involved, the responsible microorganisms and the type of histopathologic reaction observed in 23 cases of infection of the nose and paranasal sinuses found in a review of the literature.

Clinical Manifestations: For unknown reasons nasal and sinus aspergillosis affects women more frequently than men (5:1). It is most often unilateral and has a chronic course marked by periods of activity alternating with periods of quiescence. The usual symptoms are similar to those observed in other types of chronic sinusitis and include dull pain, sinus fulness, nasal obstruction, and nasal as well as postnasal secretion. Sometimes there is spontaneous discharge of tissue fragments that have a membranous appearance. If the frontal, sphenoid, or ethmoid sinuses are also involved, the sensations of fulness and pain are localized to the frontal or parietal areas.

Examination reveals tenderness on pressure over the maxillary antrum, enlargement of the inferior turbinate and opacity on transillumination of the antrum. On puncturing the maxillary sinus it may be difficult or impossible to syringe out its contents, although membranous fragments of tissue may be obtained. In the rare cases limited to the nose, a thick membrane may fill and occlude the nasal fossae (FELDERMAN, 1940). X-ray examination reveals increased density of the affected cavities.

Table 7. *Aspergillosis of nose and paranasal sinuses**

Author and Year	Age Sex	Etiol. Agent	Areas Involved	Histopath. Picture
SCHUBERT (1885) . . .	— F	—	—	—
MACKENZIE (1893) . .	— F	*Aspergillus sp.*	Rt. maxillary sinus	Chronic non-specific inflamm.
HARMER (1913) . . .	— F	*A. fumigatus*	Rt. maxillary sinus	Not described
HARMER (1914—1915)	— F	*Aspergillus*	Maxillary sinuses	Chronic non-specific inflamm.
HARMER (1914—1915)	— F	*Aspergillus*	Maxillary sinuses	?
HARMER (1914—1915)	— F	*Aspergillus*	Maxillary sinuses (bilaterally)	?
HARMER (1914—1915)	— F	*Aspergillus*	Maxillary sinuses and ethmoid sinuses (bilaterally)	?
SKILLERN (1921) . . .	50 M	*Aspergillus*	Lt. maxillary sinus	Not described
ADAMS (1933)	32 F	*A. fumigatus*	Lt. frontal, ethmoid, sphenoid, maxillary sinuses and orbit. Rt. ethmoid and maxillary sinuses	Fibrosing granulomatous inflammation
KELLY (1934)	60 F	*Aspergillus sp.*	Rt. maxillary sinus	Not described
NASH (1938)	43 F	*Monilia or Aspergillus*	Lt. nose, ethmoid sinus, orbit.	Necrotizing inflammation
FELDERMAN (1940) . .	43 F	*Aspergillus sp.*	Nose (bilaterally)	Not described
BATT (1941)	44 F	*Aspergillus sp.*	Lt. maxillary sinus	Not described
JACKSON et al. (1955)	22 F	*Aspergillus sp.*	Orbit, ethmoid, cranial cavity	Fibrosing granulomatous inflammation
MONTREUIL (1955) . .	34 F	*Aspergillus sp.*	Rt. maxillary sinus	Not described
ANDERSEN et al. (1956)	55 F	*A. fumigatus*	Lt. maxillary and ethmoid sinuses	Not described
ARNAUD et al. (1957)	29 F	*Aspergillus sp.*	Lt. maxillary and Rt. ethmoid sinuses	Not described
ARNAUD et al. (1957)	53 F	*Aspergillus sp.*	Maxillary and ethmoid sinuses (bilaterally), trachea	Not described
SAVETSKY et al. (1961)	53 F	*Aspergillus*	Lt. maxillary sinus	Chronic non-specific inflamm.
BAILEY et al. (1961)	65 M	*A. flavus*	Rt. maxillary sinus, orbit, base of brain	Granulomatous inflammation
WELLER et al. (1960)	31 M	*A. flavus*	Rt. maxillary and ethmoid sinuses, orbit, pterygo-maxillary fossa	Fibrosing granulomatous inflammation
HORA (1965)	39 M	*A. flavus*	Rt. maxillary and ethmoid sinuses	Chronic non-specific inflamm.
ARONS et al. (1966) . .	25 M	*Aspergillus sp.*	Soft tissue of Rt. preauricular and infraorbital areas and Rt. maxillary sinus	Not described

* Features of 23 cases culled from the literature.

In some instances the condition spreads into the orbit and into the cranial cavity. Symptoms referable to these organs are then predominant. There may be exophthalmos, diplopia, paralysis of cranial nerves and signs of increased intracranial pressure (JACKSON et al., 1955; WELLER et al., 1960; BAILEY and FULMER, 1961). X-ray studies reveal destruction of the bony structures such as the floor of the orbit or the walls of the maxillary sinus.

Morphologic Findings: The maxillary sinuses, as they appear during surgery, have a thickened mucosa and sometimes thickened osseous walls. They contain a material variably described as fibrinopurulent (BAILEY and FULMER, 1961), whitish and solid and resembling a cholesteatoma (SAVETSKY and WALTNER, 1961), cheesy (SKILLERN, 1921), or flaky, friable and greenish (MONTREUIL, 1955). Many writers have been impressed by the presence of friable, grumous, gray membranous material, that has been described as "pseudodiphtheritic" (MACKENZIE, 1893).

Occasionally, the mucosal lining of the sinus has a polypoid appearance (BATT, 1941; ANDERSEN et al., 1956; ARNAUD et al., 1957).

Upon microscopic examination of cases of *Aspergillus* sinusitis, two main types of inflammatory response are found. The first is a chronic non-specific inflammation, characterized by a polymorphonuclear and mononuclear leukocytic infiltrate in the lamina propria of the mucosa. Sometimes the reaction is mostly purulent. The epithelium is often ulcerated or shows focal squamous metaplasia. Typical branching *Aspergillus* hyphae are often seen infiltrating the mucosa. Sometimes the pseudomembranous material obtained at curettage contains a layer of actively proliferating hyphae and fruiting heads. In rare instances the mass filling the antrum is a compact network of hyphae, reminiscent of that observed in intrabronchial aspergillomas (ARNAUD et al., 1957).

The second type of inflammatory response is a fibrosing granulomatous process. This was first observed by ADAMS in 1933 and is characterized by numerous multinucleated foreign body giant cells, occasionally arranged in groups. Accompanying them is a fibrosing process, usually quite marked, with mononuclear cells, fibroblasts and variable degrees of collagenization. The exact nature of the process may be difficult to recognize. In conventional stains, the cytoplasm of giant cells shows clear spaces that upon staining with the appropriate techniques reveals bizarre hyphal elements with abortive branching and bulbous dilatations (ADAMS, 1933; JACKSON et al., 1955).

The bony structures of the sinuses and orbit are usually well preserved or are moderately thickened. In invasive cases however, there may be actual destruction and necrosis of the osseous plates that separate the sinuses from the orbit or from the cranial cavity (WELLER et al., 1960; BAILEY and FULMER, 1961).

The prognosis varies according to the kind of inflammatory response. The chronic non-invasive, non-specific type has a good prognosis. It may be treated with gentian-violet or crystal-violet topically applied, and systemic or topical administration of fungistatics (mycostatin). Surgical procedures of the Caldwell — Luc or De Lima type are indicated. The results are good.

The fibrosing-granulomatous type may be treated with surgical procedures, topical application of mycostatin and systemic administration of amphotericin B or griseofulvin. The results are generally poor (WELLER et al., 1960; BAILEY and FULMER, 1961).

Ear

Aspergillosis of the external ear is one of the oldest known forms of the disease. The earliest works on the subject are those of WREDEN (1867), BURNETT (1874, 1879, 1889) and SIEBENMANN (1889).

The existence of aspergillosis of the ear as an entity is highly controversial. By denying its existence, in his article entitled "The Myth of Otomycosis", KINGERY (1965) creates some suspense about this subject. Perhaps not to be outdone in matters of suspense the Hitchcock Clinic group (McGONIGLE et al., 1967) refer to otomycosis as a "distinctive clinical entity" and proceed to report 5 cases. These widely divergent opinions illustrate the controversy that surrounds the role of *Aspergillus* and other fungi as primary agents of otitis externa.

In principle, two possibilities may be outlined concerning the role of *Aspergillus* in external ear infections. First, that the fungus is the primary etiologic agent of the disease. Second, that the fungus, acting as a saprophyte, invades ears already affected by other diseases as eczema or chronic otitis media. The problem is rendered somewhat more complex by the fact that a number of fungi, including *Aspergillus*, may be cultured from normal ears (LEA et al., 1958). The weight of the evidence indicates that *Aspergillus* may indeed produce primary inflammatory changes of the ear canal (McBURNEY et al., 1936; SIMMS, 1937; STUART and BLANK, 1955). However, the fungus sometimes acts as an opportunistic element and proliferates in pre-existing inflammatory processes as otitis media and eczema (STUART and BLANK, 1955). It must be stressed that only 15—20% of cases of otitis externa are due to fungi (CONANT et al., 1954). The most important bacterial agent is *Bacillus pyocyaneus* (LESHIN, 1953).

The fungi most frequently found in otomycosis are the *Aspergilli* although other fungi as *Candida*, *Penicillium*, *Mucor* and the dermatophytes may also be encountered (WOLF, 1947; GREGSON and LA TOUCHE, 1961). The most common species is *A.niger*. Thus in STUART and BLANK's series (1955) of 29 cases of otomycosis, *A.niger* was found in 19 cases, *A.flavus* in 6, *A.fumigatus* in 2, *A.nidulans* in 1 and *A.flavipes* in 1. It is worth mentioning that in otitis externa the otherwise more prevalent *A.fumigatus* is not the most frequent species. To explain this phenomenon, SHARP and JOHN (1946) invoke the existence of a selective factor. GREGSON and LA TOUCHE (1961) believe that metabolic requirements are partially responsible for this selectivity.

Aspergillus thrives in cerumen and cellular debris that frequently accumulate in the aural canal. It is stated that otomycosis is more frequent in hot tropical climates (LESHIN, 1953) and some reports indicate a high prevalence in such regions (ISMAIL, 1962).

Clinical Manifestations: Aspergillosis of the ear attacks both sexes with the same frequency and is generally unilateral, although bilateral cases are not rare. A dermatitis of the lining membrane of the aural canal, sometimes with involvement of the pinna is the basic alteration. The disease may affect the ear drum, but rarely does it produce perforation of this membrane to involve the middle ear (WHALEN, 1938; REBOUÇAS, 1939). The clinical manifestations vary in intensity from asymptomatic eruptions of the canal to extensive, pruriginous and painful processes. The most important symptom is a dull, boring pain (PITTENGER, 1938). Other symptoms are itching, tinnitus and sensation of fulness. Wax, dried secretions and proliferating mycelium accumulate and sometimes block the ear canal with resulting impairment of hearing. Secondary bacterial infection may occur, which produces a more exudative, eczematoid and sometimes malodorous lesion. Treatment with topical applications of silver nitrate (IRIARTE, 1943), gentian violet (REBOUÇAS, 1939), Iodocloroquin (McGONIGLE et al., 1967), and nystatin (GREGSON and LA TOUCHE, 1961) has been used.

Mycotic infections of postoperative ear cavities following mastoidectomy or fenestration are a common problem. In POWELL et al's survey (1962), *A.terreus* was the fungus most commonly found (22 cases), whereas *A.niger* was rarely found (1 case). Although these data are difficult to evaluate (given the low frequency of *A. niger* infections) it is evident that fungi are usually associated with a significant quantity of debris in postoperative ear cavities, and that postoperative fungal infections of the mastoid are very common.

Morphologic Findings: On gross examination the aural canal is partially filled with a mass of wax, scales, dry secretions and a membranous substance that resembles pieces of sponge sometimes covered by tiny black dots, as if sprinkled with pepper (HOOVER, 1912; LANGERON, 1922; PITTENGER, 1938). If the material is removed, which is easily done, it leaves an inflamed congested surface. Characteristically, the material reaccumulates in 4 or 5 days (FELDERMAN, 1940). The membranous lining of the external ear may be superficially eroded, but is not deeply invaded. The pinna is oftentimes affected by a scaly dermatitis. Microscopic examination of the material obtained from the ear usually reveals cellular debris. The spongy masses are composed of an abundant mycelial growth in which fruiting heads are frequent.

Skin and Adnexa

Aspergillosis of the skin is not frequent. It may be endogenous, as a result of a blood-borne infection, or exogenous. Among the cases of generalized aspergillosis listed in Table 1, only 4 had involvement of the skin (LEDERMAN et al., 1966; ALLAN and ANDERSEN, 1960). Metastatic lesions may be suppurative encrusted papules that involve the scalp and extremities.

Exogenus aspergillosis of the skin is as a rule a chronic inflammatory process involving the dermis and subcutaneous tissues, sometimes accompanied by ulceration.

The species most frequently responsible for exogenous aspergillosis are *A. niger*, *A. fumigatus* and *A. terreus*. In contradistinction with the visceral forms of the disease *A. fumigatus* is not the most common species.

Most probably the fungus does not penetrate through an intact skin, but requires a previous break of the epidermis as exemplified by the cases of LYNCH (1923), FRANK et al. (1933), and CHEETHAM (1964). However, in some instances no history of a previous injury is apparent (LYNCH, 1925; MYERS et al., 1930; and BENNETT et al., 1962).

Clinical Manifestations: In *endogenous aspergillosis*, skin involvement is only a minor part of a generalized disease and its manifestations are superseded by the more important involvement of other organs.

Exogenous aspergillosis in the majority of cases involves only the skin and subcutaneous tissues and remains more or less localized near the portal of entry. However, in some instances, the disease spreads to more distant areas. Lesions encountered include subcutaneous abscesses (FRANK and ALTON, 1933; CHEETHAM, 1964) and raised ulcero-papular skin lesions (MYERS and DUNN, 1930; BALL, 1933; HARRELL et al., 1966). ROBINSON et al. (1935) observed a number of patients apparently with superficial infections of the skin, involving legs, hands, feet and anus, but the clinical picture was not clearly described. The case recorded by CAHILL et al. (1967) in an Egyptian farmer is unique. This patient exhibited a very large number of confluent skin nodules of the face, scalp, trunk and extremities, resembling lepromatous leprosy. However, biopsies and cultures demonstrated *A. niger*. Cases of *A. niger* infection of abdominal surgical incisions characterized by skin and suture-line abscesses are occasionally reported (FRANK and ALTON, 1933).

Morphologic Findings: In the skin as in other sites, *Aspergillus* evokes two basic types of inflammatory reaction, the purulent non-specific and the granulomatous forms. The role of *Aspergilli* as initial agents of purulent inflammation of the skin is controversial and some of these lesions may also yield pathogenic bacteria on culturing (MYERS and DUNN, 1930). However, it is the consensus that *Aspergilli* do indeed play a pathogenetic role. Grossly, the lesions are described as indurated ulcers with minute pustules along the edge (MYERS and DUNN, 1930), or as frank subcutaneous abscesses (CHEETHAM, 1964). Histopathologic findings include epithelial hyperplasia and intraepithelial or subcutaneous abscesses.

The granulomatous type of reaction may be observed in the eyelid, face, or throughout the entire skin surface. Grossly the lesions are described as an elevated skin ulcer, as large, irregular exophytic masses resembling a neoplasm or as nodular lesions. Microscopically they disclose giant cells either scattered or arranged in groups around necrotic foci, as well as an abundant leukocytic infiltration located in the dermis. Hyphae are often seen among the inflammatory infiltrate.

Nails

It is difficult to evaluate the etiologic role of *Aspergillus* in onychomycosis, although most writers believe that it can produce alterations of the nails (BERESTON et al., 1946; BERESTON, 1950; MOORE et al., 1948; MOORE, 1950). The earlier cases of *Aspergillus* onychomycosis were reported in France by EMILE-WEIL et al. (1918—1919), SARTORY (1920), and OTA (1923).

The species most frequently encountered, according to BERESTON et al. (1946) is *A. glaucus*. Other species found are *A. flavus*, *A. niger*, *A. nidulans*, *A. terreus*, *A. rubrum* and *A. onycophylus* (MOORE et al., 1948; BERESTON et al., 1941; BERESTON et al., 1945; GRECO et al., 1935; SMITH, L.M., 1934; PALDROCK et al., 1952).

The condition involves the nails of the fingers and toes approximately with the same frequency. It often starts at one of the margins. The nails are thickened, show pronounced vertical striations, crumble easily and show a greenish or whitish discoloration.

Direct examination and cultures of scrapings reveal the etiologic agent. The ultimate diagnosis rests upon the correlation of the clinical and mycological findings.

Mycetoma Pedis

Only in rare instances do *Aspergillus* infections produce mycetoma (PINOY, 1913; GAMMEL, 1927). Cases have been reported by NICOLLE and PINOY in Tunis (1906), Puestow in the United States (1929) (same case reported by FOERSTER and PUESTOW, 1927), FONSECA in Brazil (1930), and BOUFFARD (quoted by PINOY, 1913). The etiologic agents reported are *A. amstelodami* and *A. nidulans*. In general terms the clinical and anatomopathological features of this form of the disease do not differ from other types of mycetoma, although it may occasionally involve the face and upper extremities.

Aspergillosis in Children

In the young, the disease runs a faster course and disseminates more frequently. Most instances have been reported since 1947, although occasional cases had been previously recorded (PONCE, 1935). Reviews on this subject have been published by BERKEL et al. (1963), KHOO et al. (1966), STRELLING et al. (1966) and BLATTNER (1967).

As in the adult, the infection may be primary or secondary. The latter is more frequent and evolves as an acute disease. The pre-existing conditions as well as the morphologic findings are similar to those observed in adults.

Primary aspergillosis in children is rare, but a number of cases have been described (HERTZOG et al., 1949; AKKOYUNLU et al., 1957; ALLAN et al., 1960; DE LEON-ANTONI, 1962; STRELLING et al., 1966; KHOO, 1966). The disease takes the form of an acute bronchopneumonia, often causing death within a few days. Hertzog's 2 cases, a brother and a sister (one with no anatomopathologic study), and STRELLING et al's 2 cases, two sisters, are exceedingly interesting and do not leave any doubts as to the existence of an acute primary form. Each of these pairs of siblings came from farm families. In both reports, the infection of the second sibling developed shortly after that of the first and all died of an acute pneumonic process within 5—14 days after the onset of the disease. At autopsy, the lungs were very extensively infiltrated by a nodular granulomatous process, that yielded *A. fumigatus* on culture. Microscopically, the granulomas consisted of histiocytes and giant cells, many of which had a core of polymorphonuclear leukocytes.

Aspergillosis in Animals

Fungi of the genus *Aspergillus* may attack a wide variety of animals, including mammals, birds and insects. The subject has been reviewed by AINSWORTH and AUSTWICK (1959) and by AUSTWICK (1965). The three most important types of infection are: 1. mycotic abortion in cattle and mares; 2. pulmonary infection in mammals; and 3. respiratory tract infection in birds. In addition, *Aspergilli* may produce invasive disease of insects (MEDELIN, 1960) and infect incubating poultry eggs (LUCET, 1897; WRIGHT et al., 1961).

Mycotic Abortion

Mycotic abortion, a frequent disease in cows and mares, is due to inflammation and necrosis of the placenta (AINSWORTH and AUSTWICK, 1959; MAHAFFEY et al., 1964; HILLMAN, 1961). Among the various types of abortion in cattle, mycotic placentitis is found in 20% of the placentas examined histologically (HILLMAN, 1961). About 73% of these abortions are due to *Aspergillus* sp. (AUSTWICK and VENN, 1962). In cattle, most abortions occur around the 7th or 8th month of pregnancy, during the winter months. HILLMAN's experimental work (1961) in pregnant heifers suggests that the placental infection is blood-borne. In mares, the possibility of an initial endometrial infection as another portal of entry must be considered (MAHAFFEY et al., 1964).

Morphologic Findings: On gross examination the affected placentas reveal irregularly shaped areas of necrosis, sometimes with ulceration especially on the maternal surface. Microscopically there is chronic inflammation with presence of mononuclear cells, giant cells and neutrophils. Hyphae are seen among necrotic villi. Similar lesions may be present in the amnion.

Pulmonary Aspergillosis in Mammals

Pulmonary aspergillosis in mammals appears to be frequent and has been observed in horses (RANDALL, 1925), lambs (AUSTWICK et al., 1960), calves (EGGERT et al., 1960) and dairy cows (AUSTWICK, 1962b). It is probable that at least a number of infections in lambs and cows is subclinical. AUSTWICK's observations are of great importance, since they indicate that most dairy cows (66%) may harbor pulmonary *Aspergillus* infections.

Morphologic Findings: As described by AUSTWICK (1962b), the lungs of cows disclose subpleural or intraparenchymatous firm nodules measuring 0.1—2 mm and varying from very few to many hundreds. On microscopic examination, the nodules contain asteroid bodies measuring 50—80 μ, which are composed of refractile crystalline radiating structures. Hyphae are present in the majority of the asteroid bodies. Around most nodules there is a fibrous capsule and a moderate inflammatory reaction consisting of macrophages and polymorphonuclear leukocytes. Asteroids are not pathognomonic of aspergillosis and may be present in other conditions.

In lambs, pulmonary aspergillosis may be manifested by discrete purulent foci or by diffuse consolidation. Older lesions are composed of epithelioid cells and occasionally contain asteroid bodies similar to those observed in cows.

Avian Aspergillosis

Aspergillosis is encountered in chickens, ducks, turkeys, parrots (Ainsworth and Austwick, 1955), captive exotic birds (Ainsworth and Rewell, 1949) and free living wild birds (McDiarmid, 1955). The fungus species most frequently responsible is *A. fumigatus*, but *A. flavus*, *A. niger*, and *A. nidulans* are also observed. The condition may acquire the form of an epizootic pneumonia especially in chickens, pigeons and turkeys, with a mortality rate as high as 90% (Savage and Isa, 1933; Taylor, 1966; and Witter et al., 1952). Under experimental conditions leukemic chickens exhibit an enhanced susceptibility to aspergillosis (Chick, 1963).

Morphologic Findings: The organs most frequently involved are the air-sacs, followed by the lungs (Ainsworth and Rewell, 1949). Other organs involved are the trachea, central nervous system and kidneys (Ainsworth and Austwick, 1955). Grossly the surfaces of the air-sacs are covered by raised white nodules which may coalesce to form a white or green lining of the entire sac. The fungus does not tend to penetrate tissues. Microscopically the inflammatory response is characterized by proliferation of granulation tissue, on the surface of which the fungus multiplies.

The lungs are the site of scattered, small nodules with a necrotic center that resemble those observed in tuberculosis. There may be consolidation of the parenchyma or formation of large abscesses. Microscopically there is a great variability in the inflammatory response, from focal bronchopneumonia, to a granulomatous nodular infiltration with epithelioid and giant cells.

Toxins

The importance of toxic metabolites produced by fungi of the genus *Aspergillus* in human pathology remains to be settled.

The earliest works on these toxic metabolites are those of Ceni and Besta (1902), Bodin and Gautier (1906), and Henrici (1939). These investigators were able to isolate substances that exhibited neurotoxic properties for laboratory animals from *A. fumigatus*, *A. flavescens* and *A. niger*. These substances were not chemically identified.

More recently Tilden et al. (1961) have obtained endotoxins from *A. fumigatus* and *A. flavus*, that possess nephrotoxic properties. The endotoxin from *A. fumigatus* has in addition dermonecrotic and hemolytic properties. It has been partially purified and found to be a protein (Rau et al., 1961).

Forgacs et al. (1954) have observed a toxigenic strain of *A. clavatus*. It is possible that toxic substances produced by this fungus and by *A. chevalieri* and *A. fumigatus* are responsible for some cases of bovine hyperkeratosis although this is still uncertain (Forgacs, 1962).

A. fumigatus strains, under certain conditions, produce antibacterial substances known to be toxic to animals. Among them are gliotoxin, fumigacin and fumigatin. Production of these substances under natural conditions deserves further study (FORGACS and CARLL, 1962).

WILSON (1966) reviewed the toxic substances, other than aflatoxin, produced by *A. flavus*. These are more or less well defined from the chemical point of view, and include oxalic acid, kojic acid, aspergillic acid, beta-nitropropionic acid and a tremorgenic factor. Spontaneous occurrence of intoxication in man or animals due to exposure to *A. flavus* — produced toxins has not been conclusively demonstrated. Oxalic acid is a stomach irritant and kojic acid is a convulsant (WERCH et al., 1957). Aspergillic acid and related compounds have antibiotic properties. Beta-nitropropionic acid shares with chloramphenicol a nitro compound in their structure and is a vasodilator. A tremorgenic factor of unknown chemical composition, recognized by WILSON and WILSON in 1964, produces tremors that last from 1—3 days when administered to mice in doses of 0.5 mg. A new metabolite, with toxic properties for chicken embryos, has been isolated by RODRICKS et al. (1968), to which the name aspertoxin has been applied.

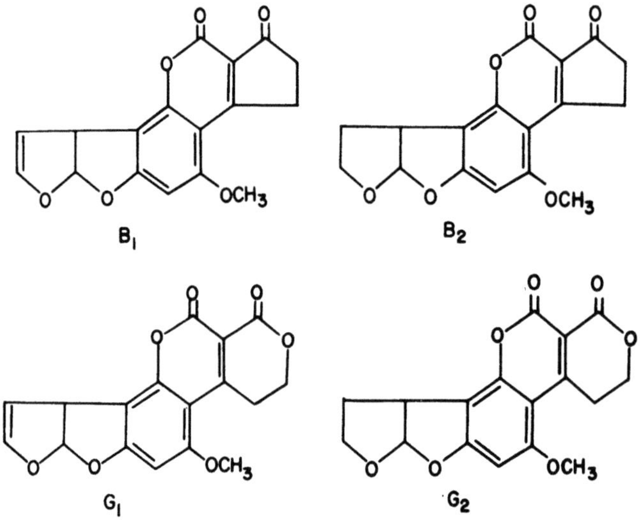

Fig. 35. Chemical structure of aflatoxins

Aflatoxins

The search for toxic metabolites produced by *Aspergilli* culminated recently with the discovery of aflatoxins (Fig. 35). WOGAN (1966) and PHELPS (1967) have published comprehensive reviews on the subject. In 1960 and 1961 a series of outbreaks of an unknown "X" disease, occurring in turkeys and other animals was observed in England. A Brazilian groundnut meal was incriminated (BLOUNT, 1961) and a toxic factor was isolated from it (ALLCROFT et al., 1961). SILLER and OSTLER (1961) studied the histopathology of the disease in turkeys.

SARGEANT et al. (1961) identified *A. flavus* as the toxin-producing agent. SMITH and MCKERNAN (1962) demonstrated that the chromatographically separated fractions of *A. flavus* extracts possessed hepatotoxic properties. VAN DER ZIJDEN et al. (1962) isolated the toxin in crystalline form. By means of chromatographic analysis two components were obtained, one showing a blue fluorescence and the other a green fluorescence under ultra-violet light, which were called aflatoxins B and G (NESBITT et al., 1962). It was further demonstrated that there are two blue (B_1 and B_2) and two green (G_1 and G_2) components. Aflatoxin B_1 is the more abundant.

These compounds possess a furanocoumarin structure (ASAO et al., 1965). Aflatoxins B_2 and G_2 are the dihydroderivatives of B_1 (CHANG et al., 1963) and G_1, respectively.

Two main groups of tests, one biological, the other chemical, are available for the detection of aflatoxins (GOLDBLATT, 1967). The most commonly used biological test consists of the administration of extracts of the suspected substance to one-day-old ducklings for 7 days. The livers are then examined microscopically for characteristic lesions such as periportal necrosis and especially for bile duct hyperplasia. The chemical tests consist of extraction of the suspected substance with a solvent, followed by chromatographic analysis and ultraviolet light examination of the chromatograms. Using this method DE IONGH et al. (1965) were able to demonstrate aflatoxins in a sample of peanuts of the year 1922. C_{14} labelled aflatoxin has recently been produced (WOGAN, 1967) by culturing *A. flavus* in media enriched with C_{14} methionine.

Biological Effects

The biological effects of the aflatoxins depend on the amount of toxin administered, on the duration of the exposure, and on the species. Young animals are more sensitive than the mature. Turkeys and ducklings are very susceptible, while chickens are comparatively resistant. Pigs and calves are also quite susceptible. Sheep are remarkably resistant (ALLCROFT, 1965).

Acute Toxicity: The LD_{50} dose of aflatoxin B_1 varies between 0.36 mg/kg (per os), for day-old ducklings (CARNAGHAN et al., 1963), and 17.9 mg/kg (per os), for female rats (BUTLER, 1964a). The LD_{50} dose for male rats is 7.2 mg/kg (per os). Aflatoxin G_1 is somewhat less toxic and aflatoxins B_2 and G_2 are much less toxic.

Morphologic Changes: The acute changes produced by aflatoxin B_1 in ducklings and rats are found mainly in the liver (BUTLER, 1964a, 1964b). In 16—24 hours after administration of an LD_{50} dose to a rat there is loss of glycogen and of cytoplasmic basophilia in the periportal areas. In 48 hours there are well developed periportal areas of necrosis, which at 72 hours are being invaded by histiocytes; proliferation of biliary channels is manifest. At 7 days there is some increase in the amount of reticulin in the portal areas, which by 14 days is more marked and accompanied by an increasing biliary proliferation. By the end of the fourth week there is proliferation of fibrous connective tissue surrounding hyperplastic liver cell nodules, with resulting distortion of the parenchyma. The toxicity response of ducklings and rainbow trout is similar to that of rats. However, hemorrhage and necrosis appear to be somewhat more severe in trout (ASHLEY, 1967). Bile duct hyperplasia is more marked in ducklings and parallels closely the amount of aflatoxin B_1 administered (WOGAN and NEWBERNE, 1964). Other acute lesions in the rat include necrosis of the proximal convoluted tubules of the kidneys, hemorrhages of the corticomedullary junction of the adrenal glands, petechial hemorrhages of the lungs and focal necrosis of the myocardium. These lesions heal in a short period of time.

In cell cultures (LEGATOR, 1966) the earliest effect of a mixture of aflatoxins is the inhibition of DNA synthesis and suppresion of mitosis.

At the ultrastructural level (BUTLER, 1966), administration of aflatoxin B_1 to rats produces early dilatation of the rough endoplasmic reticulum cisternae, dislocation of ribosomes and swelling of the mitochondria, especially in periportal liver cells. SHANK and WOGAN (1964) have observed that administration of aflatoxin B_1 to ducklings and rats produces depletion of glycogen and increase in total lipids of the liver. WOGAN and FRIEDMAN (1965) demonstrated that aflatoxin B_1 in rats causes inhibition of trytophan pyrrolase induction, indicating that the toxin alters the ability of the liver to respond to regulation of enzyme production.

Chronic Toxicity — Carcinogenesis: Aflatoxin is one of the most potent *hepatocarcinogenic substances* known (BUTLER and BARNES, 1963). BUTLER (1965) has estimated that the daily dose for induction of liver tumors in rats is 10 μg per day, whereas the dose of dimethylnitrosamine would be 750 μg/day and of butter yellow 9000 μg/day. A total dose of 2—3 mg is capable of inducing liver tumors in rats (NEWBERNE, 1967). Even before the toxic substance was identified LANCASTER et al. (1961) demonstrated the marked carcinogenic properties of samples of groundnuts in rats. These findings were confirmed by numerous other observations (LE BRETON et al., 1962; BUTLER and BARNES, 1963).

Aflatoxin is also a potent carcinogen for other species as the rainbow trout (WOLF and JACKSON, 1963; ASHLEY et al., 1965; HALVER, 1965; FALK, 1967; SINNHUBER, 1967). The incidence of carcinomas in animals seems to be directly proportional to the amount of aflatoxin and to the lenght of exposure (NEWBERNE, 1965; SINNHUBER, 1967). Continuous administration is not necessary for the induction of liver tumors (BARNES and BUTLER, 1964; WOGAN, 1966).

The cancerogenic properties of aflatoxins are not limited to the liver. DICKENS and JONES (1963) have induced sarcomas by repeated subcutaneous injection of aflatoxins in rats.

Morphologic Findings: The single non-cancerous aflatoxin induced lesion common to all species studied, perhaps with the exception of the mouse, is a proliferation of bile-duct cells in the portal spaces (NEWBERNE, 1967). A veno-occlusive reaction of the liver seems on the contrary, to be specific for bovines.

The pathology of aflatoxin-induced hepatomas has been studied among others by NEWBERNE (1965, 1967), ASHLEY (1967), WALES (1967), and SIMON et al. (1967). The species more frequently investigated are rats, ducklings, mice and trout. Although there are some individual differences, the basic changes are the same in all species. The liver tumors are multicentric. On gross examination the tumor nodules are soft, gray-yellow and hemorrhagic or necrotic. Sometimes there are metastases to the lungs, or direct extension to the peritoneum. In trout, occasional metastases may be found in the kidneys and gills (HALVER, 1967).

Microscopically, the earliest change observed is a bile-duct proliferation in the portal areas that progresses to connect with other portal tracts. This lesion does not seem to progress beyond the state of hyperplasia. Another early change is the appearance of large liver cells with large hyperchromatic nuclei and prominent nucleoli (megalocytosis), mainly in the periportal areas. This is followed by a hyperplastic stage, characterized by the appearance of liver-cell nodules (NEWBERNE, 1965).

In the fullblown state, a large portion of the liver parenchyma is replaced by anaplastic hepatocytes that may be arranged in a trabecular pattern or in solid sheets. Atypias and mitotic figures are frequent. In ducklings the induced tumors appear benign and do not produce metastasis (NEWBERNE, 1965).

At the ultrastructural level the Golgi complex and the endoplasmic reticulum of tumor cells are highly developed (SCARPELLI, 1967). The latter feature correlates with marked cytoplasmic basophilia.

The importance of the aflatoxins in human pathology, and particularly their carcinogenic potential, have not as yet been fully evaluated. The higher incidence of hepatoma in the countries of Central and Southern Africa (Mozambique, Senegal, Congo Republic, Gambia, South Africa), particularly in the male negroid population, is well known but hard to explain. Aflatoxin has naturally been held suspect (LE BRETON et al., 1962). This hypothesis seems plausible but it has still to be shown that the incidence of liver carcinoma in a population rises parallel with its degree of exposure to aflatoxin (OETTLÉ, 1965).

References

ADAMS, N.F.: Infection involving the ethmoid, maxillary and sphenoid sinuses and the orbit due to *Aspergillus fumigatus*. Arch. Surg. **26**, 999 (1933). — ADELBERG, J.L., BERKOWITZ, G., SHIELDS, T.W.: Aspergillosis complicating sarcoidosis. Successful surgical resection. Quart. Bull Northw. Univ. med. Sch. **35**, 16 (1961). — AGBAYANI, B.F., NORMAN, P.S., WINKENWERDER, W.L.: The incidence of allergic aspergillosis in chronic asthma. J. Allergy **40**, 319 (1967). — AGUILAR-CELI, P.: Aspergillosis pulmonar. An. Fac. Med. (Lima) **43**, 544 (1960). — AINSWORTH, G.C., AUSTWICK, P.K.C.: A survey of animal mycoses in Britain: General aspects. Vet. Rec. **67**, 88 (1955). ~ Fungal diseases of animals (Commonwealth Bureau of Animal Health Review Series, No. 6), 148 pp. (England): Commonwealth Agricultural Bureaux, Farnham Royal 1959. — AINSWORTH, G.C., REWEL, R.E.: The incidence fo aspergillosis in captive wild birds. J. comp. Path. **59**, 213 (1949). — AKKOYUNLU, A., YUCEL, F.A.: Aspergillose broncho-pulmonaire et encéphaloméningée chez un nouveau-né de 20 jours. Arch. franç. Pédiat **14**, 615—622 (1957). — ALLAN, G.W., ANDERSEN, D.H.: Generalized aspergillosis in an infant 18 days of age. Pediatrics **26**, 432 (1960). — ALLCROFT, R.: Aspects of aflatoxicosis in farm animals. In: G.N. WOGAN (ed.), Mycotoxins in foodstuffs, p. 153. Cambidge, Mass.: M.I.T. Press 1965. — ALLCROFT, R., CARNAGHAN, R.B.A., SARGEANT, K., O'KELLY, J.: A toxic factor in Brazilian groundnut meal. Vet. Rec. **73**, 428 (1961). — ALLEN, J.C.: Drug usage and proper diagnosis in treatment of the red eye. Wis. med. J. **65**, 480 (1966). — ALVAREZ, R., REYES, M.A., MADRIÑÁN, C.D.: Encuesta sobre hongos ambientales en la ciudad de Cali. Antioquia méd. **15**, 497 (1965). — AMBROSE, C.T., COONS, A.H.: Studies on antibody production. J. exp. Med. **117**, 1075 (1963). — ANDERSEN, H.C., STENDERUP, A.: Aspergillosis of the maxillary sinus. Acta Oto-laryng. (Stockh.) **46**, 471 (1956). — ARGEN, R.J., LESLIE, E.V., LESLIE, M.B.: Intracavitary fungus ball — Pulmonary aspergillosis. Report of a case. J. Amer. med. Ass. **179**, 944 (1962). — ARGRABITE, J.W., MORROW, M.B., MEYER, G.H.: Allergic bronchial asthma and pulmonary infection due to *Aspergillus fumigatus* treated by infections of emulsified allergen. Ann. Allergy **21**, 583 (1963). — ARKLE, C.J., HINDS, F.: A case of pneumonomycosis. Trans. path. Soc. Lond. **47**, 8 (1895—1896). — ARNAUD, G., PESLE, G.D., DE ANGELIS: L'Aspergillose des sinus (A propos de 2 cas). Ann. d'Otol. **74**, 796 (1957). — ARONS, M.S., LYNCH, J.B., LEWIS, S.R., LARSON, D.L., BLOCKER, T.G.: Hemifacial resection for rare inflammatory diseases of the parotid gland: I. Aspergillosis. II. Pseudosarcomatous fasciitis. Amer. Surg. **32**, 496 (1966). — ASAO, T., BUCHI, G., ABDEL KADER, M.C., CHANG, S.B., WICK, E.L., WOGAN, G.N.: The structures of aflatoxin B_1 & G_1. In: G.N. WOGAN (ed.), Mycotoxins in foodstuffs, p. 265. Cambridge, Mass.: M.I.T. Press 1965. — ASHLEY, L.M.: Histopathology of rainbow trout aflatoxicosis. In: HALVER, J.E., MITCHELL, I.A. (ed.), Trout hepatoma research conference papers, p. 103. Research report 70. Washington, D.C.: U.S. Govermennt Printing Off. 1967. — ASHLEY, L.M., HALVER, J.E., GARDNER, W.K., WOGAN, G.N.: Crystalline aflatoxins cause trout hepatoma. Fed. Proc. **24**, 627 (1965). — ATKINSON, J.B., ATKINSON, G.R.: Lymphosarcoma with pulmonary aspergillosis. Penn. med. J. **61**, 1373 (1958). — AUSTWICK, P.K.C.: Ecology of *Aspergillus Fumigatus* and the pathogenic *Phycomycetes*. Recent Progr. Microbiol. **8**, 644 (1962a). ~ The presence of *Aspergillus fumigatus* in the lungs of dairy cows. Lab. Invest. **11**, 1065 (1962b). ~ Chapter on Pathogenicity. In: RAPER, K.B., FENNELL, D.I.: The Genus Aspergillus, p. 82. Baltimore: Williams & Wilkins Co. 1965. — AUSTWICK, P.K.C., APPLEBY, E.C.: Unpublished data. Quoted by AUSTWICK, P.K.C. (1965). — AUSTWICK, P.K.C., GITTER, M., WATKINS, C.V.: Pulmonary aspergillosis in lambs. Vet. Rec. **72**, 19 (1960). — AUSTWICK, P.K.C., VENN, J.A.J.: Mycotic abortion in England and Wales 1954—1960. Proc. 4th International Congress Animal Reproduct. **3**, 562 (1962). — BADER, G.: Die visceralen Mykosen. Pathologie, Klinic and Therapie, 423 pp. Jena: Gustav Fischer 1965. — BAILEY, J.C., FULMER, J.M.: Aspergillosis of orbit. Report of a case treated by the newer antifungal antibiotic agents. Amer. J. Ophthal. **51**, 670 (1961). — BAKER, R.D.: The diagnosis of fungus diseases by biopsy. J. chron. Dis. **5**, 552 (1957). ~ Leukopenia and therapy in leukemia as factors predisposing to fatal mycoses. Mucormycosis, aspergillosis, and cryptococcosis. Amer. J. clin. Path. **37**, 358 (1962a). ~ Foreword to the International Symposium on Opportunistic Fungus Infections. Lab. Invest. **11**, 1017 (1962b). — BALL, H.A.: Aspergillus dermatomycosis. Report of case. Calif. west. Med. **38**, 178 (1933). — BARIÉTY, M., POULET, J., MONOD,

O., DE BRUX, J.: Aspergillose aiguë, purement pulmonaire, à forme de cancer bronchique. Bull. Soc. méd. Hôp. Paris **73**, 397 (1957). — BARNES, J.M., BUTLER, W.H.: Carcinogenic activity of aflatoxin to rats. Nature (Lond.) **202**, 1016 (1964). — BARNOLA, J., ANGULO-ORTEGA, A.: Aspergilosis pulmonar. Mycopathologia (Den Haag) **15**, 408 (1961). — BARUAH, H.K.: The air spora of a cowshed. J. gen. Microbiol. **25**, 483 (1961). — BATT, F.: Aspergillosis and other mycoses in the sinuses of the nose. Acta otolaryng. (Stockh.) **29**, 129 (1941). — BAUM, LITZMAN, EICHSTEDT: Quoted by VIRCHOW (1856). — BECH, A.O.: Diffuse broncho-pulmonary aspergillosis. Thorax **16**, 144 (1961). — BENNETT, J.E., KIRBY, E.J., BLOCKER, T.G.: Unusual inflammatory lesions of the face. Plast. reconstr. Surg. **29**, 684 (1962). — BENNETT, J.H.: On the parasitic vegetable structures found growing in living animals. Trans. roy. Soc. Edinb. **15**, 277 (1844). — BERESTON, E.S.: Further studies on *Aspergillus* infections of the nails. Sth. med. J. (Bgham, Ala.) **43**, 489 (1950). — BERESTON, E.S., KEIL, H.: Onychomycosis due to *Aspergillus flavus*. Arch. Derm. Syph. (Chic.) **44**, 420 (1941). — BERESTON, E.S., WARING, W.S.: Onychomycosis and dermatomycosis caused by *Trichophyton rubrum* and *Aspergillus nidulans*. Arch. Derm. Syph. (Chic.) **52**, 162 (1945). ~ *Aspergillus* infection of the nails. Arch. Derm. Syph. (Chic.) **54**, 552 (1946). — BERKEL, I., SAY, B., TINAZTEPE, B.: Pulmonary aspergillosis in a child with leukemia. Report of a case and a brief review of the pediatric literature. New Engl. J. Med. **269**, 893 (1963). — BERNTON, H.S.: Asthma due to mold-*Aspergillus fumigatus*. J. Amer. med. Ass. **95**, 189 (1930). — BIGUET, J., TRAN VAN KY, P., CAPRON, A., FRUIT, J.: Annalyze immunochimique des fractions antigéniques solubles d' *Aspergillus fumigatus*. C. R. Acad. Sci. (Paris) **254**, 3768 (1962). — BLATTNER, R.J.: Pulmonary aspergillosis in children. J. Pediat. **70**, 139 (1967). — BLOUNT, W.P.: Turkey "X" disease. Turkeys **9**, 52 (1961). — BODIN, E., GAUTIER, L.: Note sur toxine produite par l' *Aspergillus fumigatus*. Ann. Inst. Pasteur **20**, 209 (1906). — BOSHES, L.D., SHERMAN, I.C., HESSER, C.J., MILZER, A., MACLEAN, H.: Fungus infections of the central nervous system. Experience in treatment of cryptococcosis with cycloheximide (actidione). Arch. Neurol. Psychiat. (Chic.) **75**, 175 (1956). — BOUFFARD: Quoted by PINOY (1913). — BOYCE, R.: Remarks upon a case of aspergillar pneumomycosis. J. Path. Bact. **1**, 163 (1893). — BRAATELIEN, N.T., PERLMUTTER, H.M.: Aspergillosis of the lung. Dis. Chest **39**, 425 (1961). — British Tuberculosis Association: *Aspergillus* in persistent lung cavities after tuberculosis. Tubercle (Edinb.) **49**, 1 (1968). — BROUNST, G., ALLAMÉ, F., BROUNST, S.: Un cas primaire d' aspergillome urinaire observé au Liban. Bull. Acad. nat. Méd. (Paris) **147**, 50 (1963). — BRUNNER, A.: Das sogenannte Aspergillom. Schweiz. med. Wschr. **1958** 499—563. — BRUSKIN, S.: A comprehensive survey of the incidence of fungus spores in the New Brunswick, New Jersey, area. Ann. Allergy **11**, 15 (1953). — BURNETT, C.H.: A case of myringomycosis aspergillina. Arch. Ophthal. Otol. **4**, 121 (1874). ~ Twenty cases of the growth of *Aspergillus* in the living human ear. Amer. J. Otol. **1**, 10 and 93 (1879). ~ *Aspergillus* in the human ear. — With report of eleven cases. Med. Surg. Reporter **61**, 539 (1889). — BURSTON, J., BLACKWOOD, W.: A case of *Aspergillus* infection of the brain. J. Path. Bact. **86**, 225 (1963). — BUTLER, W.H.: Acute toxicity of aflatoxin B_1 in rats. Brit. J. Cancer **18**, 756 (1964a). ~ Acute liver injury in ducklings as a result of aflatoxin poisoning. J. Path. Bact. **88**, 189 (1964b). ~ Liver injury and aflatoxin. In: G.N. WOGAN (ed.), Mycotoxins in foodstuffs, p. 175. Cambridge, Mass.: M.I.T. Press 1965. ~ Early hepatic parenchymal changes induced in the rat by aflatoxin B_1. Amer. J. Path. **49**, 113 (1966). — BUTLER, W.H., BARNES, J.M.: Toxic effects of groundnut meal containing aflatoxin to rats and guinea pigs. Brit. J. Cancer **17**, 699 (1963). — BUTTRICK, D.D., ROBERTS, L.: Generalized cytomegalic inclusion disease. Report of two cases with associated fungal infection, one involving aspergillosis, the second with candidiasis. Amer. J. Dis. Child. **110**, 319 (1965). — CAHILL, K.M., EL MOFTY, A.M., KAWAGUCHI, T.P.: Primary cutaneous Aspergillosis. Arch. Derm. **96**, 545 (1967). — CAMPBELL, M.J., CLAYTON, Y.M.: Bronchopulmonary aspergillosis. A correlation of the clinical and laboratory findings in 272 patients investigated for bronchopulmonary aspergillosis. Amer. Rev. resp. Dis. **89**, 186 (1964). — CANNON, G.D.: Secondary aspergillosis (*Aspergillus Niger*) superimposed upon bronchiectasis. Report of a case. J. thorac. Surg. **4**, 533 (1935). — CARBONE, P.P., SABESIN, S.M., SIDRANSKY, H., FREI, E.: Secondary aspergillosis. Ann. intern. Med. **60**, 556 (1964). — CARNAGHAN, R.B.A., HARTLEY, R.D., O'KELLY, J.: Toxicity and fluorescence properties of the aflatoxins. Nature (Lond.) **200**, 1101 (1963). — CASTLEMAN, B., MCNEELY, B.U. (editors): Case records of the Massachusetts General Hospital. New Engl. J. Med. **274**, 1079 (1966). — CASTROVIEJO, R., MUÑOZ-URRA, F.: Aspergilosis ocular. Arch. Oftal. hisp.-amer. **21**, 453 (1921). — CAWLEY, E.P.: Aspergillosis and the *Aspergilli*. Report of a unique case of the disease. Arch. intern. Med. **80**, 423 (1947). — CENI, C., BESTA, C.: Über die Toxine von *Aspergillus fumigatus* und *A. flavescens* und deren Beziehungen zur Pellagra. Zbl. allg. Path. path. Anat. **13**, 930 (1902). — CHANG, S.B., ABDEL KADER, M.M., WICK, E.L., WOGAN, G.N.: Aflatoxin B_2: Chemical identity and biological activity. Science **142**, 1191 (1963). — CHARNEAU, A.: Contribution à l'étude de l'aspergillome bronchectasiant. Paris Diss. 1954, 60 S. — CHEETHAM, H.D.: Subcutaneous infection due to

Aspergillus terreus. J. clin. Path. **17**, 251 (1964). — Chick, E.W.: Enhancement of aspergillosis in leukemic chicken. Arch. Path. **75**, 81 (1963). — Chick, E.W., Conant, N.F.: Mycotic ulcerative keratitis: A review of 148 cases from the literature. Invest. Ophthal. **1**, 419 (1962). — Coe, G.E.: Primary bronchopulmonary aspergillosis, an occupational disease. Ann. intern. Med. **23**, 423 (1945). — Cogan, D.G.: Endogenous intraocular fungous infection. Report of a case. Arch. Ophthal. **42**, 666 (1949). — Cohnheim, J.: Zwei Fälle von Mykosis der Lungen. Virchows Arch. path. Anat. **33**, 157 (1865). — Collie, R.J., Figiel, L.S., Figiel, S.J., Rush, D.K.: Pulmonary aspergilloma. Report of two cases. Dis. Chest **47**, 343 (1965). — Comings, D.E., Turbow, B.A., Callahan, D.H., Waldstein, S.S.: Obstructive *Aspergillus* cast to the renal pelvis. Report of a case in a patient having diabetes mellitus and Addison's disease. Arch. intern. Med. **110**, 255 (1962). — Conant, N.F., Smith, D.T., Baker, R.D., Callaway, J.L., Martin, D.S.: Manual of Clinical Mycology. 2nd Edition, 456 pp. Philadelphia: W.B. Saunders Company 1954. — Conen, P.E., Walker, G.R., Turner, J.A., Field, P.: Invasive primary aspergillosis of the lung with cerebral metastasis and complete recovery. Dis. Chest **42**, 88 (1962). — Cooper, N.S.: Acute bronchopneumonia due to *Aspergillus fumigatus* Fresenius. Report of a case, with a description of acute and granulomatous lesions produced by the fungus in rabbits. Arch. Path. **42**, 644 (1946). — Cornet, E., Kerneis, J.P., Moigneteau, C.R., Dupon, H., Coiffard, P.: Un cas d'aspergillome bronchectasiant précocement infecté. Poumon **13**, 221—230 (1957). — Corpe, R.F., Cope, J.A.: Bronchogenic cystic disease complicated by unsuspected choleraesuis and *Aspergillus* infestation. Amer. Rev. Tuberc. **74**, 92 (1956). — Cramer, C.: Über eine neue Fadenpilzgattung: Sterigmatocystis Cramer. Vierteljahresschrift Naturforsch. Ges. Zürich. **4**, 325 (1859). — Cueva, A.: Informe preliminar sobre micosis pulmonar. Rev. méd. hondur. **33**, 250 (1965). — Darrell, R.W.: Endogenous *Aspergillus* uveitis following heart surgery. Arch. Ophthal. **78**, 354 (1967). — David, M., Charlin, A., Morice, J., Naudascher, J.: Infiltration mycosique à *Aspergillus amstelodami* du lobe temporal simulant un abcès encapsulé. Ablation en masse. Guérison operatoire. Rev. Neurol. **85**, 121 (1951). — De Iongh, H., Vles, R.O., De Vogel, P.: The occurrence and detection of aflatoxin in food. In: G.N. Wogan (ed.), Mycotoxins in foodstuffs, p. 235. Cambridge, Mass.: M.I.T.Press 1965. — De Leon-Antoni, E., Marcial-Rojas, R.A.: Pulmonary aspergillosis. Report of first case in Puerto Rico with necropsy findings. Bol. Asoc. méd. P. Rico **54**, 189 (1962). — Delikat, E., Dyke, S.C.: Acute pulmonary mycosis. Lancet **2**, 370 (1945). — Des Autels, E.J., Hoffman, O.R., Montes, M.: Invasive pulmonary aspergillosis. Difficulties in establishing the diagnosis and distinguishing primary from secondary infection. Dis. Chest **42**, 208 (1962). — Deslongchamp, E.: Note sur les moeurs du Canard Eider (Anas Mollissima Latham), et sur des moisissures developpées, pendant la vie, à la surface interne des poches aériennes d'un de ces animaux. Ann. Sci. nat. Zool. (Ser. 2) **15**, 371 (1841). — Dévé, F.: Une nouvelle forme anatomo-radiologique de mycose pulmonaire primitive. Le Mégamycétome intrabronchectasique. Arch. méd.-chir. Appar. resp. **13**, 337 (1938). — Diamond, M.A., Kirk, H.Q.: Postoperative intraocular aspergillosis. Amer. J. Ophthal. **54**, 1124 (1962). — Dickens, F., Jones, H.E.H.: The carcinogenic action of aflatoxin after its subcutaneous injection in the rat. Brit. J. Cancer **17**, 691 (1963). — Dickie, H.A., Rankin, J.: Farmer's lung. An acute granulomatous interstitial pneumonitis occurring in agricultural workers. J. Amer. med. Ass. **167**, 1069 (1958). — Dieulafoy, Chantemesse, Widal: Une pseudo-tuberculose mycosique. Gaz. Hôp. (Paris) **63**, 821 (1890). — Donahue, H.C.: Unusual mycotic infection of the lacrimal canaliculi and conjunctiva. Amer. J. Ophthal. **32**, 207 (1949). — Donaldson, J.M., Koerth, C.J., Mc Clorke, R.G.: Pulmonary aspergillosis. J. Lab. clin. Med. **27**, 740 (1942). — *Editorial:* Pulmonary Mycetoma. Lancet **2**, 439 (1968). — Edwards, J.C.: A baffling case of pulmonary carcinomatosis. New Engl. J. Med. **213**, 15 (1935). — Eger, W., Kührt, P.: Über akute Pilzencephalitis (Aspergillose) beim Menschen und im Tierexperiment. Dtsch. Z. Nervenheilk. **171**, 370—387 (1954). — Eggert, M.J., Romberg, P.F.: Pulmonary aspergillosis in a calf. J. Amer. vet. med. Ass. **137**, 595 (1960). — Ellis, R.H.: Total collapse of the lung in aspergillosis. Thorax **20**, 118 (1965). — Emile-Weil, P., Gaudin, L.: Contribution a l'étude des onychomycoses. Onychomycoses à *Penicillium*, à *Scopulariopsis*, à *Sterigmatocystis*, à Spicaria. Arch. Méd. exp. Anat. Path. **28**, 452 (1918—1919). — Emmons, C.W.: The Jekyll-Hydes of mycology. Mycologia **52**, 669 (1960). ~ Natural occurrence of opportunistic fungi. Lab. Invest. **11**, 1026 (1962). — Emmons, C.W., Binford, C.H., Utz, J.P.: Medical mycology, 380 pp. Philadelphia: Lea & Febiger, 1963. — Enjalbert, L., Ségretain, G., Eschapasse, H., Moreau, G., Bourdin, M.: Deux cas d'aspergillose pulmonaire. Etude anatomo-pathologique. Sem. Hôp. Paris **33**, 830—842 (1957). — Epstein, S.M., Miale, T.D., Moossy, J. Sidransky, H.: Experimental intracranial aspergillosis. J. Neuropath. exp. Neurol. **27**, 473 (1968). — Epstein, S.M., Verney, E., Miale, T.D., Sidransky, H.: Studies on the pathogenesis of experimental pulmonary aspergillosis. Amer. J. Path. **51**, 769 (1967). — Evans, J.H., Baker, R.D.: Treatment of experimental aspergillosis with amphotericin B. Antibiot. and Chemother. **9**, 209 (1959). — Falk, H.L.: Potential hepatocarcinogenesis for fish. In:

HALVER, J.E., MITCHELL, I.A. (ed.), Trout hepatoma research conference papers, pp. 175. Research report 70: Washington, D.C.: U.S. Government Printing Off. 1967. — FELDERMAN, L.: Infection with *Aspergillus niger*. Report of two cases. Arch. Otolaryng. **31**, 327 (1940). — FERNANDEZ-LUNA, D.: Aspergilosis pulmonar. Rev. bras. Tuberc. **26**, 431 (1958). — FETTER, B.F., KLINTWORTH, G.K., HENDRY, W.S.: Mycoses of the central nervous system, 214 pp. Baltimore: The Williams & Wilkins Co. 1967. — FINE, B.S.: Intraocular mycotic infections. Lab. Invest. **11**, 1161 (1962). ~ Mycotic keratitis. In the Cornea World Congress, p. 207. Washington: Butterworths 1965. — FINE, B.S., ZIMMERMAN, L.E.: Postoperative mycotic endophthalmitis diagnosed clinically and verified histopathologically. Brit. J. Ophthal. **43**, 753 (1959a). ~ Exogenous intraocular fungus infections with particular reference to complications of intraocular surgery. Amer. J. Ophthal. **48**, 151 (1959b). ~ Therapy of experimental intraocular *Aspergillus* infection. Arch. Ophthal. **64**, 849 (1960). — FINEGOLD, S.M., WILL, D., MURRAY, J.F.: Aspergillosis. A review and report of twelve cases. Amer. J. Med. **27**, 463 (1959). — FISHER, A.M., OSSMAN, A.G., SHAW WILGIS, E.F., KRAVITZ, S.C.: Generalized tuberculosis with pancytopenia. Report of a case with aspergillosis as a terminal event. Bull. Johns Hopk. Hosp. **119**, 355 (1966). — FOERSTER, O.H., PUESTOW, K.L.: Maduromycosis (due to *Aspergillus nidulans*). Arch. Derm. Syph. (Chic.) **16**, 91 (1927). — FONSECA, O. DA.: Mycetoma por "*Aspergillus amstelodami*". Rev. méd.-cirurg. Brasil **38**, 415 (1930). — FORGACS, J.: Mycotoxicoses in animal and human health. Proc. U.S. Livestock Sanitary Assoc., 66th Annual Meeting, p. 426, 1962. — FORGACS, J., CARLL, W.T.: Mycotoxicoses. Advanc. vet. Sci. **7**, 273 (1962). — FORGACS, J., CARLL, W.T., HERRING, A.S., MAHLANDT, B.G.: A toxic *Aspergillus clavatus* isolated from feed pellets. Amer. J. Hyg. **60**, 15 (1954). — FOUSHEE, J.H.S., NORRIS, F.G.: Pulmonary aspergillosis. A case report. J. thorac. Surg. **35**, 542 (1958). — FRANCOMBE, W.H., TOWNSEND, S.R.: Pancytopenia and disseminated aspergillosis. Canad. med. Ass. J. **92**, 81 (1965). — FRANK, L., ALTON, O.M.: Aspergillosis: A case of postoperative skin infection. J. Amer. med. Ass. **100**, 2007 (1933). — FRAUMENI, J.D., FEAR, R.E.: Purulent pericarditis in aspergillosis. Ann. intern. Med. **57**, 823 (1962). — FREIDELL, H.V., GERHART, W.F.: Rupture of abdominal aortic aneurysms complicated by acute renal failure and aspergillosis. Calif. Med. **97**, 80 (1962). — FRENKEL, J.K.: Role of corticosteroids as predisposing factors in fungal diseases. Lab. Invest. **11**, 1192 (1962). — FRESENIUS, G.: Beiträge zur Mykologie, 111 pp. Frankfurt: H.L. Bronner 1863. — FRIEDMAN, C., MISHKIN, S., LUBLINER, R.: Pulmonary resection for *Aspergillus* abscess of the lung. Dis. Chest **30**, 345 (1956). — FRIEDREICH, N.: Fall von Pneumomycosis aspergillina. Virchows Arch. path. Anat. **10**, 510 (1856). — FÜRBRINGER, P.: Beobachtungen über Lungenmycose beim Menschen. Virchows Arch. path. Anat. **66**, 330 (1876). — GALUSSIO, J.C., MOSCA, A.: Megamicetoma intracavitario (aspergiloma). A propósito de dos casos. Sem. méd. **123**, 570 (1963). — GAMMEL, J.A.: The etiology of maduromycosis. With a mycologic report on two new species observed in the U.S. Arch. Derm. Syph. (Chic.) **15**, 241 (1927). — GARCÍA-TRIVIÑO, CAMBRONERO, M., ESTEBAN, J.: Un caso interesante de aspergilosis pulmonar. Siglo méd. **95**, 30 (1935). — GAUCHER, E., SERGENT, E.: Un cas de pseudo-tuberculose aspergillaire simple chez un gaveur de pigeons. Bull. Mem. Soc. Med. Hôp. (Paris) Ser. III, **11**, 512 (1894). — GERSTL, B., WEIDMAN, W.H., NEWMANN, A.V.: Pulmonary aspergillosis: Report of two cases. Ann. intern. Med. **28**, 662 (1948). — GINGRICH, W.D.: Keratomycosis. J. Amer. med. Ass. **179**, 602 (1962). — GOLDBLATT, L.A.: Detection and elimination of aflatoxin. In: HALVER, J.E., MITCHELL, I.A. (ed.), Trout hepatoma research conference papers, p. 160. Research report 70. Washington, D.C.: U.S. Government Printing Off. 1967. — GOLDSTINE, M.T.: *Aspergillus fumigatus* vaginitis. Amer. J. Obstet. Gynec. **25**, 756 (1933). — GONZALEZ-MENDOZA, A., BRANDT, H., GONZALEZ-LICEA, A.: Coexistence of pulmonary mycosis and amebic abscess of the liver. Amer. J. trop. Med. **11**, 786 (1962). — GOWING, N.F.C., HAMLIN, I.M.E.: Tissue reactions to *Aspergillus* in cases of Hodgkin's disease and leukemia. J. clin. Path. **13**, 396 (1960). — GRAVES, C.L., MILLMAN, M.: Lobectomy for fungous abscess of the lung: effect of penicillin. J. thorac. Surg. **22**, 202 (1951). — GRAWITZ, P.: Über Schimmelvegetationen im thierischen Organismus. Experimentelle Untersuchung. Arch. Path. Anat. Physiol. Klin. Med. **81**, 355 (1880). — GRCEVIC, N., MATTHEWS, W.F.: Pathologic changes in acute disseminated aspergillosis. Amer. J. clin. Path. **32**, 536 (1959). — GRECO, N.V., BIGATTI, A.: Onicomicosis por *Aspergillus onycophylus*. Rev. Asoc. méd. argent. **49**, 509 (1935). — GREER, A.E.: Disseminating fungus diseases of the lung, 398 pp. Springfield Illinois: Charles C. Thomas 1962. — GREGORY, P.H., BUNCE, M.E.: Microfloral succession in hay. Rept. Rothamsted Exp. Sta. **1959**, 109 (1960). — GREGORY, P.H., LACEY, M.: Liberation of spores from mouldy hay. Trans. Brit. Mycol. Soc. **46**, 73 (1963). — GREGSON, A.E.W., LA TOUCHE, C.J.: Otomycosis: A neglected disease. J. Laryng. **75**, 45 (1961). — GREKIN, R.H., CAWLEY, E.P., ZHEUTLIN, B.: Generalized aspergillosis. Report of a case. Arch. Path. **49**, 387 (1950). (Case reported also by GRCEVIC and MATTHEWS, 1959). — GRUHN, J.G., SANSON, J.: Mycotic infections in leukemic patients at autopsy. Cancer **16**, 61 (1963). — GUILFOIL, P.H.: Pulmonary aspergillosis. A case report. Ohio St. med. J. **60**, 679 (1964). — GUILLAIN, G.,

Bertrand, I., Lereboullet, J.: Etude anatomo-clinique sur un abscès mycosique du lobe frontal. Rev. Neurol. **64**, 684 (1935). — Guisan, M.: Sklerosierende posttraumatische Aspergillus-Meningitis. Schweiz. Arch. Neurol. Neurochir. Psychiat. **90**, 235—254 (1962). — Hadorn, W.: Aortenruptur durch *Aspergillus*infektion nach Operation einer Aortenstenose. Endaortitis polyposa mycotica. Schweiz. med. Wschr. **90**, 929—934 (1960). — Halver, J.E.: Aflatoxicosis and rainbow trout hepatoma. In: G.N. Wogan (ed.), Mycotoxins in foodstuffs, p. 209. Cambridge, Mass.: M.I.T. Press, 1965. ~ Crystalline aflatoxin and other vectors for trout hepatoma. In: Halver, J.E., Mitchell, I.A. (ed.), Trout hepatoma research conference papers, p. 78. Research report 70. Washington, D.C.: U.S. Government Printing Off. 1967. — Harley, R.D., Mishler, J.E.: Endogenous intraocular fungus infection. Trans. Amer. Acad. Ophthal. Otolaryng. **63**, 264 (1959). — Harmer, D.: Suppuration of the antrum due to *Aspergillus fumigatus*. J. Laryng. Rhinol. Otol. **28**, 494 (1913). — Harrell, E.R., Wolter, J.R., Gutow, R.F.: Localized aspergilloma of the eyelid. Treatment with local amphotericin B. Arch. Ophthal. **76**, 322 (1966). — Hauptmann, A., Sander, C.: Ein hämatogen entstandenes solitäres Aspergillusgranulom bei einem Neunjährigen. Z. Kinderheilk. **90**, 124—133 (1964). — Hausmann, P.F.: Pulmonary aspergilloma. J. thorac. Surg. **35**, 538 (1958). — Heffernan, A.G.A., Asper, S.P.: Insidious fungal disease. A clinicopathological study of secondary aspergillosis. Bull. Johns Hopk. Hosp. **118**, 10 (1966). — Hemphill, R.A.: Mycotic lung infection. Amer. J. Med. **1**, 708 (1946). — Henderson, A.H.: Allergic aspergillosis: review of 32 cases. Thorax **23**, 501 (1968). — Henderson, A.H., English, M.P., Vecht, R.J.: Pulmonary aspergillosis. Thorax **23**, 513 (1968). — Henderson, A.H., Pearson, J.E.G.: Treatment of bronchopulmonary aspergillosis with observations on the use of natamycin. Thorax **2**, 519 (1968). — Hendrick, E.B., Conen, P.E.: Recovery from a cerebral *Aspergillus* abscess metastatic from a pulmonary focus in the lung. Canad. J. Surg. **6**, 352 (1963). (Case also reported by Conen et al., 1962). — Henrici, A.T.: An endotoxin from *Aspergillus fumigatus*. J. Immunol. **36**, 319 (1939). — Heppleston, A.G., Gloyne, S.R.: Pulmonary aspergillosis in coal workers. Thorax **4**, 168 (1949). — Hertzog, A.J., Smith, T.S., Giblin, M.: Acute pulmonary aspergillosis. Report of a case. Pediatrics **4**, 331 (1949). — Hetherington, L.H.: Primary aspergillosis of the lungs. Amer. Rev. Tuberc. **47**, 107 (1943). — Hiddlestone, H.J.H., Rosser, T.H.L., Seal, R.M.E.: Pulmonary aspergillosis. Tubercle (Edinb.) **35**, 15—18 (1954). — Hildick-Smith, G., Blank, H., Sarkany, I.: Fungus diseases and their treatment, 494 pp. Boston: Little, Brown & Co., 1964. — Hillman, R.B.: A study of mycotic placentitis and abortion in cattle, 141 pp. Thesis: Ithaca, New York: Cornell University 1961. — Hinson, K.F.W, Moon, A.J., Plummer,. N.S.: Broncho-pulmonary aspergillosis. — A review and a report of eight new cases. Thorax **7**, 317—333 (1952). — Hochberg, L.A., Griffin, E.H., Bicunas, A.D.: Segmental resection of the lung for aspergillosis. Amer. J. Surg. **80**, 364 (1950). — Höer, P.W., Schweisfurth, R.: Chronische Lungenaspergillose mit metastatischen Aspergillusgranulomen in den parenchymatösen Organen und ihren Identifizierung im Tierversuch. Verh. dtsch. Ges. Path. **1960**, 343, 346. ~ Chronische granulomatöse Form der generalisierten Aspergillose beim Menschen und ihre Identifizierung im Tierversuch. Frankfurt. Z. Path. **71**, 56—81 (1961). — Hoover, F.P.: Aspergillosis — Its diagnosis and treatment. Amer. Physician Med. Council (Phil.) **17**, 106 (1912). — Hora, J.F.: Primary aspergillosis of the paranasal sinuses and associated areas. Laryngoscope (St. Louis) **75**, 768 (1965). — Horejsi, M., Sach, J., Tomsikova, A., Mecl, A.: A syndrome resembling farmer's lung in workers inhaling spores of *Aspergillus* and *Penicillia* moulds. Thorax **15**, 212 (1960). — Hughes, W.T.: Generalized aspergillosis. A case involving the central nervous system. Amer. Dis. Child. **112**, 262 (1966). — Hunt, W., Broders, A.C., Stinson, J.C., Carabasi, R.J.: Primary pulmonary aspergillosis with invasion of the mediastinal contents and lymph nodes. Amer. Rev. resp. Dis. **83**, 886 (1961). — Husebye, K.O.: Serial studies in a case of pulmonary aspergillosis. Dis. Chest **51**, 327 (1967). — Hutson, L.R.: Bagasse litter as a contributory factor in avian aspergillosis. Canad. vet. J. **7**, 117 (1966). — Hutter, R.V.P., Collins, H.S.: The occurrence of opportunistic fungus infections in a cancer hospital. Lab. Invest. **11**, 1035 (1962). — Hutter, R.V.P., Lieberman, P.H., Collins, H.S.: Aspergillosis in a cancer hospital. Cancer (Philad.) **17**, 747 (1964). — Hyde, H.A., Richards, M., Williams, D.A.: Allergy to mould spores in Britain. Brit. med. J. **1**, 886 (1956). — Hyde, H.A., Williams, D.A.: A census of mould spores in the atmosphere. Nature (Lond.) **164**, 668 (1949). — Ikemoto, H.: Pulmonary aspergilloma or intracavitary fungus ball. Report of five cases. Sabouraudia **3**, 167 (1963—1964). — Iriarte, D.R.: Aspergilosis auricular por *Aspergillus fumigatus* (Fresenius, 1863). Arch. venez. Soc. Oto-rinolaring. **4**, 33 (1943). — Ismail, H.K.: Otomycosis. J. Laryng. **76**, 713 (1962). — Iyer, S., Dodge, P.R., Adams, R.D.: Two cases of *Aspergillus* infection of the central nervous system. J. Neurol. Neurosurg. Psychiat. **15**, 152 (1952). — Jackson, I.J., Earle, K., Kuri, J.: Solitary *Aspergillus* granuloma of the brain. Report of two cases. J. Neurosurg. **12**, 53 (1955). — Jäger, G.F.: Über die Entstehung von Schimmel im Innern des thierischen Körpers. Dtsch. Arch. Anat. Physiol. (Meckel) **2**, 354 (1816). — Janke, D., Theune, J.: Zur Kenntnis der

Aspergillose mit besonderer Berücksichtigung ihrer granulomatösen Hautmanifestation. Hautarzt **13**, 145—152 u. 193—198 (1962). — KAMPMEIER, H., BLACK, H.A.: Pulmonary aspergillosis in association with bronchial carcinoma. Amer. Rev. Tuberc. **30**, 315 (1934). — KELLY, A.B.: Aspergillosis of the nose and maxillary antrum. J. Laryng. **49**, 821 (1934). — KELMENSON, V.A.: Treatment of pulmonary aspergillosis. Dis. Chest **36**, 442 (1959). — KESHISHIAN, J.M.: Solitary pulmonary aspergilloma. Amer. Surg. **27**, 729 (1961). — KEYE, J.D., MAGEE, W.E.: Fungal diseases in a general hospital. A study of 88 patients. Amer. J clin. Path. **26**, 1235 (1956). — KHOO, T.K., SUGAI, K., LEONG, T.K.: Disseminated aspergillosis. Case report and review of the world literature. Amer. J. clin. Path. **45**, 697 (1966). — KINGERY, F.A.: The myth of otomycosis. J. Amer. med. Ass. **191**, 129 (1965). — KIRSCHSTEIN, R.L., SIDRANSKY, H.: Mycotic endocarditis of the tricuspid valve due to *Aspergillus flavus*. Report of a case. Arch. Path. **62**, 103 (1956). — KRASNITZ, A.: Broncho-pulmonary aspergillosis. N.Y. St. J. med. **57**, 3852 (1957). — LAGÈZE, P., BÉRARD, M., GALY, P., TOURAINE, R.: Le mégamycétome pulmonaire ou aspergillome intra-cavitaire à propos de trois cas. J. franc. Méd. Chri. thor. **7**, 648—662 (1953). — LANCASTER, M.C., JENKINS, F.P., PHILP, J.M.: Toxicity associated with certain samples of groundnuts. Nature (Lond.) **192**, 1095 (1961). — LANDAU, J.W., NEWCOMER, V.D., SCHULTZ, J.: Aspergillosis; report of two instances in children associated with acute leukemia and review of the pertinent literature. Mycopathologia (Den Haag) **20**, 177 (1963). — LANGERON, M.: Sur un champignon d'une otomycose brésilienne: *Sterigmatocystis Hortai*. Bull. Soc. Path. exot. **15**, 383 (1922). — LARA, A.: Algunos casos de aspergilosis observados en la península de Yucatán. Rev. Med. Yucatán **15**, 41 (1928). — LEA, W.A., SCHUSTER, D.S., HARRELL, E.R.: Mycological flora of the healthy external auditory canal: A study of 120 human subjects. J. invest. Derm. **31**, 137 (1958). — LE BRETON, E., FRAYSSINET, C., BOY, J.: Sur l'apparition d'hépatomes "spontanés" chez le Rat Wistar. Role de la toxine de l'*Aspergillus flavus*. Intérêt en pathologie humaine et cancérologie experimentale. C. R. Acad. Sci. (Paris) **255**, 784 (1962). — LEDERMAN, I.R., MADGE, G.: Endogenous intraocular aspergillosis. Arch. Ophthal. **76**, 233 (1966). — LEGATOR, M.: Biological effects of aflatoxin in cell culture. Bact. Rev. **30**, 471 (1966). — LESHIN, N.: Correlation of clinical otitis externa with mycobacteriologic studies. Arch. Otolaryng. **58**, 716 (1953). — LEVENE, N., RIVAROLA, C., BLUE, M.E.: Surgical considerations in pulmonary tuberculosis complicated by bronchopulmonary aspergillosis. Amer. Rev. resp. Dis. **91**, 262 (1965). — LEVIN, E.J.: Pulmonary intracavitary fungus ball. Radiology **66**, 9 (1956). — LEVY, E.S., COHEN, O.B.: Systemic moniliasis and aspergillosis complicating corticotropin therapy. Arch. intern. Med. **95**, 118 (1955). — LIBRACH, I.M.: Primary pulmonary aspergillosis. Report of a fatal case. Antibiot. Med. Clin. Ther. **4**, 377 (1957). — LICHTHEIM, L.: Über pathogene Schimmelpilze. I. Die Aspergillusmykosen. Berl. Klin. Wschr. **19**, 129 and 147 (1882). — LINCK, K.: Tödliche Meningitis aspergillina beim Menschen. Virchows Arch. path. Anat. **304**, 408 (1939). — LINK, H.F.: Observations in ordines plantarum natureles. Ges. Naturf. Freunde Berlin Magazin **3**, 1 (1809). — LONGBOTTOM, J.L., PEPYS, J.: Pulmonary aspergillosis: diagnostic and immunological significance of antigens and C substance in *Aspergillus fumigatus*. J. Path. Bact. **88**, 141 (1964). — LOURIA, D.B., COLLINS, H.S.: Some aspects of pulmonary mycotic infections. Tuberculology **21**, 76 (1964). — LUCET, A.: De l'*Aspergillus fumigatus* chez animaux domestiques et dans les oeufs en incubation. Etude clinique et experimentale, 108 pp. Paris: Ch. Mendel 1897. — LUKE, J.L., BOLANDE, R.P., GROSS, S.: Generalized aspergillosis and *Aspergillus* endocarditis in infancy. Report of a case. Pediatrics **31**, 115 (1963). — LYNCH, K.M.: *Aspergillus* in scalp lesions following redbug (leptus) bites. Arch. Derm. Syph. (Chic.) **7**, 599 (1923). ~ Invasion of the hair of the scalp by *Aspergillus*. Sth. med. J. (Bgham, Ala.) **28**, 119 (1925). — MCBURNEY, R., SEARCY, H.B.: Otomycosis: An investigation of effective fungicidal agents in treatment. Ann. Otol. (St. Louis) **45**, 988 (1936). — MCDIARMID, A.: Aspergillosis in free-living wild birds. J. comp. Path. **65**, 246 (1955). — MCGONIGLE, J.J., JILLSON, O.F.: Otomycosis: An entity. Arch. Derm. **95**, 45 (1967). — MCKEE, E.E.: Mycotic infection of brain with arteritis and subarachnoid hemorrhage. Report of a case. Amer. J. clin. Path. **20**, 381 (1950). — MACARTNEY, J.N.: Pulmonary aspergillosis: A review and a description of three new cases. Thorax **19**, 287 (1964). — MACKENZIE, D.: Preliminary report on *Aspergillus* mycosis of the antrum maxillare. Bull. Johns Hopk. Hosp. **4**, 9 (1893). — MACKIE, F.P.: Visceral infections due to higher fungi. Indian J. med. Res. **9**, 781 (1922). — MEDELIN, M.F.: Internal fungal parasites of insects. Endeavour **19**, 181 (1960). — MAHAFFEY, L.W., ASAM, N.M.: Abortions associated with mycotic lesions of the placenta in mares. J. Amer. vet. med. Ass. **144**, 24 (1964). — MAHVI, T.A., WEBB, H.M., DIXON, C.D., BOONE, J.A.: Systemic aspergillosis caused by *Aspergillus niger* after open heart surgery. J. Amer. med. Ass. **203**, 520 (1968). — MANGIARACINE, A.B., LIEBMAN, S.D.: Fungus keratitis (*Aspergillus fumigatus*). Treatment with Nystatin (Mycostatin). Arch. Ophthal. **58**, 695 (1957). — MARSALEK, E., ZIZKA, Z., RIHA, V., DUSEK, J., DVORACEK, D.: Plicni aspergilloza s generalizaci vyvolana druhem *Aspergillus restrictus*. Čas. Lék. čes. **99**, 1285 (1960). — MARTIN-LALANDE, J.: Aspergillose et tuberculose pulmonaire associées. (Contribution à l'étude des modalités de coexistence). Rev. Tuberc. (Paris) **25**, 1235 (1961). — MAYER, A.C.: Verschimmelung (Mucedo) im lebenden Körper. Dtsch. Arch. Anat. Physiol. (Meckel) **1**, 310 (1815). — MAYS, E.E., HAWKINS, J.A.: Cavitary bronchiolar carcinoma with an intracavitary aspergilloma. Amer. Rev. resp. Dis. **95**, 1056 (1967). — MEHROTRA, M.C.: Aspergillosis of maxilla; an unusual case. Oral Surg. **20**, 33 (1965). — MERCHANT, R.K., LOURIA, D.B., GEISLER, P.H., EDGCOMB, J.H., UTZ, J.P.: Fungal endocarditis: Review of the literature and report of three cases. Ann. intern. Med. **48**, 242 (1958). — MERKOW, L., PARDO, M., EPSTEIN, S.M., VERNEY, E., SIDRANSKY, H.: Lysosomal stability during phagocytosis of *Aspergillus flavus* spores by alveolar macrophages of cortisone treated mice. Science **160**, 79 (1968). — MICHELI, P.H.: Nova plantarum genera juxta tournefortii methodum disposita. Florence, 1729, 234 pp. — MIERES, J.F.: Caso de seudotuberculosis causada por el *Aspergillus fumigatus*. Sem. méd. **1**, 370 (1925). — MITSUI, Y., HANABUSA, J.: Corneal infections after cortisone therapy. Brit. J. Ophthal. **39**, 244(1955).— MONIZ, E., LOFF, R.: Aspergillose cerebrale. Presse méd. **39**, 273 (1931). — MONOD, O., PESLE, G., SÉGRÉTAIN, G.: L'aspergillome bronchectasiant. Presse méd. **59**, 1557—1559 (1951). — MONOD, O., PESLE, G., MEYER, A.: L'aspergillome bronchectasiant. Sem. Hôp. (Paris) 3588—3602 (1957). — MONTES, M.: Pathologic study of a case of primary pulmonary aspergillosis. Amer. Rev. resp. Dis. **87**, 409 (1963). — MONTREUIL, F.: Fungus infection of the antrum. J. Laryng. **69**, 559 (1955). — MOOLTEN, S.E.: A case of primary broncho-pulmonary aspergillosis. J. Mt Sinai Hosp. **5**, 29 (1938—1939). — MOORE, M.: Discussion following paper presented by E.S. BERESTON. Sth. med. J. (Bgham, Ala.) **43**, 492 (1950). — MOORE, M., WEISS, R.S.: Onychomycosis caused by *Aspergillus terreus*. J. invest. Derm. **11**, 215 (1948). — MOSTAFA, M.S., EL-D.: *Aspergillus niger* infection of the eye. Amer. J. Ophthal. **62**, 1204 (1966). — MUNOZ, J., GEISTER, R.: Inhibition of phagocytosis by aureomycin. Proc. Soc. exp. Biol. (N.Y.) **75**, 367 (1950). — MYERS, J.T., DUNN, A.D.: *Aspergillus* infection of the hand. J. Amer. med. Ass. **95**, 794 (1930). — NAJI, A.F.: Bronchopulmonary aspergillosis. Report of two new cases, review of literature, and suggestion for classification. Arch. Path. **68**, 282 (1959). — NASH, C.S.: Fulminating infection of nose due to *Monilia* or *Aspergillus*. Report of a case. Arch. Otolaryng. **28**, 234 (1938). — NAVARRO-BLASCO, A.: Sobre aspergilosis pulmonar. Clin. Lab. **29**, 81 (1936). — NESBITT, B.F., O'KELLY, J., SARGEANT, K., SHERIDAN, A.: Toxic metabolites of *Aspergillus flavus*. Nature (Lond.) **195**, 1062 (1962). — NEWBERNE, P.M.: Carcinogenicity of aflatoxin-contaminated peanut meals. In: G.N. WOGAN (ed.), Mycotoxins in foodstuffs, p. 187. Cambridge, Mass.: M.I.T. Press, 1965. ~ Biological activity of the aflatoxins in domestic and laboratory animals. In: HALVER, J.E., MITCHELL, I.A., (ed.), Trout hepatoma research conference papers, p. 130. Research Report 70. Washington, D.C.: U.S. Government Printing Off. 1967. — NEWBERNE, P.M., CARLTON, W.W., WOGAN, G.N.: Hepatomas in rats and hepatorenal injury in ducklings fed peanut meal or aspergillus flavus extract. Path. vet. (Basel) **1**, 105—132 (1964). ~ Hepatomas in rats and hepatorenal injury in ducklings fed peanut meal or aspergillus flavus extract. (Leberzellcarcinome bei Ratten. Leber- und Nierenparenchymschäden bei Entenküken nach Fütterung mit Erdnußmehl oder Aspergillus flavus-Extrakten.) Path. vet. (Basel) **1**, 105—132 (1964). — NEWMAN, W.H., CORDELL, A.R.: *Aspergillus* endocarditis after open-heart surgery. Report of a case and review of literature. J. thorac. cardiovasc. Surg. **48**, 652 (1964). — NICHOLSON, H.: Infección aspergilar en tuberculosis pulmonar. Tórax **13**, 89 (1964). — NICOD, J.-L., SCHLEGEL, J.: Mycose pulmonaire double à Aspergillus fumigatus Fres. et à Mucor pusillus Lindt. Schweiz. Z. allg. Path. **15**, 307—321 (1952). — NICOLLE, C., PINOY: Sur un cas de mycetome d' origine aspergillaire, observé en Tunisie. Arch. de Parasitologie **10**, 437 (1906). — NOBLE, W.C., CLAYTON, Y.M.: Fungi in the air of hospital wards. J. gen. Microbiol. **32**, 397 (1963). — OEHLERT, W.: Autoradiographische Untersuchung bei der experimentellen Aspergillose der Ratte. (Größe des Eiweiß-Umsatzes verschiedener Zellen untersucht mit S_{35}-markierten Thioaminosäuren.) Acta histochem. (Jena) **6**, 315—332 (1959). — OEHLERT, W., DÜFFEL, F.: Experimentelle Untersuchungen über die Aspergillusinfektion mit Nachweis toxisch wirkender Stoffwechselprodukte des Aspergillus fumigatus. Zbl. allg. Path. path. Anat. **98**, 41—49 (1958). — OETTLÉ, A.G.: The aetiology of primary carcinoma of the liver in Africa. A critical appraisal of previous ideas with an outline of the mycotoxin hypothesis. S. Afr. med. J. **39**, 817 (1965). — OKUDAIRA, M.: Histopathological differentiation between Candida albicans and Aspergillus fumigatus in tissue sections. Acta path. jap. **5**, 117—124 (1955). — OKUDAIRA, M., Schwarz, J.: Tracheobronchopulmonary mycoses caused by opportunistic fungi, with particular reference to aspergillosis. Lab. Invest. **11**, 1053 (1962). — OLK, W.: Bronchogene generalisierte Aspergillose. Zbl. allg. Path. **97**, 361—371 (1958). — OPPE, W.: Zur Kenntniss der Schimmelmykosen beim Menschen. Zbl. allg. Path. **8**, 301 (1897). — ORIE, N.G.M., DE VRIES, G.A., KIKSTRA, A.: Growth of *Aspergillus* in the human lung: Aspergilloma and aspergillosis. Amer. Rev. resp. Dis. **82**, 649 (1960). — OSLER, W.: *Aspergillus* from the lung. Trans. path. Soc. Philad. **13**, 108 (1887). — Ota, M.: Sur une nouvelle espèce d'*Aspergillus* pathogène: *Aspergillus jeanselmei*. Ann. Parasit. hum. comp. **1**, 137 (1923). — PALDROK, H., HOLLSTROM,

E.: Onychomycosis due to *Aspergillus terreus* Thom. Acta Dermat. Venereol. **32**, 255 (1952). — PARADIS, A.J., ROBERTS, L.: Endogenous ocular aspergillosis. Report of a case in an infant with cytomegalic inclusion disease. Arch. Ophthal. **69**, 765 (1963). — PATTERSON, R.: Hypersensitivity disease of the lung. Univ. Mich. med. Cent. J. **34**, 8 (1968). — PECORA, D.V., TOLL, M.W.: Pulmonary resection for localized aspergillosis. New Engl. J. Med. **263**, 785 (1960). — PEER, E.T.: Case of aspergillosis treated with amphothericin "B". Dis. Chest **38**, 222 (1960). — PEET, M.M.: *Aspergillus fumigatus* infection of the cerebellum. Trans. Amer. neurol. Ass. **71**, 165 (1946). (Case also reported by CAWLEY, 1947). — PEÑA, C.E., SALAZAR, H.: Aspergilosis en Colombia. Estudio clinico-patológico de 15 casos. Mycopathologia (Den Haag) **34**, 65 (1968). — PEPYS, J.: Possible role of precipitins against *Aspergillus fumigatus*. Amer. Rev. resp. Dis. **90**, 465 (1964). ~ Pulmonary hypersensitivity disease due to inhaled organic antigens. Postgrad. med. J. **42**, 698 (1966a). ~ Immunological aspects of respiratory disease. J. Laryng. **80**, 373 (1966b). ~ Pulmonary hypersensitivity disease due to inhaled organic antigens. Ann. intern. Med. **64**, 943 (1966c). — PEPYS, J., JENKINS, P.A.: Precipitin (F.L.H.) test in farmer's lung. Thorax **20**, 21 (1965). — PEPYS, J., LONGBOTTOM, J.L., JENKINS, P.A.: Vegetable dust pneumoconioses. Immunologic responses to vegetable dusts and their flora. Amer. Rev. resp. Dis. **89**, 842 (1964). — PEPYS, J., RIDELL, R.W., CITRON, K.M., CLAYTON, Y.M.: Precipitins against extracts of hay and moulds in the serum of patients with farmer's lung, aspergillosis, asthma, and sarcoidosis. Thorax **17**, 366 (1962). — PESLE, G.D., MONOD, O.: Bronchiectasis due to aspergilloma. Dis. Chest **25**, 172 (1954). — PHELPS, R.A.: Aflatoxins in feeds. A review. In: HALVER, J.E., MITCHELL, I.A. (ed.), Trout hepatoma research conference papers, p. 145. Research report 70. Washington, D.C.: U.S. Government Printing Off. 1967. — PIMENTEL, J.C.: Pulmonary calcification in the tumor-like form of pulmonary aspergillosis: pulmonary aspergilloma. Amer. Rev. resp. Dis. **94**, 208 (1966). — PINOY, E.: Actynomycoses et mycetomes. Bull. Inst. Pasteur **11**, 929 and 977 (1913). — PITTENGER, B.N.: Case reports of aspergillosis. Kentucky med. J. **36**, 161 (1938). — PLIHAL, V., JEDLICKOVA, Z., VICKLIKY, J., TOMANEK, A.: Multiple bilateral pulmonary aspergillomata. Thorax (Lond.) **19**, 104—111 (1964). — PONCE, G. de O.: Forma broncho-pneumonica da aspergillose pulmonar pura e primitiva? Rev. bras. Med. Pharm. **11**, 72 (1935). — PONS, E.R., BELAVAL, M.E.: A one-year aeroallergen survey of Puerto Rico. J. Allergy **32**, 195 (1961). — PORTA, E.A.: Endocarditis micótica por *Aspergillus* en un caso de estenosis subaórtica. Rev. Hosp. Niños (B. Aires) **1**, 37 (1959). — POWELL, D.E.B., ENGLISH, M.D.: Clinical bacteriological and mycological findings in post-operative ear cavities. J. Laryng. **76**, 12 (1962). — PROCKNOW, J.J.: Treatment of opportunistic fungus infections. Lab. Invest. **11**, 1217 (1962). — PROCKNOW, J.J., LOEWEN, D.F.: Pulmonary aspergillosis with cavitation secondary to histoplasmosis. Amer. Rev. resp. Dis. **82**, 101 (1960). — PUESTOW, K.L.: A contribution to the study of maduromycosis, with report of a case of infection with *Aspergillus nidulans*. Arch. Derm. Syph. (Chic.) **20**, 642 (1929). — RAMIREZ, J.: Pulmonary aspergilloma. Endobronchial treatment. New Engl. J. Med. **271**, 1281 (1964). — RAMSEY, B.H.: The upper lobe lesion — old or new, with reference to a case of aspergillosis (mycetoma). Dis. Chest **38**, 625 (1960). — RANDALL, R.: Aspergillosis. Vet. Bull. (Washington) **16**, 103 (1925). — RANKIN, N.E.: Disseminated aspergillosis and moniliasis associated with agranulocytosis and antibiotic therapy. Brit. med. J. **4816**, 918—919 (1953). — RAPER, K.B., FENNELL, D.I.: The genus *Aspergillus*, 686 pp. Baltimore: Williams & Wilkins Co. 1965. — RAU, E.M., TILDEN, E.B., KOENIG, V.L.: Partial purification and characterization of the endotoxin from *Aspergillus fumigatus*. Mycopathologia (Den Haag) **14**, 346 (1961). — RAYER, MONTAGNE: Mycose aspergillaire dans les poches aeriennes d'un bouvreuil. J. Inst. (Paris) In-**4**, 270 (1842). (Quoted by ROBIN, C., 1853). — REBOUÇAS, J.: Aspergilose niger do conduto. Rev. bras. Oto-rino-laring. **7**, 169 (1939). — REDMOND, A., CARRE, I.J., BIGGART, J.D., MACKENZIE, D.W.R.: Aspergillosis (*Aspergillus nidulans*) involving bone. J. Path. Bact. **89**, 391 (1965). — REED, J.A.: Aspergillosis and moniliosis. Sth. med. J. (Bgham., Ala.) **28**, 729 (1935). — RÉNON, L.: Recherches cliniques et experimentales sur la pseudotuberculose aspergillaire. Thèse de Paris, 1893. ~ Deux cas familiaux de tuberculose aspergillaire simple chez des peigneurs de cheveux. C. R. Soc. Biol. (Paris) **47**, 694 (1895). ~ Etude sur l' Aspergillose chez les animaux et chez l' homme, 301 pp. Paris: Masson et Cie. 1897. — RIFKIND, D., MARCHIORO, T.L., SCHNECK, S.A., HILL, R.B.: Systemic fungal infections complicating renal transplantation and immunosuppressive therapy. Clinical, microbiologic, neurologic and pathologic features. Amer. J. Med. **43**, 28 (1967). — RILEY, E.A., TENNENBAUM, J.: Pulmonary aspergilloma or intracavitary fungus ball. Report of five cases. Ann. intern. Med. **56**, 896 (1962). — ROBIN, C.: Histoire naturelle des vegetaux parasites qui croissent sur l'homme et sur les animaux vivants, 702 pp. Paris: Baillière 1853. — ROBINSON, C.L.N., MCPHERSON, A.R.: Bronchopulmonary aspergilloma. Canad. J. Surg. **5**, 411 (1962). — ROBINSON, G.H., GRAUER, R.C.: Use of autogenous fungus extracts in treatment of mycotic infections. Arch. Derm. Syph. (Chic.) **32**, 787 (1935). — RODIN, A.E., HNATKO, S.I.: Nonreactive tuberculosis. Canad. med. Ass. J. **89**, 817 (1963). — RODRICKS, J.V., HENERY-

LOGAN, K.R., CAMPBELL, A.D., STOLOFF, L., VERRETT, M.J.: Isolation of a new toxin from cultures of *Aspergillus flavus*. Nature (Lond.) **217**, 668 (1968). — RODRÍGUEZ-VILLEGAS, R., SCHENA, A.T.: Micosis pulmonar en una cavidad residual hidatídica. Sem. méd. **2**, 93 (1941). — ROMINGER, L., BÖHM, F.: Über einen Fall von generalisierender pyämischer Aspergillose. Beitr. Klin. Tuberk. **113**, 221—233 (1955). — ROSENVOLD, L.K.: Dacryocystitis and blepharitis due to infection by *Aspergillus niger:* report of cases. Amer. J. Ophthal. **25**, 588 (1942). — ROSS, C.F.: A case of pulmonary aspergillosis. J. Path. Bact. **63**, 409 (1951). — RUBINSTEIN, P.: Micosis broncopulmonares consecutivas al tratamiento antibiótico o corticoide. Rev. Asoc. méd. argent. **77**, 53 (1963). — RUEDA, G., MEJÍA, A.: Micetoma pulmonar. (Revisión de literatura, presentación de un caso.) Trib. méd. (Bogotá) **1**, 1 (1962). — SALFELDER, K., CAPRETTI, C., HARTUNG, M., MONCADA-REYES, F.: Dos casos poco comunes de aspergilosis pulmonar y pleural. Mycopathologia (Den Haag) **14**, 78 (1961). — SANCHEZ-FREIJO, C.: La aspergilosis pulmonar primitiva (Un caso de forma aguda). Siglo méd. **95**, 222 (1935). — SANERKIN, N.G., SEAL, R.M.E., LEOPOLD, J.G.: Plastic bronchitis, mucoid impaction of the bronchi and allergic broncho-pulmonary aspergillosis, and their relationship to bronchial asthma. Ann. Allergy **24**, 586 (1966). — SARGEANT, K.A., SHERIDAN, A., O'KELLY, J., CARNAGHAN, R.B.A.: Toxicity associated with certain samples of groundnuts. Nature (Lond.) **192**, 1096 (1961). — SARTORY, A.: Sur un champignon nouveau du genre *Aspergillus* isolé dans un cas d'onychomycose. C. R. Acad. Sci. (Paris) **170**, 523 (1920). — SATHMARY, M.N.: *Bacillus subtilis* septicemia and generalized aspergillosis in a patient with acute myeloblastic leukemia. N.Y. St. J. Med. **58**, 1870 (1958). — SAUBERMANN, G., SCHOLER H.J.: Aspergillose der Hornhaut. I. Kasuistischer und experimenteller Beitrag zur Diagnose und Therapie, p. 1—22. Basel: S. KARGER 1959. — SAUNDERS, A.M., BIEBER, C.: Pathologic findings in a case of cardiac transplantation. J. Amer. med. Ass. **206**, 815 (1968). — SAVAGE, A., ISA, J.M.: A note on mycotic pneumonia in chickens. Sci. Agr. **13**, 341 (1933). — SAVETSKY, L., WALTNER, J.: Aspergillosis of the maxillary antrum. Report of a case and review of the available literature. Arch. Otolaryng. **74**, 695 (1961). — SAYLE, B.A., LANG, P.A., GREEN, W.O., BOSWORTH, W.C., GREGORY, R.: Cushing's syndrome due to islet cell carcinoma of the pancreas. Report of two cases: One with elevated 5-hydroxyindole acetic acid and complicated by aspergillosis. Ann. intern. Med. **63**, 58 (1965). — SCARPELLI, D.G.: Ultrastructural and biochemical observations on trout hepatoma. In: HALVER, J.E., MITCHELL, I.A. (ed.), Trout hepatoma research conference papers, p. 60. Research report 70. Washington, D.C.: U.S. Government Printing Off. 1967. — SCHECHTER, F.R., LEIVY, F.E., NIEDZWIEDZ, A.: Pulmonary aspergillosis complicating therapy for lymphoma with reversible uric acid uremia. J. Einstein Med. Cent. **8**, 153 (1960). — SCHEIDEGGER, S.: Pilzerkrankungen. In: HENKE, F., LUBARSCH, O. (ed.), Handbuch der speziellen pathologischen Anatomie und Histologie, Vol. 13, part 2, section A. Berlin-Göttingen-Heidelberg: Springer 1958. — SCHMIDT, E.L., BANKOLE, R.O.: Detection of *Aspergillus flavus* in soil by immuno-fluorescent staining. Science **136**, 776 (1962). — SCHNEIDER, L.V.: Primary aspergillosis of the lungs. Amer. Rev. Tuberc. **22**, 267 (1930). — SCHNYDER, H.K.: Aspergillose der Schädelbasis. Pract. oto-rhino-laryng. (Basel) **10**, 402—421 (1948). — SCHOLER, H.J., SAUBERMANN, G.: Aspergillose der Hornhaut. II. Diskussion von 90 Fällen der Literatur, p. 23—58. Basel: S. Karger 1959. — SCHUBERT, P.: Zur Casuistik der Aspergillusmykosen. Dtsch. Arch. Klin. Med. **36**, 162 (1885). — SCHUMACHER, H.R., GINNS, D.A., WARREN, W.J.: Fungus infection complicating leukemia. Amer. J. med. Sci. **247**, 313 (1964). — SCHWARZ, J., BAUM, G.L., STRAUB, M.: Cavitary histoplasmosis complicated by fungus ball. Amer. J. Med. **31**, 692 (1961). — SEABURY, J.H., SAMUELS, M.: The pathogenic spectrum of aspergillosis. Amer. J. clin. Path. **40**, 21 (1963). — SÉGRETAIN, G.: Pulmonary aspergillosis. Lab. Invest. **11**, 1046 (1962). — SÉGRETAIN, G., VIEU, M.: Formes parasitarires des aspergillus dans l'aspergillome bronchique: Diagnostic biologique des aspergilloses broncho-pulmonaires. Sem. Hôp. Paris/Path. Biol./Arch. Biol. méd. **1957**, 421—429. — SHANK, R.C., WOGAN, G.N.: Effects of aflatoxin B_1 on some aspects of liver metabolism. Fed. Proc. **23**, 200 (1964). — SHARP, W.B., JOHN, M.B.: Pathogenicity of the *Aspergilli* of otomycosis. Tex. Rep. Biol. Med. **4**, 353 (1946). — SHAW, F.W., WARTHEN, H.J.: Aspergillosis of bone. Sth. med. J. (Bgham, Ala) **29**, 1070 (1936). — SHERMAN, I.C., BOSHES, L.D.: Some fungus infections of the central nervous system: experience with recent treatments. Bol. Asoc. méd. P. Rico **48**, 225 (1956). — SIDRANSKY, H., FRIEDMAN, L.: The effect of cortisone and antibiotic agents on experimental pulmonary aspergillosis. Amer. J. Path. **35**, 169 (1959). — SIDRANSKY, H., PEARL, M.A.: Pulmonary fungus infections associated with steroid and antibiotic therapy. Dis. Chest. **39**, 630 (1961). — SIEBENMANN, F.: Die Schimmelmykosen des menschlichen Ohres, 118 pp. Wiesbaden: J. F. Bergmann 1889. — SILLER, W.G., OSTLER, D.C.: The histopathology of an entero-hepatic syndrome of turkey poults. Vet. Rec. **73**, 134 (1961). — SILVER, M.D., MASON, W.E.H., ROBINSON, C.L.N., BLANK, F.: Pulmonary aspergillosis. Canad. med. Ass. J. **87**, 579 (1962). — SIMMS, R.F.: *Aspergillus* infection as the cause of external ear diseases. Sth. med. J. (Bgham, Ala.) **30**, 1224 (1937). — SIMON, R.C., DOLLAR, A.M., SMUCKLER, E.A.: Descriptive classification on normal and altered histology of trout livers. In: HALVER, J. E., MITCHELL, I. A. (ed.), Trout hepatoma research conference papers, p. 18. Research report 70. Washington, D.C.: U.S. Government Printing Off. 1967. — SINNHUBER, R. O.: Aflatoxin in cottonseed meal and liver cancer in rainbow trout. In: HALVER, J. E., MITCHELL, I. A. (ed.), Trout hepatoma research conference papers, p. 48. Research report 70. Washington, D.C.: U.S. Government Printing Off. 1967. — SKILLERN, R. H.: Aspergillos of the maxillary sinus. Laryngoscope (St. Louis) **31**, 946 (1921). — SLUYTER, F. T.: De vegetabilibus organismi animalis parasitis, ac de novo epiphyto in pitryiasi versicolore obvio. Diss. Inaug. Berlin 1847. Quoted by VIRCHOW (1856). — SMITH, A. G., SCHULTZ, R. B.: Observersations in the disposition of *Aspergillus fumigatus* in the respiratory tract. Amer. J. clin. Path. **44**, 271 (1965). — SMITH, L. M.: *Aspergillus* infection of the nails. Urol. cutan. Rev. **38**, 783 (1934). — SMITH, R. H., MCKERNAN, W.: Hepatotoxic action of chromatographically separated fractions of *Aspergillus flavus* extracts. Nature (Lond.) **195**, 1301 (1962). — SMITH, W. R.: Aspergillosis. J. Tenn. med. Ass. **27**, 407 (1934). — SNELLING, M. R. J., MCGLADDERY, H. M., PONNAMPALAM, J. T.: Bronchopulmonary aspergillosis. Dis. Chest **44**, 100 (1963). — SOCHOCKY, S.: Infection of pneumonectomy space with *Aspergillus fumigatus* treated by "nystatin". Dis. Chest **36**, 554 (1959). — SOERGEL, K. H., COTE, R. A., TAUGHER, P. J.: Felty's syndrome and generalized aspergillosis. Marquette med. Rev. **30**, 33 (1964). — SPOTNITZ, M., OVERHOLT, E. L.: Mucoid impaction of the bronchi associated with *Aspergillus*. Report of a case. Dis. Chest **52**, 92 (1967). — STANLEY, P.: Pulmonary aspergillosis occurring in a patient receiving large doses of prednisolone. Tubercle (Edinb.) **46**, 227 (1965). — STARK, J. E.: Allergic pulmonary aspergillosis successfully treated with inhalations. of nystatin. Report of a case. Dis. Chest **51**, 96 (1967). — STEELE, A. E.: A case of infection with *Aspergillus versicolor*. Boston med. surg. J. **195**, 536 (1926). — STEVENS, K. M.: The effect of antibiotics upon the immune response. J. Immunol. **71**, 119 (1953). — STERN, S. G., KULVIN, M. M.: Aspergillosis of the cornea. Amer. J. Ophthal. **33**, 111 (1950). — STOLOW, A. J.: Primary broncho-pulmonary aspergillosis. J. med. Soc. N. J. **36**, 484 (1939). — STRELLING, M. K., RHANEY, K., SIMMONS, D. A. R., THOMSON, J.: Fatal acute pulmonary aspergillosis in two children of one family. Arch. Dis. Childh. **41**, 34 (1966). — STUART, E. A., BLANK, F.: Aspergillosis of the ear. A report of twenty-nine cases. Canad. med. Ass. J. **72**, 334 (1955). — SUIE, T., HAVENER, W. H.: Mycology of the eye: a review. Amer. J. Ophthal. **56**, 63 (1963). — SWAEBLY, M. A., CHRISTENSEN, C. M.: Molds in house dust, furniture stuffing, and in the air within homes. J. Allergy **23**, 370 (1952). — SYMMERS, W. St. C.: Histopathologic aspects of the pathogenesis of some opportunistic fungal infections, as exemplified in the pathology of aspergillosis and phycomycosis. Lab. Invest. **11**, 1073 (1962). — TAKANO, J., CUELLO, C., HOFFMANN, E., CORREA, P.: Estudio de lesiones residuales pulmonares. Rev. lat.-amer. Anat. pat. **6**, 63 (1962). — TAYLOR, P. A.: Mycotic pneumonia in pigeons. Canad. vet. J. **7**, 262 (1966). — TEIXEIRA, L.: Um caso de aspergillose pulmonar. Rev. Flum. de Med. **6**, 147 (1941). — THOM, C., CHURCH, M. B.: The *Aspergilli*, 272 pp. Baltimore: Williams & Wilkins Co. 1926. — THOM, C., RAPER, K. B.: A manual of the *Aspergilli*, 373 pp. Baltimore: Williams & Wilkins Co. 1945. — TILDEN, E. B., HATTON, E. H., FREEMAN, S., WILLIAMSON, W. M., KOENIG, V. L.: Preparation and properties of the endotoxins of *Aspergillus fumigatus* and *Aspergillus flavus*. Mycopathologia (Den Haag) **14**, 325 (1961). — TILLEY, H.: Aspergillosis of the maxillary antrum. Proc. roy. Soc. Med. **8**, (pt. 2), 14 (1914—1915). — TOBLER, W., MINDER, W.: Generalisierte chronische Aspergillose beim Kind und ihre Beziehung zur antibiotischen Therapie. Helv. paediatr. Acta, Ser. C., **9**, 209—230 (1954). — TOIGO, A.: Pulmonary aspergillosis. Amer. Rev. resp. Dis. **81**, 392 (1960). — TORACK, R. M.: Fungus infections associated with antibiotic and steroid therapy. Amer. J. Med. **22**, 872 (1957). — TURHAN, B.: Aspergillosis der Lunge. Istanbul. Üniv. Tip. Fak. Meck. **15**, 748—757 (mit engl. Zusammenfassung 1952) (Türkisch). — TVETEN, L.: Cerebral mycosis. A clinic-pathological report of four cases. Acta Neurol. Scand. **41**, 19 (1965). — UTZ, J. P.: The spectrum of opportunistic fungus infections. Lab. Invest. **11**, 1018 (1962). ~ Chemotherapeutic agents for the systemic mycoses. New Engl. J. Med. **268**, 938 (1963a). ~ Systemic fungal infections amenable to chemotherapy. D. M. p. 1—52, September 1963b. — UTZ, J. P., ANDRIOLE, V. T., EMMONS, C. W.: Chemotherapeutic activity of X-5079 C in systemic mycoses of man. Amer. Rev. resp. Dis. **84**, 514 (1961). — UTZ, J. P., GERMAN, J. L., LOURIA, D. B., EMMONS, C. W., BARTTER, F. C.: Pulmonary aspergillosis with cavitation — Iodide therapy associated with an unusual electrolyte disturbance. New Engl. J. Med. **260**, 264 (1959). — VADALA, A. J.: Mycotic infection of the bronchopulmonary tract. Ann. Otol. (St. Louis) **49**, 291 (1940). — VAITHIANATHAN, T., BOLONIK, S. J., GRUHN, J. G.: Leukemic reticuloendotheliosis. Amer. J. clin. Path. **38**, 605 (1962). — VALLERY-RADOT, P., HALPERN, B. N., SÉGRETAIN, A., DOMART, A.: Etude de la nature et de la densité de la flore mycologique dans l'atmosphere de Paris durant l' année 1948. Acta allerg. (Kbh.) **3**, 179 (1950). — VAN DER ZIJDEN, A. S. M., KOELENSMID, W. A. A. B., BOLDINGH, J., BARRETT, C. B., ORD, W. O., PHILIP, J.: *Aspergillus flavus* and turkey X disease. Nature (Lond.) **195**, 1060 (1962). — VAN ORDSTRAND, H. S.: Pulmonary aspergillosis. Report of a case. Cleveland Clin. Quart. **7**, 66 (1940). — VANTRAPPEN, G., SIMONS, C., VAN de Woestijne, K.P., Gyselen, A.: Intracavitary and intrabronchial fungus masses. Dis. Chest **35**, 528 (1959). — Vedder, J.S., Schorn, W.F.: Pulmonary aspergillosis with metastatic skin nodules. J. Amer. med. Ass. **209**, 1191 (1969). — Vellios, F., Crawford, A.S., Gatzimos, C.D., Haynes, E.: Bronchial aspergillosis occurring as intracavitary "fungus ball". Amer. J. clin. Path. **27**, 68 (1957). — Villar, T.G.: The diagnosis of pulmonary aspergilloma. Geriatrics **18**, 883 (1963). — Villar, T.G., Cortez-Pimentel, J., Costa, M.F.E.: The tumour-like forms of aspergillosis of the lung (pulmonary aspergilloma). A report of 5 new cases and a review of the Portuguese literature. Thorax **17**, 22 (1962). — Virchow, R.: Beiträge zur Lehre von den beim Menschen Vorkommenden pflanzlichen Parasiten. Virchows Arch. path. Anat. **9**, 557 (1856), — Wahl, E.F.: Pneumomycosis. Trans. Amer. ther. Soc. **35**, 36 (1936). — Wahl, E.F., Erickson, M.J.: Primary pulmonary aspergillosis. J.M.A. Georgia **17**, 341 (1928). — Wahner, H.W., Hepper, N.G.G., Andersen, H.A., Weed, L.A.: Pulmonary aspergillosis. Ann. intern. Med. **58**, 472 (1963). — Wales, J.H.: Degeneration and regeneration of liver parenchyma accompanying hepatomagenesis. In: Halver, J.E., Mitchell, I.A. (ed.), Trout hepatoma research conference papers, p. 56. Research report 70. Washington, D.C.: U.S. Government Printing Off. 1967. — Wätjen, J.: Zur Kenntnis der Gewebsreaktionen bei Schimmelmykosen. Virchows Arch. path. Anat. **268**, 665 (1928). (with complete reference of the older European Lit.). ~ Durch Schimmel- und Sproßpilze bedingte Erkrankungen der Lungen. In: Henke, F., Lubarsch, O. (ed.), Handbuch der speziellen pathologischen Anatomie und Histologie, vol. 3, part 3. Berlin: Springer 1931. — Weens, H.S., Thompson, E.A.: The pulmonary air meniscus. Radiology **54**, 700 (1950). — Weiner, A.: Occupational bronchial asthma in a baker due to *Aspergillus*. A case report. Ann. Allergy **18**, 1004 (1960). — Weinstein, L., Lewis, R.M.: A case of fungus infection of the vagina. Yale J. Biol. Med. **11**, 85 (1938—1939). — Weller, W.A., Joseph, D.J., Hora, J.F.: Deep mycotic involvement of the right maxillary and ethmoid sinuses, the orbit and adjacent structures. Laryngoscope (St. Louis) **70**, 999 (1960). — Welsh, R.A., Buchness, J.M.: *Aspergillus* endocarditis, myocarditis and lung abscesses. Report of a case. Amer. J. clin. Path. **25**, 782 (1955). — Welsh, R.A., McClinton, L.T.: Aspergillosis of lungs and duodenum with fatal intestinal hemorrhage. Arch. Path. **57**, 379 (1954). — Werch, S.C., Oester, Y.T., Friedemann, T.E.: Kojic acid — a convulsant. Science **126**, 450 (1957). — Whalen, E.J.: Fungous infections of the external ear. J. Amer. med. Ass. **111**, 502 (1938). — Wheaton, S.W.: Case primarily of tubercle, in which a fungus (*Aspergillus*) grew in the bronchi and lung, simulating actinomycosis. Trans. path. Soc. Lond. **41**, 34 (1890). — Whitaker, L.W., Jackson, B.F.: Buccal infection with *Aspergillus niger*. Med. Bull. Veterans' Adm. (Wash.) **10**, 162 (1933). — Wilhite, J.L., Cole, F.H.: Invasion of pulmonary cavities by *Nocardia asteroides*. Amer. Surg. **32**, 107 (1966). — Wilson, B.J.: Toxins other than aflatoxins produced by *Aspergillus flavus*. Bact. Rev. **30**, 478 (1966). — Wilson, B.J., Wilson, C.H.: Toxin from *Aspergillus flavus:* Production on food materials of a substance causing tremors in mice. Science **144**, 177 (1964). — Wilson, J.A.S., Munro, D.D.: Surgical treatment of mycotic infections of the lung. Canad. med. Ass. J. **97**, 166 (1967). — Witter, J.F., Chute, H.L.: Aspergillosis in turkeys. J. Amer. vet. med. Ass. **121**, 387 (1952). — Wogan, G.N.: Chemical nature and biological effects of the aflatoxins. Bact. Rev. **30**, 460 (1966). ~ Isolation, identification, and some biological effects of aflatoxins. In: Halver, J.E., Mitchell, I.A. (ed.), Trout hepatoma research conference papers, P. 121. Research report 70. Washington, D.C.: U.S. Government Printing Off. 1967. — Wogan, G.N., Friedman, M.A.: Effects of aflatoxin B_1 on tryptophan pyrrolase induction in rat liver. Fed. Proc. **24**, 627 (1965). — Wogan, G.N., Newberne, P.M.: Characteristics of the bile duct hyperplastic response to aflatoxin in ducklings. Fed. Proc. **23**, 200 (1964). — Wolf, F.T.: Relation of various fungi to otomycosis. Arch. Otolaryng. **46**, 361 (1947). — Wolf, H., Jackson, E.W.: Hepatomas in rainbow trout: Descriptive and experimental epidemiology. Science **142**, 676 (1963). — Wreden, R.: Recherches sur deux nouvelles especes de vegetaux parasites (*Aspergillus flavescens* et *Aspergillus nigricans*) de l'home. C. R. Acad. Sci. (Paris) **65**, 368 (1867). — Wright, M.L., Anderson, G.W., McConachie, J.D.: Transmission of aspergillosis during incubation. Poultry Sci. **40**, 727 (1961). — Wright, R.E.: Two cases of granuloma invading the orbit due to an *Aspergillus*. Brit. J. Ophthal. **11**, 545 (1927). — Wybel, E.R.: Mycosis of cervical spinal cord following intrathecal penicillin therapy. Report of a case simulating cord tumor. Arch. Path. **53**, 167 (1952). — Yesner, R., Hurwitz, A.: A report of a case of localized broncho-pulmonary aspergillosis successfully treated by surgery. J. thorac. Surg. **20**, 310 (1950). — Young, F.H.: Two cases of neoplasms of the mediastinum with unusual complications. Lancet **210**, 1196 (1926). — Zimmerman, L.E.: *Candida* and *Aspergillus* endocarditis. With comments on the role of antibiotics in dissemination of fungus disease. Arch. Path. **50**, 591 (1950). ~ Fatal fungus infections complicating other diseases. Amer. J. clin. Path. **25**, 46 (1955). ~ Mycotic keratitis. Lab. Invest. **11**, 1151 (1962). — Ziskind, J., Pizzolato, P., Buff, E.E.: Aspergillosis of the brain: Report of a case. Amer. J. clin. Path. **29**, 554 (1958).

Mucormycosis

(Opportunistic Phycomycosis)

ROGER D. BAKER, New Brunswick/New Jersey, USA

With 43 Figures

Definition

Mucormycosis is an acute, frequently fatal, fungus disease characterized by the occurrence of broad, non-septate (coenocytic) hyphae in tissues which tend to grow into arteries and produce thrombosis and infarction. The infection nearly always develops in a person whose resistance is lowered by a metabolic disorder, a blood dyscrasia, corticosteroid therapy, or malnutrition. Mucormycosis complicates diabetic acidosis and leukemia.

Mucormycosis assumes cranial, pulmonary, gastrointestinal, disseminated, cutaneous and focal forms.

In the *cranial form* the fungus enters the nose and spreads rapidly, producing vascular thrombosis and infarcts of the nasal mucous membrane, the hard palate, the orbit, the tissues of the face, and the brain and meninges. In the *pulmonary form* the fungus proliferates in the main bronchi and then enters the hilar tissues and pulmonary arteries to produce infarcts of the lung. In *gastrointestinal mucormycosis* the fungus enters from the lumen and causes gangrene (infarction) of portions of the digestive tract. In *disseminated mucormycosis* the infection spreads from the lungs or head and causes lesions in many organs. Usually in *primary cutaneous mucormycosis* an ulcer develops on the leg of a diabetic patient. In *focal mucormycosis*, a single lesion occurs, as, for example, mucormycosis of the kidney.

The organism most frequently causing mucormycosis is a species of the genus *Rhizopus* (Fig. 1). This organism is ubiquitous in nature and a saprophyte, growing on bread and fruit and causing spoilage (Fig. 2). As it grows in nature or on culture it produces stolons (runners), rhizoids (stick-tights), sporangiaphores (bearers of sporangia), and sporangia (sacs full of red-blood-cell-sized spores). Species of the genera *Mucor* and *Absidia* have also been reported as causal agents of mucormycosis. *Rhizopus*, *Mucor* and *Absidia*, belonging to the family Mucoraceae of the order Mucorales, cause mucormycosis. Because of the component Mucor- in these designations of family and order, the term mucormycosis is appropriate, and should be retained. However, the order Mucorales belongs to the class Phycomycetes characterized by non-septate hyphae, and mucormycosis might be termed opportunistic phycomycosis to indicate that it is a phycomycosis usually found in a person of lowered bodily resistance. The unmodified name phycomycosis includes not only mucormycosis but also two other phycomycosis, subcutaneous phycomycosis and rhinoentomophthoromycosis which are found in persons without predisposing diseases, in Indonesia, Africa and Latin-America (see Chapter by CLARK and EDINGTON 15).

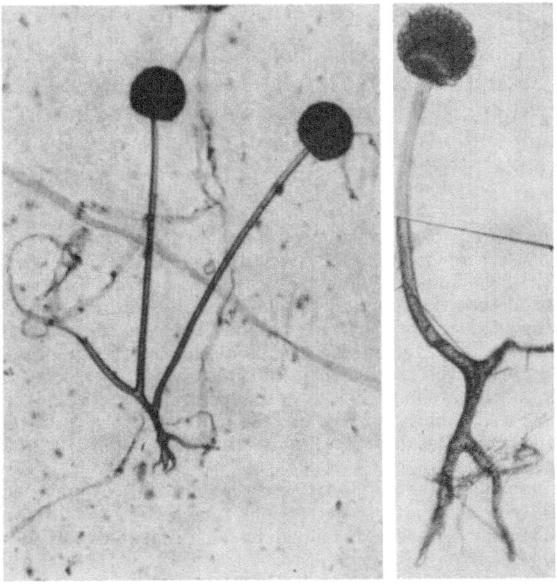

Fig. 1. To left, *Rhizopus oryzae* in a wet preparation, unstained, photographed at 200×. Note the stolon and rhizoids below, the sporangiophores and the sporangia. To right, at 300×, the columella is clearly visible and the spores can just be seen. The rhizoids show clearly. (From BAKER, 1957a)

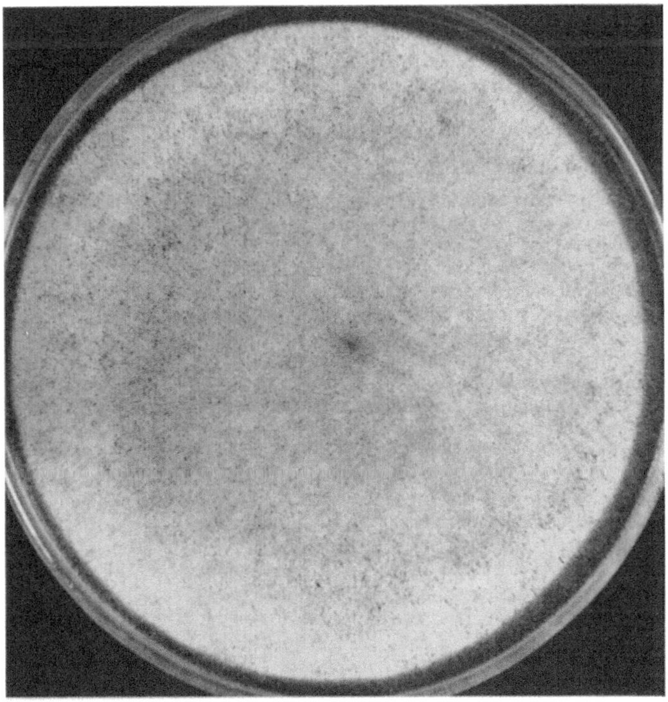

Fig. 2. Culture of *Rhizopus oryzae* two days after inoculation of Sabouraud's glucose agar. The white cottony growth is developing a brownish tinge due to the pigment in the sporangia and endospores

The History of Mucormycosis

BARTHELAT published an account of the pathogenic *Mucors* and the mucormycoses of animals and man in 1903, and reviewed the literature and the published cases up to that time. He illustrated his account with line-drawings of the filamentous mycelium without cross-septa in the hyphae and the spores in sporangia or sacs on the ends of stalks or sporangiophores. He pictured the rhizoids of *Rhizopus*.

These organisms did not frequently infect animal or man, Barthelat wrote. SLUYTER may have described a *Mucor* in a pulmonary cavity in 1847. In 1855 KÜCHENMEISTER described a *Mucor* in a cancer of the lung. His figure shows a sporangium with a non-septate mycelium. There is no indication that the fungus was more than a contaminant. Many of the cases of "mucormycosis" reported before 1900 are dubious because of failure to demonstrate the fungus in tissue sections. The Mucoraceae are ubiquitous saprophytes and frequently contaminate pathologic lesions. Two cases of COHNHEIM had been mentioned by some as mucormycosis, but the organisms are more like those of *Aspergillus*, according to BARTHELAT.

In 1876, FÜRBRINGER published 2 new cases of mucormycosis. In the first the patient died of disseminated cancer and the right lung showed a hemorrhagic infarct with fungus hyphae and a few sporangia.

In the second case there was pulmonary emphysema and chronic gastro-enteritis. At the apex of each lung there was an infarct the size of a nut. Behind the infarct situated in the left apex there was a similar nodule surrounded by an infiltrated and indurated zone which contained mycelium.

In the two cases, the fungus showed the same features, except that in the second the sporangia seemed to be less mature. FÜRBRINGER considered the two fungi as being *Mucor mucedo*, but he added that they looked also like *M. circinelloides*. LINDT thought it was M. *corymbifer*.

I am willing to accept Fürbringer's cases as the earliest reported examples of pulmonary mucormycosis (See Table 3).

In 1885, PALTAUF published the case of a 52-year-old man who died after having presented the symptoms of an infectious disease. Histologically PALTAUF found fungus filaments in the pus of an abscess of the pharynx and larynx, and in lesions of the brain, lung and intestine. There was no culture but PALTAUF thought the fungus was *Mucor corymbifer*.

In 1886 BOSTROEM demonstrated 2 preparations from the tuberculous lungs of a 30-year-old subject with the simultaneous presence of *Aspergillus* and a *Mucor*. These were probably contaminants.

OBRASZOW and PETROV, in 1890, reported the case of a girl who died with lesions of actinomycosis. In her right lung there were foci with fungus filaments with the appearance of *Mucor* or *Penicillium*. No cultures of the parasite were made.

HERLA in 1895 observed a *Mucor* in a pulmonary cavity in a woman who died with a cancer of the liver.

In 1899 PODACK published a treatise on mucormycosis. He reported a case of endothelioma of the right pleural cavity, associated with which there were numerous fungus filaments, sporangia, and sporangiophores in the subpleural pulmonary tissue. These filaments were seen not only in the alveoli but also in the interstitial tissue and in the walls of veins. There were lesions rich in mycelium on the surface of the pleura. PODACK called the fungus *Mucor corymbifer*. I have accepted Podack's case as the third example of pulmonary mucormycosis (See table of pulmonary mucormycosis).

In 1901, LUCET and CONSTANTIN reported a case observed by LAMBRY which came from a woman with a pulmonary affliction. In the sputum of the patient these authors found mycelial filaments which were cultured and characterized as a new *Mucor* to which they gave the name *Rhizomucor parasiticus*. Subjected to treatment with arsenic and potassium iodide, as recommended for aspergillosis, the patient improved and, 2 months later, her condition was considered satisfactory. I have not included this case in the table of pulmonary mucormycosis because there were no tissue studies.

In summary, according to BARTHELAT, if one viewed all the publications relative to mucormycosis in animals and man before 1903, they could be arranged in 4 groups: (1) Those in which a fungus was mentioned and named arbitrarily. The majority were insufficiently described and without scientific value. (2) Those publications concerning tumors from which a fungus had been isolated and considered to be a species of *Mucor*. In these cases the parasite had not been proved to be the cause of the neoplasm. (3) Those reporting mucormycoses of the respiratory tract. The fungus grew on a pre-existing lesion. Nearly all were in autopsies and it was impossible to determine whether the fungus was a saprophyte or a parasite. (4)

The fourth group of publications concerned the cases of primary mycoses (FÜRBRINGER, PALTAUF, LUCET and CONSTANTIN, and PODACK). These cases demonstrated that *Mucors* could play an active infectious role like other living agents of disease.

In his review, BARTHELAT devoted much attention to otomycosis and nasopharyngeal lesions "caused" by *Mucors*. He cited 53 cases of otomycosis of which 2 were attributed to *Mucors*. My own review of these cases suggests that the fungus was growing as a saprophyte in the ear wax and was not causing a lesion.

EMMONS (1964) believes that PALTAUF'S (1885) case of generalized mucormycosis should be accepted as the first genuine example of the disease. He refers to earlier cases, but believes that the diagnosis of these was based upon the growth of an organism in culture rather than the demonstration of a fungus in the tissues. *Mucor*, *Absidia* and *Rhizopus* are common saprophytes which grow rapidly and their spores may contaminate ulcers, collected sputum, exudates and cultures made from clinical specimens.

If Fürbringer's cases of pulmonary mucormycosis (1876) were the first, then in 1969 the disease would be 93 years old. Throughout this time mucormycosis has been recognized as a very rare disease of sporadic occurrence. Though still rare, it is appearing more frequently since 1943.

My attention was first drawn to the disease by an account of 3 fatal cases in diabetic patients, reported in the Johns Hopkins Hospital Bulletin by GREGORY et al., in 1943. It was an amazing story for a fungus disease. The course of the disease was rapid, only two to five days to death, and the lesions of the head and brain were infarcts caused by the growth of the broad, coenocytic (non-septate), hyphae into arteries where they formed thrombi.

The first case of pulmonary mucormycosis in the United States was reported in 1948 by BAKER and SEVERANCE. This was in a white female child, 3 years old, with diabetes, who at postmortem had hemorrhagic lesions in the lungs and infarcts in the spleen and brain, all due to vascular thromboses caused by broad coenocytic hyphae. It appeared that the pulmonary lesions were secondary to a primary cranial mucormycosis.

ZIMMERMAN in 1955 called attention to the manner in which mucormycosis complicates other diseases. MOORE et al. in 1949 published the first well-documented case of mucormycosis of the digestive tract. In 1959 HUTTER published a comprehensive review of mucormycosis and emphasized this mycosis as a complication of the therapy of leukemia.

Mycology

Among the 255 reported cases of mucormycosis the results of cultures were positive in 43 cases (Table 1). *Rhizopus* was reported as having been cultured in 32 instances, *Mucor* in 10, *Absidia* in 2, and a fungus of the Mucorales order in 1. *Rhizopus* predominated in the cranial and gastrointestinal cases, and *Mucor* in the pulmonary cases.

The genera *Rhizopus*, *Mucor* and *Absidia* belong to the family Mucoraceae, the order Mucorales, the class Phycomycetes (non-septate hyphae), the superclass Eumycetes (true fungi), the subphylum Fungi (absence of chlorophyll) and the phylum Thallophyta (comparatively simple plants) (PROCKOP and SILVA-HUTNER).

The class Phycomycetes contains the orders Mucorales and Entomophthorales. This chapter on mucormycosis deals with infection by the family Mucoraceae of the order Mucorales. The chapter on subcutaneous phycomycosis and rhino-entomophthoromycosis deals with infections of the family Entomophthoraceae of the order Entomophthorales.

The class Phycomycetes is characterized by coenocytic (non-septate) hyphae, and by zygospores.

The genus *Rhizopus* is characterized by sporangiophores which arise singly or in clusters opposite rhizoids and bear spherical sporangia with amber spores between sporangial wall and columella (Fig. 1).

The small size and shape of the rhizoids, sporangia (30 μ—70 μ), sporangiophores (300 μ—600 μ long by 5 μ—12 μ wide), and the small columellae (10 μ—25 μ) and spores (4 μ—8 $\mu \times$ 3 μ—7 μ), and their ridged markings, identified the species of the fungus derived from the case of cephalic mucormycosis of PROCKOP and SILVA-HUTNER as *Rhizopus arrhizus*. This was confirmed by Dr. H. HESSEL-

Table 1. *Fungi cultured from 43 cases of mucormycosis diagnosed on the basis of characteristic non-septate hyphae in tissues. From 1 gastrointestinal case 3 cultures were obtained*

	Total	Cranial	Pul-monary	G. I.	Disse-minated	Cu-taneous	Focal
Rhizopus cultures (32)							
Rhizopus oryzae	9	8	1	—	—	—	—
Rhizopus arrhizus	5	4	—	—	1	—	—
Rhizopus nigricans	5	2	—	3	—	—	—
Rhizopus rhizopodiformis	1	—	—	—	—	1	—
Rhizopus microsporus	4	—	—	4	—	—	—
Rhizopus	8	8	—	—	—	—	—
Mucor cultures (10)							
Mucor corymbifer(a)	2	—	2	—	—	—	—
Mucor hiemalis	1	—	—	1	—	—	—
Mucor ramosissimus	1	—	—	—	—	1	—
Mucor	6	2	3	—	1	—	—
Absidia cultures (2)							
Absidia lichtheimi	1	1	—	—	—	—	—
Absidia ramosa	1	—	—	—	—	—	1
Non-specified cultures (1)							
Fungus of Mucorales order	1	1	—	—	—	—	—
Total	45	26	6	8	2	2	1

TINE of the Northern Regional Research Laboratories, Peoria, Illinois, where the culture was entered as No. A-11, 715 in his collection.

Rhizopus grows rapidly and fills the petri dish in 5 days with a dense, cottony aerial mycelium which is at first white, then dark gray (Figs. 2 and 3). *Rhizopus arrhizus* was cultured from the case of HARRIS which is presented in detail later in this chapter. In case 2 of BAUER et al. *Rhizopus oryzae* was cultured and identified as such by Dr. HESSELTINE.

In the case of BAKER et al. (1962) *Rhizopus rhizopodiformis* was the species identification given by Dr. HESSELTINE. In his thesis, Hesseltine cites two instances in which this species was recovered from a mycosis in animals, in one case from a dog and in another from a horse. *R. rhizopodiformis* has very close affinities with the genus *Mucor* and is distinguished from other species of *Rhizopus* chiefly by the branched sporangiophores, the presence of occasional pyriform columellae, and the lack of striations in the walls of the sporangiospores.

It is apparent that the species identification of *Rhizopus* is beyond the scope of the usual hospital or even medical school microbiologist or mycologist.

The species terminology of *Rhizopus* apparently varies according to the expert who makes the identification, for none of the species mentioned above are included in the list of species recovered by NEAME and RAYNER from cases of mucormycosis involving the digestive system. The Commonwealth Mycological Institute of England reported *Rhizopus microsporus* van Tiegh from 3 of their cases. *Rhizopus stolonifer* (Syn. *R. nigricans*) from 2 cases, and *Rhizopus stolonifer*, *Rhizopus microsporus* van Tiegh, and *Mucor hiemalis* Wehm from 1 case.

The genus *Mucor* is also a rapidly growing fungus which fills the petri dish in days with an abundant growth of floccose aerial mycelium which is at first white, then dark gray (CONANT et al., 1954). The non-septate vegetative mycelium of *Mucor* gives rise to numerous unequalled length sporangiophores which branch irregularly and bear terminal globose, spore-filled sporangia. The wall of the

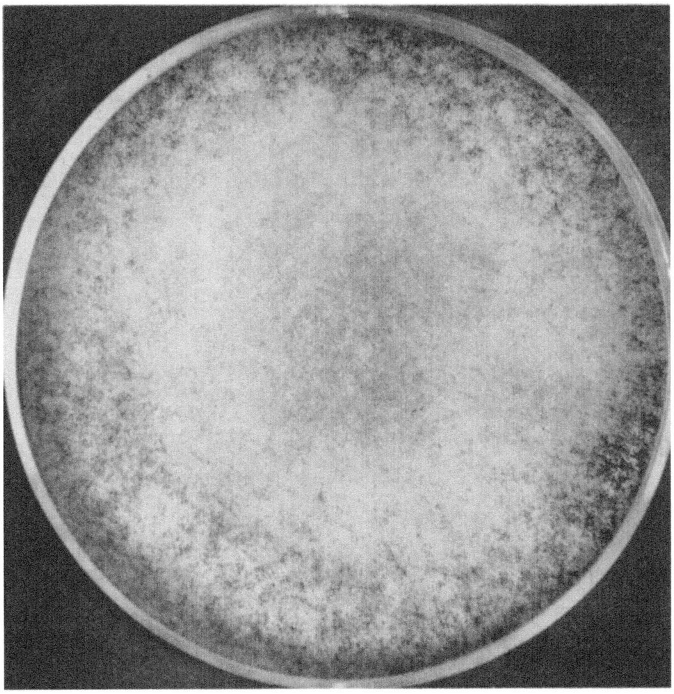

Fig. 3. Culture of *Rhizopus oryzae* 4 days after inoculation of Sabouraud's glucose agar. The aerial mycelium has become browner. *Rhizopus* will grow on to the cover of the Petridish attaching itself by the rhizoids

sporangium is easily broken, scattering the elliptical spores, leaving a fragment of the sporangial wall (collarette) at the base of the spherical columella. This structure is the swollen end of the sporangiophore which extends into the sporangium and is seen only when the sporangial wall ruptures and the mass of spores are dispersed.

The genus *Absidia* has rhizoids which resemble those of *Rhizopus*, but come from the stolons between the sporangiophores, rather than opposite them as is characteristic of *Rhizopus*.

EMMONS et al. give no description of *Mucor* but include a description of *Absidia corymbifera* (Cohn) SACCARDO and TROTTER (1912), and give as the synonymy: *Mucor corymbifer* 1884; *Lichtheimia corymbifera* VUILLEMIN (1903). *A. corymbifer* grows rapidly, they write, covering a Petri dish in a few days. The colony is a coarse, gray, wooly, coenocytic mycelium with numerous upright branched sporangiophores. Sporangia are pearshaped and are borne on aerial hyphae *between* systems of branching root-like hyphae which are called rhizoids. With a lens the rhizoids can be seen attached to the inner wall of a culture tube. The origins of the sporangiophores *between* these rhizoids are more difficult to find.

I wonder, as I record the diagnosis of *Rhizopus*, *Mucor*, and *Absidia* in Tables 2—7, whether the distinctions between the three genera have always been clear to the person who diagnosed the fungus. Certainly it would be easy to diagnose as *Rhizopus* a fungus with rhizoids, without regard to whether these "rootlets" came from an origin opposite the sporangiophores or between them.

In the subsequent reporting of new cases of mucormycosis, I hope that cultures may be obtained more frequently, that the fungus will be fully described in the

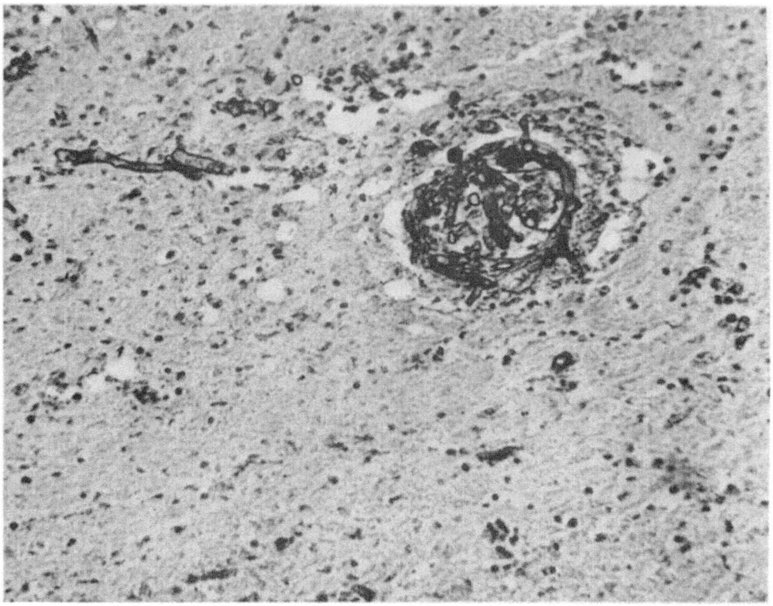

Fig. 4. The coenocytic (non-septate) hyphae of mucormycosis in the brain of one of the cases of GREGORY et al. (1943). Note the growth in the vessel wall and lumen, the right-angle branching and the devitalized (infarcted) cerebral tissue. The hyphae can be confused with capillaries (H & E, × 200)

case report, and that the identification of the fungus will be corroborated by an expert in the field of the Phycomycetes.

EMMONS mentions another problem in connection with the identification of Phycomycetes and in relating them to disease in man and animals. The order Mucorales contains not only the family Mucoraceae (which includes *Rhizopus*, *Mucor* and *Absidia*), but also the families Cunninghamellaceae and Mortierellaceae; and phycomycete diseases in addition to mucormycosis, subcutaneous phycomycosis and rhinoentomophthoromycosis have been described in animals. For man, Emmons mentions that one case of lymphosarcoma reported by HUTTER (1959) had disseminated mycosis and *Cunninghamella elegans* was cultured. *Mortierella* has also been isolated from cutaneous ulcers, according to EMMONS.

Appearance of Mucoraceae in Tissues

The members of the family Mucoraseae are seen in paraffin sections stained with hematoxylin and eosin as broad (6—50 μ) coenocytic (non-septate) hyphae which branch at right angles or nearly at right angles (Fig. 4). *Aspergillus* is easily confused with the Mucoraceae but the hyphae have cross septa, are narrower, and branch at an acute angle. It may grow in colonies in the tissues. *Candida* may be confused with the Mucoraceae, but the filaments are much narrower, are often composed of pseudohyphae, and frequently accompanied by small oval spores.

The Mucoraceae can be distinguished from the Entomophthoraceae (*Basidiobolus meristosporus* and *Entomophthora coronata*) because the latter have eosin-staining hyaline material around the hyphae in tissue.

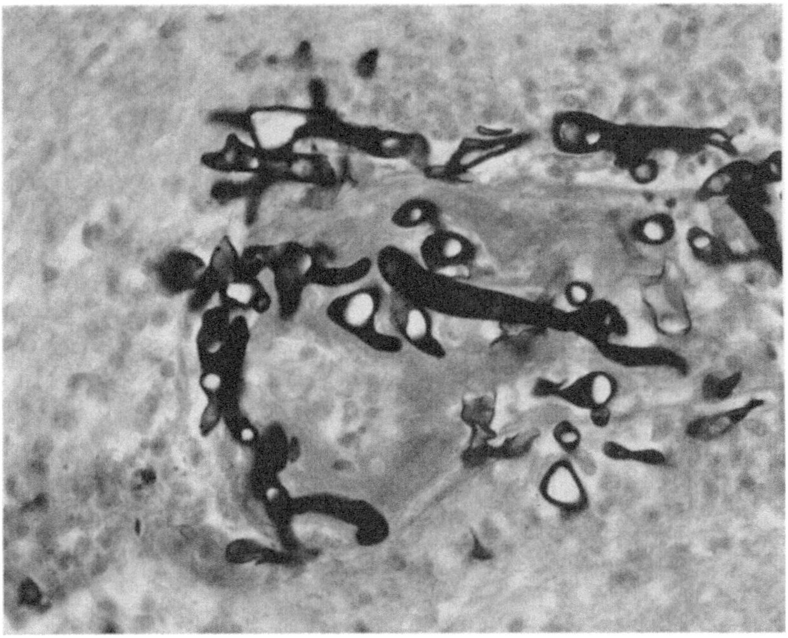

Fig. 5. Hyphae of mucormycosis in and about a thrombosed cerebral blood vessel. Grocott-Gomori methenamine silver stain (G.M.S.), × 500. (From BAKER, 1957a, Fig. 5)

The circumstances of the case also assist in the differential diagnosis. The Mucoraceae are usually found in vessel walls or infarcts within the tissues of a person suffering from a debilitating disease such as diabetes or leukemia. They are found in the internal organs, and cause fatalities. The Entomophthoraceae may be found in an otherwise healthy individual who has a localized subcutaneous lesion or one which involves the air sinuses or bony structure of the face. These infections are usually non-fatal. Infections with the Entomophthoraceae have been observed only in Indonesia, Africa and Latin America, with a few exceptions, whereas the infections with the Mucoraceae have been observed in all parts of the world and where predisposing diseases and conditions exist favoring the invasion of a usually non-pathogenic or opportunistic fungus.

The organisms of mucormycosis are fairly well demonstrated in tissue with hematoxylin and eosin. The Gomori methenamine silver method is much better (BAKER, 1957b). If study with hematoxylin and eosin suggests an organism of mucormycosis, the Gomori methenamine silver stain with a green or blue counterstain should be obtained (Fig. 5). The indigo-carmine-picric acid stain (BAKER, 1956) stains the hyphae of mucormycosis in tissues, but is less valuable than the G. M. S. stain. The periodic acid Schiff and hematoxylin stain (P. A. S. H.) does not color the organisms of mucormycosis well, nor does the Gram stain.

The hyphae of mucormycosis often present an internal structure which is separated slightly from the delicate wall. Rarely the hyphae present swellings or enlargements. The hyphae when cut crosswise may suggest spores to the inexperienced observer.

Usually only hyphae are seen in tissues, but sporangia have been reported in several cases. BAUER et al. found sporangia in an air sinus at postmortem and LA TOUCHE et al. demonstrated sporangia in the mucosa of the left middle tur-

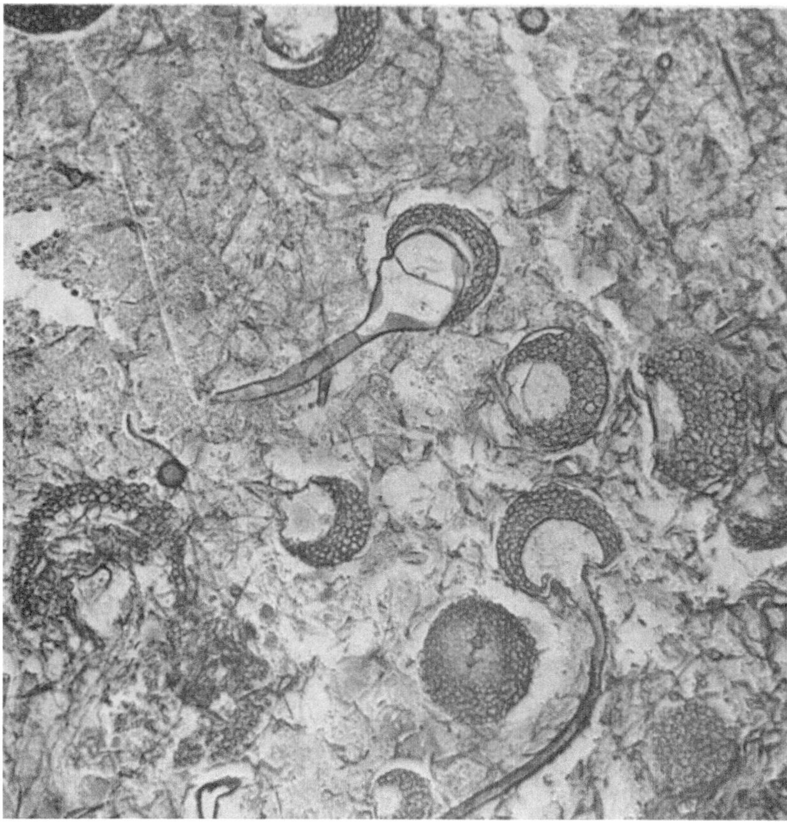

Fig. 6. Phycomycete organism causing mucormycosis in the left middle turbinate of the case of LA TOUCHE et al. (1963) (Fig. 2). Note the sporangia, the contained spores, the columellae and the sporangiophores (H & E, × 400)

binate (Fig. 6). Possibly the formation of sporangia is dependent on proximity to a cavity containing air.

The tables show that the disease is world-wide, with a preponderance of cases reported from the United States. Whether the *Rhizopus* organism is more prevalent in the United States or not, I cannot say, for the greater frequency in the United States may be due to better diagnosis, to more reports in medical journals, to more diabetic and treated leukemic patients.

Epidemiology

Mucormycosis is clearly not transmissible from person to person. The organisms live on the ground, in houses, and are abundant on old bread and spoiling fruit. The organisms are saprophytes which must be inspired frequently by normal persons. In the person with a predisposing condition, such as diabetic acidosis or leukemia, the fungus proliferates in the mucosa of the nose, air sinuses or bronchi and spreads locally. *In vitro* studies indicate that the serum from a normal person inhibits the growth of *Rhizopus* (GALE and WELCH). In the several cases of secondary infection of cutaneous burns, the burn, for an unknown reason, acted like the abnormally receptive mucosa of the nose or bronchus.

Predisposing Diseases and Conditions

Mucormycosis is usually a secondary or opportunistic mycosis. The normal human organism resists infection with the ubiquitous *Rhizopus*, *Mucor* and *Absidia*. How the normal body effects this, and what the essence of the change is which permits the fungus to enter, is not understood.

Diabetes is the most frequent antecedent disease in cranial and focal mucormycosis, leukemia in the pulmonary and disseminated forms, and malnutrition in the gastrointestinal form. Diabetes may also predispose to the pulmonary, gastrointestinal and disseminated forms of the disease. Diabetes and blood diseases are the antecedent conditions in most of the reported cases of human mucormycosis. Why was mucormycosis not reported more frequently before 1943, the date of publication of the paper by GREGORY et al. concerning diabetic patients? Why was the pulmonary and disseminated form not reported more frequently before the papers of BAKER, in 1957, and of HUTTER in 1959, which emphasized diseases of the hematopoetic system, especially leukemia, as antecedent conditions?

Apparently mucormycosis was not a frequent disease before the publication dates just cited. The organisms in the tissues are readily seen in hematoxylin and eosin sections of postmortem or surgically removed tissues, and competent histopathologists have been active since 1885.

At Johns Hopkins Hospital the autopsy reports of 1902 are on file with protocols by OPIE, WHIPPLE and MCCALLUM. At Duke Hospital the files of autopsy protocols date from 1930 and there was minute review of each case by FORBUS, SPRUNT and BAKER. In Boston, Chicago, Philadelphia, and in medical centers of Europe, particularly in Germany, mucormycosis could not have gone unrecognized at postmortem examination as a complication of diabetes or leukemia.

In the three cases of cranial mucormycosis reported by GREGORY et al. an individual with diabetes mellitus developed diabetic acidosis, then mucormycosis of the face and head, and died with cerebral mucormycosis within two to five days.

In my tabulation of 255 reported cases of mucormycosis I have listed diabetes 46 times and diabetic acidosis 46 times. Therefore diabetes was present in 36% of the cases of mucormycosis. Certainly the well-regulated diabetic is less likely to develop this severe mycosis than the poorly-regulated one. Probably the acidotic complication of diabetes was necessary to permit the infection but had not been recognized in the patients. HUTTER (1959) speaks of steroid diabetes in connection with patients dead of pulmonary mucormycosis who had received corticosteroids in the therapy of leukemia.

It is a curious observation that *blood diseases* are complicated by the pulmonary and disseminated forms of mucormycosis while diabetes mellitus plays a minor role in these forms of the fungus infection. The blood dyscrasias predisposing to mucormycosis were as follows: leukemia, 39; neutropenia, 8; aplastic anemia, 3; lymphosarcoma and Hodgkin's disease, 4; myeloma, 2. The leukemias were acute or chronic, myelogenous or lymphogenous. Corticosteroids were listed as predisposing factors in 30 of these disorders of the hematopoetic system.

In my study of fatal mycoses complicating leukemia (BAKER, 1962), I concluded that neutropenia, drug therapy, hypogammaglobulinemia and splenectomy were probably factors accessory to the leukemia in permitting the development of the mycosis. In leukemia a neutropenia may develop as a phase in the course of the disease or as a result of anti-leukemic drugs, and the mycosis becomes established after the neutropenia has been severe and prolonged (Fig. 7). *Corticosteroid therapy*, as I have demonstrated in normal monkeys, has a powerful spreading factor for *Rhizopus*, and it is probable that leukemia, neutropenia and

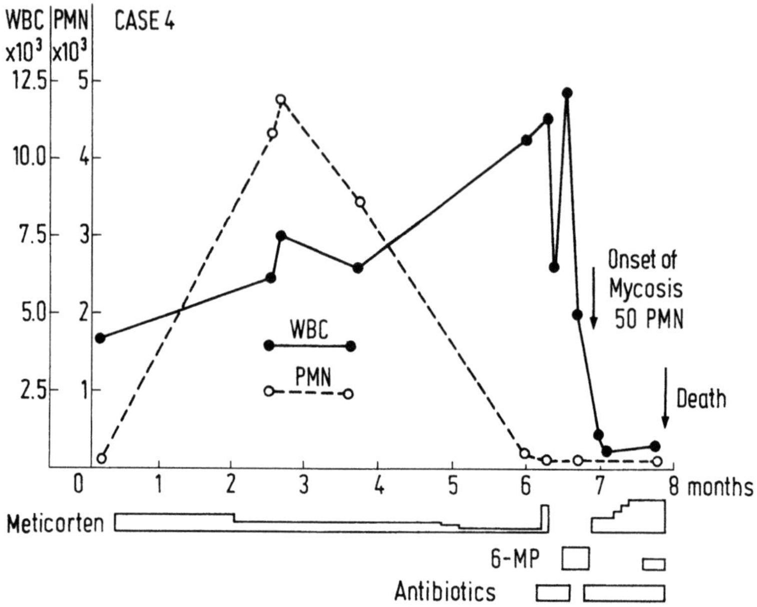

Fig. 7. Clinical plot of mucormycosis complicating a case of acute lymphocytic leukemia. The onset of the mycosis occurred when there was severe neutropenia and after corticosteroids had been administered for a long time

corticosteroid therapy provide an ideal triad of circumstances for the development of complicating mucormycosis (Baker, 1964).

Renal disease or renal insufficiency were mentioned as possible predisposing conditions in 16 cases: glomerulonephritis 3 times, acute renal failure and renal insufficiency 5 times, hydronephrosis with uremia twice; renal cortical necrosis once, the nephrotic syndrome once and uremia alone in 4 cases. Uremia also complicated some of the cases with renal diseases. Renal disease appears to lower resistance by a metabolic change induced by uremia.

Malnutrition in the form of kwarshiorkor was given as a predisposing factor in the development of the gastro-intestinal form of mucormycosis in many of the cases reported from South Africa by Neame and Raynor.

Diarrhea, or diarrhea and vomiting, with acidosis, was thought to have been the forerunner of some of the cranial and gastrointestinal cases of mucormycosis.

Extensive *body burns* preceded the development of disseminated mucormycosis in 4 cases (Baker, 1956; Rabin et al.; Straatsma et al.) and focal mucormycosis, of the hand only, in 1 case (Foley and Shuck). In the cases of Rabin et al. the bones of the face became necrotic where the fungus had proliferated.

Of the miscellaneous preceding or associated conditions mention was made of carcinomas, postoperative status, congenital heart disease, amebiasis, cirrhosis, prematurity, hepatic failure, infectious hepatitis, pancreatitis, septic abortion, and fracture.

In all of the anatomical forms of mucormycosis there were between 5 and 50% of the cases in which no predisposing condition or associated disease appeared to be present. Apparently *Rhizopus* and possibly *Mucor* and *Absidia* can sometimes act as primary pathogens rather than secondary invaders or opportunists.

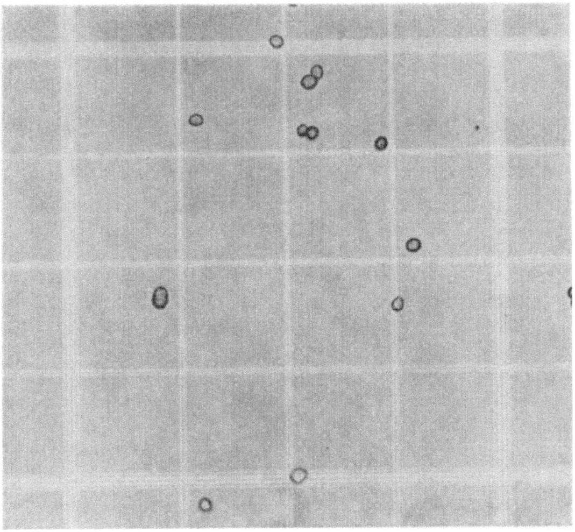

Fig. 8. Endospores of *Rhizopus oryzae* from a ruptured sporangium, in a blood-counting chamber, being enumerated for intravenous animal inoculation. They resemble erythrocytes in size and shape

Experimental Mucormycosis

A large number of reports on experimental mucormycosis have appeared since 1955 (Fig. 8). These have been directed toward the hypothesis that the predisposition to mucormycosis is a metabolic disturbance (Bauer et al., 1956).

Acute alloxan diabetes permitted a fulminating mucormycosis to develop in rabbits (Bauer et al., 1955a, b; Elder and Baker, 1956; Baker et al., 1956). However, chronic alloxan diabetes, like human non-acidotic diabetes, failed to modify host resistance (Schofield and Baker, 1956). In acute alloxan diabetes the inflammatory response is by no means paralyzed (Schauble and Baker, 1957) though phagocytosis may be diminished (Wertman and Henney, 1962). The induction of acute alloxan diabetes activated quiescent mucormycotic granulomata in rabbits (Sheldon and Bauer, 1958).

The role of predisposing factors in experimental fungus infections is reviewed by Sheldon and Bauer (1962). Other papers on the subject are by Bauer et al. (1956) and Bauer and Sheldon (1957), Bauer et al. (1957), Osswald and Seeliger (1958), Sheldon and Bauer (1959), and Paplanus and Sheldon (1965). But experimental mucormycosis can be produced in the healthy rat (Josefiak and Foushee, 1958).

The most spectacular indication of the spread of *Rhizopus* in the monkey pretreated with the corticosteroid prednisolone was displayed in the form of kidneys swollen by infection and gastric ulcers due to vascular thromboses. This contrasted with the absence of these findings in the monkeys injected with *Rhizopus* intravenously but not pretreated with prednisolone (Baker et al., unpublished).

To round out the subject of experimental mucormycosis, mention is made here of the preventive and curative effect of Amphotericin B in experimental *Rhizopus oryzae* infection (Chick et al., 1958a, b).

Cranial Mucormycosis

Cranial mucormycosis begins in the nasal mucosa and extends to the palate, the paranasal sinuses, the orbit, the face and the brain (Table 2). The manifestations of this progression of the disease have been termed the classical central nervous system syndrome of mucormycosis. This was a uniformly fatal condition prior to 1955, as observed in the 3 cases of Gregory et al., Le Compte and Meissner, Baker and Severance, Wolf and Cowen, Stratemeier, Kurrein, Martin et al., the 2 cases of Bauer et al., and the case of Gunson and Bowden, 13 cases reported from 1943 to 1955.

Table 2. *Cranial Mucormycosis (107 cases)*

Authors and Case Number	Location of Case Age, Sex, Race	Predisposing Factors	Anatomic Involvement	Cultures	Duration	Treatment	Status of Patient
OPPE (1897)	Dresden/Germany 37, M, W	None	Base of brain	No	?	None	Dead (nodule found at autopsy)
HAFSTROM et al. (1941)	Stockholm/Sweden 69, M, W	None	Brain, left central	No	10 years	Surgical removal	Survived
GREGORY et al. (1943) Case 1	Baltimore/USA 43, F, N	Diabetic coma	Orbit, face, brain, carotid artery, cavernous sinus	No	4 days	For shock and acidosis	Dead, mucormycosis
— Case 2	Baltimore/USA 52, F, N	Diabetic coma Chronic sinusitis	Nose, ethmoid sinus, eye, brain	No	5 days	For diabetic coma. Enucleation of eye	Dead, mucormycosis
— Case 3	Baltimore/USA 75, M, W	Diabetes	Eye, brain, cervical cord	No	2 days		Dead, mucormycosis
LECOMPTE and MEISSNER (1947)	Boston/USA 57, M, W	Diabetes Hemochromatosis	Orbit, brain	No	2 days	Insulin	Dead, mucormycosis
BAKER, SEVERANCE (1948) BAKER (1956), Case 4 BAKER (1957), Case 2	San Antonio/USA 3, F, W	Diabetic coma	Left orbit, brain, left internal carotid artery, lung	No	10 days	Insulin	Dead, mucormycosis
WOLF, COWEN (1949) BURNS (1959), Case 1	New York/USA 42, M, N	Diabetes	Left ethmoid, orbit, eye, brain, cavernous sinus	No	10 days	Antibiotics, anticoagulants	Dead, mucormycosis
STRATEMEIER (1950)	Philadelphia/USA 32, F, N	Diabetic coma	Nose, left turbinate, orbit, brain	No	2 days		Dead, mucormycosis
WADSWORTH (1951)	New York/USA 10, M	None	Right bulbus oculi	No	4 years and 8 m.	Enucleation	Well
KURREIN (1954)	Worcester/Engl. 5, M	Glomerulonephritis, pneumonia, diarrhea	Left frontal lobe of brain, meninges	No	14 days	Penicillin	Dead, mucormycosis
MARTIN et al. (1954)	Texas/USA 2.5 mos., M, W	Diarrhea, acidosis	Left orbit, face, brain, internal carotid artery, esophagus, lungs, pancreas, marrow	No	8 days	Antibiotics	Dead, mucormycosis

BAUER et al. (1955) Case 1	Atlanta/USA 25, F, W	Diabetes	Left orbit, brain	No	5 days	Intravenous fluids, insulin	Dead, mucormycosis
— Case 2	Atlanta/USA 40, M, N	Diabetes	Ethmoid, right orbit, brain, pituitary, carotid artery	*Rhizopus oryzae*	2 days	Insulin, antibiotics	Dead, mucormycosis
GUNSON and BOWDEN (1955)	Toronto, Canada 8, F, W	Chronic glo-merulonephr., uremia	Brain, left internal carotid artery, left middle cerebral artery	No	10 days	Not stated	Dead, mucormycosis
HARRIS (1955) GEORGIADE et al. (1956), Case 1 BAKER (1957), Case 2 DILLON and SEALY (1962), Case 3	Durham/USA 14, F, N	Diabetes	Palate, left paranasal sinuses, left orbit	*Rhizopus arrhizus*	2 mos.	Insulin, iodides, desensitization	Well in 1968
FOUSHEE and BECK (1956)	Winston-S./USA 65, F, W	Cholecystect. Carc. of kidney	Right cheek, brain, meninges	No	4 days	Aureomycin	Dead, mucormycosis
GEORGIADE et al. (1956), Case 2	Durham/USA 16 mos., M, W	Diabetes	Left side of palate	No	6 mos.	Plastic repair	Living
KEYE and MAGEE (1956), Case 1	St. Louis, Mo./USA Infant	—	Brain, duodenum	—	—	—	Dead, mucormycosis
— Case 2	St. Louis, Mo./USA Infant	—	Brain	—	—	—	Dead, mucormycosis
— Case 3	St. Louis, Mo./USA Adult	—	Brain	—	—	—	Dead, mucormycosis
Clinicopathologic conference (1956)	Sydney/Australia 53, F	Diabetic acidosis	Right eye, brain, lung	No	10 days	Insulin, penicillin, fluids	Dead, mucormycosis
BAKER (1957) Case 1	Durham/USA 62, M, W	Cirrhosis	Sphenoid and ethmoid sinuses, left eye, brain, internal carotid artery	No	14 days	Penicillin	Dead, mucormycosis
— Case 3	Durham/USA Adult, M, W	Diabetic acidosis	Left ethmoid, orbit	*Rhizo-pus*	—	Insulin	Cured
JACKSON and KARNAUCHOW (1957)	Ottawa/Canada 20 days, F	Congenital heart disease	Brain, meninges, ileum	No	13 days	Descrysticin, chloramphenicol	Dead, mucormycosis

Table 2. Continuation

Authors and Case Number	Location of Case Age, Sex, Race	Predisposing Factors	Anatomic Involvement	Cultures	Duration	Treatment	Status of Patient
McCALL and STROBOS (1957)	Winston-S./USA 18, F, W	Diabetes	Left eye, face	*Rhizopus oryzae*	—	Enucleation of left eye	Living, with hemiparesis and aphasia caused by mucormycosis
MERRIAM and TEDESCHI (1957)	Boston/USA 57, M, W	Diabetic acidosis	Orbit, left eye, cavernous sinus	*Rhizopus oryzae*	14 weeks	Insulin, enucleation, antibiotics	Cured
BRYAN et al. (1958)	Iowa City/USA 34 months, F	Diabetic acidosis	Right orbit, palate, face, ear, internal carotid artery, brain, intestines, lungs, aorta	No	6 days	Insulin, hydrocortisone	Dead, mucormycosis
DWYER and CHANGUS (1958)	Norwalk/USA 46, F, W	Diabetic acidosis	Left facial gangrene, nose, palate, sinuses, orbit, brain, cavernous sin.	No	9 days	Insulin, antibiotics	Dead, mucormycosis
SMITH et al. (1958)	San Francisco/USA 38, M, W	Diabetes	Right orbit, both internal carotid arteries, cavernous sinuses, sphenoid sinus	No	9 days	Antibiotics	Dead, mucormycosis
SMITH and KIRCHNER (1958), Case 1 GABRIELE (1960), Case 2	New Haven/USA 72, M, W	Diabetes	Right facial gangrene, turbinates, sinuses, orbit, internal carotid, brain	No	20 days	Antibiotics	Dead, mucormycosis
SMITH and KIRCHNER (1958), Case 2 GABRIELE (1960), Case 2	New Haven/USA 6.5 years	Diabetes	Turbinates, sinuses, left orbit, internal carotid artery, brain, pituitary	No	4 days	Antibiotics	Dead, mucormycosis
SMITH and KIRCHNER (1958), Case 3 GABRIELE (1960), Case 3	New Haven/USA 50, F, W	Diabetes	Turbinates, oral cavity, maxillary sinus and bone, right orbit	No	44 days	Antibiotics, mycostatin, Amphotericin B	Dead, mucormycosis
BORLAND (1959)	Kansas City/USA 7 mos., M, N	Diarrhea, acidosis	Face, ethmoid, right orbit, brain	No	7 days	Hydration	Dead, mucormycosis

Burns (1959) Case 2	New York/USA 60, M, N	Diabetic acidosis	Ethmoid, left orbit, brain	*Rhizopus*	13 days	Intravenous Amphotericin B	Dead, mucormycosis
Dolman and Herd (1959)	Vancouver/Canada 15, F	Acute pancreatitis, renal cortical necrosis, diabetes	Left face, orbit, cavernous sinus, internal carotid artery, brain, pituitary	No	10 days	Insulin, Levophed, hydrocortisone, antibiotics	Dead, mucormycosis
Faillo et al. (1959)	Fort Benning/USA 50, M, W	Diabetes	Left face, maxillary sinus, orbit	No	2 mos.	Operative	Living
Lie-Kian-Joe et al. (1959)	Indonesia 15, F	Diabetes	Orbit, left nostril, turbinate, brain	No	16 days	—	Dead, mucormycosis
Smith and Yanagasawa (1959) Gabriele (1960), Case 4	New Haven/USA 39, M, N	Diabetes	Right eye and sinus, turbinate, carotid artery, brain	*Rhizopus*	6 days	Opening of air sinuses	Dead, mucormycosis
Long and Weiss (1959)	Washington/USA 24, F	Diabetes	Left orbit, brain, pituitary, vessels of lung	No	5 days	Insulin, intravenous fluids, Levarterenol	Dead, mucormycosis
Muresan (1960)	Bucharest/R. 18, M	None	"Tumor" of left parietal lobe of brain	No	Many weeks	Surgical removal	Well 2.5 years postoperative
Neame (1960), Case 7	Durban/S. Africa 3 weeks, M	Gastroenteritis	Inferior surface, frontal lobe, brain	No	7 days	—	Dead, mucormycosis
— Case 11	Durban/S. Africa 27, M	Uremia, chron. glomerulonephritis	Inferior surface, frontal lobe, brain	No	2 days	—	Dead, mucormycosis
Banker (1961), Case 2	Boston/USA 3.5 years, M	Nephrotic syndrome	Orbit, sinuses, internal carotid artery, brain	*Mucor*	1 month	Not stated	Dead, mucormycosis
Berk et al. (1961)	Oakland/USA 49, M	Chronic pancreatitis with diabetes	Left nasal cavity, orbit, sinuses, internal carotid artery, cavernous sinus	*Rhizopus*	16 days	U-0178	Dead, mucormycosis
Gass (1961) Case 1	Baltimore/USA 44, M	Diabetic acidosis	Nose, sinus, left orbit	No	2 months	Amphotericin B, exenteration of left eye and sinuses	Well
— Case 2	Baltimore/USA 13, M, N	Diabetic acidosis	Nasal tissues, frontal sinuses, orbit, brain	No	—	—	Dead, mucormycosis

Table 2. Continuation

Authors and Case Number	Location of Case Age, Sex, Race	Predisposing Factors	Anatomic Involvement	Cultures	Duration	Treatment	Status of Patient
HOAGLAND et al. (1961), Case 1	Fort Benning/USA 50, M	Diabetes, pancreatitis	Palate, right orbit, antrum, sphenoid	No	1 month	Extirpation of right face	Well
— Case 2	Fort Benning/USA 33, M Amer. Indian	Diabetic acidosis	Left palate, maxillary and ethmoid sinuses	*Rhizopus nigricans*	18 days	Extirpation of left face; Amphotericin B	Well
TURPIN et al. (1961)	Paris/France 13, M	Acute leukemia	Brain	No	10 days	Transfusions	Dead, mucormycosis
WASSERMAN et al. (1961), Case 1	Richmond/USA 59, M, N	Diabetic acidosis	Palate, sinuses, orbits, internal carotid artery, brain (no autopsy)	*Rhizopus*	7 days	Insulin, Amphotericin B	Dead, mucormycosis
— Case 2	Richmond/USA 45, M, N	Diabetic acidosis	Palate, sinuses, left orbit, internal carotid artery, brain	No	4 days	Insulin, fluids, Amphotericin B	Dead, mucormycosis
— Case 3	Richmond/USA 65, F, N	Diabetic acidosis	Nose, palate, sinuses, right orbit	No	42 days	Fluids, insulin, iodides	Living, with defect
BANK et al. (1962)	Israel 60, M	Diabetic acidosis	Right orbit	*Rhizopus arrhizus*	—	Insulin, antibiotics	Survived
EGGENSCHWILER (1962)	Basel/Switzerland 48, M	None	Right antrum with bony involvement, ethmoids	*Absidia lichtheimi*	Chronic	Operation: Caldwell-Luc	Well
LANDAU and NEWCOMER (1962)	Los Angeles/USA 15, F, W	Diabetic acidosis	Palate, left paranasal sinuses, orbit	*Rhizopus arrhizus*	44 days	Amphotericin B	Cured
RINALDI and ASHBY (1962)	Newport News/USA 53, M, N	Diabetes	Face, right orbit, palate (no autopsy)	No	12 days	Insulin	Dead, mucormycosis
STRAATSMA et al. (1962), Case 1	AFIP Accession No. 575574; F	Diabetic acidosis	Classic C.N.S. syndrome	—	—	—	
— Case 2	AFIP Accession No. 580752 21, F, N	Hodgkin's disease	Classic C.N.S. syndrome	—	—	—	

— Case 3							
Reeves et al. (1965)	AFIP Accession No. 694863 St. Barbara/USA 30, F, N	Diabetic acidosis, renal insufficiency	Left orbit, face, ethmoids, internal carotid artery, meninges	No	6 days	Antibiotics	Dead, mucormycosis
Straatsma et al. (1962), Case 4	AFIP Accession No. 864638 26, F, N	Diabetic acidosis, renal insufficiency	Classic C.N.S. syndrome	—	—	—	
— Case 5	AFIP Accession No. 989662 39, M, N	Diabetic acidosis, renal insufficiency	Classic C.N.S. syndrome	—	—	—	
— Case 25	AFIP Accession No. 216244 30, M, N	Brain tumor and crainiotomy	Brain	—	—	—	
— Case 26	AFIP Accession No. 532008 25, M, N	None	Nasal cavity	—	—	—	
— Case 27	AFIP Accession No. 542127 47, F, N	Hypertension	Brain	—	—	—	
— Case 28	AFIP Accession No. 682762 55, M, N	Renal insufficiency	Antrum, face	—	—	—	
— Case 30	AFIP Accession No. 734644 38, F, N	Diabetic acidosis	Antrum, orbit	—	—	—	
— Case 32	AFIP Accession No. 909312 65, F, Race ?	Renal insufficiency, disseminated cryptococcosis	Brain	—	—	—	
— Case 33	AFIP Accession No. 982789; 32, M (Amer. Ind.)	Diabetic acidosis	Nose, ethmoid, antrum, orbit	—	—	—	
Suprun (1962)	Tel Aviv/Israel 46, F	Diabetic acidosis	Left orbit, skin of nose, brain	No	2 days	—	Dead, mucormycosis

Table 2. Continuation

Authors and Case Number	Location of Case Age, Sex, Race	Predisposing Factors	Anatomic Involvement	Cultures	Duration	Treatment	Status of Patient
Agrest et al. (1963) Case 1	B. Aires/Argentina 19, F	Abortion, acute renal failure, hyperglycemia	Right orbit, cerebral arteries, brain	No	7 days	Insulin, Levaphed	Dead, mucormycosis
— Case 2	B. Aires/Argentina 59, F	Diabetic acidosis	Brain, middle and anterior cerebral arteries	*Rhizopus*	5 days	Insulin, dialysis	Dead, mucormycosis
Burrow et al. (1963)	New Haven/USA 47, M	Diabetic acidosis	Turbinate, right face, orbit	*Rhizopus nigric.*	60 days	Amphotericin B	Cured
Fuentes (1963), Case 1	Mexico City/Mex. 9, M	Diabetic acidosis	Left naris, orbit, brain, lung	No	9 days	Corticoids and antibiotics	Dead, mucormycosis
— Case 2	Mexico City/Mex. 10, M	Diabetes	Right orbit, palate, brain	No	4 days	Insulin, parametasona	Dead, mucormycosis and candidosis
— Case 3	Mexico City/Mex. 22 months, F	Amebiasis	Left orbit, brain	No	3 days	Antibiotics	Dead, mucormycosis and amebiasis
— Case 4	Mexico City/Mex. 11, F	Infectious diarrhea	Brain and meninges	No	—	Antibiotics	Dead, mucormycosis, and Klebsiella septicemia
La Touche et al. (1963, 1964)	Leeds/England 58, F	Diabetic acidosis	Orbit, palate, left antrum, pons	*Rhizopus oryzae*	54 days	Amphotericin B	Cured
Suga (1963)	Keio/Japan 38, F	Agranulocytosis	Paranasal sinuses, orbit, bulbus oculi, brain, lungs, endocardium, colon	No	78 days	—	Dead, mucormycosis
Frágner and Rokas (1964)	Czechoslovakia 55, M	Hyperglycemia, pulmonary tuberculosis	Ethmoids, internal carotid artery, pituitary	*Rhizopus oryzae*	—	—	Dead, mucormycosis

LUBBE and PENNINGTON (1964)	Melbourne/Austr. 41, F	Renal failure	Nasal cavity, palate, right face, orbit, brain	*Mucor*	11 days	Antibiotics	Dead, mucormycosis
ROWE and PAYNE (1964)	Sydney/Austr. 48, F	Diabetic acidosis	Right face, palate, ethmoids, orbit	No	7 days	Insulin, antibiotics	Dead, mucormycosis
TAYLOR et al. (1964)	USA 40, F	Diabetes	Left palate, left maxillary sinus	*Rhizopus*	—	Amphotericin B, regulation of diab.	Well
WEISSKOPF (1964) Case 1	San Francisco/USA 68, F	Diabetes	Left nostril, turbinates, orbit, face, internal carotid artery	No	14 days	Antibiotics, Amphotericin B	Dead, mucormycosis
— Case 2	San Francisco/USA 45, M	Diabetes	Face, arteries	No	Not stated	Antibiotics, cortisone	Dead, mucormycosis
— Case 3	San Francisco/USA 23, F	Diabetes	Palate, sinuses, left orbit, internal carotid artery, cavernous sinus	No	25 days	Antibiotics, ACTH	Dead, mucormycosis
DE WEESE et al. (1965), Case 1	Portland/USA 1.5, M	Acidosis from diarrhea	Nares, right orbit, ethmoids	No	3 days	Amphotericin B	Well
— Case 2	Portland/USA 25, F	Diabetes	Nose, right eye, palate	No	3 mos.	Amphotericin B	Well
GORDON and LITTLE (1965), Case 1	Albany/USA 40, M, N	Diabetic acidosis	Nose, left orbit, sinuses, brain (no autopsy)	No	4 mos.	Amphotericin B Surgical explorat.	Dead, mucormycosis
— Case 2	Rochester/USA 44, M, W	Diabetes	Sinus, right orbit, brain, cavernous sinus	*Rhizopus oryzae*	5 days	—	Dead, mucormycosis
HURTADO et al. (1965)	Santiago/Chile 58, F	Diabetic acidosis	Right orbit, face, palate, ethmoids, brain	Fungus of *Mucorales* order	8 days	Insulin, antibiotics	Dead, mucormycosis
PARMENTIER et al. (1965)	Brussels/Belgium 75, F	Diabetes	Right orbit, ethmoids	No	3.5 weeks	Insulin	Dead, mucormycosis
PEÑA and DORADO (1965), Case 2	Palmira/Colombia 5 months, M	Diarrhea, acidosis	Orbit, meninges, brain	No	—	Amphotericin B	Dead, mucormycosis
STEHBENS (1965) Case 2	Sydney/Australia 35, F	Acute leukemia, cortis.	Branch of right internal carotid artery	No	—	Antibiotics, purinethol	Dead, subarachnoid hemorrhage

Table 2. Continuation

Authors and Case Number	Location of Case Age, Sex, Race	Predisposing Factors	Anatomic Involvement	Cultures	Duration	Treatment	Status of Patient
Sun et al. (1965) Case 11	Peking/China 11, M	Uremia, glomerulonephr.	Cerebellum	No	—	Antibiotics	Dead, uremia
Tomiyasu and Baker (1965)	Los Angeles/USA 65, M, W	Diabetic acidosis	Right orbit, arteries, brain	No	—	Amphotericin B	Dead, mucormycosis
Ginsberg et al. (1966)	Cincinnati/USA 76, M, W	Diabetic acidosis	Nose, sinuses, orbit, palate, brain	*Rhizopus oryzae*	14 days	Antibiotics	Dead, mucormycosis
Grover et al. (1966)	Nagpur/India 16, M, Hindu	Diabetic acidosis	Nasal mucosa, ethmoids, orbit, internal carotid artery, brain	*Rhizopus oryzae*	18 days	Insulin, penicillin	Dead, mucormycosis
Perez et al. (1966)	Ponce/P. Rico 43, F, W	Diabetes	Orbit, brain	*Rhizopus*	5 days	Amphotericin B	Dead, mucormycosis
Tinaztepe and Tinaztepe (1966)	Ankara/Turkey 4.5, M	Hydronephrosis, uremia, acidosis	Orbit, brain	No	23 days	Gantrisin	Dead, mucormycosis
Abramson et al. (1967), Case 1	Boston/USA 38, M, N	Diabetes	Right orbit, sphenoid	No	60 days	Amphotericin B	Dead, ruptured internal carotid artery
— Case 2	Boston/USA 40, F, N	Diabetic acidosis	Palate, right orbit	No	55 days	Amphotericin B	Living
Baum (1967)	New York/USA 53, M, W	Diabetes	Left turbinate, left orbit	—	7 weeks	Amphotericin B	Living
Green et al. (1967)	Philadelphia/USA 45, M, N	Diabetic acidosis	Turbinates, palate, ethmoids, right orbit, brain	No	18 days	Control of diabetes	Dead, mucormycosis
Miyake and Okudaira (1967), Japan; Claim 3 cases							
Pastore (1967)	Richmond/USA 59, M, W	Diabetes	Left maxillary sinus	No	27 days	Surgical removal of mucous membrane	Survived
Prokop and Silva-Hutner (1967)	New York/USA 46, F	Diabetic acidosis	Right antrum, orbit, right carotid, cranial nerves	*Rhizopus arrhizus*	2 mos.	Insulin, antibiotics, nystatin, Amphotericin B	Survived

The case of HARRIS was the first survival of a patient with all of the features of the classical central nervous system syndrome of mucormycosis, except for a pronounced involvement of the central nervous system itself. That is, there was involvement of the palate, the paranasal sinuses, the orbit and several of the cranial nerves, but the advance of the fungus ceased before it reached the brain and meninges and the patient survived. There were two other cases of survival of patients with cranial mucormycosis before the report of HARRIS (Table 2), but they did not show the syndrome of classical central nervous system mucormycosis. In one, the case of HAFSTROM et al., there was an inactive nodule in the brain of 10 years duration removed surgically and in the other, the case of WADSWORTH, there was an inactive mucormycosis in an enucleated eye.

As the case of HARRIS was the first survival in the classic central nervous system syndrome of mucormycosis, as the patient was thoroughly studied clinically and followed after the cure, and as the case demonstrated the lesions of mucormycosis so well, I am recording the clinical story here, in full detail.

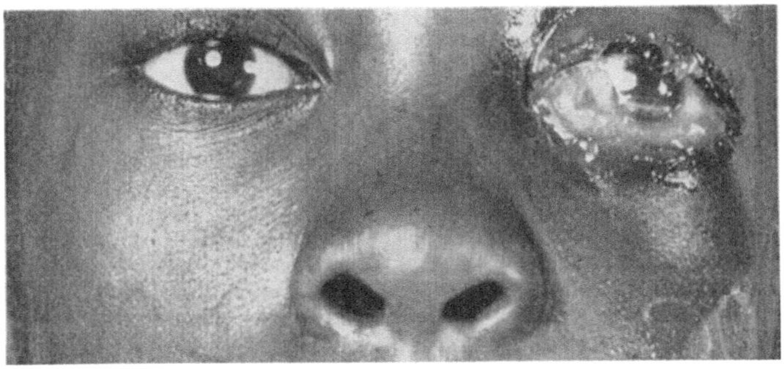

Fig. 9. Cranial mucormycosis. Note the proptosis of the left eye 1 week after admission to hospital of this 14-year-old diabetic patient of HARRIS (1955) (Fig. 1)

Case Report (HARRIS). History: Constance M., a 14-year-old colored girl, was admitted to Duke Hospital, Durham, N.C./USA, January 29, 1954 (Fig. 9). For a month she had noted polyuria, polydypsia, weakness and weight loss. For a week she had had nasopharyngitis, soreness of the roof of her mouth and pain in the left side of her head. She entered another hospital 6 days previously in impending diabetic coma. The blood sugar concentration was 500 mg/100 ml, and the carbon dioxide combining power 24 volumes per cent. Treatment of diabetic acidosis and diabetes was begun. The next day the left eye become injected, proptosed and fixed. Her temperature rose to 103° F and her leukocyte count to 18,000. Penicillin and dihydrostreptomycin were prescribed. She developed paralysis of the left side of the face, weakness of the left side of the soft palate, weakness on swallowing and drooling from the left side of her mouth.

Physical Findings: There was fever (39.2° C). The left side of the face was swollen. The left eye was proptosed and there was a complete internal and external ophthalmoplegia on that side. The conjunctivae on the left were edematous and injected. Corneal clouding and ulceration were present and there was some greenish-yellow purulent exudate. The left fundus showed severe venous congestion. The left ear canal was macerated and contained white cheesy-looking material. The left ear drum was white and thickened, and the landmarks were obscured. The nasal turbinates on the left were swollen. The breath was foul.

At the junction of the soft and hard palate on the left side, there was an ulceration 3 by 3 cm which was partly covered by a brownish-yellow cheesy material (Fig. 10). In the center of the ulcer there was a perforation (1 by 1 cm) of the palate which revealed the nasal cavity. The bare pterygoid plate on the lateral side of the nasal cavity could be seen through the perforation. The entire area was covered with a whitish cheesy necrotic material and a serosanguinous exudate oozed from the perforation. On the left lateral aspect of the tongue there was a firm, white mass (0.5 by 0.5 cm). The neck resisted passive movement and was painful, particularly on flexion.

There was almost complete blindness on the left. The left third, fourth and sixth nerves were completely paralyzed, the left pupil was moderately dilated and fixed. There was complete anesthesia of the maxillary division of the left fifth nerve and partial anesthesia of its orbital division with absent corneal sensation. There was weakness of the left side of the face but this might have been more apparent than real because of the swelling. The left side of the tongue showed diffuse atrophy. The tongue deviated to the left on protrusion and the palate deviated to the right on gagging. The lower jaw deviated to the left on opening the mouth and there was weakness of the muscles of mastication on the left, and of the left sternocleidomastoid and trapezoid muscles.

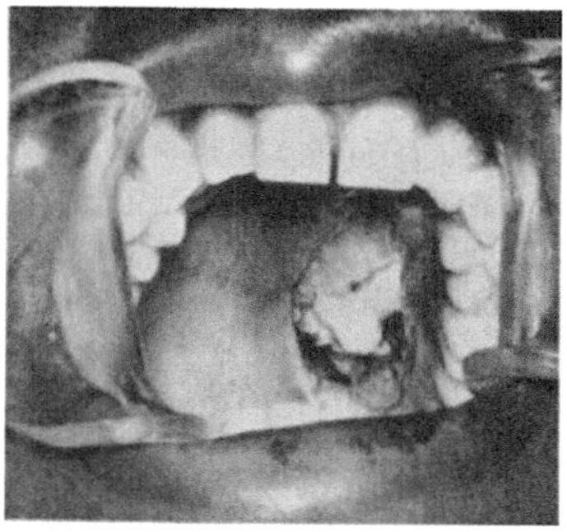

Fig. 10. Cranial mucormycosis. Note the large perforation of the left side of the palate. Patient of Harris (1955) (Fig. 1)

Laboratory Findings: The hemoglobin was 12 g. The total white count and the proportion of neutrophils were increased during the first 4 days. On admission the urine showed ketone bodies, a trace of protein, 3 plus sugar, 12—15 leucocytes, and rare erythrocytes. The fasting blood sugar was 225 mg/100 ml; the carbon dioxide combining power, 66 vol.-%; the serum chloride, 95 m Eq./l; serum sodium, 143 m Eq./l; and serum potassium, 4.4 m Eq./l. The total protein was 7.4 g/100 ml with A/G ratio of 1.0. Intradermal tuberculin skin test was negative in 1:1000 dilution. The histoplasmin, Frei and coccidioidin skin tests were negative but that for blastomycosis showed a slight transient erythema. There was a mildly positive reaction to intradermal Rhizopus vaccine in 1:1000 dilution and a very intense, 3 plus, indurated reaction to the 1:100 dilution.

A biopsy from the ulcer of the palate on February third showed necrotic fatty areolar tissue without nuclear staining but with maintenance of tissue structure.

Broad, branching, non-septate hyphae, 4—14 μ in thickness and up to 180 μ in length, occurred in the walls and lumina of blood vessels and lymphatics, and in nerves and other tissues. The hyphae were those of the coenocytic phycomycetes. No bacteria were stainable deep in the tissue though gram-positive cocci, gram-positive and gram-negative bacilli and a yeast suggestive of *Candida* were adherent to the surface. A biopsy on February eighth showed healthy granulation tissue free of organisms, but with broad branching hyphae in the pus on the surface.

Cultures of these biopsied lesions showed *Rhizopus arrhizus*. Cultures for acid-fast organisms were negative. On March 11, the nasal mucosa was biopsied. One fragment showed lymphoid tissue covered on all sides by stratified squamous epithelium. The other was a fragment of skeletal muscle heavily infiltrated by acute inflammatory cells but without fungi. On March 18, *Rhizopus arrhizus* was again cultured from the palatal lesion.

Electroencephalographic findings were interpreted as suggestive of a fronto-temporal dysfunction.

Roentgenograms of the chest on January 29 showed an area of increased density in the left lower lung field associated with shift of the heart to the left and elevation of the left diaphragm, indicating atelectasis of segments of the left lower lobe.

Roentgenograms of the skull, sinuses and orbits showed cloudiness of all sinuses on the left side but no evidence of bone destruction (Fig. 11). On February 6 the lungs were normal. On February 20 the sinuses on the left side still appeared clouded without bone destruction. On March 25 the left antrum, ethmoid and frontal sinuses were still markedly clouded. Similar dense clouding of the left sinuses and thickening of the membrane on the left side of the sphenoidal sinus were noted on April 14. On April 22 haziness but no definite bone destruction was found in the left mastoid cells.

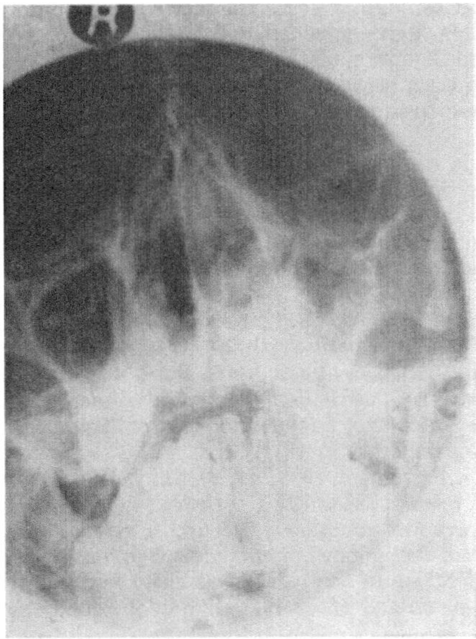

Fig. 11. Cranial mucormycosis. Note the cloudiness of the sinuses on the left. Patient of HARRIS (1955), shown in Figs. 9 and 10. Photograph of X-ray from BAKER (1957a)

Course: The admission diagnosis was diabetes complicated by either tumor or infection causing bronchopneumonia and extensive involvement of the left side of the head and face. The child was thought to have left pansinusitis, cavernous and left lateral sinus thrombosis with extension to the jugular vein and inflammation around the hypoglossal foramen. The diagnosis of mucormycosis was suggested shortly after admission and was verified by biopsy and culture.

Therapy was directed toward rigid control of the diabetes, elimination of secondary bacterial contamination, symptomatic relief and the possibility of specific action against the invading fungus.

The diabetes was controlled with diet and insulin. Therapy against the fungus consisted of iodides and desensitization. After March 16, the patient was afebrile save for 2 spikes to 39°C on March 28 and April 4. A very marked improvement in general condition occurred during the first week probably due to the control of the diabetes and bacterial infection. She was ambulatory after 10 days and showed slow, steady improvement during the remainder of her hospital stay. There were occasional and fairly severe headaches referred to the left ear with lancinating pains deep in the left ear.

Shortly after admission, loss of vision in the left eye progressed to complete blindness. The conjunctivitis and corneal ulceration responded rapidly to hot saline compresses and medication with chlortetracycline eye drops and ointment. A left tarsorrhaphy was done on February 13 because of the child's proptosis, corneal anesthesia and inability to close the eye. The proptosis slowly receded. Towards the end of the hospital stay, slight movement of the eyeball on attempted lateral gaze could be detected. There was no light perception through the closed lids.

A left myringotomy was done on February 9 and a small amount of purulent material obtained. No organisms were seen and cultures did not show growth of fungi or bacteria. Another culture for fungi on April 3 was negative. The otitis externa slowly subsided and the canal was clear by April 16. At that time the left drum appeared thick, edematous and white, particularly in the inferior portions, an appearance ascribed to chronic otitis media possibly of fungal origin.

Under local therapy the white cheesy necrotic material disappeared in 3 weeks, leaving a healthy edge to the palatal perforation. The ulceration filled in from the periphery and by the end of hospitalization was only one-half the size on admission. The perforation into the left nasopharynx remained unchanged. The medial pterygoid plate was exposed and the structures distorted.

On admission there was swelling of the left orbit, left cheek and area in front of the left ear. The left anterolateral aspect of the neck was also swollen with moderate tenderness along the course of the left internal jugular vein. The latter was firm and cordlike to palpation. As the left-sided facial swelling slowly subsided, two small, firm, non-tender subcutaneous nodules (2 by 3 cm and 1 by 1 cm in size) could be felt beneath the left eye. These nodules slowly diminished in size and had disappeared by the time of discharge. The swelling around the left ear and in the neck likewise disappeared, but the left internal jugular vein remained thickened to palpation.

Dysarthria, dysphagia and difficulty in chewing diminished. The left facial paresis improved. Weakness and atrophy of the left tongue, palate and masseters remained unchanged. However, there was a return of sensation to portions of the left side of the face. The complete anesthesia which originally extended from forehead to jaw gradually receded so that, by discharge, the area extended from the lower forehead to the upper lip and from the front of the ear to the lateral side of the nose.

At the time of discharge, June 2, 1954, the child felt well. The perforation remained unchanged. Many neurological abnormalities persisted: blindness of the left eye, left ophthalmoplegia, deafness on left, left facial weakness, left palatal weakness, atrophy and paresis of the left side of the tongue, weakness of the left mastication muscles, diminished to absent sensation of the left side of the face and weakness of the left trapezius. Roentgenograms showed clouding of the sinuses on the left.

The child was seen on November 19, 1954, at which time she was feeling well and going to school. The urine had been consistently free from sugar on the same dose of insulin. The exophthalmos of the left eye had completely receded. External ophthalmoplegia was still present. The area of anesthesia was somewhat smaller but extended from the lips to the forehead on the left face. The perforation of the palate was clean and less than one-half of its original size. The patient was re-examined on January 26, 1955, one year after admission. She has continued to do well.

Discussion of Case of Dr. HARRIS: The case demonstrated an extensive invasion of *R. arrhizus* in an individual predisposed to infection by uncontrolled diabetes mellitus. The portal of entry of the fungus was apparently the nasal cavity or the accesory nasal sinuses on the left. The fungus then invaded the palate where a perforation developed. Invasion of blood vessels, lymphatics and nerves together with necrosis and inflammation were demonstrated in biopsies. The necrosis was probably in part infarction, as thrombosis was undoubtedly present. The dense clouding of all the nasal sinuses on the left side was probably due to mucosal invasion by hyphae as has been seen in other cases. The fungus organisms extended from the palate and sinuses directly through the tissues and along the vessels and nerves. The proptosis and ophthalmoplegia were due to retro-orbital inflammation and the involvement of vessels and nerves entering and leaving the orbit. The involvement of the fifth, seventh, ninth, eleventh and twelfth nerves suggested an extensive invasion along the left side of the base of the skull through blood vessels so that the left posterior jugular outflow was involved. The cord-like thickening of the left internal jugular vein indicated invasion and thrombosis of that vein. The nodules in the cheek demonstrated local invasion with abscess or with infarcted foci.

In the light of the frequency of fatal cerebral involvement in the literature it was amazing that the child recovered from the acute phase of her disease and a year later seemed to be well except for the neurological residue and a relatively clean palatal perforation. The vascular thrombi organized and recanalized, the organisms died and necrotic tissue became fibrotic.

This case of HARRIS cannot be tabulated exactly as one of classical central nervous system mucormycosis, for the brain was largely spared. The facial features place it within the cranial category and it may be termed rhinoorbital mucormycosis.

Nasal Mucormycosis

In cranial mucormycosis there is usually a lesion of the nasal cavity on physical examination or at postmortem or a history of bloody discharge from the nose. The involvement is most frequently unilateral, though it may be bilateral, and palatal, sinus, orbital or facial involvement tend to be on the same side as the unilateral nasal involvement.

Examination of the interior of the nostril may reveal a bloody discharge and a black, dry-looking infarcted region. The turbinates may be black and gangrenous, due to vascular obstruction. In most case reports it has apparently been taken for granted that the fungus entered via the nose and the nasal mucous membranes, but the appearances suggest that it could be those of infarction secondary to vascular obstruction. Presumably the fungus could enter a paranasal sinus or the conjunctival sac of the eye and produce the nasal lesions.

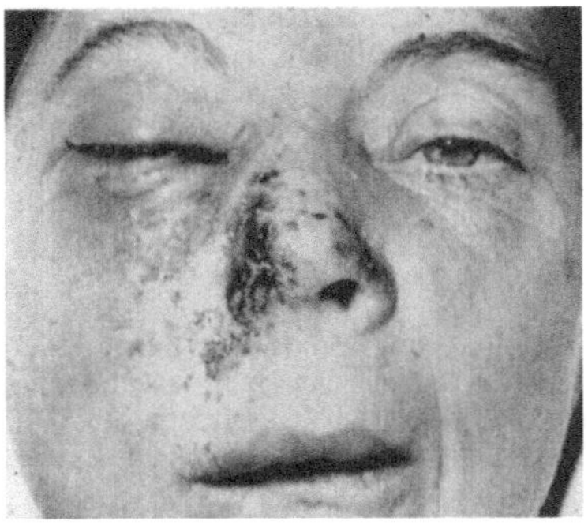

Fig. 12. Mucormycosis of the face. Note the proptosis of the right eye, the infarction of the nose and the effect of nerve paralyses on the eyelids and mouth. There was infarction of the right side of the palate and necrotizing ethmoiditis due to the fungus. From LUBBE and PENNINGTON (Fig. 1). Courtesy of authors and the Medical Journal of Australia

SMITH and YANAGISAWA say that the most important physical sign of rhinomucormycosis is the gray-black appearance of the nasal mucosa, which resembles dried clotted blood; and that the gray-black discoloration of the nasal mucosa is unlike that seen in any other nasal disease.

Two of the cases of GREGORY et al. showed nasal lesions. In one patient there was a dark, discolored mucosa and necrosis of the nasal septum.

BAUM described a swollen middle turbinate with overlying easily bleeding friable tissue. A biopsy of the turbinate revealed the hyphae of mucormycosis in areas of necrosis with scant neutrophils. The hyphae were invading blood vessels.

In GASS' patient there had been epistaxis from the left nares two days before admission. Dried blood was present in the left nares when the patient was brought to the hospital. On the third hospital day an otolaryngological consultant described peculiar, almost black mucous membrane in the left nares. The nasal septum was ulcerated and perforated.

In the case of DWYER and CHANGUS the left nostril contained hemorrhagic exudate, and the left side of the nose was gangrenous.

BERK et al. found a black and soft nasal septum at autopsy and microscopically there were hyphae in the nasal septum. In Case 1 of FUENTES there was, at autopsy, a congested and hemorrhagic inflammatory area with a gangrenous aspect, in the left nostril. WEISSKOPF mentions a crust in the left nostril and red, dry and atrophic turbinates.

DE WEESE and ROBINSON report 2 cases of mucormycosis of the nose and sinuses with recovery following treatment. In the first case nosebleed was the first event and a perforation of the septum was discovered. Biopsies of the nasal lesions showed hyphae. In their second case there was a depression of the cartilaginous part of the nasal septum, a foul nasal discharge, necrotic dehiscence of the right lateral wall of the nose and posterior nasal septum.

SMITH and KIRCHNER illustrate the autopsy specimen of black thrombosed turbinate from their Case 2 (Fig. 15).

Review of these case reports impresses one with the gangrenous nature of the nasal lesions, based on ischemia. Bone and cartilage share in the process and become soft, and the nasal septum may be perforated.

Palatal Mucormycosis

HARRIS, as already mentioned, reported a case of mucormycosis with a palatal perforation on the left at the junction of the hard and soft palate. The perforation, 1 × 1 cm, was in the center of an ulcer 3 × 3 cm in diameter, partially covered by a brownish-yellow, cheesy material. The bare pterygoid plate of the lateral side of the nasal cavity could be seen through the perforation. With healing the ulcer filled in from the periphery, but the perforation into the nasopharynx remained unchanged.

GEORGIADE et al. described and illustrated the palatal slough in the case of HARRIS and reported an infant case of palatal slough which they thought might have been mucormycotic in origin.

DWYER and CHANGUS described the left side of the palate of their case as escharotic. HOAGLAND et al. described as gray the right half of the palate in a non-fatal case. In their second case, also non-fatal, they mention an abnormality of the palate on the left. In case 1 of WASSERMAN et al. there was a black eschar in the midline of the palate. RINALDI and ASHBY mention a lesion of the hard palate in their case. LUBBE and PENNINGTON have an illustration of a large, right-sided lesion of the palate. This was probably gangrenous, as there was gangrene of the face in this case. TAYLOR et al. report the case of a 40-year-old diabetic woman with necrosis of the palatal bone proved to be mucormycotic in cultures and tissue sections. Successful treatment consisted of the use of Amphotericin B and regulation of the diabetes. WEISSKOPF describes ulceration of the palate in his case 3. DE WEESE et al. describe ulceration of the roof of the mouth in their second case. It was located posteriorly and to the right. In the autopsy report of their case, HURTADO et al. observed incipient gangrene of the palatal arch on the right. In the nonfatal case of ABRAMSON et al., the necrotic palatal bone was removed and later a prosthesis was adapted to the defect.

In Case 3 of SMITH and KIRCHNER there was a sinus tract from the hard palate and alveolus of the maxilla leading to the maxillary sinus.

In summary, the palatal lesions of cranial mucormycosis tend to be unilateral in location and ischemic in nature, that is, they are regions of gangrene due to infarction. Hyphae were demonstrated in the tissues and blood vessels of the palate. The affected palate bone became necrotic and softened and in a few cases perforations developed with the formation of a permanent communication between the oral and nasal cavity.

Paranasal Mucormycosis

The case reports of cranial mucormycosis indicate that the maxillary, ethmoid, and occasionally sphenoidal and frontal sinuses are involved and act as way-stations on the route of the fungus to the orbital cavity and the brain.

The sinusitis is usually unilateral and on the side on which the nose, palate and eye are affected. The left side is said to be affected more frequently than the right.

In one of the cases of GREGORY et al. there was purulent ethmoiditis with necrosis of bone, and perforation into the orbit and beneath the dura, and into the nose. STRATEMEIER's case had chronic empyema of the anterior ethmoids. BAUER et al. found the mucosa of the ethmoids and frontal sinuses red and necrotic. There was necrosis of bone, and the fungus was demonstrated in it. MERRIAM et al. describe dark brown contents of the sphenoid and ethmoid sinuses with demonstrable fungi in the contents. In HARRIS' case the antrum had a thickened mucous membrane, as seen in X-ray, and there was no bony involvement.

In most cases the sinuses do not contain fluid and the majority of X-ray studies fail to show a fluid level.

In one of GASS' cases, however, there was clouding of the left ethmoidal sinuses and a questionable fluid level in the left antrum was found on X-ray examination. On the fifth hospital day the ethmoidal bone biopsy was reported as showing mucormycosis. On the 33rd hospital day a rhinotomy, debridement of the left nasal cavity, exenteration of the left sinus complex, and left eye enucleation were done. At operation much of the frontal process of the maxillary bone, the entire left ethmoidal labyrinth, nasal septum, the turbinates, the anterior wall of the sphenoid sinus, the bone comprising the optic foramen, and the floor and medial wall of the orbit were necrotic. The necrosis extended to the cribriform plate and was thought to extend through the plate into the anterior cranial fossa. The patient was discharged from the hospital 27 days following his admission, and 3 months after the onset of his disease the activity of the fungus had ceased and the patient appeared to be "well."

In Case 1 of BURNS the sinuses on the affected side were cloudy in X-ray. In his Case 2 there were mycelia in tissue from the ethmoid sinus at operation and *Rhizopus sp.* was cultured from the contents. In Fig. 1 of ROWE and PAYNE hyphae are pictured in the necrotic wall of an ethmoid sinus. REEVES et al., in an autopsy report, describe ethmoiditis and sphenoiditis.

SMITH and YANAGISAWA described and illustrated a fluffy growth of white fungus on the skin of the face at the operative site following an external operation on a mucormycotic sinusitis. The fungus was growing at the inner angle of the eye close to the nose and resembled an early culture of *Rhizopus*. Apparently the organism had extended from the sinus and had spread to the skin surface adjacent to the incision.

In the case of SMITH and KIRCHNER there was a peculiar perforation extending from the right maxillary alveolar ridge and hard palate to the maxillary sinus.

In Case 2 of BAUER et al. (1955a) the postmortem examination revealed, in addition to hyphae in the tissues of the mucosa of the right ethmoid sinus, sporangia attached to their sporangiophores and containing spores. Scattered spores were present in the tissues.

Mucormycotic Otitis Media

HARRIS described otitis media during the course of the mucormycosis of his patient, but the fungus could not be cultured from pus from the middle ear or from the contents of the external ear canal. Later, the left ear drum appeared thick as though from chronic otitis media, possibly of fungal origin.

REEVES et al. reported a suppurative left otitis media observed at autopsy in a case of mucormycosis of the central nervous system. There was also ethmoiditis and sphenoiditis.

In the case of BRYAN et al., that of a 34-month-old infant, a moderate inflammation of the right tympanic membrane was seen clinically. At autopsy the right middle ear contained red-brown fluid. Microscopically, hyphae of mucormycosis were observed in the Haversian system and intertrabecular spaces of the bone around the right middle ear.

Orbital Mucormycosis

Orbital or rhinoorbital mucormycosis consists of unilateral orbital cellulitis with proptosis, ptosis, internal and external ophthalmoplegia, optic neuritis, posterior uveitis and obstruction of the central retinal artery (GINSBERG et al.). Marked visual impairment is usual and the ocular signs tend to be permanent.

This dramatic disease usually presents as a well-defined clinical syndrome (BAUM). The typical story is that of an adult with uncontrolled diabetes, usually in acidosis, who suddenly develops facial pain and headache, rhinitis with epistaxis, lid edema, and the other features mentioned above (Fig. 12). The patient lapses into coma, and death commonly ensues within two weeks. To date, almost all cases of orbital mucormycosis have occurred in patients with some type of debilitating disease, usually diabetic acidosis.

GASS and FERRY, and STRAATSMA et al. have written excellent reviews of rhinoorbital mucormycosis.

There are 77 cases (68%) of orbital and ocular mucormycosis among the 107 cases of cranial mucormycosis listed in Table 2. (The 4 cases listed as classic CNS mucormycosis are presumed to have had orbital disease and are included in the figure of 77.)

The predisposing disease in 68 (88%) of the 77 cases of orbital mucormycosis was diabetes, almost always in an acidotic state. The other predisposing conditions were hydronephrosis, uremia, and acidosis; diarrhea and acidosis (2 cases); the nephrotic syndrome; renal failure; cirrhosis; Hodgkin's disease; agranulocytosis; and amebiasis. Orbital mucormycosis, therefore, can be considered an important and frequently lethal complication of diabetes mellitus. The importance of keeping the diabetic patient from becoming acidotic cannot be overemphasized. Particularly, the diabetic patient should avoid extended overindulgence in alcohol, as this has led to acidosis and mucormycosis in several of the reported cases.

The causative fungus, usually a *Rhizopus* species, probably reaches the tissues of the orbital cavity from the nasal cavity or the paranasal sinuses, especially the

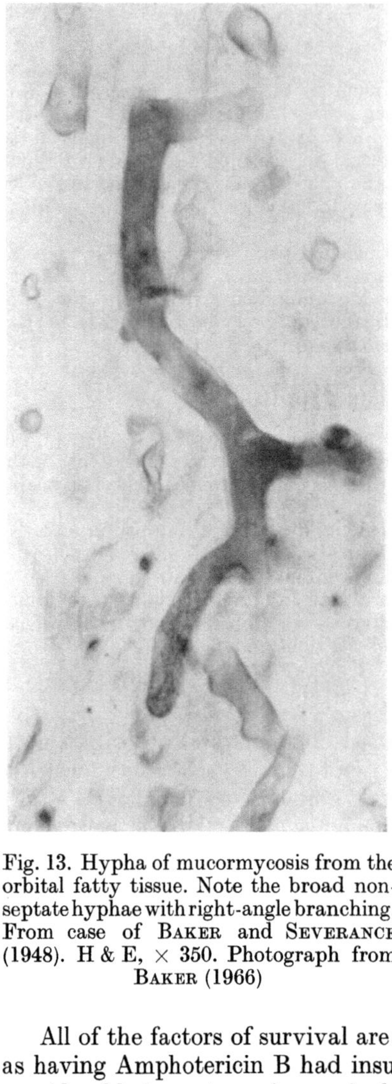

Fig. 13. Hypha of mucormycosis from the orbital fatty tissue. Note the broad nonseptate hyphae with right-angle branching. From case of BAKER and SEVERANCE (1948). H & E, × 350. Photograph from BAKER (1966)

ethmoid air cells or the maxillary sinus. A direct infection of the conjunctival sac or cornea cannot be excluded. Infection of the soft tissues of the orbital cavity may be complicated by infection of the bulbus oculi, the bony orbit and the tissues of the face and nose. The infection may extend through the roof of the orbit to the anterior fossa of the cranial cavity and to the meninges and parenchyma of a frontal lobe of the brain producing mucormycotic meningitis and mucormycosis of the brain.

The infection of the orbit may involve the ophthalmic artery causing it to become thrombosed. The mucormycotic thromboarteritis may extend to the artery of origin, the internal carotid artery and to the circle of Willis and its branches.

Of the 77 cases of orbital mucormycosis death has ensued in the majority (77%). The 18 which are listed as survivals (23%) together with the factor apparently responsible for survival are listed as follows:

WADSWORTH, self-limited; HARRIS, insulin; BAKER (1957), insulin; McCALL and STROBOS, enucleation of eye; MERRIAM and TEDESCHI, insulin, enucleation; FAILLO et al., operation; GASS (case 1), Amphotericin B and exenteration of eye and sinuses; HOAGLAND et al. (case 1), extirpation of right face; WASSERMAN et al. (case 3), insulin and iodides; BANK et al., insulin; LANDAU and NEWCOMER, Amphotericin B; BURROW et al., Amphotericin B; LA TOUCHE et al., Amphotericin B; DE WEESE et al. (case 1), Amphotericin B; DE WEESE et al. (case 2), Amphotericin B; ABRAMSON et al. (case 2), Amphotericin B; BAUM, Amphotericin B; PROKOP and SILVA-HUTNER, insulin and Amphotericin B.

All of the factors of survival are not included in the above listing. Some listed as having Amphotericin B had insulin and other therapeutic agents. A factor of considerable importance in survival may be the shorter period of diabetic acidosis occasioned by prompt diagnosis and therapy of the uncontrolled diabetes. The spread of the infection to the brain is halted.

Apparently Amphotericin B has been a factor in improving the prognosis of orbital mucormycosis.

Pathologic Findings. Examination of the orbital contents at postmortem from within the cranial cavity discloses hemorrhagic, grayish, infarcted tissues. Numerous hyphae are free in the tissues and within arteries and nerves.

Ocular Mucormycosis

GINSBERG et al. in 1966 described the eyes removed at autopsy from a fatal case of cerebral mucormycosis, making clear why there is blindness and ophthalmoplegia in these cases.

The partially collapsed left globe, the mucormycotic one, had numerous fragments of adherent muscle and fatty tissue. A horizontal section revealed the vitreous to be diffusely

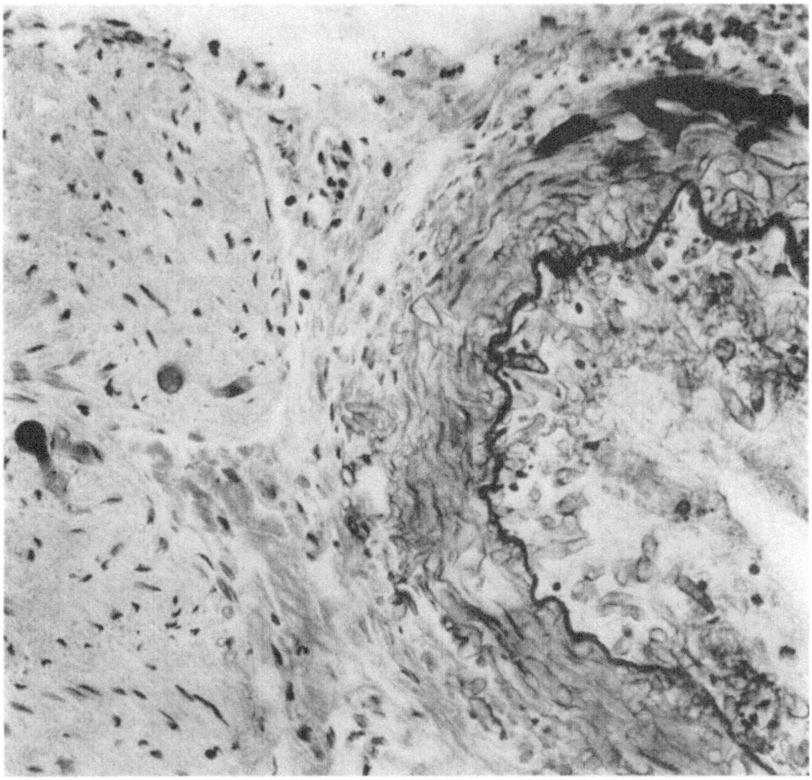

Fig. 14. Mucormycosis, with hyphae in a nerve (left) and in the wall and lumen of an artery (right) (H & E, × 145). Hyphae also may travel along lymphatic vessels. Photograph from BAKER (1966)

cloudy. Microscopically a disseminated cellular infiltrate, predominantly neutrophilic, was present in the posterior choroid, sclera, episclera, extraocular muscles and orbital tissues. Anteriorly, the inflammation extended to the edge of the choroid, the tendons of the horizontal rectus muscles, the limbal episclera and the bulbar conjunctiva. The amount of cellular exudation varied considerably, being moderate within the eye and quite severe in the orbit where there was considerable necrosis with moderate edema.

Leukocytic infiltration was mainly perivascular but also surrounded foci of extravascular hyphae in necrotic tissue. Although some areas of suppuration appeared to reflect the direct effect of the fungus, most of the necrosis was attributable to infarction. The inflammatory reaction appeared most intense in relation to small or medium-sized arteries, especially the choroidal, short and long ciliary vessels and those of the extraocular vessels. The infiltrate in extensively necrotic extraocular muscle with intense reaction around a thrombosed vessel is sufficient to explain the ophthalmoplegia.

As for the loss of vision, although the optic nerve was unaffected, there was vascular invasion, chiefly arterial, with and without thrombosis, and hyphae were identified in blood vessels of the uveal tract, optic meninges, conjunctiva and orbit.

STRAATSMA et al. give particular attention to ocular disorders in 5 cases with the classic central nervous system syndrome of mucormycosis. Three patients had proptosis, with edema and limitation of mobility indicative of orbital cellulitis. Three patients demonstrated extraocular muscle paralysis, and one showed retinal hemorrhages. Three developed unilateral or bilateral occlusion of the central retinal artery as shown by pallor of the fundus, a cherry-red spot in the macula, and a fixed, dilated pupil.

Histologically, the hyphae were present throughout the soft tissues of the orbit and showed a predilection for invasion of the major vessels with resultant thrombosis and infarction (Fig. 14). Involvement of the optic and ciliary nerves was frequent, and in a few cases panophthalmitis was observed.

In WADSWORTH's case of ocular mucormycosis (1951) the eye was removed surgically. There was massive fibrosis of the retina in the posterior portion of the globe and the retina was completely detached. Under high magnification one could see bone formation in the region of fibrosis, and old hemorrhage with cholesterol crystals. Incorporated in scar near the optic nerve were branching, non-septate, phycomycete hyphae. There was none of the surrounding eosin-staining material characteristic of infection by Entomophthora and infiltrations of eosinophils were not present. The walls of the hyphae are so sharp, and in places so thickened, in Fig. 3 of WADSWORTH's article, that one wonders whether calcification has not occurred.

The active infection must have occurred months or years before the enucleation of the eye, because the fungi were embedded in scar tissue. The clinical story indicates increase in size of the lesion by ophthalmologic examination and the development of acute secondary glaucoma about 4 months before enucleation. Probably there was a subclinical orbital and ocular infection from the nose at a former time or, less likely, a blood-born deposit in the eye from pulmonary mucormycosis. There was no associated diabetes or other predisposing condition.

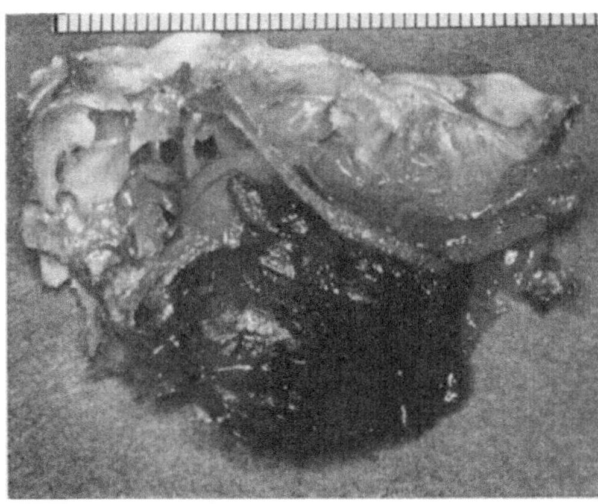

Fig. 15. Autopsy specimen of black thrombosed turbinate from case 2 of SMITH and KIRCHNER (1958) (Fig. 4)

Facial Mucormycosis: The first case of facial mucormycosis was reported by GREGORY et al. in 1943.

It was that of a diabetic patient with acidosis who died of cerebral mucormycosis. Two days before death the face became edematous, especially about the eyes, and then dark blue over the bridge of the nose. There were no microscopic sections of the affected skin, but undoubtedly fungi would have been found in the subcutaneous or cutaneous tissues and blood vessels.

The second case, of MARTIN et al., was one in which sections were obtained (Fig. 16). This was the case of a $2^1/_2$ month old diabetic infant who died of cerebral mucormycosis with thrombosis of the carotid artery. Four days before death, gangrene of the left side of the face developed and hyphae were observed in tissue sections of the left side of the face and its underlying tissues.

McCALL and STROBOS, DWYER and CHANGUS, FAILLO et al. and BORLAND report similar cases. Of the 6 patients just mentioned, 4 died and 2 recovered.

The location of the facial lesion varied from case to case. For example, in the case of SMITH and KIRCHNER (Fig. 17), the gangrene was a round region, about 6 cm in diameter, just

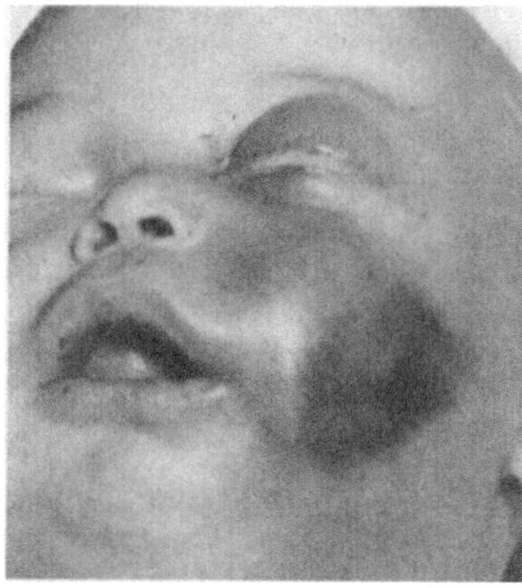

Fig. 16. Cranial mucormycosis. Infarction (gangrene) of left orbit and cheek. Case of MARTIN et al. (1954). Photograph by courtesy of Dr. J.M. LUKEMAN

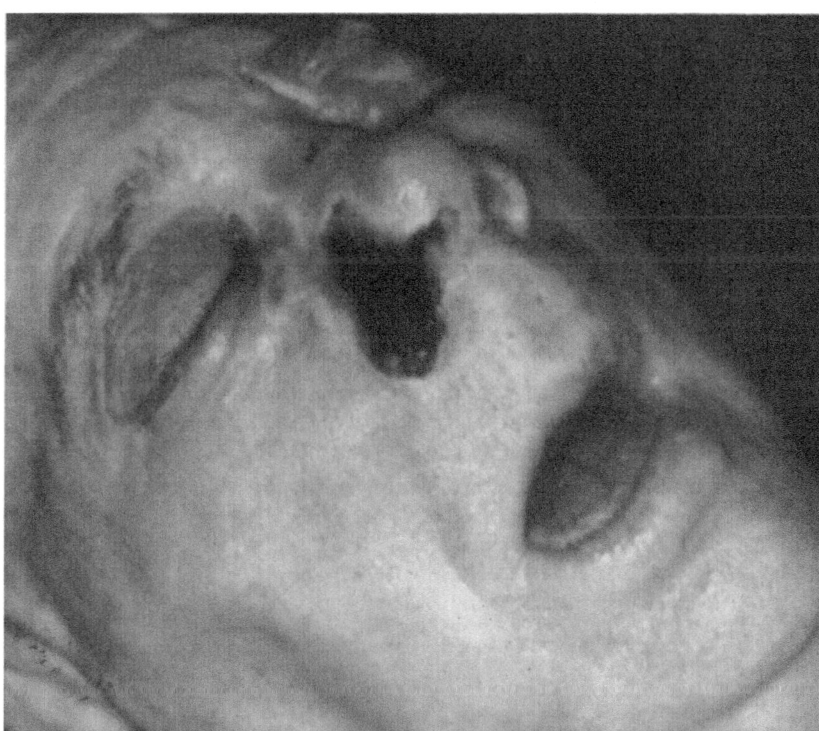

Fig. 17. Cranial mucormycosis. Note the black necrotic region of infarction of the cheek and nose adjacent to the right nostril, and also the periorbital swelling with discoloration. Case 1 of SMITH and KIRCHNER (1958) (Fig. 1)

to the right of the right nostril. In the case of LUBBE and PENNINGTON (Fig. 12), it was a little larger and elongated, running parallel to, and just to the right of the nose. In the case of HURTADO et al., the area of reddish induration extended from the nose well out on the right cheek.

Other cases are described by RINALDI and ASHBY, FOUSHEE and BECK, HOAGLAND et al., WEISSKOPF, and DE WEESE and ROBINSON.

In summary, mucormycosis of the face, of a gangrenous, infarcted nature, occurs in about a fourth of the cases of cranial mucormycosis. It is not as common as orbital edema and proptosis, but occurs on the same side and is usually unilateral, though in a few cases it has spread across the upper part of the nose to involve both sides of the nose and adjacent cheeks. A whole cheek may be red, indurated and swollen, or livid regions may occur next to a nostril or between the eye and the nose, fanning out onto the cheek. The majority of the lesions are not ulcerated. One lesion was reported as an abscess of the right naso-labial fold (GABRIELE). The coldness of the infarcted area of the cheek during the life of the patient was mentioned by one observer.

Mucormycosis of the Brain: The tables include 64 cases of mucormycosis of the central nervous system, 1 of which had spinal cord as well as cerebral lesions and 2 of which had mucormycosis of the spinal cord without involvement of the brain.

My Table 2 of cranial mucormycosis includes all cases of mucormycosis with lesions in the head or face, except 5 cases listed as disseminated. In the 5 tables of mucormycosis cases there is no duplication of listing.

If the classic central nervous system syndrome is defined as combined orbital and cerebral involvement, the following cases qualify for this designation:

Cases 1, 2 and 3 of GREGORY et al., LECOMPTE and MEISSNER, BAKER and SEVERANCE, WOLF and COWAN, STRATEMEIER, MARTIN et al. (Fig. 16); cases 1 and 2 of BAUER et al., Sydney Hospital (1956a, b), BRYAN et al., DWYER and CHANGUS; cases 1 and 2 of SMITH and KIRCHNER (Figs. 15, 17), BORLAND, DOLMAN and HEARD, LIE KIAN-JOE et al., SMITH and YANAGASAWA, LONG and WEISS, BANKER; case 2 of GASS; cases 1 and 2 of WASSERMAN et al.; cases 1, 2 and 3 of STRAATSMA et al., SUPRUN, AGREST et al.; cases 1 and 2 of FUENTES, SUGA, LA TOUCHE et al., LUBBE and PENNINGTON (Figs. 12, 27); case 1 of GORDON and LITTLE, HURTADO et al.; case 2 of PEÑA and DORADO, TOMIYASU and BAKER, GINSBERG et al., GROVER et al., PÉREZ et al., and TINAZTEPE and TINAZTEPE (Figs. 22—26).

In a small number of cases, the orbit was not affected, but there were *lesions* of the *nose*, *paranasal sinuses* or other portions of the face which suggested that the infection of the brain had come from the region of the face.

In the remaining group of cases of cranial mucormycosis, with lesions of the brain or meninges and sometimes of the internal carotid artery, it is not clear how the cerebral lesion developed, whether from the face region or through the blood stream, or from the lungs or elsewhere. Some of these cases have a single, solitary tumor-like mass in the brain.

This group comprises the cases of OPPE, HAFSTROM et al., KURREIN, GUNSON and BOWDEN, FOUSHEE and BECK, KEYE and MAGEE (3 cases), JACKSON and KARNAUCHOW, MURESAN, TURPIN et al., STRAATSMA et al. (cases 5, 27 and 32), AGREST et al., and FUENTES (cases 3 and 4).

Allied to this last group, but with a high probability that the brain lesion has developed as part of a disseminated mucormycosis with the primary lesion in the lung or elsewhere, are the brain lesions listed in Table 5 of disseminated mucormycosis.

These are the cases of PALTAUF, TORACK (case 7), HUTTER (case 4), RABIN et al. (cases 1 and 2) and BAKER (1962) (case 11).

In 32 cases of the classical central nervous system syndrome of mucormycosis the prosector examined the brain in detail and his observations have found their way into print and into published illustrations.

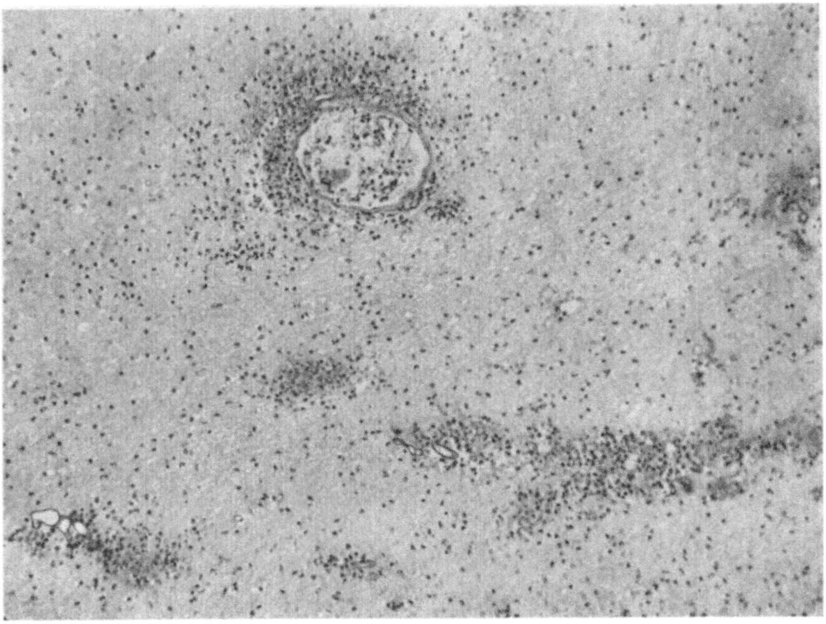

Fig. 18. Cerebral mucormycosis, to show cellular infiltrate, mostly neutrophilic, about blood vessels. Hyphae are just visible in the wall of the largest vessel. Swollen hyphae are seen elsewhere (H & E, × 100)

The classical central nervous system lesions of mucormycosis were described by GREGORY et al. in 1943 (Figs. 4, 18). A patient in diabetic acidosis developed nasal, paranasal, or orbital mucormycosis and the infection spread to infect the brain.

In 29 of the 32 cases the *Rhizopus* hyphae appeared to enter the brain directly from the nasopharynx, paranasal sinuses or orbit by invading the bone of the floor of the anterior fossa of the skull, the dura, the leptomeninges, and finally the cortex and white matter of a frontal lobe of the brain. During the advance the fungus invaded vessels and caused thrombosis and infarction. Bone and brain were softened. Microscopically, hyphae, thrombosis, necrosis, and cells of acute inflammation were described.

I will review the findings in the first case of GREGORY et al., and then those of the subsequent cases:

This historic case of medical mycology was that of a 43-year-old colored woman who developed diabetic acidosis and then pain and loss of vision in the left eye. She soon showed gangrene of the left orbit and of the bridge of the nose, and died only 4 days after the onset of this peculiar infection. There was infarction (gangrene) of the left side of the nasal septum and the turbinates, mycosis of the eyeball, thrombosis of the left cavernous sinus, the left internal carotid artery, the anterior cerebral artery, the middle cerebral, the posterior cerebral and the inferior cerebellar. There was meningovascular mucormycosis of the base of the brain as shown by dull, opaque and blood-stained leptomeninges. This appearance extended over the tips of the frontal lobes.

There was infarction of the inferior portions of the frontal lobes, with minimal inflammatory reaction, and infarction, early, of the left cerebral hemisphere. Coronal sections of the frontal lobes showed softening of the orbital surfaces to a depth of 2—3 cm. There was mucormycosis of the pituitary gland.

The authors suggested that the infection started in the nose and extended through the angular vein and its branches which reach the orbit and empty into the cavernous sinus by way of the ophthalmic vein.

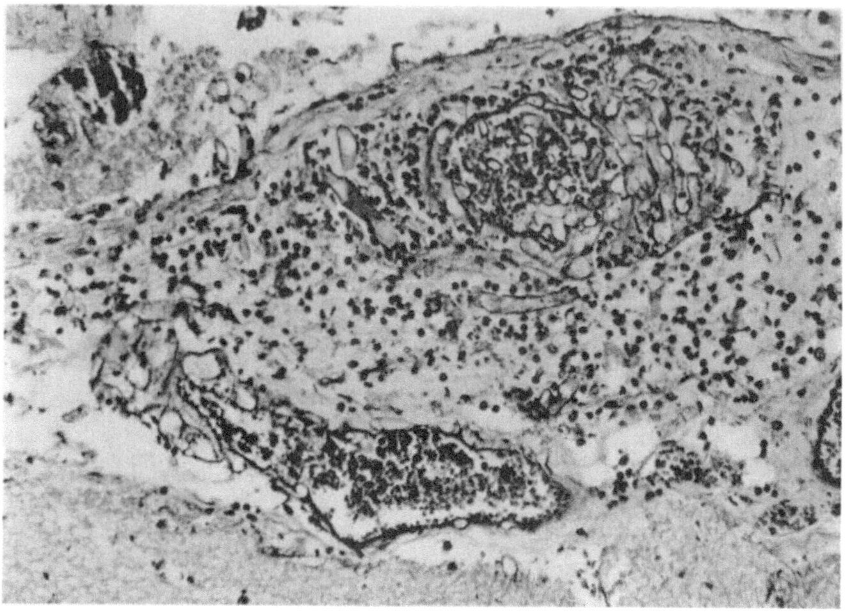

Fig. 19. Mucormycotic meningitis. Note the polymorphonuclear inflammatory edudate in the edematous leptomeninges and the hyphae both free and within vessels (H & E, × 130)

This is a possibility, and there has been cavernous sinus thrombosis in at least six of the 32 reported cases.

In the second case of GREGORY et al. there was no thrombus in any dural sinus. The patient, a diabetic who had been on an alcoholic bout for several weeks, was brought to the hospital with a proptosis, i.e. forward projection, of the right eye. Apparently the disease began with mucormycotic ethmoiditis and there was direct extension to the right orbit and to the floor of the anterior cranial fossa to form an abscess beneath the dura and to involve the frontal lobes of the brain. The protocol states that after the brain was removed, purulent material was seen beneath the dura on both sides of the crista galli. On the inferior surfaces of the frontal lobes of the brain the tissue was destroyed and ragged. On coronal sectioning the necrosis was found to extend 4 cm inward from the orbital surfaces and posteriorly to the corpus callosum. There was osteomyelitis of the bony plates forming the medial border of the right orbital cavity.

In the third case of GREGORY et al. the right orbit and the pons of the brain showed severe mucormycosis at autopsy. Could the process have begun in the right conjunctival sac? It is not clear that the nose and paranasal sinuses were examined at autopsy. Here was a case apparently without inferior frontal lobe necrosis.

There are only two other cases in which the evidence for a direct spread from below to the frontal lobes is not adequate. I will give the findings in these two cases, LE COMPTE and MEISSNER, and STRAATSMA et al., Case 3, and then the findings in the remainder of the cases, in which the evidence for direct spread was good.

In the case of LE COMPTE and MEISSNER the right eye was swollen and inflamed and the pupil was larger than on the left. There was no movement in the left arm or leg. At autopsy, note was made of the great tendency of the fungus to grow in vessels. The brain showed petechial hemorrhages of the cerebellum, pons and medulla. There was acute mucormycotic meningoencephalitis with an intense polymorphonuclear reaction in many places. The orbital surfaces of the frontal lobes seemed to be no more involved than any other portion of the brain. My comment was that the disease appeared to begin in the conjunctival sac of the right orbit, but that the left hemiplegia indicated that the fungus was already in the right motor tracts of the brain affecting the left motor tracts of the cord and that both the eye and central nervous system involvement might have developed from an undiscovered focus, as from the ethmoid air cells.

In the third case of STRAATSMA et al., which is the case later reported in more detail by REEVES et al., there was, in a diabetic woman of 30, gangrene of the eyelids and left side of the bridge of the nose. Suppurative ethmoiditis and sphenoiditis developed. There was bilateral cavernous sinus thrombosis, and thrombosis of the left anterior and middle cerebral arteries, the intracranial portion of the left internal carotid artery, and both ophthalmic arteries. There was basilar fibrinopurulent meningitis. An interesting finding was an acute suppurative left otitis media.

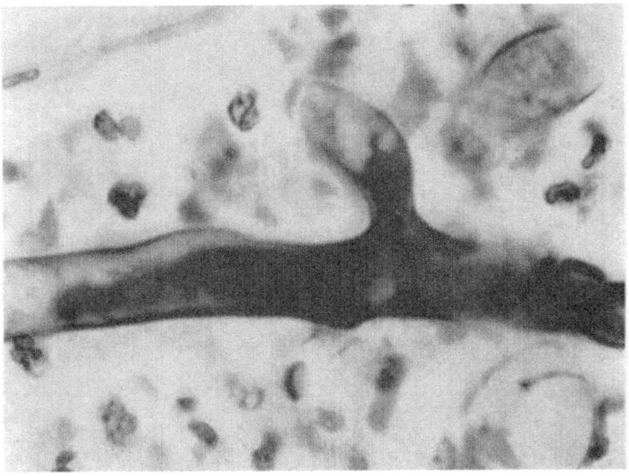

Fig. 20. Hypha in meningeal mucormycosis. Note the right-angle branching and the neutrophils and red cells in the leptomeninges (H & E, × 800)

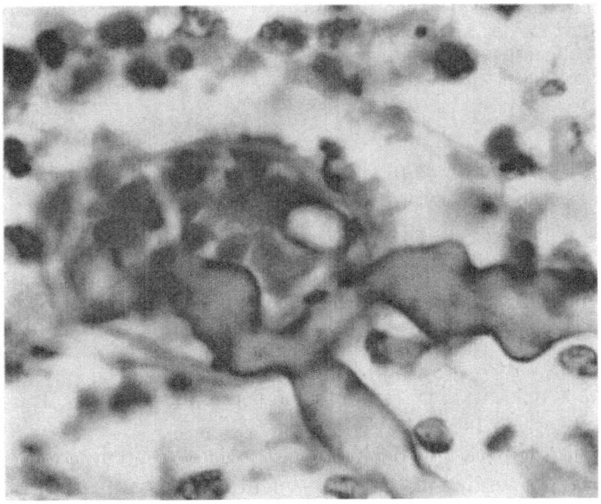

Fig. 21. Meningeal mucormycosis. Note the hyphae penetrating a minute vessel and stimulating the conglomeration of platelets and the agglutination of blood cells as the process of thrombosis is initiated (H & E, × 1000)

As nothing was said about the orbital surface of the left frontal lobe of the brain, I conclude that a venous spread was possible as was described in the first case of GREGORY et al.

The abstracts of the cases in which there is evidence for a direct spread of infection from the nose, paranasal sinuses or orbit to the brain, through bone and

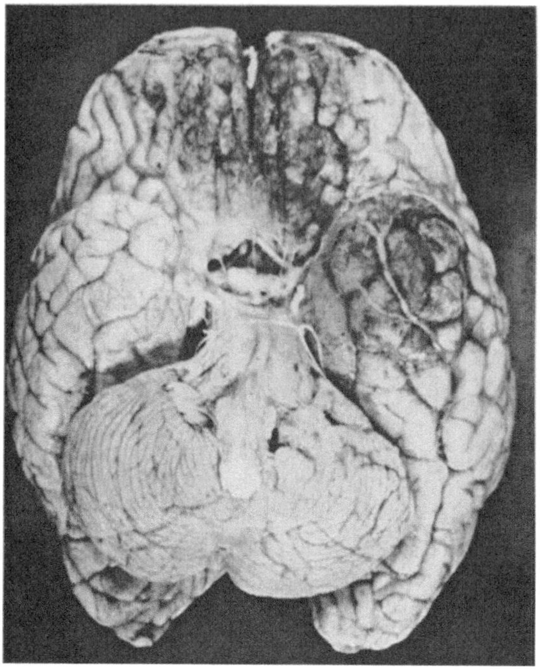

Fig. 22. Meningeal and cerebral mucormycosis. "There was an opaque, red-grey thickening of the meninges of the inferior surface of the frontal and temporal lobes," wrote the Doctors TINAZTEPE, and they continued: "The lepto-meninges over the cerebral convexities appeared thin and transparent except in the depth of the cerebral sulci where they had a slight gray-white appearance, suggesting exudate, around the blood vessels. Basilar and anterior cerebral arteries showed thrombosis." Previously unpublished photograph. Courtesy of Drs. TINAZTEPE and TINAZTEPE

dura will now follow. The neuropathologic findings are included in these abstracts, and the cases are arranged chronologically.

Case of BAKER and SEVERANCE (1948). A 3-year-old girl with diabetes had hemorrhage in the left eye, and the left pupil was larger than the right. At postmortem examination the fat of the left orbit was congested and red. Microscopically there were hyphae in the fat, arteries, veins, perineurial lymphatics, edematous muscle and bone. The thrombus in the left internal carotid artery showed numerous hyphae. On the orbital surfaces of the frontal lobes of the brain, more pronounced on the left, were regions of injection and softening. Frontal sections through these areas showed necrosis of brain extending through the gray matter and into the white matter. Microscopically there was minimal acute inflammation. The bony wall of the orbit and the capsule of the pituitary contained hyphae.

My comment on this case was that the mycosis first appeared in the orbit and conjunctiva, that the under surface of the brain was infected, probably by the fungus coming up from the orbits or paranasal sinuses below and that the fungus disseminated from the cranium to the lung.

Case of WOLF and COWEN (1949). A 42-year-old colored male diabetic developed left proptosis with redness and swelling of the lids. The left pupil was larger than the right. The left nasal cavity was edematous and full of discharge. An external right ethmoidectomy showed hyphae. There were hyphae in the eyeball. There was left cavernous sinus thrombosis, lepto-meningitis over the orbital surfaces of the frontal lobes, more on the left, and necrosis of the frontal lobes far into the white matter.

In this case the infection could have entered the cranial cavity via veins to the cavernous sinus or it could have penetrated upward through the orbital roofs to the frontal lobes of the brain.

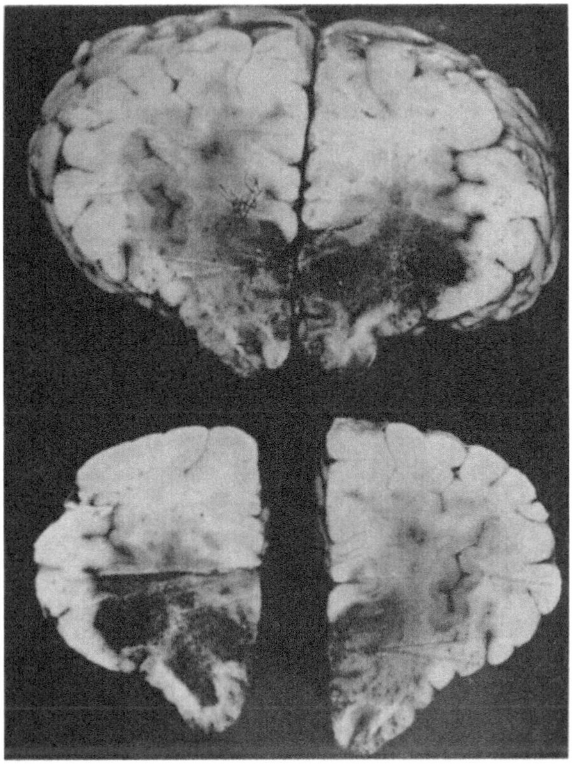

Fig. 23. Cerebral mucormycosis. Coronal section of the frontal lobes of the brain shown in the previous figure. There is bilateral hemorrhagic necrosis and infarction. The fungus has come up through the floor of the anterior fossa of the skull from the orbits and paranasal sinuses invading the meninges and brain, and causing thromboses of arteries. Case of TINAZTEPE and TINAZTEPE (1966) (Fig. 2)

Case of STRATEMEIER (1950). A 33-year-old colored woman developed diabetic coma after a heavy alcoholic bout. The left eyelid was swollen and ptosed.

At postmortem examination there was chronic empyema of the anterior ethmoid cells. The left posterior nares showed a thickened, necrotic and hemorrhagic membrane. The left turbinate bone and the nasal septum were necrotic. Microscopically the mucosa of the middle turbinate on the left side was deeply infiltrated with lymphocytes and occasional fungi and spores.

There were thrombosed vessels on the orbital surfaces of the frontal lobes and microscopically these vessels were blocked by masses of leucocytes, fibrin and hyphae. There was infarction of the orbital surface of the frontal lobes. Coronal cuts showed this infarction to extend into the frontal lobes a short distance.

The author's interpretation was that the fungus had spread from the nares to the brain.

Case of MARTIN et al. (1954). An infant of two and one half months with diarrhoea and acidosis developed necrosis of the left orbit including the globe, and gangrene of the left cheek. At autopsy there was thrombosis of the left ophthalmic and left internal carotid arteries. There was meningovascular mucormycosis. The orbital surface of the left frontal lobe was necrotic. The fungus disseminated to the lungs, pancreas and bone marrow.

Case 1 of BAUER et al. (1955). A 25-year-old white woman developed diabetic acidosis and left periorbital edema with a fixed pupil, larger than that on the right. At autopsy the under surfaces of the frontal lobes of the brain were adherent to the dura and the leptomeninges here and at the base of the brain were more opaque than normal. The inferior portions of both frontal lobes and the tips of both temporal lobes were softer than normal. A coronal section through the anterior half of the brain showed an infarct of the superior and lateral portion of the left cerebral hemisphere extending to include the basal ganglia on the left, Microscopically sections

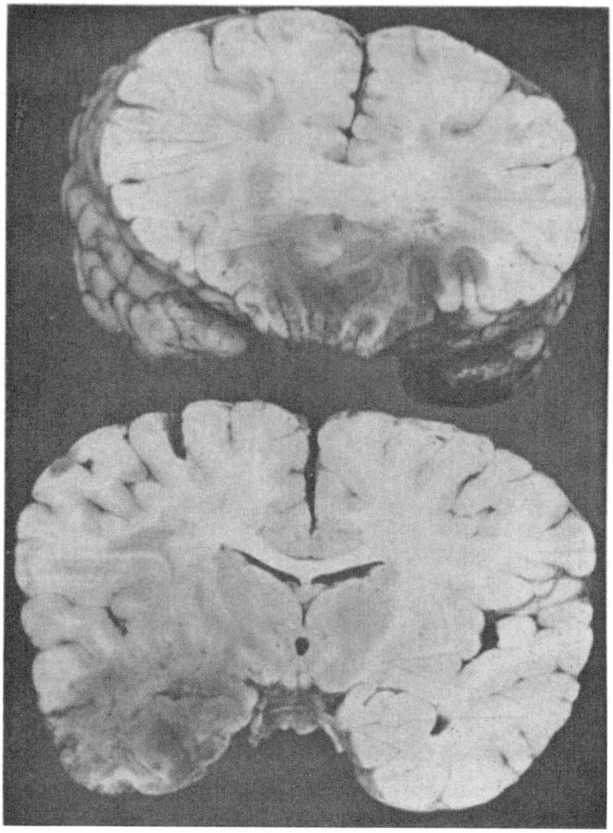

Fig. 24. Cerebral mucormycosis. Coronal section of the brain showing extension of the hemorrhagic necrosis. Case of TINAZTEPE and TINAZTEPE (Fig. 3)

from the infarct showed hyphae, necrosis, polymorphonuclear neutrophils and gitter cells. There were a few hyphae in the pons. There were hyphae in the dura about the pituitary gland but the dura itself was free.

Interpretation: The infection spread from the orbits through the orbital roofs to the dura of the anterior cranial fossa and into the leptomeninges. The infarct suggests a thrombosis of the middle cerebral artery, and the arteries and veins of the brain and meninges were involved.

Case 2 of BAUER et al. A 40-year-old colored person with diabetic acidosis had mucormycosis of the paranasal sinuses and orbit on the right. The dura overlying the roof of the right orbit was roughened and adherent to the leptomeninges of the right frontal lobe. There was necrotic bone in the walls of the ethmoid air cells and the right frontal sinus. The diaphragma sellae and the coverings of the pituitary stalk contained hyphae.

The area of necrosis at the base of the right frontal lobe extended into the adjoining white matter. There were numerous minute mucormycotic infarcts throughout the cortex and particularly in the floor of sulci. This showed grossly as areas of discoloration 0.5—1.2 cm in diameter. They had a red pinpoint center surrounded by a ring of gray tissue with a pink halo. The right internal carotid artery was thrombosed.

In this case mucormycosis developed in the right nostril and paranasal sinuses and orbit and passed upward to involve the dura and right frontal lobe of the brain. The leptomeningitis and vasculitis became general over the brain and caused minute infarcts.

The thrombosis of the internal carotid artery can be explained on the basis of a probable thrombosis of the ophthalmic artery, via such tributaries as the anterior and posterior ethmoidal arteries and frontal arteries. Also it is well to realize that thrombosis of the inferior orbital branch of the anterior cerebral could produce infarction of the inferior portion of the frontal lobe, perhaps similar to what is observed in these cases of mucormycosis.

Case at Clinical Pathological Conference of the Sydney Hospital, Australia, 1956. This was a case of orbito-cerebral mucormycosis with frontal lobe involvement and metastatic lesions in the lungs. On the medial surface of the anterior pole of each frontal lobe there was an area of congestion with underlying petechial hemorrhages and softening. Conspicuous petechial hemorrhages were present throughout both cerebral hemispheres and in the brain stem and pons. Microscopically in the frontal lobe there was acute hemorrhagic meningoencephalitis, with hyphae in blood vessels and gray and white matter. Many were surrounded with well-preserved polymorphonuclear neutrophils indicating that the infection occurred during life and was of short duration. In other parts of the brain there were petechial hemorrhages but no mycelium.

In the middle and lower lobes of the right lung and in the left lung were round, pale brown regions 1—3 cm in diameter. Near one lesion a firm thrombus was found in a vessel.

Case of BRYAN et al. (1958). This 3-year-old girl had orbito-cerebral mucormycosis. Clinically there was muscular weakness of the left side of the body.

The infection appeared to begin in the nose and involve the right-sided structures of the face, the palate, the ear and then the frontal and temporal lobes of the brain. The fungus spread to the lungs and intestines. The author called the case disseminated. It is a cranial case with dissemination.

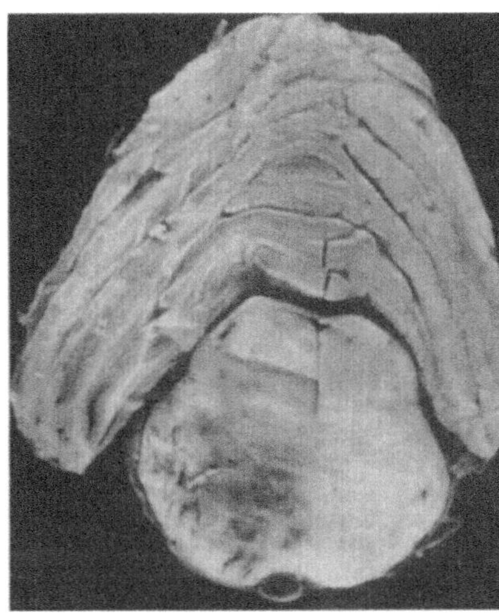

Fig. 25. Mucormycosis of the brain. Section of the pons showing thrombosis of the basilar artery with hemorrhagic necrosis (infarction) of the left side. Case of TINAZTEPE and TINAZTEPE (Fig. 4)

The meningeal vessels were congested. The inferior portions of both frontal lobes and of the tip of the right temporal lobe were red and soft. These changes were more extensive in the right frontal lobe than in the left. The opposing surfaces of the dura mater were soft and red.

The pituitary gland was red and soft. The right middle ear contained red brown fluid. Fungi were found in the Haversian system and intertrabecular spaces of the bone around the right middle ear, in the pituitary gland and in the lungs. There was patchy infarction of the small intestine.

The right internal carotid artery was occluded by thrombus. There was recent thrombotic occlusion of the aorta below the origin of the inferior mesenteric artery and into the common iliac artery.

Case of DWYER and CHANGUS (1958). The authors write: "The route of spread was demonstrated from the maxillary sinus into adjacent bone, along vessels, especially through the perineural vessels of the branches of the maxillary nerve, into the sphenopalatal ganglion involving the periorbital tissues, both cavernous sinuses, meninges, brain."

The frontal lobes were not involved but the left temporal lobe was.

The above should be a sufficient number of case reports to indicate the usual route of access of the fungus to the brain and the types of cerebral pathology produced.

There appeared to be evidence for the direct spread of the fungus from the paranasal sinuses and orbit, through the softened bone of the floor of the anterior fossa of the brain and

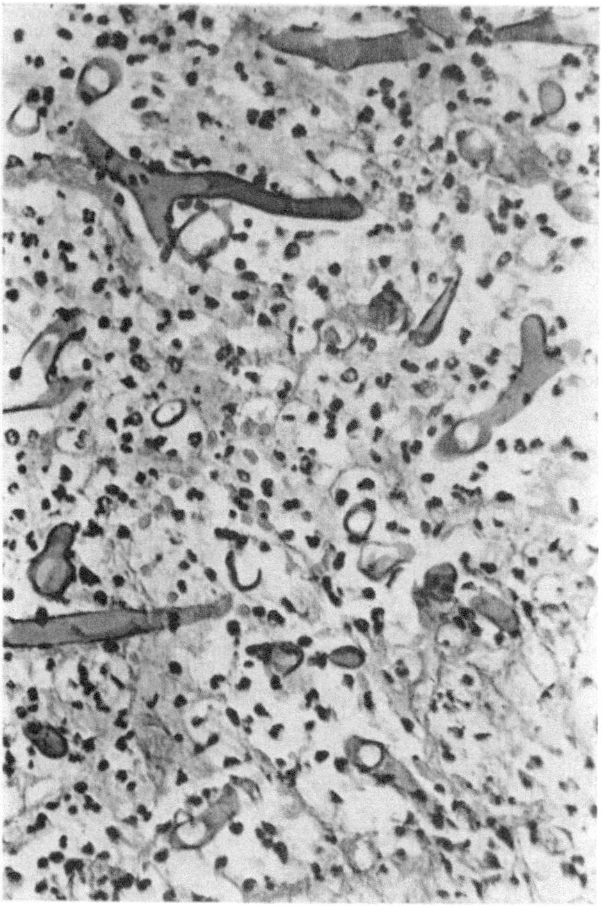

Fig. 26. Mucormycosis of the brain. High power microscopy showing hemotoxophilic, large nonseptate fungi (Fig. 7). Case of TINAZTEPE and TINAZTEPE, a $4^1/_2$-year-old boy with uremia and acidosis complicating bilateral hydronephrosis

through dura to the orbital surfaces of the frontal lobe or lobes of the brain in the case of KURREIN (1954), the cases 1 and 2 of SMITH and KIRCHNER (1958), BORLAND (1959), case 2 of BURNS (1959), DOLMAN and HERD (1959), LIE-KIAN-JOE (1959), LONG and WEISS (1959), GASS' case 2 (1961), case 1 of WASSERMAN (1961), SUPRUN (1962), AGREST (1963), case 2 of FUENTES (1963), LA TOUCHE et al. (1963), case 1 and 2 of GORDON (1965), HURTADO et al. (1965), PEREZ et al. (1966) and TINAZTEPE and TINAZTEPE (1966) (Figs. 22—26).

In summary, the changes in the brain frequently encountered at autopsy in cases of the classical central nervous system syndrome of mucormycosis may be:

1. Softening of the orbital surface of one of the frontal lobes corresponding to a region of localized pachymeningitis and necrosis of bone above a paranasal sinus or the orbit (Figs. 22, 23). At times the process is bilateral.
2. A diffuse acute meningovascular mucormycosis (Figs. 19, 20, 21).
3. Multiple minute infarcts throughout the brain.
4. Infarcts of the pons, the temporal lobes or even an entire hemisphere (Figs. 24, 25, 26, 27).

The general change is an acute meningo-encephalitis with infarcts of various sizes secondary to mucormycotic arteritis and thrombosis. Abscesses scarcely

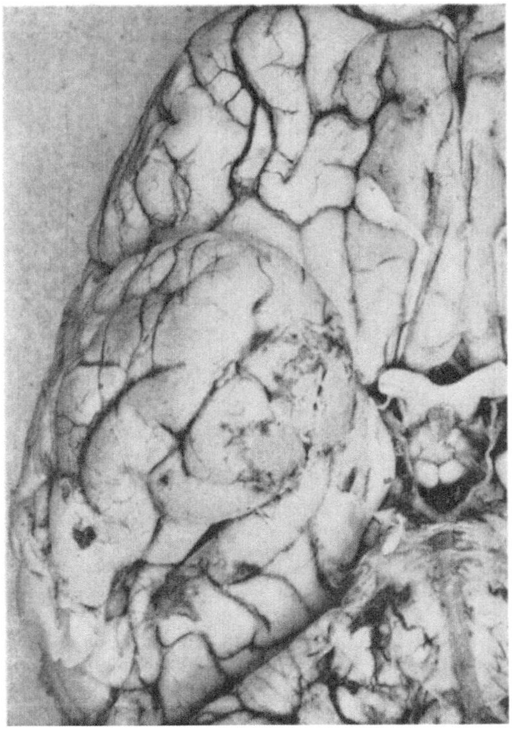

Fig. 27. Mucormycosis of the base of the right temporal lobe. The lesion corresponded to a region of thickened dura in the right middle fossa. This plaque of thick dura was firm, adherent to the meninges and brain, 3 cm in diameter, 3 mm in height and sharply outlined. Hyphae were demonstrated in the temporal lobe lesion Fig. 3 of LUBBE and PENNINGTON, courtesy of authors and Medical Journal of Australia

occur. Areas of encephalomalacia may simulate abscesses. The inflammatory response to the presence of the hyphae of *Rhizopus* is at times nil or minimal. The hyphae may stimulate microabscess formation very rarely. Much of the inflammatory reaction in the brain is secondary to the necrotic cerebral tissue caused by ischemia. Accompanying bacterial infecton is reported, but rarely.

Cerebral Lesions from the Systemic Blood Stream

Not all mucormycosis of the brain originates in the nose, paranasal sinuses or the orbit. In some cases the infection begins in the lung and there is secondary hematogenous dissemination to the brain. Two cases illustrate this occurrence.

Case 11 of BAKER (1962), was one of acute lymphoblastic leukemia with pulmonary mucormycosis. The brain contained a swollen red region, 3 cm in diameter, which was softer than normal, and which occupied the internal capsule, the caudate and lentiform nuclei, and the thalamus (Fig. 28). This region broke away easily from the surrounding cerebral tissue.

Case 23 of GRUHN and SANSON is remarkably similar to BAKER's case. They picture a coronal brain slice with a rounded lesion in the corpus striatum and basal ganglia.

Mucormycotic lesions of the brain which are secondary to primary pulmonary infection are rounded and deep in the brain, contrasting with mucormycotic lesions which have developed on the inferior aspect of the frontal lobe as a result of extension of the infection upward.

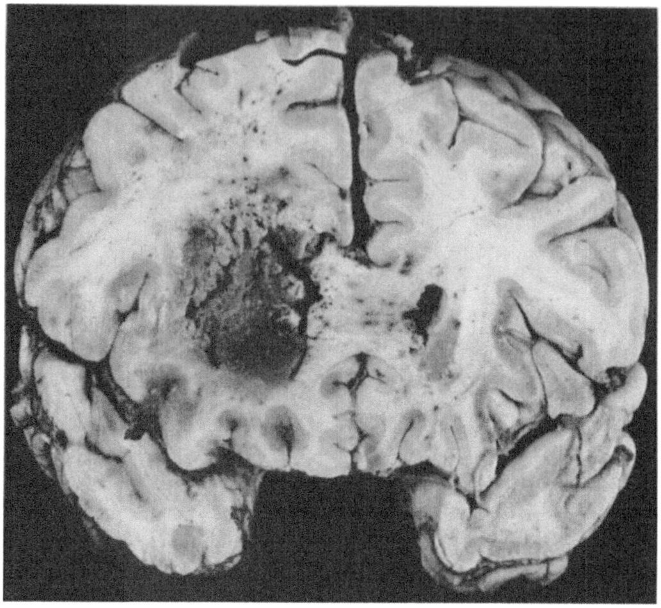

Fig. 28. Mucormycotic infarct in acute lymphoblastic leukemia. The primary lesion in this case 11 of Baker (1962) was in the lower lobe of the right lung; and the cerebral lesion is metastatic from the lung

Mucormycotic Mycotic Aneurysms

In 3 cases of cranial mucormycosis in the literature mucormycotic aneurysms have developed. As with other infectious agents such as the tubercle bacillus or *Streptococcus viridans*, the organism grows into the wall of an artery and weakens it. An aneurysm or swelling may develop, and may rupture, causing hemorrhage. This is a mycotic aneurysm no matter what the living agent causing the disease may be. Here the term mycotic does not refer to a fungus infection but to any living agent. Three illustrative cases are pertinent.

A white male of 62 years, a non-diabetic, had had a portacaval shunt operation performed 2 years antemortem (Table 2, Baker, 1957a, case 1). Three days before death he became comatose and spinal tap revealed xanthochromic, bloody spinal fluid. The carotid angiograms showed obstruction of the left internal carotid artery, and there was an aneurysmal dilatation of this artery just proximal to the origin of the left anterior cerebral artery, and the aneurysm filled from the right. At autopsy, a massive subarachnoid hematoma was found in the left middle fossa, extending through the comminuted tissue of the left temporal lobe into the lateral ventricle. Embedded in the hematoma was the left internal carotid artery, the free basal end of which was expanded into a ragged, saccular dilatation 1 cm in diameter, suggesting the distal half of an aneurysm. Obstruction of the remaining extracranial portion of the artery in the neck was demonstrated by the injection of water into the aortic end of the common carotid and by noting that water did not escape from the internal carotid artery inside the skull.

Microscopically, fungus invasion of cerebral arteries was strikingly evident in sections of the left internal carotid, both middle cerebrals, both posterior communicating arteries, and the basilar artery. The fungus extended along small blood vessels and nerves into the left orbit where it was limited to those trunks, not being found free in the areolar tissue. The mycelium was most abundant in the lumina and subendothelial tissues of arteries, frequently obstructing them with a tangle of filaments rich in neutrophils and embedded in fibrin. Occlusion of the left internal carotid artery was due to such a mass. A satisfactory section of the aneurysm was not obtained, but sections of this segment of the internal carotid artery showed ulcerations of the artery wall extending from the luminal surface through to the adventitia.

The wall of the left cavernous sinus was invaded by fungus, as was the pituitary capsule in its entirety, with some extension of the mycelium into the posterior lobe of the gland. The remaining dural sinuses were uninvolved. Intra-arterial extension of the fungus into the brain was evident, with involvement of small areas of the left frontal cortex, left insular cortex and left caudate nucleus. Hyphae penetrated freely into the brain substance and evoked very little inflammatory response. Pial vessels overlying the left temporal lobe, cerebellum, and pyramids also bore their load of fungus and had excited remarkably little meningeal reaction.

The immediate cause of death was considered to be massive intracranial hemorrhage due to rupture of an aneurysm of the left internal carotid artery, the origin of which is still not clear. From its location, it may well have been congenital rather than mycotic. Rupture may have been precipitated by invasion of the wall of the aneurysm by the fungus.

In brief, there was an aneurysm, which may have been a congenital aneurysm to begin with and then became a mycotic aneurysm and ruptured. Or, possibly the aneurysm was entirely fungal in origin. In any event, the case illustrates the affinity the fungus has for arteries and its ability to ulcerate the inner surfaces and to penetrate and weaken the walls of the artery.

In an article entitled "atypical cerebral aneurysms", STEHBENS (1965) of Australia reported the case of a woman of 35 years who had developed acute lymphoblastic leukemia and had received cortisone, antibiotics, purinethol and multiple transfusions. The cortisone therapy favored the entrance and spread of the fungus. Autopsy confirmed the diagnosis of acute leukemia with widespread hemorrhages. A large subdural hematoma was found over the anterior half of the left cerebral hemisphere, with an extensive subarachnoid hemorrhage at the base of the brain. Blood clot was firmly adherent to the right internal carotid artery at the origin of the posterior communicating artery. An aneurysmal sac was not seen, but was believed to be embedded in blood clot and fibrin. Another small artery embedded in clot nearby had a fusiform dilatation over a length of 4 mm. The second, third, fourth, and fifth cranial nerves on the right side were incorporated in the organizing blood clot.

Serial paraffin sections were cut of the fusiform aneurysm and of the right internal carotid artery. On microscopic examination, the architecture of their walls was grossly disturbed by arteritis with extensive infiltration of polymorphonuclear leucocytes, many of which were necrotic. Some loose fibroblastic reaction had occurred, and giant cells were present. Numerous broad branching hyphae, mostly non-septate, had infiltrated the arterial wall and were present in thrombus and inflammatory exudate in the lumen. No sporangia were seen. The diagnosis was made on histological grounds, and meningeal involvement was also found.

In this case there was a fusiform unruptured mucormycotic mycotic aneurysm, and possibly another one which had ruptured.

ABRAMSON et al. (1967), reported from Boston, Massachusetts, a terminal hemorrhage arising from a ruptured mycotic aneurysm of the internal carotid artery, in the carotid canal of the petrous portion of the temporal bone. There were hemorrhagic infarctions in the right parietal and occipital areas. Extensive granulation tissue was present around the cranial nerves at the apex of the right orbit, but no fungi were present in that region. In this case at autopsy there was also necrosis involving the sphenoid bone and adjacent temporal, occipital and frontal bones and superior nasopharynx. Typical large, rarely septate hyphae were present in the necrotic material. The patient was a 38-year-old Negro man with diabetes for many years who developed acidosis and orbital mucormycosis and a mucormycotic mass in the nasopharynx.

Further information than that given in the published report is obtained through the courtesy of Dr. RONALD A. ARKY and Dr. KAROLY BOLOGH of Boston (ABRAMSON et al.). Dr. V. SANGALANG diagnosed thrombosis with aneurysmal rupture of the petrous portion of the right internal carotid artery, secondary to direct extension from mucormycotic infection of the nasopharynx. Longitudinal sections of the thrombosed petrous segment of the right internal carotid artery showed an abrupt transition from the intact to the degenerated segment, especially well demonstrated by elastic tissue stains. The wall was totally replaced by necrotic inflammatory debris with mural thrombosis. Micro-organisms were not seen.

The failure to find hyphae in the wall of the vessel suggested that the necrosis was of ischemic origin due to obstruction of vessels by the mucormycotic process elsewhere.

The above three cases indicate that infiltration and weakening of the walls of arteries by hyphae of mucormycosis may produce aneurysms and rupture of arteries.

Table 3. *Pulmonary Mucormycosis (46 cases)*

Authors and Case Number	Location of Case Age, Sex, Race	Predisposing Factors	Anatomic Involvement	Cultures	Duration	Treatment	Status of Patient
FÜRBRINGER (1876) Case 1	Europe 66, M	Metastatic carcinoma	2 walnut-sized foci, right upper lobe	No	—	—	Dead, mycosis incidental
— Case 2	Europe 31, M	Gastrointestinal catarrh	3 apical walnut-sized nodules	No	—	—	Dead, mycosis incidental
PODACK (1899)	Europe 39, M	Carcinoma of pleura	Subpleural nodules	No	—	—	Dead, mycosis incidental
LANG and GRUBAUER (1923)	Europe 56, F	Tuberculous pulmonary cavity	Wall of pulmonary cavity	*Mucor corymbifer*	—	—	Dead, mucormycotic and coccal pneum.
WÄTGEN (1929)	Graz/Austria 34, F	Sarcoma	Hemorrhagic infarcts of lung	No	Not stated	Not stated	Dead, mycosis incidental
GUKELBERGER (1938)	Bern/Switzerland 46, M	Pneumonia, mild diabetes	Lung	No	2 mos.	Iodides	Living
LLOYD et al. (1949)	Boston/USA 26. F, N	Pregnancy, macerated fetus, dehydration	Lung: infarct of right upper lobe	No	13 days	For dehydration	Dead, mucormycosis
BAKER (1956), Case 1 BAKER (1962), Case 2	Durham, N.C./USA 47, M, W	Leukemia, cortisone, ACTH	Right lung	No	30 days	Antibiotics	Dead, mucormycosis
— Case 2	Chapel Hill/USA 23, M, W	Leukemia, cortisone	Right lung, right pleura, spleen	No	19 days	6-MP-antibiotics, cortis.	Dead, mucormycosis
— Case 3	Greenwood/USA 51, M, W	Diabetes	Hilus and base of left lung	No	13 days	Antibiotics, insulin	Dead, mucormycosis
— Case 5	Durham/USA 50, M, N	Diabetic acidosis, syphilis	Upper lobes of both lungs	No	8 days	—	Dead, mucormycosis
KEYE and MCGEE (1956), Case 1	St. Louis/USA 38	Hodgkin's disease	Lungs	—	—	—	Dead, mucormycosis

— Case 2	St. Louis/USA 23	Leukemia	Lung	—	—	—	Dead, mucormycosis
— Case 3	St. Louis/USA 1.5	Leukemia	Lung	—	—	—	Dead, mucormycosis
— Case 4	St. Louis/USA 38	Hodgkin's disease	Lung and esophagus	—	—	—	Dead, mucormycosis
— Case 6	St. Louis/USA Age unknown ?	Leukemia	Lungs	—	—	—	Dead, mucormycosis
Mayfield and Condie (1957)	Seattle/USA 24, M, W	Leukemia, cortisone	Left lung, pulmonary artery	No	23 days	6MP, amethopterin, cortisone	Dead, mucormycosis
Stefanini (1957)	Boston, USA 35, M	Leukemia, ACTH, steroids	Lungs, pulmonary arteries	No	2 weeks	ACTH, corticosteroids, penicillin	Dead, mucormycosis
Torack (1957) Case 11	New York/USA 50, F	Leukemia, meticorten	Lung	No	18 days	Urethane, meticorten	Dead, mucormycosis
Dillon et al. (1958) Dillon and Sealy (1962), Case 1	Durham/USA 20, F, N	Diabetic acidosis	Right bronchus intermedius	No	30 days	Resection of right, middle and lower lobes	Cured
Blankenberg and Verhoeff (1959)	Huntersville/USA 23, F, N	None (Diabetes appeared 1 year after operation)	Right, middle, and lower lobes	No	3 mos.	Operative removal	Living
Burns (1959) Case 3)	New York/USA 36, M, N	Diabetic acidosis, prednisone, splenectomy	Right lung, pulmonary arteries, meninges	No	3 days	Prednisone	Dead, mucormycosis
Hutter (1959) Case 1	New York/USA 13, M, W	Leukemia, Meticorten	Left lower lobe of lungs, heart	No	23 days	6-MP, antibiotics	Dead, mucormycosis
— Case 2	New York/USA 62, M, W	Leukemia, Meticorten	Right upper and lower, and left lower lobes, spleen	No	10 days	6-Chloropurine, antibiotics	Dead, mucormycosis

Table 3. Continuation

Authors and Case Number	Location of Case Age, Sex, Race	Predisposing Factors	Anatomic Involvement	Cultures	Duration	Treatment	Status of Patient
— Case 3	New York/USA 8, M	Lymphosarcoma, Meticorten	Lungs, larynx, heart, esophagus, stomach, ileum	*Mucor*	21 days	Antibiotics, mycostatin	Dead, mucormycosis
SHANKLIN (1959)	Syracuse, USA 54, M, W	Cushing's syndrome	Left upper and lower, and right upper lobes of lungs, thyroid	No	34 days	Antibiotics	Dead, mucormycosis
CUADRADO et al. (1961), Case 2	Dallas/USA 65, F, W	Leukemia	Upper lobes of lungs, main pulmonary artery	No	10 days	6-MP, cortico-tropin, antibiotics	Dead, mucor-mycosis and leukemia
GLOOR et al. (1961) Case 1	Basel/Switzerland 63, M	Leukopenia, corticosteroids	Right upper lobe, entire left lung	*Rhizopus oryzae*	17 days	Antibiotics, corticosteroids	Dead, mucormycosis
BAKER (1962) Case 4	Durham/USA 20, M, W	Leukemia, neutro-penia, Meticorten	Lung, liver	No	4 weeks	Meticorten	Dead, mucormycosis
— Case 9	Durham/USA 70, M, W	Leukemia, neutropenia, prednisone	Right middle lobe of lungs	*Mucor*	6 weeks	Antileukemics, prednisone, antibiotics	Dead, mucormycosis and aspergill.
DILLON and SEALY (1962), Case 2	Durham/USA 15, F	Diabetic acidosis	Right middle lobe	*Mucor*	3 mos.	Lobectomy	Cured
STRAATSMA et al. (1962), Case 15	AFIP Accession No. 545672 55, M, W	Carcinoma and renal insuffi-ciency	Pulmonary	—	—	—	
— Case 16	AFIP Acc. No. 561012; 67, M, W	Leukopenia	Pulmonary	—	—	—	
— Case 17	AFIP Acc. No. 732623; 25, M, W	Leukemia	Pulmonary	—	—	—	
— Case 18	AFIP Accession No. 755869 69, M, W	Carcinoma, dia-betes, renal in-sufficiency	Pulmonary	—	—	—	

— Case 19	AFIP Acc. No. 758019; 67, M, N	Carcinoma	Pulmonary	—	—	—	
— Case 20	AFIP Acc. No. 846871; 32, M, W	Leukemia	Pulmonary	—	—	—	
— Case 21	AFIP Acc. No. 915355; 45, M, W	Leukemia	Pulmonary	—	—	—	
Balasubrahmanyan, Chaudhuri (1963)	Pondicherry/India 5, M	Uremia of hydronephrosis	Both lungs	No	7 days	Antibiotics	Dead, mucormycosis
Darja and Davy (1963)	Saskatoon/Canada 72, M, W	Leukemia, steroids	Upper left lobe, right upper and middle lobes of lungs	*Mucor corymbifera*	21 days	Antibiotics, total body irradiation	Dead, mucormycosis
Dourov and Dustin (1964)	Belgium 65, F	Leukemia	Vessels of lungs	No	Not stated	Not stated	Dead, mucormycosis
Hutter et al. (1964) Case 22	New York/USA 62, M, W	Leukemia, hypogammaglobulinemia, steroid-induced diabetes	Right middle lobe of lungs	No	9 days	Transfusions, X-ray to spleen and prednisone	Dead, aspergillosis and mucormycosis
Sun et al. (1965) Case 8	Peking/China 42, F	Uremia, hydrocortisone	Right upper lobe of lungs, bronchi and vessels	No	Not stated	Antibiotics, hydrocortisone	Dead, uremia, mucormycosis
Wall and Madison (1965)	Winston-Salem/USA; 36, M, W	Leukemia, prednisone	Right lung and right pulmonary artery	No	3 weeks	Hydroxy urea, prednisone	Dead, mucormycosis, hemoptysis
Winston (1965)	England 29, M	Diabetes	Left main bronchus, pulmonary artery	No	35 days	—	Dead, mucormycosis, hemorrhage
Ellis (1966)	London/England 55, M	Diabetes	Right upper lobe of lungs	No	3 days	Insulin, intravenous fluids, aramine	Dead, mucormycosis

Mucormycosis of the Spinal Cord

There are 3 cases in which spinal cord lesions are mentioned.

In case 3 of GREGORY et al. the mucormycosis affected the cervical cord.

In the postmortem examination of case 3 of BANKER the hyphae of mucormycosis were identified in the walls and lumens of thrombosed arteries and veins surrounding the thoracic, lumbar and sacral segments of the spinal cord and in the subarachnoid space. The spinal cord was infarcted and had probably been so for 5 days. There was no inflammation or glial reaction. This case was that of a 12-year-old boy with erythroleukemia of a year's duration treated with steroids for 6 months. One month before death he developed spastic paraplegia and a sensory level at the mid-thoracic region. Hyphae were identified in the lungs and kidneys at autopsy as well as in the spinal canal. The infection originated in the lung, and the spinal cord involvement is the result of dissemination of the organisms. The case is listed in Table 6 of disseminated mucormycosis (BANKER, 1961).

In case 5 of PARKHURST and VLAHIDES the spinal cord from T-1 to T-4 was infarcted, with thrombi and hyphae present in adjacent vessels. The autopsy also presented a large hemorrhagic infarct of the left lung with infarction of the pulmonary arteries supplying the infarct. Moreover, in the heart there was an infarct in the right anterior ventricular wall with an adherent mural thrombus. The thrombus and many vessels contained numerous nonseptate hyphae (see Table 6, PARKHURST and VLAHIDES, 1967). The patient had leukemia and had been treated with steroids and cytotoxic drugs.

Thus, mucormycosis does invade the spinal cord. In one instance it was an extension of the process from the brain in a cranial case, and in two instances a metastatic lesion from pulmonary mucormycosis.

Primary Pulmonary Mucormycosis

Table 3 lists 49 reported cases of pulmonary mucormycosis, but does not indicate 29 cases of disseminated mucormycosis which originated in the lung and displayed primary pulmonary mucormycosis (Table 5). In Table 4 of gastrointestinal mucormycosis there are two cases with pulmonary lesions, but it appears more probable that the infection passed from the digestive tract to the lung, rather than vice versa. Thus there are 78 cases of primary pulmonary mucormycosis in the world literature. This compares with 107 cases of cranial mucormycosis or about 70% as many.

In slightly less than half of the cases of primary pulmonary mucormycosis there was antecedent leukemia and many of the leukemic patients had received corticosteroids. About 10% of the cases developed in diabetic patients and 4% in patients with Hodgkin's disease. There was a variety of other predisposing conditions, and in a few cases, specifically in those of the early German literature, no predisposing condition was cited.

Cultures in the pulmonary cases were reported by LANG and GRUBAUER (1923) as *Mucor corymbifer*, by HUTTER, case 3 (1962) as *Mucor*, by GLOOR et al. (1961) as *Rhizopus oryzae*, by BAKER (1962) as *Mucor*, by DILLON and SEALY (1962) as *Mucor*, and by DARJA and DAVY (1963) as *Mucor corymbifera*.

Among the disseminated cases which may have begun in the lungs, cultures were reported by MCBRIDE et al. (1960) as *Rhizopus arrhizus*, and by BAKER (1962) as *Mucor*.

In summary, *Mucor* was reported 5 times, *Rhizopus* twice, and *Absidia* once. This is in contrast to cranial mucormycosis, in which the reports of cultures listed *Rhizopus* as first and *Mucor* and *Absidia* less frequently. It appears that *Mucor* is the predominant organism in pulmonary cases and *Rhizopus* in cranial cases. However, it would be interesting to analyze the evidence for the diagnosis of *Mucor* and *Absidia* in all of these cases. In many reports there is simply the statement that such and such an organism was cultured and there is no description, illustration, or indication that the organism was submitted to a mycologist with a special interest in the Mucoraceae.

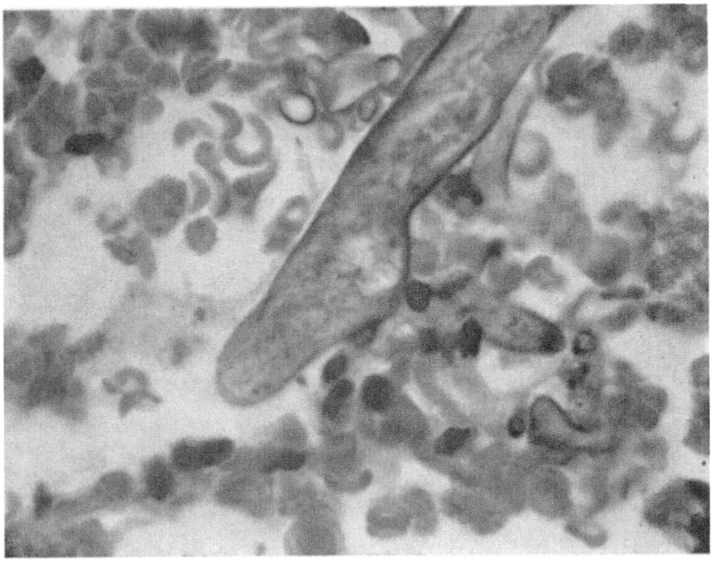

Fig. 29. Pulmonary mucormycosis. Growing tip of nonseptate hypha entering an alveolus (H & E, × 1,000)

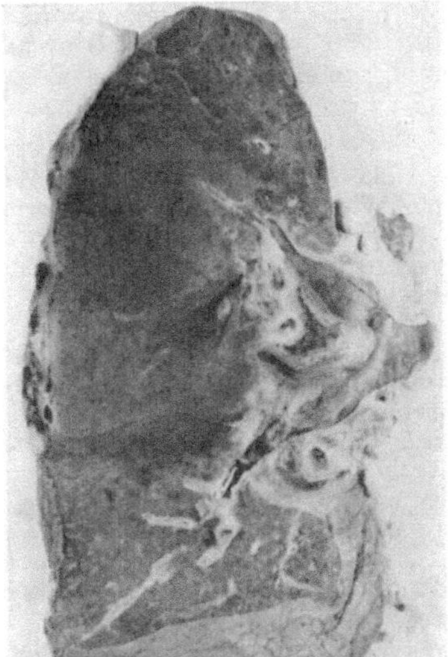

Fig. 30. Pulmonary mucormycosis. Massive mucormycotic infarct of the upper lobe and the upper portion of the lower lobe of the right lung. Note the enormous mucormycotic thrombus of the pulmonary artery and its main branches, and the pleurisy on the lateral aspect of the upper lobe. Duke Hospital autopsy No. 8775

Fürbringer (2 cases) (1876), and Podack (1899) described walnut-sized pulmonary foci and subpleural nodules as instances of mucormycosis of the lung. Lang and Grubauer (1923) found the fungus in the wall of a tuberculous cavity and cultured *Mucor corymbifer*.

Wätjen (1929) described infarcts of the lung.

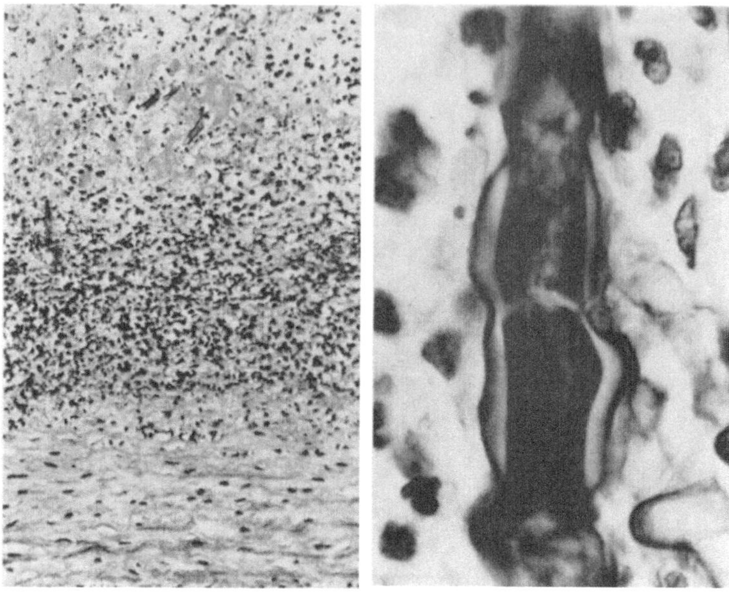

Fig. 31 (left). Mucormycotic thrombosis of an artery of the lung. Media of artery below, neutrophilic infiltrate at periphery of thrombus, and hyaline thrombus with fungus hyphae, at the top. H & E, × 200. Figure 3 of Case 1 of BAKER (1956)
(Right). Segment of hypha in the wall of a pulmonary blood vessel. Iron hematoxylin. H & E, × 1487. Figure 4 of Case 1 of BAKER (1956)

In the case of pulmonary mucormycosis described by LLOYD et al. (1949), the pleural surfaces of the right lung were pinkish red to purple, with fibrin on the upper lobe, especially posteriorly. This upper lobe was rubbery in consistency. On the pleural surface there were scattered, fine cystic blebs measuring 0.1 cm in diameter, none of which was ruptured. The cut surface of the right upper lobe showed a red, homogeneous, wet surface which oozed large amounts of fluid on pressure. The medial aspect of the right upper lobe just above the interlobar fissue revealed a black, homogeneous-looking area of softening occupying about one third of the lobe, which was in contrast to the firm surrounding red area. The right lower and middle lobes were of a dark red color and had a wet red surface which oozed large amounts of fluid on pressure. The whole lung was subcrepitant to firm in consistency, the upper lobe giving an impression of early consolidation, the middle and lower lobes giving an impression more consistent with atelectasis, congestion, and edema. The pulmonary vessels were dissected to their third and fourth branches. A questionable firm reddish-yellow thrombus was noted at the second bifurcation of the right lower lobe vessel. After fixation, a moderate sized, gray, friable, adherent thrombus was found in the artery leading to the softened area in the right upper lobe. The left lung was congested.

The bronchial mucosa was a deep reddish pink and edematous. The lumen contained tenacious, reddish gray, mucopurulent material. The hilar nodes showed moderate anthracosis.

Microscopically, LLOYD et al. examined 18 sections of lung. Sections from the soft, black, homogeneous area of the right upper lobe showed broad fungal hyphae devoid of cross septa. In the lung tissue where the mycelia were numerous there was preservation of the alveolar pattern but the cellular detail was blurred. There was abundance of cellular debris and edema fluid in the alveolar spaces. The pleura was markedly thickened and edematous. There was a fibrinopurulent exudate on the surface containing many polymorphonuclear neutrophils. One section showed a definite, recent, pulmonary infarct with the characteristic microscopic findings. Section of the pulmonary artery supplying the area of the infarct showed a large, antemortem thrombus with massive invasion of the thrombus and of the vessel walls by mycelial hyphae.

This careful examination and description of LLOYD et al. indicates clearly that the lesion of the right upper lobe was an antemortem infarct dependent upon vascular thrombosis stimulated by the intraluminal growth of the fungus. This pulmonary picture is presented in the majority of the cases subsequently reported.

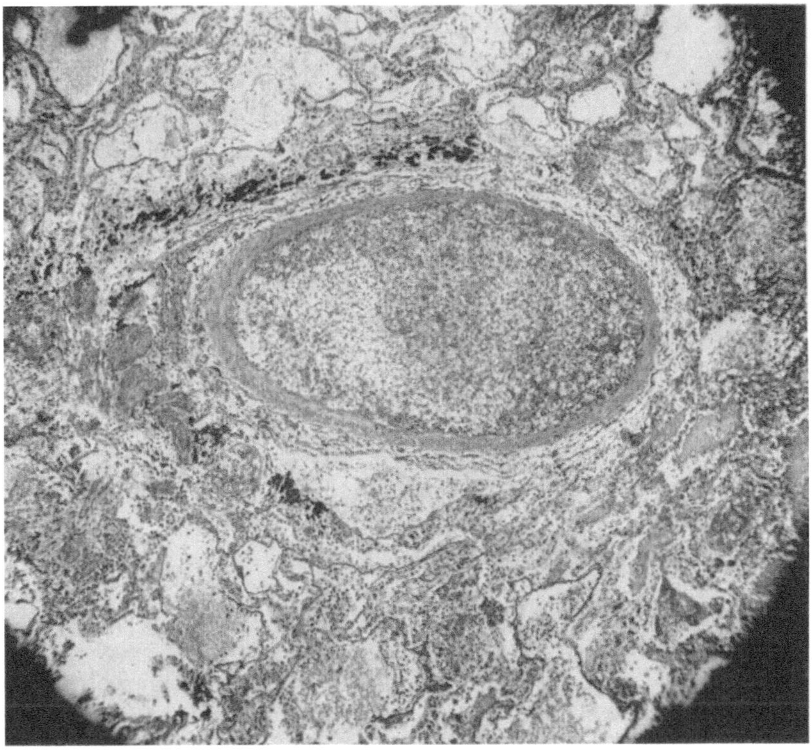

Fig. 32. Mucormycotic thrombus of an artery in the mucormycotic infarct of the lower lobe of the right lung in case 11 of BAKER (1962). The pulmonary infection gave rise to the metastatic infarct of the brain illustrated previously. H & E, × 60. Hyphae are hardly visible in the thrombus. In G.M.S. stain they would be easily visible. Figure 14, case 11 of BAKER (1962)

MURPHY and BORNSTEIN (1950) obtained a calcified nodule of the lung at operation and reported it as an example of mucormycosis. The cross section of the excised specimen shows concentric rings and there is a thin peripheral capsule. The appearance suggests histoplasmosis. The photomicrographs are not sharp, and the appearances of the organisms of mucormycosis are not clearly shown. A *Mucor* was grown from the calcified material and injected into a rabbit. The photomicrograph of the section of kidney from the rabbit shows non-septate hyphae like those of mucormycosis. The cultured fungus was identified as *Absidia italiana* (COST and PERIN) Dodge. The case must be listed as doubtful because of the uncharacteristic appearance of the fungus in the human tissue preparations.

The case of NICOD et al. (1952), reported as a double pulmonary infection with *Aspergillus fumigatus* and *Mucor pusillus*, also lacks documentation with respect to histopathology.

BAKER (1956) reported four cases of primary pulmonary mucormycosis, two in patients with leukemia and two in patients with diabetes (Figs. 30—36). The pulmonary lesions were as follows:

Case 1: Thrombosis, massive infarct, pneumonia, pleurisy, right lung; mycotic bronchitis.

Case 2: Thrombosis, infarct, right lower lobe; mycotic bronchitis.

Case 3: Thrombo-arteritis, thrombo-phlebitis, mycotic pneumonia, left lung; mycotic bronchitis.

Case 4: Thrombo-arteritis, mycotic pneumonia, right upper lobe.

MAYFIELD and CONDIE (1957) reported mucormycotic thrombosis of intrapulmonic arteries with pulmonary infarcts. An interesting feature of their case was the presence of a mucormycotic mural thrombus of the main pulmonary artery close to the pulmonary valve.

DILLON et al. (1958) reported a case, in a diabetic patient, of mucormycosis of the bronchus successfully treated by lobectomy. Biopsies of the distal portion of the right bronchus intermedius had shown the broad nonseptate hyphae in the necrotic wall of the bronchus. There

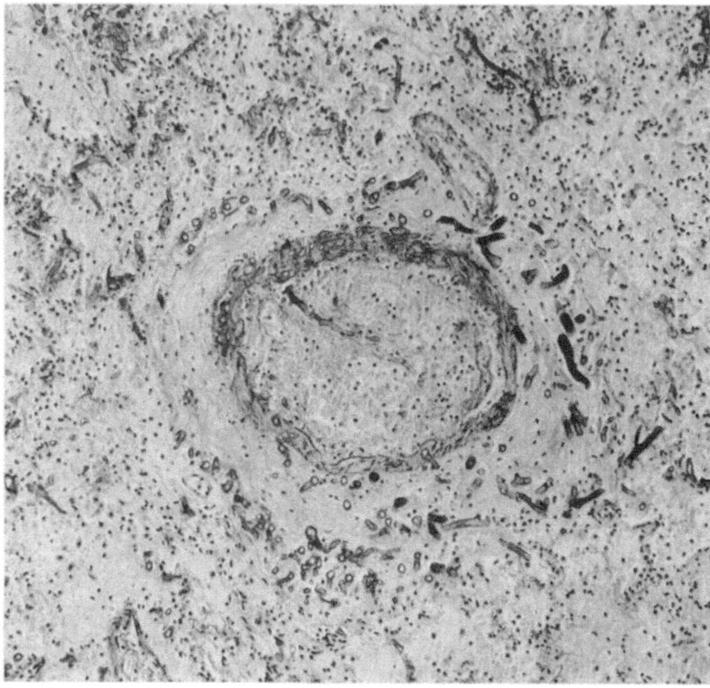

Fig. 33. Pulmonary mucormycosis. Hyphae in and about a thrombosed blood vessel within a pulmonary infarct. H & E, × 150. From CONANT et al. (1954) (Fig. 105)

was an abscess in the removed right lower lobe, but without fungus organisms. The mucormycosis of the bronchus apparently had led to the development of the pulmonary abscess of other cause. Or, the abscess could have been a softened infarct.

In the case of BLANKENBERG and VERHOEFF (1959) there was a cavity measuring 6 by 5 by 4 cm in the resected right lower lobe and another cavity measuring 4 by 4 by 4 cm in the resected right middle lobe. The cavities were surrounded by thick layers of grayish-brown fibrinous hemorrhagic material and thick layers of indurated whitish-gray necrotic tissue. Many branches of the medium-sized bronchi were completely obliterated and filled with necrotic whitish-brown firm material. Cross sections of the lung showed many foci of yellowish-gray, firm, and soft friable areas near the large cavity. Microscopically there were areas of liquifaction necrosis surrounded by thick layers of eosinophilic, amorphous, or granular material with mild infiltration of neutrophils, lymphocytes and plasma cells. Many multinucleated foreign body giant cells were seen. In the necrotic area numerous basophilic, irregularly shaped structures were observed. Around the necrotic spaces many large blood vessels and bronchi showed almost complete necrosis of the wall. Other blood vessels showed hyalinization of the wall or recent thrombus of the lumen, characterized by obliteration with spindle shaped cells with hyperchromatic large nuclei. Occasionally even small bronchi contained much foreign material. The foreign material was characterized by broad branching coenocytic hyphae of varying thickness. When the fungus appeared to be growing luxuriantly, it was thick; in regions in which infarction had occurred or the tissues were necrotic, hyphae were shrunken and irregular. A marked tendency to invade vessels was noted.

It is my opinion that the cavities represented liquifaction of infarcts. The clinical history suggested that the process in the lungs might have begun more than three months prior to lobectomy when a non-productive cough began. Nearly six weeks before lobectomy she developed what she described as a chest cold. Nineteen days before lobectomy she began to produce about one ounce of blood-streaked sputum daily and developed a fever. There was some evidence at this time that there was an associated bacterial infection, and there was x-ray evidence of some form of pneumonia in the left lung. However, I would think it likely that the cavities in this case, and in the case reported by DILLON et al., represented liquifaction necrosis of *pulmonary infarcts* caused by mucormycosis. These would represent late stages of pulmonary infarcts.

In Case 3 of BURNS (1959) there was *thrombosis of the pulmonary artery* branches to the left upper and lower lobes with pulmonary infarction. The thrombosis was due to invasion of the pulmonary artery by mycelia of mucormycosis. In addition, mycelia were in the pulmonary alveoli. In this case there was a history of slight left ptosis, and it may be that the case is primarily of cranial type. I have, however, classified it as primarily pulmonary because the pulmonary findings were so striking, but may be in error in so doing.

Case 1 of HUTTER (1959) was that of a boy with acute leukemia. In each upper lobe of the lungs there were yellowish-gray bulging zones, 1 cm in diameter, surrounded by congestion. The artery to the posterior basal segment of the left lower lobe was thrombosed and was surrounded by congestion. Microscopically this vessel contained a fungus with large, coenocytic (nonseptate) hyphae with branching at obtuse angles. These hyphae were present in other smaller vessels and extended through their walls and into the parenchyma, where there was infarction. Alveoli contained hyaline membranes, and smaller vessels had pale hyaline material appearing like loose white thrombi. Aspiration pneumonitis, clumps of bacteria, and leukemic infiltrate were also noted in the lungs. Dr. HUTTER's comment was: "This child had pulmonary mucormycosis with mycotic embolization to the myocardium. There was also histological evidence of bacterial infection in the lungs. Although the infecting fungi may have been airborne, the possibility of infection from draining maxillary sinuses exists. Possible predisposing factors were leukemia, chemotherapy (steroids and antibiotics) and metabolic alterations (steroid-induced diabetes).

In HUTTER's case 2, at autopsy, the infarct in the right upper lobe measured 5 by 5 by 3 cm, the one in the right lower lobe, 3 cm in diameter. An infarct in the left lower lobe was small and was not measured. The vessels adjacent to the hemorrhagic areas contained antemortem clots. Microscopically there were branching, nonseptate hyphae in these clots as well as in smaller arteries and venules. The fungi extended through vessel walls and were plentiful in the alveoli as well as in the bronchioles. Pulmonary mucormycosis with mycotic embolization to the spleen were the essential features of this case.

In Case 3 of HUTTER the lungs (the right 600, and the left 300 g in weight) had numerous nodular hemorrhagic infarcts that on section showed central yellow pinhead-sized foci. There were also ulcerations of the major branches of the right main stem bronchus and granular reddish tan exudate on the vocal cords. Microscopically, there was marked bronchial and bronchiolar infestation with the organisms of mucormycosis. Two observations were of note: (1) the hyphae appeared to penetrate intact mucosal epithelium in places; and (2) sporangia were noted in the lumina of the bronchi. There were mucormycotic thrombi of large and small arteries and veins; red and white cells, and fibrin and platelets were present. The fungi penetrated the walls of bronchi and vessels and were present in alveoli in the surrounding zone of hemorrhagic infarction. Alveoli at the periphery of the infarct contained fibrin deposits. The vocal cords were ulcerated, and numerous hyphae of mucormycosis were penetrating into the tissue. In the aerated lumen sporangia were forming. In addition, there were numerous blastospores and pseudomycelia, characteristic of *Candida*, present on the surface of the vocal cords.

In the case of pulmonary *mucormycosis* complicating *Cushing's syndrome* reported by SHANKLIN, his Fig. 1 shows the posterior halves of the lungs with a cavity 5—7 cm in diameter in the left lower lobe, containing soft, friable, caseous material. In other parts of the left lower lobe there were several smaller cavities, 1—2 cm in diameter of a similar type. Other similar areas were found in the left upper lobe. In the apex of the right upper lobe there was a similar area, sharply demarcated by scar tissue. Microscopically, he describes foci of infarction, abscesses with hyphae and thrombosed vessels. Grossly, the lesions in the lungs are not wedge-shaped and do not extend to the pleura at the level of section, but are rounded and look like abscesses. The microscopic description suggests that they are truly infarcts with disintegration of the infarcted tissue, simulating abscesses.

In case 2 of CUADRADO et al. (1961) there were numerous wedge-shaped, dark red infarcts in the upper lobes of the lungs. The main pulmonary artery contained a premortem thrombus that occluded the main branch to the right upper lobe. Microscopically, organisms of mucormycosis were numerous in thrombi and infarcts.

In case 1 of GLOOR et al. (1961) there were, in both lungs, numerous large and small arteries occluded with fresh thrombi. In both upper lobes and in the left lower lobe there were larger hemorrhagic infarcts. In addition, there were foci of terminal bronchopneumonia in the lower lobes.

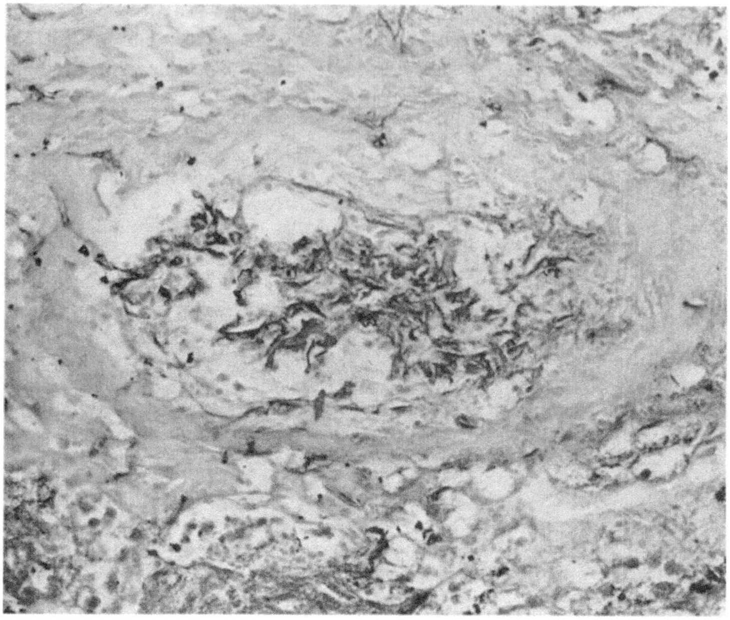

Fig. 34. Shrunken hyphae in the thrombosed lumen of a pulmonary vessel in an infarct of long duration. H & E, × 100

In illustration of his case 4, BAKER (1962) pictures a gray, firm, infarct of the right middle lobe, of 4 weeks' duration. There was a brown, thick exudate in the right pleural cavity, and fibrous adhesions overlay the right pleural cavity. Microscopically, the vessels and pulmonary parenchyma of the infarct contained thrombi and nonseptate hyphae. In the peripheral region there was granulation tissue and fibrosis.

At the postmortem examination of case 9 of BAKER there was hemorrhagic consolidation of the upper portion of the right lower lobe and the lower portion of the left lower lobe. In these consolidated areas there were yellow nodules, measuring from 0.5—1.5 cm in diameter. Microscopically, thrombi with nonseptate hyphae were observed in the lesions of the lung.

In DILLON and SEALY's case 2 a right middle lobe was removed surgically. Mucormycotic organisms were found in sections of the wall of an abscess in this middle lobe. One wonders if this abscess represented disintegration of an infarct.

In the case of BALASUBRAHMANYAN and CHAUDHURI (1963) there were small, firm areas scattered throughout the middle and upper lobes of the right lung and both lobes of the left lung.

On section, these were firm and congested and looked like small areas of necrosis. Microscopically, these nodules showed edema of the alveoli with small areas of necrosis around which dense collections of inflammatory cells were visible. The lesions were close to blood vessels which showed invasion of their walls and lumina by non-segmented mycelia.

These foci apparently represented scattered nodules of infarction dependent upon thrombosis.

In DARJA and DAVY's case there was a raised, purple area 7 cm in diameter in the base of the upper lobe of the left lung, and another such area 11 cm in diameter in the base of the right middle lobe. The pulmonary artery divisions leading to the upper lobe of each lung and to the right middle lobe were occluded by thrombi. On section, the indurated areas were well-circum-

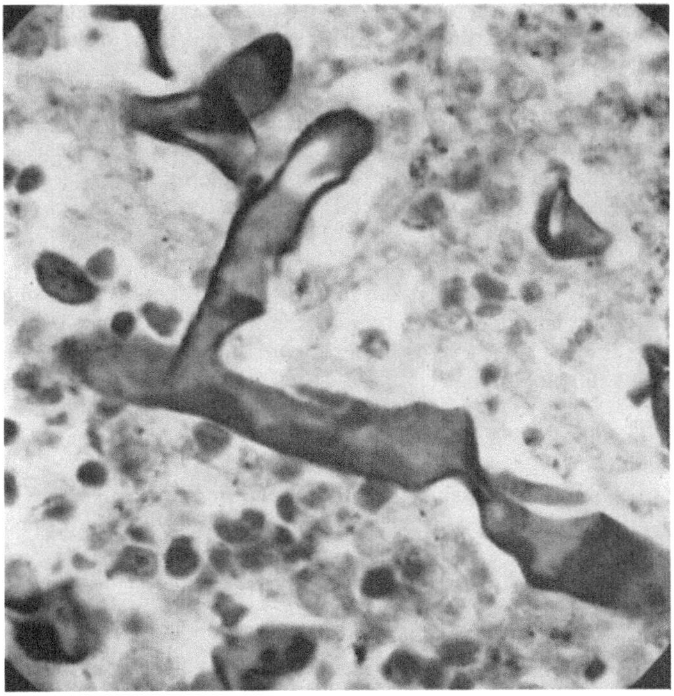

Fig. 35. Thrombus of a pulmonary blood vessel containing hyphae of mucormycosis. From a case of leukemia complicated by the cenocytic fungus infection. G.M.S., × 800

scribed, purplish-red and centrally necrotic. Multiple blood vessels within these hemorrhagic areas were thrombosed. Small amounts of mucus were present in the bronchi. Microscopically tangles of broad non-septate hyphae occupied many air-spaces and infiltrated the alveolar septa. Penetration of hyphae into the walls of arteries with thrombosis and infarction was conspicuous. Hyphae were incorporated in the thrombi. Tissue necrosis was extensive.

In case 8 of Sun et al. (1965) the lesion was in the right upper lobe. Hyphae of mucormycosis were seen in bronchial lumina and blood vessels of this region. There was acute suppurative inflammation and the formation of small abscesses in the surrounding pulmonary tissue.

This description is different in two respects from that in the previous cases. Infarction is not mentioned and abscess formation is a prominent feature. However, there were hyphae in bronchi and in blood vessels. If there were hyphae in blood vessels, there were certainly thrombi and probably infarcts. The author does not illustrate the appearances. Infarction may be attended with pronounced neutrophilic infiltrates.

In the case of Wall and Madison (1965), fatal pulmonary phycomycosis complicating leukemia, there was adherent, occlusive, dark red thrombotic material 3 cm proximal to the hilar bifurcation of the right pulmonary artery. Distally, the lung was mottled dark red. Microscopically, hematoxylinophilic coenocytic hyphae permeated and apparently colonized the thrombotic material and were scattered through the disrupted internal elastic membrane of the artery. In addition, there was associated arteritis and early organization of the thrombus. Focal ischemic necrosis was present in the upper and lower right lung lobes; in viable tissue there was pulmonary edema.

Winston (1965) reported a case of mucormycosis in a 29-year-old diabetic man in whom death was due to hemorrhage from a lesion of the right main bronchus and adjacent pulmonary artery. The right main brochus was almost completely blocked by a soft grayish-purple mass. On section a spherical mass $1^1/_2$ cm in diameter infiltrated the bronchus and the pulmonary artery. The lesion was an ischemic infarct of the bronchus and pulmonary artery surrounded by polymorphonuclear neutrophils. Rupture had occurred at the edge of the infarct and here there was fibrosis. Hyphae were plentiful in the infarct.

This rupture of an artery of the lung, with hemorrhage, is reminiscent of the similar process in the cranial cavity when the internal carotid becomes weakened by the mycotic arteritis and forms an aneurysm and ruptures.

Summary of Primary Pulmonary Mucormycosis

From this review of the findings in the lungs in reported cases of pulmonary mucormycosis there emerges a picture of the pathology of this condition.

It becomes clear that the usual initial process is the proliferation of the fungus in the bronchi at the hilus of the lung. These *penetrate* into the hilar structures but especially into the *arteries*. It is not clear why the fungus prefers arteries to veins, but it is clear that arteritis is much more frequent than phlebitis. The fungus, once within the artery, produces thrombosis. There may be huge thrombi in the main pulmonary artery to a lung, or in the smaller branches to capillary size. An infarct develops, possibly because of concomitant involvement of veins and lymphatics. This thought is interjected because, in general, infarcts of the lung do not develop except in the presence of chronic passive congestion due to congestive failure.

The infarcts may be massive, involving two or more lobes or they may be small and multiple, only a centimeter or so in diameter. They are not always wedge-shaped.

The early infarcts are hemorrhagic and they may remain this type for, certainly, a week or so. At a later stage they are gray and homogeneous, and often there is scar tissue about them and thickening of the pleura.

A later stage is softening of the infarct to form a cavity often erroneously interpreted as an abscess cavity. Possibly some of these cavities are abscess cavities due to secondary bacterial infection.

The lesions of the lung almost always have the vascular obstructive features mentioned. There is certainly an element of acute edematous and fibrinous bronchopneumonia, accompanying or complicating the infarction.

The only other form of primary involvement is ulceration of a main bronchus by the fungus. There is usually also involvement of the larger arteries. The last case cited was of this type, and fatal hemorrhage ensued.

Hematogenous Pulmonary Mucormycosis

We have seen, in our consideration of cranial mucormycosis, that the fungus can gain a foothold in the orbit or paranasal sinuses and disseminate through the systemic circulation to the lungs. The case of BAKER and SEVERANCE, reported in more detail in BAKER (1956) is pertinent:

On the day before the death of this child of 3 years, the left pupil was much larger than the right pupil, and there was hemorrhage in the left eye. At autopsy the left orbital fat was congested and red in contrast to the ordinary yellow fat of the right orbit. On the inferior surface of the left frontal lobe of the brain there was an area of injection and softening which measured 4 by 2.5 cm. The left internal carotid artery presented a thrombus at the level where it was severed in removing the brain and there were thrombi in vessels of the left middle fossa.

These findings indicate that the infection began in the cranial region of this diabetic child. At the *postmortem examination* the lower lobe of the left lung was congested and edematous on cut section, and there was a dark purple area, 4 cm in diameter, in the middle of the lobe with thrombi in branches of the pulmonary artery. The bronchi contained thick, brown, mucoid material. The upper lobe on the left was pink and air-containing. The upper right lobe was congested and edematous on cut section, with a severely congested area 2 cm in diameter. There was a thrombosed vein leading from this lobe. The middle lobe was dark purple, but air-containing. The lower lobe was pink, edematous, and rubbery. The bronchi on the right contained more of the thick, brown, mucinoid material noted on the left.

The *microscopic findings* supported the gross observations and clinical story indicating that the case was primarily cranial. Branching hyphae were numerous in the orbital tissues.

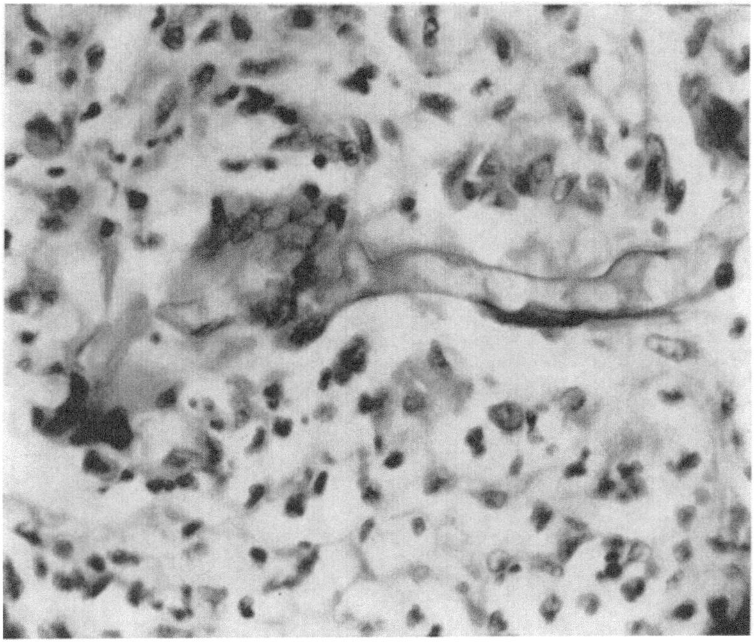

Fig. 36. Giant cell response to a hypha in case 7 of BAKER (1956), a disseminated case of 3 to 4 months duration. Giant cell response is very uncommon (H & E, × 700)

Microscopically, sections of bronchi of all sizes contained branching hyphae, 4—18 μ thick and up to 100 μ long, with occasional neutrophils and red cells. The wall of the largest bronchus was necrotic. The pulmonary parenchyma was not necrotic and contained numerous hyphae like those in the bronchi. Alveolar exudate consisted of edema fluid, hemorrhage, fibrin, and small numbers of neutrophils. The larger hilar blood vessels presented hyphae everywhere, including, especially, permeation of walls. Thrombi estimated to be of several day's duration, also showed hyphae. They occluded all but an axial pathway for fluid blood. Organization of thrombi was not observed. Many of the smaller vessels contained fluid blood. Lymph vessels in edematous pulmonary fibrous septa contained hyphae and neutrophils. Since all of the numerous sections of lung showed hyphae, it was presumed that the infection involved all lobes of both lungs (Fig. 29). There were fungi in hilar lymph nodes, fat, connective tissue, perineurial lymphatics, nerve trunks and pleura.

Thus we see how completely the metastatic fungus permeated the tissues of the lungs, including the bronchi and pleura. The pulmonary lesions were early infarcts and edematous pneumonia.

Similar cases of pulmonary lesions resulting from hematogenous dissemination from primarily cranial mucormycosis are reported by MARTIN et al., BRYAN et al., LONG and WEISS, FUENTES (case 1), and SUGA. The ages of their patients were $2^1/_2$ months, 34 months, 24 years, 9 years, and 38 years. Including the case of BAKER and SEVERANCE, 4 of the cases were in the pediatric age group and 2 in the adult group. This suggests that hematogenous, systemic, dissemination from cranial mucormycosis may be more frequent in children and infants than in adults.

In the 34-month-old diabetic girl reported by BRYAN et al. (1958) the metastatic lesions of the lung were described as follows: The upper and lower lobes of the right lung and the lower lobe of the left lung were bluish, firm and non-aerated; the remaining lobes were congested. Histologically, fibrin thrombi occluded the vessels of the lungs and these thrombi showed fungi. There was an associated fibrinous, hemorrhagic, fluid exudate in the air spaces. This was thought to be more an effect of infarction than of the direct toxic effect of the fungi.

Table 4. *Esophageal and Gastrointestinal Mucormycosis*

Authors and Case Number	Location of Case Age, Sex, Race	Predisposing Factors	Anatomic Involvement	Cultures	Duration	Treatment	Status of Patient
MARCHAND (1910)	Leipzig, Germany 18, F, W	Mushroom poisoning	Stomach, ulcer	No	15 days	Not stated	Dead
BENEKE (1911)	Marburg/Germ. 33, M, W	None	Stomach, ulcer	No	Few days	Not stated	Dead, pneumonia
LJUBIMOWA (1913)	St. Petersburg/ Russia; 46, F, W	None	Stomach	No	1 day	Exploratory operation	Dead, peritonitis
TEUTSCHLÄNDER (1917)	Heidelberg/Germ. 64, M, W	None	Stomach	No	7 days	Gastroenter-ostomy	Dead, pneumonia
LÖHLEIN (1920) Case 2	Germany Male	War prisoner. Malnutrition ?	Stomach	No	—	—	Dead
MOORE et al. (1949)	Milwaukee/USA 37, F, W	None	Large intestine with perforation of cecum	No	15 days	Sulfonamide	Dead, generalized peritonitis
LIE KIAN JOE and N. ENG (1956)	Java 48, F, Chinese	None	Stomach; ulcer of lesser curvature	No	6 months	Resection	Well
BAKER et al. (1957) Case 1	Durham/USA 43, M, W	Cortisone, Toxic hepatitis	Small and large intestines, stomach and liver	No	3 days	Cortisone, tetracycline	Dead, gangrene of intestine and intestinal hemorrhage
— Case 2	Jackson/USA 75, M, W	None	Cecum and ascending colon	No	12 days	Intravenous fluids	Dead, mucormycosis and bronchopneum.
CLARK (1957)	Cleveland/USA 59, M, W	Diabetes, leukemia (?)	Duodenum with extension to left lobe of liver and cecum	No	5 weeks	Antibiotics	Dead, mucormycosis; terminal pneumonia
TORACK (1957) Case 10	New York/USA 65, M, W	Aplastic anemia	Stomach	No	4 days	Corticosteroids, antibiot.	Dead, mucormycosis
WATSON (1957)	Natal/Afr. 26 mos., M, N	Kwashiorkor	Anterior stomach wall, perforation	No	4 weeks	High-protein diet	Dead, mucormycosis

GATLING (1959)	Roanoke/USA 16 days, F, W	Otitis media	Entire stomach wall, extension to adjacent organs	No	11 days	Antibiotics, nystatin	Dead, mucormycosis
MONTENEGRO et al. (1959) Case 1	São Paulo/Brazil 10 months, F	Malnutrition	Small intestine	—	3 days	Nystatin	Dead, bronchopneumonia
— Case 2	São Paulo/Brazil 1 F	Malnutrition	Colon	—	30 days	—	Dead, intestinal hemorrhage
LEVIN and ISAACSON (1960)	Johannesburg/Afr. Premature, N	Prematurity	Descending colon with perforation	No	5 days	Antibiotics	Dead, meconium peritonitis
NEAME, RAYNER (1960), Case 1	Durban/S. Afr. 20, M	Amebic colitis, hepatitis	Stomach	—	—	—	—
— Case 2	Durban/S. Afr. 52, M	Pellagra	Esophagus, colon	—	—	—	—
— Case 3	Durban/S. Afr. 47, M	Uremia	Stomach	—	11 days	—	—
— Case 4	Durban/S. Afr. 23, F	Uremia, peritonitis	Stomach, colon, liver	—	13 days	—	—
— Case 5	Durban/S. Afr. 36, F	Uremia	Stomach	—	27 days	—	—
— Case 6	Durban/S. Afr. 2, F	Amebic colitis and liver abscess	Colon	—	28 days	—	—
— Case 8	Durban/S. Afr. 30, M	Amebic colitis, renal tuberculosis	Colon	—	14 days	—	—
— Case 9	Durban/S. Afr. 3, F	Kwashiorkor, pulm. tuberculosis	Stomach	—	15 days	—	—
— Case 10	Durban/S. Afr. 1, F	Kwashiorkor	Colon, stomach, liver, gall bladder	—	70 days	—	—
— Case 12	Durban/S. Afr. 2, M	Kwashiorkor, empyema	Stomach, colon	—	9 days	—	—
— Case 13	Durban/S. Afr. 11 months, F	Kwashiorkor	Stomach	*Rhizopus microsporus* van Tiegh	15 days	—	—

Table 4. Continuation

Authors and Case Number	Location of Case Age, Sex, Race	Predisposing Factors	Anatomic Involvement	Cultures	Duration	Treatment	Status of Patient
— Case 14	Durban/S. Afr. 6 days, M	Prematurity	Stomach, colon. liver	—	3 days	—	—
— Case 15	Durban/S. Afr. 2, M	Kwashiorkor	Esophagus, rectum	*Rhizopus stolonifer* (Syn. *R. nigricans*)	14 days	—	—
— Case 16	Durban/S. Afr. 27, F	Typhoid, peritonitis	Colon	*Rhizopus microsporus* van Tiegh	8 days	—	—
— Case 17	Durban/S. Afr. 11 months, M	Kwashiorkor	Colon	*Rhizopus stolonifer* (Syn. *R. nigricans*)	35 days	—	—
— Case 18	Durban/S. Afr. 26, M	Amebic dysentery with peritonitis	Colon	—	12 days	—	—
— Case 19	Durban/S. Afr. 50, F	Amebic dysentery with peritonitis	Stomach	*Rhizopus stolonifer*, *Rhiz. microsporus*, *Mucor hiemalis*	4 days	—	—
— Case 20	Durban/S. Afr. 2, F	Kwashiorkor	Stomach	*Rhizopus microsporus*	56 days	—	—
— Case 21	Durban/S. Afr. 35, M	Typhoid	Ileum	—	12 days	—	—
— Case 22	Durban/S. Afr. 37, F	Hepatic failure	Stomach	—	—	—	—
SUTHERLAND and JONES (1960)	Acornhoek/Transv. 50, M, N	None	Greater curvature of stomach near cardia, 5×8 cm. "tumor"	No	1 month	Penicillin	Dead, terminal pneumonia

CUADRADO et al. (1961), Case 3	Dallas/USA 8, M, W	Leukemia, steroids	Right colon and lung	No	Not stated	6-MP, aminop-terin, steroids	Dead, mucormycosis
DE FEO (1961)	Schenectady/USA 51, M	Leukemia, hydrocortisone	Descending colon, branch of pulmonary artery	No	3 weeks	Antibiotics, hydrocortisone	Dead, peritonitis, mucormycotic perforation
TRUJILLO (1961)	Medellin/Colombia 7 months, M, W	Enteritis, dehydration	Stomach and duodenum	No	—	Operative exploration	—
STRAATSMA et al. (1962), Case 22	AFIP Accession No. 571372 36, M, Mongolian	Gastroenteritis	Gastrointestinal mucormycosis	—	—	—	—
— Case 23	AFIP Acc. No. 955878; 37, M, N	Diabetic acidosis and renal failure	Gastrointestinal mucormycosis	—	—	—	—
— Case 24	AFIP Acc. No. 996826; 70, M, N	Renal insufficiency	Gastrointestinal mucormycosis	—	—	—	—
KAHN (1963)	South Africa 17, F, N	Diabetes, puerperium	Stomach and esophagus	No	—	—	Dead
ABRAMOWITZ (1964)	South Africa 35, M, N	None	Stomach and colon	No	13 days	Hydrocortis., antibiotics	Dead, peritonitis
KUBO (1965)	Japan	—	Stomach and duodenum	—	—	—	—
STEIN and SCHMAMAN (1965), Case 1	Johannesburg/SA 35, M, N	Malnutrition	Stomach and duodenum	No	9 days	Antibiotics, operation	Dead, peritonitis
— Case 2	Johannesburg/SA 35, M, N	—	Stomach with pyloric ulcer, 3 cm diameter	No	4 days	Surgical	Dead, peritonitis
SUN et al. (1965) Case 6	Peking/China 23 days, M, Chin.	Prematurity, Hydrocortisone	Pyloric antrum of stomach	No	24 days	Surgical	Dead, post-operative
CALLE and KLATSKY (1966)	Baltimore/USA 23, F, N	None	Infarcts, ulcers, terminal ileum and cecum	No	28 days	Antibiotics	Dead, generalized peritonitis
PEÑA (1967)	Colombia Newborn	Prematurity	Small intestine	No	15 days	—	Dead, mucormycosis

In the 24-year-old pregnant diabetic patient of LONG and WEISS (1959) the metastatic pulmonary lesions were in the posterosuperior portion of the lower right lobe where there was a poorly demarcated pyramidal acrepetant lesion. There was a thrombus in an artery leading to this infarct. Microscopically there were phycomycetes in vessels of the lung and in the pulmonary tissue. Their Fig. 1 shows the fungus in the pulmonary vessel with destruction of the vessel wall and thrombosis.

In the 9-year-old diabetic boy reported by FUENTES the thoracic cavities showed no macroscopic alterations of importance except for severe congestion of organs. Microscopically colonies of fungi were observed with the production of focal chronic inflammation. This disposition of organisms and this type of inflammation is extraordinary and peculiar in a case with a short clinical course.

These are sufficient examples to indicate that the metastatic pulmonary lesions from cranial mucormycosis are less striking and well-defined than in primary pulmonary mucormycosis, probably because they develop late in the course of the infection.

In summary, pulmonary lesions of mucormycosis may be encountered in postmortem examination of cases dying of primary cranial infection. Such dissemination appears to be more frequent in infancy and childhood than in adult life. The fungus comes to lie in the vascular bed of the lungs. It produces thromboses of vessels, and swarms into the pulmonary parenchyma and hilar structures including the main bronchi. Infarction develops and there is an edematous pneumonia as well. The lesions are like those of primary pulmonary mucormycosis, but less advanced.

In some of the cases of gastrointestinal mucormycosis (Table 4), there were also lesions in the lung, and the question is whether the fungus infected the digestive tract first and spread to the lung, or vice versa. In some of the cases of disseminated mucormycosis (Table 5). the infection may have begun in the gastrointestinal tract and spread to the lung.

Esophageal Mucormycosis

Esophageal mucormycosis is much less common than esophageal candidosis (Table 4). The instances of mucormycosis of the esophagus mentioned in the literature follow:

In Case 3 of HUTTER (1959) the lower end of the esophagus was eroded, and hemorrhage extended throughout all layers. Histologically there were numerous hyphae and sporangia of "*Mucor*" in the lumen. Hyphae extended interstitially through all walls and penetrated vessels. Blastospores and pseudomycelia of *Candida* were present superficially without deep penetration. This then, was a double mycotic infection of the esophagus.

In case 4 of HUTTER there were gross ulcers of the esophagus. On microscopic examination, the esophageal ulcers contained submucosal "*Mucor*" thrombi with infarction of the overlying mucosa. There were fibrin clots and hemorrhagic dissection through the muscle wall, with intramural *Mucor* thrombi in small vessels. Hyphae were also present interstitially. On the surface of the mucosa there were large clusters of blastospores of *Candida* (no pseudohyphae) and colonies of bacteria.

In both of HUTTER's cases the esophageal mucormycosis was probably blood-borne because the lesions were deep in the wall of the esophagus and because the lesions in the body were disseminated, indicating mycemia. However, in the first case, with hyphae of "*Mucor*" in the esophageal lumen, the fungus may have spread directly into the wall.

In case 1 of McBRIDE et al. the distal third of the esophagus was black and ulcerated. The necrotic membranous inflammation of the distal esophagus was sharply separated from the uninvolved esophagus above and the stomach below, as shown in Fig. 2 of the article. Microscopically in the wall of the distal segment of the esophagus there were many hyphae and extensive areas of necrosis, hemorrhage, neutrophilic infiltration and vascular thrombosis.

Two of the cases of NEAME and RAYNER (1960) had esophageal lesions. One case had multiple small ulcers and the other large ulcers.

KAHN (1963) reported an esophageal and gastric mucormycosis.

MIYAKE and OKUDAIRA (1967) tabulate mucormycosis of the esophagus in one or two cases in which the infection was in multiple organs.

Mucormycosis of the Stomach

For cases in which the mucormycosis of the stomach is the only mucormycotic lesion in the body, or when it is clearly the primary lesion, the descriptions of the gross and microscopic features follow in chronologic sequence (See Table 4).

BENEKE (1911) described a small, elongated, gray-white ulcer close to the pylorus of the stomach. Microscopically, thick, largely non-septate hyphae were growing into blood vessels and there was thrombosis. Bacilli were present. Death in this case was not due to mucormycosis but to a terminal pneumonia and the lesion of the stomach was an incidental finding. It was not unlike a chronic peptic ulcer.

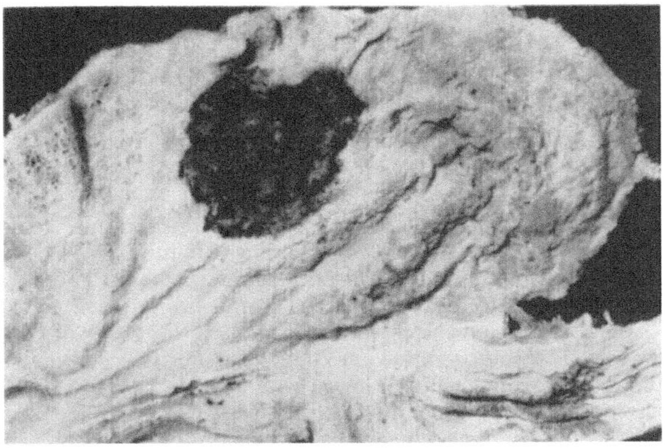

Fig. 37. Mucormycosis of stomach in chronic lymphocytic leukemia. The black slough is an infarct caused by fungal thrombosis. From case 5 of BAKER (1962), which also had lesions in the lung, liver, and kidney. The gastric lesion is secondary

LIE KIAN JOE and NJO-INJO TJOIE ENG (1956) described a surgical specimen of gastric mucormycosis. Roentgenograms of the stomach had showed a large defect filled with barium on the lesser curvature. It was noted during the operation that the stomach was thickened along the lesser curvature. Partial gastrectomy was performed. The removed tissue showed an ulcer. Eighty sections were cut of a block from the stomach wall. The slides showed an ulcer with overhanging edges. The ulcer extended to the muscular layer. The base of the ulcer consisted of three zones, a thin one of necrotic tissue, a zone of granulation tissue with many dilated blood capillaries and a broad zone of fibrous or scar tissue which contained blood vessels with thick walls. This scar tissue reached to the peritoneum. It had almost entirely destroyed the muscular coat under the ulcer. Small pieces of muscular tissue, remnants of the muscular coat, were seen in the scar tissue. Numerous neutrophils were present in the zone of granulation tissue. The inflammatory reaction in the scar tissue was chronic. A feature was the presence of many eosinophils diffusely scattered throughout the granulation tissue, giving it a reddish appearance. In the scar tissue the eosinophils were around blood vessels with plasma cells and round cells. The submucosa at the margin of the ulcer was thickened by the same fibrous tissue as that found at the base of the ulcer. Cross and longitudinal sections of large hyphae were found in the necrotic zone and in the zone of granulation tissue at the base of the ulcer, not in the scar tissue. The hyphae had a thin wall and were blue-violet in the H and E slides. They contained a granular mass, reddish brown in color. No septate divisions were seen. Their diameter varied from 3.5—34 μ and they were present in the necrotic zone singly or in small groups. From here they grew into the zone of living granulation tissue where they were often found in small abscesses. One doubtful giant cell was found in one of the slides. The hyphae showed no tendency to grow into the blood vessels. Groups of bacteria were seen in the necrotic zone. No culture could be made.

In discussion of their case, LIE KIAN JOE and NJO-INJO TJOEI ENG refer to the presence of many eosinophils and say that as far as they know there is only one fungus among the Phycomycetes which provokes a prominent tissue eosinophilia in man, i. e. *Basidiobolus ranarum*, a saprophytic fungus found in the digestive tract of frogs and lizards and in beetles which feed upon the excrement of these animals and are in turn eaten by them. This fungus was found to cause induration and swelling of the subcutaneous tissue in Indonesian boys. Marked tissue eosinophilia was also found in pigs which were infected with *Absidia racemosa* or *Rhizopus equinus*, both members of the family Mucoraceae (CHRISTIANSEN, 1929). However, it seemed unlikely to the authors that the fungus in their case was identical with one of these, or with *Basidiobolus ranarum*, because of a marked difference in diameter of the hyphae.

In addition to this reason against the fungus being *Basidiobolus ranarum* is the fact that there is no hyaline substance adjacent to the fungus, as is so characteristic of cases of subcutaneous phycomycosis caused by *Basidiobolus ranarum*. The tissue response is rather like that of a case of mucormycosis of the arm reported by BAKER et al. (1962), from which *Rhizopus rhizopodiformis* was grown, in that in both cases the fungus lies within microabscesses. However, the failure of the fungus to invade blood vessels is peculiar for a Mucoraceae. From all the evidence, I am willing to accept this case as one of mucormycosis.

WATSON (1957) described a mucormycotic gastric perforation in a child with kwashiorkor. When the abdomen was opened at postmortem there was found to be a black, necrotic-looking piece of omentum adherent to the anterior aspect of the stomach. There was no general peritonitis. On the omentum being removed from the stomach a large perforation of the anterior wall of the viscus was found. This consisted of a complete sloughing of an area about 5 cm in diameter. The lower edge of the perforation extended as far as the greater curvature and the upper edge was about 1 cm from the lesser curvature. The margins of the perforation consisted of black, ragged, necrotic tissue, surrounding which was a zone of reaction with small hemorrhagic spots. The stomach wall was edematous and exhibited a red flare about 1 cm in width.

Microscopic sections through the stomach wall showed a wide-spread inflammatory exudate with numerous polymorphonuclear and round cells. There was destruction of the mucosa in the necrotic edge of the perforated area, and the adjacent mucosa was edematous. Numerous fungal hyphae were present in the stomach wall. These were particularly noticeable in the walls of the blood vessels in the submucosa, and in places the hyphae were found encroaching on the lumen of the vessels. The hyphae were non-septate and showed considerable variation in size. The average diameter was 10 μ but some were as large as 20 μ. Many of the blood vessels appeared to be thrombosed and this probably accounted for the extensive necrosis. The appearances were typical of those described in mucormycotic infections. No cultures were made before the viscus was fixed with formalin. Since no lesions were found in the body anywhere other than in the stomach, it is reasonable to assume that the fungus entered the stomach wall from the lumen of this viscus because the tissues were devitalized by malnutrition. The location of the lesion was not at all that of a chronic peptic ulcer.

GATLING (1959) reported a case of gastric mucormycosis occurring in an otherwise apparently healthy newborn infant who died of the infection on the 16th day. At autopsy, the wall of the stomach, from the cardia to within 0.5 cm proximal to the pylorus, was greatly thickened and indurated. The mucosa of the stomach was ragged and tan-brown and opaque. The serosal surface was covered with fibrin. Microscopically large numbers of nonseptate hyphae were present within the mucosa of the stomach and extended through the entire stomach wall. There was extensive vascular invasion. The infection had extended within the wall beneath the intact mucosa at the pyloric end. Many gram-negative bacilli and occasional gram-positive cocci were demonstrated in the superficial portion of the necrotic mucosa. Cultures of the gastric contents and a subdiaphragmatic abscess were overgrown by Friedländer's bacillus. The anatomical diagnosis was acute necrotizing phlegmonous, mucormycotic gastritis, with extension to the liver, spleen, pancreas, colon, diaphragm and left lung.

The report of NEAME (1960), concerning gastric mucormycosis is the most comprehensive of the entire literature. He reports 11 cases from Durban, South Africa, eight of which are limited to the stomach. He gives a description of the histopathology which applies to any part of the digestive tract, as follows:

Under low magnification, the most striking feature was necrosis and hemorrhage. In less advanced lesions, only the mucosa and part of the submucosa were affected, while, in more advanced cases, the remainder of the submucosa, the muscle layer, and even the serosa, were involved. There was perforation in 5 cases. In most cases, thrombosis was apparent in submucosal and serosal vessels. Under higher magnification, numerous pleomorphic hyphae could be seen in the necrotic tissue, invading the lumina and vessel walls, where thrombi were often present. The hyphae were nonseptate, averaged 15—20 μ in width, and had a well-defined edge and a cytoplasm showing areas of lighter, vacuolated and darker, granular staining, which could, on occasion, give the impression that the hyphae were septate. They showed numerous lateral branches, while sporangia were seen in a number of sections. In his case 9 what was presumed to be a columella of *Rhizopus* was found. Neutrophils were numerous in most sections, many around vessels and in hemorrhagic and necrotic areas.

NEAME described the morbid anatomy applicable to the entire digestive system as follows: The disease presented in three ways: (1) as small ulcers approximately 1 cm in diameter; (2) as larger ulcers with well-circumscribed borders; and (3) as more diffuse and often perforated necrotic lesions. Each is a further stage in the progression of the disease.

1. The small ulcers were superficial and consisted of a central portion of yellow slough, surrounded by a zone of erythema. A greenish-black discoloration was noted on the undersurface of the serosa. The lesions could easily be mistaken for amebic ulceration. Ulcers of this type were found in the descending or sigmoid colon of three cases and the ileum of one case, while multiple ulcers were found in the colon and esophagus of another.

2. Larger ulcers averaged 3—4 cm in diameter, the central 1.5 cm being black and necrotic. The peripheral 1 cm was raised, edematous, and gray in color. Surrounding this was a zone of intense erythema. On the serosal surface, blackish discoloration could be seen, deep to the peritoneum. This type of lesion was found most frequently at the pyloric antrum, as in 6 of the gastric cases, but occurred in the colon of 3 cases and the rectum and esophagus of 1 case.

3. A typical example of the more diffuse lesion occurred in a 6-day-old child who had necrosis of the whole thickness of the stomach, with sloughing and perforation, and the peripheral part of the lesion showed a zone of erythema. Another case showed a diffuse hemorrhagic necrotic lesion involving 8 cm of the transverse colon. On the mucosal surface, there were multiple areas of ulceration, one of which had perforated. On one occasion ulceration was diffuse but superficial with a velvety surface. Diffuse lesions of the stomach were present in four cases and of the colon in two.

The lesions of the digestive tract, excluding the small-ulcer type, were easily discernible, and of the last 15 cases, 12 were recognized at autopsy.

Because NEAME was remarkably successful in demonstrating the causative organism in his cases, data concerning the mycology and his methods of taking materials are included here.

Mycology: Swabs were taken from all lesions where fungal invasion was suspected and were submitted to the Bacteriology Department for direct examination and culture. 1. Direct Examination: A wet preparation was made in 20% KOH solution, allowed to stand overnight, and examined the next day. By this method, nonseptate hyphae of the family Mucoraceae were found in nine cases. 2. Culture: No positive swab cultures occurred until a medium in a Petri dish was used, consisting of Sabouraud's glucose agar on which white bread, which had been sterilized by autoclaving at 15 lb. for 15 min, and by dry heat in a hot air oven at 160 C. was placed. In eight cases there was a profuse growth of *Rhizopus*, while two cases were overgrown with various molds. It is of interest that all the cases of mucormycosis from which culture was obtained were of *Rhizopus* species.

Pathogenesis: Following ingestion and given the right environment, the fungus invaded the alimentary wall, causing local infarction and subsequent ulceration. Contiguous spread from stomach to liver and transverse colon was noted. "Kissing ulcers" on opposing walls of the stomach developed in two cases. In one case spread via the portal vein to the liver occurred, judging from the appearance at autopsy and on microscopic sectioning.

Site: An outstanding feature of Neame's series was the predominance of mucormycosis of the digestive tract. Two of his cases involved the central nervous system and the remainder the digestive system.

Associated Disease: Among the associated diseases of the 12 cases of NEAME involving the stomach there were the following: Kwashiorkor in 6 cases, amebiasis in 2, hepatic disease in 2, uremia in 3.

Fig. 38. Intestinal mucormycosis. Hemorrhagic necrosis (infarction) of ileum with central area of ulceration of necrotic mucosa. Case 1 of BAKER et al. (1957)

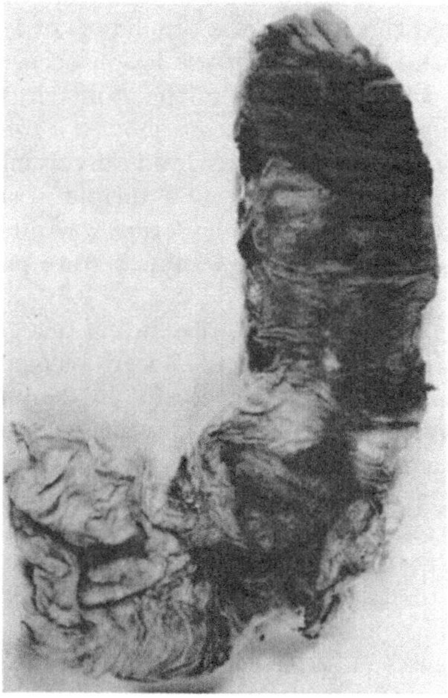

Fig. 39. Intestinal mucormycosis. Terminal ileum (above), ileocecal valve, and proximal cecum (lower left), showing areas of hemorrhagic necrosis. Case 1 of BAKER et al. (1957)

It is interesting that there was so much digestive tract mucormycosis in South Africa and one wonders if *Rhizopus* is particularly prevalent there, possibly growing on ingested food. There are other areas of the globe where malnutrition is prevalent, as in Latin America and India, but no such incidence of mucormycosis of the digestive tract.

NEAME noted that *cortisone* has been mentioned in the literature as a predisposing factor, and stated that no cortisone had been used prior to the development of the infection in the patients of his series. On the other hand, antibiotics had been given to all.

The author noted, further, the difficulty of diagnosis in digestive tract mucormycosis. None of his cases was diagnosed clinically. However, diagnosis could be made on the basis of knowledge of the predisposing factors and pathogenesis, sensitivity tests and biopsy, and the finding of numerous non-septate hyphae in the gastric aspirate or stool, -though the last could be merely saprophytic.

SUTHERLAND and JONES (1960), of Eastern Transvaal, Africa, reported a case of gastric mucormycosis in a Swazi of about 50 years. At postmortem examination there was a grayish white mass 8 by 5 cm along the greater curvature near the cardia. The surface of the mass was elevated one centimeter above the surrounding mucosa and was covered with a gelatinous exudate. Microscopically the mass was composed of necrotic and hemorrhagic material which was diffusely infiltrated with numerous non-septate hyphae of mucormycosis. Vascular invasion was prominent. The presence of a tumor-like mass in gastro-intestinal mucormycosis may signify an infection of longer duration than usually occurs. The authors suggested that the fermented foods and beer of the Bantu may possibly be factors in gastro-intestinal fungus infections by the introduction of excessive numbers of spores.

TRUJILLO's (1961) case was in an infant of 7 months. At operation the pyloric portion of the stomach was enlarged and palpable. A biopsy included the pylorus and a part of the duodenum. Microscopically there was ulceration of the mucosa of the pylorus, and the entire gastric and duodenal wall was invaded by thick, non-septate hyphae.

KAHN (1963) reported a case in South Africa involving the stomach and esophagus, found at the postmortem examination of a 17-year-old, puerperal, diabetic Negro girl.

ABRAMOWITZ (1964) reported fatal mucormycotic perforations of the stomach in a 35-year-old Bantu. There was peritonitis at postmortem examination. Two ulcers of the stomach had perforated, one into the greater sac through the anterior stomach wall, the other into the gall-bladder. There were areas of superficial ulceration throughout the colon. Histological sections of the stomach at the sites of the perforations and of the bowel at the areas of colonic ulceration showed a marked hemorrhagic and inflammatory reaction, with large branching non-septate fungal hyphae.

In case 1 of STEIN and SCHMAMAN there was, at surgical exploration, a rent along the lesser curvature of the stomach extending from the cardia to the pyloric sphincter. The posterior wall was adherent to the pancreas and completely covered by an exudate. The duodenum was edematous and had a perforation 0.7 cm in diameter in the anterior wall. The posterior wall of the stomach was freed, and found to have some very hard areas in it. A biopsy was taken. The rest of the mucosa appeared normal. The stomach was repaired. The duodenum was extremely friable and could not be sutured. The patient died and there was no postmortem examination. The biopsies taken from the stomach and duodenum, however, showed the characteristic features of mucormycosis. There were varying degrees of necrosis with only slight infiltration with neutrophils. Numerous broad, non-septate, branching, fungal hyphae were present. Several blood vessels in the sections showed penetration of their walls by these hyphae, which were also present in the lumen accompanied by thrombosis. A remarkable feature was the virtual absence of inflammatory reaction in the walls of the vessels where they were infiltrated by the fungal hyphae. As mucormycosis was not suspected during the surgical procedure, specimens were not taken for culture.

In case 2 of STEIN and SCHMAMAN (1965) there was a perforated ulcer about 1 cm in diameter on the lesser curvature of the stomach. A biopsy was taken and the perforation sutured. Microscopic examination of the gastric biopsy showed acutely inflamed granulation tissue in which there were numerous broad, non-septate, branching hyphae. At autopsy on the sixth postoperative day there was generalized peritonitis. The stomach was distended and contained altered blood. At the pyloric end of the lesser curvature there was an ulcer measuring 3 cm in diameter, to which omentum had been sutured. Numerous fungal hyphae were present even on the peritoneal surface.

BAKER (1962) has an excellent photograph of mucormycosis of the stomach in chronic lymphocytic leukemia (Fig. 37). Figure 3 of that article shows the black slough, the size of a silver dollar towards the greater curvature and about half way between the cardia and pylorus. It was an infarct caused by fungus thrombosis. This case 5 of this article had mucormycotic lesions also in the lung, liver, and kidney, and represented a disseminated mucormycosis. The infarct of the stomach was a hematogenous lesion, the primary lesion being in the right lung.

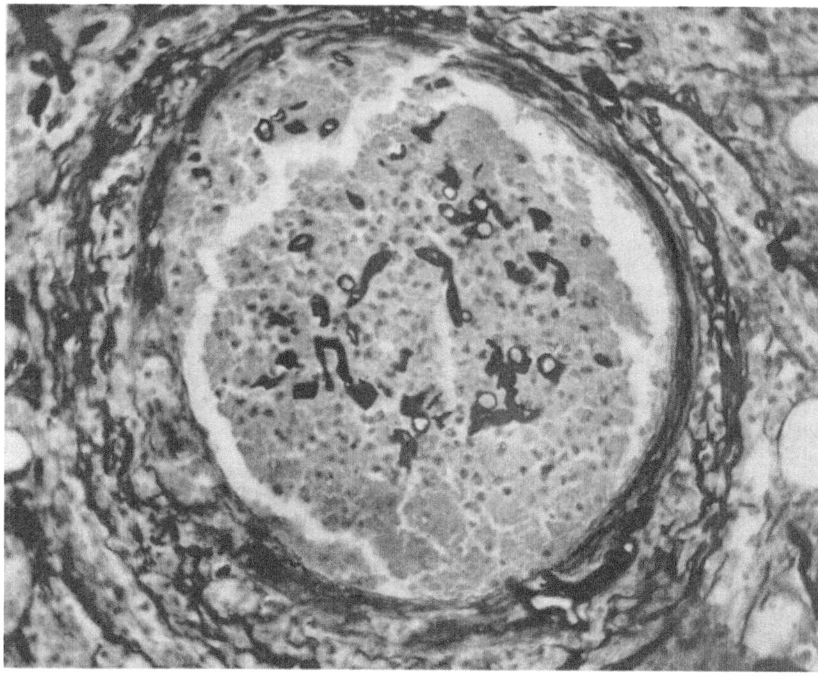

Fig. 40. A thrombosed mesenteric vessel showing hyphae in the lumen, wall, and surrounding tissue. Methenamine silver nitrate stain, × 160. Case 1 of BAKER et al. (1957)

In case 7 of this same article a similar, but less conspicuous gastric mucormycosis, is described. There was a hemorrhagic indurated region, 2.5 by 2 cm in the fundus of the stomach involving the mucosa and extending to the serosa. Hyphae hematogenously borne from the primary pulmonary lesion had stimulated the formation of thrombi in vessels of the stomach, producing the hemorrhagic infarct.

From the foregoing extensive descriptive anatomic pathology, one obtains a concept of the characteristic changes in gastric mucormycosis.

Most of the reported cases have been primary infections of the stomach and most have been reported from South Africa and in subjects with kwarshiorkor. This may devitalize the gastric mucosa and permit the fungus to enter. Possibly the gastric mucosa exhibits nutritional edema and develops fissures permitting access of the fungus. Possibly the food eaten by these subjects is overgrown with *Rhizopus*. Neame showed that *Rhizopus* is the Mucoraceae usually involved in gastric mucormycosis.

Rhizopus is especially important as a cause of spoilage of fruits and stored potatoes, especially sweet potatoes (SKINNER et al., 1947). The small fruits, especially strawberries, are particularly susceptible. The disease in strawberries is known as leak, because of the softening and dripping of the fruit. It is a source of considerable loss in shipment and is prevented by refrigeration. In sweet potatoes, a characteristic soft rot is produced.

The fungus entering the mucosa of the stomach penetrates the blood vessels and stimulates thrombosis and consequent infarction (gangrene). With the necrosis of the mucosa an ulcer is produced. Later there may be gangrene of the wall and a perforation.

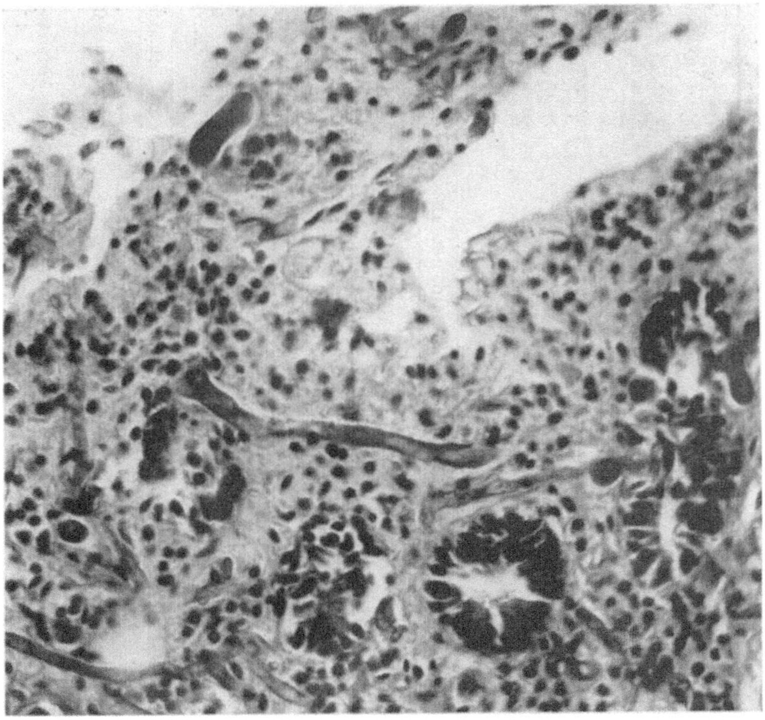

Fig. 41. Hyphae in intestinal mucosa. Case 2 of BAKER et al. (1957). H & E, × 160

The lesions occur in various parts of the stomach. The entire stomach wall may be thickened and necrotic. Ulcers may develop along the lesser curvature of the stomach in the prepyloric region, suggesting chronic peptic ulcers, but it is not clear that any mucormycotic lesions in this region begin as chronic peptic ulcers.

Finally, attention is called to the possibility of regions of gastric infarction or gangrene developing from organisms reaching the stomach in the blood stream. The lesions would not be ulcerated at first, but would soon become so.

Intestinal Mucormycosis

BAKER et al. (1957) reported two cases of intestinal mucormycosis (Figs. 38, 39, 40, 41) and tabulated two previously reported cases as follows: (1) PALTAUF (1885). A 52-year-old white male in Austria had the mycosis for 17 days, died with mycotic septicemia and showed lesions in the terminal ileum; MOORE et al. (1949): A 37-year-old white woman in Wisconsin, U.S.A., had mucormycosis for 15 days, dying of peritonitis, and showed lesions in the large bowel (See Table 4).

Case 1 of BAKER et al. (1957) was that of a 43-year-old white man who developed hepatitis and jaundice while under therapy for cavitary tuberculosis of the right lung. At postmortem examination there was extensive mucormycosis of the gastrointestinal tract, minimal mucormycosis of the liver, and none elsewhere in the body.

Apparently the fungus invaded the intestinal wall from the lumen and produced thrombosis of small mesenteric vessels of the serosa. The fungus reached the liver by the portal blood stream and produced a small focus of infection there. The absence of fungus infection in other parts of the body indicated that the fungus was not disseminated to the gastrointestinal tract via the mesenteric arteries. Intramural hemorrhage of the intestine due to hepatic insufficiency

Table 5. *Disseminated Mucormycosis (37 cases)*

Authors and Case Number	Location of Case Age, Sex, Race	Predisposing Factors	Anatomic Involvement	Cultures	Duration	Treatment	Status of Patient
PALTAUF (1885)	Graz/Austria 52, M	None stated	Brain, lungs, pharynx, ileum	No	8 days	None stated	Dead, mucormycosis
CRAIG and FARBER (1953)	Boston/USA	Leukemia	Disseminated visceral	—	—	—	Dead, mucormycosis
ZIMMERMAN (1955) Case 6	Washington/USA 53, N	Myeloma, hemosiderosis, hepatoma	Lung, heart, kidney	No	—	Urethane, cortisone, X-ray	Dead, mucormycosis
BAKER (1956) Case 6	Durham/USA 3, F, N	75% of body burned	Lung, spleen, liver	No	3 days	Penicillin	Dead, mucormycosis
— Case 7	Phoenix/USA 4, M, W	—	Lung, liver, kidney	No	30 days	—	Dead, mucormycosis
TORACK (1957) Case 7	New York/USA 64, M, W	Myeloma, anemia, corticosteroids	Lungs, brain, stomach, spleen, heart	No	20 days	Corticosteroids, antibiotics	Dead, mucormycosis
HUTTER (1959) Case 4	New York/USA 8, M, W	Leukemia, steroid diabetes	Brain, lungs, spleen, digestive tract, thyroid, diaphragm	No	21 days	Antileukemics, antibiotics, steroids	Dead, mucormycosis
MCBRIDE et al. (1960) Case 1	Nova Scotia/ Canada 32, F, W	Septic abortion, renal cort. necrosis, anuria and uremia	Intestines, lungs, liver, kidneys, spleen, larynx	No	7 days	Antibiotics	Dead, mucormycosis
— Case 2	Boston/USA 26, M, W	Hepatitis, uremia	Lungs, esophagus, stomach, ileum, colon	*Rhizopus arrhizus*	10 days	Antibiotics	Dead, mucormycosis
BANKER (1961) Case 3	Boston/USA 12, M	Erythroleukemia, steroids	Lung, kidney, spinal cord	No	1 month	Steroids	Dead, mucormycosis
CUADRADO et al. (1961), Case 1	Dallas/USA 31, M, W	Diabetic acidosis, uremia	Lung, liver, spleen, pleura	No	30 days	Heparin, femoral vein ligation	Dead, mucormycosis
GLOOR et al. (1961), Case 2	Basel/Switzerland 69, M	Myeloblastic leukemia, corticosteroids	Aortic valve, myocardium, lung, kidney, spleen	No	21 days	Ilotycin, corticosteroids	Dead, mucormycosis
BAKER (1962) Case 5	Durham/USA 49, M, N	Chron. lymphocytic leukemia, neutrop., steroids	Right lung, stomach, kidney, liver	*Mucor*	4 weeks	Antileukemics, steroids, antibiotics	Dead, mucormycosis, aspergillosis

— Case 7	Durham/USA 45, M, W	Acute myeloblastic leukemia, neutrop.	Both lungs, stomach, right kidney	No	1 week	Antileukemics, antibiotics	Dead, mucormycosis
— Case 11	Durham/USA 58, M, W	Acute lymphoblastic leukemia, neutrop., steroids	Lungs, brain, thyroid	No	1 week	BW 57-323, steroids	Dead, mucormycosis
STRAATSMA et al. (1962), Case 6	AFIP Acc. No. 336851; 55 M, W	Aleukemic leukem., renal insufficiency	Disseminated mucormycosis	—	—	—	—
— Case 7	AFIP Acc. No. 486865; 66 M, W	Subleukemic myelogen. leukemia	Disseminated mucormycosis	—	—	—	—
— Case 8	AFIP Acc. No. 557484; 68 M, W	Chronic myelogenous leukemia	Disseminated mucormycosis	—	—	—	—
— Case 9	AFIP Acc. No. 565089; 28 M, N	Cutaneous burns	Disseminated mucormycosis	—	—	—	—
— Case 10	AFIP Acc. No. 729137; 62 M, W	Renal insufficiency	Disseminated mucormycosis	—	—	—	—
— Case 11	AFIP Acc. No. 801945; 62 M, W	Diabetes, agranulocytosis	Disseminated mucormycosis	—	—	—	—
— Case 12	AFIP Acc. No. 840049; 64 M, N	Postoperative status	Disseminated mucormycosis	—	—	—	—
— Case 13	AFIP Acc. No. 847623; 28 M, W	Chronic pulmonary disease	Disseminated mucormycosis	—	—	—	—
— Case 14	AFIP Acc. No. 856793; 43 M, W	Chronic myeloid leukemia	Disseminated mucormycosis	—	—	—	—
ERIKSEN (1963)	Copenhagen/Denmark; 23, M	Lymphogranulomatosis	Lungs, cerebrum, kidneys, myocardium	No	—	Antitumor drugs, prednisone, antibiotics	Dead, mucormycosis
GRUHN and SANSON (1963), Case 23	Chicago/USA 50, F	Chronic lymphocytic leukemia	Lung and brain	No	—	Steroids, antibiotics	Dead, mucormycosis
BRUNSON (1964)	Jackson/USA 63, M, N	Leucopenia and neutropenia	Lungs, liver, stomach, kidney, brain	No	—	Antibiotics	Dead, mucormycosis

Table 5. Continuation

Authors and Case Number	Location of Case Age, Sex. Race	Predisposing Factors	Anatomic Involvement	Cultures	Duration	Treatment	Status of Patient
Simon et al. (1964)	Evanston/USA 76, M	Splenectomy, steroids, anemia	Lungs, liver, kidneys, duodenum, brain	No	4 days	Stilbestrol, steroids	Dead, mucormycosis
Cussen (1965)	Melbourne/Aust. 13, F, W	Hypopituitary dwarfism, Fanconi's syndrome, renal hypertension, prednisolone	Brain, thyroid, lungs	No	1 week	Prednisolone, antibiotics	Dead, mucormycosis
Sun et al. (1965) Case 6	Peking/China 23 days, M	Hydrocortisone	Pylorus, jejunum, liver, gall bladder, pancreas, lungs	No	21 days	Antibiotics, hydrocortisone	Dead, mucormycosis
— Case 7	Peking/China 56, M	Infectious hepatitis	Heart, kidney, lungs	No	—	Antibiotics, steroids	Dead, mucormycosis and bacterial infection
McMillan (1967) CPC Ohio	Columbus/USA 49, M, W	Acute leukemia, prednisone	Pulmonary vessels, appendix, kidneys	No	—	Transfusions, cytoxine, prednisone	Dead, mucormycosis
Parkhurst and Vlahides (1967), Case 1	Schenectady/USA 62, F	Aplastic anemia, steroids	Disseminated mucormycosis	—	—	Antibiotics, steroids	Dead, mucormycosis
— Case 2	Schenectady/USA 51, M	Lymphoblastic leukemia, steroids	Disseminated mucormycosis	—	—	Antibiotics, steroids	Dead, mucormycosis
— Case 3	Schenectady/USA 84, F	Acute stem cell leukemia, steroids	Disseminated mucormycosis	—	—	Antibiotics, steroids, cytotoxics	Dead, mucormycosis
— Case 4	Schenectady/USA 28, F	Acute myeloblastic leukemia, steroids	Disseminated mucormycosis	—	—	Antibiotics, steroids, cytotoxics	Dead, mucormycosis
— Case 5	Schenectady/USA 67, F	Chronic myelocytic leukemia, steroids	Lung, heart, spinal cord	No	1 week	Antibiotics, steroids, cytotoxics	Dead, mucormycosis

may have been the first event, followed by secondary infection of devitalized tissues. The extension of the infection to small mesenteric vessels then led to intestinal infarction. Death was due to massive intraluminal hemorrhage. Figure 1 of the paper of BAKER et al. (1957) is that of a segment of ileum showing a focal area of hemorrhagic necrosis with a central area of ulceration of the necrotic mucosa. Figure 2 shows the intestinal mucormycosis just above and below the ileocecal valve, with areas of hemorrhagic necrosis.

CALLE and KLATSKY (1966), report another case of mucormycosis of the terminal ileum and cecum with death due to generalized peritonitis. An interesting feature was the presence, at the margin of a lesion, of giant cells containing the non-septate hyphae of mucormycosis, presumably *Rhizopus*. This is probably to be correlated with the relatively long duration of the mycosis, estimated at 28 days.

CALLE and KLATSKY also present an excellent tabulation of the 14 cases of intestinal mucormycosis in the literature, including their own case. Not included in their table are cases in which involvement of the intestine was only part of a disseminated mucormycosis. The table includes the case of MOORE et al., the two cases of BAKER et al. (1957), the case of CLARK, the two cases of MONTENEGRO, five cases of NEAME and RAYNER, a case of LEVIN and ISAACSON and finally the case of CALLE and KLATSKY. Half of these 14 cases were from South Africa, 2 from Brazil and 5 from the United States of America. This paper also contains a valuable table showing the subdivisions of the Phycomycetes and indicating the *Rhizopus* and *Absidia* species isolated from cases of human phycomycosis and the species of *Rhizopus* and *Absidia* used to produce experimental phycomycosis in animals. This data refers to mucormycosis in general, and is not restricted to intestinal mucormycosis. It is interesting that no *Mucors* are listed.

The case of intestinal mucormycosis by CLARK, listed in the table of CALLE and KLATSKY, deserves special mention, because of the manner in which the fungus extended from the duodenum into the liver forming a tumor-like mass in the liver. There was also a mucormycotic carcinoma-like mass in the cecum. The case illustrates that occasionally mucormycosis of the gastrointestinal tract may take the form of tumor-like masses, in addition to infarction and ulceration.

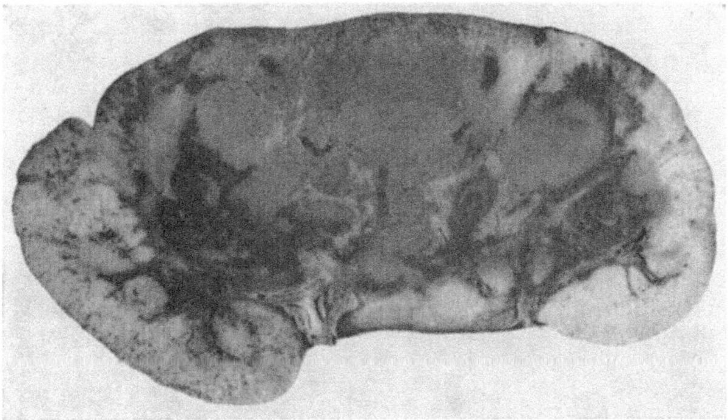

Fig. 42. Hematogenous mucormycotic infarct of kidney in myeloblastic leukemia. Mucormycosis also in lung (the primary site) and stomach. Case 7 of BAKER (1962)

Disseminated Mucormycosis

Table 5 includes 40 cases of disseminated mucormycosis which complicated leukemia, myeloma, septic abortion, body burns, diabetic acidosis, cortisone therapy, and renal insufficiency (Fig. 42). Leukemia, treated with modern drugs, is the most frequent antecedent of disseminated mucormycosis.

Each case of mucormycosis reported in this chapter is entered in only one of the five tables. In several of the cases of cranial mucormycosis there was dissemination to the lungs and elsewhere. Moreover, some of the cases listed as gastrointestinal and esophageal disseminated or were part of a dissemination from the cranial region or lung. In gastrointestinal mucormycosis the infection may spread centrifugally to the diaphragm, liver and omentum without regard to anatomic boundaries. A similar spreading, not along vascular channels but directly through the tissues, occurs in cranial mucormycosis and at the hilus of the lung in pulmonary mucormycosis. However, the fungus has such an attraction to blood vessels that these are invariably involved in the instances of direct spread of *Rhizopus*.

Pulmonary mucormycosis and mucormycosis of body burns constitute important primary sites for disseminated mucormycosis.

Cutaneous and Subcutaneous Mucormycosis

Secondary cutaneous and subcutaneous mucormycosis has already been discussed in the section on the cranial form of the disease. The infarct or gangrene of the skin of the face or nose is secondary to vascular obstruction by thrombi of arteries, veins and lymphatics caused by hyphae growing in the lumina of these vessels.

Cases of this type were reported by Bryan et al., Smith and Kirchner, Dolman and Herd, Rinaldi and Ashby, Suprun, Burrow et al., Reeves et al., Straatsma et al. (case 28), Lubbe and Pennington, Weisskopf (case 1 and 2), and Hurtado et al.

The eight cases of primary cutaneous and subcutaneous mucormycosis are listed in Table 6. The predisposing factors were diabetes in 3 cases, cutaneous burns in 3, and there was no predisposing factor in two. The infection was on the leg in 3 cases, on an upper extremity in two, and on the face in 3. Abstracts of the eight cases follow.

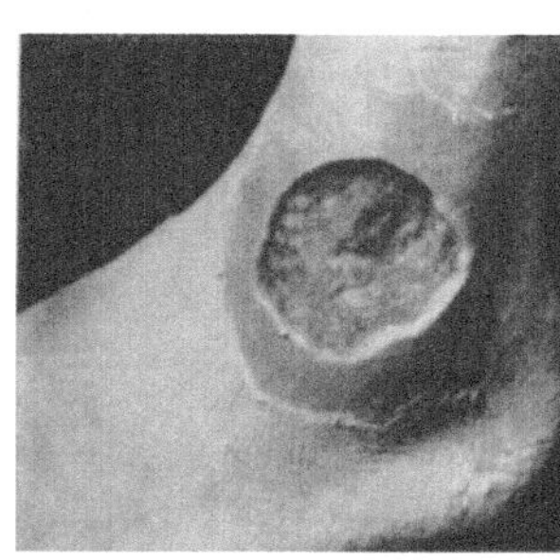

Fig. 43. Cutaneous mucormycosis. An ulcer which developed spontaneously in a diabetic patient. The ulcer healed in $4^1/_2$ months. (From Josefiak et al.)

The lesion of the case of Josefiak et al. developed *de novo* in a diabetic of 26 years (Fig. 43). A region of black gangrene appeared above the malleolus of the leg. The necrotic tissues sloughed out leaving a deep ulcer. Biopsies showed the characteristic broad hyphae in abscesses and in the walls of vessels. With the control of the diabetes, and the administration of potassium iodide, the ulcer healed in $4^1/_2$ months. It was not clear whether the fungus was implanted locally or reached the leg *via* the blood stream.

The case of Knoth-Born involved both legs of a 50-year-old farm woman. The illustrations in the report show organisms which are coenocytic and compatible with those of mucormycosis.

In the two cases reported by Rabin et al. the patients had sustained extensive thermal cutaneous burns involving the face, and the facial tissues became infected with an organism of mucormycosis. This led to gangrenous necrosis of the central and midportion of the face and to systemic mucormycosis. The fungus caused softening of the nasal cartilages and bones and

Table 6. *Primary cutaneous and subcutaneous Mucormycosis (8 cases)*

Authors and Case Number	Location of Case Age, Sex, Race	Predisposing Factors	Anatomic Involvement	Culture	Duration	Treatment	Status of Patient
JOSEFIAK et al. (1958)	Winst.-Salem/ USA; 26, M	Diabetes	Lower left leg	No	$4^1/_2$ mos.	Diabetic control	Well
KNOTH-BORN (1959)	Gießen/Germany 50, F, W	—	Skin of legs	No	—	—	—
RABIN et al. (1961) Case 1	Ft. Sam Houston/ USA; 22, M	Body burns, 64%	Face, nasal cavity, brain, lungs, kidney, spleen, heart	No	12 days	Antibiotics	Dead, mucormycosis
RABIN et al. (1962) Case 2	Ft. Sam Houston/ USA; 20, M, N	Body burns, 45%	Face, nasal cavity, left orbit, meninges, heart, kidneys	No	8 days	Antibiotics	Dead, mucormycosis
BAKER et al. (1962)	New Orleans/USA 79, F, W	Diabetes	Right arm	*Rhizopus rizopodiformis*	8 days	Antibiotics	Dead, mucormycosis
ROBERTS (1962)	West Palm Beach/ USA; 24, M, N	Diabetic acidosis	Right leg	No	17 days	Control of diabetes	Recovery
VIGNALE and MACKINNON (1964)	Bagé/Brazil 39, F	None	Face, neck, nasal septum	*Mucor ramosissimus*	24 days	Amphotericin B	Cured
FOLEY and SHUCK (1968)	Ft. Sam Houston/ USA; 20, M, N	Burn	Hand	No	7 days	Amputation	Cured

these structures were removed by debridement. In these two cases the burn represented the predisposing factor and there was no diabetes.

The case of BAKER et al. (1962) was that of a 79-year-old white woman admitted to the Louisiana State University Medical Service of Charity Hospital on September 22, 1961, because of fever and stupor. A diabetic for many years, she had been hospitalized at Charity Hospital for 2 weeks for the same complaints 1 month previously.

During the first admission she was treated with penicillin, chloramphenicol, diet and insulin. She was discharged on an 1800-calorie diabetic diet and 1 tablet of chlorpropamide daily.

The patient's course at home was satisfactory for 1 week only. She then developed temporary loss of motor power and sensation in the left leg. She became lethargic, unstable in her movements, and intermittently refused food and fluids. A carbuncle appeared over the right scapula on September 15, and was incised and drained the next day. Intramuscular penicillin was administered for several days.

From September 17 until the day of admission she refused her chlorpropamide, food, and most fluids. She complained of pain, stiffness of the neck, and headache, and became unresponsive on September 21.

Significant findings at admission on September 22 were: brachial blood pressure, 190 mm of mercury, systolic, and 80, diastolic; dehydration; bilateral cataracts; stiff neck which appeared painful on attempted movement; draining carbuncle over the right scapula; absence of abnormal reflexes; and fever of 102° F. Acetone was absent from the urine and serum, and only 1 plus glycosuria was present. The hematocrit was 44.5%, and the white blood cell count and differential were normal. Examination of the cerebrospinal fluid was normal chemically and cytologically.

Although the patient was intermittently responsive, treatment did not alter her fever or progressive deterioration. It was necessary to cannulate several veins in order to administer intravenous fluids and antibiotics. Transfusions with whole blood, regulation of electrolytes, digitalization, and insulin did not improve her status.

She was treated as if she had staphylococcal and gram-negative rod septicemia, because these organisms were grown from the carbuncle. The absence of positive blood cultures may have been due to antibiotic treatment which was instituted on the night of admission.

Multiple decubiti developed. On October 3 the right forearm above the venous cannula was swollen, purplish, and firmly nodular. The arm became progressively more dusky and edematous. On October 5 the epidermis became elevated by a large vesicle which contained serosanguineous fluid. The material from the unbroken vesicle revealed *Rhizopus rhizopodiformis* and *Aerobacter aerogenes*. Melena was observed on the same day and was intermittently present thereafter.

The right arm was incised and packed on October 8, and material from the subcutaneous tissues was obtained for laboratory study. The tissue contained hyphae which were recognized as belonging to the phycomycetes. Cultures of the tissue and of deep exudate the following day grew *A. aerogenes* and *Rhizopus rhizopodiformis*.

At postmortem examination the remnants of the carbuncle were represented by an ulcer of the back, 4 by 4 cm in diameter, with healthy-looking granulation tissue in its base.

On the dorsum of the right arm there were incisions over an area 9 by 3.5 cm. There were foci of yellow and white opacity in the subcutaneous tissue which represented fat necrosis and minute abscesses. There was no enlargement of lymph nodes at the elbow or axilla.

In the right globus pallidus of the brain there were minute foci of old encephalomalacia, each 3—4 mm across. The heart (340 g) had no infarct or severe coronary arteriosclerosis. The aorta was rigid and calcified in its terminal third. The left pleural cavity contained 600 cc of clear yellow fluid, and the right, 400 cc. There was atelectasis of the lower lobe of each lung. The liver weighed 1980 g and the spleen 220 g.

There was an ulcer, 2 by 1 cm in diameter and 1.5 cm deep, in the first portion of the duodenum, with hemorrhagic foci in the base. The terminal ileum contained small amounts of tarry blood, and there were brown feces in the colon. Five pedunculated polypi were found along the large bowel. The right kidney weighed 110 g and the left, 120 g. The trigone of the bladder was red and swollen. The pancreas was of normal size. The thyroid was atrophic (7 g) and contained colloid nodules. The right adrenal weighed 10 g, the left 11 g.

Microscopically, there were broad (8—15 μ), branching, coenocytic hyphae in the skin and underlying tissues of the right forearm. They were in abscesses and necrotic fat, bordered in some instances by giant cells, and in the walls of blood vessels. Bacilli were found in the incised necrotic fat close to the hyphae but not in the abscesses or giant cells. There was mild fibroblastic response near the muscle. Fungi were not found in other tissues of the body.

Microbiologically, an antemortem culture from the arm grew rapidly at room temperature as a fluffy growth, at first white and later dark brown or black. Dr. C. W. HESSELTINE of Peoria, Illinois, identified it as *Rhizopus rhizopodiformis*. The following description appears in HESSELTINE's thesis, 1950:

"Colonies at first white, then black-plumbeus, nearly as black in reverse, little or no sterile mycelium, less than 3 mm in height; odor sometimes strong; stolons hyaline; rhizoids poorly developed, hyaline or brown-colored; sporangiophores branched, especially near the surface of the medium, from stolens, or from swellings of the hyphae or opposite the rhizoids; septations only above a branch, up to 800 μ in height (mostly less than 500 μ), up to 15 μ in width; sporangia spherical, white, then a glistening black, finally a gray-black, averaging 90—120 μ, exceptionally to 140 μ, columellae hyaline or colored, variable in shape from pyriform or oblong to globose, reaching 72 μ in diameter. Sporangiospores non-striate, smooth, heavy-walled, spherical, nearly hyaline singly but black in mass, averaging 4—5 μ; chlamydospores cylindrical or spherical or oval, up to 25 μ in diameter; zygospores unknown."

R.rhizopodiformis is rarely encountered as a saprophytic contaminant and only a few isolates have been described in the literature. In his thesis Hesseltine cites two instances in which the species was recovered from a mycosis in animals, in one case from a dog and in another from a horse. The organism has very close affinities with the genus *Mucor* and is distinguished from other species of *Rhizopus* chiefly by the branched sporangiophores, the presence of occasional pyriform columellae, and the lack of striations in the walls of the sporangiospores.

Undoubtedly *diabetes* predisposed to the subcutaneous infection of the arm by *R.rhizopodiformis*. Also, it is reasonably certain that the fungus entered the tissues around the cannula used in intravenous therapy. The fungus must have been present in the arm for at least a week, as there was giant cell and localized abscess production in response to its presence. The *Rhizopus* penetrated vessel walls, as in cerebral and pulmonary mucormycosis. The presence of microabscesses, however, indicated that the organism exerted its effect as a direct inflammatory agent as well as by vessel invasion, thrombosis, and ischemia. The necrosis of fat may have been due to ischemia or to the presence of bacteria.

In the local tissue reaction in the arm, *A.aerogenes* appeared to play a minor role, as no bacteria could be found in the abscesses or giant cells. However, a bacillus was visible in necrotic fat in the neighborhood of hyphae, and this may have represented a *Proteus* organism.

The fungus infection of the arm, possibly reinforced by a bacillary infection, may have been responsible for the fever which the patient had during her hospital stay. No other satisfactory cause of fever was demonstrated at postmortem and, while Gram-negative rod septicemia was postulated, it was never proved by blood culture. There was evidence of mild gastrointestinal bleeding at autopsy, not sufficient to be the main cause of death. It is reasonable to conclude, therefore, that the fungus infection of the arm was a major contributing factor in the causation of the death of this patient.

The case of Roberts was similar to that of Josefiak et al. An ulcer developed in the right lower leg of a 24-year-old Negro diabetic. He survived following excision of the ulcer, rigid diabetic management, and supportive nutritional measures.

The case of Vignale et al. is unique in the annals of mucormycosis. The case differs from classical cranial, pulmonary, digestive tract and disseminated mucormycosis with respect to the type of lesion, the length of the course of the disease, and the absence of diabetes or other predisposing factor. The case resembles somewhat the African cases of entomophthoromycosis. If it were not that the fungus agent was a *Mucor*, I would have been inclined to regard the case as one of entomophthoromycosis or subcutaneous phycomycosis. Vignale et al. note that subcutaneous phycomycosis described in horses by Bridges and Emmons (1961) and Bridges, Romaine and Emmons (1962) offers a number of features common to those described in this case: extreme chronicity, extensive destruction and failure to cure spontaneously.

At the age of 15 years the girl from South Brazil noticed a flat, faintly erythematous lesion on the left cheek; it measured 1 cm in diameter and had undefined boundaries and a soft consistence. There was no history of trauma or local infection. After a few months the lesion became slightly raised and remained stationary for about 5 years. By the end of 10 years, the lesions had spread centrifugally forming a papulose, congestive, infiltrated plaque.

When the patient was 39 the process had extended to involve the cartilage and bone of the nose and face. The nasal bridge was absent, the nasal orifices appearing at the same level as the remaining surrounding skin. At the level of the hard and soft palates there was a 15 by 20 mm perforation, slightly towards the left. The uvula had disappeared. In the neck, as on some portions of the face there were papules from 2—4 mm in diameter.

Numerous biopsies showed a fibrosing and granulomatous inflammatory process made up of histiocytes, lymphocytes and a few giant cells and irregularly distributed mycelial filaments.

Cultures of biopsy material were made on Sabouraud glucose agar. In every case, within a few days, yellowish white mycelium developed forming little raised colonies. The microscopic features were characteristic of a phycomycete and sporangia indicative of the genera *Mortierella* or *Mucor* were present. Repeated isolation of the same species as well as morphological similarities in tissues and in cultures were indications of its pathogenicity. On four occasions cultures of biopsy samples yielded the same fungus which HESSELTINE and ELLIS (1964) regard as *Mucor ramosissimus* Samutsevitsch. Treatment with amphotericin B was apparently successful.

FOLEY and SHUCK (1968) report the case of a 20-year-old Negro man whose right hand, together with 65% of his body surface, was burned after a mine explosion in Vietnam. On the eleventh post burn day his right hand became swollen and tender over the thenar area. Topical chemotherapy had been used, and in some way this may have contributed to the infection. At exploratory operation the hand appeared to be gangrenous and was disarticulated. Histopathologic examination revealed ischemic necrosis of the subcutaneous, muscle and connective tissue at all levels sampled. There was invasion through unhealed burn by broad, non-septate, irregularly branching hyphae characteristic of Phycomycetes. The hyphae infiltrated the wall and lumen of small vessels that were infected and thrombosed. There was a cellulitis and abscess of the thenar area with necrosis, and hyphae invading blood vessels.

Focal Mucormycosis

This table of 6 cases is a miscellany of isolated lesions of mucormycosis. There is one lesion of the tibia, one of a fracture site, one of the larynx, two of the kidney, and one of the aorta (See Table 7).

The lesions of the larynx and of the fracture site in the leg were reported in tabular form by STRAATSMA et al., and no detailed data are available. ERIKSEN et al. (1962) reported mucormycosis of a fracture of a tibia and the recovery of *Absidia ramosa* from the lesion.

DIRIART et al. (1963) reported renal mucormycosis. There are no illustrations.

PROUT and GODDARD (1960) reported a case of renal mucormycosis in a surgically removed kidney. There were non-septate hyphae and thrombosed arteries. Giant cells formed part of the cellular response. Amphotericin B was used in treatment and the patient lived.

FIENBERG and RISLEY (1959) reported the first case of mucormycosis limited to the aorta. The fungus infection occurred in a 45-year-old woman with controlled diabetes. The source of the infection on the basis of autopsy data was considered to be a ruptured appendix with extruded fecal material. Complete occlusion of the lumen of the abdominal aorta was the result of the fungal infection.

Spontaneous Mucormycosis in Animals

GISLER and PITCOCK described 2 cases of intestinal mucormycosis in monkeys (*Macaca mulatta*). In the first case there were ulcers of the cecum and ascending colon. Hyphae were demonstrated in them and in a focus in the liver. These were thrombosed vessels. The case resembles those of intestinal mucormycosis in man. In the second case there was neutropenia of 1,330 cells per cubic millimeter. There was ulcerative colitis of the ascending and transverse portions of the large intestine. There were hyphae of mucormycosis and also *Endameba histolytica* in the lesions. The mucormycosis seemed to be incidental.

LUCKE and LINTON reported the case of an 11-year-old mandrill which had been in the Bristol Zoological Gardens for 4 years. It died after 3 days of illness. At autopsy ulcerative lesions were found in the stomach and intestines and granulomatous lesions were found involving the lungs, heart, kidneys, liver, spleen and

Table 7. *Focal Mucormycosis (6 cases)*

Authors and Case Number	Location of Case Age, Sex, Race	Predisposing Factors	Anatomic Involvement	Cultures	Duration	Treatment	Status of Patient
FIENBERG and RISLEY (1959)	Beverly/USA 45, F	Diabetes, abscess of appendix	Abdominal aorta	No	5 days	Penicillin, cortisone	Dead, mucormycosis
PROUT and GODDARD (1960)	Miami/USA 51, M	Alcoholism, pulmonary tuberculosis	Kidney	No	—	Nephrectomy, Amphotericin B	Survived
ERIKSEN et al. (1962)	Copenhagen/Denmark	Fracture	Tibia	*Absidia ramosa*	—	—	Well, fungus disappeared
STRAATSMA et al. (1962) Case 29	AFIP Acc. No. 705571 40, F, N		Larynx	—	—	—	—
STRAATSMA et al. (1962) Case 31	AFIP Acc. No. 787760 45, F, N		Fracture site, leg	—	—	—	—
DIRIART et al. (1963)	Paris/France 8 months	Mucoviscidosis, azotemia	Kidney	No	5 weeks	Antibiotics	Dead

occasional lymph nodes, and there was a discharging ulcer in the skin of the right shoulder. Large non-septate hyphae were seen in all the lesions but the fungus was not isolated.

HESSLER et al. described mucormycosis in a rhesus monkey.

Thus the occurrence of intestinal lesions in primates is as in man. The disseminated granulomatous lesions of LUCKE and LINTEN are of a different nature than the disseminated lesions of man except that one of BAKER'S (1956) cases had giant cell response to the fungus.

AINSWORTH and AUSTWICK published a book entitled *Fungal Diseases of Animals* (1959). They discuss mucormycosis occurring in pigs, cattle, horses, dogs, mink, guinea pigs, fowls, and mice, and state that the lesions may be ulcerative and confined to the gastric and intestinal mucosa. A granulomatous form involves lymph nodes and is thus unlike mucormycosis in man. Occasionally the granulomatous lesions become generalized. In their section on bovine mycotic abortion they mention many cases of infection of the placenta with fungi of the *Mucoraceae*.

AUSTWICK and VENN (1957, 1961) found 12.3% of 638 cases of mycotic abortion associated with mucoraceous fungi, and 64% with *Aspergillus*.

NICOLET et al., reporting on bovine abortion, found 14 cases of mucormycosis and 19 of aspergillosis. *Absidia corymbifera* was the most prevalent species.

MAHAFFEY and ADAMS reported phycomycetous mycotic lesions of the placenta in mares.

AINSWORTH and AUSTWICK, GITTER and AUSTWICK, and SHIRLY report abomasal ulcers containing phycomycete hyphae.

CHRISTIANSEN (1929) described chronic granulomatous lesions in cattle. MOMBERG-JORGENSEN described subcutaneous lesions in the mink. GLEISER reported a renal lesion in a dog, a stomach lesion in another dog, and lymph node lesions in a heifer.

DAVIS et al. described pulmonary and lymph-nodal lesions in cattle and pigs. GITTER and AUSTWICK (1959) described mucormycosis in a litter of sucking pigs. KAPLAN et al. described mucormycosis in a harp seal, MARCATO and DIMON in a gazelle, MORQUER et al. in cattle and swine, TURNER in bovine foetuses, MEHNERT in young pigs, MOSTAFA et al. in goats, TEWARI in buffaloes, HEWER in an okapi, SAUER in a squirrel.

A case of mucormycosis in the uterus and rumen of a cow was considered in a clinicopathologic conference at the New York State Veterinary College.

Diagnosis: The clinical and postmortem diagnosis of the cranial form of mucormycosis should offer no difficulty to the physician or pathologist who knows that unilateral proptosis, gangrene of the face, black mucosa of the nose, and gangrene of the palate proclaim mucormycosis. The diagnosis of the other forms of the disease are, in contrast, exceedingly difficult, as a mucormycotic infarct of the lung might pass for pneumonia in X-ray and gangrene of the bowel for mesenteric thrombosis.

Mucormycosis should be kept in mind as a possible complicating infection in diabetic and leukemic patients.

Biopsy has proved to be the most effectice means of making a diagnosis in the reported cases of mucormycosis. Smears are of less value than wet preparations, i. e., placing material in a drop of normal saline and placing a cover-slip on the fluid, and examining in the microscope with the light cut down. The hematoxylin and eosin stain of the biopsy is usually sufficient for a diagnosis.

To grow the fungus in *culture* is difficult, and many reports indicate failure even though the samples of material seemed ample. The method of Neame should be tried. He simply took material on a swab and inoculated it into Sabouraud glucose agar to which had been added autoclaved and heat-dried white bread. By this method he grew *Rhizopus* from several gastro-intestinal lesions. If pathologists are to obtain cultures of the Mucoraceae at postmortem, they must resist the all-too-frequent practise of fixing tissues in formalin without saving material for culture. A mucormycotic infarct of the lung or bowel may resemble infarcts due to thrombosis or embolism of other nature.

There is usually little difficulty in differential histologic diagnosis. *Aspergillus*, however, may be confused with *Rhizopus*.

Prevention: Mucormycosis might be prevented by keeping diabetics in optimal regulation, by avoiding the use of corticosteroids in leukemic patients, and by avoiding the use of antileukemic drugs which induce neutropenia. There is probably no means of keeping the infecting fungus from the patient, but studies along these lines are indicated.

Treatment: Regulation of the acidotic diabetic patient, the use of Amphotericin B and the surgical resection of gangrenous tissue appear to have been curative measures in surviving patients (Procknow, Utz).

References

Abramowitz, M.B.: Fatal perforations of the stomach due to mucormycosis of the gastrointestinal tract. S. Afr. med. J. **38**, 93 (1964). — Abramson, E., Wilson, D., Arky, R.A.: Rhinocerebral phycomycosis in association with diabetic ketoacidosis. Ann. intern. Med. **66**, 735 (1967). — Agrest, A., Manuel, A., Barcat, J.A., Ejden, J.: Cerebral mucormycosis. Medicina (B. Aires) **23**, 102 (1963). — Ainsworth, G.C., Austwick, P.K.C.: Fungal diseases of animals. Commonwealth Agricultural Bureaux (Farnham Royal, Bucks, 1959). — Austwick, P.K.C., Venn, J.A.J.: Mycotic abortion in England and Wales 1954—1960. Proceedings of the IV International Congress on Animal Reproduction, p. 562. The Hague 1961. ~ Routine investigation into mycotic abortion. Vet. Rec. **69**, 488 (1957). — Bader, G.: Die visceralen Mykosen (Pathologie, Klinik und Therapie), **423** pp. Jena: Gustav Fischer 1965. ~ Bader, G., Stiller, D., Ruffert, K.: Fluorochrome stains for histological diagnosis of visceral mycoses. Nature (Lond.) **208**, 796 (1965). — Baker, R.D.: Pulmonary mucormycosis. Amer. J. Path. **32**, 287 (1956). ~ Mucormycosis, a new disease? J. Amer. med. Ass. **163**, 805 (1957a). ~ The diagnosis of fungus diseases by biopsy. J. chron. Dis. **5**, 552 (1957b). ~ Diabetes and mucormycosis. Diabetes **9**, 143 (1960). ~ Essential pathology. Baltimore: Williams & Wilkins Co. 1961. ~ Leukopenia and therapy in leukemia as factors predisposing to fatal mycoses. Mucormycosis, aspergillosis, and cryptococcosis. Amer. J. clin. Path. **37**, 358 (1962). ~ Drug-induced mycoses. In: Proceedings of the Second Symposium on Drug-Induced Diseases, State University of Leyden, October 1964, pp. 50—60. Excerpta Medica International Congress Series No. 85. ~ Fungus infections. Chapter in Pathology. Ed. by W.A.D. Anderson. St. Louis: C.V. Mosby Co. 1966. — Baker, R.D., Bassert, D.E., Ferrington, E.: Mucormycosis of the digestive tract. Arch. Path. **63**, 176 (1957). — Baker, R.D., Schofield, R.A., Elder, T.D., Spoto, A.P.: Alloxan diabetes and cortisone as modifying factors in experimental mucormycosis (Rhizopus infection). Fed. Proc. **15**, 506 (1956). — Baker, R.D., Seabury, J.H., Schneidau, J.D.: Subcutaneous and cutaneous phycomycosis. Lab. Invest. **11**, 1091 (1962). — Baker, R.D., Severance, A.O.: Mucormycosis with report of acute mycotic pneumonia. Amer. J. Path. **24**, 716 (1948). — Balasubrahmanyan, M., Chaudhuri, S.: A case of pulmonary mucormycosis. Indian J. Path. Bact. **6**, 60 (1963). — Bank, H., Shibolet, S., Gilat, T., Altmann, G., Heller, H.: Mucormycosis of head and neck structures. A case with survival. Brit. med. J. **1**, 766 (1962). — Banker, B.Q.: Cerebral vascular disease in infancy and childhood. I. Occlusive vascular diseases. J. Neuropath. exp. Neurol. **20**, 127 (1961). — Barthelat, G.-J.: Les mucorinées pathogénes et les mucormycoses chez les animaux et chez l'homme. Arch. parasitol. **7**, 1 (1903). — Bauer, H., Ajello, L., Adams, E., Hernandez, D.: Cerebral mucormycosis. Pathogenesis of the disease. Amer. J. Med. **18**, 822 (1955a). — Bauer, H., Flanagan, J.F., Sheldon, W.H.: Experimental cerebral mucormycosis in rabbits with alloxan diabetes. Yale J. Biol. Med. **28**, 29 (1955a). ~ Experimental cerebral mucormycosis in diabetic rabbits. Amer. J. Path. **31**, 600 (1955). ~ The effects of metabolic alterations on experimental Rhizopus oryzae (mucormycosis) infection. Yale J. Biol. Med. **29**, 23 (1956). — Bauer, H., Sheldon, W.H.: Leukopenia with granulocytopenia in experimental mucormycosis (Rhizopus oryzae infection). J. exp. Med. **106**, 501 (1957). — Bauer, H., Wallace, G.L., Sheldon, W.H.: The effects of cortisone and chemical inflammation on experimental mucormycosis (Rhizopus oryzae infection). Yale J. Biol. Med. **29**, 389 (1957). — BAUM, J.L.: Rhino-orbital mucormycosis occurring in an otherwise apparently healthy individual. Amer. J. Ophthal. **63**, 335 (1967). — BENEKE, R.: Ein Fall von Schimmelpilzgeschwür in der Magenschleimhaut. Frankfurt. Z. Path. **7**, 1 (1911). — BERK, M., FINK, C.I., UYEDA, C.T.: Rhinomucormycosis. J. Amer. med. Ass. **177**, 511 (1961). — BERNARD, P.N.: Sur un Rhizopus pathogene de l'homme: Rhizopus equinus (1) LUCET et CONSTANTIN 1903, variété annamensis (2) NOEL BERNARD 1914. Bull. Soc. Path. exot. **7**, 430 (1914). — BIANCHI, L., DELLA TORRE, B.: A fatal case of phycomycosis. Mycopathologia (Den Haag) **19**, 145 (1963). ~ Rara forma di ficomicosi umana a decorso piemico. Minn. Med. **53**, 2462 (1962). — BIANCHI, L., DELLA TORRE, B., MARTINAZZI, M.: Fatal pancreatic necrosis in human phycomycosis. Path. et Microbiol. (Basel) **30**, 15 (1967). — BIANCHI, L., MAGRASSI, B.: Osserrazioni su una rara forma di ficomicosi piemica con localizzazioni scheletriche multiple. Chir. Organi Mov. **53**, 160 (1964). — BLANKENBERG, H.W., VERHOEFF, D.: Mucormycosis of the lung. Amer. Rev. Tuberc. **70**, 357 (1959). — BORLAND, D.S.: Mucormycosis of the central nervous system. Amer. J. Dis. Child. **97**, 852 (1959). — BOSTROEM: Demonstration mikroskopischer Präparate von Schimmelpilzen. Berl. klin. Wschr. **23**, 232 (1886). — BOYD, W.: A textbook of pathology. Structure and function in diseases. Seventh edition, 1378 pp. Philadelphia: Lea and Febiger 1961. — BRUNSON, J.G.: Clinicopathologic conference conducted by the Department of Pathology, University of Mississippi School of Medicine, Jackson, Mississippi. J. Miss. med. Ass. **5**, 95 (1964). — BROWN-THOMSEN, J.: Phycomycosis vulneris. Ugeskr. Laeg. **128**, 17 (1966). — BRYAN, G.T., READ, C.H., ZIMMERMANN, G.R.: Disseminated mucormycosis in a child with diabetes mellitus. A case report. J. Iowa St. med. Soc. **48**, 193 (1958). — BURNS, R.P.: Mucormycosis of the sinuses, orbit and central nervous system. Trans. Pacif. Cst otoophthal. Soc. **40**, 83 (1959). — BURROW, G.N., SALMON, R.B., NOLAN, J.P.: Successful treatment of cerebral mucormycosis with Amphotericin B. J. Amer. med. Ass. **183**, 370 (1963). — CALLE, S., KLATSKY, S.: Intestinal phycomycosis (mucormycosis). Amer. J. clin. Path. **45**, 264 (1966). — CHICK, E.W., EVANS, J., BAKER, R.D.: Treatment of experimental mucormycosis (Rhizopus oryzae infection) in rabbits with Amphotericin B. Antibiot. et Chemother. (Basel) **8**, 394 (1958). ~ The inhibitory effect of amphotericin B on localized Rhizopus oryzae infection (mucormycosis) utilizing the pneumoderma pouch of the rat. Antibiot. et Chemother. (Basel) **8**, 506 (1958). — CHRISTIANSEN, M.: Chronic lesions in cattle-granulomatous. Virchows Arch. path. Anat. **273**, 879 (1929). — CLARK, R.M.: A case of mucormycosis of the duodenum, liver, and cecum. Gastroenterology **33**, 985 (1957). — *Clinical Pathological Conference. New York State Veterinary College*, Cornell vet. **57**, 308 (1967). — *Clinicopathological Conference* from the Ohio State Univ. Hosp., Columbus/Ohio. Ohio St. med. J. **63**, 494 (1967). — COHNHEIM: Zwei Fälle von Mykosis der Lungen. Virchows Arch. path. Anat. **33**, 157 (1865). — CONANT, N.F., SMITH, D.T., MARTIN, D.S., CALLOWAY, J.L., BAKER, R.D.: Manual of clinical mycology. Philadelphia and London: W.B. Saunders Co. 1954. — CRAIG, J.M., FARBER, S.: Development of disseminated visceral mucormycosis during therapy for acute leukemia. Amer. J. Path. **29**, 601 (1953). — CUADRADO, S.P., HABERMAN, S., RACE, G.J.: Visceral mucormycosis (phycomycosis). Tex. St. J. Med. **57**, 712 (1961). — CUSSEN, I.J.: Primary hypopituitary dwarfism with Franconi's hypoplastic anemia syndrome, renal hypertension and phycomycosis. Report of a case. Med. J. Aust. **2**, 367 (1965). — DARJA, M., DAVY, M.I.: Pulmonary mucormycosis with cultural identification. Canad. med. Ass. J. **89**, 1235 (1963). — DAVIS, C.L., ANDERSON, W.A., MCCRORY, B.R.: Mucormycosis in food-producing animals. J. Amer. vet. med. Ass. **126**, 261 (1955). — DEFEO, E.: Mucormycosis of the colon. Amer. J. Roentgenol. **86**, 86 (1961). — DEWEESE, D.D., ROBINSON, L.B.: Mucormycosis of the nose and paranasal sinuses. Trans. Amer. laryng. rhin. otol. Soc., p. 452 (1965). — DEWEESE, D.D., SCHEUNING, A.J., ROBINSON, L.B.: Mucormycosis of the nose and paranasal sinuses. Laryngoscope (St Louis) **75**, 1398 (1965). — DILLON, M.L., SEALY, W.C.: Surgical aspects of opportunistic fungus infections. Lab. Invest. **11**, 1231 (1962). — DILLON, M.L., SEALY, W.C., FETTER, B.F.: Mucormycosis of the bronchus successfully treated by lobectomy. J. thorac. Surg. **35**, 464 (1958). — DIRIART, H., POUJOL, J., SCHNEIDER, J., ISIDOR, P.: Mucoviscidose à localisation rénale dominante parasitée par une mucoracée. Arch. franç. Pédiat. **20**, 220 (1963). — DOLMAN, C.L.A., HERD, J.A.: Acute pancreatitis in pregnancy complicated by renal cortical necrosis and cerebral mucormycosis. Canad. med. Ass. J. **81**, 562 (1959). — DOUROV, N., DUSTIN, P.: Fréquence des mycoses dans les hemopathies malignes traitées. Intl. Colloquium on Medical Mycology. Ann. Soc. belge Méd. trop. **44**, 909 (1964). — DWYER, G.K., CHANGUS, G.W.: Rhinomucormycosis resulting in fatal cerebral mucormycosis. Arch. Otolaryng. **67**, 619 (1958). — EGGENSCHWILER, E.: Die Mucormykose der Nasennebenhöhlen und ihre Komplikationen (sog. craniale Form der Mucormykose). Pract. oto-rhino-laryng. (Basel) **24**, 166 (1962). — ELDER, T.D., BAKER, R.D.: Pulmonary mucormycosis in rabbits with alloxan diabetes. Arch. Path. **61**, 159 (1956). — ELLIS, P.A.: Mucormycosis. A cause of abscess of the hilum of the lung. Brit. J. Dis. Chest **60**, 203 (1966). — EMMONS, C.W., BINFORD, C.H., UTZ, J.P.: Medical mycology, 380 pp. Philadelphia: Lea & Febiger 1963. — EMMONS, C.W.: Phycomycosis in man and animals. Riv. Pat. Veg. (Pavia) **4**, 329 (1964). — ERIKSEN, K.R.: Dissemineret mucormykose. Ugeskr. Laeg. **125**, 1849 (1963). — ERIKSEN, K.R., CHRISTENSEN, A.M., HORNBAK, H.: Localized mucormycosis. Ugeskr. Laeg. **124**, 1700 (1962). — ERNST, H.C.: Case of mucor infection. J. med. Res. **39**, 143 (1918, 1919). — FAILLO, P.S., SUBE, H.P., ANDERSON, N.H.: Mucormycosis of the paranasal sinuses and the maxilla. Oral Surg. **12**, 304 (1959). — FANKHAUSER, R., KELLER, H., LANZ, E.: Granulomatose Entzündung von Zwischenwirbelscheiben beim Rind. Morphologische Mucormykose. Schweiz. Arch. Tierheilk. **108**, 699 (1966). — FEGELER, F.: Medizinische Mykologie in Praxis und Klinik, 125 pp. Berlin-Heidelberg-New York: Springer 1967. — FEJER, E.: Medizinische Mykologie und Pilzkrankheiten, 987 pp. Budapest: Akadémai Kiadó 1966. — FERRY, A.P.: Cerebral mucormycosis (phycomycosis). Surv. Ophthal. **6**, 1 (1961). — FETTER, B.F., KLINTWORTH, G.K., HENDRY, W.S.: Mycoses of the central nervous system, 214 pp. Baltimore: Williams and Wilkins 1967. — FIENBERG, R., RISLEY, T.S.: Mucormycotic infection of arteriosclerotic thrombus of the abdominal aorta. New Engl. J. Med. **260**, 626 (1959). — FOLEY, F.D., SHUCK, J.M.: Burn-wound infection with Phycomycetes requiring amputation of hand. J. Amer. med. Ass. **203**, 596 (1968). — FOUSHEE, S., BECK, W.C.: Mucormycosis of central nervous system. N.C. med. J. **17**, 26 (1956). — FRÁGNER. P., ROKOS, J.: Prípad mukormykózy (*Rhizopus oryzae*). Čas. Lék. čes. **103**, 1084 (1964). — FRANKS, A.G., GUIDUCCI, A.: Mucormycosis. In: Medical mycology, ed. by R.D.G. PH. SIMONS, chapter 36. Amsterdam: Elsevier Publ. Co. 1954. — FRENKEL, J.K.: Role of corticosteroids as predisposing factors in fungal diseases. Lab. Invest. **11**, 1192 (1962). — FREYCON, M.T.: La mucormycose ou phycomycose. Pédiatrie **18**, 109 (1963). — FUENTES OLANO, C.: Cuatro casos de mucormicosis cerebral. Bol. méd. Hosp. infant. (Méx.) **20**, 571 (1963). — FÜRBRINGER, P.: Beobachtungen über Lungenmycose beim Menschen. Virchows Arch. path. Anat. **66**, 330 (1876). — GABRIELE, O.F.: Mucormycosis. Amer. J. Roentgenol. **83**, 227 (1960). — GALE, G.R., WELCH, A.M.: Studies of opportunistic fungi. I, Inhibition of Rhizopus oryzae by human serum. Amer. J. med. Sci. **241**, 604 (1961). — GASS. J.D.M.: Ocular manifestations of acute mucormycosis. Arch. Ophthal. **64**, 226 (1961). ~ Acute orbital mucormycosis. Report of 2 cases. Arch. Ophthal. **65**, 214 (1961). — GATLING, R.R.: Gastric mucormycosis in a newborn infant. Arch. Path. **67**, 249 (1959). — GEORGIADE, M., MAGUIRE, C., CRAWFORD, H., PICKRELL, K.: Mucormycosis and palatal sloughs in diabetics. Plast. reconstr. Surg. **17**, 473 (1956). — GINSBERG, J., SPAULDING, A.G., LAING, V.O.: Cerebral phycomycosis (mucormycosis) with ocular involvement. Amer. J. Ophthal. **62**, 900 (1966). — GISLER, D.B., PITCOCK, J.A.: Intestinal mucormycosis in the monkey (*Macaca mulatta*). Amer. J. vet. Res. **23**, 365 (1962). — GITTER, M., AUSTWICK, P.K.C.: Mucormycosis and Moniliasis in a litter of sucking pigs. Vet. Rec. **71**, 6 (1959). ~ The presence of fungi in abomasal ulcers of young calves. A report of 7 cases. Vet. Rec. **69**, 924 (1957). — GLEISER, C.A.: Mucormycosis in animals. A report of three cases. J. Amer. vet. med. Ass. **123**, 441 (1953). — GLOOR, F., LÖFFLER, A., SCHOLER, H.J.: Mucormykosen. Path. et Microbiol. (Basel) **24**, 1043 (1961). — GORDON, M.A., LITTLE, G.N.: Annual report of the division of laboratories and research for 1964. Albany, New York State Dep. of Health, p. 98 (1965). — GREEN, W.H., GOLDBERG, H.I., WOHL, G.T.: Mucormycosis infection of the craniofacial structures. Amer. J. Roentgenol. **101**, 802 (1967). — GREER, A.E.: Disseminated Fungus diseases of the Lung. Springfield/Ill.: Charles C. Thomas 1962. — GREGORY, J.E., GOLDEN, A., HAYMAKER, W.: Mucormycosis of the central nervous system. A report of three cases. Bull. Johns Hopk. Hosp. **73**, 405 (1943). — GROCOTT, R.G.: A stain for fungi in tissue sections and smears using Gomori's methenamine-silver nitrate technic. Amer. J. clin. Path. **25**, 975 (1955). — GROVER, S., NAIDU, A., JUNNARKAR, R.V.: Rhinocerebral phycomycosis. A case report. Indian J. Path. Bact. **9**, 264 (1966). — GRUHN, J.G., SANSON, J.: Mycotic infections in leukemic patients at autopsy. Cancer (Philad.) **16**, 61 (1963). — GUENIOT, M.: La mucormycose chez les diabetiques. Sem. Hôp. Path. Biol. (Paris) **7**, 1077 (1959). — GUKELBERGER, M.: Pneumomykosis Mucarina als Sekundärinfektion einer Bronchopneumonie. Dtsch. Arch. klin. Med. **182**, 28 (1938). — GUNSON, H.H., BOWDEN, B.H.: Cerebral mucormycosis. Report of a case. Arch. Path. **60**, 440 (1955). — HAFSTRÖM, T., SJÖQUIST, O., HENSCHEN, F.: Zur Kenntnis der mykotischen Veränderungen des Gehirns. Acta chir. scand. **85**, 115 (1941). — HARRIS, J.S.: Mucormycosis. Report of a case. J. Pediat. **16**, 857 (1955). — HAWKER, L.E., ABBOTT, P.M.: Fine structure of vegetative hyphae of Rhizopus. J. gen. Microbiol. **30**, 401 (1963). — HERLA, V.: Note sur un cas de pneumomycose chez l'homme. Bull. Acad. roy. Med. Belg. **9**, 1021 (1895). — HESSELTINE, C.W.: Revision of the Mucorales based especially upon a study of the representatives of this order in Wisconsin. Thesis. University of Wisconsin, Madison, Wisconsin, 1950. ~ Survey of the mucorales. Trans. Acad. Sci. **14**, 210 (1952). — HESSLER, J.R.: Mucormycosis in a Rhesus monkey. J. Amer. vet. med. Ass. **151**, 909 (1967). — HEWER, T.F.: Aspergillosis and Mucormycosis in an okapi. A clinical-pathological conference. Bristol med.-chir. J. **79**, 121 (1964). — HILDICK-SMITH, G., BLANK, H., SARKANY, I.: Fungus diseases and their treatment. Boston: Little, Brown & Co. 1964. — HOAGLAND, R.J., SUBE, J., BISHOP, R.H., jr., HOLDING, B.F., jr.: Mucormycosis. Amer. J. med. Sci. **242**, 415 (1961). — HÖER, P.W., BLÄTTNER, G.: Pulmonale Mucormykose

(Phycomykose) beim Neugeborenen. Frankfurt. Z. Path. 74, 13 (1964). — HÜCKEL, A.: Zur Kenntnis der Biologie des Mucor corymbifer. Beitr. path. Anat. 1, 115 (1886). — HURTADO, R., INNER, R. S., POZO, S., VEIT, Y. O.: Mucormycosis cerebral. Rev. méd. Chile 93, 179 (1965). — HUTTER, R. V. P.: Phycomycetous infection (mucormycosis) in cancer patients. A complication of therapy. Cancer (Philad.) 12, 330 (1959). — HUTTER, R. V. P., COLLINS, H. S.: The occurrence of opportunistic fungus infections in a cancer hospital. Lab. Invest. 11, 1035 (1962). — HUTTER, R. V. P., LIEBERMAN, P. H., COLLINS, H. S.: Aspergillus in a cancer hospital. Cancer (Philad.) 17, 747 (1964). — JACKSON, J. R., KARNAUCHOW, P. N.: Mucormycosis of central nervous system. Canad. med. Ass. J. 76, 130 (1957). — JOSEFIAK, E. J., FOUSHEE, J. H. S.: Experimental mucormycosis in the healthy rat. Science 127, 1442 (1958). — JOSEFIAK, E. J., FOUSHEE, J. H. S., SMITH, L. C.: Cutaneous mucormycosis. Amer. J. clin. Path. 30, 547 (1958). — KAHN, L. B.: Gastric mucormycosis. Report of a case with a review of the literature. S. Afr. med. J. 37, 1265 (1963). — KAPLAN, W., GOSS, L. J., AJELLO, L., IVANS, S.: Pulmonary mucormycosis in a harp seal caused by mucor pusillus. Mycopathologia (Den Haag) 12, 101 (1959, 1960). — KEENEY, E. L.: Practical medical mycology. Springfield/Ill.: Charles C. Thomas 1955. — KEYE, J. D., MAGEE, W. E.: Fungal diseases in general hospital. A study of 88 patients. Amer. J. clin. Path. 26, 1235 (1956). — KNOTH-BORN, R. C.: Cutane Mucorinfektion bei einer 50-jährigen Bäuerin. Z. Haut- u. Geschl.-Kr. 26, 348 (1959). — KUBO, T.: Tumor of small intestine (mucormycosis). Jap. J. clin. Path. 13, 710 (1965). — KÜCHENMEISTER, G. F. H.: Die in und an dem Körper des lebenden Menschen vorkommenden Parasiten. Ein Lehr- und Handbuch der Diagnose und Behandlung der tierischen und pflanzlichen Parasiten des Menschen. Leipzig (Germany): B.G. Teubner 1855. — KURREIN, F.: Cerebral mucormycosis. J. clin. Path. 7, 141 (1954). — LANDAU, J. W., NEWCOMER, V. D.: Acute cerebral phycomycosis (mucormycosis). A report of a pediatric patient successfully treated with amphotericin B and cycloheximide and review of the pertinent literature. J. Pediat. 61, 363 (1963). — LANG, F. J., GRUBAUER, F.: Über Mucor- und Aspergillusmykose der Lunge. Virchows Arch. path. Anat. 245, 480 (1923). — LATOUCHE, C. J., SUTHERLAND, T. W., TELLING, M.: Histopathological and mycological features of a case of rhinocerebral mucormycosis (phycomycosis) in Britain. Sabouraudia 3 (pt. 2), 148 (1964). ~ Rhinocerebral mucormycosis. Lancet 1963 II, 811. — LECOMPTE, P. M., MEISSNER, W. A.: Mucormycosis of the central nervous system associated with hemochromatosis. Amer. J. Path. 23, 673 (1947). — LEVIN, S. E., ISAACSON, C.: Spontaneous perforation of the colon in the newborn infant. Arch. Dis. Childh. 35, 378 (1960). — LICHTHEIM, L.: Über pathogene Mucorineen und die durch sie erzeugten Mykosen des Kaninchens. Z. klin. Med. 7, 140 (1884). — LIE-KIAN-JOE, ENG, N. T., TJOKRONEGORO, S., SCHAAFMA, S., EMMONS, C. W.: Phycomycosis of the central nervous system, associated with diabetes mellitus in Indonesia. Amer. J. clin. Path. 32, 62 (1959). — LJUBIMOWA, W. J.: Ein Fall von Ulcus ventriculi verursacht durch Schimmelpilze. Virchows Arch. path. Anat. 214, 432 (1913). — LLOYD, J. B., SEXTON, L. I., HERTIG, A. T.: Pulmonary mucormycosis complicating pregnancy. Amer. J. Obstet. Gynec. 58, 548 (1949). — LÖHLEIN, M.: Über Schimmelmykosen des Magens. Virchows Arch. path. Anat. 227, 86 (1920). — LONG, E. L., WEISS, D. L.: Cerebral mucormycosis. Amer. J. Med. 26, 625 (1959). — LOPEZ, H., HURTADO, H., CORREA, E.: Las micosis profundas en el Hospital de San Juan de Dios (Anos 1959—1962). Cali, Colombia: Publ. Hosp. S. Juan de Dios 16, 1 (1964). — LUBBE, T. R., PENNINGTON, Orbital mucormicosis. Med. J. Aust. 1, 681 (1964). — LUCET, A., CONSTANTIN, J.: Contributiona à l'étude des mucorinées pathogénes. Arch. parasit. 4, 362 (1901). — LUCKE, M., LINTON, A. H.: Phycomycosis in a Mandrill (*Mandrillus sphinx*). Vet. Rec. 77, 1306 (1965). — MAHAFFEY, L. W., ADAM, N. M.: Abortions associated with mycotic lesions of the placenta in mares. J. Amer. vet. med. Ass. 144, 24 (1964). — MARCATO, P. S., DIMOU, D.: Su un caso di mucormycosi del rumine in una gazzella. Nuova Vet. 38, 279 (1962). — MARCHAND, F.: Ein eigentümlicher Magenbefund (Ulcus gangraenosum durch Fadenpilzwucherung) nach Vergiftung durch verdorbene Steinpilze. Verh. dtsch. Ges. Path. 14, 183 (1910). — MARTIN, F. P., LUKEMAN, J. M., RANSON, R. F., GEPPERT, L. J.: Mucormycosis of the central nervous system associated with thrombosis of the internal carotid artery. J. Pediat. 44, 437 (1954). — MAYFIELD, G. R., CONDIE, F.: Paradoxical mucorthrombosis in thrombocytopenic purpura. Arch. Path. 63, 260 (1957). — MCBRIDE, R. A., CORSON, J. M., DAMMIN, J. J.: Mucormycosis. Two cases of disseminated disease with cultural identification. Review of literature. Amer. J. Med. 28, 832 (1960). — MCCALL, W., STROBOS, R. R. J.: Survival of a patient with central nervous system mucormycosis. Neurology (Minneap.) 7, 290 (1957). — MEHNERT, B.: The role of fungi in diseases of young pigs. Zbl. Vet.-Med. (B) 13, 201 (1966). — MERRIAM, J. C., jr., TEDESCHI, C. G.: Cerebral mucormycosis, a fatal fungus infection complicating other diseases. Neurology (Minneap.) 7, 510 (1957). — MIYAKE, M., OKUDAIRA, M.: A statistical survey of deep fungus infection in Japan. Acta path. jap. 17, 401 (1967). — MOMBERG-JORGENSEN, A. C.: Enzoötic mycosis in mink. Amer. J. vet. Res. 11, 334 (1950). — MONTENEGRO, M. R., DEBRITO, T., LOMBARDI, J., LACAZ, C. S.: Mucormicose intestinal; Registro de dois casos. Rev. Hosp. Clin. Fac. Med. S. Paulo 14, 59 (1959). — MOORE, M., ANDERSON, W. A. D., EVERETT, H. H.: Mucormycosis of large bowel. Amer. J. Path. **25**, 559 (1949). — MORQUER, R., LOMBARD, C., BERTHELON, M., LACOSTE, L.: Pathogénie de quelques Mucorales pour les animaux. Une nouvelle mucormycose chez les Bovidés. Bull. trim. Soc. mycol. Fr. **81**, 421 (1965). — MOSS, E.S., MCQUOWN, A.L.: Atlas of medical mycology, 335 pp. Baltimore: Williams & Wilkins Co. 1960. — MOSTAFA, I.E., CERNA, J., CERNY, L.: Mucormycosis in goats. Report of two cases. Ceylon vet. J. **14**, 79 (1966). — MURESAN, A.: A case of cerebral mucormycosis diagnosed in life, with eventual recovery. J. clin. Path. **13**, 34 (1960). — MURPHY, J.D., BORNSTEIN, S.: Mucormycosis of the lung. Ann. intern. Med. **33**, 44 (1959). — NARASIMHAN, M.J., GANLA, V.G., DEODHAR NAR N.S.: Epidemic polyuria in man caused by a phycomycetous fungus (the Sassoon hospital syndrome). Lancet 1967**I**, 760. — NEAME, P., RAYNER, D.: Mucormycosis. Report on 22 cases. Arch. Path. **70**, 261 (1960). — NICOD, J.L., FLEURY, C., SCHLEGEL, J.: Mycose pulmonaire double à Aspergillus fumigatus Fres. et à mucor pusillus. Lindt .Schweiz. Z. allg. Path. **15**, 307 (1952). — NICOLET, J., LINDT, S., SCHOLER, H.J.: L'avotemente mycosique de la vache. Considérations sur le diagnostic de routine. Path. et Microbiol. (Basel) **29**, 644 (1966). — OBRASZOW, E.S., PETROV, N.V.: Fall gleichzeitiger Aktinomykose und Schimmelmykose. Russ. Med. **7**, 457 (1889). — OPPE: Zur Kenntnis der Schimmelmykosen beim Menschen. Zbl. allg. Path. path. Anat. **8**, 301 (1897). — OSSWALD, H., SEELIGER, H.: Tierexperimentelle Untersuchungen mit antimycotischen Mitteln. Arzneimittel-Forsch. **8**, 370 (1958). — PALTAUF, A.: Mycosis mucorina. Virchows Arch. path. Anat. **102**, 543 (1885). — PAPLANUS, S.H., SHELDON, W.H.: The inflammatory response of hypothyroid rats to subcutaneous *Rhizopus oryzae* infection. Bull. Johns Hopk. Hosp. **117**, 140 (1965). — PARKHURST, G.F., VLAHIDES, G.D.: Fatal opportunistic fungus diseases. J. Amer. med. Ass. **202**, 279 (1967). — PARMENTIER, N., BALASSE, E., MEUR, G., PIRART, J., VANDERHAEGEN, J.J.: Etude anatomo-clinique d'un cas de phycomycose de l'étage craniofacial. Bull. Soc. belge Ophthal. **139**, 304 (1965). — PASTORE, P.N.: Mucormycosis of the maxillary sinus and diabetes mellitus. Report of a case with recovery. Sth. med. J. (Bgham, Ala.) **60**, 1164 (1967). — PEÑA, C.E.: Deep mycotic infections in Colombia. A clinicopathologic study of 162 cases. Amer. J. clin. Path. **47**, 505 (1967). — PEÑA, C.E., DORADO, J.A.: Mucormycosis (ficomicosis) en Colombia. Presentación de dos casos. Rev. Fac. Med. (Bogotá) **33**, 2 (1965). — PEREZ ARZOLA, M., GARCÍA SORIA, M., RIVIERA, E.: Cerebral mucormycosis. Bol. Asoc. méd. P. Rico **58**, 374 (1966). — *Phycomycosis* (Editorial): Actualid. pediát. **24**, 501 (1963). — PIRILA, P.: Eine Mucormykose der äußeren Genitalien. Acta derm.-venereol. (Stockh.) **22**, 377 (1941). — PLAUT, H.C., GRÜTZ, O.: Mucoraceen, pp. 159—163, and Die Schimmelpilze, pp. 156. In: W. KOLLE and A.V. WASSERMANN **5**, 1 (1928). — PODACK, M.: Zur Kenntnis des sog. Endothelkrebses der Pleura und die Mucormycosen im menschlichen Respirationsapparat. Dtsch. Arch. klin. Med. **63**, 1 (1899). — POLEMANN, G.: Klinik und Therapie der Pilzkrankheiten, 396 pp. Stuttgart: Georg Thieme 1961. — PROCKNOW, J.J.: Treatment of opportunistic fungus infections. Lab. Invest. **11**, 1217 (1962). — PROCKOP, L.D., SILVA-HUTNER, M.: Cephalic mucormycosis (phycomycosis). A case with survival. Arch. Neurol. (Chic.) **17**, 379 (1967). — PROUT, G.R., jr., GODDARD, R.: Renal mucormycosis. New Engl. J. Med. **263**, 1246 (1960). — RABIN, E.R., LUNDBERG, G.D., MITCHELL, E.T.: Mucormycosis in severely burned patients. Report of 2 cases with extensive destruction of face and nasal cavity. New Engl. J. Med. **264**, 1286 (1961). — REEVES, D.L., DICKSON, D.A., BENJAMIN, E.L.: Phycomycosis (mucormycosis) of the central nervous system. Report of a case. J. Neurosurg. **23**, 82 (1965). — RINALDI, I., ASHBY, S.F.: Facio-cranial mucormycosis. Report of a case. Virginia med. Mth. **89**, 595 (1962). — ROBERTS, H.J.: Cutaneous mucormycosis. Report of a case, with survival. Arch. intern. Med. **110**, 108 (1962). — ROWE, P.B., PAYNE, W.H.: Rhino-cerebral mucormycosis. Med. J. Aust. **2**, 960 (1964). — SAUER, R.M.: Cutaneous mucormycosis (phycomycosis) in a squirrel (*Sciureus carolinensis*). Amer. J. vet. Res. **27**, 380 (1966). — SCHAUBLE, M.K., BAKER, R.D.: The inflammatory response in acute alloxan diabetes. Arch. Path. **64**, 563 (1957). — SCHOFIELD, R.A., BAKER, R.D.: Experimental mucormycosis (Rhizopus infections) in mice. Arch. Path. **61**, 407 (1956). — SHANKLIN, D.R.: Pulmonary mucormycosis complicating Cushing's syndrome. Arch. Path. **68**, 262 (1959). — SHELDON, W.H., BAUER, H.: Activation of quiescent mucormycotic granulomata in rabbits by induction of acute alloxan diabetes. J. exp. Med. **108**, 171 (1958). ~ The development of the acute inflammatory response to experimental cutaneous mucormycosis in normal and diabetic rabbits. J. exp. Med. **110**, 845 (1959). ~ The role of predisposing factors in experimental fungus infections. Lab. Invest. **11**, 1184 (1962). — SHIRLEY, A.G.H.: Two cases of phycomycotic ulceration in sheep. Vet. Rec. **77**, 675 (1965). — SIMON, R., HOFFMAN, G.G., HARDING, H.B.: Phycomycosis. Aerospace Med. **35**, 668 (1964). — SLUYTER: De vegetabilibus organismi animalis parasitis ac de novo epiphyto in pityriasi versicolore obvio. Inaug. Diss. Berlin (Germany) 1847. — SMITH, D.T.: Miscellaneous fungus diseases. J. chron. Dis. **5**, 528 (1957). — STEVENSON, H.N.: Mucormycosis of maxillary sinus. Arch. Otolaryng. **18**, 775 (1933). — SMITH, H.W., KIRCHNER, J.A.: Cerebral mucormycosis. Report of 3 cases. Arch. Otolaryng. **68**, 715 (1958). — SMITH, H.W., YANAGISAWA, E.: Rhinomucormycosis. Report of a fatal case. New Engl. J. Med. **260**, 1007 (1959). — SMITH, M.E., BURNHAM, DEW.K., BLACK,

M. B.: Cerebral mucormycosis. Arch. Path. **66**, 468 (1958). — STEHBENS, W. E.: Atypical cerebral aneurysms. Med. J. Aust. **1**, 765 (1965). — STEFANINI, M., ALLEGRA, S.: Pulmonary mucormycosis in acute histocytic leukemia. New Engl. J. Med. **256**, 1026 (1957). — STEIN, A., SCHMAMAN, A.: Rupture of the stomach due to mucormycosis. S. Afr. J. Surg. **3**, 123 (1965). — STRAATSMA, B. R., ZIMMERMANN, L. E., GASS, J. D.: Phycomycosis. A clinicopathologic study of 51 cases. Lab. Invest. **11**, 963 (1962). — STRATEMEIER, W. P.: Mucormycosis of central nervous system. Report of a case. Arch. Neurol. Psychiat. (Chic.) **63**, 179 (1950). — SUGA, J., HAGAL, A., KASHIMA, H.: Autopsy case of disseminated mucormycosis with ocular involvement. J. clin. Ophthal. (Tokyo) **17**, 365 (1963). — SUIE, T., HAVENER, W. H.: Mycology of the eye. A review. Amer. J. Ophthal. **56**, 63 (1963). — SUN, H. L., WANG, T., MA, W., KAO, H.: A study of the pathology and morphology of deep mycosis. China med. J. **84**, 125 (1965). — SUPRUN, H.: Mucormycosis infection. A post-mortem case report with a summarized survey of the literature. Israel med. J. **21**, 40 (1962). — SUTHERLAND, J. C., JONES, T. H.: Gastric mucormycosis. Report of a case in a Swazi. S. Afr. med. J. **34**, 161 (1960). — SUTHERLAND-CAMPBELL, H., PLUNKETT, O. A.: Mucor paronychia. Syph. **30**, 651 (1934). — *Sydney Hospital:* Clinicopathological conference. Med. J. Aust. **2**, 98 (1956a). ~ Clinicopathological conference. Med. J. Aust. **11**, 30 (1956b). — SYMMERS, W. ST. C.: Histopathologic aspects of the pathogenesis of some opportunistic fungal infections, as exemplified in the pathology of aspergillosis and phycomycetoses. Lab. Invest. **11**, 1073 (1962). ~ The tissue reactions in deep-seated fungal infections. The role of histological examination in mycological diagnosis. Ann. Soc. belge Méd. trop. **44**, 869 (1964). ~ The occurrence of deep-seated fungal infections in general hospital practice in Britain today. Proc. roy. Soc. Med. **57**, 405 (1964). — TAKAHASHI, Y.: A critical survey of medical mycology for the years 1946—1956 in Japan. Mycopathologia (Den Haag) **19**, 105 (1963). — TAYLOR, R., SHKLAR, G., BUDSON, R., HACKETT, R.: Mucormycosis of the oral mucosa. Arch. Derm. **89**, 419 (1964). — TEUTSCHLAENDER, O.: Mucormykose des Magens. Mitt. Grenzgeb. Med. Chir. **29**, 127 (1916, 1917). — TEWARI, R. P.: Studies on some mycotic infections of domestic animals and poultry. Agra Univ. J. Res. **12**, 163 (1963). — TINAZTEPE, B., TINAZTEPE, K.: Cerebral mucormycosis in a child. Turk. J. Pediat. **8**, 207 (1966). — TOMIK, F.: Histologische und histochemische Diagnose der Systemmykosen. Zbl. allg. Path. path. Anat. **104**, 189 (1963). — TOMIYASU, U., BAKER, R. N.: Phycomycosis (mucormycosis) of the orbital apex and cavernous sinus. Bull. Los Angeles neurol. Soc. **31**, 177 (1966). — TORACK, R. M.: Fungus infections associated with antibiotic and steroid therapy. Amer. J. Med. **22**, 872 (1957). — TRUJILLO, S. H.: Monografia sobre la mucormicosis en pediatrica. Anot. pediát. **4**, 296 (1961). — TURNER, P. D.: Simultaneous infection of a bovine foetus by 2 fungi. Nature (Lond.) **205**, 300 (1965). — TURPIN, R., BOCQUET, L., CAILLE, B., DEFRANOUX, A.: Mucormycose cérébral. Épisode terminale d'une leucose aiguë. Sem. Hôp. Ann. Péd. (Paris) **37**, 2134 (1961). — UTZ, J. P.: The spectrum of opportunistic fungus infections. Lab. Invest. **11**, 1018 (1962). ~ Systemic fungal infections amenable to chemotherapy. DM (Chic.) **1**, 1963. — VIGNALE, R., MACKINNON, J. E., CASELLA DE VILA BOA, E., BURGOA, F.: Chronic destructive, mucocutaneous phycomycosis in man. Sabouraudia **3**, 143 (1964). — WADE, J. L., MATTHEWS, A. R. K.: Cutaneous mucor infection of face. J. Amer. med. Ass. **114**, 410 (1940). — WADSWORTH, J. A. C.: Ocular mucormycosis. Report of a case. Amer. J. Ophthal. **34**, 405 (1951). — WALL, G. H., MADISON, W. N., jr.: Fatal pulmonary phycomycosis complicating leukemia. A case report. N. C. med. J. **26**, 66 (1965). — WASSERMAN, A. J., SNIELS, W. S., SPORN, I. N.: Cerebral mucormycosis. Sth. med. J. (Bgham, Ala.) **54**, 403 (1961). — WÄTJEN, J.: Pathologisch-anatomische Demonstrationen. Klin. Wschr. **8**, 280 (1929). ~ Durch Schimmel und Sproßpilze bedingte Erkrankungen der Lungen. In: Handbuch der speziellen pathologischen Anatomie und Histologie Hrsg. von HENKE-LUBARSCH. Vol. 3, part 3, p. 397. Berlin: Springer 1931. — WATSON, K. C.: Gastric perforation due to fungus *Mucor* in a child with kwashiorkor. S. Afr. med. J. **31**, 99 (1957). — WATSON, K. C., NEAME. P. B.: In vitro activity of Amphotericin B on strains of Mucoraceae pathogenic to man. J. Lab. clin. Med. **56**, 251 (1960). — WEISSKOPF, A.: Mucormycosis, a rhinologic disease. Ann. Otol (St. Louis) **73**, 16 (1964). — WELLER, W. A., JOSEPH, D. J., HORA, J. F.: Deep mycotic involvement of the right maxillary and ethmoid sinuses. Laryngoscope (St. Louis) **70**, (2), 999 (1960), — WERNER, H. J., WRIGHT, P., BAKER, R. D.: Electron microscope observations of *Rhizopus rhizopodiformis*. J. gen. Microbiol. **37**, 205 (1964). — WERTMAN, K. F., HENNEY, M. R.: The effects of alloxan diabetes on phagocytosis and suceptibility to infection. J. Immunol. **89**, 314 (1962). — WILSON, J. W., PLUNKETT, O. A.: The fungus diseases in man. Berkeley and Los Angeles: University of California Press 1965. — WINSTON, R. M.: Phycomycosis of the bronchus. J. clin. Path. **18**, 729 (1965). — WOLF, A., COWEN, D.: Mucormycosis of central nervous system (abstr.). J. Neuropath. exp. Neurol. **8**, 107 (1949). — WRIGHT, P., SYMMERS, W. ST. C.: Systemic pathology. New York: Elsevier Press 1967. — ZIMMERMAN, L. E.: Fatal fungus infections complicating other diseases. Amer. J. clin. Path. **25**, 46 (1955).

Geotrichosis

J. MORENZ, Magdeburg, Germany (DDR)

With 12 Figures

A. Definition

Geotrichosis is a world-wide mycosis caused by *Geotrichum candidum*. The oral, bronchial and bronchopulmonary geotrichoses are endogenous in origin and have little tendency to hematogenous dissemination. The cutaneous geotrichoses are mostly exogenous in origin, limited to the skin, chronic, occur in superficial and deep forms, and sometimes are accompanied by allergic skin reactions in the form of geotrichids. Mycetemia without lesions in the organs, ulcerative keratitides, and a mycetoma have been observed.

B. Geographical Extent and Epidemiology

Geotrichum candidum is world-wide and ubiquitous. It is in soil and sewage, on damaged or dying plants and almost always present in sour milk and sour milk products. It grows especially well in sour vegetable liquid, moist food for farm animals, mash, spoiled wine, manure, and other animal excreta.

Although geotrichosis is of world-wide occurrence most of the cases have been observed in Europe and North America, probably because of the better diagnostic possibilities in these regions. SMITH (1949) gave data on the frequency of geotrichosis in the United States. He reported that among 250,000 admissions to Duke Hospital (Durham, North Carolina, USA) there were 207 patients with systemic mycoses and among them 4 with geotrichosis.

With regard to the age incidence of the cases of geotrichosis, there appears to be no predilection of the disease as a whole, or of the various forms, for any particular age period. The extremes of occurrence are in a 2-month-old infant and an 80-year-old man. The ratio of men to women is about 2:1. This difference may be fortuitous because of the small number of cases. Occupational predispositions are not evident.

As *G. candidum* can be isolated from the sputum and stool of healthy persons, that is, it is normally already present on the mucous membranes of the respiratory and digestive tract, therefore bronchopulmonary, bronchial and oral geotrichoses are endogenous infections. There are also exogenous infections. These are illustrated by the two corneal infections described by PERZ (1964) after mud, evidently containing *G. candidum* splashed into the eye, and injured the cornea. The mycetoma of the foot described by NICOLAU et al. (1957) apparently developed after a wound from a nail in the sole of a shoe. In cutaneous geotrichosis the manner of infection is, to some extent, still unclear. In the superficial forms, and in the deep forms which begin with an epidermal lesion, the concept of an exogenous infection (an injury mycosis) can be accepted. In contrast to this, in the patient of THIERS et al. (1953) the infection began with subcutaneous nodules, and the epidermis was implicated later.

C. Geotrichum Candidum - Occurrence in Humans

1. Intestinal Tract

G. candidum can be demonstrated most frequently in the stool, in 25—30% of persons without intestinal disorders, and in 50—60% of patients with intestinal disorders, especially in the region of the colon. In the latter group the quantity of excreted fungus is significantly greater.

SCHNOOR (1939) isolated *G. candidum (G. sp.)* from 29% of 314 normal stools. SWARTZ and JANKELSON (1941) found *G. candidum* (*G. versiforme*) in 67% of 24 cases of nonspecific ulcerative colitis and in only 21% of the control group of 24 cases of healthy persons and patients with diarrhea due to other causes. Quantitatively, almost pure cultures were obtained from the colitis group, while in the control group there were generally only a few colonies. GOLAY and WYSS-CHODAT (1942) isolated *G. candidum (G. sp.)* in only 3% of 102 stools of healthy persons and of patients with various skin lesions, but they obtained 54% positive stools in 28 patients with psoriasis. As there were 83% positives in the more severe psoriasis cases, the authors thought there might be a relationship between psoriasis and the presence of *G. candidum* in the stool. This observation was corroborated by ETTIG (1965) who was able to demonstrate *G. candidum* in 98% of 125 psoriasis patients and in only 41% of 138 healthy persons and persons with other skin diseases. *G. candidum* grew luxuriantly in 22% of the stool cultures of the psoriasis patients in contrast to 4% of the control group. ETTIG related the increased occurrence of the fungus in psoriasis to a dysfunction of the intestinal mucous membrane. The psoriasis was not influenced in 5 patients treated for 3 weeks with nystatin. The stools were fungus free in 5 days. Also, the geotrichin test was negative in the psoriasis patients. A larger research series comes from FELSENFELD (1944). He isolated *G. candidum (G. sp.)* in 27% of 900 cases without intestinal disturbances and in 26% of 100 carriers of typhoid bacteria. However, he isolated the fungus in 56% of 103 cases with diarrhea of recent onset. A study of 47 untreated bacterial diarrheas yielded 53% with *G. candidum* while in 51 cases treated with sulfathiazole only 6% were positive for *G. candidum*. MARSELOU and PASCHALI-PAVLATOU (1956) demonstrated *G. candidum (G. sp.)* in 6% of rectal smears from 105 women. SAËZ (1957) found *G. candidum* in 30% of 62 persons (both ill and healthy), KÄRCHER (1953) in 15% of 124 patients with various illnesses and GALINOVIC-WEISGLASS (1959) in 56% of 100 patients with various intestinal disorders and in 33% of 100 healthy subjects. In the patients the fungus was present in larger quantities than in the healthy subjects. PÉTER and HORVÁTH (1959) established an increasing case frequency with age. In 119 children under a year of age, *G. candidum (G. sp.)* could be isolated in only 5% and these were bottle-fed infants. In the age period of 1—3 years 15% of 168 infants and children had isolates of *G. candidum* from their stools; in the age period of 6—12 years 37% of 184, and in the age period above 20 years 53% of 176 were positive. The average frequency of demonstration was 22% in 342 healthy subjects in contrast to 39% in 305 patients treated with antibiotics. CARETTA (1961) isolated *G. candidum* from the stools in 10% of 135 patients more than 15 years old, TZAMOURANIS and CRIMBITHIS (1962) in 13% of 155 patients, HEIDENBLUTH (1963) in 40% of 50 subjects without skin diseases, OTT (1963) in 31% of 300 patients with various skin diseases, ARIYEVICH et al. (1963) in 28% of 107 patients suspected of having candidosis, and MORENZ (1963) in 38% of 577 stools submitted for the investigation of pathogenic intestinal bacteria. PÉTER et al. (1967) could demonstrate *G. candidum* in 30% of 674 stool specimens from healthy persons and in 52% of 802 stools from patients with various diseases, such as epidemic hepatitis, etc.

Because of the frequent occurrence of *Geotrichum candidum* in the stool and its increased elimination in many intestinal disturbances, the relationship of this fungus to afflictions of the digestive tract, especially to diarrheal disorders, is discussed.

G. candidum was isolated by CASTELLANI (1914; *Monilia asteroides*) in enteritis and sprue, by MATTLET (1926; *Mycoderma issavi*, *M. muyaga* and *M. kieta*) in enteritis and dysentery, and viewed as a possible etiological agent. JANNIN (1913) found *G. candidum* twice in hyperacidic gastric juice in gastric dilatation. Although he did not accept the idea of an invasion of the mucosa, he considered the fungi to foster chronic bacterial processes. CIFERRI et al. (1938) considered a *G. candidum* strain (*G. matalense*) isolated from an enterocolitis of 3 years duration to be not the cause but to play a secondary role. According to SWARTZ and JANKELSON (1941), in spite of the frequent and abundant occurrence of *G. candidum* in the stool in colitis cases, there is no reason to consider it causal. COLONNELLO (1944) rejects an etiologic role for *G. candidum* in intestinal diseases. He found no difference between 30 healthy persons and 30 patients with various intestinal disorders with respect to the occurrence of *G. candidum*. *G. candidum* was isolated from rectal ulcerations by ALMEIDA and LACAZ (1940) and COLONNELLO (1944). Two patients of COUDERT et al. (1958) developed a feeling of abdominal fullness and anal itching in the course of tetracycline therapy, and also there were large numbers of the fungus in smears of the stool. As there was a diminution of the number of fungi with the disappearance of symptoms, the authors ascribed a causal relation to the fungus. In the 2 enterocolitis cases of KÄRCHER (1963) that had been treated with aureomycin, *G. candidum* was present in the stool in large quantity. After oral therapy with gentian violet there was a rapid improvement and a decrease in the numbers of *G. candidum* in the stool.

ARIYEVICH et al. (1963) reported 14 patients with various underlying lesions (5 cases of dysentery, 5 of chronic colitis, 2 of ulcerative colitis, one of pleuropneumonia and one of virus-grippe) who had been treated unsuccessfully with antibiotics concurrently or successively. Microscopically and culturally abundant *G. candidum (G. sp.)* was demonstrated as the only fungus or in combination with *Candida*. Therapy with *Candida*-vaccine and an acid-milk-bacteria preparation and later with nystatin was relatively successful. An intestinal geotrichosis or an infection with a combination of *Geotrichum* and *Candida* was postulated. The fungus infections were regarded as super-infections of a condition of disturbed bacterial flora due to antibiotics. The intestinal symptoms corresponded in general to the symptoms of the underlying disease. In some patients there were inconstant bowel movements, in others a chronic obstipation with moderate periodic abdominal pain. In all patients, after high dosages of antibiotics there was a worsened condition, especially with meteorism, severe girdle pains, increased bowel movements and occasionally with subfebrile temperatures and pruritus ani. In these cases the pathogenetic significance of *G. candidum* is difficult to appraise and can not be considered as completely certain. ABRAMOVA et al. (1964) described very completely a case regarded as intestinal geotrichosis (in relation to antibiotic therapy), in a 3 months old, male infant. The child was treated, shortly after birth, with various antibiotics for pneumonia, otitis and gastroenteritis. There was subnormal temperature; the infant was restless and the stools (4 to 17 a day) were pasty and mucoid. Pathogenic intestinal bacteria could not be demonstrated, but there was abundant *G. candidum (G. sp.)* microscopically and culturally. After the giving of colistin the condition was worse, whereas two courses of nystatin over 3 week periods and an intensive treatment with blood transfusions lead to healing and the elimination of the fungus. Also in this case the original significance of the *G. candidum* is not clear as the related general treatment could have accounted for the improvement.

G. and M. NEAGOE (1967) reported the cases of two 22—23-year-old patients, who, in the course of glutamic acid treatment of asthenia, developed a chronic diarrhea with weight loss of 8—10 kg and a large content of *G. candidum* in their stools. After discontinuing the glutamic acid therapy and after the giving of iodine, gentian violet or methylene blue, the intestinal symptoms improved and the content of *G. candidum* in the stools decreased. A renewal of the glutamic acid therapy lead to relapses. The authors believe that the increased growth of fungus caused by glutamic acid may lead to enterocolitis.

From a study of these cases, a conclusive judgment concerning the occurrence of intestinal geotrichosis is still not possible. Apparently there are geotrichoses of the mucous membranes of the respiratory and upper digestive tracts. These develop when the fungus proliferates in an altered terrain, and infection ensues. As the increase in the amount of *G. candidum* in the stool might be only an indication of an altered condition in the intestine, the success of therapy with fungistatic agents comes to have great diagnostic significance. However, one must bear in mind the possibility of an underlying disease or predisposing condition.

2. Respiratory Tract

Geotrichum candidum can be isolated from the sputum of from 1—6% of the population and of 25% of patients with pulmonary tuberculosis.

ALMEIDA and LACAZ (1940) isolated 4 strains of *G. candidum (G. sp.)* from 422 sputa. Among 124 sputum examinations KÄRCHER (1953) found *G. candidum (G. sp.)* once. COUDERT et al. (1957) was able to demonstrate *G. candidum* in 25% of 151 cases of pulmonary tuberculosis in a sanatorium. The presence of the fungus could not be correlated with the therapy of the tuberculosis or with the presence of the tubercle bacillus in the sputum. CARETTA (1961) found *G. candidum* in the sputum of 2 of 135 patients. VÖRÖS-FELKAI and NOVÁK (1961) demonstrated *G.candidum (G.sp.)* 17 times in the pharyngeal, laryngeal, and bronchial secretions and gastric juice of 1200 patients suspected of having mycoses. Among 143 isolates of fungi from the sputum and oral cavity MONTEMAYOR and GAMERO (1961) found *G.candidum (G.sp.)* 10 times. MORENZ (1963) isolated *G.candidum* in 14% of 262 sputa submitted for a search for the tubercle bacillus, and in only 6% of 326 other sputa. ARIYEVICH et al. demonstrated *G.candidum (G.sp.)* three times in 105 sputa and twice in 100 tongue smears of patients suspected of having candidosis. PÉTER et al. (1967) were able to culture *G.candidum* 6 times in 120 sputa and 3 times in 279 throat swabs from healthy persons, and 34 times in 279 sputa, 6 times in 121 throat swabs and once in 33 bronchial washings from patients, mainly patients with tuberculosis. REIERSÖL (1953) found *G.candidum (G.sp.)* twice in 250 laryngeal smears, TZAMOURANIS and CRIMBITHIS (1962) once in 155 pharyngeal smears, and OTT (1963) once in tongue smears from 300 dermatologic patients, while KOURKOUMELI-KONTOMICHALOU et al. (1961) were able to demonstrate *G.candidum* in 6% of the pharyngeal specimens from 240 healthy persons (hospital personnel). Among 6 fungi isolated from bronchial washings by MONTEMAYOR and GAMERO (1961) *G.candidum (G.sp.)* was isolated once. Among 580 bronchoscopic specimens GERNEZ-RIEUX (1960) found *G.candidum (G.sp.)* 5 times, twice in patients with pulmonary tuberculosis, twice in patients with tumors, and once in a patient with mega-esophagus. In 4 of these cases *G.candidum* was visible in direct preparations. From the lungs of 71 autopsies HALEY and MCCABE isolated *G.candidum* once.

3. Skin

Geotrichum candidum occurs on the skin only accidentally. However, GUÉGUEN (1913) mentioned the more frequent occurrence of *G.candidum (Oidium lactis)* on the skin of persons working in milk plants.

CIFERRI and REDAELLI (1935) describe a strain of *G.candidum* isolated from eczema. REDAELLI and CIFERRI (1943) cite two additional strains, one from normal skin and one, with another fungus, from a lesion of the foot. FRÁGNER and SVATEK (1956) found *G.candidum* twice among 170 fungi isolated from the interdigital spaces of miners. PINKERTON et al. (1957) were able to demonstrate *G.candidum (G.sp.)* more often with *Trichophyton mentagrophytes* or *T.rubrum* in tinea pedis and tinea cruris. COUDERT et al. (1958) isolated *G.candidum* five times from scrapings and nail clipping among 128 specimens, but only from moist skin areas. MONTEMAYOR and GAMERO (1961) among 12 fungi isolated from intertrigo and 10 from onychomycosis found in each group a single strain of *G.candidum (G.sp.)*. DVOŘÁK and OTČENÁŠEK (1966) were able to demonstrate *G.candidum* 8 times culturally (4 times together with *Candida albicans*) and 3 times by direct microscopic observation among 2500 lesions of skin and nails. The material came from nails four times, four times from skin scrapings from between the toes, and from between the fingers and from the back of the hand. As *G.candidum* was demonstrated microscopically and in heavy growth of pure culture in a finger nail and in an interdigital lesion in these two cases, a *G.candidum* infection of a previously pathologically altered area was conjectured. The clinical picture and the course of the condition are unknown.

4. Urogenital Tract

G.candidum is only occasionally present in the urogenital tract and may be considered a nonsignificant contaminant.

FRÁGNER et al. (1955) isolated *G.candidum* from the vaginal secretions of 2 patients without giving the finding a clinical meaning. SAËZ (1957) was unable to isolate *G.candidum* in 64 samples from the vagina. PÉTER et al. (1967) isolated the fungus from 2 of 19 vaginal smears from patients with various disorders, but none from 50 smears from healthy subjects.

In urine *G.candidum* was present with bacteria once in the experience of DUNCAN et al. (1951) and COUDERT et al. (1958), but SAËZ (1957) could not demonstrate *G.candidum* in urines obtained aseptically. PÉTER et al. (1967) found *G.candidum* once in 43 urines and MORENZ (1963) twice in 877 urines. Both urines were, however, not taken aseptically nor were the majority of investigated specimens. ARIYEVICH et al. (1963) isolated *G.candidum (G.sp.)* four times from 168 urines from patients suspected of having candidosis but without disease of the urinary tract. On the other hand, *G.candidum (G.sp.)* in great quantity was demonstrated microscopically and culturally in the urine of a 32-year-old patient (from Moscow) with agranulocytosis and symptoms of cystitis who had been treated with penicillin, streptomycin, and other antibiotics. Cystoscopically a white coating was seen on the mucosa of the bladder. Also there were similar plaques on the oral mucosa, but they consisted of *Candida* elements.

D. Mycology

The mycologic diagnosis is relatively simple, as *G. candidum* is the only *Geotrichum* which occurs in man, apart from some insignificant exceptions (MORENZ, 1963—1964). There may be confusion with *Trichosporon* BEHREND (Synonyms: *Geotrichoides, Neogeotrichum*) which is mentioned because even in recent times mycoses caused by *Trichosporon* have been called geotrichoses.

The causal agents of the mycoses described by VRIJMAN (1933), CASTELLANI (1942), VOLK and CAÑAS (1942), PÉTOURAUD and COUDERT (1953), GAUVREAU and GRANDBOIS (1954), LOFGREN (1955) and BREDNOW (1958, Case O. BU.) do not belong to the genus *Geotrichum.*

Geotrichum candidum Link is known by about 100 synonyms (CARMICHAEL, 1957; MORENZ, 1963). Most often this fungus is listed among the genera *Oidium, Oospora,* and *Mycoderma.* The synonyms of the strains found in cases are given in Tables 1—5. The synonyms used in the text are placed in parentheses.

1. Micromorphology

Geotrichum candidum is a filamentous fungus with a hyaline, ramifying mycelium which breaks up into arthrospores (Fig. 1). They are 1.5 to 11×2 to 40 μ in size, and of a rectangular shape. From the arthrospores originate germ tubes, which grow out to form mycelium. In contrast to the *Trichosporon*, reproduction takes place by fission only. *Geotrichum candidum* belongs to the fungi imperfecti, as a sexual development is not known. The micromorphology can be observed best in slide cultures. The medium is poured on to a slide and the fungus is inoculated in streaks. Then a coverglass is put on and the preparation is incubated in a moist chamber for 3 days at room temperature. *Trichosporon* and *Coccidioides* may be mistaken for *Geotrichum.* However, *Trichosporon* propagates by forming arthrospores and blastospores that can be clearly demonstrated by slide culture and in fluid media. The mycelium formed by *Coccidioides immitis* fragments differently from that of *Geotrichum candidum,* not into arthrospores but into thick-walled chlamydospores, which are separated from one another in a characteristic way by empty hyphae. *C. immitis,* moreover, grows as a mold.

2. Direct Preparations

Sputum, bronchus aspirate, pus, ulcer scrapings or material from mucus membranes or from biopsies is placed directly on a glass slide. When there is much cellular material, a drop of physiological NaCl solution or of 20% sodium hydroxide solution is added and a coverglass is applied. The preparation can be seen by reduced illumination, unstained, for the mycelium of the fungus is relatively strongly light refracting. *Geotrichum candidum* stains very well with methylene blue, fuchsin, gentian violet, or PAS. Young mycelium is Gram positive, older mycelium is partly Gram negative. In sputum, feces and mucus there are rectangular arthrospores and mycelial fragments (Fig. 2) while in pus and tissue sections the fungus elements are frequently rounded off (4—12 μ in diameter) and are found in small numbers only. Round forms are not to be mistaken for yeast or *Blastomyces dermatitidis,* in which, in contrast to *Geotrichum candidum,* budding can always be demonstrated. Moreover, in *B. dermatitidis* there are never arthrospore-like structures in the tissue. However, with *Geotrichum candidum* peculiar shaped formations may occur, which look like budding cells. Cases in doubt can be easily clarified by culture. The strongly rounded or irregular forms, as well as

1
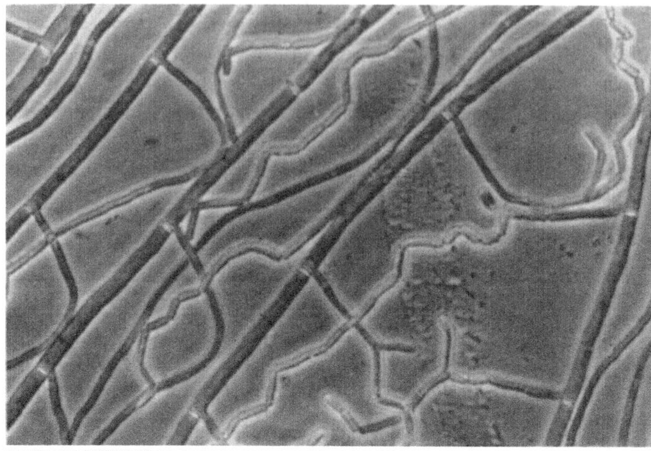

2
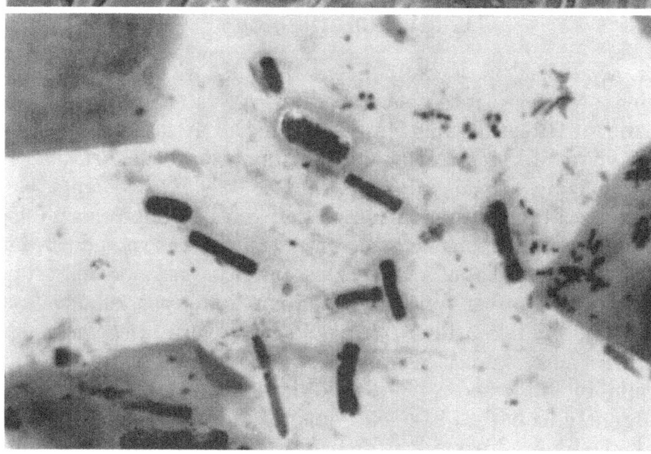

Fig. 1. *Geotrichum candidum*. Branching and septate mycelium with hyphae breaking up into arthrospores. Slide culture. Phase contrast, × 200

Fig. 2. Smear of sputum. *Geotrichum candidum* arthrospores, bacteria and epithelial cells. Gram stain, × 1000

the structures hard to stain, are involution changes in the fungus, determined by an inadequate environment in the interior of the tissue and by unfavorable cultural conditions.

3. Culture

The unpretentious fungus grows on all the usual media at room temperature, and not as well at 37°C (Fig. 3). Old specimens at times do not grow at 37°C, but can be adapted again to this temperature. The growth is strongly promoted by 2% glucose. Vitamins have no influence. The most commonly used medium is Sabouraud's glucose agar. Antibiotics are recommended to suppress bacterial growth. A mold fungus inhibitor for this relatively fast growing fungus is not necessary. In 2—4 days, at the latest, relatively large, shallow, yeast-like, greyish-white to cream colored colonies form with a predominantly flour-like, irregular surface. On liquid media *Geotrichum candidum* produces thick pellicles. Under the microscope one sees typical arthrospores and non-fragmented hyphae (Fig. 4).

3
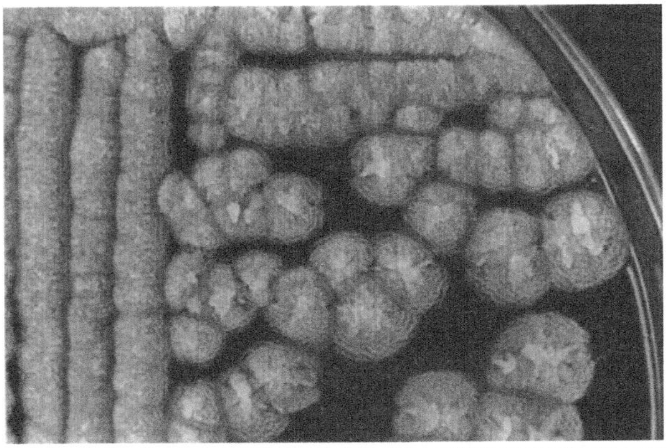

4
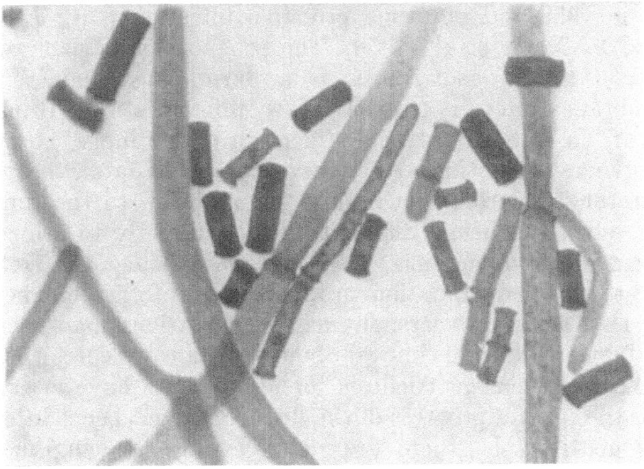

Fig. 3. *Geotrichum candidum.* Colonies on Sabouraud-glucose agar after 3 days at room temperature

Fig. 4. *Geotrichum candidum.* Septate hyphae and arthrospores. Smear of culture. PAS stain, × 1000

4. Physiology

Fermentation and assimilation are tested with the media commonly used in yeast diagnosis (LODDER and KREGER-VAN RIJ, 1952). The assimilation test can always be read clearly, when the strain is inoculated on to the surface of the test agar. The reading is done after 5—10 days of incubation at room temperature. Testing methods which prove the carbohydrate splitting by formation of acids are not appropriate, since the acids are only intermediate products.

Geotrichum candidum assimilates (without gas formation) glucose and galactose. However, erythritol, xylose, maltose, cellobiose, lactose, melibiose and saccharose are not assimilated. Sometimes strains are found which possess a low proportion of population, which can utilize sugars listed above as not assimilated. The biochemical behavior is an important feature for the differentiation of the *Geotrichum* species. Here they are serving only for the confirmation of the *Geotrichum candidum* diagnosis, which is determined, for all practical purposes, for isolates from human beings by means of micromorphology.

E. Pathogenesis and Immunology

1. Pathogenesis

As in other endogenous infections, factors which reduce resistance can initiate infection with *Geotrichum*. Bronchopulmonary geotrichosis has appeared with tuberculosis of the lungs, with typhoid fever, with bronchiectasis, and after acute febrile infections of the upper respiratory tract. Mycetemia has also been observed with miliary tuberculosis and diabetes. A membrane of *G. candidum* has appeared in the mouth cavity with pertussis, with gastroenteritis and with tonsillitis. Cases of geotrichosis have also been described without any predisposing disease. Disturbance of the intestinal flora caused by antibiotic treatment does not seem to play a role in the pathogenesis of geotrichosis. Only the cases reported by WEGMANN (1954) and GRIMMER (1954) can be explained in this way. Apart from tubercle bacilli, only one genuine mixed infection (with *Klebsiella*) was reported by WEBSTER in 1959 and only one probable infection (with *Proteus* and *Coli*) were described by NAGY et al. (1958). The prognosis is variable with individual manifestations. An oral geotrichosis is a harmless secondary infection. The mucous membranes may not even show inflammation. Bronchopulmonary geotrichosis can, in spite of extensive infection of the lungs, be present without affecting the general condition. Other cases have been fatal, the patients having died of respiratory or circulatory failure. In addition to the functional disturbances of the lungs and heart, due directly or indirectly to injury to the lungs, animal experimentation suggests an additional pathogenic factor, a chronic intoxication by fungal degradation products, the intoxication being recognized by cachexia. Histologically parenchymal degenerations may occur in the liver and in the kidneys. Nothing is known about spontaneous cure in cases of bronchopulmonary and bronchial geotrichosis for all patients have either been treated with iodine or the further progress of the disease has not been stated. Of the cases of cutaneous geotrichosis, there are no resistance-reducing factors reported, though doubtless there are unknown predisposing factors. Attempts to infect human beings have been futile. In 1901, RICKETTS scarified the skin and rubbed in *G. candidum (Oidium aceticum)*. However, the skin healed promptly. In 1884, HUEPPE was unsuccessful when he inoculated *G. candidum (Oidium lactis)* repeatedly into his own skin. TZAMOURANIS, in 1962, reported that the infant he was observing, three months prior to the development of geotrichosis, was suffering from intertrigo due to *Candida*. In 1964, PERZ (Case 2) reported that in the course of geotrichosis of the cornea, on the face, and on the hands and feet, there were desquamate skin changes with rhagades, from which *Candida albicans* could be isolated. In 1952, BENDOVE and ASHE reported that in a geotrichosis case *C. albicans* was found in the urine. These findings point to the presence of resistance-reducing factors which are of importance for the genesis of geotrichosis as well as candidosis. Favorable external factors were assumed for the patient of BALZER et al. (1912), as the localization of the skin lesions partly corresponded to the pressure spots on the patient's body from carrying a basket. Although spontaneous healing has not been observed, the prognosis of cutaneous geotrichosis is good, as the internal organs are not affected.

2. Immunology

a) Agglutination and Complement Fixation Reaction: In 1912, BALZER et al. determined an agglutination titer of 1:100 in their patient with cutaneous geotrichosis. In 1958, SEELIGER found among 47 healthy persons: a titer of 1:40 or less 36 times, a titer of 1:80 4 times, and a titer of 1:160 7 times. Out of 200 sera examined by SCHABINSKI in 1960, 0.5% had an agglutinin titer of 1:100. MORENZ, in 1963, in an examination of 2, 284 sera, which were sent for the diagnosis of lues, found a normal distribution of the agglutinin titer. In a serum dilution of 1:85, 1%, and in a serum dilution of 1:120, 0.5% were positive. In 1935, MARTIN examined 3 blastomycosis patients with respect to the complement fixation reaction, using *Geotrichum* antigen (fungus suspension). The result was negative. The complement fixation test was positive with culture autolysate antigen, in the case of CHIALE (1937), a patient with geotrichids, but negative in the case of WEGMANN (1954). THJÖTTA and URDAL (1949) found the complement fixation test positive in a dilution of 1:60 in one of their cases and negative in another. Of ten patients, thought by ARIYEVICH et al. (1963) to have intestinal geotrichosis, 5 showed complement fixation.

SEELIGER (1958) studied 52 normal persons, in whom only 4 showed complement fixation in a serum dilution of 1: 2 and only one in a serum dilution of 1: 4. In more recent research SEELIGER (1962) tested 1, 334 sera, finding 16 positive in a dilution of 1:2 to 1:10 and 2 positive in greater dilutions. These 18 persons showed no clinical indication of geotrichosis, nor could *G. candidum* be demonstrated in the feces among 14 of them. MORENZ (1963) had negative tests on 565 sera at a dilution of 1: 5.

In summary, from the foregoing results, positive agglutination tests in titers of 1:100 to 1: 200, and positive complement-fixation tests in titers of 1:5 or higher point to geotrichosis in the presence of clinical symptoms, although a negative reaction in these dilutions does not rule out geotrichosis.

b) The Geotrichin Test: CHIALE (1937) was the first to have succesfully applied the intracutaneous injection of a culture autolysate of *G. candidum* in a case of geotrichosis of the skin. KUNSTADTER et al. (1950—1951) found the geotrichin test strongly positive in two patients with *Geotrichum* pneumonia, using bouillon culture filtrates of 1:10 and 1:100 dilutions, but also in 3 other members of the family who were not ill. Furthermore, two of 63 children with pneumonia gave a positive geotrichin skin test. The bronchopulmonary and bronchial cases of geotrichosis of WEGMANN (1954) and WEBSTER (1957/1959) were geotrichin-positive. However the two patients of SUNDGAARD et al. (1950) reacted negatively. SKOBEL et al. (1956) found the geotrichin test positive in 2% of adults. Their antigen consisted of an autoclaved suspension in physiological salt solution of 100,000 fungus cells per milliliter. A positive reaction was an erythema of at least 5 mm diameter of the tuberculin type. CANITROT-ARAUJO (1957) found a similar percentage of positives (2.8%) in more than 100 children tested with a culture filtrate of *G. candidum (G. pulmoneum)*.

From these scanty observations it appears that only about 2—3% of healthy persons or persons with unrelated diseases have a positive geotrichin test. If, on the other hand, the test should prove positive in every case of all or some disease types of geotrichosis, it will have value in differential diagnosis.

F. Symptomatology, Diagnosis, and Therapy

Because of the small number of described cases of geotrichosis, generalizations are possible only to a limited extent. As there are only sketchy pathologic-anatomic and histologic studies of geotrichosis, a consideration directed chiefly toward the morphological aspects of the disease is not possible. Complete studies by biopsy and autopsy are necessary in future cases. However, the casuistic representation of the various forms of geotrichosis and the symptomatology and diagnostic features will be given. Here let it be stated that, especially with bronchopulmonary and bronchial geotrichosis, a sure diagnosis requires more than the demonstration of the fungus. Moreover, not in every case cited in this chapter have all of the criteria for the diagnosis of geotrichosis been fulfilled yet considering all the evidence geotrichosis appears to be the appropriate diagnosis.

1. Bronchopulmonary Geotrichosis

The x-ray usually shows widespread, disseminated, partly confluent, irregularly delimited, soft, small spotty infiltrates in all or in some lung fields, without especial preference for particular segments. Less commonly there are large circumscribed tumor-like, non-homogeneous shadows in a lung field. In a quarter of the cases there are one or several thin-walled cavities. In children there may also be an enlargement of the hilar lymph nodes. There is slight, or no, tendency to fibrosis. Calcification is not observed.

The physical findings may not be prominent even with considerable x-ray changes. The percussion note is, at the most, slightly dulled, and the vocal fremitus is infrequently increased. Usually bronchial sounds are to be heard, and increased vesicular breathing, with dry rales, and also moist, rattling rales. Less frequently musical rales and pronounced bronchial breathing are indicative of pulmonary infiltrates. Especially annoying to the patients is a regularly present, often severe cough, even tormenting, and productive. The sputum, which may amount to 80 ml daily, is white to gray, thick and mucoid, and in some cases purulent. In a third of the cases it is tinged with blood. Occasionally there are hemoptyses. The temperature may be normal, subfebrile, or febrile (38—39° C, seldom higher). The sedimentation-rate is normal or slightly elevated (25—45 mm in an hour), seldom greatly increased (over 100 mm). The leucocyte count is normal or moderately elevated (up to 15,000 white cells per cubic mm, rarely higher). The red blood cell count and the hemoglobin are normal; at times there is a mild anemia.

The disease continues chronic or subacute for months, even for years, occasionally with acute phases, which possibly are caused by bacterial superinfection. With an acute onset the geotrichosis may be related to another acute infection, especially of the upper respiratory tract. The patients often complain of weakness, shortness of breath on exertion, loss of appetite and loss of weight. The prognosis can be characterized as good, especially when there is treatment with iodides. In spite of occasional mycetemias, lesions in other organs are seldom observed. However, there are 4 cases (not treated with iodides), with a fatal outcome. Postmortem examinations were performed on the cases of ŞERBAN and TAŞCĂ (1962) and PENINOU-CASTAING et al. (1964).

Bronchopulmonary geotrichosis, like the other forms of geotrichosis, has no typical clinical symptoms. In differential diagnosis pulmonary tuberculosis must be excluded and the pulmonary mycoses, especially candidosis, coccidiodomycosis, histoplasmosis, cryptococcosis and North American blastomycosis. Moreover, pulmonary neoplasms may be suspected, although in geotrichosis the infiltrate is usually non-homogeneous and irregularly demarcated.

A consideration of the following points of view permits a secure diagnosis.

1. *G.candidum* is demonstrable constantly and in large quantity in fresh sputum from a patient not being treated with iodine or nystatin. In healthy persons *G.candidum* is often present in the sputum, but microscopically it is present only in an isolated fashion and only a few colonies grow. Also in other pulmonary mycoses, in contrast to pulmonary tuberculosis, an increase in the

Table 1. *Bronchopulmonary Geotrichosis*

Author	Age and Sex	Occupation	Location	Organism
1. Bennett (1842)	M		Great Britain	Cryptogamic plant
2. Linossier (1916)	26 M	Soldier	France	*Oidium lactis A*
3. Martins (1928)	13 M		Portugal	*Zymonema pulmonalis membranogenes*
4. Smith (1934)	25 M	Student	Puerto Rico and North Carolina, USA	*Geotrichum* species
5. Smith (1934)	18 F		South Carolina, USA	*Geotrichum* species
6. Moore (1934)	22 M	Chemist	Missouri, USA	*Geotrichum* species
7. Smith (1945, Fig. 10)				*Geotrichum* species
8. Kunstadter et al. (1946)	22 M	Soldier	Texas, USA and North Africa	*Geotrichum* species
9. Thjötta and Urdal (1949)	37 F	Farm housewife	Norway	*Geotrichum* species
10. Thjötta and Urdal (1949)	31 F	Farm housewife	Norway	*Geotrichum* species
11. Kunstadter et al. (1950, Case 1)	6 M		Illinois, USA	
12. Kunstadter et al. (1950, Case 2)	4 F		Illinois, USA	
13. Good (1951, Fig. 4c)			Minnesota, USA	
14. Dávalos and Arcos (1954)	47 M	Civil official	Ecuador	*Geotrichum candidum*
15. Hauser (1954)	67 M	Farmer	Germany	*Geotrichum candidum*
16. Wegmann (1954)	55 F		Switzerland	*Geotrichum* species
17. Mahoudeau et al. (1955)	47 M		France	*Geotrichum candidum*
18. Webster (1957)	46 M	Truck driver	Tennessee, USA	*Geotrichum candidum*
19. Webster (1959, Case 2)	64 M	Locomotive engineer in retirement	Tennessee, USA	*Geotrichum candidum*
20. Webster (1959, Case 3)	66 M	Merchant in retirement	Tennessee, USA	*Geotrichum candidum*
21. Webster (1959, Case 4)	52 M	Farmer	Tennessee, USA	*Geotrichum candidum*
22. Bell et al. (1962)	57 M	Laborer	Great Britain	*Geotrichum* species
23. Nicolescu (1962)	23		Rumania	*Geotrichum* species
24. Peninou-Castaing et al. (1964)	54 M	Mason	France	*Geotrichum candidum*

occurrence of *G. candidum* is unknown. In association with *G. candidum* other fungi may occasionally be present, but only in small numbers. The sputum should be examined shortly after it is expectorated, as *G. candidum* can greatly increase in sputum preserved for a long time.

2. The geotrichin test is presumably positive in all pulmonary geotrichoses. As only 2—3% of healthy persons react positively, the negative geotrichin test permits the exclusion of geotrichosis.

3. A successful therapy with iodides or nystatin permits a non-mycotic pulmonary condition to be excluded.

4. Evidence against the diagnosis of tuberculosis is based on the failure to demonstrate tubercle bacilli, a negative tuberculin test and lack of improvement following treatment with anti-tuberculosis drugs. However, there are descriptions of cases of pulmonary geotrichosis combined with pulmonary tuberculosis.

5. Evidence against other pulmonary mycoses consist of negative results with skin tests, for example the coccidiodin and histoplasmin tests.

A plentiful growth of *G. candidum* from the sputum, a positive geotrichin test and successful iodide therapy are diagnostic of bronchopulmonary geotrichosis.

The first time *G. candidum* was demonstrated in a patient was in 1842 by BENNETT. The patient suffered from phthisis with pneumothorax. The fungus was so plentiful in the thick, purulent sputum that it was macroscopically visible on the sides of the sputum container. At postmortem examination the left lung contained numerous cavities of various sizes. Some were filled with a soft tuberculous material in which the fungus was present as abundantly as in the sputum. Apparently there was secondary infection of cavitary pulmonary tuberculosis.

LINOSSIER (1916) described a case thought to be tuberculous. However, tubercle bacilli could not be demonstrated in the sputum. Instead, *G. candidum* was present in large amounts, along with staphylococci and pneumococci. The patient died after the rapid development of cavities. There was no autopsy.

MARTINS (1928) found *G. candidum* microscopically and culturally in every examination of sputum from a 13-year-old youth who had suffered from a pneumonic condition and had frequent and severe hemoptyses. His general condition was poor and his temperature varied between 37 and 39 degrees centigrade. There was no support for the diagnosis of tuberculosis.

SMITH (1934, 1947; see also CONANT et al., 1954, Fig. 97) reported two cases of pulmonary geotrichosis. The first was that of a 25-year-old student who had two small cavities in the upper portion of the right lung and a large one in the left upper lung field (Fig. 5). The sputum was mucopurulent, frequently blood-tinged and without fetid odor, and totalled 15—30 ml per day. Tubercle bacilli could not be demonstrated, but present microscopically and culturally were numerous *G. candidum* and some Friedländer bacilli. Two weeks after the beginning of iodine therapy the cavities became smaller and one disappeared. After 7 months only residual fibrosis was visible in x-ray. The other case was that of an 18-year-old girl with small infiltrates in the lower two-thirds of both lungs. *G. candidum* was demonstrated in the sputum. After a year of iodine therapy the condition was not completely cured. The x-ray picture of another case (SMITH, 1945, Fig. 10) showed numerous small thinwalled cavities in the lower halves of both lungs. No further details were given.

MOORE (1934) reported the case of a man of 22 years who had had chronic pulmonary disease in the form of bronchiectasis for 7 years. He developed fever and increased expectoration and infiltrates in the left lung. *G. candidum* was demonstrated in relatively large amounts in the sputum. With three months of iodine therapy there was much improvement in the condition of the patient and it was thought there had been a bronchopulmonary geotrichosis complicating the bronchiectasis.

KUNSTADTER et al. (1946) reported the case of a 22-year-old solider who developed left-sided pleurisy and then bilateral spotty infiltrates (Fig. 6). On three consecutive occasions *G. candidum* was demonstrated in the sputum and gastric juice. With iodine therapy for three months the pulmonary infiltrates disappeared. The *G. candidum* disappeared from the sputum after the first week of iodine therapy.

SUNDGAARD et al. (1950; see also THJÖTTA and URDAL, 1949) isolated *G. candidum* from the sputum of 8 of 11 members of a farm family. Three sisters of this family gave evidence of pulmonary disease. A 37-year-old sister had apparently suffered from tuberculosis of the hip from age 14 to 26, although tubercle bacilli had not been found in discharge from the fistulae of the hip and the tuberculin reaction was negative. There were delicate, small, spotty, partly

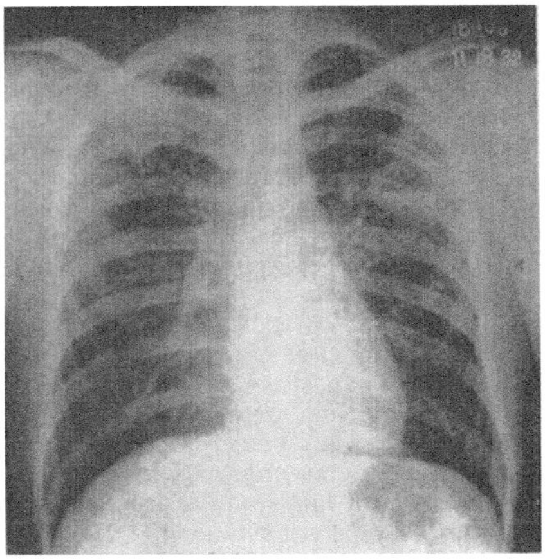

Fig. 5. Bronchopulmonary geotrichosis with a large cavity in the left upper lung field and two smaller cavities in the right upper lung field. (From SMITH, 1934)

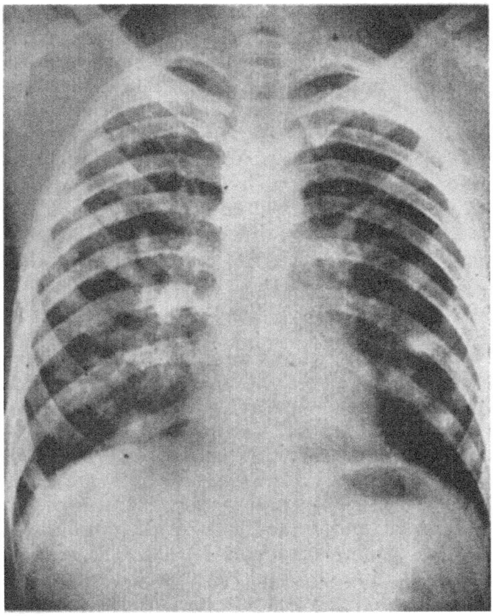

Fig. 6. Bronchopulmonary geotrichosis with disseminated, soft, infiltrates in all lung fields. (From KUNSTADTER et al., 1946)

confluent infiltrates in both lungs, particularly in the middle and subclavicular regions. The serum of this patient gave a positive complement-fixation reaction in a dilution of 1:60 with a suspension of a *G. candidum* from the patient, and also with the *G. candidum* of Case 2, although the serum of Case 2 reacted negatively with both strains. As the geotrichin test in the two patients was negative, in contrast to all other tested cases of bronchopulmonary geotrichosis, and as no iodine treatment was carried out, the diagnosis rests only on the demonstration of *Geotrichum candidum* in the sputum. One can not be sure of the diagnosis of geotrichosis in these two cases, particularly in Case 2 which had a negative complement fixation reaction. The 31-year-old sister (Case 2) had a similar pulmonary picture with soft, indistinct, partly confluent, infiltrations in both lungs from the apices to the bases, especially in the central fields. The hilar lymphnodes were not enlarged, the tuberculin test was negative and tubercle bacilli were not demonstrable. The blood picture was normal, the sedimentation-rate

24 mm. Both sisters were in good general condition and had no fever. For silicosis and carcinomatosis there was no support. The clinical symptoms, which were rather mild, consisted of moderate cough with a rather watery expectoration in which white firm flecks were present. Moreover, *G. candidum* was found in large quantities microscopically. The findings in the third sister were not characteristic of geotrichosis.

The author described 3 further cases (Cases 4, 5, and 6) in which the patients were found to have *G. candidum* in the sputum. In Case 4 there was pulmonary tuberculosis. For a pathognomonic meaning of the moderate quantity of fungus in the sputum, even as a secondary infection, there was little diagnostic proof except the positive geotrichin test. Nor is the evidence sufficient in the two cases with negative geotrichin tests to accept them as instances of geotrichosis.

KUNSTADTER et al. (1950; see also WHITCOMB et al., 1949) describe 2 cases in whom the demonstration of *G. candidum* was not successful, apparently because of the early giving of iodine therapy. The geotrichin test was positive, although the test with histoplasmin, blastomycin, coccidiodin, torulin, and tuberculin were negative. In favor of geotrichosis was the success of iodine therapy. A 6-year-old boy became acutely ill with sore throat, rhinitis and fever. Except for low fever, all symptoms improved in a few days. Three weeks later the x-ray showed spotty infiltrates in both lungs and a pronounced hilar adenopathy. In spite of streptomycin treatment the fever persisted and in three months a severely productive cough had developed. The sputum was whitish, mucoid and often flaky though not bloody. The patient lost considerable weight and tired easily. Leucocytes 9,900, sedimentation-rate 32 mm, temperature 37.1 °C. Tubercle bacilli and other pathogenic bacteria could not be demonstrated. The x-ray after six months still showed extensive, soft irregularly spotty infiltrates. With iodine therapy the cough disappeared in a few weeks. The x-ray showed a distinct clearing of the infiltrate with regression of the hilar lymphadenopathy. The patient felt well and put on weight. After a year, in spite of subjective freedom from complaints, the x-ray abnormalities had not completely faded out. However the iodine therapy was discontinued.

The 4-year-old sister of the above patient became ill shortly after her brother did, and had a mild acute upper respiratory infection, which lasted only 2 or 3 days. The x-ray, except for a mild enlargement of the hilar lymph nodes, was not remarkable. Although she had normal laboratory findings and only a mild cough, she showed after 6 months, as did her brother, a definite hilar adenopathy and a diffuse, spotty infiltrate, which affected the hilar regions and extended into the lower lobes. After 6 months of iodine therapy the lung fields were again clear and the enlargement of the hilar lymph nodes had subsided. In the other 3 members of this family, in whom the geotrichin reaction was likewise positive, no abnormal changes were noted in x-rays.

GOOD (1951) cited a case of geotrichosis which was limited to the right pulmonary apex when first studied. The continuing investigation showed a disseminated infiltrate of the left lower lobe field, and a hilar lymphadenopathy. After 2 months of iodides and antibiotics the x-ray changes and the clinical findings had disappeared.

DÁVALOS and ARCOS (1954) reported the case of a 47-year-old man who complained of weakness and weight loss of 4 months duration, and of cough and whitish expectoration of 2 months duration. Roentgenologically there were infiltrates of both lung bases suggestive of tuberculosis. Tubercle bacilli and tumor cells could not be demonstrated. The general condition of the patient was good, leukocytes 12,000. The sputum cultures yielded *Penicillium*, *Aspergillus*, *Geotrichum* and *Acladium* at first, later only *G. candidum*. Although the iodine treatment was continued for only 3 weeks because the patient developed an intolerance for the drug, within a week the cough and expectoration had improved, and the laboratory findings became normal. A year later the x-rays showed only areas of fibrosis in the lower lung fields.

HAUSER (1954) reported the case of a 67-year-old farmer who had suffered from a dermatitis herpetiformis Duhring for 4 years. In addition, he had cough and white, mucoid expectoration for years. His temperature was occasionally subfebrile. Leucocytes were 11,000 and sedimentation-rate 110 mm. In x-ray there was the picture of a "bilateral productive, cirrhotic, pulmonary tuberculosis" in the apical and subapical regions. In both middle lung fields there were small, punctate, infiltrates, and in the lower lung fields moderately dense shadows with clearing toward the periphery. In the tomogram a cavity was identified in the right upper lung field. In sputum, *G. candidum* could be demonstrated microscopically and culturally in large quantities and tubercle bacilli could not be found. Therapy with INH and streptomycin was unsuccessful.

WEGMANN (1954/1957; see also ROSSIER, 1952) described a fatal geotrichosis of both lungs. The 55-year-old female patient had chronic cough with episodes of choking, and gelatinous expectoration for two years. Because of the suspicion of bronchiectasis she was treated continuously with penicillin and aureomycin. Moist musical and non-musical rales were

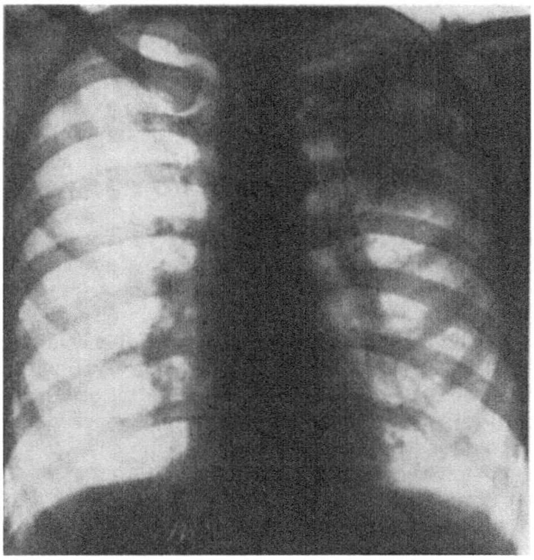

Fig. 7. Bronchopulmonary geotrichosis with a large round shadow in the left upper lung field. (From MAHOUDEAU et al., 1955)

audible in both lung bases. In x-ray there were pinhead-sized, thickly beset, rather sharply demarcated foci, increasing in number in the inferior portions of the lungs. Leukocytes 4,900; tuberculin and blastomycin tests negative; tubercle bacilli not demonstrable; oidiomycin and geotrichin test positive; *Geotrichum* complement fixation and precipitin test negative. In the sputum *G. candidum* was demonstrable in large quantities microscopically and culturally. Later, a spontaneous pneumothorax formed on the right, and this was treated with suction. Death from increasing circulatory insufficiency. No postmortem examination.

WEGMANN (1961; Fig. 103) shows, without further explanation, the x-ray of another case with pneumoperitoneum and a large thin-walled cavity.

MAHOUDEAU et al. (1955) described a tumor-like geotrichosis of the lung. The 47-year-old patient became ill with pain on the left side of the chest, cough, and bloody sputum. He lost weight and then had low grade fever. In x-ray there was a large, non-homogenous shadow in the upper left lung field (Fig. 7). Tubercle bacilli and tumor cells were not demonstrable. However, material obtained by bronchoscopy grew out *G. candidum* in pure culture. Iodine therapy produced rapid improvement. The x-ray changes disappeared completely and the patient was discharged cured.

WEBSTER (1957) described a pulmonary geotrichosis with cavity formation and mycetemia in a 46-year-old truck driver. The patient became ill three weeks previously with a shaking chill and high fever. Twenty-four hours before admission he had a relapse with persisting cough and thick, mucoid, rusty, earthy-smelling sputum. Shortly thereafter he coughed up bright red blood. The patient was acutely ill and very dyspneic. Temperature: 40.8° C. Leukocytes 24,600. X-ray showed a thin walled cavity and infiltrates in the upper portion of the left lower lobe. *G. candidum* was demonstrated in the sputum, gastric contents, stool, bronchial aspirate, and twice in the blood. The geotrichin reaction was strongly positive. Tubercle bacilli and other pathogenic bacteria were not demonstrable. Skin tests with tuberculin, coccidioidin, and histoplasmin were negative. He was first treated with tetracycline, and after clarification of the diagnosis, with iodine. Thereupon there was an involution of the infiltrates and a stepwise closing of the cavity. After six months only some peribronchial infiltrates could be seen.

WEBSTER (1959) described three brochopulmonary cases in addition to one bronchial geotrichosis. A 64-year-old locomotive engineer, who had suffered from Parkinsonism for five years, complained of cough with gray mucoid expectoration, shortness of breath and fever of three weeks duration. Temperature 38.8° C. Leukocytes 13,400. The x-ray showed a lobar pneumonia of the right upper lobe. In the sputum there were Friedländer bacilli, and *G. candidum*. There were no tubercle bacilli demonstrable. The tuberculin, histoplasmin, and coccidioidin tests were negative. Bronchoscopically there was no support for the diagnosis of neoplasm. In the exudate from the right upper lobe there were large quantities of the fungus and *Klebsiella*. After 9 months of treatment with streptomycin, aureomycin, sulphadiazine, and iodine, the pulmonary changes disappeared, and no bacteria or fungi were demonstrable. According to WEBSTER, it was a double infection, with both organisms playing an important role.

A 66-year-old merchant in retirement suffered with a chronic cough, with gray gelatinous sputum, and dyspnea. Temperature 36.9° C, and leukocytes 6,050. In x-ray there was emphysema, and infiltrates in both lungs. In the sputum there were *G. candidum* and no tubercle bacilli. The geotrichin test was strongly positive, the tuberculin and histoplasmin tests were negative. Treatment with iodine improved the condition. An x-ray after 6 months showed less infiltrate, although *G. candidum* could still be cultured.

A 52-year-old farmer complained of increasing cough of two months duration, with thick gray gelatinous sputum, evening fever, night sweats, weakness and weight loss. For two weeks he had vomited blood. Temperature 38° C and leukocytes 12,000. In x-ray, the left upper lobe of the lungs showed a spotty infiltrate and cavity formation. Tubercle bacilli could not be demonstrated. The tuberculin, histoplasmin, and coccidioidin tests were negative, but the geotrichin test was positive. *G. candidum* was demonstrable in all cultures of sputum, in bronchial aspirates, and once in the blood. After oral iodine and intravenous gentian violet treatment the symptoms disappeared. In a later x-ray, no changes were recognizable. However *G. candidum* could still be demonstrated in the sputum.

BELL et al. (1962) reported the case of a 57-year-old laborer who, for four months, had complained of shortness of breath, weight loss, and a productive cough with 50—80 ml daily of a purulent sputum. Temperature normal, leukocytes 12,500. In the sputum *G. candidum* was constantly present in large quantities. *B. coli* and staphylococci were present in some examinations. Tubercle bacilli were never found. With nystatin and iodine the condition improved considerably and the fungus disappeared from the sputum.

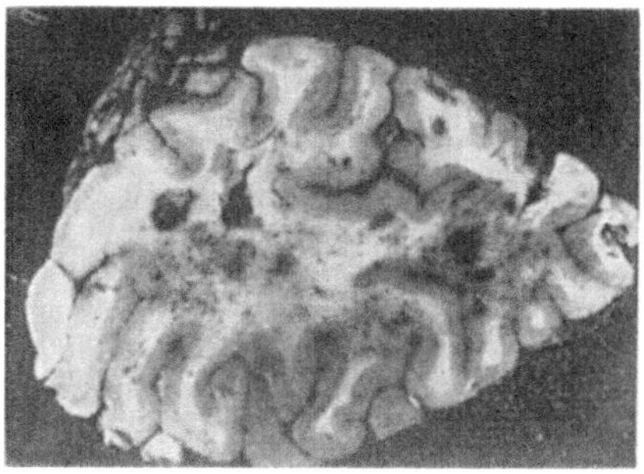

Fig. 8. Bronchopulmonary, metastasizing geotrichosis. Foci of softening containing *Geotrichum*, in the white substance of the brain. (From ŞERBAN and TAŞCĂ, 1962)

NICOLESCU (1962) reported the case of a patient with chronic hepatitis who, in the course of cortisone treatment, developed fever, dyspnea and cough. The x-ray showed irregular and nodular foci in both lungs. The patient died in 6 days of a generalized infection. The pathologic-anatomical findings were described by ŞERBAN and TAŞCĂ (1962). The grayish-white bronchopneumonic foci were rounded and sharply defined. The centers of these nodules were purulent or granulomatous with giant cells which sometimes contained fungi. At the periphery of the necrotic zones, there were clusters of the fungus. The thickness of hyphae was 6—12 μ and the diameter of the free-lying arthrospores, 12—15 μ. *G. candidum* was cultured. The white substance of the brain contained softenings with a diameter up to one cm (Fig. 8). Necrotic infarcted nodules were found in the kidneys and spleen. In the myocardium of the heart there were microinfarcts of 2—3 mm diameter. Microscopically the microabscesses of the myocardium showed the fungus (Fig. 9).

PENINOU-CASTAING et al. (1964) described a fatal cavitary pulmonary geotrichosis, which at first was associated with pulmonary tuberculosis. The 54-year-old brick layer became ill 18 months before his death, lost weight and developed a cough with mucoid and purulent sputum, in which tubercle bacilli and much *G. candidum* were demonstrated. The x-ray showed bilateral infiltrates with a large thick walled cavity at the right apex and another cavity in the left subapical region. There was a shadow on the right next to the pericardium. In the aspirate from the bronchus there were tubercle bacilli and nummerous *G. candidum*. With antituberculous treatment with drugs for six months the tubercle bacilli disappeared from the sputum. The cavity at the right apex closed and the general condition improved. However, in spite of an extensive treatment with nystatin, there was a relapse with more cavities. The patient became cachectic and died. At autopsy, the right upper lobe was scarred, without a cavity. The cavities in the lower lobes had no granulation tissue wall. There was only fibrin containing fungus and a mucoid fluid with arthrospores in necrotic collections of neutrophils (Fig. 10). The tuberculosis was histologically cured. However *G. candidum* was isolated in culture in large quantities. The shadow next to the pericardium turned out to be large cysts of the emphysematous lung containing pus in which the fungus was also visible.

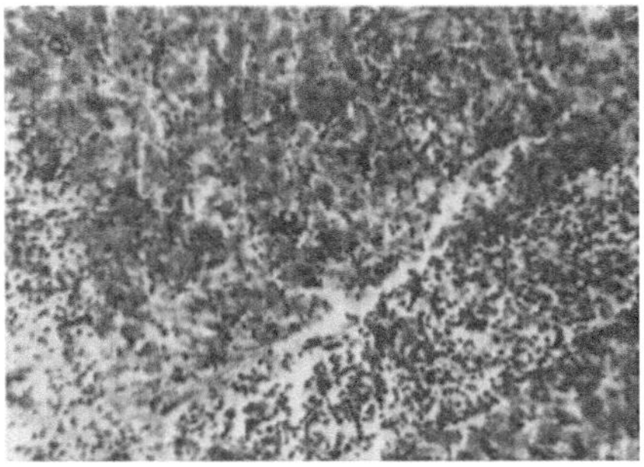

Fig. 9. Bronchopulmonary, metastasizing geotrichosis. Interstitial myocarditis with fungus containing microabscesses. H&E, × 56. (From Şerban and Taşcă, 1962)

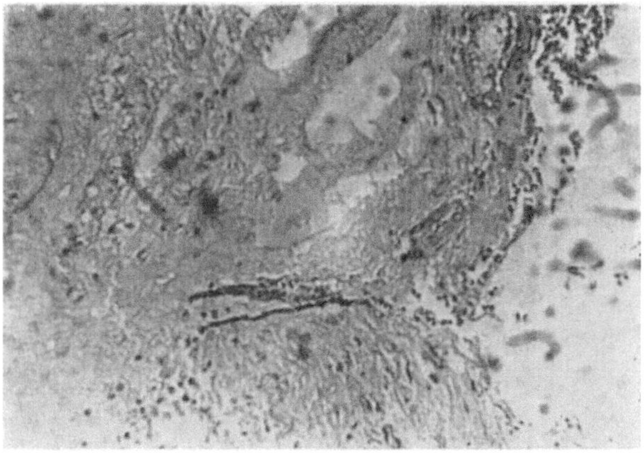

Fig. 10. Bronchopulmonary geotrichosis. Bizarre forms of arthrospores in necrotic masses of fibrin and polymorphonuclear cells at the edge of a pulmonary cavity. (From Peninou-Castaing et al., 1964)

2. Bronchial Geotrichosis

Bronchial geotrichosis has essentially no involvement of the lung itself, and is chronic or subacute. Persistent cough is a prominant feature. The diagnosis is made as in bronchopulmonary geotrichosis by demonstrating the fungus in the sputum. Especially convincing is the demonstration of *G. candidum* in material obtained by bronchoscopy, and improvement with iodine therapy. Concerning the geotrichin test in bronchial geotrichosis nothing is known.

The first report of the presence of *G. candidum* (*Monilia asteroides, Oidium matalense*) in bronchitis came from Castellani and Chalmers (1919). The next mention was that of Mattlet in 1926. In his case *G. candidum* was isolated repeatedly. Smith (1934) reported two cases in which *G. candidum* could be demonstrated in the sputum. One case was that of a

Table 2. *Bronchial Geotrichosis*

Author	Age and Sex	Occupation	Location	Organism
1. MATTLET (1926)	F		Urundi, Africa	*Mycoderma nyabisi*
2. SMITH (1934)	35 F		New York, USA	*Geotrichum* species
3. SMITH (1934)	30 M		New York, USA	*Geotrichum* species
4. REEVES (1941)	25 F		North Carolina, USA	
5. REEVES (1941)	25 M		North Carolina, USA	
6. MINTON et al. (1954, Case 1)	53 M	Painter	Panama	*Geotrichum* species
7. MINTON et al. (1954, Case 2)	54 M		Panama	*Geotrichum* species
8. MÁNDI and BERENCSI (1954, Case 1)	39 M		Hungary	*Geotrichum* species
9. WEBSTER (1959, Case 1)	22 F	Housewife	Tennessee, USA	*Geotrichum candidum*

35-year-old woman without changes in the pulmonary parenchyma. After iodine therapy there was rapid cure. The other case was similar, with improvement of the condition in a man of 30 with the use of iodine. REEVES (1941) found two cases of geotrichosis amoung 79 bronchomycoses. There were symptoms of bronchitis for two to three years in both patients.

MINTON et al. (1954) described two cases of endobronchial geotrichosis. A 53-year-old painter had had, for 4 years, several spontaneously self-limited episodes of coughing with bright red sputum. During a relapse there was coughing and hemoptyses. The daily amount of the dark or light red sputum was 60—120 ml. In bronchoscopic examination white exudative easily bleeding foci were observed in both main bronchi. Material from this region revealed *G. candidum*, microscopically and culturally. With iodine for two weeks, the blood disappeared from the sputum and the fungus was no longer present in the material from bronchoscopy. The second case was that of a 54-year-old man who had countless arthrospores in his sputum with positive *G. candidum* cultures. With iodine therapy the fungus disappeared from the sputum. The patient died of a heart attack. At autopsy the lungs and bronchi appeared normal. MÁNDI and BERENCSI (1954) described an endobronchial geotrichosis in a 39-year-old man who had cavernous pulmonary tuberculosis of three years duration and a tuberculous pleurisy of one year. A pneumothorax was performed and INH and PAS were given. The mucosa of the trachea and the main bronchi showed numerous yellowish white foci relatively well attached which culturally gave *G. candidum*. With the thought that *G. candidum* was related to *Candida albicans*, the patient was treated for two months with a vaccine of *C. albicans*, with which the skin test was positive. During this period the coughing stopped. Bronchoscopically only a few gray white fungus foci were seen in the right bronchi. The tuberculosis had greatly improved during this time.

WEBSTER in 1959 reported the case of a 22-year-old housewife who complained of weakness and cough of 4 months duration, and a yellow mucoid sputum with small gray flecks. There was no hemoptysis. In the mouth and throat, there was a white exudate suggestive of thrush. The x-ray of the lung showed increase in the peribronchial markings. From the exudate of the mouth, and from the sputum and stool, *G. candidum* was repeatedly cultured. The blood culture was negative. Tubercle bacilli were not demonstrable. Tuberculin, histoplasmin and coccidioidin tests were negative. With iodine therapy the symptoms improved. Two years later there was exacerbation of the bronchitis. Bronchoscopically, an exudate was seen bilaterally from which *G. candidum* was grown, but no tubercle bacilli. A year later the patient developed a pulmonary tuberculous cavity. Both tubercle bacilli and *G. candidum* were isolated from the sputum. As this fungus also occurs in the sputum of healthy people, the relationship between the two infections was not clear.

3. Oral Geotrichosis

In oral geotrichosis a membrane occurs on the mucosa of the oral cavity or pharynx, and is composed of arthrospores and fungus mycelium. A definite differention from the membranes of other causes (for example, those due to

Table 3. *Oral Geotrichosis**

Author	Age and Sex	Location	Organism
a) Thrush			
1. CASTELLANI and CHALMERS (1919)		Tropics	*Monilia asteroides*, *Oidium matalense*
2. BACHMANN (1921)	1	Germany	*Oospora* species
3. CASTELLANI (1928)	25 M	Ceylon	*Oidium matalense*
4. COLONNELLO (1944)	4	Italy	*Geotrichum candidum*
b) Tonsillar mycosis			
1. CASTELLANI (1928)			*Oidium matalense*, *Oidium asteroides*
2. MOTTA (1931)	F	Italy	*Oospora lactis*
3. CORTESE (1933)	37 F	Italy	*Geotrichum candidum*
c) Black Hairy Tongue			
1. SCHAEDE (1934)	21 M	Germany	*Oospora catenata*, *Oospora fragilis*
2. GRIMMER (1954)		Germany	*Endomyces lactis*
3. CARTEAUD and DROUHET (1955, Case 2)	79 M	France	*Geotrichum* species
4. CARTEAUD et al. (1957)		France	*Geotrichum* species
5. ARRIGHI and DROUHET (1957)	80 M	France	*Geotrichum* species

* The occupations of the patients were not known.

Candida albicans) is possible only microscopically or culturally. In thrush caused by *G. candidum*, grayish yellow, mostly loosly adherent patches develop on the reddened mucous membrane of the mouth, throat and esophagus. Tonsillar mycosis (in the narrower sense) is characterized by small gray-white or brown fungal foci on the tonsils without there being an inflammatory reaction. In black hairy tongue there is an easily removed, dark brown or blackish coating on the back of the tongue. This coating consists of a thick layer of arthrospores and mycelium. Except for a hypertrophy of the papillae of the tongue there are no signs of inflammation. Black hairy tongue is caused more frequently by bacteria or other fungi than by *G. candidum*. All types of coatings with *G. candidum* appear to develop on the basis of some unrelated previous pathologic change in the mucosa. The diagnosis is made by microscopic and cultural investigation of the membrane. Concerning the geotrichin test in oral geotrichosis nothing is known.

Thrush: The first indication that oral thrush could be caused by *G. candidum* came from CASTELLANI and CHALMERS (1919), who had observed it in the tropics. BACHMANN (1921) found at the autopsy of a year-old child with whooping cough pneumonia a gray white or gray yellow coating on the tip of the tongue, on the entire pharynx, throughout the esophagus and in the stomach, which histologically and culturally consisted of *G. candidum*. Under the coating, which was in part firmly and in part loosly attached, was a grayish-white, granularity on a reddish base. The tongue and pharynx were also reddened.

CASTELLANI (1928) mentioned a 25-year-old man with acute tonsillitis, severe sore throat, difficulty in swallowing and temperature up to 40°C. After 12 hours both tonsils were covered with a white exudate from which *G. candidum* and not diphtheria bacilli were isolated. COLONNELLO (1944) mentioned the appearance of thrush membranes on the tonsils and the mucous membranes of the cheeks as a complication of an acute gastroenteritis in a 4-year-old child. This coating consisted of *G. candidum*. The fungus was also in the stool in large quantities as long as the membranes persisted.

ARIYEVICH et al. (1963) reported the appearance of membranes on the tongue and in the pharynx of a 23-year-old female patient who had been treated with penicillin and ACTH for lupus erythematosus disseminatus. As both *G.candidum (G.sp.)* and *Candida* were isolated from the membranes, and the numbers of organisms were not stated, it is not clear whether the condition was oral geotrichosis or candidosis. Genuine mixed infections of *G.candidum* and other fungi are not known.

Tonsillar Geotrichosis: CASTELLANI (1928) cited *G. candidum* as the cause of a lesion which he termed spinous tonsillar mycosis. On the tonsils there were numerous white, gray or brown, erect spicules generally originating in the crypts. These spicules consisted of fungal mycelia. CORTESE (1933) described a similar case in a 37-year-old woman who complained of slight pain on swallowing. On the mucous membranes of the enlarged tonsils were several dozen grayish-white flecks firmly attached to the tonsils, and these consisted of *G.candidum*. Histologic examination showed fungal filaments among epithelial cells extending to the lymph follicles, but without appreciable inflammation. Also, in the case described by MOTTA (1931), there were similar, white, nodules on the gums and lingual tonsils. The histologic picture was like that described by CORTESE.

Black Hairy Tongue: SCHAEDE (1934) published a case in the lingual membrane of which two strains of *G.candidum* were isolated. In a healthy 21-year-old man there developed a black color of the tongue. A year later almost the entire tongue was covered with a 3 mm thick, dark brown coating. The superficial part of this covering could be easily removed. The edges of the tongue were red and moist. There was fetid odor. Microscopically the papillae of the tongue were covered with arthrospores. Also, a mycelium could be recognized, and this clung to the papillae and extended downward or upward from the surface. There were, also, numerous bacteria and occasional *Penicillium* mycelium. The prescribed iodine therapy was not regularly followed by the patient. A half year later the coating suddenly disappeared. The region of the former coating could be recognized by a slight elevation and a velvety pale white character of the mucous membrane.

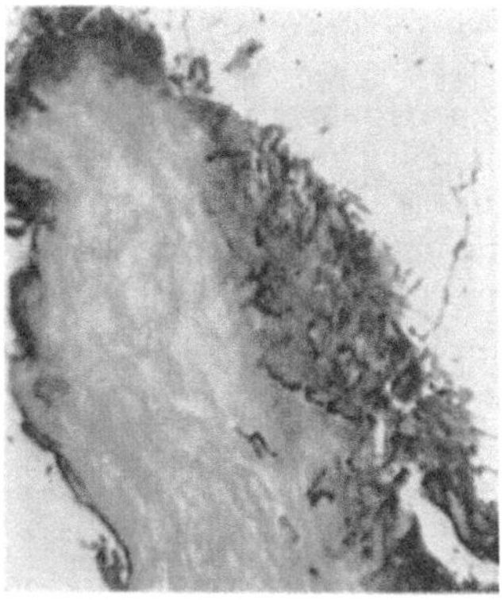

Fig. 11. Black hairy tongue. Coating of arthrospores and mycelium on a papilla of the tongue. PAS stain, × 350. (From CARTEAUD et al., 1957)

GRIMMER (1954) reported the case of a black hairy tongue which developed after prolonged penicillin treatment of a secondarily infected tinea barbae profunda. *G.candidum* was microscopically and culturally demonstrable in great quantity. CARTEAUD et al. (1955/1957) in 15 cases of black hairy tongue were able to isolate *Candida* 7 times, *Geotrichum* 3 times, and both fungi together 5 times. *G.candidum* without *Candida* was demonstrated in a 79-year-old man with symptoms of two months duration. In the other cases no clinical data were given.

A histological preparation showed a coating of hyphae and arthrospores on a hypertrophied papilla (Fig. 11). In another histologic preparation the surface of the papilla bears a mixture of bacteria, *Candida* and *Geotrichum*. The papillae were not edematous or infiltrated with cells.

ARRIGHI and DROUHET (1957) reported the case of an 80-year-old man in whom a black hairy-tongue had appeared two weeks before without obvious cause. It consisted of a considerable papillary hypertrophy with a shaggy coating which could be easily scraped off. *G.candidum* was demonstrated culturally.

4. Cutaneous Geotrichosis

Cutaneous geotrichosis is a chronic fungus infection of the skin which untreated may last for many years. The skin lesions appear gradually, and may be disseminated, and may be accompanied by geotrichids. Areas of predilection are not recognized. The single or confluent lesions consist either of superficial erythematous, pustular, and squamous eruptions or as deeper infiltrates, mostly with miliary abscesses or small nodules. In some patients the infiltrated regions ulcerate superficially and later form scars. In others, extensive lesions occur, with central scarring, but with ulcerative, verrucous vegetations at the periphery. Geotrichids are polymorphous, erythematous-squamous, or eczematoid lesions. The infection is confined to the skin. The internal organs, draining lymph vessels and the lymph nodes are not affected. There are no large subcutaneous abscesses. The general condition is not affected. The skin lesions are indolent or slightly painful on pressure.

In the deeper forms, one or another of the following reactions may be observed. 1. A proliferation of the connective tissue tending toward fibrosis; 2. Diffuse, histiocytic and lymphocytic infiltrates; 3. Microabscesses with polymorphonuclear neutrophiles; 4. Granulomas with central necrosis surrounded by giant cells and histiocytes.

The epidermis tends to be hyperplastic. Fungus elements are found, in pus microscopically even if in small numbers. In tissue sections the fungi are hard to find, but special staining has been little used. Usually *G.candidum* is markedly rounded in tissue, and differentiation from yeast-like fungi may be difficult. Yet the regularly positive culture leaves no doubt as to the etiology of the skin lesions. The demonstration of the causative agent, preferably together with a biopsy, also permits a differential diagnosis, especially in ruling out verrucous skin tuberculosis, nodular and ulcerative syphilids, fungating epitheliomas, epidermal sporotrichosis, cutaneous North American blastomycosis and primary cutaneous coccidioidomycosis. In the superficial forms, *G.candidum* can be culturally demonstrated in the pus of the pustules and probably also microscopically in tissue sections. As *G.candidum* occasionally occurs on the normal skin, it is important to demonstrate it repeatedly and to rule out other causal agents, especially dermatophytes, yeasts, and staphylococci. The geotrichin test was positive in the one case of geotrichosis of the skin in which it was applied.

Deep Forms: RICKETTS (1901, Case 3; see also HYDE and RICKETTS, 1901, Case 12) described the case of a 33-year-old farmer with a skin lesion of the left lower eyelid and the adjacent portions of the cheek and temple. The condition had begun 2 years earlier as a "carbuncle". After it was opened, it failed to heal. The process became larger and twice it was scraped but without success. The clearly defined, reddish-brown, painless lesion consistend of irregular, raised, in part excoriated nodules with minor areas between, of non-involved skin. Histologic examination showed a pronounced epithelial hyperplasia and distinct, interepithelial and subcutaneous microabscesses with polymorphonuclear neutrophils and occasional giant and plasma cells.

Table 4. *Cutaneous Geotrichosis*

Author	Age and Sex	Occupation	Location	Organism
a) Deep Forms				
1. Ricketts (1901, Case 3)	33 M	Farmer	Illinois, USA	*Oidium aceticum*
2. Balzer et al. (1912)	37 M	Vegetable carrier	France	*Mycoderma pulmoneum*
3. Gougerot (1920)	24 M		France	*Mycoderma pulmoneum*
4. Gougerot (1936)			France	*Mycoderma pulmoneum*
5. Gougerot (1936)			France	*Mycoderma pulmoneum*
6. Thiers et al. (1953)	63 F	Teacher in retirement	France	*Geotrichum candidum*
7. Rousset et al. (1957)	54 M	Colonial official	France	*Geotrichum candidum*
8. Tzamouranis (1962)	4/12 F		Greece	*Geotrichum candidum*
b) Superficial Forms				
1. Chiale (1937, Case 3)	54 F		Italy	*Geotrichum candidum*
2. Canellis (1959)	50 M		Greece	*Geotrichum candidum*
3. Heidenbluth (1963)			Germany	*Geotrichum candidum*

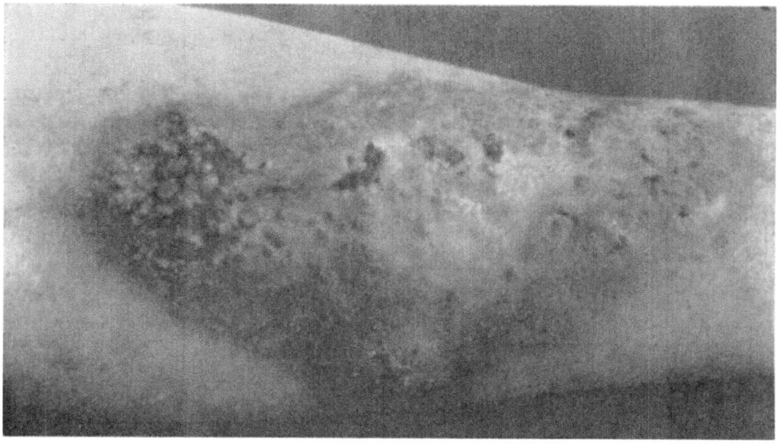

Fig. 12. Cutaneous geotrichosis. Ulcerous-verrucous, proliferating and centrally irregular, scarred lesions on the right lower arm. (From Balzer et al., 1912)

The surrounding tissue was diffusely infiltrated with the same cells. The sparse fungal elements in and about the abscesses were mostly round (12 μ in diameter), at times elongated, and with a double-contoured wall. They were stainable with methylene blue, carbol toluidin blue and acid orcein, except for old cells without recognizable protoplasm. From pus and from tissue scrapings *G. candidum* could be cultured, and in pus it could also be demonstrated microscopically. Tubercle bacilli were not demonstrable. Iodine treatment brought prompt improvement. The process was not completely healed at the time of the report.

Balzer et al. (1910/1912) described as "dermatomycose végétante disséminée" a geotrichosis of the skin in a 37-year-old male patient (Fig. 12).

The first ulcerations, which after treatment with calomel powder for 1—2 months became scarred, appeared 7 and 4 years before on the right lower arm and right lower leg. A year previously there was recurrence in the form of a disseminated ulcerative-vegetative process. Ulcers now appeared on the back, in the lumbar region and on the right shoulder. With a one month treatment with potassium iodide and arsenic, the patient was almost completely cured. After 9 months the lesions were again active and now also affected the left foot and the right temple. The general condition of the patient was always good. The internal organs were not affected, and there was no support for the diagnosis of lues or tuberculosis. The affected skin regions consisted of isolated lesions, "1 to 5 franc piece sized" and of confluent ulcers which in part formed long bands.

The lesions began as pustules which spread, ulcerated and became crusted. They finally developed into plaques. In the center of the plaque there was smooth, firm, irregular scarring. At the borders there were red-violet, firm infiltrates, which gave rise to drops of pus when pressed upon. With renewed iodine and arsenic medication the ulcers were largely healed in two months, but a month later they were again partly ulcerated. The complete scarring lasted more than three months. The further course is unknown. Histological examination gave a mixture of the following reaction types with all intermediate steps. There was, in one type, an inflammation of the connective tissue, with macrophages, plasma cells and lymphocytes which tended toward fibrosis. In another type there were tubercle-like structures and giant cells. In still another type there were micro-abscesses with neutrophils, macrophages and some eosinophils. The fungus was seen with difficulty and in small numbers in sections and smears. The fungus cells were round, oval or pear-shaped (5×7 μ) or longer filaments (2—3 by 9—17 μ). *G. candidum* could be regularly isolated, often in pure culture, but at other times associated with bacteria.

Gougerot (1920) reported a second case of dermatomycosis vegetans with verrucous-ulcerative and gummous lesions, in which *G. candidum* was demonstrated microscopically and culturally. The disease began during the first year of life of the 24-year-old male patient. As the lesions were not influenced by iodine, they were excised and cauterized. In spite of almost complete healing, recurrences occurred. The right foot had two ulcerosquamous lesions, and the popliteal space had verrucous lesions. On the hands were scars which appeared to arise from abscesses. Gougerot (1936) reported two additional cases with numerous disseminated foci on all four extremities.

Thiers et al. (1953) wrote concerning disseminated, nodular lesions of geotrichosis with geotrichids in a 63-year-old woman teacher. The cutaneous changes developed gradually for a year, first on the face, then on the extremities and trunk. The firm, indolent, non-itching, pea-sized nodules arose subcutaneously and later affected the dermis which was slightly reddened. Over the nodules there formed small, dry, yellowish crusts which came off spontaneously, leaving a sunken, white scar. Between these nodules and scars there were polymorphous, erythematous and squamous changes with slight subcutaneous edema which in places resembled a moist eczema. These lesions, from which the fungus could not be cultured, were regarded as geotrichids, and the nodules in which *G. candidum* was demonstrated microscopically and culturally were regarded as the related foci of infection. In biopsy, the nodules consisted of granulation tissue, one zone with prominent histiocytes and giant cells and a central, necrotic focus with polymorphonuclear leukocytes. The giant cells contained rounded, yeast-like, elements 4—8 μ in size which could be stained with toluidin blue and erythrosin. The patient was treated with INH for 45 days with good results, although the fungus was resistant to INH *in vitro*. No new nodules appeared. The geotrichids healed. Later the patient was completely cured. Preceding therapy with penicillin, streptomycin, tetracycline, chloramphenicol and fongeryl were ineffective.

Rousset et al. (1957) reported the case of a 54-year-old colonial official, in whom the disease began with a conjunctivitis (reddened eye, lacrimation), and a small, white, epithelioma-like induration on the mucous membrane of the lower lid, which later ulcerated superficially. After a month there appeared on the mucous membrane of the lower lip (later on the upper lip), a vesicle, which was accompanied by a progressing leukoplakia-like infiltrate, with central superficial ulceration. Except for the sensation of foreign body, the patient was free of pain. The biopsy showed a granuloma with foreign-body giant cells and yeast-like elements, which were embedded in thick histiocytic-plasma cell infiltrates and less frequently (together with filamentous elements) in the interstitial tissues. *G. candidum* could be cultured from the lesions in pure form. Although a 15-day treatment with intramuscular lipiodol brought only slight improvement, there developed, after a 10-day oral nystatin therapy, considerable improvement, but with scarring. Local treatment with argyrol, methylene blue and aureomycin was unsuccessful.

TZAMOURANIS (1962) described a cutaneous geotrichosis in a 4-month-old infant. On the right cheek was a small reddening and infiltrate, which spread in a few days to nearly the entire cheek. At the periphery there appeared new isolated papules which soon became confluent. Small lesions appeared on the forehead, the neck and the left cheek. Little scaling and itching occurred. In direct preparations, on two occasions, there were single, round to oval, double-contoured, non-budding elements and culturally numerous *G. candidum* colonies and some streptococcal and aerobic spore forms. After treatment with nystatin ointment, the lesions healed rapidly, as did a recurrence which appeared a month later. The first attempted treatment with chloramphenicol and penicillin was unsuccessful.

Superficial Forms: CHIALE (1937) viewed the foot lesions of the following case as foci from which the hypersensitivity of the organism developed and led to a geotrichid. In a 54-year-old woman, who suffered from an erythematous, vesicular and squamous exzematoid lesion of the foot for 10 years, there suddenly developed on both legs and on the lower arms, small, diffusely scattered, spotty, slightly scaling, herpetic, somewhat itchy lesions. From the old foci on the foot and, rarely, from the new cutaneous lesions, *G. candidum* could be isolated, but not from the blood. The geotrichin test and the *Geotrichum* complement fixation reaction were positive.

CANELLIS (1959) described a folliculitis decalvans of the scalp due to *G. candidum*. In a 50-year-old man there developed, over the course of 5 years, several disseminated, pin-head sized, follicular pustules and crusts, and small, erythematous and squamous foci in the region of the temples, and in the region of the hair-part, several sharply circumscribed, bald places with smooth, whitish or erythematous surfaces and atrophy of the skin. At the edge of the bald areas and in the hairy scalp there were also numerous pustules related to the follicles. In the pus *G. candidum* was demonstrated microscopically and culturally. The serologic test for syphilis and the blood sugar test were negative. After 25 days of local and oral nystatin treatment the skin lesions were cured. Nine months later the patient was free of lesions.

HEIDENBLUTH (1963) was able to demonstrate *G. candidum* microscopically and culturally in a series of patients who suffered from persistent anal eczema. The fungus was recovered from the stool and from scrapings from the perianal region. The cutaneous appearances and the subjective symptoms of the patients disappeared rapidly under nystatin therapy.

5. Other Geotrichoses

In this section cases of geotrichosis will be described which do not fit into the categories of bronchopulmonary, bronchial, oral, or cutaneous geotrichosis.

Mycetemia due to Geotrichum: In the two cases to be described there was a mycetemia due to *G. candidum* without a demonstrable geotrichosis of other organs. The portal of entry of the fungus was not ascertainable. Apparently the invasion, without appreciable local proliferation of the fungus, was via the digestive or upper respiratory tracts.

BENDOVE and ASHE (1952) reported on a 79-year-old man with diabetes who, because of dietary negligence, developed hyperglycemia and glycosuria and a greatly reduced general condition. Furthermore he acquired pharyngitis. With the use of penicillin, chloramphenicol and aureomycin he became free of fever. Later, fever of 37.8—39.2° C. appeared which could not be influenced. The blood sugar reached 380 mg%, the glycosuria 70 grams per day. Red blood cells were 3.4 millions, leukocytes 12,150, sedimentation-rate 90 mm. By X-ray the lungs showed no lesions. Although the glycosuria was controlled, the patient lost weight and appeared cachectic. In the blood *G. candidum* was demonstrated four times culturally, and also in the sputum. Thereupon the patient was treated with neomycin, which, however, had to be abandoned after 12 days, as difficulty in hearing developed which, moreover, was enduring. Although the *in vitro* effect of neomycin on the strain isolated was slight, the temperature slowly returned to normal. *G. candidum* could no longer be demonstrated in the blood or the sputum, and the patient soon recovered.

KALISKI et al. (1952) reported the case of a 2-month-old infant with miliary tuberculosis, in whom shortly before death abundant *G. candidum* could be cultured from blood samples. Even in smears of the blood, typical rectangular arthrospores were visible. The portal of entry could not be determined. Also, in sections of the lung no fungi were found.

Table 5. *Other Geotrichoses*

Author	Age and Sex	Occupation	Location	Organism
a) *Geotrichum* mycetemia				
1. Bendove and Ashe (1952)	79 M		New York, USA	*Geotrichum* species
2. Kaliski et al. (1952)	2/12 M		Texas, USA	*Geotrichum* species
b) *Geotrichum* mycetoma				
1. Nicolau et al. (1957)	37 F	Farmwife	Rumania	*Geotrichum candidum*
c) *Geotrichum* wound infection				
1. Nagy et al. (1958)	64 M		Hungary	*Geotrichum* species
d) Geotrichosis of the cornea				
1. Perz (1964, Case 1)	53 M	Mechanic	Poland	*Geotrichum candidum*
2. Perz (1964, Case 2)	60 M	Farmer	Poland	*Geotrichum candidum*

Geotrichum Mycetoma: Nicolau et al. (1957, 1958, 1959, 1964) described a mycetoma of the foot with bony changes caused by *G. candidum*. The lesion was cured with INH.

A 37-year-old farm woman developed a firm, plum-sized swelling on the sole of the left foot which was operatively removed, as was a recurrence which developed two years later. After a year there was again a recurrence, which, in the course of another year, led to the development of a painful, unwieldy swelling of the foot. The reddish-brown, dry skin, without ulcers or fistulae, showed numerous reddish nodules of millet-seed size, partly confluent. Nodules which were opened mechanically were found to consist of an oily mass with numerous white pinhead sized granules in which *G. candidum* was demonstrated microscopically and culturally. Roentgenologically the posterior half of the 5th metatarsal bone and the posterior end of the 4th metatarsal were thickened, the bony structure deformed and the cortex dentated. The general condition of the patient was good. Erythrocytes were 3,600,000, leukocytes 4,600, sedimentation-rate 60 mm. The biopsy showed a nodular infiltrate in the corium and partly also in the upper layers of the subcutaneous tissue, which were separated from one another by young connective tissue. The infiltrate in the edematous connective tissue consisted largely of plasma cells. Between them, isolated or in clusters, there were polymorphonuclear neutrophils, lymphocytes, histiocytes and individual giant cells. Furthermore, there were some interstitial, hemorrhagic foci. Capillaries were rich in the granulation tissue. In the infiltrate there were numerous round or oval fungus granules composed of an amorphous or slightly granulated mass without clubshaped projections. In the granule, by Gram stain, one could see relatively thin fungal hyphae (1 μ in diameter), but not in the PAS stain, because the matrix of the granule was PAS positive. In the infiltrate and especially in the neighborhood of the granules there were numerous cells with PAS-positive Russell bodies. With oral and local treatment with INH there was healing in four months and this was complete at the end of a year. Two months after the beginning of therapy no granules could be demonstrated. The bony lesions showed no change at four months but after a year they had healed, with calcification. In the following six years there was no recurrence.

Geotrichum Wound Infection. Nagy et al. (1958) described a suppurative and metastasizing wound infection in a 64-year-old man who, because of left-sided renal tuberculosis, had a nephrectomy. Three weeks later a thrombophlebitis appeared in both legs. From the region of the operation thick yellow pus escaped and *Proteus* and colon bacilli were cultured from it. There was no fever, but the white cells numbered 17,400 and the sedimentation-rate 102 mm in one hour. On the extremities appeared bean-sized abscesses which contained hemorrhagic pus in some instances. A week later the patient died. He had received streptomycin, INH, penicillin and aureomycin. At the postmortem examination, numerous microabscesses were found, in the corium, lungs, pancreas and retroperitoneal tissues. *G. candidum* could be demonstrated culturally in pus from the pustules of the skin and from the retroperitoneal abscesses. Microscopically there were Gram-positive, round or oval elements, single or in groups, in the region of the micro abscesses, in the lymphocytic infiltrate of the mucosa of the air passages, in the wall of the abscess cavity, and in the liver. The abscess wall was infiltrated with lymphocytes, leukocytes and giant cells in many places, but the picture was not that of tuberculous granulation tissue. Nor were tubercle bacilli demonstrable.

Apparently in this case there was a secondary infection of a bacterial process in the form of disseminated geotrichosis.

Geotrichosis of the Cornea. PERZ (1964) described two local infections of the cornea. A 53-year-old locksmith received a squirt of mud in his right eye. In spite of thorough washing, pain and lacrimation developed the following day. A chronic, ulcerative keratitis ensued. The edges of the grayish yellow ulcers were undermined and the neighboring cornea was slightly dull. The conjunctiva was more or less injected, the iris severely injected, the corneal reflex diminished and precipitates were visible on Descemets membrane. Temporarily there was a hypopyon. Local treatment with antibiotics, iodine and Nipagin were unsuccessful. A smear of the ulcer showed *G. candidum* in pure culture. In the eye itself there spread a network of fungus mycelia from the ulcer into the neighboring corneal tissues. After three months vascularization began. Further ulceration had occurred. In six months there was healing with corneal scars. The patient could appreciate only strong light through this eye.

In the second case, a 60-year-old farmer got hay dust into his eyes. The next day an acute inflammation of the left eye developed, with severe pain and lacrimation. At first, serpigenous corneal ulcer with hypopyon was diagnosed. The conjunctiva was injected, and the lower half of the cornea presented grayish-yellow ulcers with undermined edges. The iris had a washed-out appearance and showed posterior synechiae. Local treatment was with atropine, sulfathiazole and silver nitrate. After a month the edge of the ulceration began to vascularize. After three months the whole process healed. The light sensitivity of the eye was maintained. From two excisional biopsies of the infiltrated cornea *G. candidum* could be grown in pure culture.

6. Specific Therapy

In the treatment of bronchopulmonary geotrichosis iodine is still the drug of choice. After several weeks of iodine therapy there is a lessening of the amount of sputum, a stopping of the troublesome cough and subjective feeling of well-being with increase in weight. Also dense pulmonary shadows clear in a few months and hilar lymph nodes become smaller. Even cavities disappear but some residual fibrosis remains. The fungus is no longer demonstrable in stool or sputum, or it occurs in smaller quantity. As a rule, oral potassium iodide is effective (CONANT et al., 1954, page 71). Additionally or by itself sodium iodide can be given intravenously. Also ethyl-iodide inhalations are possible (SWARTZ and JANKELSON, 1941). Acute reactions or disseminations which occasionally can develop as an expression of strong hypersensitivity to the iodine therapy in North American blastomycosis are not seen in the iodine therapy of geotrichosis. For complete cure, iodine therapy may be continued some months, or a year or longer. In bronchial geotrichosis one can count on cure or much improvement after several weeks of iodine therapy.

Nystatin was used in two cases of pulmonary geotrichosis (3—4 months by mouth, daily 3—9 mill. units and as an aerosol) by BELL et al. (1962) with success, and by PENINOU-CASTAING et al. (1954) without success. With cutaneous geotrichosis treated with nystatin in ointment and/or by mouth (2—5 mill. units daily) for 10—25 days, the lesions healed with scar formation. With cutaneous geotrichosis treated with iodine, recurrences appeared in the case of BALZER et al. (1912), in spite of extensive healing, and in the case of ROUSSET et al. (1957) there was, only with the following nystatin treatment, a significant improvement. The fungistatic dose of nystatin according to DROUHET (1955) and DOBINS and HAZEN (1961) is 6—16 units/ml. Also with nystatin, after a few days, there is a cessation or considerable diminution in the excretion of the fungus. As DUNCAN et al. (1951) found *G. candidum* moderately sensitive to neomycin, BENDOVE and ASHE (1952) treated their patient intramuscularly for three days with daily doses of 1000 units per kg and 10 days more with half doses. The infection could be dominated with that. There was, however, a persisting deafness. NICOLAU et al. (1958) were able to cure completely a mycetoma of the foot with bony change by the use of INH. By mouth daily 500—600 mg (altogether 35 grams) of INH were given and in addition 33 local injections of 5—10 ml of a 2% INH solution. The fungistatic INH dose for *G. candidum* is 50—100 μg/ml. Also in the patient of THIERS et al. (1953) there was improvement, with 250 mg of INH daily for 45 days, of the cutaneous lesions and they later healed completely, although the *G. candidum* strain isolated was, *in vitro*, resistant to INH. The pulmonary geotrichosis of HAUSER (1954) and PENINOU-CASTAING et al. (1964) was not influenced by INH. Concerning the therapy of oral geotrichosis there is only one example, that of CARTEAUD and DROUHET (1955), who saw the coating of a black hairy tongue disappear after the use of sodium bicarbonate mouth washings. If a therapy is necessary, oral geotrichosis can be treated with gentian violet etc. as is oral candidosis. The customary antibiotics, like penicillin, streptomycin, chloramphenicol and tetracycline, are ineffective against *G. candidum*. According to HEIDENBLUTH (1963) *G. candidum* is resistent to griseofulvin. Treatment with Amphotericin B was recommended by PROCKNOW and LOOSLI (1958) and WEGMANN (1961). The therapeutic use of *G. candidum* vaccine has not been studied.

G. Geotrichum Candidum Infections in Animals

Natural Infections: In spontaneously dying white mice which during life displayed weakness of the extremities, AGOSTINI (1932) found the lungs beset with small white pustules. In the lung and liver there were microscopically fungus filaments and culturally a bacteria-free pure culture of *G. candidum*. To test the pathogenicity of the isolated strain, healthy mice were injected intravenously, and they died in 3—4 months. *G. candidum* was demonstrable in their lungs and liver. Then a second passage was carried out, and the animals died after 1—2 months and finally, with the third passage, after 1—3 days. The last mice also showed changes in the intestines. TORRES et al. (1942) reported a gastritis in normally nourished white mice. The histologic investigation of 19 animals showed a thin coating on the stomach mucous membrane, which in seven mice extended into the crypts and gland lumina. From the membrane *G. candidum (G. sp.)* could be isolated. The mucosa itself showed a mild acute catarrhal inflammation.

MORQUER et al. (1955) described a caseous adenitis in pigs, which for ten years was repeatedly observed in the slaughter house of Oran. From the caseous, but not calcified, pinhead-sized to pea-sized lesions of lymph nodes *G. candidum* could be isolated. The spleen showed no changes and there was no lymphangitis. With the isolated strain, guinea pigs, white mice and piglets were experimentally inoculated. An acute, non-caseous adenitis was produced by subcutaneous injection and inoculation of the inguinal lymph nodes. Oral infections were not obtained in the pigs, nor were infections obtained from intratracheal and nasal instillations with another strain (*G. versiforme*).

AINSWORTH and AUSTWICK (1955) mentioned three mastitis cases in cows, in whom *G. candidum* was the only agent isolated, and two abortions, in which as a cause a *G. candidum* infection was considered possible. *G. candidum* could be found in abnormally large quantity in the feces of a pig and of a dog, both of whom had severe diarrhea. HEYN (1958) described in a 6-month aborted fetus of a cow a purulent and necrotizing dermatitis with pronounced parakeratosis. The skin changes consisted of foci the size of 5-Mark pieces, extending 1 cm above the surface of the skin, the slimy coating of which could be easily removed. A collection of lymphocytes with occasional leukocytes and histiocytes extended to the subcutaneous tissues. Especially intense were the infiltrates in and about the hair follicles. Within the hair follicles were isolated, segmented and branching hyphae stained by the method of Giemsa. On the basis of the histologic picture an infection with *G. candidum (Oospora lactis)* was predicated. A cultural investigation was not possible.

Experimental Infections: HUEPPE (1884) reported that in cats, rabbits, rats, and mice, after giving them *G. candidum* (Oidium lactis) by mouth, or intracutaneously, subcutaneously or intravenously, no disease process appeared. LANG and FREUDENREICH (1893) and LINOSSIER (1916) got the same results with the intravenous infection of rabbits. Also, THJÖTTA and URDAL (1949) could see no pathological change after intravenous, intraperitoneal and intramuscular injections in rabbits and guinea pigs. The *G. candidum* strain isolated by NICOLAU et al. (1957) from a mycetoma was not pathogenic for mice, guinea pigs and rats. Also the extensive research of COUDERT et al. (1958) with different strains of *G. candidum* in rats, guinea pigs and mice was negative. They infected the animals intraperitoneally and by mouth. The lowering of the resistence by antibiotics, cortisone and streptokinase did not lead to infection. No changes were seen in animals, macroscopically or microscopically, in the period between three weeks and two months after the injections. Nor could the organism be cultured from the internal organs. VIEU and SEGRETAIN (1959) injected mice intraperitoneally with various strains of *G. candidum (G. candidum, G. versiforme* and *G. asteroides).* The infections did not cause the death of the animals, nor were there changes in the organs at autopsy 45 days after injection. CARETTA and MARTINAZZI (1960) injected 2, 10, and 20 mill. *G. candidum* arthrospores in rabbits without there being demonstrable clinical or histological changes after 30 days. Only in one further animal, which had received a considerably higher dose could there be found in the lung a diffuse, interstitial, microinfiltrate of polymorphonuclear neutrophils, eosinophils, lymphocytes and epithelioid cells. There were scattered foreign-body giant cells and foci of remnants of mycelium, bordered by connective tissue, to be seen. The fungus could, however, not be isolated. Brain, heart, liver, spleen, kidneys and adrenals were histologically unchanged. CARETTA (1960) obtained the same result with a broader series of experiments. He injected eight strains of *G. candidum* in three different doses intravenously into 24 rabbits. According to BADER (1965) whose experimental infections likewise were unsuccessful, the fungus cells were taken up by macrophages and rendered harmless, after intravenous injections. In the liver, spleen, bone marrow and lymph nodes there appeared numerous cells whose cytoplasm contained large PAS-positive inclusions consisting of fungus break-down products.

CAO (1900) injected *G. candidum (Oidium lactis)* intravenously. A pregnant animal passed a dead fetus ten days later, and from its kidneys *G. candidum* in large quantity could be cultured, although visible pathologic changes were not present. The mother rabbit died after 25 days having been in very poor condition. There were no macroscopic lesions to be seen. However, the fungus could be demonstrated microscopically in the kidneys and culturally from the brain and kidneys. Another animal died, cachectic, after 42 days. Except for atrophy of the renal cortex there was macroscopically no finding of lesions. The cultural evidence was successful only from the brain. A third animal, which was injected with the strain isolated from the first animal died, somewhat emaciated, after 14 days. The left kidney showed a white cortical infarct, the right one 5 white nodules with purulent contents. The culture from kidney and brain was positive. MARTINS (1928) injected his strain *(Zymonema pulmonalis membranogenes)* into rabbits, dogs and guinea pigs. The intravenously injected rabbit and the intraperitoneally injected guinea pig showed no changes. An intravenously inoculated dog lost weight and a subcutaneously inoculated guinea pig developed a knotty infiltrate at the site of injection. CIFERRI and REDAELLI (1935) found a mild, generalized, miliary infection with prominent renal changes, after they injected a strain of *G. candidum* (Colombo Pasini Nr. 4) in a rabbit. CIFERRI et al. (1938), in four rabbits, injected a thick suspension of *G. candidum (G. matalense)* intravenously. Only the animal injected with 10 ml of suspension died after 23 days. The other animals survived and those receiving 3 and 5 ml showed no changes while those receiving 8 ml were emaciated. The dead, severely emaciated animal showed somewhat hyperemic lungs, a slightly enlarged liver and spleen of decreased consistency, numerous enlarged abdominal lymph nodes and distinctly enlarged kidneys. The microscopic renal finding was that of well-maintained glomeruli. The capsular spaces were, in some cases, filled with coagulated protein. The vessels of medium caliber contained numerous fungus elements, mostly in the form of foci of various size. Fungal foci were also found in the renal parenchyma. The tissue surrounding the fungal elements was partially necrotic. In part there was a slight, nodular, non-proliferating inflammation present (slight exudate with few cells). The tubular epithelium showed a diffuse albuminoid degeneration. Fungus elements could be found only in the kidneys. The other organs were normal or showed signs of a slight inflammation with parenchymatous degeneration. The rabbit injected with 8 ml and sacrificed two months later, also showed the same changes, even if in lesser degree. Here, also, the changes in the kidneys were prominent. But not every strain of *Geotrichum* produced lesions, as two other rabbits, injected with other *G. candidum* strains (*Pseudomycoderma matalense* and *Oidium* sp. CHAPMAN) in the same quantities as above did not become ill. These strains were old ones, from the culture collection, while the successfully injected strain had been freshly isolated from the stool of a patient ill with enteritis.

VÖRÖS-FELKAI and NOVÁK (1961) reported, without details, fatal intraperitoneal infections (within 7—10 days) in mice injected with 17 strains of *G. candidum (G. sp.)* isolated from humans. ARIYEVICH et al. (1963) injected rats. With the intravenous injection of 75 mill. arthrospores the animals, sacrificed after three days, showed grayish white nodules on the surface of the kidneys and histologically numerous renal abscesses with fungus elements. Also in the lungs there were small nodules with central, fungus-containing giant cells which were surrounded with lymphocytes and leukocytes. Also in the brain fungus elements and mild infiltrates near vessels could be demonstrated. With the subcutaneous injection of 50 million arthrospores abscesses formed after one or two days and these contained numerous hyphae. Intranasally injected fungus cells, 50 million of them, under light ether anesthesia, led, after one or two days, to exudative, leukocyte-rich foci in the lungs having fungus elements with right angled and rounded corners. In 3-day-old rats a part of the tongue surface was chemically traumatized and then brushed with a suspension of arthrospores. After one or two days one could see, histologically, numerous microabscesses with fungus filaments. The process later involved the non-traumatized portion of the tongue and also the esophagus. PERZ (1964) injected intraperitoneally two guinea pigs with 1 ml each and two mice with 0.5 ml each, using a strain of *G. candidum* isolated from a corneal ulceration in a suspension of 1 million arthrospores per ml. In the guinea pigs infection was unsuccessful. The mice died after 9 days, after they had been apathetic and without appetite for a day. At autopsy only a hyperemia of the liver and dilatation of the small intestine could be seen grossly but microscopically there were small thrombi and arthrospores in the coronary vessels. The fungus could not be cultured. SCHIEFER (1967) described, in mice, arthrospore-containing abscesses and epithelioid cell granulomas with a tendency toward caseation, and there were collections of foam cells with arthrospores.

In the following investigations of RICKETTS, BALZER et al., and NAGY et al., a bacterial contamination of the *G. candidum* strain, sometimes difficult to recognize, cannot be definitely ruled out. RICKETTS (1901) injected several animals with his strain (*Oidium aceticum*) isolated from cutaneous abscesses. In guinea pigs the intraperitoneal infection was negative. After subcutaneous injection into the skin of the belly there developed, after 5 days, abscesses with spontaneous rupture (Staphylococcal ?). With injection into the back there was elevation of temperature, but no abscess formation. Subcutaneous injection in mice and intraperitoneal and subcutaneous injection in white rats had negative results. Also there were no effects following the subcutaneous injection of a dog. Another dog injected intravenously and sacrificed after a month showed only miliary, subcapsular, renal nodules consisting of young connective tissue and plasma cells. The causal agent was not demonstrable. An intravenously injected rabbit died, after 5 min, of a pulmonary embolus. Another died after two days without the fungus being demonstrated culturally. The kidneys were enlarged and pale with the features of fatty degeneration, as was also the liver. In the lungs there were small foci with epithelioid cell, lymphocytic and leukocytic infiltrates. The other organs were normal.

BALZER et al. (1912) carried out investigations with a strain isolated from a skin lesion (*Mycoderma pulmoneum*). The intraperitoneally and subcutaneously injected guinea pigs died of septicemia after 2 days. An intravenously injected rabbit died of septicemia after 8 days, having suffered rapid emaciation. Two subcutaneously injected rats died in very poor condition after 3 months, one with a large ulcer of the anus.

NAGY et al. (1958) injected two mice intraperitoneally with the *G. candidum* (*G.* sp.) isolated from their patient. Both animals succumbed to purulent peritonitis at 6—7 weeks. In smears of the peritoneal exudate there were numerous arthrospores with rounded ends and short hyphae. The internal organs also contained fungi.

The investigations described were carried out with living cultures. But also from intravenous injection of dead suspensions, for example in the immunizing of rabbits, fatal cases have been observed. This occurs not only after *G. candidum* inoculations, but often when large quantities of relatively bulky fungus material is injected intravenously (THJÖTTA et al., 1951; SEELIGER, 1958; TORHEIM, 1963; MORENZ, 1963). The long thick hyphae can block small vessels and induce micro-infarcts.

There appears to be a special concentration of the fungus elements in the kidney, perhaps related to the excretion function of the kidneys. If the pulmonary capillaries are rapidly blocked, there can be the so-called rapid death situation with the picture of a pulmonary embolism. The organ changes are not dependent on the particulate nature of the fungus alone, but also on the toxic effect of the fungus. With homogenized *G. candidum* material immunized animals after several injections may become emaciated and finally succumb. In this way the degradation products of the fungus may lead to parenchymal degenerations, especially in the kidneys and the liver.

In summary, it is possible to say that under certain unknown circumstances *G. candidum* may also be pathogenic for animals, as the already described natural infections show. Moreover, an increase in virulence with animal passage may occur. Experimental infections were possible, as a rule, only with very large doses of infectious material. As the same symptoms also appeared with the killed fungus, the reactions are related to the large size of mycelial fragments (emboli, microinfarcts) and to the toxic fungal degradation products (microinfiltrates, foreign body granulomas, parenchymal degenerations) without an increase in the number of fungi being necessary. For saprophytic and pathogenic types within the *Geotrichum candidum* species there is no support. The "pathogenic differences" in experimental infections with different strains can be explained by differences in fragmentation of the mycelium, as immunization with completely arthrospore-containing cultures or mechanically fragmentated hyphae is better tolerated by the animals.

References

ABRAMOVA, E.V., KRYLOV, L.M., SOKOLOVA, T.S.: Intestinal geotrichosis in a three-month-old child. (Russian). Vopr. Ochr. Mat. (Moskva) **9/2**, 87 (1964). — ACOSTINI, A.: *Geotrichum candidum* Link, patogeno del mus. Boll. Soc. ital. Biol. sper. **7**, 727 (1932). — AINSWORTH, G.C., AUSTWICK, P.K.C.: A survey of animal mycosis in Britain: general aspects. Vet. Rec. **67**, 88 (1955). ~ A survey of animal mycosis in Britain: mycological aspects. Trans. Brit. Mycol. Soc. **38**, 369 (1955). — ALMEIDA, F. de, LACAZ, C.S.: Considerações micólogicas sobre 4 amostras de *Geotrichum* isoladas do escarro. Folia clín. biol. (S. Paulo) **12**, 41 (1940). ~ Cogumelo do genero *Geotrichum* isolado de lesoes ulcerativas do reto. Folia clín. biol. (S. Paulo) **12**, 57 (1940). — ARIEVICH, A. M., KRYLOV, L. M., STEPANISHCHEVA, Z.G.: Geotrichosis of mucous membranes and internal organs (clinical and experimental investigation). (Russian). Klin. Med. (Moskva) **41**, 14 (1963). — ARRIGHI, F., DROUHET, E.: Langue noire villeuse à *Geotrichum*. Bull. Soc. franç. Derm. Syph. **64**, 753 (1957). — BACHMANN, W.: Ein Fall von Soorvarietät. Zbl. Bakt., I. Abt. Orig. **86**, 129 (1921). — BADER, G.: Die viszeralen Mycosen (Pathologie, Klinik und Therapie). Jena: Gustav Fischer 1965. — BALZER, F., BURNIER, R., GOUGEROT, H.: Dermatite végétante et ulcéreuse due un champignon filamenteux constaté dans le pus, isolé par la culture et encore indéterminé. Bull. Soc. franç. Derm. Syph. **21**, 342 (1910). — BALZER, F., GOUGEROT, H., BURNIER, R.: Dermatomycose végétante disséminée due au *Mycoderma pulmoneum*. Ann. Derm. Syph. (Paris), Sér. V, **3**, 461 (1912). — BELL, D., BRODIE, J., HENDERSON, A.: A case of pulmonary geotrichosis. Brit. J. Dis. Chest **56**, 26 (1962). — BENDOVE, R.A., ASHE, B.I.: *Geotrichum* septicemia. Arch. intern. Med. **89**, 107 (1952). — BENNETT, J.H.: On the parasitic fungi found growing on the bodies of living animals. Proc. roy. Soc. Edinb. **1**, 356 (1832—1844). ~ On the parasitic vegetable structures found growing in living animals. Trans. roy. Soc. Edinb. **15**/II, 277 (1842). — BREDNOW, W.: Zur Klinik der Lungenmykosen. Med. Klin. **53**, 622 (1958). — CANELLIS, P.: Geotrichosis der Kopfhaut unter dem Bilde einer Folliculitis decalvans. Hautarzt **10**, 43 (1959). — CANITROT-ARAUJO, M.: Aportación al estudio de la alergia cutánea a diversos hongos patógenos. An. Fac. Med. Santiago de Compostela **1**, 453 (1957). (Quoted from Rev. Med. vet. Myc. [Kew] III/1958). — CAO, G.: Oiden und Oidiomykose. Z. Hyg. **34**, 282 (1900). — CARETTA, G.: Isolamento comparativo da escreti umani del *Geotrichum candidum* e saggi di patogenicità. Atti Ist. Bot. Lab. Critt. Pavia. Ser, V, **17**, 293 (1960). ~ Reperti simultanei di *Candide e Geotrichi* in sputi e in feci umane. Nuovi Ann. Ig. **12**, 309 (1961). — CARETTA, G., MARTINAZZI, M.: Aspetti morfologici delle candidosi, aspergillosi e geotricosi sperimentali. Pathologica **52**, 245 (1960). — CARMICHAEL, J.W.: *Geotrichum candidum*. Mycologia **49**, 820 (1957). — CARTEAUD, A.J.-P., DROUHET, E.: Étude mycologique à propos de six observations typiques de langue noire. Bull. Soc. franç. Derm. Syph. **62**, 162 (1955). — CARTEAUD, A.J.-P., DROUHET, E., VIEU, M.: Les langues noires villeuses. Presse méd. **65**, 966 (1957). — CASTELLANI, A.: Notes on the hyphomycetes found in sprue: with remarks on the classification of fungi of the genus "*Monilia Gmelin* 1791". J. trop. Med. Hyg. **17**, 305 (1914). ~ Fungi and fungous diseases. Lecture II. Arch. Derm. Syph. (Chic.) **17**, 61 (1928). ~ Dermatite eczematose micotica delle dita dei piedi dovuta probabilmente al *Geotrichum rotundatum* CASTELLANI 1911 ed al *Geotrichum rotundatum var. gallicum n. v.* Arch. ital. Sci. med. colon. **23**, 1 (1942). — CASTELLANI, A., CHALMERS, A.J.: Manual of Tropical Medicine. London: Baillière, Tindall & Cox 1919. — CHIALE, G.F.: In tema di "foci" cutanei micotici. G. ital. Derm. Sif. **78**, 771 (1937). — CIFERRI, R., REDAELLI, P.: Contribuzioni alla sistematica delle Torulopsidaceae. XV—XXXIII. Arch. Mikrobiol. **6**, 9 (1935). — CIFERRI, R., VERONA, O., SAGGESE, V.: Reisolamento della *Pseudomycoderma matalense* e revisione del gruppo. Mycopathologia (Den Haag) **1**, 212 (1938). — COLONNELLO, F.: Il *Geotrichum candidum* quale ospite dell' organismo umano. Atti Ist. Bot. Lab. Critt. Pavia, Ser. V, **3**, 197 (1944). — CONANT, N.F., SMITH, D.T., BAKER, R.D., CALLAWAY, J.L., MARTIN, D.S.: Manual of Clinical Mycology. Philadelphia-London: W. B. Saunders Comp. 1954. — CORTESE, F.: Tonsillomicosi da "*Geotrichum candidum.*" Valsalva **9**, 149 (1933). (Quoted from COLONNELLO, 1944.) — COUdert, J., Despierres, G., Saëz, H., Hollard, J.: Recherches sur la flore levuriforme bronchique en milieu sanatorial. Sem. Hôp. Paris 33, 2978 (1957). — Coudert, J., Garin, J.-P., Saëz, H.: Quelques aspects de la géotrichose humaine. J. méd. Lyon 38, 189 (1958). — Dávalos, R., Arcos, L.: Geotricosis pulmonar. Gac. méd. (Guayaquil) 9, 155 (1954). — Dobias, B., Hazen, E.L.: Nystatin. Chemotherapia 3, 108 (1961). — Drouhet, E.: Action de la nystatine (fungicidine) in vitro et in vivo sur *Candida albicans* et autres champignons levuriformes. Ann. Inst. Pasteur (Paris) 88, 298 (1955). — Duncan, G.G., Clancy, C.F., Wolgamot, J.R., Beidleman, B.: Neomycin: results of clinical use in ten cases. J. Amer. med. Ass. 145, 75 (1951). — Dvořák, J., Otčenášek, M.: *Geotrichum candidum* Link 1809 in einer Haut- und Nagelläsion. Derm. Wschr. 152, 1183 (1966). — Ettig, B.: Das Vorkommen von *Geotrichum candidum* Link im Stuhl Psoriasiskranker. Hautarzt 16, 491 (1965). — Felsenfeld, O.: Yeast-like fungi in the intestinal tract of chronically institutionalized patients. Amer. J. med. Sci. 207, 60 (1944). — Frágner, P., Petrů, M., Vojtěchovská, M.: *Geotrichum candidum* při vaginálním fluoru. Čsl. Hyg. Epid. Mikrobiol. 4, 434 (1955). — Frágner, P., Svatek, Z.: Mykosy kladenských havířů. Čsl. Hyg. Epid. Mikrobiol. 5, 75 (1956). — Galinovic-Weisglass, M.: *Geotrichum* u intestinalnom traktu čovjeka. Higijena (Beograd) 11, 316 (1959). — Gauvreau, L., Grandbois, J.: Un cas de mycose cutanée causée par un champignon de l'ordre des arthrospores. Laval méd. 19, 59 (1954). — Gernez-Rieux, C.: Les pneumopathies fongiques. Rev. Tuberc. (Paris) 24, 1088 (1960). — Golay, J., Wyss-Chodat, F.: De la présence fréquente d'un champignon du genre *géotrichum* (Link) dans l'intestin des psoriasiques. Rev. méd. Suisse rom. 62, 961 (1942). — Good, C.A.: Fungus diseases of the lungs. Review of roentgenologic manifestations. Tex. St. J. Med. 47, 817 (1951). — Gougerot, H.: Mycodermose cutanée due au "*Mycoderma pulmoneum.*" Bull. Soc. franç. Derm. Syph. 27, 185 (1920). ~ Mycodermoses (ou géotrichoses). In: Darier, J. et al.: Nouvelle Pratique Dermatologique, T. II, p. 569—579. Paris: Masson 1936. — Grimmer, H.: Antibiotika und Pilzerkrankungen der Haut und Schleimhaut. Antibiot. et Chemother. (Basel) 1, 180 (1954). — Guéguen, F.: Méconnaissance fréquente de *l'Oidium lactis* Fresenius, saprophyte facilement identifiable de l'homme et des animaux. C. R. Soc. Biol. (Paris) 74, 943 (1913). — Haley, L.D., McCabe, A.: A mycologic study of seventy-one autopsies. Amer. J. clin. Path. 20, 35 (1950). — Hauser, W.: Geotrichose der Lunge. Ärztl. Wschr. 9, 244 (1954). — Heidenbluth, J.: Über *Geotrichum candidum* und *Oospora lactis* und ihr Vorkommen beim Analekzem. Z. Haut- u. Geschl.-Kr. 34, 222 (1963). — Heyn, W.: Hautmykose bei einem Rinderfetus. Dtsch. tierärztl. Wschr. 65, 561 (1958). — Hueppe, F.: Untersuchungen über die Zersetzung der Milch durch Mikroorganismen. Mitt. kaiserl. Gesundheitsamt 2, 309 (1884). — Hyde, J.N., Ricketts, H.T.: A report of two cases of blastomycosis of the skin in man, with a survey of the literature of human blastomycosis. J. cutan. Dis. 19, 44 (1901). — Jannin, L.: Les "Mycoderma". Leur rôle en pathologie. Thése Fac. Méd. No. 1001, Nancy, 1913. — Kaliski, S.R., Beene, L.M., Mattman, L.: *Geotrichum* in blood stream of an infant. J. Amer. med. Ass. 148, 1207 (1952). — Kärcher, K.H.: Neue Gesichtspunkte zur Klinik und Pathogenese der Hefeerkrankungen. Arch. Derm. Syph. (Berl.) 197, 51 (1953). ~ Die Geotrichose. In: Jadassohn, J.: Handbuch der Haut- und Geschlechtskrankheiten, Ergänzungswerk Bd. IV/4, p. 197—203. Berlin-Göttingen-Heidelberg: Springer 1963. — Kourkoumeli-Kontomichalou, P., Papadakis, E., Daikos, G.: Nasopharyngeal flora of personnel of "Alexandra" hospital. (Greek). Acta microbiol. Hellen. 6, 150 (1961). — Kunstadter, R.H., Milzer, A.: Incidence of mycotic infections in children with acute respiratory tract disease. Amer. J. Dis. Child. 81, 306 (1951). — Kunstadter, R.H., Milzer, A., Whitcomb, F.: Bronchopulmonary geotrichosis in children. Amer. J. Dis. Child. 79, 82 (1950). — Kunstadter, R.H., Pendergrass, R.C., Schubert, J.H.: Bronchopulmonary geotrichosis. Amer. J. med. Sci. 211, 583 (1946). — Lang, M., Freudenreich, E. v.: Über *Oidium lactis*. Landwirtsch. Jahrb. Schweiz 7, 229 (1893). — Linossier, G.: Contribution à l'étude des mycoses bronchopulmonaires. Étude biologique d'un "*Oidium lactis*" parasite de l'homme. Bull. Soc. méd. Hôp. Paris, Sér. III, 40, 575 (1916). — Lodder, J., Kreger-Van Rij, N.J.W.: The Yeasts: A Taxonomic Study. North-Holland, Amsterdam, 1952. — Lofgren, R.C.: Oospora-like fungus found pathogenic to man. Arch. Derm. Syph. (Chic.) 71, 56 (1955). — Mahoudeau, D., Lemoine, J.M., Poulet, J., Dubrisay, J.: Mycoses respiratoires pseudo-tumorales. (Aspergillose et géotrichose). J. franç. Méd. Chir. thor. 9, 53 (1955). — Mándi, L., Berencsi, G.: Adatok a tüdömykozis kérdéséhez két eset kapcsán. Orv. Hetil. 27, 742 (1954). — Marselou, U., Paschali-Pavlatou, M.: Fréquence des levures du genre *Candida* dans le vagin, le rectum et le pharynx. (Greek). Acta microbiol. Hellen. 1, 239 (1956). — Martin, D.S.: Complement-fixation in blastomycosis. J. infect. Dis. 57, 291 (1935). — Martins, C.: *Zymonema pulmonalis membranogenes* isolé d'un crachat de pneumopathie grave et mortelle. C. R. Soc. Biol. (Paris) 98, 1162 (1928). — Mattlet, G.: Mycoses dans l'Urundi. Ann. Soc. belge Méd. trop. 6, 1 (1926). — Minton, R., Young, R.V., Shanbrom, E.: Endobronchial geotrichosis. Ann. intern. Med. 40, 340 (1954). — Montemayor, L. de, Gamero, B.H. de: Aislamiento y determinacion de 300 cepas del genero *Candida.*

Mycopathologia (Den Haag) **15**, 343 (1961). — MOORE, M.: A new *Geotrichum* from a bronchial and pulmonary infection, *Geotrichum versiforme* MOORE, *n. sp.* Ann. Missouri Botan. Garden **21**, 349 (1934). — MORENZ, J.: *Geotrichum candidum* Link. Taxonomie, Diagnose und medizinische Bedeutung. Mykol. Schriftenreihe (Leipzig) **1**, 1 (1963). ~ Taxonomische Untersuchungen zur Gattung *Geotrichum* Link. Mykol. Schriftenreihe (Leipzig) **2**, 33 (1964). — MORQUER, R., LOMBARD, C., BERTHELON, M.: Pouvoir pathogène de quelques espèces de *Geotrichum*. C. R. Acad. Sci. (Paris) **240**, 378 (1955). — MOTTA, R.: Micosi benigna della tonsilla linguale da *Oospora lactis*. Valsalva **7**, 599 (1931). (Quoted from COLONNELLO, 1944). — NAGY, L., MOLNÁR, L., FLÓRIÁN, E.: Geotrichosis. Zbl. allg. Path. **98**, 374 (1958). — NEAGOE, G., NEAGOE, M.: Enterokolitis durch *Geotrichum candidum* nach Therapie mit Glutaminsäure. Dtsch. Z. Verdau.- u. Stoffwechselkr. **27**, 205 (1967). — NICOLAU, S.G., AVRAM, A.: Der gegenwärtige Stand der Behandlung des Myzetoms im Lichte eigener Erfahrung. Derm. Wschr. **140**, 792 (1959). ~ Cas de mycétomes observés en Roumanie. Arch. Union Méd. Balkan. (Bucureşti) **2**, 597 (1964). — NICOLAU, S.G., AVRAM, A., COLINTINEANU, L., BĂLUŞ, L., STOIAN, M.: Efect favorabil al hidrazidei acidului izonicotinic asupra leziunilor osoase dintr-un caz de micetom al piciorului cu grăunţi albi, provocat de *Geotrichum candidum*. Derm.-Vener. (Buc.) **3**, 263 (1958). — NICOLAU, S.G., HULEA, A., AVRAM, A., COLINTINEANU, L., BĂLUŞ, L.: Studiu clinic şi micologic asupra unei forme tuberoase de micetom cu grăunţi albi al piciorului produs de *Geotrichum candidum*. Derm.-Vener. (Buc.) **2**, 217 (1957). — NICOLESCU, N.: Aspecte radiologice în micozele pulmonare. Viaţa med. **9**, 707 (1962). — OTT, E.: Über das Vorkommen von Hefen im Magen-Darm-Trakt unter besonderer Berücksichtigung quantitativer Unterschiede. Mykosen **6**, 7 (1963). — PENINOU-CASTAING, J., COLLAS, R., BATTESTI, M.R.: Géotrichose pulmonaire d'évolution fatale associée a une tuberculose pulmonaire. Poumon **20**, 287 (1964). — PERZ, M.: Drożdżyca ragówki spowodowana przez *Geotrichum candidum*. Klin. oczna **34**, 273 (1964). — PÉTER, M., HORVÁTH, G.: Cercetări clinico-micologice privind frecvenţa genului *Geotrichum*. Rev. Medicală **5**, 430 (1959). — PÉTER, M., HORVÁTH, G., DOMOKOS, L.: Date referitoare la frecvenţa genului *Geotrichum* în diferite produse biologice umane. Med. Internă **19**, 875 (1967). — PÉTOURAUD, C., COUDERT, J.: Blastomycose cutanée à "*Geotrichum cutaneum*" de Beurmann et Gougerot (1909) localisée, chez un employé dans une fromagerie. Bull. Soc. franç. Derm. Syph. **60**, 299 (1953). — PINKERTON, M.E., MULLINS, J.F., SHAPIRO, E.M.: The ecology of superficial fungus infections in Galveston, Texas. A five year survey. Tex. Rep. Biol. Med. **15**, 26 (1957). — PROCKNOW, J.J., LOOSLI, C.G.: Treatment of the deep mycoses. Arch. intern. Med. **101**, 765 (1958). — REDAELLI, P., CIFERRI, R.: Relazione sul primo quinquennio (1938—1943) di attività del Centro Micologia Umana e Comparata della R. Università di Pavia. Atti Ist. Bot. Lab. Critt. Pavia, Ser. V, **3**, 1 (1943). — REEVES, R.J.: The incidence of bronchomycosis in the South. Amer. J. Roentgenol. **45**, 513 (1941). — REIERSÖL, S.: An investigation of the occurrence of fungi in 250 laryngeal swabs from tuberculous patients. Acta path. microbiol. scand. **32**, 500 (1953). — RICKETTS, H.T.: Oidiomycosis (blastomycosis) of the skin and its fungi. J. med. Res. **6**, 373 (1901). — ROSSIER, P.H.: Antibiotica und Mykosen. Helv. med. Acta **19**, 261 (1952). — ROUSSET, J., COUDERT, J., GARIN, J.P.: Blastomycose superficielle des muqueuses à foyers multiples. Bull. Soc. franç. Derm. Syph. **64**, 316 (1957). — SAËZ, H.E.: Le *Geotrichum candidum* Link. Caractéristiques morpho-biologiques; fréquence chez l'homme. Bull. Soc. Myc. Fr. **73**, 343 (1957). — SCHABINSKI, G.: Grundriß der medizinischen Mykologie. Jena: Gustav Fischer 1960. — SCHAEDE, R.: Über zwei neue Oosporaarten bei „schwarzer Haarzunge". Derm. Wschr. **98**, 521 (1934). — SCHIEFER, B.: Pathomorphologie der Systemmykosen des Tieres. Jena: Gustav Fischer 1967. — SCHNOOR, T.G.: The occurrence of *Monilia* in normal stools. Amer. J. trop. Med. **19**, 163 (1939). — SEELIGER, H.P.R.: Mykologische Serodiagnostik. Leipzig: J.A. Barth 1958. ~ Serodiagnostik der Pilze und mykotischen Infektionen. Zbl. Bakt., I. Abt. Orig. **184**, 203 (1962). — ŞERBAN, P., TAŞCĂ, C.: Aspecte privind diagnosticul morfologic al unor micoze generalizate, în practica prosecturală. Morfol. norm. si pat. (Buc.) **7**, 257 (1962). — SKOBEL, P., SCHABINSKI, G., ESSIGKE, G.: Untersuchungen über die Immunitätslage und den Wert der Serodiagnostik bei *Candida*-Mykosen. Ärztl. Wschr. **11**, 317 (1956). — SMITH, D.T.: Oidiomycosis of the lungs. Report of a case due to a species of *Geotrichum*. J. thorac. Surg. **3**, 241 (1934). ~ Pulmonary mycoses. Clinics **4**, 994 (1945). ~ Fungus diseases of the lungs. Springfield: Thomas 1947. ~ Fungous infections in the United States. J. Amer. med. Ass. **141**, 1223 (1949). — SUNDGAARD, G., THJÖTTA, T., URDAL, K.: Familiaer opptreden av geotrichosis pulmonum. Nord. Med. **43**, 434 (1950). — SWARTZ, J.H., JANKELSON, I.R.: Incidence of fungi in the stools of non-specific ulcerative colitis. Amer. J. dig. Dis. **8**, 211 (1941). — THIERS, H., COUDERT, J., COLOMB, D.: Lésions nodulaires disséminées (*geothricum candidum*) avec allergides. Efficacité remarquable de l'Isoniazide. Bull. Soc. franç. Derm. Syph. **60**, 165 (1953). — THJÖTTA, T., RASCH, S., URDAL, K.: Preparation of fungous antigens for immunization and for serological reactions. Acta path. microbiol. scand. **28**, 132 (1951). — THJÖTTA, T., URDAL, K.: A family endemic of geotrichosis pulmonum. Acta path. microbiol. scand. **26**, 673 (1949).

— TORHEIM, B.J.: Immunochemical investigations in *Geotrichum* and certain related fungi. III. Gel-precipitation studies of reactions between rabbit immune sera and polysaccharides extracted from fungi. Sabouraudia **2**, 292 (1963). — TORRES, C.M., LEÃO, A.E.A., SALLES, J.F.: Gastrite espontânea do camondongo e cogumelos do gênero *Geotrichum* Link. Mem. Inst. Osw. Cruz **39**, 97 (1942). — TZAMOURANIS, N.: Sur un cas de géotrichose cutanée. Arch. Inst. Pasteur hellén. **8**, 49 (1962). — TZAMOURANIS, N., CRIMBITHIS, E.: La fréquence de *Geotrichum* et des levures dans les selles et la cavité buccopharyngée de l'homme en Grèce. Arch. Inst. Pasteur hellén. **8**, 119 (1962). — VIEU, M., SEGRETAIN, G.: Contribution à l'étude de *Geotrichum* et *Trichosporum* d'orgine humaine. Ann. Inst. Pasteur (Paris) **96**, 421 (1959). — VOLK, R., CAÑAS, E.: Un caso de blastomicosis. Medicina (Méx.) 22, 615 (1942). — VÖRÖS-FELKAI, G., NOVÁK, E.K.: Incidence of yeasts in human material. Acta microbiol. Acad. Sci. hung. **8**, 89 (1961). — VRIJMAN, L.H.: *Geotrichum*-infectie. Ned. T. Geneesk. **77**/II, 2129 (1933). — WEBSTER, B.H.: Pulmonary geotrichosis. Amer. Rev. Tuberc. **76**, 286 (1957). ~ Bronchopulmonary geotrichosis: A review with report of four cases. Dis. Chest **35**, 273 (1959). — WEGMANN, T.: Pilzerkrankungen der inneren Organe als Folge von Behandlung mit Antibiotica, unter besonderer Berücksichtigung des Respirationstraktes. Antibiot. et Chemother. (Basel) **1**, 235 (1954). ~ Mykosen der inneren Organe. Ergebn. inn. Med. **8**, 457 (1957). ~ Pilzkrankheiten der inneren Organe. In: POLEMANN, G.: Klinik und Therapie der Pilzkrankheiten. Stuttgart: Georg Thieme 1961. — WHITCOMB, F.C., MILZER, A., KUNSTADTER, R.H.: Incidence of mycotic infections in children with acute respiratory disease. J. Pediat. **35**, 715 (1949).

Miscellaneous Uncommon Diseases Attributed to Fungi and Actinomycetes

JOSEPH C. PARKER JR., Boston/Mass., USA and GORDON K. KLINTWORTH, Durham/N. C., USA

With 60 Figures

Adiaspiromycosis

(Haplomycosis)

In 1942, a thick-walled, spherical fungus was discovered in the lungs of wild rodents in Arizona, USA, and designated *Haplosporangium parvum* (EMMONS, 1942; EMMONS and ASHBURN, 1942). The genus *Emmonsia* was proposed for this organism by CIFERRI and MONTEMARTINI (1959), but CARMICHAEL (1951, 1962) considers that the fungus should be included in the genus *Chrysosporium* Corda. Four species have been described (Table 1).

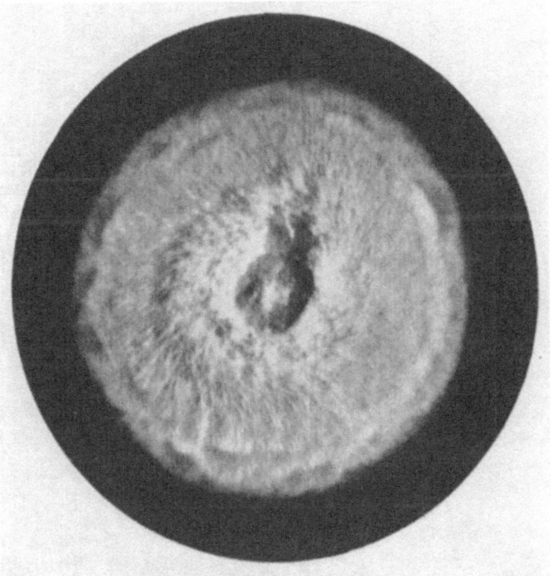

Fig. 1. Colony of *Emmonsia parva* on modified Sabouraud's glucose agar after two weeks at 30°C. (After EMMONS, C.W., and JELLISON, W.L.: Ann. N.Y. Acad. Sci. 89, 91—101, 1960)

The fungus grows slowly on Sabouraud's agar at room temperature producing a white colony, whose appearance varies with the species and strain (Fig. 1). Conidia (aleurospores) that are almost spherical (2—14 μ in diameter) are borne on simple or compound conidiophores, which arise at right angles from delicate branching septate hyphae (0.5—2 μ in diameter). The spores are borne singly or rarely in chains of two. When the fungus is incubated at 37°C, the conidia and

Table 1. Comparison between different species of *Emmonsia* (Modified from THIRUMALACHAR et al., 1965)

	E. brasiliensis *	E. ciferrina *	E. crescens	E. parva
Conidia	(1) 7—10 × 5—7 μ (2) Hyaline smooth	(1) 7—14 × 4.8 μ (2) Hyaline smooth	(1) 2—4 × 2.5—4.5 μ (2) Hyaline smooth	(1) 2.4 × 2.5—4.5 μ (2) Hyaline smooth
Adiaspores in culture	(1) Produced at 37° C	(1) Produced on blood agar and several other media at 28 to 37° C	(1) Produced at 37° C on blood agar	(1) Produced at 40° C on blood agar
	(2) 20—25 μ in diameter	(2) 20—45 μ in diameter	(2) 250—480 μ in diameter	(2) 15—25 μ in diameter
	(3) Wall 1.5—2.7 μ thick	(3) Wall 1.5—2 μ thick	(3) Wall up to 70 μ in thickness	(3) Wall 2 μ thick
	(4) Multinucleate		(4) Multinucleate	(4) Uninucleate
Adiaspores in lung tissue	20—25 μ in diameter	(1) 15—25 μ in diameter (2) Wall thickness 1.5—2 μ	250—480 μ in diameter	15—40 μ in diameter
Mycelium		Septate and branching, in later stages forming arthrospores in alternate cells as in *Coccidioides immitis*	Septate and branching	Septate and branching

* PADHYE and CARMICHAEL (1962) have drawn attention to the fact that *E. brasiliensis* and *E. ciferrina* are almost identical in colonial and microscopic morphology to *Chrysosporium pruinosum*.

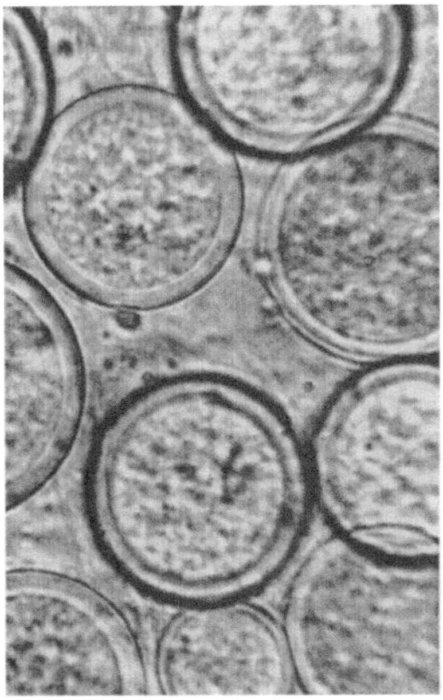

Fig. 2. Adiaspores of *Emmonsia ciferrina* on oatmeal tomato paste agar. × 1000. (After THIRUMALACHAR, M.J., PADHYE, A.A., and SRINIVASAN, M.C.: Mycopathologia (Den Haag) 26, 323—332, 1965)

some of the hyphal cells enlarge and form large spherules (adiaspores) comparable to those that develop in the host (Fig. 2). These adiaspores like other fungal spores do not reproduce by budding or endosporulation. In some species such as *E. crescens* they are multinucleated, while in others, such as *E. parva*, they possess a single nucleus. Following isolation of the fungus from the host at room temperature, or after the removal of colonies from 37° C to room temperature, hyphae develop from the adiaspores. The number of emerging hyphae is directly proportional to the number of nuclei in the adiaspores. For example, a single germ tube arises with *E. parva*, while there are multiple germ tubes with *E. crescens* (Fig. 3). Arthrospores develop in the mycelium of at least some species, including *E. ciferrina* (THIRUMALACHAR et al., 1965). At culture the fungus resembles *Histoplasma capsulatum* but differs in not producing macroconidia. In tissue *Emmonsia* is not only larger than *Histoplasma*, but does not reproduce by budding. Though the fungus was originally discovered while investigating the natural history, host range, and ecology of *Coccidioides immitis*, and was often associated with coccidioidomycosis, the fungus differs in morphology and colonial characteristics from *C. immitis*.

EMMONS and JELLISON (1960) designated the large spherules in the lungs as "adiaspores" and proposed the term "adiaspiromycosis" for the resultant pulmonary disease.

Since the condition first became recognized, it has been detected in numerous small mammals throughout the world (BAKERSPIGEL, 1956, 1957, 1965; DOWDING, 1947a, b; DVǑRAK et al., 1965, 1967; ERICKSON, 1949; JELLISON, 1947, 1950, 1954, 1956, 1958; JELLISON and LORD, 1964; JELLISON and PETERSON, 1964; MCDIARMID and AUSTWICK, 1954; PROKOPIČ et al., 1965; SMITH and LANCASTER, 1965; TAYLOR et al., 1967; TEVIS, 1956). The fungus has been isolated not only from animals, but also from soil and air (BATISTA and SHOME, 1963; MENGES and HABERMANN, 1954; THIRUMALACHAR et al., 1965).

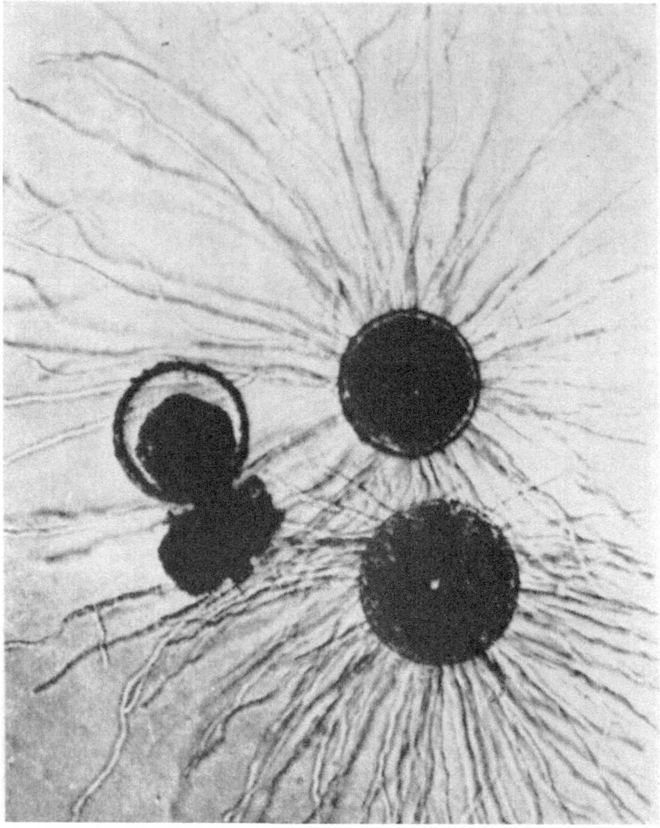

Fig. 3. Multiple germ tubes have arisen from adiaspores of *Emmonsia crescens* that were incubated for three weeks at 37°C on blood agar and then removed to room temperature. × 337. (After EMMONS, C.W., and JELLISON, W.L.: Ann. N.Y. Acad. Sci. 89, 91—101, 1960)

Reports of the organism in man have appeared in Brazil, France, and India (BATISTA et al., 1963a, b; CHEVREL et al., 1964; DOBY-DUBOIS et al., 1964; MISRA et al., 1966). Though the fungus has been associated with pulmonary disease in man, further documentation is necessary in order to evaluate its role in the genesis of the lesions.

Following inhalation, the individual fungal spores increase in size from the original diameter of 2—4.5 μ to enormous thick-walled cells (the adiaspores) that are many times their initial dimensions. These may become 500 μ in diameter, with cell walls reaching a thickness of 70 μ in *E. crescens*. The walls stain well with the periodic acid Schiff technique. The organism appears to be incapable of reproduction in the host and apparently dies without budding or undergoing endosporulation. The giant fungal spores are surrounded by a minimal, but definite, cellular reaction (Fig. 4).

Allescheriosis

(Monosporiosis)

The fungus *Allescheria boydii* belongs to the class Ascomycetes. In the perfect or ascocarpic stage the organism forms brown, thin-walled, spherical bodies (cleistothecia) which contain transparent, ovoid asci with eight yellow-brown

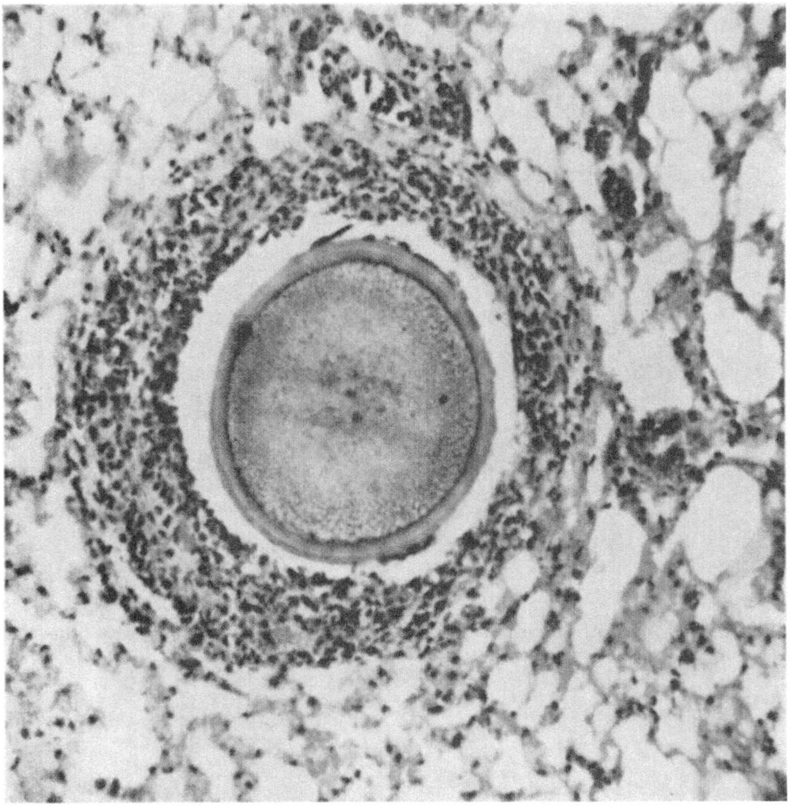

Fig. 4. Adiaspore of *Emmonsia parva* in the lung of a wild rodent (*Sorex cinereus*) × 300. (After BAKERSPIGEL, A.: Canad. J. Microbiol. 7, 676, 1961)

ascospores (Fig. 5). The binomium *Monosporium apiospermum* refers to the conidial or imperfect form of the fungus. This form possesses conidiophores with single, terminal conidia (Fig. 6), and lacks cleistothecia, characterizing the sexual stage. The fungus has a worldwide distribution, and is a common soil saprophyte (AJELLO, 1952). It grows rapidly on artificial media at both room temperature and 37° C and forms a white cottony colony which later turns dark smoky gray (Fig. 7). The reverse side of the colony is usually black. The majority of isolates do not produce ascocarps in culture but form the imperfect stage (EMMONS, 1944).

Allescheria boydii was first isolated in 1921 by BOYD and CRUTCHFIELD from a mycetoma in man. Since that time, the organism has been cultured from similar lesions on numerous occasions. In most instances the imperfect form of the same fungus has been described, perhaps because most isolates do not produce the perfect stage in culture. Most cases of mycetoma due to *A. boydii* have been reported from the United States, though the organism has been isolated from such lesions in many parts of the world. The fungus is the commonest isolate from mycetoma in the *United States* and continental *Europe:*

(ARVEIRA-NEVES, 1942; AVRAM, 1967; AVRAM et al., 1968; BATTISTINI et al., 1958; BOL'SHAKOVA et al., 1960; BORELLI, 1957; BOYD and CRUTCHFIELD, 1921; BRICENOMAAZ, 1967; CAMAIN et al., 1957; CARRION and KNOTT, 1944; CIFERRI and REDAELLI, 1950; COCKSHOTT, 1957; CONVIT et al., 1961; COURTOIS et al., 1954; DE ALMEIDA and SIMOES BARBOSA,

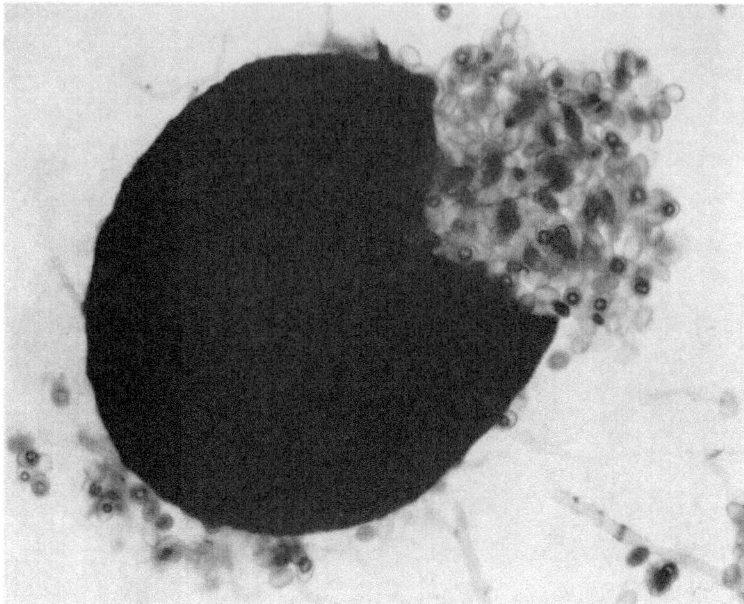

Fig. 5. A ruptured cleistothecium with ascospores (perfect stage of *Allescheria boydii*). Lacto-phenol cotton blue preparation. × 475

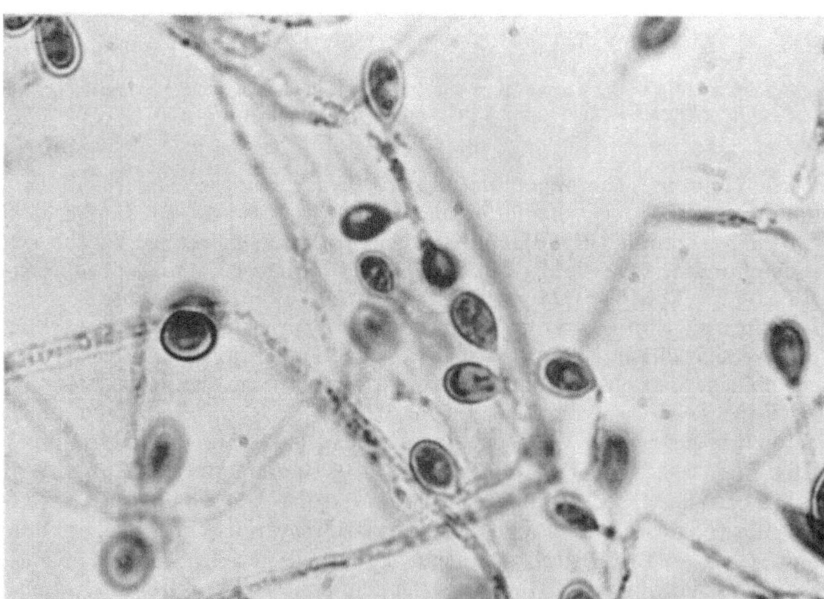

Fig. 6. Imperfect form of *Allescheria boydii* showing single, terminal conidia. Lacto-phenol cotton blue preparation. × 1000

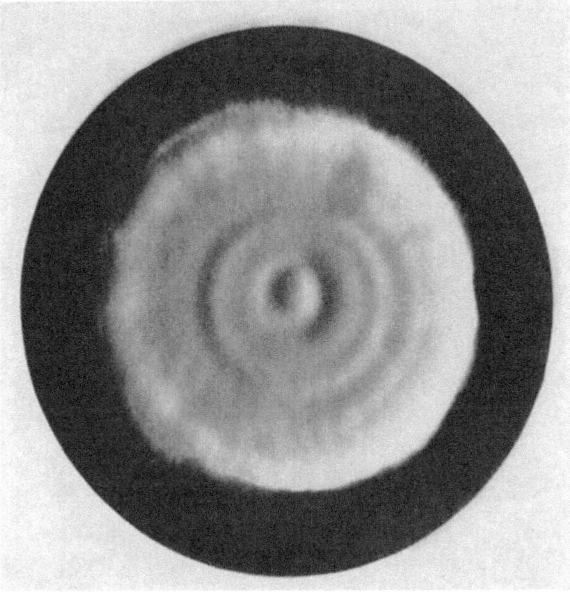

Fig. 7. Colony of *Allescheria boydii* on Sabouraud's glucose agar

1940; DESTOMBES et al., 1966; DOWDING, 1935; DROUHET, 1955; EL-MOFTY et al., 1965; EMMONS and GREENHALL, 1962; FIENBERG, 1944; GAY and BIGELOW, 1930; GELLMAN and GAMMEL, 1933; GREEN and ADAMS, 1964; HANLON et al., 1955; HAUKOHL and SADOFF, 1954; HERRERO et al., 1955; JONES and ALDEN, 1931; JORGE et al., 1941; JOSEFIAK and KOKITO, 1959; KATSNEL'SON et al., 1962; LATAPÍ, 1963; LAPATÍ and ORTIZ, 1963; MACKINNON, 1963; MAHGOUB, 1964; MARIAT, 1963; MEYERDING and EVERT, 1947; MOHR and MUCHMORE, 1958; MONTPELLIER and GUILLON, 1921; NIÑO, 1941, 1949, 1953; NIÑO et al., 1962; NIÑO and FREIRE, 1966; REES, 1962; SEABURY et al., 1959; SHAW and MACGREGOR, 1935; TAYLOR et al., 1964; THEODORESCU et al., 1959; TWINING et al., 1946; VANBREUSEGHEM, 1967; VERGHESE and KLOKKE, 1966).

As with mycetoma produced by other fungi (see Chapter 12, Mycetoma), granules consisting of matted mycotic elements are common in both the lesions and the purulent discharge. The granules are up to 1 cm in diameter and possess a white to yellow color. In tissue sections they are eosinophilic and readily demonstrated with Grocott's methenamine silver and the periodic acid Schiff stains. The granules are composed of a central mass of septate hyphae (4—5 μ) and chlamydospores (Figs. 8, 9). The cellular infiltrate consists predominantly of polymorphonuclear leukocytes, lymphocytes, and plasma cells. Granulomas and multinucleated giant cells are rarely observed. Mycetomata with granules have been produced in guinea pigs, hamsters, and mice following inoculation with *Allescheria boydii* (AVRAM, 1967; SCHMITT et al., 1963).

Several examples of pulmonary infection with *Allescheria* have been described (CREITZ and HARRIS, 1955; DROUHET, 1955; LOURIA et al., 1966; MISRA et al., 1966; SCHARYJ et al., 1960; STOECKEL et al., 1960; TRAVIS et al., 1961). The organism has not only been isolated from sputa and pulmonary lesions, but also has been identified in tissue from the lung (CREITZ and HARRIS, 1955; TRAVIS et al., 1961). In most instances there has been an associated disease in the respiratory system. The latter has included sarcoidosis, chronic bronchitis, emphysema, and a benign bronchogenic cyst. The pulmonary lesions resemble those of myce-

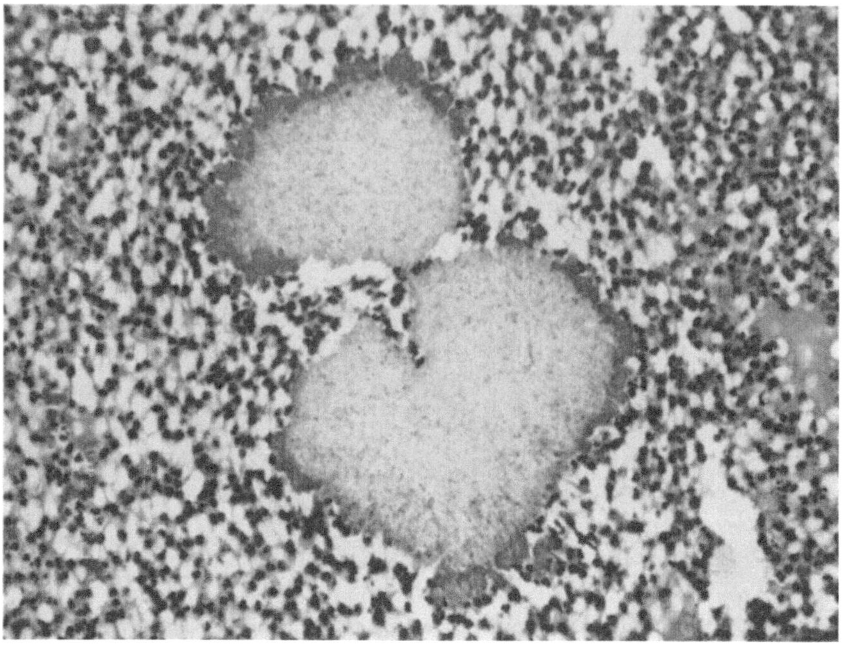

Fig. 8. Granule in subcutaneous tissue from a mycetoma due to *Allescheria boydii*. Hematoxylin and eosin stain. × 240

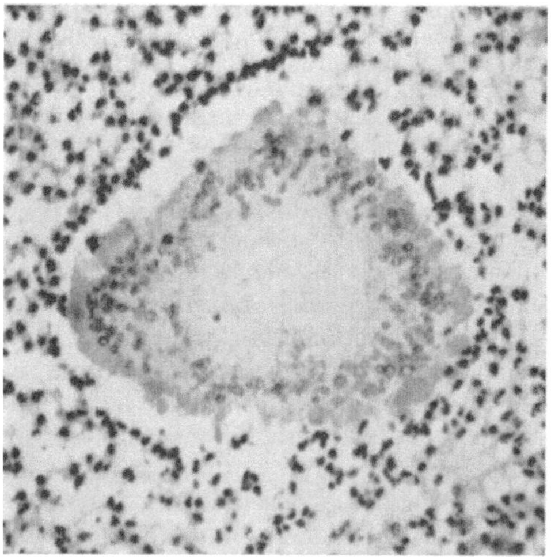

Fig. 9. Granule from a mycetoma due to *Allescheria boydii* showing fungal elements in its periphery. Modified periodic acid Schiff stain. × 240

toma in the extremities and are characterized by a purulent exudate and yellow to white granules.

Allescheria boydii has been isolated from corneal ulcers (ERNEST and RIPPON, 1966; GORDON et al., 1959; LEY, 1956; NAUMANN et al., 1967; PAULTER et al., 1955). In some such instances fungal elements have also been identified in corneal

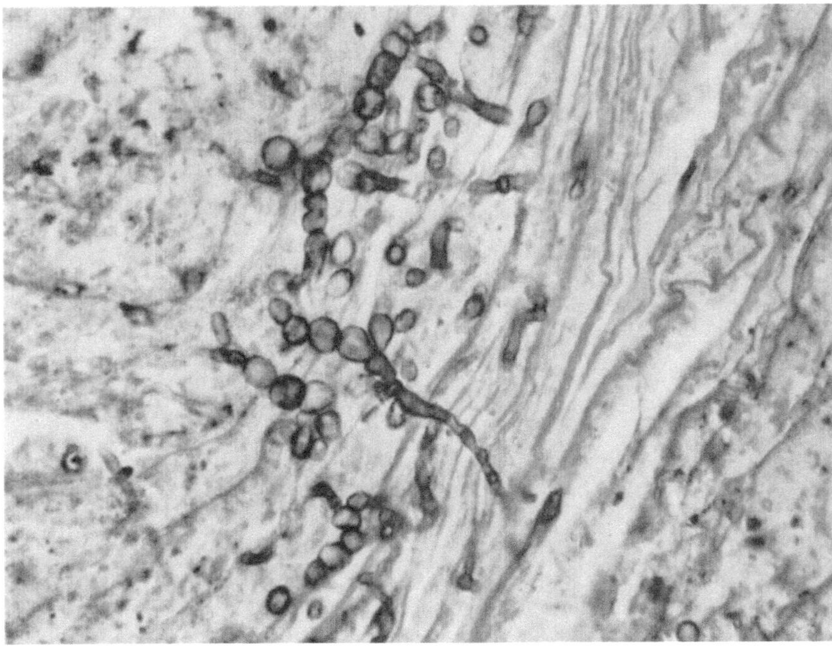

Fig. 10. Hyphal elements and chlamydospores of *Allescheria boydii* in tissue from the eye. Grocott's methenamine silver stain. × 550

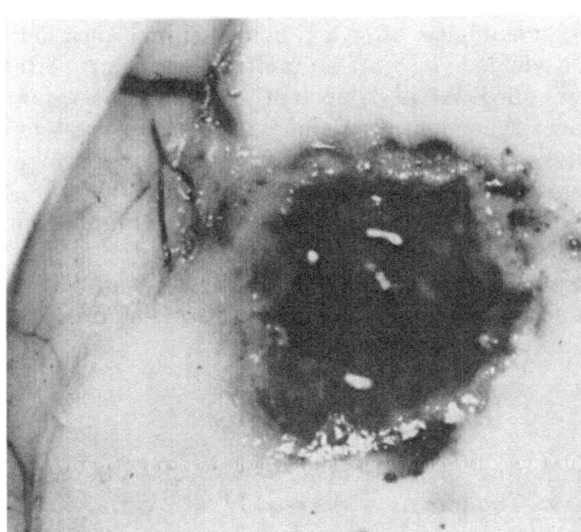

Fig. 11. A cerebral abscess due to *Allescheria boydii*. (After ROSEN, F., DECK, J. H. N., and REWCASTLE, N. B.: Canad. med. Ass. J. 93, 1125—1127, 1965)

scrapings or histologic sections (ERNEST and RIPPON, 1966) (Fig. 10). In most cases there has been a past history of corneal injury and antecedent therapy with steroids and/or antibiotics.

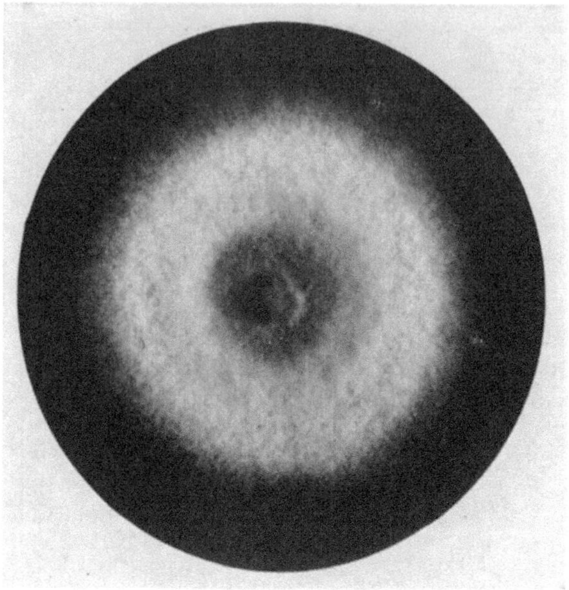

Fig. 12. Colony of *Alternaria sp.* on Sabouraud's glucose agar

Other sites have rarely been infected by *Allescheria boydii*. The organism accompanied a chronic meningitis in a middle-aged woman from Trinidad (ARONSON et al., 1953; BENHAM and GEORG, 1948; WOLF et al., 1947). The patient developed signs of meningitis after a previous spinal anesthetic. Cultures of the cerebrospinal fluid yielded *A.boydii* on multiple occasions. After a course of eight months the patient died. At post-mortem examination a granulomatous meningitis was found over the spinal cord and around the brainstem and cerebellum.

The fungus has rarely produced disseminated lesions in man. ROSEN et al. (1955) documented abscesses in the brain and thyroid due to *A.boydii* in a patient who had been treated with both steroids and immunosuppressive agents for subacute glomerulonephritis (Fig. 11).

Allescheria has occasionally been cultured from urine and prostatic secretions (MEYER and HERROLD, 1961), a crust in the external auditory canal (BELDING and UMANZIO, 1935), and blood (ZAFFIRO, 1938).

Alternariosis

One of the ubiquitous plant pathogens which has been associated with human disease is *Alternaria*. The fungus grows readily on a variety of media which it may contaminate. The round colonies may be white early in their growth, but later become dark olive-green or brown to gray-black. With age they are covered by a loose, light colored, aerial mycelium (Fig. 12). The dark hyphae are septate and the conidiophores contain simple or branched chains of inverted, clavate, olivaceous to brown macroconidia with an elongated, lighter tip and with longitudinal and/or transverse septa. These muriform conidia possess apices that are often attenuated in long simple chains (Fig. 13).

As early as 1930, HENRICI suggested that *Alternaria* may cause suppurative lesions, but the evidence was inconclusive. Since then, the fungus has been isolated

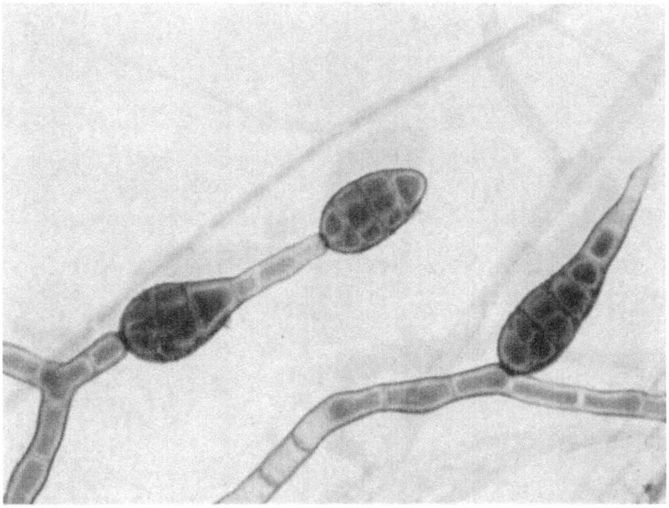

Fig. 13. *Alternaria tenuis* with attenuated macroconidia. Lacto-phenol cotton blue preparation. × 1000

from many different sources in man, including the bile, cerebrospinal fluid, conjunctival sac, and skin (AINSLEY and SMITH, 1965; BORSOOK, 1933; BOTTICHER, 1966; OHASHI, 1960). As *Alternaria* is a common contaminant of laboratory media, it has generally been discarded as insignificant when cultured from clinical material. However, it is questionable whether the organism is always non-pathogenic to man. For example, *A.tenuis* has been repeatedly cultured from a lesion in the hand produced by a splinter (BORSOOK, 1933). BOTTICHER (1966) also raised doubt about the non-pathogenic nature of *Alternaria* when she isolated *Alternaria sp.* on 121 occasions from 2,381 specimens of suspected superficial mycoses of the skin, hair, and nails. Broad pigmented septate hyphae and/or muriform spores resembling those of *Alternaria* were identified in most potassium hydroxide preparations from the lesions. In six skin scrapings of typical mycotic lesions, *Alternaria* was the only fungus isolated and identified in the tissue. At the Duke University Medical Center *Alternaria sp.* was cultured on four separate occasions from an ovoid 8.5 cm × 5 cm, denuded, cutaneous lesion on the lower calf of a woman with discoid lupus erythematosus and leukopenia (MILLER and TINDALL, 1967). This lesion, which was studded with black granules, followed a scratch with a tree limb. Rare fungal elements characterized by hyphal structures and oval bodies unlike those usually seen were identified in the patient's tissue. They were associated with an inflammatory reaction that was composed of both abscesses and discrete granulomas (Figs. 14, 15).

Besides invading tissue, *Alternaria* has been implicated as an allergen in some individuals (BOTTICHER, 1966; HARRIS, 1941; HOPKINS et al., 1930; HYDE and WILLIAMS, 1949; PRATT et al., 1954; PRINCE et al., 1949, 1961; WALTON, 1949).

Arthrodermosis

Several investigators have described a cutaneous disease due to *Arthroderma simii (Epidermophyte simii, Trichophyton simii)* in dogs, monkeys, poultry, and even man, who may become infected through contact with diseased animals

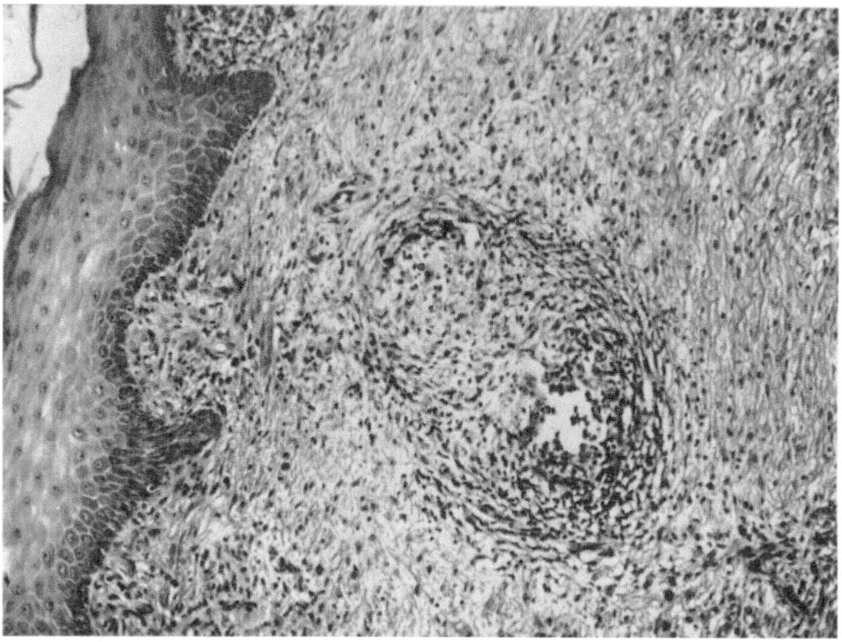

Fig. 15

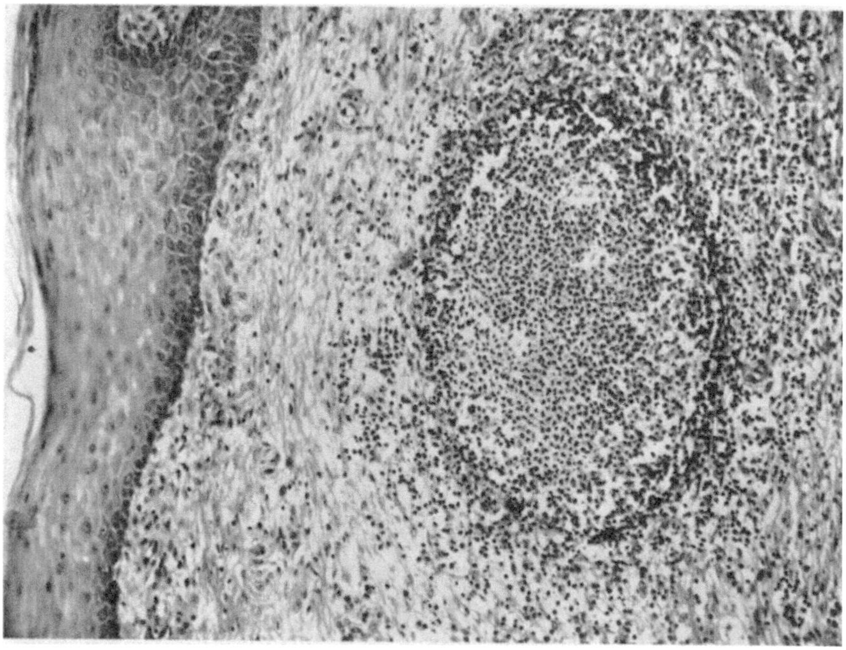

Fig. 14

Figs. 14, 15. Cutaneous lesions, which yielded *Alternaria* on culture, contained abscesses (Fig. 14) and granulomas (Fig. 15). Hematoxylin and eosin stain. × 120

(EMMONS, 1940; GUGNANI et al., 1967a; KLOKKE and DURAIRAJ, 1967; PINOY, 1912a, b; RIPPON et al., 1968; SINGH, 1963; STOCKDALE et al., 1965). This heterothallic fungus, called *Trichophyton simii* in its imperfect stage and *Arthroderma simii* in its perfect stage, appears to have a restricted geographic distribution. All infections, for which a locality is known, have originated from India, where the

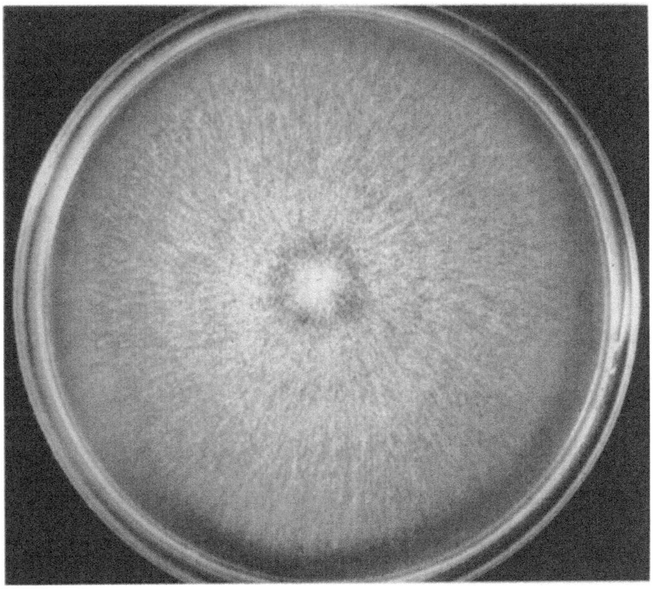

Fig. 16. *Arthroderma simii* on malt agar extract. (After STOCKDALE, P.M., MACKENZIE, D.W.R., and AUSTWICK, P.K.C.: Sabouraudia 4, 112—123, 1965)

fungus has been isolated from the soil and hair of several, healthy small mammals (GUGNANI et al., 1967b; RIPPON et al., 1968; STOCKDALE et al., 1965).

The morphology of the fungus and its colonial characteristics have been thoroughly described by STOCKDALE et al. (1965). The organism grows rapidly on numerous media and produces colonies that are white to very pale buff or rose-buff in color. They are uniformly and finely granular and possess a diffuse margin and central, white, conical, velvety umbo. The reverse side of the colony is non-pigmented at first, but later has a red-yellow discoloration (Fig. 16). Numerous

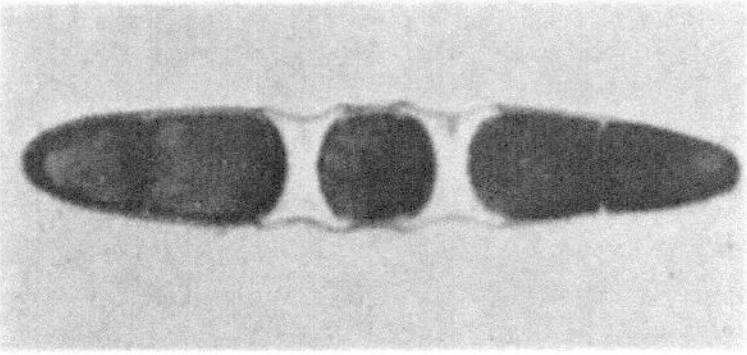

Fig. 17. Chlamydospores of *Arthroderma simii*. × 900. (After STOCKDALE, P.M., MACKENZIE, D.W.R., and AUSTWICK, P.K.C.: Sabouraudia 4, 112—123, 1965)

macroconidia, borne terminally on complexly branched hyphae, are hyaline and cylindrical to fusiform. They measure 6—11 $\mu \times$ 35—85 μ with up to 10 septa. Thick-walled spores formed by one or more of the cells may be contained by empty intercalary cells (Fig. 17). Clavate to pyriform (1.5—3 $\mu \times$ 2—6.5 μ) microconidia are rare in young cultures, but more abundant in old ones. They are borne along the sides of simple hyphae. Cleistothecia (closed sporebearing structures) characteristic of the genus *Arthroderma* have been produced on soil and other media (STOCKDALE et al., 1965).

In the monkey the lesions consist of small, superficial, silvery desquamating scales with some inflammation of the underlying dermis (EMMONS, 1940; STOCKDALE et al., 1965) (Fig. 18). Hair loss is not a feature in the monkey, and though some hair follicles may be invaded, the hair shaft is not penetrated (EMMONS, 1940; PINOY, 1912a, b). Alopecia does, however, occur in the guinea pig and mouse in whom the disease can readily be produced experimentally. Infection can also be transmitted to chickens, monkeys, and rabbits, and the resulting lesions are indistinguishable from those occurring naturally (EMMONS, 1940; GUGNANI et al., 1967b; STOCKDALE et al., 1965).

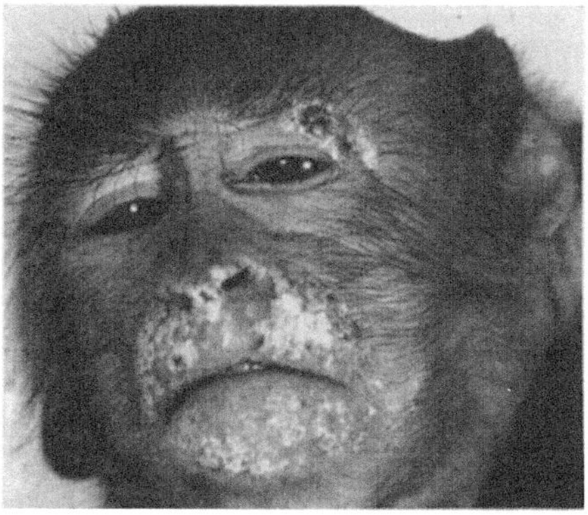

Fig. 18. Rhesus monkey with desquamating lesions on the face due to *Arthroderma simii* (After STOCKDALE, P.M., MACKENZIE, D.W.R., and AUSTWICK, P.K.C.: Sabouraudia 4, 112—123, 1965)

Fig. 19. Human skin lesions due to *Arthroderma simii*. (After KLOKKE, A.H., and DURAIRAJ, P.: Sabouraudia 5, 153—158, 1967)

The pathogenicity of the fungus for man was demonstrated experimentally by EMMONS in 1940. Later, the organism was isolated from cutaneous lesions in a laboratory technician, who handled infected guinea pigs, and in a poultry attendant (STOCKDALE et al., 1965). Since then, the fungus has also been isolated from numerous superficial cutaneous lesions in individuals from India (GUGNANI et al., 1967a; KLOKKE and DURAIRAJ, 1967; RIPPON et al., 1968) (Fig. 19).

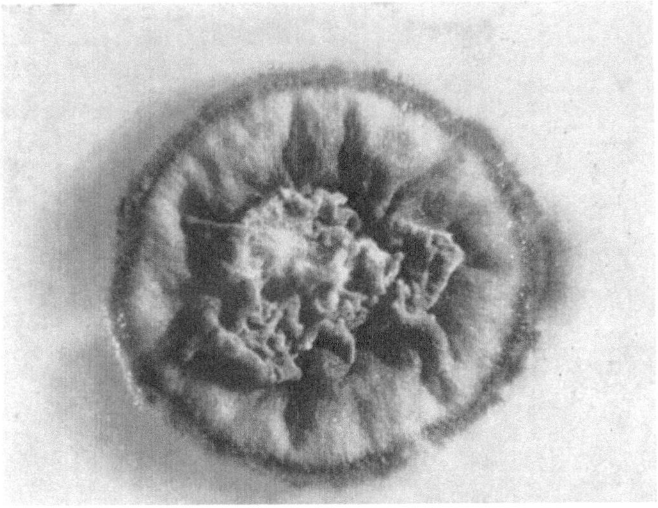

a

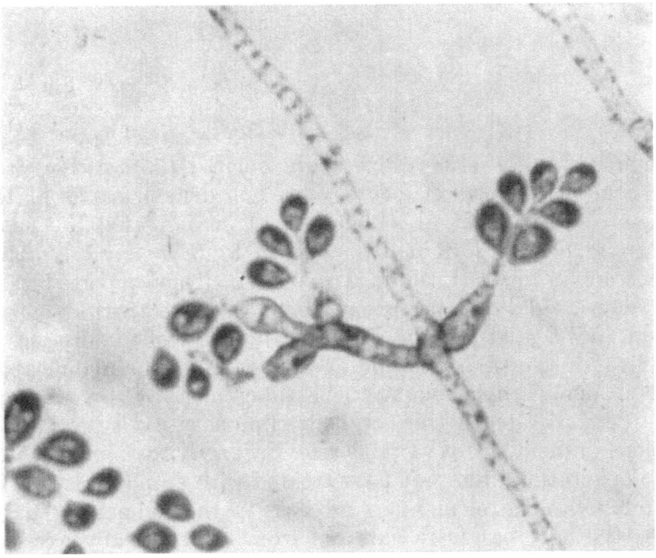

b

Fig. 20. a) Colony of *Beauveria tayeaui* on Sabouraud's agar (Courtesy of Dr. M. LAHOURCADE). b) and c) Filaments and spores of *Beauveria tayeaui*. (After FRÉOUR, P., LAHOURCADE, M., and CHOMY, P.: Presse med. 74, 2317—2320, 1966)

Beauveriosis

In 1835 a micro-organism first became recognized as a cause of disease. This occurred when the fungus *Beauveria bassiana* (formerly termed *Botrytis bassiana*) was found to produce muscardine, a contagious disease of silkworms (BASSI, 1835). Subsequently, this fungus was found to be widespread in nature and pathogenic to many insects (ELLINGBOE et al., 1957; MADELIN, 1963).

For more than a century *Beauveria sp.* was generally regarded as non-pathogenic for man and other vertebrates. However, pulmonary infection by *B. bassiana* has been documented in giant tortoises that were kept in captivity (GEORG et al.,

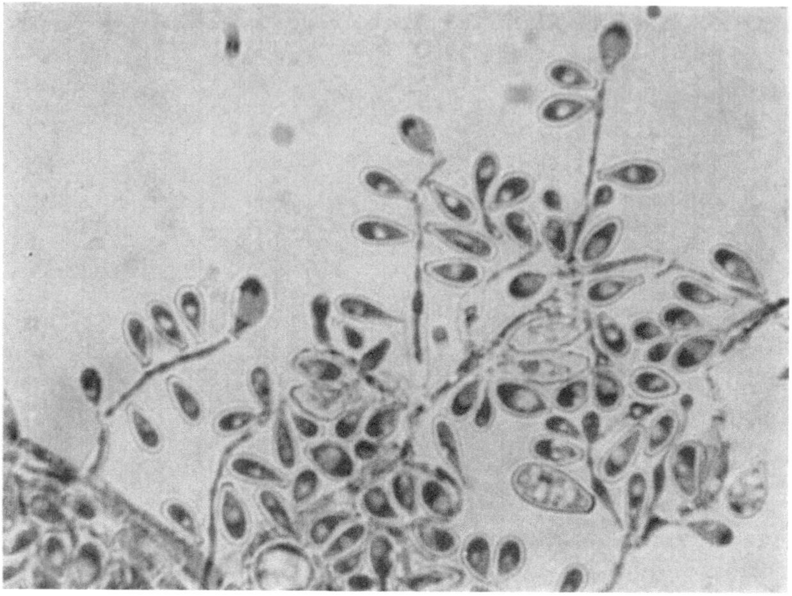

Fig. 20c

1962). Though LANGERON (1934) repeatedly isolated *Beauveria brumpti* from an infected human eye, fungal elements were not demonstrated in the affected tissues. Recently FRÉOUR et al. (1966a, b, c) incriminated the entomogenous fungus *Beauveria tayeaui* (originally described as *Beauveria bassiana*) in a pulmonary infection in man (LAHOURCADE, 1966). They reported a 22-year-old woman who developed a dry, persistent cough associated with debility, anorexia, loss of weight, and a mild pyrexia. Prior to the onset of the disease, the patient had always lived in North Africa and at the age of 11 she manifested ulcerative cervical lymphadenopathy thought to be tuberculosis. Subsequent radiological examination demonstrated a thick-walled pulmonary cavity and discrete disseminated lesions in the lungs of variable size. *Beauveria bassiana* was isolated from her bronchial aspirations. She was later treated with amphotericin B followed by pneumonectomy. Fungal elements were seen in the pulmonary tissue, and *B. bassiana* was the only organism cultured from the surgically excised specimen (Fig. 20). Following surgery, the patient's health was stable.

Cephalosporiosis

The cephalosporia are widely dispersed throughout the world, and have been isolated from soil, air, and sewage. Some species have been associated with disease in insects and a variety of plants, including trees, wheat, corn, and sugar cane (PISANO, 1963). In addition *Cephalosporium spp.* are also important to man, as a source of antibiotics.

Most species of *Cephalosporium* are readily grown on many artificial media, including Sabouraud's dextrose agar (Fig. 21). At the tips of slender or swollen conidiophores, the fungus produces conidia that are usually hyaline and non-septate. The spores vary in size with the species and usually measure up to 8 μ wide and 15 μ long (Fig. 22).

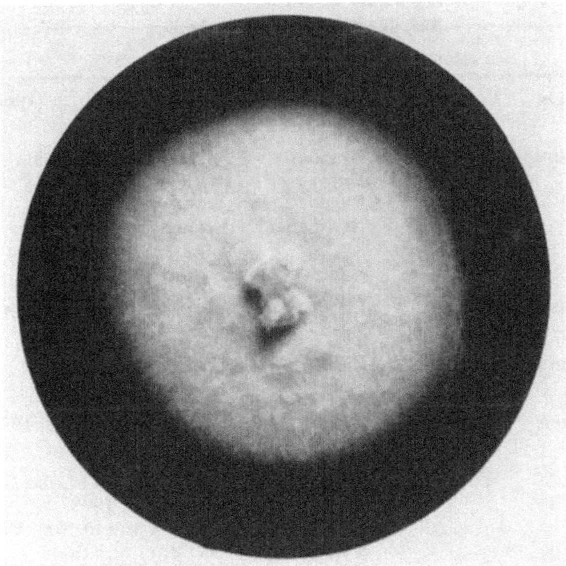

Fig. 21. *Cephalosporium sp.* grown on Sabouraud's glucose agar

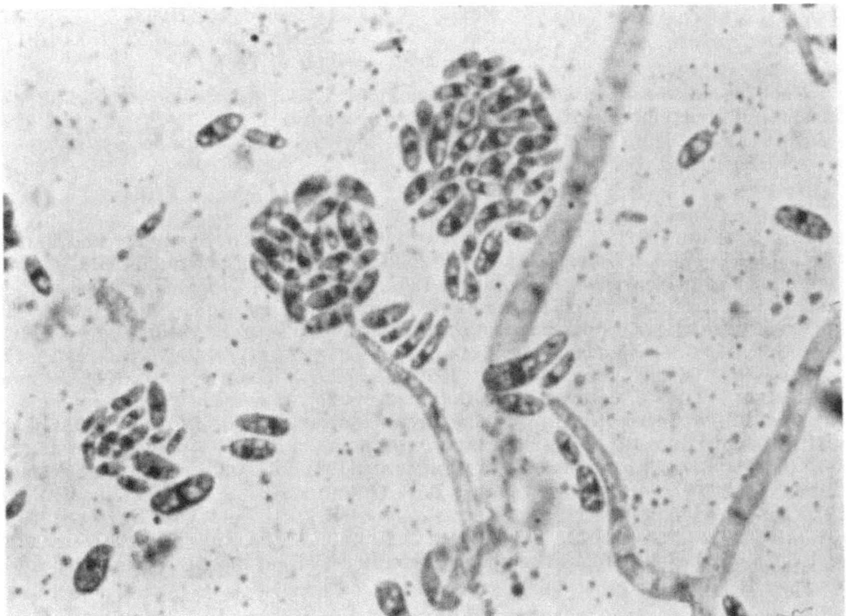

Fig. 22. *Cephalosporium sp.* showing masses of oval conidia at the ends of conidiophores. Lacto-phenol cotton blue preparation. × 1000

A wide variety of species have been implicated in human disease.

These include *C. acremonium* (ARIEVITCH et al., 1966; BATISTA et al., 1960; BOMMER et al., 1961; CABRINI and REDAELLI, 1929; CATANEI, 1944; COUTELEN and COCHET, 1945; COUTELEN et al., 1948; DEBUSMANN, 1939; GRÜTZ, 1925; MURRAY and HOLT, 1964; PITOTTI, 1932),

Table 2. *Mycetoma Produced by Cephalosporium sp.*

Species	Color of granules	Country	References
I. *Cephalosporium acremonium*	Colorless	West India	MURRAY and HOLT (1964)
II. *Cephalosporium falciforme*	White	Brazil	ALMEIDA et al. (1948)
		Malaya	PONNAMPALAM (1964)
		Puerto-Rico	CARRIÓN (1939, 1951)
		Roumania	AVRAM (1964a, 1966)
			AVRAM et al. (1968)
		Senegal	BAYLET et al. (1961)
III. *Cephalosporium*[1] *granulomatis*	Not mentioned	USA	WEIDMAN and KLIGMAN (1945)
IV. *Cephalosporium infestans*	Black	India	GAIND et al. (1962)
			PADHYE et al. (1963)
V. *Cephalosporium madurae*[2]	Black	India	PADHYE et al. (1962, 1963)
VI. *Cephalosporium recifei*	White	Brazil	ARÊA-LEÃO and LÔBO (1934a, b)
			LÔBO (1943, 1963)
			ALMEIDA and SIMÕES-BARBOSA (1940b)
		Senegal	BAYLET et al. (1961)
		Venezuela	BASTARDO (1963/1964)
VII. *Cephalosporium* sp.		Cameroon	GAMET et al. (1964)
		Chad	MARIAT (1963)
		Mexico	LATAPÍ and ORTIZ (1963)
			GONZALEZ-OCHOA and RUITOBA (1944)
		Venezuela	BORELLI (1962)

1 This species was studied by MACKINNON (1951, 1954) who considered it to be a contaminant and synonymous with *Cephalosporium acremonium*.

2 MURRAY and HOLT (1964) pointed out that the description of the species *C. madurae* is not precise enough to be substantiated.

C. ballagii (BALLAGI, 1932), *C. cinnabarinum* (ARIEVITCH et al., 1966; KUKOLEVA and MALKUHA 1967), *C. cordoniformis* (SIMÕES-BARBOSA, 1941), *C. doukourei* (BOUCHER, 1918), *C. falciforme* (ALMEIDA et al., 1948; AVRAM 1964, 1966a, 1967; AVRAM et al., 1968; BAYLET et al., 1961; BORELLI, 1962; CARRION, 1939; DESTOMBES and SEGRETAIN, 1962; MACKINNON, 1963; PANNAMPALAM, 1964), *C. granulomatosis* (WEIDMAN and KLIGMAN, 1945), *C. griseum* (GOUGEROT et al., 1933), *C. infestans* (GAIND et al., 1962; PADHYE et al., 1963), *C. keratoplasticum* (BATISTA et al., 1960; HARADA and USUI, 1960), *C. madurae* (PADHYE et al., 1962), *C. nigrum* (KAMBAYASHI, 1937), *C. niveolanosum* (BENEDEK, 1927b), *C. onychophilum* (BATISTA et al., 1960), *C. pseudofermentum* (CIFERRI, 1932), *C. recifei* (ALMEIDA and SIMÕES-BARBOSA, 1940a; ARÊA-LEÃO and LÔBO, 1934a, b; BASTARDO DE ALBORNOZ, 1963/1964; BAYLET et al., 1961; LÔBO, 1943, 1963), *C. roseo-griseum* (WARD et al., 1961; ZAIAS, 1966), *C. rubrobrunneum* (HARTMANN, 1926), *C. serrae* (BATISTA et al., 1960; GINGRICH, 1962; FOCOSI, 1932a, b; MAFFEI, 1929), *C. spinosus* (NEGRONI, 1933a, b), *C. stühmeri* (SCHMIDT and VAN BEYMA, 1933), *C. tomskianum* (MATRUCHOT, 1948). In most instances where *Cephalosporium* has been cultured from lesions in man, the fungus has not been identified within lesions, so the significance of the isolates is equivocal.

The association of *Cephalosporium* with mycetoma has been reported by numerous investigators including:

ALMEIDA and SIMÕES-BARBOSA (1940a, b), ALMEIDA et al., (1948) ARÊA-LEÃO and LÔBO (1934a, b), AVRAM (1966a), BASTARDO DE ALBORNOZ (1963/1964), BAYLET et al. (1961), BORELLI (1962), CARRIÓN (1939, 1951), COUTELEN et al. (1948), DESTOMBES and SEGRETAIN (1962), GAIND et al. (1962), GAMET et al. (1964), GONZALEZ-OCHOA and RUILOBA (1944), LACAZ and FAVA NETTO (1954), LATAPÍ and ORTIZ (1963), LÔBO (1943, 1963), MACKINNON (1951, 1954, 1963), MARIAT (1963), MURRAY and HOLT (1964), NAKAMURA (1933), PONNAMPALAM (1964), SEGRETAIN (1962), VERGHESE and KLOKKE (1966), and WEIDMAN and KLIGMAN (1945).

Six species have been isolated from mycetomas (*C. acremonium* CORDA, *C. falciforme* CARRIÓN, *C. granulomatis* WEIDMAN and KLIGMAN, *C. infestans* GAIND and THIRUM, *C. madurae* PADHYE, SUKA, and THIRUM, and *C. recifei* LEÃO and LÔBO)[1]. The color of the granules in the draining sinuses have varied with the isolated species (Table 2) (Fig. 23). AVRAM (1966b, 1967) has produced comparable granules in mice, hamsters, and guinea pigs following inoculation with *Cephalosporium falciforme*.

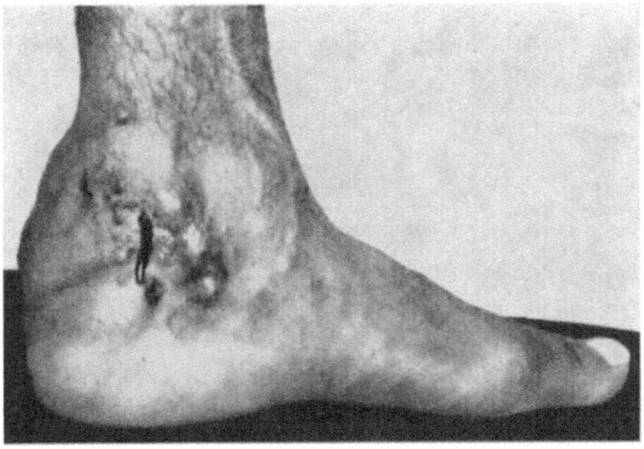

Fig. 23. Multiple draining sinuses due to *Cephalosporium madurae* in an ankle. (After PADHYE, A. A., and THIRUMALACHAR, M. J.: Sabouraudia 1, 230—233, 1961/1962)

In addition to mycetomas, species of *Cephalosporium* have been implicated as etiologic agents of gummato-ulcerative and superficial cutaneous lesions and onychomycosis.

(ASCHIERI, 1932; ARIEVITCH et al., 1964, 1966; BALLAGI, 1932; BENEDEK, 1927a, b; BOMMER et al., 1961; BOUCHER, 1918; CATANEI, 1944; COUTELEN and COCHET, 1945; COUTELEN et al., 1948; DUBOIS, 1942; GOUGEROT et al., 1933; GRÜTZ, 1925; HAENSCHE, 1957; HARADA and USUI, 1960; HARTMANN, 1926; JANKE, 1949; KAMBAYASHI, 1937; KESTEVEN, 1939; LEHNER, 1932; MALE and TAPPEINER, 1965; MILLER and MORROW, 1932; MOORE, 1955; NEGRONI, 1933a, b; PITOTTI, 1932; SCHMIDT and VAN BEYMA, 1933; SIMÕES-BARBOSA, 1941; WALSHE and ENGLISH, 1966; ZAIAS, 1966).

Occasionally, cases of cutaneous cephalosporiosis may spread to the regional lymphatics and produce a lymphangitis reminiscent of sporotrichosis (BALLAGI, 1932). In some cases, it has been suggested that the organism may become disseminated by auto-inoculation to other parts of the body. This may have occurred in a patient with cephalosporiosis at the knee in whom the infection spread to other parts of the skin (JANKE and ROHRSCHNEIDER, 1951). It has been suggested that *Cephalosporium* may occasionally spread to other parts of the body by the blood stream. Although this is rare, systemic cephalosporiosis has been reported (BOMMER et al., 1961; HAENSCH, 1957).

Besides cutaneous disease, *Cephalosporium* has been implicated as a pathogen of the mouth, tonsils, and pharynx (CABRINI and REDAELLI, 1929; CIFERRI, 1932; COUTELEN et al., 1948; COWEN, 1965; COWEN and DINES, 1964; COWEN et al., 1965; DEBUSMANN, 1939). A midline granuloma eroding the hard palate, and resulting in destruction of the maxilla and mandible was described by COWEN and

1 The isolate designated *C. granulomatosis* by WEIDMAN and KLIGMAN (1945) was studied by MACKINNON (1951, 1954) who regarded it as *C. acremonium*. MURRAY and HOLT (1964) have pointed out that the description of *C. madurae* is not precise enough to be substantiated.

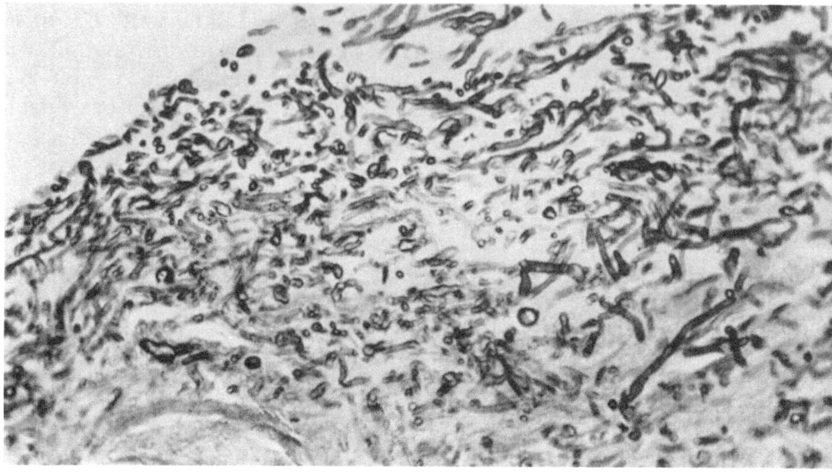

Fig. 24. *Cephalosporium sp.* in the corneal tissue. Periodic acid Schiff stain. × 305. (After ZIMMERMAN, L.E.: Survey Ophthal. 8, 1—25, 1963)

his colleagues (COWEN, 1965; COWEN and DINES, 1964; COWEN et al., 1965). Fungal hyphae and spores were identified in the latter lesion and *Cephalosporium* was cultured from it. *Cephalosporium* has also been associated with cystitis (MÜHLENS, 1938).

Ocular infection with *Cephalosporium* has been reported by several investigators. In most instances, the fungus was cultured from corneal ulcers (BEDELL, 1946, 1947; BYERS et al., 1960; FOCOSI, 1932b; GINGRICH, 1962; HOFFMANN and NAUMANN, 1963; KÜPER, 1962; MAFFEI, 1928; NAUMANN et al., 1967; ZIMMERMAN, 1962, 1963). In some of these cases, the organism was not only isolated from the corneal lesion, but demonstrated within tissue removed at keratoplasty (ZIMMERMAN, 1962) (Fig. 24). *Cephalosporium serrae* has most often been implicated in corneal lesions (GINGRICH, 1962; FOCOSI, 1932b; MAFFEI, 1928). The fungus is reported to have a potent proteolytic enzyme, which produces corneal ulceration and opacification in rabbits in 2—4 h (BURDA and FISHER, 1960). In addition to the keratitis, endophthalmitis following cataract extractions and other ocular surgery has been attributed to *Cephalosporium* (FINE, 1962; FINE and ZIMMERMAN, 1959; GREEN et al., 1965; KÜPER, 1962; THEODORE et al., 1961). The fungus has also been implicated in chronic dacryocystitis (FOCOSI, 1932a) and dacryocanaliculitis (JANKE and ROHRSCHNEIDER, 1951).

A chronic meningitis due to a *Cephalosporium* was reported by DROUCHET et al. (1965). While living in central Africa, a 33-year-old woman underwent a caesarean section under a spinal anesthetic. A left sciatica subsequently ensued and was treated with corticosteroids. A second pregnancy followed soon thereafter, during and after which the patient developed neurologic and psychiatric disturbances. At a surgical exploration a granulomatous meningitis was found, and hyphae were identified in the surgically excised meninges. *Cephalosporium sp.* was cultured from the ventricular and spinal fluids. The patient, who failed to respond to therapy (including actidione and amphotericin B), died approximately 15 months after the onset of neurological symptoms and about 4 months after surgical intervention. At autopsy examination hyphae were visualized in the meninges, but not in other parts of the body. *Cephalosporium sp.* was isolated from the brain and antibodies to the organism were detected in the blood after death.

Arthritis involving a major joint has been attributed to *Cephalosporium roseo-griseum* Saksena. Ward et al. (1961) isolated this species from the synovium and fluid from an arthritic knee following two traumatic injuries and treatment with corticosteroids.

Cephalosporium has been cultured from the bile, blood, gastric juice, pleural fluid, and sputum of human patients in the absence of tissue diagnoses (Batista et al., 1960; Debusmann, 1939; Douglass and Simpson, 1943; Matruchot, 1948; Mühlens, 1938; Verkholomov, 1961).

In most instances of cephalosporiosis the organism appears to have been inoculated into the tissues. This has been described, for example, following an injury with palm leaves (Ward et al., 1961), a cow's tail (Bedell, 1946, 1947), and a cotton stalk (Zimmerman, 1963). Infection has also followed ocular surgery (Fine, 1962; Fine and Zimmerman, 1959; Green et al., 1965; Theodore et al., 1961), a tooth extraction (Grütz, 1925), a compound fracture contaminated with soil (Arievitch et al., 1966), an incised wound (Gaind et al., 1962), and a spinal anesthetic (Drouchet et al., 1965).

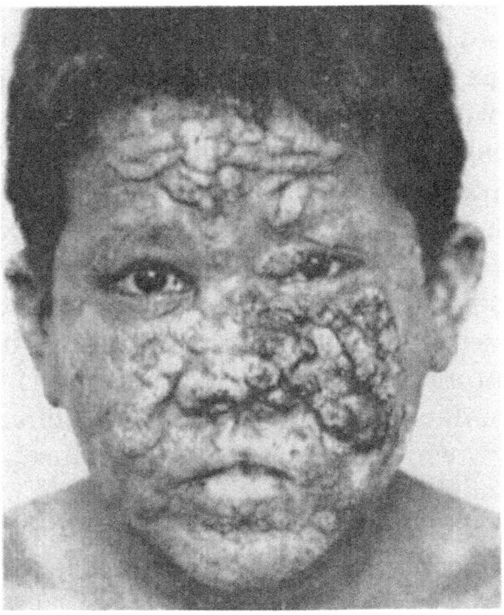

Fig. 25. Facial lesion which contained an unusual dematiaceous fungus. (After Emmons, C. W., Lie-Kian-Joe, Njo-Injo Tjoei Eng, Pohan, A., Kertopati, S., and van der Meulen, A.: Mycologia 49, 1—10, 1957)

Cercosporamycosis

The possibility that human disease may result from a fungus which causes leaf spots on plants was raised following the investigation of a patient in Indonesia (Emmons et al., 1957, 1963; Lie-Kian-Joe et al., 1957). The patient, a 12-year-old boy, suffered from extensive, indurated, verrucous and ulcerated, cutaneous and subcutaneous lesions of the face (Fig. 25). The lesions began many years before as a small nodule on the cheek, that could not be related to injury by thorns or insect bites. The nasal mucosa was also infected, and the nasal septum was perforated with a resultant saddle nose. The condition was not associated with any fever or regional lymphadenopathy. The ulcerated areas became crusted and the lesions subsequently extended to other areas, such as the thorax and thigh. Affected regions increased in extent and severity until the patient died, possibly from the mycosis. Unfortunately, an autopsy was not performed.

During life multiple biopsies of the lesions revealed hyperkeratosis, epithelial hyperplasia, keratin pearl formation and intraepithelial abscesses. Subcutaneous granulomas extended beyond the ulcerated regions. Brown septate hyphae (4—8 μ in diameter) were observed in unstained sections of all surgically excised lesions and were present throughout the skin and within subcutaneous granulomas. A fungus which did not resemble any known human pathogen was recovered repeatedly from multiple biopsy specimens, nasal secretions, and crusts from nasal mucosal ulcers. The cultured organism grew slowly at room temperature and at 30°C. An olive-gray colony (15 mm in diameter) possessed a luxuriant growth, dome-shaped tuft, and very short aerial hyphae after only 10-days growth on corn meal agar. On Sabouraud's medium the dark colony had a higher central dome. The fungus did not grow at 37°C. Sporulation occurred on corn meal agar and Sabouraud's agar. The conidiophores and conidia were brown and varied markedly in size and shape. Many conidiophores, most of which had poorly-differentiated hyphal elements, had barely discernible lateral and terminal scars from which conidia were readily detached. The conidia were clavate or acicular and possessed a truncated (usually convex) base and a rounded or greatly enlarged whip-like tip. Some immature conidia were unicellular, whereas the mature conidia (4—6 μ by 26—120 μ) had three to several septa. No muriform spores occurred. The fungus was identified as *Cercospora apii* Fresenius, a phytopathogen found throughout the world where celery is grown (LIE-KIAN-JOE et al., 1957; EMMONS et al., 1957, 1963). The isolate produced leaf spots on lettuce, tomatoes, and potatoes following its inoculation onto these plants.

The experimental production of lesions in animals using spore suspensions of this fungus was attempted, but without success. However, the fungus could be recovered in culture from the peritoneum of mice inoculated some weeks previously. Several attempts at self-inoculation by intradermal injection of conidia were made by some investigators (LIE-KIAN-JOE et al., 1957). Most of the intradermal injections produced a small nodule which resolved within a few days. One injection (using the inoculum from which leaf spots were produced in lettuce) was followed by the appearance of a slightly raised, brown papule (1.5 mm in diameter) which persisted for about 4 months. Mild pressure and friction produced an area of edema extending about 2 mm beyond the papule. The lesion slowly regressed and disappeared without any sequelae.

CHUPP (1957), who has extensive experience with the genus *Cercospora*, studied the isolate. Although Chupp could not specifically identify the fungus, he disagreed with the identification of the causal agent as *Cercospora apii* for the following reasons: (1) the colonies and the shape of the conidia did not resemble any of the species of *Cercospora* which he had seen; (2) the conidia of the fungus were pale brown whereas *Cercospora apii* has colorless conidia; (3) the bases of the conidia described by EMMONS et al. (1957) were convex whereas in *C. apii* they are clearly truncate; (4) EMMONS et al. (1957) did not report the inoculation of celery, which is maintained to be the only plant infected by *C. apii*; and (5) *C. apii* is worldwide in distribution and yet its pathogenicity for man has never been incriminated before.

Cryptostromosis

(Coniosporiosis or Maple-Bark Disease)

In 1932 TOWEY et al. described an acute respiratory disease in 35 bark peelers in Northern Michigan, USA. The condition was characterized by cough, dyspnea, fever, night sweats, and substernal pain. The disease was considered to result from the inhalation of spores of the fungus *Coniosporium corticale* (currently termed *Cryptostroma corticale*), which could be identified in the dust on maple logs from which the bark was peeled. Affected individuals had positive skin tests to extracts of the spore, whereas unaffected individuals gave negative results. They recovered following their removal from the contaminated environment.

The fungus, *Cryptostroma corticale (Coniosporium corticale)*, which may be pathogenic to trees (GREGORY and WALLER, 1951), grows on logs, especially hard

Fig. 26. Log from a maple tree with a pigmented area containing numerous spores of *Cryptostroma corticale*. (After WENZEL, F. J., and EMANUEL, D.A.: Arch. environm. Hlth 14, 385—389, 1967)

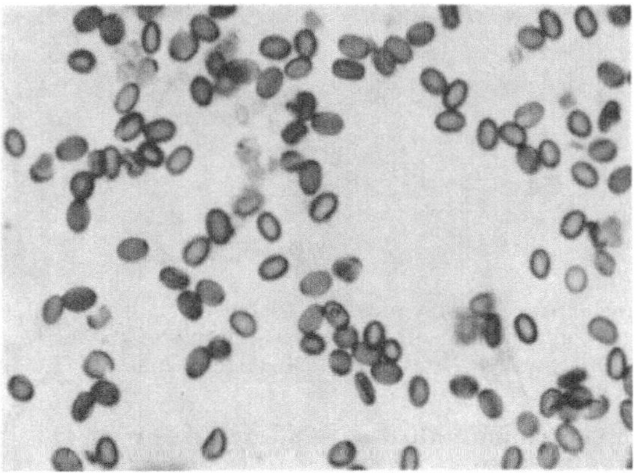

Fig. 27. Spores of *Cryptostroma corticale*. × 1225. (Courtesy of DR. D.A. EMANUEL)

wood such as maple, elm, ash, oak, and hickory. Maple logs appear to be more frequently involved than the other wood (Fig. 26). Although the wood and bark of infected trees appears healthy, a fine black dust is evident after the outer bark is removed.

Cryptostroma corticale grows on Sabouraud's medium at room temperature, but not at 37 °C. Within three days, a white pleomorphic mycelium which is hyaline, septate, and profusely branched develops. The spores are ovoid and measure 4—5 μ in their greatest diameter (Fig. 27). In large masses, these spores

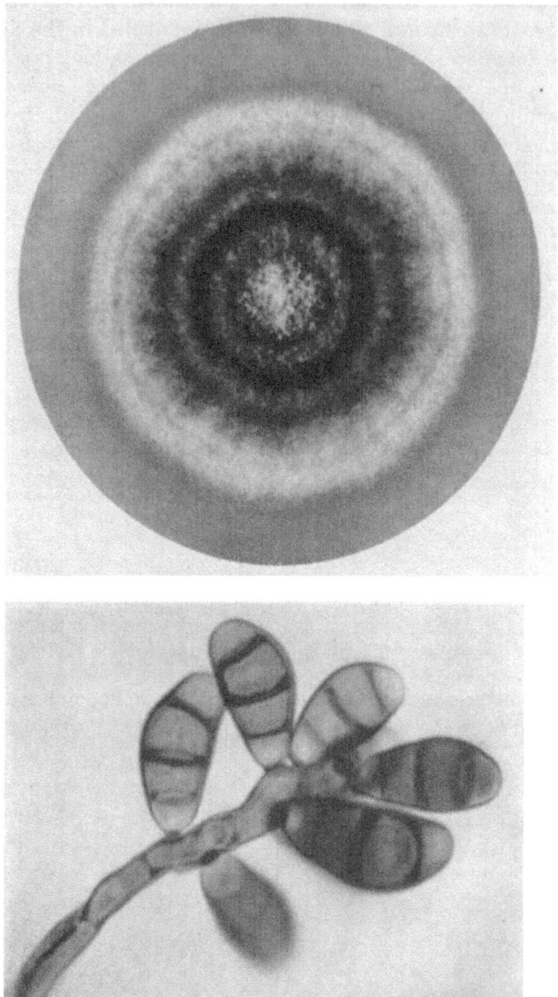

a

b

Fig. 28. (a) Colony of *Curvularia geniculata.* (After GEORG, L.K.: J. Med. Ass. Ala. 33, 234—236, 1964). (b) Conidiophore with conidia of *Curvularia geniculata.* (After NITYANANDA, K., SIVASUBRAMANIAM, P., and AJELLO, L.: Arch. Ophthal. 71, 456—458, 1964)

are black, although each individual one appears red-brown (EMANUEL et al., 1962, 1966; WENZEL and EMANUEL, 1967).

Thirty years after the fungus was first incriminated in human disease, other cases were recognized in a paper mill in Wisconsin, USA (EMANUEL et al., 1962, 1966). Surgically excised lung tissue in these cases, which contained small oval bodies, revealed prominent pulmonary changes with interstitial fibrosis, inflammatory exudates, and occasional granulomas with multinucleated giant cells. *C. corticale* was isolated from this tissue. Those patients with overt respiratory disease had positive skin reactions and circulating antibodies to an extract from the spores of *C. corticale*. In addition, some of the other workers in the same paper mill manifested radiological alterations, positive serological tests or positive skin tests (WENZEL and EMANUEL, 1967).

In unstained tissue sections the ovoid, thick-walled spores (3—5 μ in diameter) appear red-brown and are neither associated with hyphal elements nor evidence of replication (EMANUEL et al., 1966). The organism resembles *Histoplasma capsulatum* not only in its size and shape but also in the host's reaction to the fungus (EMANUEL et al., 1966).

It has been suggested that this disease is a hypersensitivity reaction to the inhaled spores of *Cryptostroma corticale* (EMANUEL et al., 1962, 1966; TOWEY et al., 1932).

Curvulariosis

Curvularia is a common fungal saprophyte of soil, and some species are phytopathogens of grasses (SPRAGUE, 1950). Two species have been implicated in human disease, namely *C. geniculata* and *C. lunata*.

The colony of *Curvularia* is dark brown to black and has a fluffy surface with many aerial hyphae and a reverse black pigmentation (Fig. 28a). The hyphae, conidiophores, and conidia have a natural light brown discoloration. *Curvularia* closely resembles *Helminthosporium*, but differs in possessing curved conidia with a prominent distal medial cell. In *C. geniculata* the conidiophore contains several cells with a diameter of 4—5 μ (Fig. 28b). The flat scars, which mark the sites of attachment of the spores to the conidiophore, project from the latter

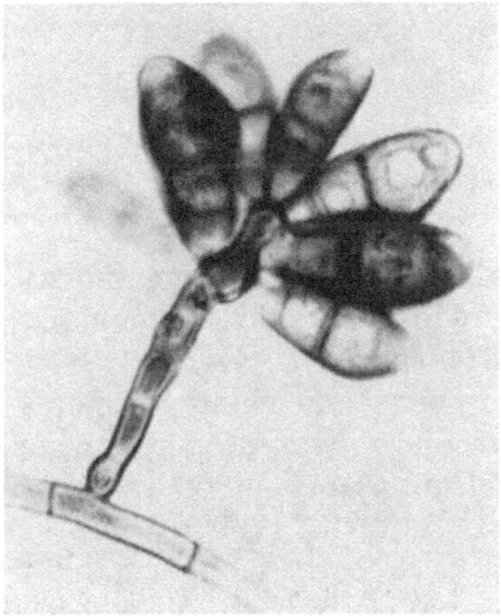

Fig. 29

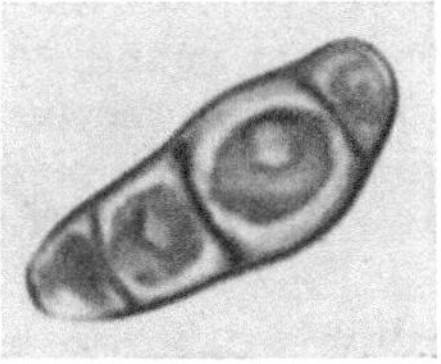

Fig. 30

Fig. 29. Conidiophore with conida of *Curvularia lunata.* $\times$ 600. (After NITYANANDA, K., SIVASUBRAMANIAM, P., and AJELLO, L.: Sabouraudia 2, 35—39, 1962)

Fig. 30. Curved septate conidium of *Curvularia lunata.* $\times$ 1150. (After NITYANANDA, K., SIVASUBRAMANIAM, P., and AJELLO, L.: Sabouraudia 2, 35—39, 1962)

structure and give it the appearance of a knobby stick. The septate conidia vary in shape and size (21—50 μ $\times$ 10—15 μ in diameter), possess thick walls, and usually consist of 3—5 cells of which the distal medial cell is enlarged. The conidial attachment protrudes from the basal cell when the spore is released from the conidiophore (GEORG, 1964). *C. lunata* is microscopically similar to *C. geniculata*, except the point of conidial attachment is inserted into the conidiophore and does not protrude from it. Its conidia measure 19—25 μ by 8—14 μ (NITYANANDA et al., 1962; PARMELEE, 1956) (Figs. 29, 30).

Curvularia has been isolated from mycetomas of man and other animals. BAYLET et al. (1959) described a Senegalese patient with a mycetoma that dischar-

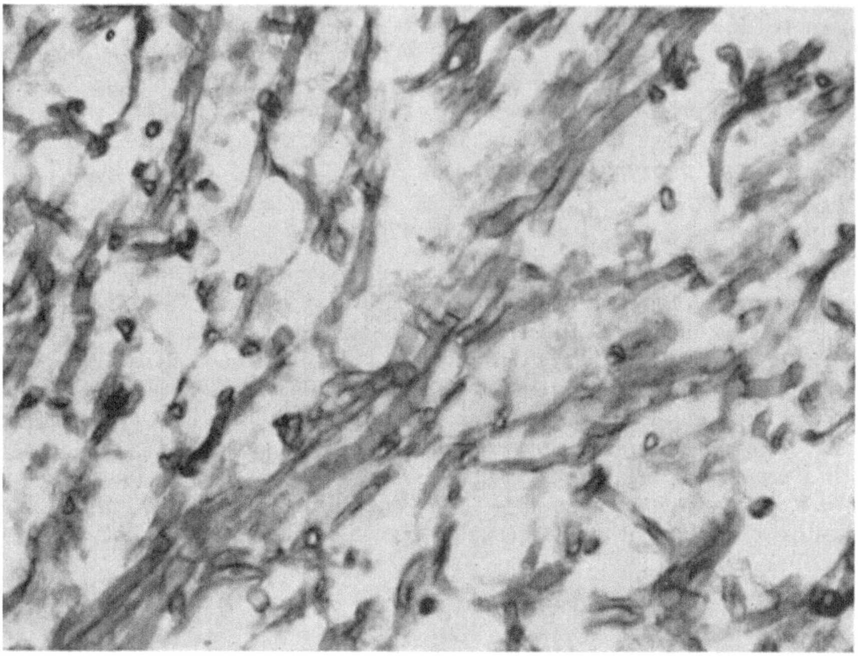

Fig. 31. *Curvularia geniculata* in corneal scrapings. Gridley stain. × 400. (After NITYANANDA, K., SIVASUBRAMANIAM, P., and AJELLO, L.: Arch. Ophthal. 71, 456—458, 1964)

ged black grains in the purulent exudates from its draining sinuses. The grains contained pigmented hyphae and chlamydospores and cultures of them yielded pure colonies of *C. lunata*. BRIDGES (1957) isolated *C. geniculata* from mycetomas of a dog's feet that contained brown hyphae and chlamydospores in granules within the tissue. Dematiaceous hyphae and chlamydospores, which may have been *Curvularia*, have also been observed in nasal granulomata and mycetoma in cattle, but were not culturally identified (BRIDGES, 1960; ROBERTS et al., 1963).

Curvularia geniculata and *C. lunata* have both been isolated from corneal ulcers that contained brown, branched, septate hyphae (3—5 μ in diameter) in scrapings and smears (NITYANANDA et al., 1962, 1964; WARREN, 1964) (Fig. 31). *C. lunata* was also repeatedly isolated, but not identified in scrapings, from another patient with a corneal ulcer (ANDERSON et al., 1959).

Dermatophilosis

(Cutaneous Streptothricosis)

In 1915 VAN SACEGHEM described a disease in cattle in the Republic of the Congo (formerly known as the Belgium Congo) due to an organism that he designated *Dermatophilus congolensis*. Since then, it has become apparent that the pathogen has a world-wide distribution, (DEAN et al., 1961; KAPLAN, 1966; SMITH et al., 1967; VANDERMAELE, 1961), and the resulting disease (dermatophilosis or cutaneous streptothricosis), is of considerable economic importance in countries where livestock raising is an important industry.

The organism, a facultative anaerobe, is an actinomycete and requires a medium which is rich in nutrients, such as brain-heart infusion agar or blood agar

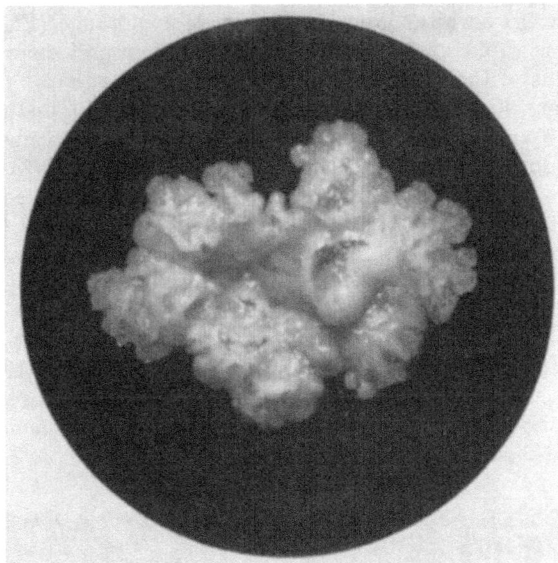

Fig. 32. Colony of *Dermatophilus congolensis* on Sabouraud's glucose agar

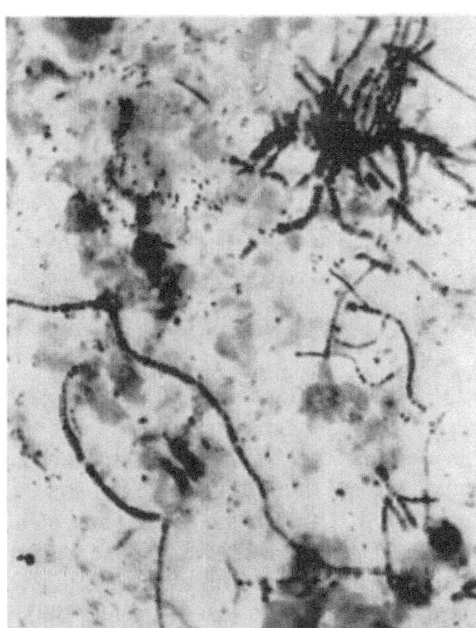

Fig. 33. Beaded filaments of *Dermatophilus congolensis*. Giemsa stain. × 1250. (After KAPLAN, W.: Southwest. Vet. 20, 14—19, 1966)

for growth. *D. congolensis* does not grow on common mycological media, such as Sabouraud's dextrose agar, and numerous antibiotics inhibit its growth (KAPLAN, 1966). On solid media, growth is apparent within 24—48 h, depending upon the temperature of incubation. The organism grows more rapidly at 37 °C than at room temperature. Colonies vary in appearance depending upon numerous environmental factors. They may possess rough or smooth, shiny or dull surfaces, and are initially grey-white, later becoming yellow-orange (Fig. 32). In liquid media, growth is rapid at 37° C and small, fuzzy, round colonies are evident within the depths of the medium. Cultures that are less than 2 days old possess branched, septate hyphae which measure 0.5—1.5 μ in diameter, and later increase to 3—5 μ in width. These filaments divide transversely into narrow segments which form clusters of coccoid spores up to eight cells in width (Figs. 33, 34). Mature hyphae break up and release flagellated, mobile, coccoid zoospores (GORDON and EDWARDS, 1963; GORDON, 1964; KAPLAN, 1966) (Fig. 35). These zoospores do not replicate, but germinate to form mycelium that later produces more zoospores. Formerly, AUSTWICK (1958) suggested that three distinct species of *Dermatophilus* existed (*D. congolensis*, *D. dermatonomus*, and *D. pedis*), but after a morphological and biochemical investigation GORDON (1964) concluded that there was only one species, namely *D. congolensis*. ROBERTS (1965c) later confirmed this on serological evidence. Synonyms for *Dermatophilus congolensis* include: *Actinomyces dermatonomus*, *Dermatophilus dermatonomus*, *Dermatophilus pedis*, and *Nocardia dermatonomus* (HART, 1967).

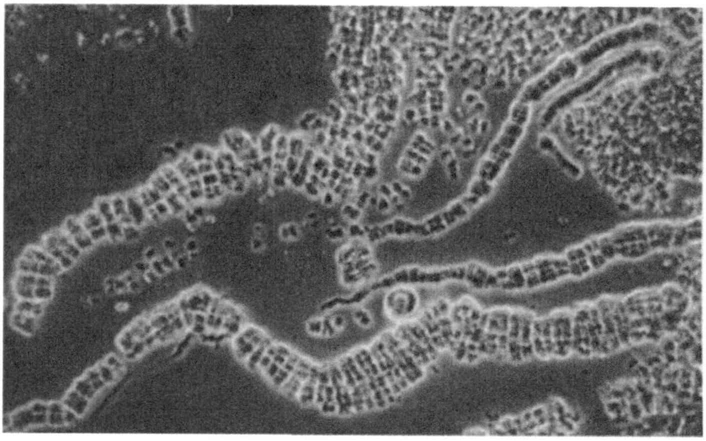

Fig. 34. Wet mount and dark phase contrast of *Dermatophilus congolensis* showing broad, beaded hyphae. × 500. (After GORDON, M.A.: J. Bact. 88, 509—522, 1964)

The cutaneous disease caused by *D. congolensis* has been reported in several animals, including cattle, deer, horses, goats (HUDSON, 1937; KAPLAN, 1966; KAPLAN and JOHNSTON, 1966; LE RICHE, 1968; PIER et al., 1963; SEARCY and HULLAND, 1968a, b; VANDEMAELE, 1961; VAN SACEGHEM, 1915). This actinomycete causes superficial cutaneous lesions with cellular infiltration and serous exudate after invasion of the epidermis (DEAN et al., 1961, ROBERTS, 1965a—c; ROBERTS, 1967). The organism readily infects hair follicles, but does not invade hair or wool.

A number of possible modes of transmission of dermatophilosis have been suspected. Trauma and ticks have been implicated in the spread of this disease in livestock (LE RICHE, 1968; MACADAM, 1964). In sheep, the transmission of the actinomycete occurs especially by contact between wet animals (LE RICHE, 1968; ROBERTS, 1963; ZLOTNIK, 1955). *Dermatophilus congolensis* has been transmitted from infected to normal rabbits by both stable flies (*Stomoxys calcitrans*) and by house flies (*Musca domestica*). No disruption of the skin was necessary to infect

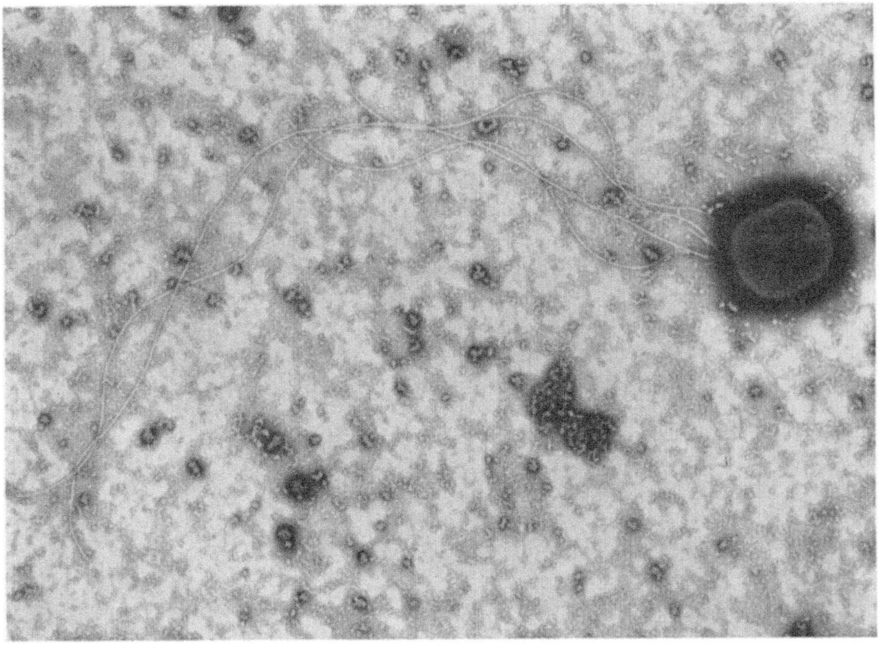

Fig. 35. Electron micrograph of an encapsulated, flagellated zoospore of *Dermatophilus congolensis*. × 30,000. (After GORDON, M.A.: J. Bact. 88, 509—522, 1964)

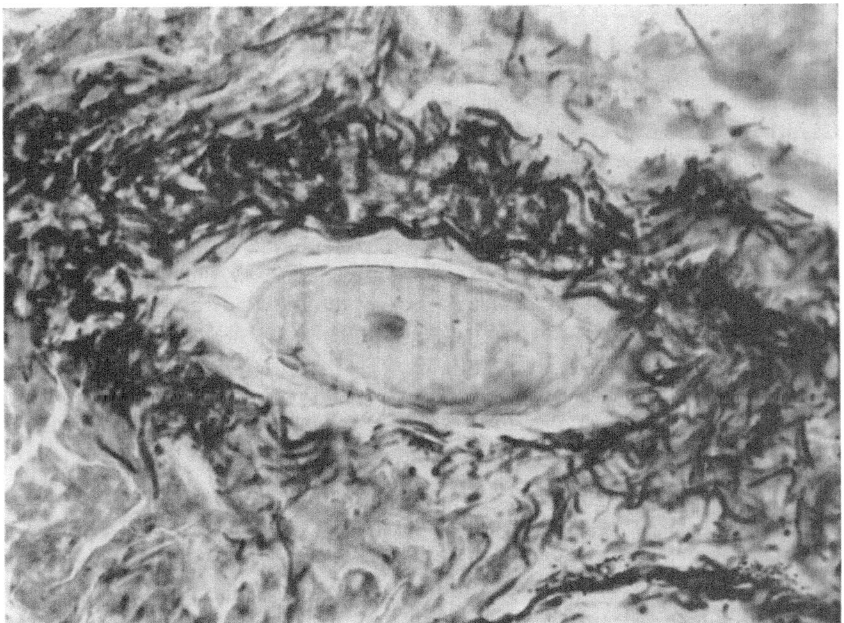

Fig. 36. *Dermatophilus congolensis* in skin showing numerous black filamentous organisms around the hair (center) but not involving it. Grocott's methenamine silver stain. × 220

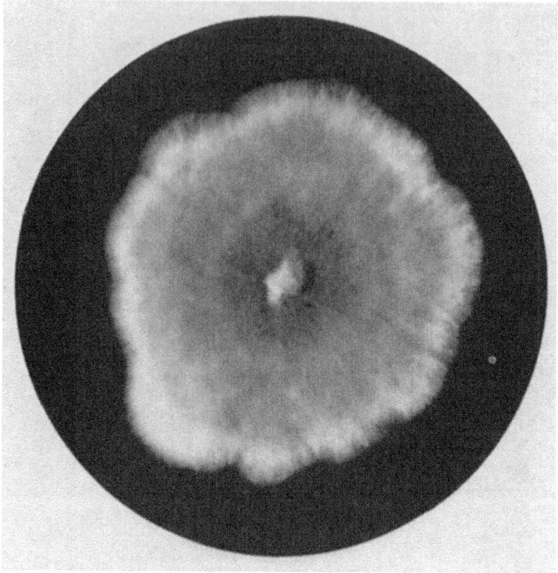

Fig. 37. Colony of *Fusarium sp.* on Sabouraud's glucose agar

them, but moistening of the animal's skin enhanced the transmission (RICHARD and PIER, 1966).

Dermatophilosis has been reported in four humans who handled an infected deer (DEAN et al., 1961). Two to seven days after exposure to the organism, multiple, non-painful vesicles developed on their hands. The lesions were 2—5 mm in diameter, contained a serous or purulent exudate, and those which remained active for less than two weeks, healed spontaneously with brown scabs and later red-purple scars. The patients had no systemic effects from the infection. No human to human transmission was evident.

Dermatophilosis can be readily diagnosed in the host by microscopic examination of stained smears of exudates, scabs, or crusts. In tissue *D. congolensis* has a similar morphology to the cultured organism and appears as branching thin filaments. The organism is Gram-positive, not acid fast, and stains well with the Giemsa and Grocott's methenamine silver stains (Figs. 33, 36). In specimens from acute lesions, elements of *D. congolensis* are abundant and can be readily noticed (KAPLAN, 1966). In material from healing lesions, the actinomycete may be sparse and careful search may be necessary to detect them. The fluorescent antibody technique may be of value in observing the organism in clinical material (PIER et al., 1964). The use of rabbits for recovery of *D. congolensis* has also been recommended (KAPLAN, 1966).

Fusariosis

The ubiquitous *Fusarium* which has antibiotic properties (ARNSTEIN et al., 1946; BOISSEVAIN, 1946; CAJORI et al., 1954; COOK et al., 1947; LACEY, 1950) contains numerous species that are phytopathogens (WOLLENWEBER et al., 1925; WOLLENWEBER and REINKING, 1935). The organism grows rapidly on Sabouraud's dextrose agar at room temperature, but only slowly at 37°C. The colony, which appears fuzzy and diffuse, is initially white and later becomes gray or brightly colored. It may be slightly granular and elevated. The reverse side may have an orange-yellow to brown-gray pigment (Fig. 37). Several species of *Fusarium* produce pigment, and these particularly resemble *Trichophyton rosaceum* (EMMONS, 1944). On blood agar the colonies manifest a clear zone of hemolysis around each colony (beta hemolysis). Spider-like flocculi appear in broth. The colorless, multiseptate macroconidia, which may occur in clusters, are 30—70 μ by 4—10 μ and sickle-shaped (Fig. 38). Smaller fusiform microconidia (6—9 μ by 3—5 μ) are occasionally present, too. In young cultures the conidiophores are not branched, but in older ones they are. Many smooth or tuberculate layers of spores are produ-

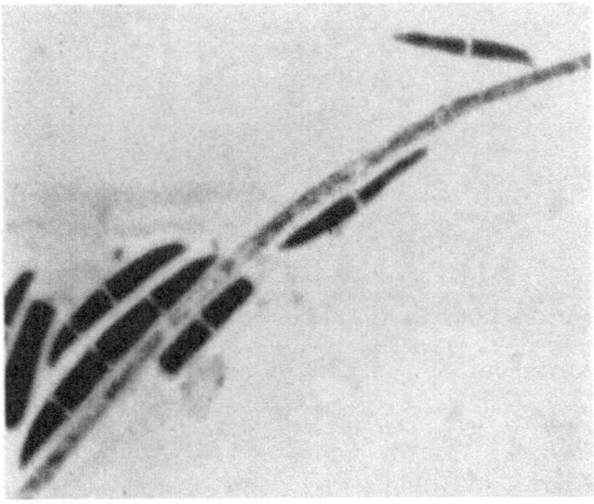

Fig. 38. *Fusarium sp.* with its elongated conidia. Lacto-phenol cotton blue preparation. × 1000

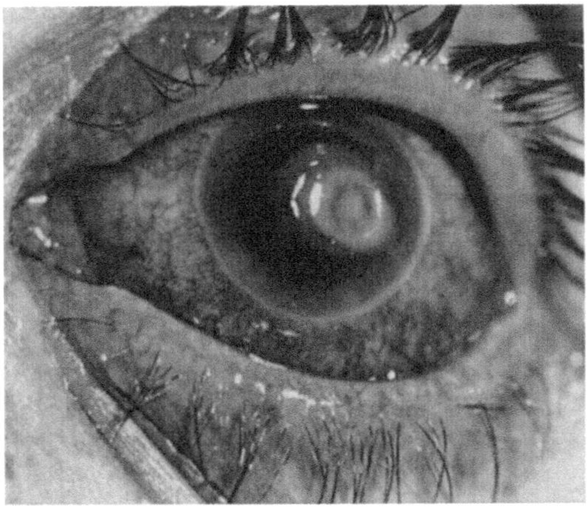

Fig. 39. Corneal ulcer due to *Fusarium sp.* (After ZIMMERMAN, L. E.: Surv. Ophthal. 8, 1—25, 1963)

ced (pionnotes). The widest septate hyaline hyphae (2—6 μ in width) are found in the older cultures. They are branched and contain intercalary chlamydospores.

The relatively few isolates of *Fusarium* in human disease usually have been identified as *Fusarium oxysporum*. The fungus has been associated with cutaneous, corneal, and urinary tract infections. Occasionally, the organism has been cultured from corneal ulcers, within which nonpigmented septate hyphae have sometimes been identified (ANDERSON and CHICK, 1963; ANDERSON et al., 1959; CHICK and CONANT, 1962; GILLESPIE, 1963; GINGRICH, 1962; LYNN, 1964;

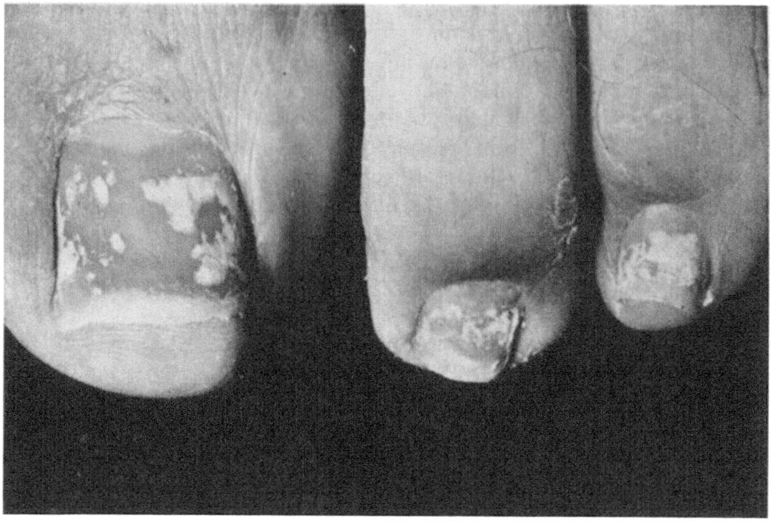

Fig. 40. Extensive involvement of the nail by *Fusarium oxysporum*. (Courtesy of Dr. N. ZAIAS)

MIKAMI and STEMMERMANN, 1958; PERZ, 1966; SIGTENHORST and GINGRICH, 1957; STOKES, 1959; ZIMMERMAN, 1963) (Fig. 39).

Fusarium has been cultured from cutaneous lesions in man (BOTTICHER, 1966; GAMET et al., 1964; HOLZEGEL, 1964; MIKAMI and STEMMERMANN, 1958; MING and YU, 1964; PETERSON and BAKER, 1959). The fungus has also been isolated from and observed in diseased nails (ZAIAS, 1966) (Figs. 40, 41).

On several occasions the fungus has been cultured from the urine of different patients. In some instances banana-shaped cells were observed in urine that

Fig. 41. Cross-section of a toe nail showing superficial fungal elements of *Fusarium oxysporum*. Periodic acid Schiff stain. × 120.(After ZAIAS, N.: Sabouraudia 5, 99—103, 1966)

yielded *Fusarium* on culture (FREI, 1925; LAZARUS and SWARTZ, 1948). The presence of *Fusarium* in the urine has usually been accompanied by leukocytes and often bacteria. FREI (1925) described a patient with small, white, star-shaped lesions in the posterior urethra from which *Fusarium sp.* was isolated.

Besides the above, toxins of *Fusarium* have been implicated in human disease. Outbreaks of a condition have been described in man and domesticated animals in rural populations of the U.S.S.R.. The disease, which has been referred to as alimentary toxic aleukia, has been attributed to a mycotoxin produced by *Fusarium sporotrichoides*. This species was identified in overwintered grain which affected individuals had ingested (JOFFE, 1960, 1962, 1963; SARKISOV and KVASHNINA, 1948). The clinical course was characterized by three distinct phases (FORGACS, 1962a, b; MAYER, 1953a). Initially, there was an acute onset of diarrhea, vomiting, and an epigastric burning sensation. In several days, even if the subject continued to ingest the contaminated grain the initial symptoms subsided. Subsequently, the peripheral blood demonstrated a pancytopenia yet the patient felt well. About 2—8 weeks later, bone marrow aplasia was evident and necrotic skin lesions appeared (FORGACS and CARLL, 1962). Large populations died from secondary infections or hemorrhages due to this disease. Disease has been reproduced experimentally with this fungus in cats, domestic swine, and other animals (JOFFE, 1963, 1965; MAYER, 1953b). *Fusarium sporotrichoides* is capable of both proliferating and forming toxin at 0—10 °C (MAYER, 1953b). A toxin, *sporofusariogenin*, with a steroid configuration has been isolated from the fungus and found to cause the condition (JOFFE, 1965). This toxin withstands, like many other mycotoxins, storage for a prolonged period (2—6 years), acid and alkaline denaturation, and autoclaving at 125°C for 30 min or 110°C for 18 h (MAYER, 1953b). An additional species of *Fusarium* has also been implicated in food poisoning in the U.S.S.R.. Food prepared from grains that are contaminated with *Fusarium graminearium* produces abdominal pain, nausea, diarrhea, and ataxia. The disease is commonly called "Drunken Bread". Once ingestion of the toxin is discontinued, the symptoms subside (FORGACS, 1962a, b).

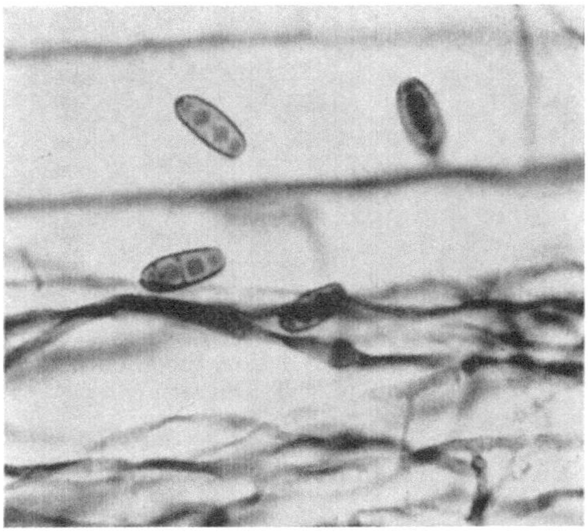

Fig. 42. *Helminthosporium sp.* with straight conidia. Polyvinyl alcohol cotton blue stain. × 320. (Courtesy of DR. C.T. DOLAN)

Helminthosporiosis

The ubiquitous species of *Helminthosporium* grow readily on standard mycological media, including Sabouraud's. The colonies become brown-gray and filamentous and then gray-brown to black with a velvety surface. The reverse side of the colony is brown-black. The occasionally racquet-shaped, septate hyphae (2.5—5.5 μ in width) possess branched conidiophores with terminal or laterally arranged black-brown, smooth-walled, cylindrical conidia (10 μ by 30 μ). Three or four septa are evident on these conidia which are straight (Fig. 42) and, therefore, readily distinguished from the curved conidia of *Curvularia* with their prominent distal medial cell.

Helminthosporium, a well known phytopathogen, has rarely been identified in mammalian tissue. The organism has been observed in and isolated from lesions in bovine nasal mucosa (ROBERTS et al., 1963). It has also been cultured from a mycetoma on a dog's foot, that contained numerous black granules of brown hyphae and chlamydospores (GEORG, 1964).

In man the fungus is frequently isolated from the sputum and skin, and may produce hypersensitivity reactions (WALTON, 1949). DOLAN et al. (1968, 1970) have evidence of *Helminthosporium* in human pulmonary tissue. They cultured *Helminthosporium sp.* from the lungs of two patients with chronic respiratory disease manifested by purulent sputum, hemoptysis, and fever. Multiple abscesses in pulmonary tissue were surrounded by lymphocytes, histiocytes, plasma cells, granulocytes, and fibrosis. Septate branching hyphae were observed in the lesions as well as the bronchi, bronchioles, and alveoli. The fungal elements appeared light brown in unstained sections and were readily demonstrated after silver impregnation (Fig. 43).

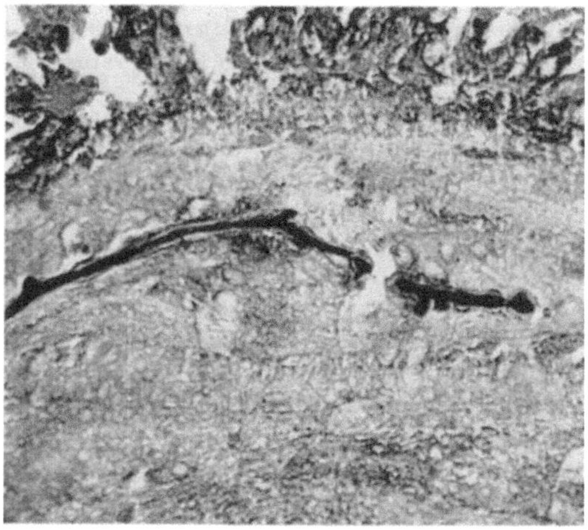

Fig. 43. *Helminthosporium sp.* in a pulmonary lesion. Silver chromate stain. × 350. (Courtesy of DR. C. T. DOLAN)

Keratomycosis

The first known case of mycotic keratitis is that of LEBER (1879). Since this report numerous fungi, as well as the actinomycetes, *Actinomyces bovis*, *Nocardia asteroides*, and *Streptomyces sp.*, have been cultured in many parts of the world from patients with keratitis. The fungal isolates have included *Acremonium sp.*, *Acrostalagmus cinnabarensis*, *Allescheria boydii*, *Alternaria sp.*, *Aspergillus spp.*, *Blastomyces dermatitidis*, *Botrytis sp.*, *Candida spp.*, *Cephalosporium spp.*, *Cryptococcus neoformans*, *Curvularia spp.*, *Fusarium spp.*, *Fusidium spp.*, *Gibberella fujikurae*, *Glenosporium spp.*, *Hormodendrum sp.*, *Mucor sp.*, *Penicillium spp.*, *Periconia keratitidis*, *Rhizopus sp.*, *Scopulariopsis spp.*, *Sporotrichum schenckii*, *Sterygmatocystis nigra*, *Trichosporon sp.*, *Verticillium spp.* (AINLEY and SMITH, 1965; ANDERSON and CHICK, 1963; ANDERSON et al., 1959; BEDELL, 1946, 1947; BYERS, et al., 1960; CHICK and CONANT, 1962; ERNEST and RIPPON, 1966; FAZAKAS, 1959; FOCOSI, 1932; GEORG, 1964; GILLESPIE, 1963; GINGRICH, 1962;

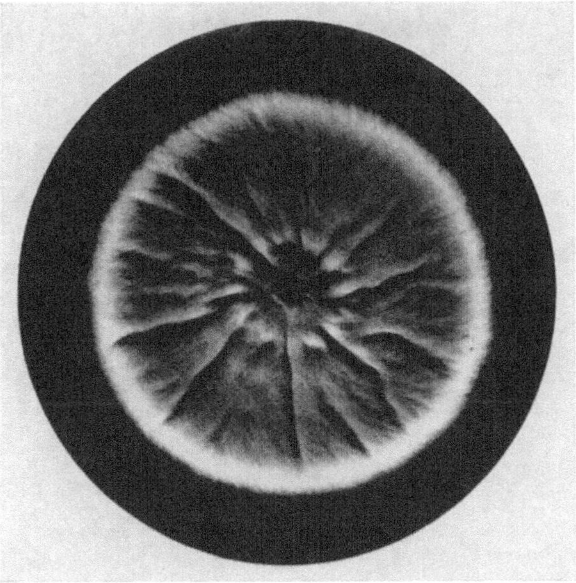

Fig. 44. Colony of *Paecilomyces* on Sabouraud's glucose agar

GORDON et al., 1959; HAGGERTY and ZIMMERMAN, 1958; HILDICK-SMITH et al., 1964; HOFFMAN and NAUMANN, 1963; KINNAS, 1965; LEY and SAUNDERS, 1956; LYNN, 1964; MAFFEI, 1928; MANCHESTER and GEORG, 1959; MANGIARACINE and LIEBMAN, 1957; MENEDELBLATT, 1953; MIKAMI and STEMMERMANN, 1958; MINODA, 1957; MITSUI and HANABRUSA, 1955; NAUMANN et al., 1967; NITYANANDA et al., 1962; PAUTLER et al., 1955; PERZ, 1966; ZIMMERMAN, 1962, 1963). Most isolates have not been accompanied by the identification of fungal elements within the excised tissue, and as many of these fungi are ubiquitous, their role in the genesis of the corneal lesions is often open to question. However, in many instances fungi such as *Allescheria boydii*, *Aspergillus spp.*, *Candida spp.*, *Cephalosporium sp.*, and *Curvularia spp.*, have been cultured from the involved tissue, which contained fungal elements.

Previous corneal trauma and the topical application of broad spectrum antibiotics and/or corticosteroids have been significantly associated with mycotic keratitis in man. Experimental evidence supports the clinical impression that corticosteroids and broad-spectrum antibiotics not only alter the normal flora of the eye, but also enhance fungal infection of the cornea (AGARWAL et al., 1963; SELIGMANN, 1952, 1953). Corticosteriods by altering the host's inflammatory response not only increase the relative virulence of recognized pathogens, but may allow other species that are general commensals to become pathogenic (AGARWAL et al., 1963; SUIE and HAVENER, 1963). Some fungi may cause corneal lesions in the absence of predisposing factors, for example *Cephalosporium sp.* has been reported to produce corneal ulceration and opacification in rabbits within 2—4 h (BURDA and FISHER, 1960).

Otomycosis

A wide variety of fungi have been cultured from the external ear of man in most parts of the world, but especially in the tropics (BRICEÑO-MAAZ and BRICEÑO-MAAZ, 1964; LAKSHMIPATHI and MURTI, 1960; SHARP et al., 1946). Some cases

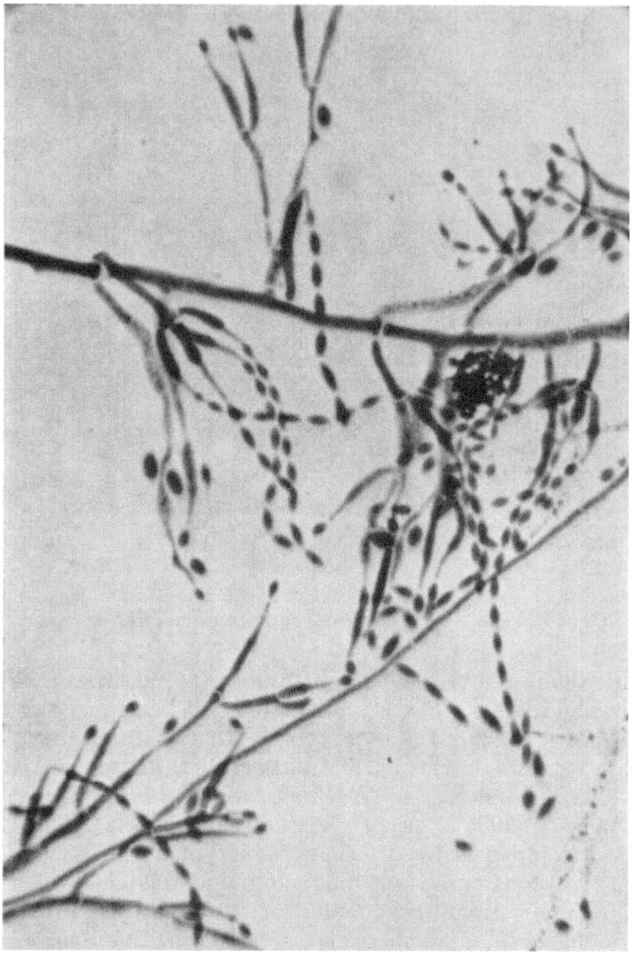

Fig. 45. *Paecilomyces sp.* showing chains of elliptical conidia arising from the tip of flask-shaped phialides. Lacto-phenol cotton blue preparation. × 1000. (After FETTER, B.F., KLINTWORTH, G.K., and HENDRY, W.S.: *Mycoses of the Central Nervous System*. Baltimore: Williams and Wilkins 1967)

have been associated with itching, pain, redness, scaling, and discharges of the external auricle, meatus, and auditory canal, but other isolates have been from seemingly normal external ears (HALEY, 1950; LEA et al., 1958; SOOD et al., 1967). The fungi that have been cultured have included numerous dermatophytes.

(ENGLISH, 1957; STUART and BLANK, 1955), as well as *Allescheria boydii* (BELDING and UMANZIO, 1935; BLANK and STUART, 1955), *Aspergillus spp.* (ANANTHANARAYAN, 1951; BEANEY and BROUGHTON, 1967; DURCAN et al., 1968; SOOD et al., 1967), *Candida albicans* (TOMIC-KAROVIC and POPADIC, 1953), *Cryptococcus neoformans* (GILL, 1947; WOLF, 1947), *Mucor spp.* (MCBURNEY and GULLEDGE, 1955; SOOD et al., 1967), *Penicillium spp.* (ARIEVITCH and STEPANISCHEVA, 1964; KIRK, 1959; KUNEL'SKAIA, 1964; KUNEL'SKAIA and STEPANISCHEVA, 1964; MCBURNEY and GULLEDGE, 1955; MOHAPATRA, 1961; SMYTH, 1961; SOOD et al., 1967), *Scopulariopsis spp.* (SMYTH, 1961), and *Waldemaria pernambucensis* (BATISTA et al., 1960; SMYTH, 1961).

As most of these fungi have only been cultured from the external ear and not identified in lesions, they simply may be saprophytes. However, fungi have

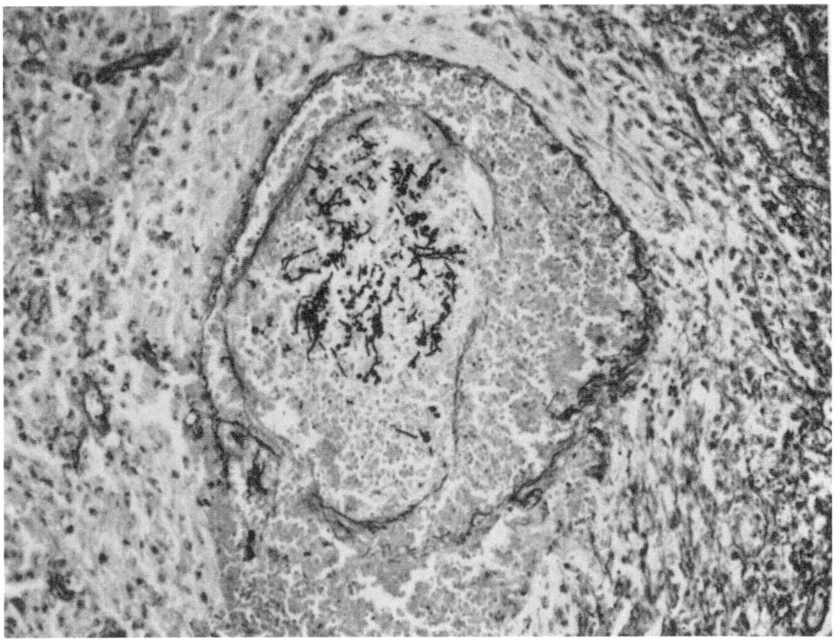

Fig. 46. Fungal elements of *Paecilomyces* in the lumen of a blood vessel. Grocott's methenamine silver stain. × 110

sometimes been observed in tissue scrapings and discharges and may be important in the genesis of the aural lesion. Moreover, lesions have been produced in the ear canals of man with fungi isolated from patients with otomycosis (SOOD et al., 1967).

Paecilomycosis

Paecilomyces has a worldwide distribution and is a common saprophyte of fruits and vegetables. The fungus, which consists of several species, grows readily on numerous culture media including Sabouraud's medium. It produces a thin colony which becomes yellow-brown and powdery following the production of conidia (Fig. 44). The mycelium has broad, branching septate hyphae. Chains of ovoid spores (2—4 μ by 6—8 μ) arise from flask-shaped phialides that have long tapering necks at an angle to the main axis of the sterigmata (Fig. 45).

The fungus has rarely caused human disease. UYS et al. (1963) described a patient who died a year after the surgical correction of an incompetent mitral valve. The terminal clinical manifestations included emaciation, pyrexia without leucocytosis, and signs of cardiac failure and embolization. *Paecilomyces varioti*, which was isolated from several blood cultures three weeks before death, was also cultured from a thrombus overlying the mitral valve and from an embolus in an iliac artery after death. On post-mortem examination the fungus was demonstrated microscopically within thrombi in the region of the mitral valve and emboli in the iliac and cerebral arteries. An area similar to a tuberculoma with caseous necrosis, epithelioid cells, and multinucleated giant cells was present in the vessel wall overlying a thrombus in the iliac artery. This lesion as well as others contained hyphae (1.5—2 μ in diameter) with and without rounded bodies

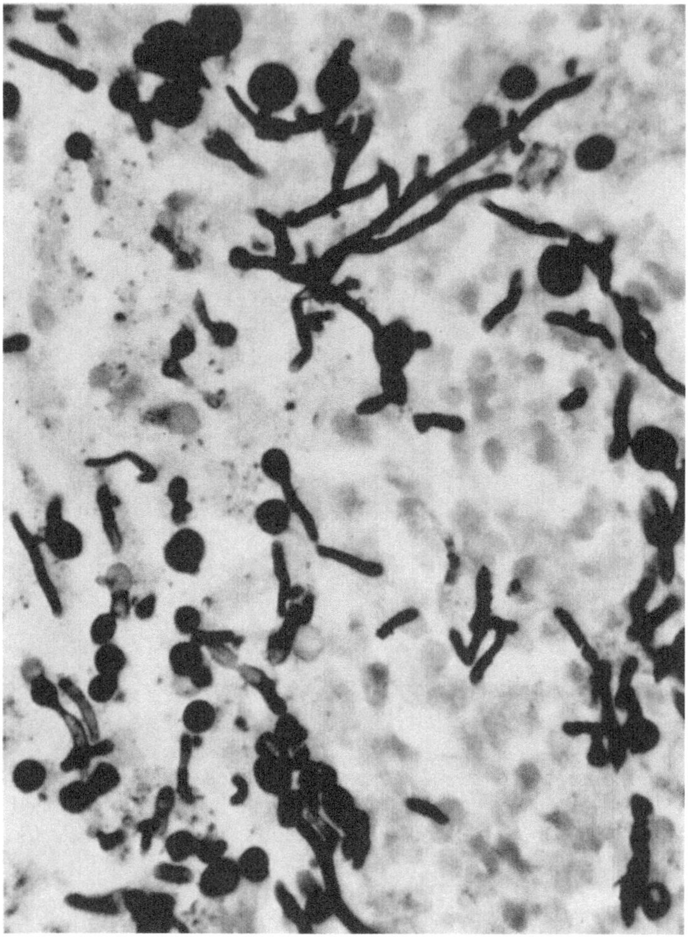

Fig. 47. *Paecilomyces varioti* in the brain. Grocott's methenamine silver stain. × 1000. (After FETTER, B.F., KLINTWORTH, G.K., and HENDRY, W.S.: *Mycoses of the Central Nervous System*. Baltimore: Williams and Wilkins 1967)

connected to them. The fungus was readily demonstrated with Grocott's methenamine silver technique (Figs. 46, 47).

Species of *Paecilomyces* have been cultured from a postoperative scleral lesion (PODEDWORNY and SUIE, 1964) and infected nails (ŠAULOV, 1965). The fungus was not identified in the tissue in these cases, and therefore they remain equivocal examples of paecilomycosis.

Penicilliosis

The genus *Penicillium*, commonly known as the blue or green mold, contains more than 200 different species that are widespread in nature. The fungus is often observed on decaying organic matter, and frequently contaminates laboratory media. Numerous species of *Penicillium* produce the antibiotic penicillin.

All species of *Penicillium* grow rapidly at room temperature on Sabouraud's glucose agar, producing a variety of pigmented colonies (Fig. 48). The aerial

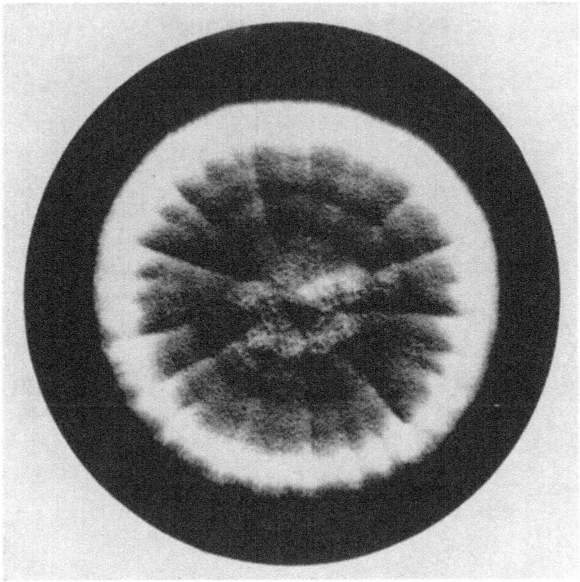

Fig. 48. *Penicillium sp.* on Sabouraud's glucose agar

mycelium forms a complex system of branches (metulae), from which the flask-shaped phialides radiate, like the bristles of a brush (*penicillus*). The conidia occur in unbranched chains from the tips of the latter structures (Fig. 49). The various species differ in the size, texture, and color of the colonies.

Penicillium has rarely been found to be pathogenic to plants or animals. It is commonly cultured from man in the abscene of disease. Several species have been implicated in disease, including *P. bertai*, *P. bicolor*, *P. citrinum*, *P. commune*, *P.crustaceum*, *P.glaucom*, *P.marneffei*, *P.mycetogenum*, and *P.spinulosum*. Most of the isolates are, however, of uncertain significance, as the fungus has usually not been observed in tissue from which it was cultured. Indeed, *Penicillium* is a frequent constituent of the conjunctival flora of healthy individuals (AINLEY and SMITH, 1965), and a common isolate of bronchial secretions (GERNEZ-RIEUX, 1964). Moreover, as the fungus is so widely dispersed in nature, even the repeated isolation of the *Penicillium* is of equivocal status.

Penicillium spp. have been isolated from patients with broncho-pulmonary disease.

(AIME et al., 1933; BUGYI, 1958; CASTELLANI, 1920; DASILVA LACAZ, 1937; DELORE et al., 1955; GARRETÓN, 1945; GIORDANO, 1918; HOŘEJŠÍ et al., 1960; HUANG and HARRIS, 1963; JACKSON and YOW, 1961; NIÑO, 1932; NUSSBAUM and BENEDEK, 1927; PEZZALI, 1921; SOUCHERAY, 1954; TALICE and MACKINNON, 1929), keratitis (AINLEY and SMITH, 1965; ANDERSON et al., 1959; CHICK and CONANT, 1962; FAZAKAS, 1959; GINGRICH, 1962; MINODA, 1957; MITSUI and HANABUSA, 1955), chronic ear infections (ARIEVITCH and STEPANISCHEVA, 1964; KIRK, 1959; KUNEL'SKAIA, 1964; KUNEL'SKAIA and STEPANISCHEVA, 1964; MCBURNEY and GULLEDGE, 1955; MOHAPATRA, 1961; SMYTH, 1961; SOOD et al., 1967), urinary tract infection (CHUTE, 1911; GILLIAM and VEST, 1951; SALISBURY, 1868), vaginitis and vulvo-vaginitis (CASTELLANI, 1920), lesions of nails and skin (ANDREEV, 1964; KESTEN et al., 1932; ŠAULOV, 1965), a post-operative mastoid cavity (SMYTH, 1962), black hairy tongue (LINEBACK, 1959), and a temporal lobe abscess (POLYANSKIY, 1938).

When *Penicillium* has been isolated from patients with respiratory disease there has usually been an underlying pulmonary disorder such as bronchiectasis,

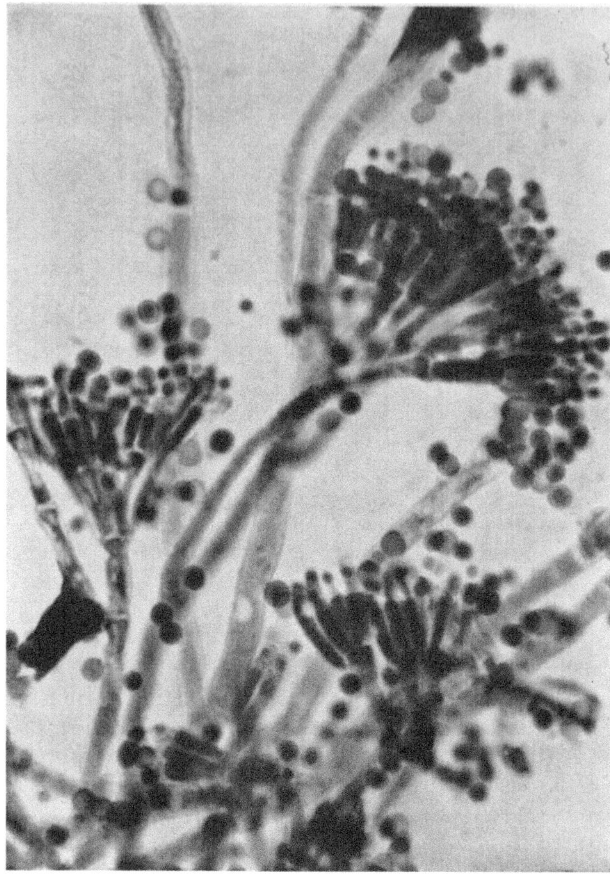

Fig. 49. *Penicillium sp.* showing typical brush-like conidia. Lacto-phenol cotton blue preparation. × 1000

tuberculosis, and lung abscesses. In most of these cases adequate proof of disease being caused by *Penicillium* is lacking. Sometimes a large mass of matted mycelia accumulates in a pulmonary cyst or cavity caused by an associated disease. Such lesions are occasionally characterized roentgenographically by a narrow zone of radiolucency between radio-opaque zones. Histologically, there is generally little or no reaction to the fungus and the organisms seem to be saprophytes. Occasionally, when the isolates have been associated with non-specific manifestations such as cough, hemoptysis, weight loss, and pyrexia, the etiology of the respiratory disease has not been clear.

P.mycetogenum has been isolated from a mycetoma that contained black granules. This species was found to be pathogenic to the pigeon, but not to the guinea pig (MANTELLI and NEGRI, 1915).

HUANG and HARRIS (1963) described a rare example of disseminated penicilliosis. The patient had acute leukemia complicated by pulmonary and cerebral penicilliosis and gastrointestinal candidosis. Antibiotics and steroid therapy anteceded the onset of the mycotic infection. *P.commune* was cultured from the pulmonary lesions at autopsy and observed within and around blood vessels. The

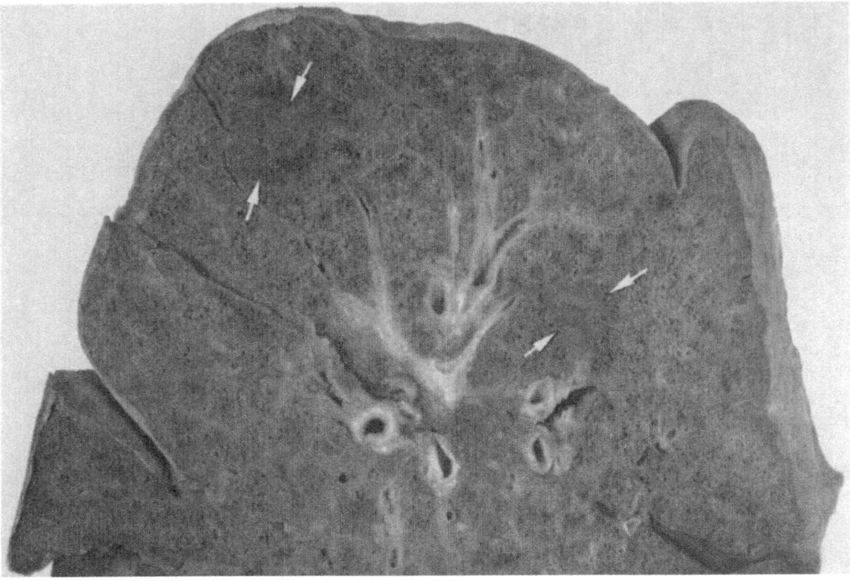

Fig. 50. Lung with hemorrhagic nodular lesions containing *Penicillium commune* (arrows) and secondary to a *Penicillium* thromboangiitis. (After HUANG, S.N., and HARRIS, L.S.: Amer. J. clin. Path. 39, 167—174, 1963)

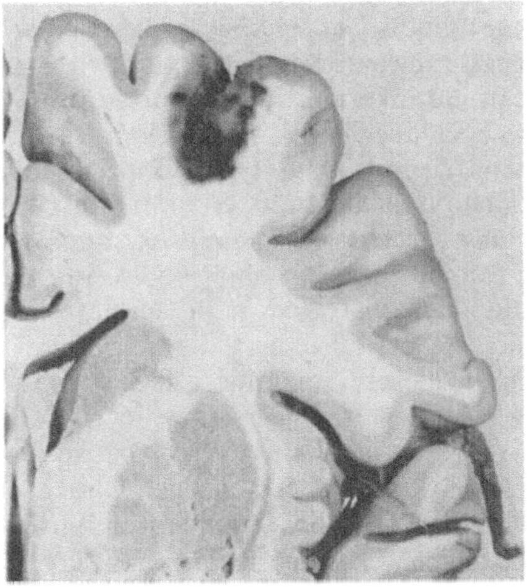

Fig. 51. Infarct in the brain caused by *Penicillium commune.* (After HUANG, S.N., and HARRIS, L.S.: Amer. J. clin. Path. 39, 167—174, 1963)

fungus had also invaded the adjacent tissue around infarcts in the lungs and brain (Figs. 50—52).

It has been suggested that sensitization to *Penicillium* may cause asthma and allergic rhinitis (DE MONTEMAYOR et al., 1965; SHELLEY, 1964; WALTON, 1949).

One species of *Penicillium (P. marneffei)* is known to cause spontaneous disease in the Bamboo rat (*Rhizomys sinensis*) that lives in the high plateaus of Vietnam. In this rodent the fungus produces a systemic infection manifested by

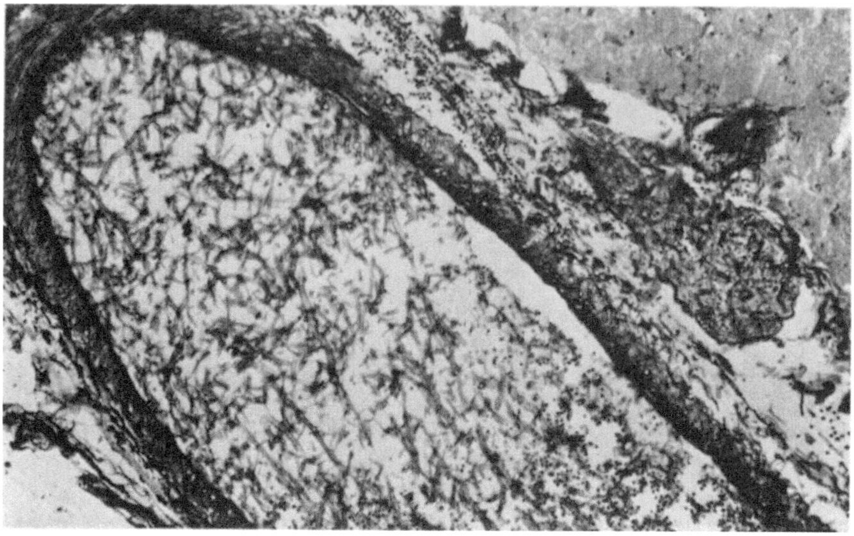

Fig. 52. Numerous hyphae of *Penicillium commune* within cerebral blood vessels. Grocott's methenamine silver stain. × 120

ascites, splenomegaly, hepatomegaly, lymphadenopathy, and intestinal lesions involving the Peyer's patches. The fungus is evident within macrophages and to a lesser extent in polymorphonuclear leukocytes, particularly in the liver, spleen, and lymph nodes. Though a clear case of infection with *P. marneffei* has not yet been documented in man, an investigator is reported to have developed a nodule in a finger 10 days after its accidental inoculation. The lesion was followed by palpably enlarged regional lymph nodes. All symptoms disappeared after treatment with Nystatin (CAPPONI et al., 1956; SEGRETAIN, 1959, 1962).

In tissue sections most species of *Penicillium* appear as branching septate hyphae. Though they can be seen in hematoxylin and eosin preparations, the hyphae are more readily demonstrated with special stains, such as Grocott's methenamine silver and the modified periodic acid Schiff stains (Figs. 53, 54). Histologically, *Penicillium* closely resembles *Aspergillus* and these two genera usually cannot be differentiated with certainty in the absence of culture. The flask-shaped phialides may aid in distinguishing between *Aspergillus* and *Penicillium*, but these are rarely observed in tissue. Unlike other species of *Penicillium*, *P. marneffei* does not produce prominent hyphae in lesions, but intracytoplasmic rounded bodies which superfically resemble *Histoplasma capsulatum* and *Leishmania donovanii*. However, with *P. marneffei* elongated fungal elements (20 μ), often with septate hyphae are sometimes evident.

Different toxins produced by *Penicillium spp.* cause disease in cattle, rodents, and other animals (KOBAYASHI et al., 1958, 1959a, b; MIYAKE et al., 1955, 1959, 1960; OHMORI et al., 1954; SAKAI, 1955; SAKAI and URAGUCHI, 1955; SAKAKI, 1891; TSUNODA, 1953, 1963; URAGUCHI et al., 1961a—c).

There has been evidence to suggest that toxins for *Penicillium* sp. produce gastrointestinal distubances in man. Lesions have not been adequately documented in humans, but the toxins produced by *Penicillium spp.* (Table 3) have been associated in rodents with such variable features as cirrhosis, hepatomas, nephrosis, and respiratory paralysis (ENOMOTO, 1959; ISHIKO, 1957; KOBAYASHI et al., 1958,

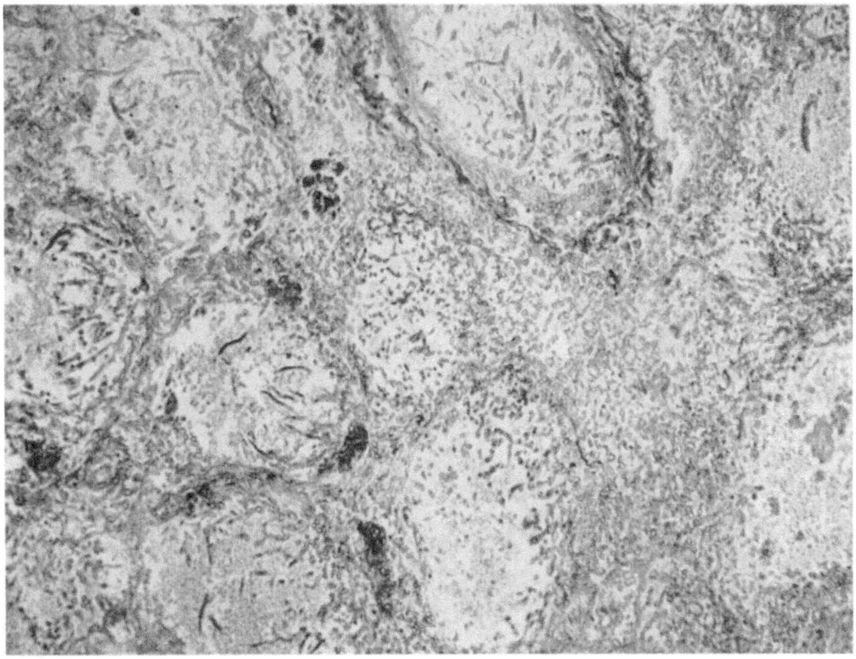

Fig. 53. Numerous hyphae of *Penicillium commune* in the lung. Grocott's methenamine silver stain. × 120

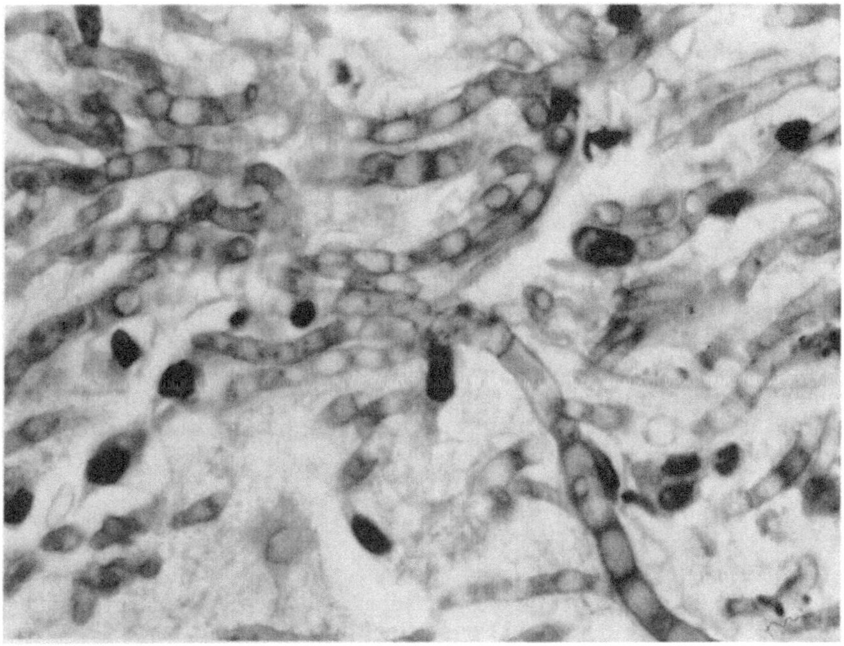

Fig. 54. Hyphae of *Penicillium commune*. Modified periodic acid Schiff stain × 1000

Table 3. Toxins of *Penicillium*

Species	Toxins
Penicillium brunneum	rugulosin, emodin, and skyrin
P. citreo-viride	citreoviridin
P. citrinum	citrinin
P. frequentans	frequentic acid
P. islandicum	islanditoxin luteoskyrin
P. notatum	notatin
P. ochrosalmoneum	citreoviridin
P. rugulosum	rugulosin
P. tardum	rugulosin
P. urticae	patulin
P. viridicatum	viridiatin

1959a, b; MIYAKE et al., 1955, 1959, 1960; OHMORI et al., 1954; SAITO, 1959; SAKAKI and URAGUCHI, 1955; TSUNODA, 1953, 1963; URAGUCHI et al., 1961a—c). Death of cattle has also been attributed to the toxins, patulin and viridiatin.

Rhodotorulosis

Rhodotorula, a red yeast-like organism, grows readily on Sabouraud's glucose agar and other common microbiological media at both room temperature and 37°C (Fig. 55). The colony possesses a prominent orange-red color and a dull, smooth surface. Though the fungus contains a red carotenoid pigment, the individual organisms generally appear colorless. The yeast-like cells are ovoid in shape, possess thin capsules, and replicate by budding. Pseudohyphae are produced in some instances. Like *Cryptococcus*, the organism hydrolyzes urea. Several species of *Rhodotorula* have been described, and include: *R. flava*, *R. glutinis*, *R. mucilaginosa*, and *R. saccharomyces*.

The organism has generally been regarded as nonpathogenic, but has been cultured from numerous sources in human patients including skin and nails, bronchial secretions, urine, blood, aortic valve, ascitic fluid, vulvovaginal area, cerebrospinal fluid, pharynx, and feces (AHEARN et al., 1966; BERGMANN and LIPSKY, 1964; CARETTA, 1961; CHAKRAVARTY, 1966; GERNEZ-RIEUX et al., 1964; LEEBER and SCHER, 1969; LINEBACK, 1959; LOURIA et al., 1960, 1967; MIGUENS, 1960; MÜLLER et al., 1967; SHELBOURNE and CAREY, 1962; SILVEIRA, 1959; VIEIRA and BATISTA, 1962; WILSON, 1965). Though the pathogenicity of *Rhodotorula* for man and other animals has yet to be established, several investigators have suggested that the organism may have adverse effects in man. They have drawn attention to the importance of not discarding isolates of this fungus as contaminants of culture media. The organism has not only been repeatedly isolated from multiple sources, including the cerebrospinal fluid, pharynx, feces, urine, and skin in the same patient (RIOPEDRE et al., 1960), but has also been cultured on several separate occasions from the blood and/or the urine in other patients (LEEBER and SCHER, 1969; LOURIA et al., 1967; SHELBOURNE and CAREY, 1962). Whether the fungus lives in man as a saprophyte or parasite remains to be established. It has usually been isolated in patients with some serious disease and has not been associated with any specific clinical features. Intravenous therapy appears to have been an important predisposing factor in the development of fungemia due to *Rhodotorula* (LEEBER and SCHER, 1969; LOURIA et al., 1967).

The organism has not been adequately demonstrated in tissue sections. Relatively few post-mortem examinations have been performed on patients from

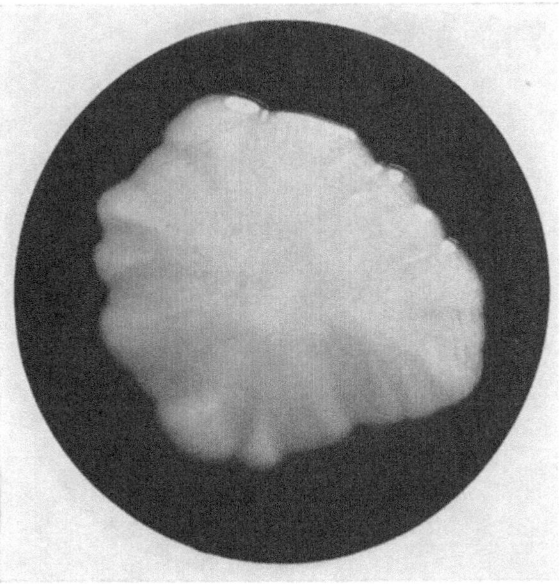

Fig. 55. Colony of *Rhodotorula sp.* on Sabouraud's glucose agar

whom *Rhodotorula* was cultured during life. In these cases the fungus could not be identified in the tissues (LOURIA et al., 1960, 1967). However, the patients had been treated with amphotericin B, and this may have accounted for the lack of histologic confirmation (LOURIA et al., 1960, 1967). Attempts to produce disease in laboratory animals with *Rhodotorula* have been inconclusive (LOURIA et al., 1960, 1967; RIOPEDRE et al., 1960).

Streptomycosis

Of all the genera of the Actinomycetes, *Streptomyces* is represented in nature by the largest number of species and varieties. Many of these play an extremely important role in medicine, not because of their pathogenicity, but on account of their importance in chemotherapy. Thus far, more than 350 species of *Streptomyces* produce numerous antibiotics, that are used in the treatment of infectious diseases, in the prevention of plant diseases, in the preservation of food, and for the promotion of animal growth (UMEZAWA et al., 1967).

Streptomyces spp. are aerobic, and grow readily at room temperature in ordinary media and at the ordinary atmosphere. The colonies of most *Streptomyces* on artificial media are smooth or lichenoid, hard and tensely textured, raised and adherent to the medium. They are usually covered completely or partially (in the form of spots or concentric rings) by an aerial mycelium, which may be pigmented (Fig. 56). The nature and intensity of the pigment depend on the species and on the composition of the media. The hyphae vary greatly in length; some are long with limited branching, others are short and profusely branched. They usually have a diameter of 0.7—0.8 μ (Fig. 57). The mycelium does not segment into bacillary and coccoid forms. Reproduction occurs by the formation of chains of spores (conidia) in spore-bearing hyphae (sporophores) that arise from the aerial mycelium either monopodially or in the form of small clusters of whorls. The sporophores may be straight or curved with curvatures ranging from waves to

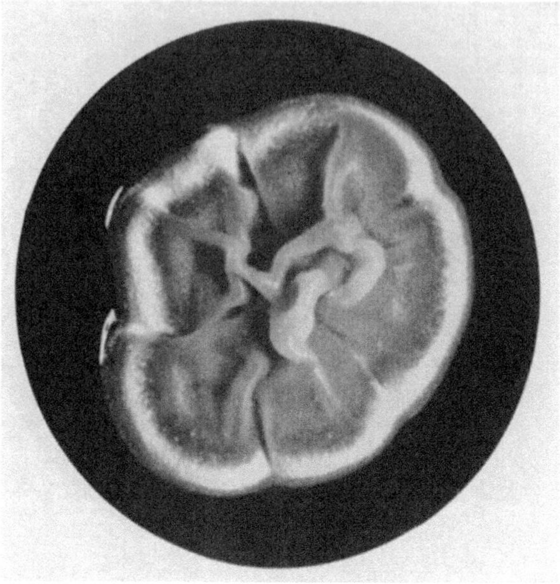

Fig. 56. Colony of *Streptomyces sp.* on Sabouraud's glucose agar

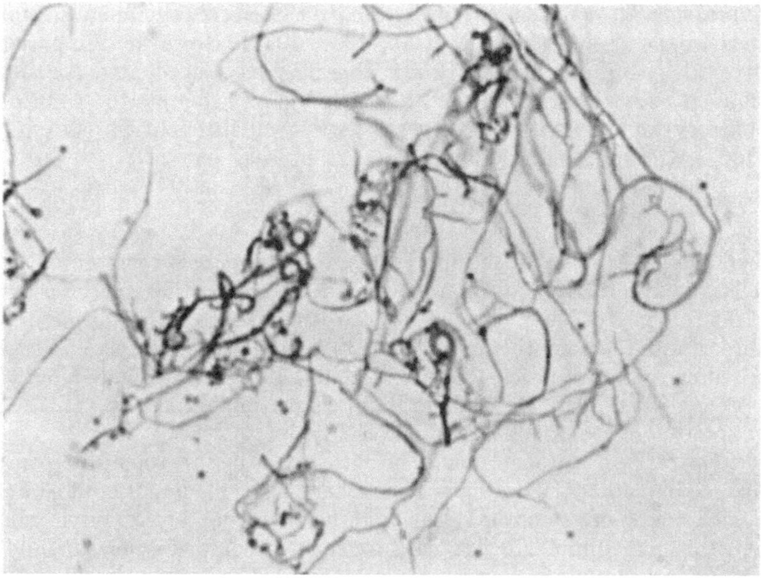

Fig. 57. Filamentous network of *Streptomyces sp.* × 625

spirals (Fig. 58). The marked variation in spore size and shape is of limited usefulness in taxonomical differentiation.

The *Streptomyces spp.* grow in soil or upon organic residues and often contaminate culture media. They are sometimes present on the skin, or in feces, saliva or sputum. Very few are pathogenic to either animals or plants. Some

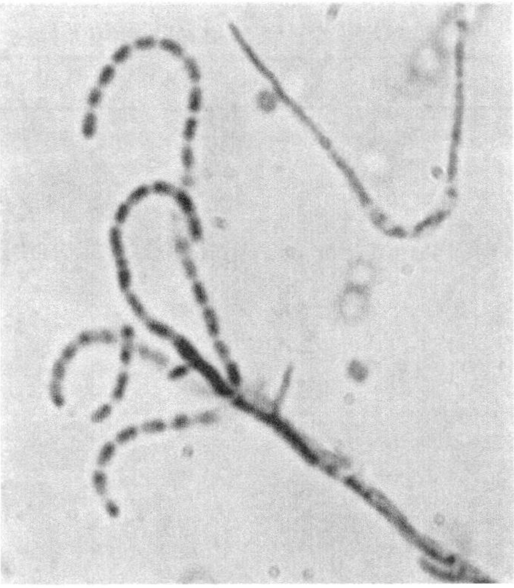

Fig. 58. *Streptomyces sp.* showing chains of conidia arising from a terminal sporophore. Lactophenol cotton blue preparation. × 2500

species of *Streptomyces*, including *S. scabies*, produce the common potato scab, the most important plant disease caused by the actinomycetes (Waksman, 1967).

Three species of *Streptomyces* (*S. madurae* Vincent, *S. pelletieri* Laveran, and *S. somaliensis* Brumpt) cause mycetomas (see Chapter 12, Mycetoma). In some parts of the world the resulting disease is endemic with the pathogenic species differing in the various countries.

(Abbott, 1956; Adami and Kirkpatrick, 1895; André, 1958; Blanc and Brun, 1919; Brault, 1906; Brumpt, 1960; Carter, 1860, 1861; Catanei, 1934, 1942, 1943; Catanei and Goinard, 1934; Catanei et al., 1927; Catanei and LeGroux, 1931; Chadli et al., 1963; Convit et al., 1961; Cornwall and La Frenais, 1922; Destombes and André, 1958; Destombes et al., 1965, 1966; Dey, 1962; Diouf et al., 1965; Gamet et al., 1964; Guichard and Jausion, 1923; Gemy and Vincent, 1892, 1896; Gonzalez-Ochoa and Ahumada Pidalla, 1958; Halde and Ringrose, 1958; Homez, 1963; Hyde, 1896; Latapí, 1959; Latapí et al., 1961; Latapí and Ortiz, 1963; Le Dantec, 1894; Legrain, 1898; Lincoln et al., 1965; Lobo, 1963; Lynch, 1964; Lynch and Moghraby, 1961; Mackinnon, 1954; Mackinnon and Artagaveytia-Allende, 1956; Maegraith, 1963; Mariat, 1963; Montpellier and Lacroix, 1921; Nicolau and Avram, 1959; Orio et al., 1963; Porritt, 1962; Proctor, 1966; Raynaud et al., 1922; Remlinger, 1912, 1913; Rey, 1961; Saul, 1961; Segretain and Destombes, 1963; Segretain and Mariat, 1958; Symmers and Sporer, 1944; Vanbreuseghem, 1958; Verghese and Klokke, 1966; Victoria et al., 1967; Vincent, 1894) (Table 4). In countries where *S. madurae*, *S. pelletieri*, and *S. somaliensis* have all been isolated, the incidence of disease produced by each of the species is not the same. For example, *Streptomyces somaliensis* is the commonest species of the actinomycetes that has been cultured from mycetoma in the Sudan, being more than ten times as common as *S. madurae* and *S. pelletieri* (Lynch, 1964).

The mycetomata produced by *Streptomyces*, like those due to other organisms, are characterized by sinus formation, a granulomatous reaction, and grains. The granules, whose color varies with the species (Table 4), are dispersed throughout the diffuse and ill-defined lesions. *S. pelletieri* has the greatest polymorphonuclear leukocytic response. Multinucleated giant cells are generally evident in older lesions due to *S. somaliensis*, but rarely with infections by *S. madurae* and *S. pelletieri* (Camain et al., 1957; Lynch, 1964). After several years regional lymph node involvement may accompany the streptomycotic mycetoma (Lynch, 1964). The rate of progression of mycetoma due to *Streptomyces* is stated to be more rapid

Table 4. *Geographic Distribution of Mycetomas due to S. madurae, S. pelletieri, and S. somaliensis and the Characteristics of their Granules*

Geographic Distribution	S. madurae	S. pelletieri	S. somaliensis
Algeria	+	—	—
Arabia	—	—	+
Brazil	+	+	—
Cameroon	—	+	—
Chad	+	+	+
Democratic Republic of the Congo	+	—	—
India	+	+	—
Israel	+	—	+
Mauritania	+	+	+
Morocco	+	—	—
Mexico	+	+	+
Nigeria	+	+	+
Roumania	+	—	—
Senegal	+	+	+
Somaliland	+	+	+
Sudan	+	+	+
Tanzania	—	+	+
Tunisia	+	—	—
United States of America	+	+	+
Upper Volta	+	+	—
Venezuela	+	+	+
Vietnam	+	—	—
Yemen	—	—	+

\+ = organism identified
— = organism not identified

Characteristics of Granules:

	S. madurae	S. pelletieri	S. somaliensis
Color:	Usually white or yellowish, occasionally reddish	Red	White or yellowish
Size:	1—5 mm	0.1—0.5 mm	0.5—2 mm
Staining:	Margins stain deeply with hematoxylin	Stains deeply and uniformly with hematoxylin	Eosinophilic

than those due to *Madurella*, and infections caused by *S. pelletieri* seem to be the most virulent of the actinomycetes (LYNCH, 1964).

Extra-pedal sites may be involved with any of the *Streptomyces* that cause mycetoma. Even the testis may be affected (CLARKE, 1953). In an extensive investigation of mycetoma, REY (1961) found that *S. pelletieri* affected sites other than the feet more frequently than other organisms.

There is some evidence to suggest that different species of *Streptomyces* vary in their affinity for bone. It is generally agreed that osseous lesions are common, with infection by *S. pelletieri* (DELAHAYE et al., 1962; KLÜKEN et al., 1964; LYNCH, 1964; REY, 1961). Lesions of the bone commonly accompany mycetomas that have been present for more than one year. The time that elapses between infection and the production of bone lesions depends, aside from the species, on the distance of the lesion from the bone (KLÜKEN et al., 1965). The bone undergoes osteolysis and develops a reticulated appearance. Sometimes the periosteum is primarily involved. Under such circumstances there may be swelling without sinus formation, and the disease may simulate an osteogenic sarcoma, both clinically and radiologically.

The roentgenographic "sun ray" appearance that commonly occurs with osteogenic sarcomas may be present. Even a "sabre" tibia may be produced (LYNCH, 1964).

LYNCH (1964) inoculated *Streptomyces somaliensis* subcutaneously into guinea pigs, rabbits, rats, and monkeys, with or without acacia thorns and Freund's adjuvant, but was unable to produce lesions comparable to those in man. Collagenase activity has been demonstrated with *S. madurae* and the enzyme has been implicated in the pathogenicity of the species (RIPPON and LORINCZ, 1964; RIPPON and PECK, 1967).

Besides mycetoma, species of *Streptomyces* have rarely been incriminated in other human lesions. KOHN et al. (1951) described a patient with a pulmonary infection from whose blood *Streptomyces sp.* was cultured three times. Species of *Streptomyces* have been cultured from ulcerative keratitis (CHICK and CONANT, 1962). CLARK et al. (1964) isolated *Streptomyces griseus* from a cerebellar abscess of a patient who had a past history of frontal sinusitis, but no other associated disease. A pre-operative lumbar cerebrospinal fluid analysis revealed 2,240 neutrophils per cu-mm, 43 mg sugar per 100 ml and 195 mg protein per 100 ml with an increased globulin level. Culture of the fluid revealed no organisms. The ventricular fluid contained 650 erythrocytes per cu-mm, 4 lymphocytes per cu-mm, 80 mg protein per 100 ml, 70 mg sugar per 100 ml and 203 mEq/L chloride. An apparent cure followed treatment with antibiotics and aspiration of the abscess.

Streptomyces is readily demonstrated in tissue sections with bacterial stains, such as the Brown and Brenn or MacCallum-Goodpasture, and Grocott's methenamine silver stain, but are poorly visualized in hematoxylin and eosin stained preparations. The fungus forms a delicate Gram-positive mycelium which is morphologically indistinguishable from *Actinomyces* and *Nocardia*. Granules are generally present in tissue with infection by *S. madurae*, *S. pelletieri*, and *S. somaliensis*. Fibrin may resemble *Streptomyces* in tissue sections, but it can be readily differentiated from the actinomycete with suitable stains. For example, fibrin is Gram-negative, but *Streptomyces* is Gram-positive with the Brown and Brenn bacterial stain. Fibrin and *Streptomyces* are both Gram-positive with the McCallum-Goodpasture bacterial stain. *Streptomyces* is not acid-fast and does not tend to break up into rod-shaped and coccus-like bodies as species of *Actinomyces* and *Nocardia* sometimes do.

Torulopsiosis

Torulopsis, a genus belonging to the Cryptococcaceae, is closely related to *Cryptococcus* and *Candida* and reproduces by budding, but does not form pseudohyphae (SILVA-HUFNER, 1970; WICKERHAM, 1957). The fungus which is widely distributed in nature has been cultured from the sputum, stools and urine of apparently healthy humans, but the only species of this generally saprophytic genus that has been incriminated consistently in human disease is *T. glabrata*. Its nonhemolytic colonies, which are pasty and colorless or white to cream colored, become gray-brown with age. The organism is very sensitive to elevated pH and grows only in an acid medium (pH <6.5) (EDEBO and SPETZ, 1965).

Torulopsis has been recovered from the bile, blood bone marrow, peritoneal fluid and pleural fluid as well as wound exudates and lesions of the oral cavity, fallopian tubes, vagina, nails, skin and subcutaneous tissue.

(AHEARN et al., 1966; ARTAGAVEYTIA-ALLENDE et al., 1961; BATISTA and PEREIRA, 1960; BATISTA et al., 1958, 1961; BERGMAN and LIPSKY, 1964; BERNHARDT, 1960; BLACK and FISHER, 1937; BRUNNER, 1964; CARETTA, 1961; CERNIK and DOANHOAG-HOA, 1963; COHEN et al., 1963; DAGLIO and MARTINEZ-MONTES, 1957; EDEBO et al., 1966; EDEBO and SPETZ, 1965; FÖLDVARI and FLORIAN, 1965; GUZE and HALEY, 1958; HALEY, 1961; HERMANEK, 1964; HURLEY and MORRIS, 1964; HOLMSTRÖM et al., 1959; KATZ and PICKARD, 1967; KEARNS

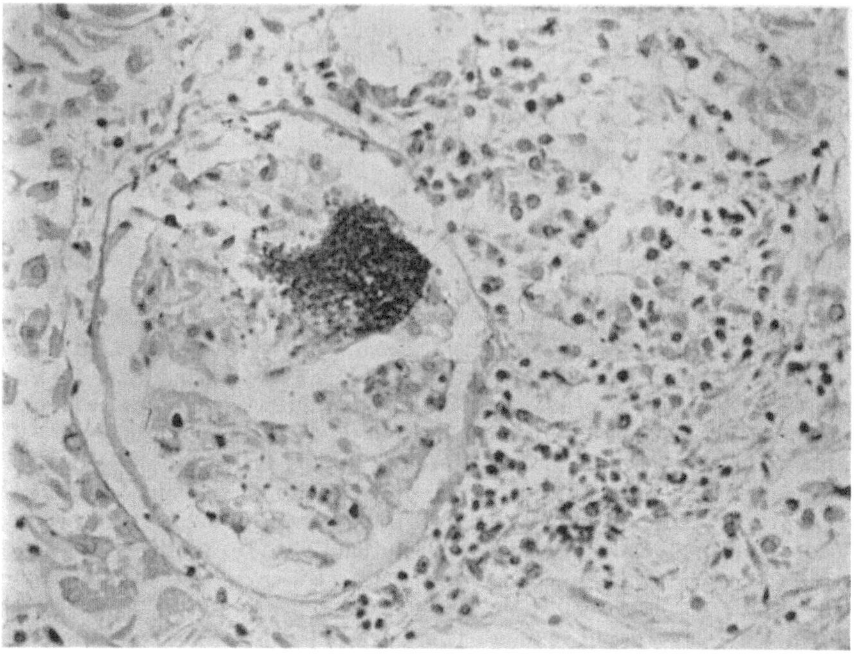

Fig. 59. Focal accumulation of *Torulopsis glabrata* in a glomerulus. Periodic acid Schiff stain. × 82. (After EDEBO, L., JÖNSSON, L. E., HALLÉN, A., and LINBOM, G.: Acta chir. scand. 131, 473—480, 1966)

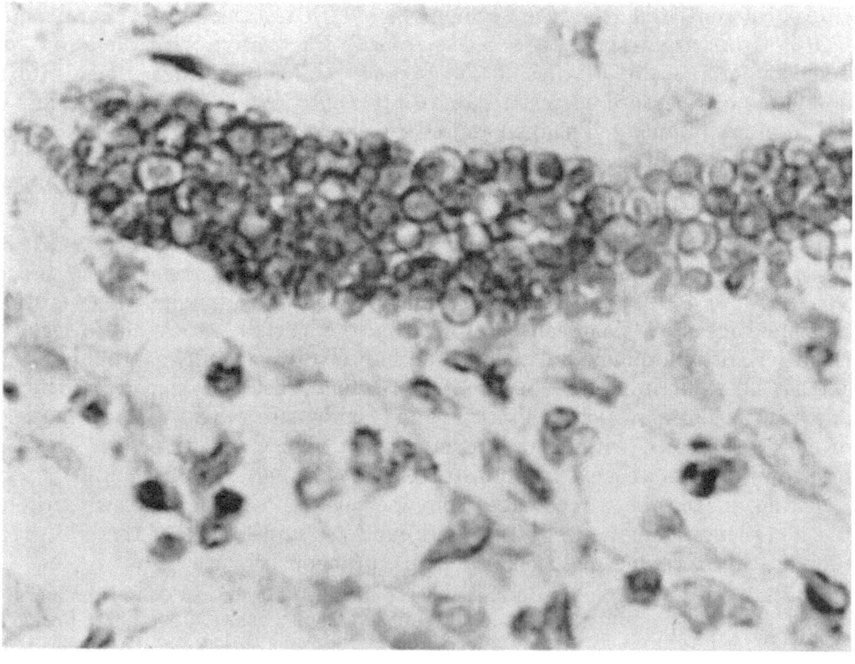

Fig. 60. Rounded yeast organisms of *Torulopsis glabrata* in the kidney. Periodic acid Schiff stain. × 400. (Courtesy of Dr. L. EDEBO)

and GRAY, 1963; LOURIA, 1960; LOURIA et al., 1960, 1967; MALICKE, 1964; MARKS et al., 1970; MIQUENS, 1960; MINKOWITZ et al., 1963; MÜLLER et al., 1967; NEGRONI et al., 1966; OLDFIELD et al., 1968; PASSOS, 1961; PEREZ and GIL, 1960; PERRY, 1964; PLAUT, 1950; REIERSÖL, 1958; RIDDELL and CLAYTON, 1958; ŠAULOV, 1965; SILVEIRA, 1959; SMITH et al., 1963; VIEIRA and BATISTA, 1962). In some instances palely eosinophilic, 2—3 μ by 1—2 μ, yeast-like cells with thin refractile capsules were also evident in tissue sections (Figs. 59, 60) (EDEBO et al., 1966) and were associated with variable inflammatory patterns, most commonly mononuclear (MARKS et al., 1970).

Underlying debilitating diseases and chemotherapeutic agents like antibiotics, immunosuppressive drugs, and steroids appear to predispose to infections with this fungus (GRIMLEY et al., 1965; LOURIA et al., 1967; MARKS et al., 1970; MINKOWITZ et al., 1963; OLDFIELD et al., 1968). The possibility of contaminated intravenous catheters contributing to the entry of *T. glabrata* has been raised by LOURIA et al. (1960) and ROSE and HECKMAN (1970).

The pathogenicity of the organism has been investigated in guinea pigs, mice, rabbits, and rats (HASENCLEVER and MITCHELL, 1962; LÓPEZ-FERDINÁNDEZ, 1952). Following intravenous or intraperitoneal injections of *T. glabrata*, abscesses and lesions consisting predominantly of yeast-like cells in macrophages occurred in these animals.

Ustilagosis

(Ustilagomycosis)

An important phytopathogen affecting the corn industry is the smut fungus (*Ustilago zeae*), which parasitizes corn and closely related teosinte. The fungus affects the various parts of corn plants that are above the ground and characteristically causes excrescences and black smutty leaves on the host. The excrescences contain black chlamydospores and may attain a diameter of several inches. The fungus only causes local infection and does not pervade the entire plant. It is not seedborne and does not infect the germinating seedlings. Eventually, the black lesions release myriads of dark spores, which remain viable for some time in the soil. With favorable conditions the chlamydospores germinate and produce sporidia, which multiply by budding. The wind and other agents spread the spores to young, susceptible corn plants.

For some time *Ustilago* has been of interest to investigators of human disease, because it occasionally contaminates culture media. Though the pathogenicity of the fungus for man has yet to be established, it has been associated with human disease on rare occasions. For example, PREININGER (1937) described a 31-year-old farmer who developed cutaneous lesions at sites of contact with wet clothing after spending the night in a cornfield in a drizzling rain. Spores of *Ustilago* were identified in scrapings from the cutaneous lesions on the patient. Moreover, *Ustilago* has been cultured from the cornea of a patient with keratitis (ANDERSON et al., 1959). In neither of these instances was the fungus identified in the tissue, and its role in the genesis of the lesions is questionable. MOORE et al. (1946) believed that *U. zeae (U. maydis)* was the etiologic agent of a chonic granulomatous meningitis and ependymitis. Lesions with multinucleated giant cells and macrophages contained and surrounded many peculiar structures that resembled the sprout mycelium and spiny or echinulate spores of *U. zeae*. However, the fungus was not cultured and the identity of the pathogen remains speculative. In sensitized individuals spores of *Ustilago* may cause allergic reactions such as asthma (LE COULANT and LOPES, 1957).

Acknowledgements: This chapter could not have been completed without the assistance of numerous individuals, and especially DRS. L. AJELLO, P. K. C. AUSTWICK, A. BAKERSPIGEL, E. H. BOSSEN, N. F. CONANT, C. T. DOLAN, L. EDEBO, D. A. EMANUEL, C.W. EMMONS, B. F. FETTER, L. K. GEORG, M. A. GORDON, W. L. JELLISON, W. KAPLAN, A. H. KLOKKE, M. LAHOURCADE, H. S.

Neilsen, Jr., N.B. Rewcastle, P.M. Stockdale, J.E. Tindall, C.J. Uys, N. Zais, and L.E. Zimmerman, who supplied us with information, tissue sections, cultures, and illustrative material; Mr. Carl Bishop for photography; Miss Margaret F. Jones and the Duke Medical Center Library for reference service; the Williams and Wilkins publishing company; and the editors of the *American Journal of Clinical Pathology, Annals of the New York Academy of Science, Archives of Environmental Health, Archives of Ophthalmology, Canadian Journal of Microbiology, Journal of Bacteriology, Journal of the Medical Association of Alabama, Mycologia, Mycopathologia et Mycologia Applicata, La Presse Medicale, Sabouraudia, Southwestern Veterinarian,* and the *Survey of Ophthalmology* for permission to reproduce previously published illustrations.

References

Adiaspiromycosis

Bakerspigel, A.: *Haplosporangium* in Saskatchewan rodents. Mycologia **48**, 568 (1956). ~ *Haplosporangium* in an additional rodent host *Microtus pennsylvanicus* Drumondi. Nature (Lond.) **179**, 875 (1957). ~ Additional cases of adiaspiromycosis in Canadian rodents. Sabouraudia **4**, 176 (1965). — Batista, A.C., De Lima, J.A., Pessoa, F.P., Shome, S.K.: Adiaspiromycose humana doença pulmonar causada por *Emmonsia brasiliensis* n. sp. Rev. Fac. Med. Univ. Ceara **3**, 24 (1963a). ~ *Emmonsia brasiliensis* n. sp. um Hifomiceto de interesse para a mycopatologia humana. Rev. Fac. Med. Univ. Ceara **3**, 45 (1963b). — Batista, A.C., Shome, S.K.: Isolation of *Emmonsia brasiliensis* from soil. Rev. Fac. Med. Univ. Ceara **3**, 42 (1963). — Carmichael, J.W.: The pulmonary fungus *Haplosporangium parvum*. II. Strain and generic relationships. Mycologia **43**, 604 (1951). ~ *Chrysosporium* and some other aleuriosporic Hyphomycetes. Canad. J. Botany **40**, 1137 (1962). — Chevrel, M.L., Souquet, R., Ferrand, B., Richier, J.L., Doby, J.M., Doby-DuBois, M., Louvet, L.: Polyparasitisme pulmonaire mycosique: premier cas humain d'adiaspiromycose associée à une aspergillose. Ann. Anat. path. **9**, 463 (1964). — Ciferri, R., Montemartini, A.: Taxonomy of *Haplosporangium parvum*. Mycopathologia (Den Haag) **10**, 303 (1959). — Doby-Dubois, M., Chevrel, M.L., Doby, J.M., Louvet, M.: Premier cas humain d'adiaspiromycose, par *Emmonsia crescens* Emmons et Jellison 1960. Bull. Soc. Path. exot. **57**, 240 (1964). — Dowding, E.S.: *Haplosporangium* in Canadian rodents. Mycologia **39**, 372 (1947). ~ The pulmonary fungus *Haplosporangium parvum*, and its relationship with some human pathogens. Canad. J. Res., Sect. E **25**, 195 (1947b). — Dvǒrák, J., Otčenášek, M., Prokopič, J.: The distribution of adiaspiromycosis. J. Hyg. Epidem. (Praha) **9**, 510 (1965). ~ Seasonal incidence of adiaspores of *Emmonsia crescens* Emmons and Jellison 1960 in wildly living animals. Mycopathologia (Den Haag) **31**, 71 (1967). — Emmons, C.W.: Coccidioidomycosis. Mycologia **34**, 452 (1942). — Emmons, C.W., Ashburn, L.L.: The isolation of *Haplosporangium parvum* n. sp. and *Coccidioides immitis* from wild rodents: their relationship with coccidioidomycosis. Publ. Hlth. Rep. (Wash.) **57**, 1715 (1942). — Emmons, C.W., Jellison, W.L.: *Emmonsia crescens* n. sp. and adiaspiromycosis (haplomycosis) in mammals. Ann. N.Y. Acad. Sci. **89**, 91 (1960). — Erickson, A.B.: The fungus (*Haplosporangium parvum*) in the lungs of the beaver (*Castor canadensis*). J. Wildlife Management **13**, 419 (1949). — Jellison, W.L.: An undetermined parasite in the lungs of a rock rabbit, *Ochotona princeps* Richardson (Lagomorpha: Ochotonidae). Proc. helminth. Soc. Wash. **14**, 75 (1947). ~ Haplomycosis in Montana rabbits, rodents and carnivores. Publ. Hlth. Rep. (Wash.) **65**, 1057 (1950). ~ The presence of pulmonary fungus in Korean rodents. Publ. Hlth. Rep. (Wash.) **69**, 996 (1954). ~ Haplomycosis in Sweden. Nord. Vet.-Med. **8**, 504 (1956). ~ Haplomycosis in Japan and Africa. Mycologia **50**, 580 (1958). — Jellison, W.L., Lord, R.D.: *Adiaspiromycosis* in Argentine mammals. Mycologia **56**, 374 (1964). — Jellison, W.L., Peterson, R.S.: *Emmonsia* a fungus, and *Besnoitia* a protozoan, reported from S. America. Bot. Chileno-Parasitol. **15**, 46 (1964). — McDiarmid, A., Austwick, P.K.C.: Occurrence of *Haplosporangium parvum* in the lungs of the mole (*Talpa eutopaea*). Nature (Lond.) **174**, 843 (1954). — Menges, R.W., Habermann, R.T.: Isolation of *Haplosporangium parvum* from soil and results of experimental inoculations. Amer. J. Hyg. **60**, 106 (1954). — Misra, S.P., Shende, G.Y., Yerawadekar, S.N., Padhye, A.A., Thirumalachar, M.J.: *Allescheria boydii* and *Emmonsia ciferrina* isolated from patients with chronic pulmonary infections. Hindustan Antibiot. Bull. **9**, 99 (1966). — Padhye, A.A., Carmichael, J.W.: *Emmonsia brasiliensis* and *Emmonsia ciferrina* are *Chrysosporium pruinosum*. Mycologia **60**, 445 (1968). — Prokopič, J., Dvořák, J., Otčenášek, M.: Adiaspiromycosis (haplomycosis) in Czechoslovakia: preliminary report. Sabouraudia **4**, 35 (1965). — Smith, J.M.B., Lancaster, M.C.: Adiaspiromycosis in the brush opossum, *Trichosurus vulpecula*, in New Zealand, Sabouraudia **4**, 146 (1965). — Taylor, R.L., Miller, B.E., Rust, J.H., Jr.: Adiaspiromycosis in small mammals of New Mexico. Mycologia **59**, 513 (1967). — Tevis, L.: Additional records of *Haplosporangium parvum* in mammals in Britain. Nature (Lond.) **177**, 437 (1956). — Thirumalachar, M.J., Padhye, A.A., Srinivasan, M.C.: *Emmonsia ciferrina*, a new species from India. Mycopathologia (Den Haag) **26**, 323 (1965).

Allescheriosis

AJELLO, L.: The isolation of *Allescheria boydii* Shear: an etiologic agent of mycetomas from soil. Amer. J. trop. Med. **1**, 227 (1952). — AROEINA-NEVES, J.: Contribucae ao estudo dos micetomas em Minas Georis, Brasil, *Monosporium apiospermum* Saccardo, 1911. Rev. bras. Biol. **2**, 305 (1942). — ARONSON, S.M., BENHAM, R., WOLF, A.: Maduromycosis of the central nervous system. J. Neuropath. exp. Neurol. **12**, 158 (1953). — AVRAM, A.: Grains expérimentaux maduromycosiques et actinomycosiques à *Cephalosporium falciforme, Monosporium apiospermum, Madurella mycetomi et Nocardia asteroides.* Mycopathologia (Den Haag) **32**, 319 (1967). — AVRAM, A., COJOCARU, I., DULĂMITĂ, L.: Le pied de madura à *Monosporium apiospermum.* A propos de 3 cas roumains. Dermatologica (Basel) **136**, 176 (1968). — BATTISTINI, F., BRICEÑO-MAAZ, T., DEBRICEÑO-MAAZ, C.: Caso de pie de madura producido por el *Monosporium apiospermum.* Gac. méd. Caracas **67**, 181 (1958). — BAXTER, M., MURRAY, I.G., TAYLOR, J.J.: A case of mycetoma with serological diagnosis of *Allescheria boydii.* Sabouraudia **5**, 138 (1966). — BELDING, D.L., UMANZIO, C.B.: A new species of the genus *Monosporium* associated with chronic otomycosis. Amer. J. Path. **11**, 856 (1935). — BENHAM, R.W., GEORG, L.K.: *Allescheria boydii,* causative agent in a case of meningitis. J. invest. Derm. **10**, 99 (1948). — BLANK, F., STUART, E.A.: *Monosporium apiospermum* Sacc., 1911, associated with otomycosis. Canad. med. Ass. J. **72**, 601 (1955). — BOL'SHAKOVA, G.M., STEPANISHCHEVA, Z.G.: [On Madura foot]. Vestn. Derm. Vener. **34**, 40 (1960) (Russian). — BORELLI, D.: Madurella mycetomi: fialides, fialosporos, inoculacion al raton. Bol. Venez. de Lab. Clin. **2**, 1 (1957). — BOYD, M.F., CRUTCHFIELD, E.D.: A contribution to the study o mycetoma in North America. Amer. J. trop. Med. **1**, 215 (1921). — BRICENOMAAZ, T.: [Mycetoma caused by *Monosporium apiospermum.*]. Derm. Venez. **6**, 155 (1967). — Camain, R., SEGRETAIN, G., NAZIMOFF, O.: Les mycétomes du Sénégal et de la Mauritanie: apercu épidémiologique et étude histopathologique. Sem. Hôp. Paris **33**, 771 (1957). — CARRION, A.L., KNOTT, J.: Mycetoma by *Monosporium apiospermum* in St. Croix, Virgin Islands. Puerto Rico J. publ. Hlth. **20**, 84 (1944). — CIFERRI, R., REDAELLI, P.: Probabilo sinonimi di *Allescheria boydii (Monosporium apiospermum).* Mycopathologia (Den Haag) **5**, 120 (1950). — COCKSHOTT, W.P.: The therapy of mycetoma. W. Afr. med. J. **6**, 101 (1957). — CONANT, N.F., SMITH, D.T., BAKER, R.D., CALLAWAY, J.L., MARTIN, D.S.: *Manual of Clinical Mycology,* Ed. 2, pp. 240—252. Philadelphia: W.B. Saunders Company 1962. — CONVIT, J. BORELLI, D., ALBORNOZ, R., RODRÍQUEZ, G., HÓMEZ, J.: Micetomas, Cromomicosis, esporotricosis y enfermedad de Jorge Lobo. Mycopathologia (Den Haag) **15**, 394 (1961). — COURTOIS, G., DELOOF, C., THYS, A., VANBREUSEGHEM, R., BURNETTE, A.: Neuf cas de pied de Madura congoleus par *Allescheria boydii, Monosporium apiospermum et Nocardia madurae.* Ann. Soc. belge Méd. trop. **34**, 371 (1954). — CREITZ, J., HARRIS, H.W.: Isolation of *Allescheria boydii* from sputum. Amer. Rev. Tuberc. **71**, 126 (1955). — DE ALMEIDA, F., SIMOES BARBOSA, F.A.: Mycetomas brasileiros. An. Fac. Med. S. Paulo 235 (1940). — DESTOMBES, P., MARIAT, F., ROSATI, L., SEGRETAIN, G.: *Mycetomata in the Republic of Somaliland.* C. R. Acad. Sci. (Paris) Ser. D. **263**, 2062 (1966). — DOWDING, E.S.: *Monosporium apiospermum,* a fungus causing madura foot in Canada. Canad. med. Ass. J. **33**, 23 (1935). — DROUHET, M.: The status of fungus diseases in France. In: *Therapy of Fungus Diseases: An International Symposium,* p. 43. Boston: Little, Brown, and Company 1955. — EL-MOFTY, A.M., ISKANDER, I.D., NADA, M., ZAKI, S.M.: "Madura foot" in Egypt. Brit. J. Derm. **77**, 365 (1965). — EMMONS, C.W.: *Allescheria boydii* and *Monosporium apiospermum.* Mycologia, **36**, 188 (1944). — EMMONS, C.W., GREENHALL, A.M.: *Histoplasma capsulatum* and house bats in Trinidad. Sabouraudia **2**, 18 (1962). — ERNEST, J.T., RIPPON, J.W.: Keratitis due to *Allescheria boydii (Monosporium apiospermum).* Amer. J. Ophthal. **62**, 1202 (1966). — FIENBERG, R.: Madura foot in a native American. Amer. J. clin. Path. **14**, 239 (1944). — GAY, D.M., BIGELOW, J.B.: Madura foot due to *Monosporium apiospermum* in a native American. Amer. J. Path. **6**, 325 (1930). — GELLMAN, M., GAMMEL, J.A.: Madura foot: a third case of monosporiosis in a native American. Arch. Surg. **26**, 295 (1933). — GORDON, M.A., VALLOTTON, W.W., CROFFEAD, G.S.: Corneal allescheriosis: a case of keratomycosis treated successfully with nystatin and amphotericin B. Arch. Ophthal. **62**, 758 (1959). — GREEN, W.O., ADAMS, T.E.: Mycetoma in the United States: a review and report of seven additional cases. Amer. J. clin. Path. **42**, 75 (1964). — HANLON, T.J., GEPHARDT, M.C., HOPPS, H.C.: Mycetoma in the U.S.: review with report of a case from new area. J. Okla. med. Ass. **48**, 299 (1955). — HAUKOHL, R.S., SADOFF, H.B.: Mycetoma: report of case due to *Monosporium apiospermum* in a native of Minnesota. Wis. med. J. **53**, 477 (1954). — HERRERO, F.J., USABEL, E.J., SIRENA, A.: Primer case en Tucumán (Argentina) de micetoma podal por *Monosporium apiospermum.* Pren. méd. argent. **42**, 3652 (1955). — JONES, J.W., ALDEN, H.S.: Maduromycotic mycetoma (Madura foot): report of case occurring in American Negro. J. Amer. med. Ass. **96**, 256 (1931). — JORGE, J.M., INTROZZI, A.S., NIÑO, F.L.: Micetoma podal maduromicósico con granos blancos por *Monosporium apiospermum.* Bol. Trab. Acad. Argent. Cir. **25**, 508 (1941). — JOSEFIAK, E.J., KOKIKO, G.V.: Mycetoma of the hand: report of a case. Arch. Path. **67**, 55 (1959). — KATSNEL' son, I.I., Levkov, A.A.: [A deep mycosis with mycetoma type skin damage]. Vestn. Derm. Vener. **36**, 78 (1962). (Russian). — Lapatí, F.: Das Mycetom. In: *Die Pilzkrankheiten der Haut durch Hefen. Schimmel, Aktinomyceten und verwandte Erreger*, p. 479. Marchionni, A., and Götz, H. (Eds.). Berlin-Göttingen-Heidelberg: Springer 1963. — Lapatí, F., Ortiz, Y.: Los micetomas en Mexico. Memorias del primer congreso de dermatologia. 126 (1963). — Ley, A.P.: Experimental fungus infections of the cornea: Preliminary report. Amer. J. Ophthal. **42**, 59 (1956). — Louria, D.B., Lieberman, P.H., Collins, H.S., Blevins, A.: Pulmonary mycetoma due to *Allescheria boydii*. Arch. intern. Med. **117**, 748 (1966). — Pillay, V.K.G., Wilson, D.M., Ing, T.S., Kark, R.M.: Fungus infection in steroid treated systemic lupus erythematosus. J. Amer. med. Ass. **205**, 261 (1968). — Mackinnon, J.E.: Causal agents of maduromycosis in the neotropical region. An. Fac. Med. Montevideo **48**, 453 (1963). — Mahgoub, E.S.: The value of gel diffusion in the diagnosis of mycetoma. Trans. roy. Soc. trop. Med. Hyg. **38**, 560 (1964). — Mariat, F.: On the geographic distribution and spread of the agents of mycetomas. Bull. Soc. Path. exot. **56**, 35 (1963). — Meyer, E., Herrold, R.D.: *Allescheria boydii* isolated from a patient with chronic prostatitis. Amer. J. clin. Path. **35**, 155 (1961). — Meyerding, H.W., Evert, J.A.: Mycetoma or Madura foot: report of cases including one case of maduromycosis of the hand. Minn. Med. **30**, 407 (1947). — Misra, S.P., Shende, G.Y., Yernadekar, S.N., Padhye, A.A., Thirumalachar, M.J.: *Allescheria boydii* and *Emmonsia ciferrina* isolated from patients with chronic pulmonary infections. Hindustan Antibiot. Bull. **9**, 99 (1966). — Mohr, J.A., Muchmore, H.G.: Maduromycosis due to *Allescheria boydii*. J. Amer. med. Ass. **204**, 335 (1968). — Montpellier, J., Guillon, P.: Mycétome du pied (type pied de Madura) du à *l'Aleuvisma apiospermum*. Bull. Soc. Path. exot. **14**, 285 (1921). — Naumann, G., Green, W.R., Zimmerman, L.E.: Mycotic keratitis. Amer. J. Ophthal. **64**, 668 (1967). — Niño, F.L.: Micetoma podal maduromicósico con granos blancos por *Monosporium apiospermum* en la Republica Argentina. Bol. Inst. Clín. quir. (B. Aires) **17**, 483 (1941). ~ *Allescheria boydii* Shear, 1921, agents etiológico del micetoma maduromicósico a granos blancos en la Argentina. Pren. méd. argent. **36**, 314 (1949). ~ Estudio micologico de la segunda observación argentina de micetoma podal por, *Monosporium apiospermum*. Pren. méd. argent. **40**, 764 (1953). — Niño, F.L., Freire, R.S.: Maduromycotic mycetoma in Chaco province. Mycopathologia (Den Haag) **28**, 95 (1966). — Niño, F.L., Freire, R.S., Salica, P.: El mícetoma maduromicosico en la provincia del Chaco. Rev. Asoc. méd. argent. **76**, 359 (1962). — Pautler, E.E., Roberts, R.W., Beamer, P.R.: Mycotic infection of the eye: *Monosporium apiospermum* associated with corneal ulcer. Arch. Ophthal. **53**, 385 (1955). — Rees. W.H.: A case of mycetoma in the New Hebrides., Trans. roy. Soc. trop. Med. Hyg. **56**, 538 (1962). — Rosen, F., Deck, J.H.N., Rewcastle, N.B.: *Allescheria boydii*: unique systemic dissemination to thyroid and brain. Canad. med. Ass. J. **93**, 1125 (1965). — Scharyj, M., Levine, N., Gordon, H.: Primary pulmonary infection with *Monosporium apiospermum*. J. infect. Dis. **106**, 141 (1960). — Schmitt, J.A., Zabransky, R.J., Janidlo, A.S., Pearsons, J.E.: Experimental maduromycosis in the laboratory mouse. Mycopathologia (Den Haag) **18**, 164 (1963). — Seabury, J.H., Kroll, V.R., Landreneau, R.: A conservative method of treatment for maduromycosis. South med. J. (Birmingham, Ala.) **52**, 1176 (1959). — Shaw, R.M., MacGregor, J.W.: Maduromycosis: with the report of a case due to *Monosporium apiospermum*. Canad. med. Ass. J. **33**, 23 (1935). — Stoeckel, H., Ermer, C.: Ein Fall von *Monosporium*-mycetom der Lunge. Beitr. Klin. Tuberk. **122**, 130 (1960). — Taylor, W.W., Radcliffe, F., Van Peenen, P.F.D.: The isolation of pathogenic fungi from the soils of Egypt, the Sudan, and Ethiopia. Sabouraudia **3**, 235 (1964). — Theodorescu, St., Vulcan, P., Avram, A., Stoica, I.: Micetom cugraunti albi al piciorului determinat *Monosporium apiospermum*. Derm. Vener. (Buc.) **1**, 114 (1956). — Tong, J.L., Valentine, E.H., Durrance, J.R., Wilson, G.M., Fisher, D.A.: Pulmonary infection with *Allescheria boydii:* report of a fatal case. Amer. Rev. Tuberc. **78**, 604 (1958). — Travis, R.E., Ulrich, E.W., Phillips, S.: Pulmonary allescheriasis. Ann. intern. Med. **54**, 141 (1961). — Twining, H.E., Dixon, H.M., Weidman, F.D.: Penicillin in the treatment of Madura foot: report of 2 cases. U.S. nav. med. Bull. **46**, 417 (1946). — Vanbreuseghem, R.: Early diagnosis, treatment, and epidemiology of mycetoma. Rev. Med. Vet. Mycol. **6**, 49 (1967). — Verghese, A., Klokke, A.H.: Histologic diagnosis of species of fungus causing mycetoma. Indian J. med. Res. **54**, 524 (1966). — Wolf, A., Benham, R., Mount, L.: Maduromycotic meningitis. J. Neuropath. exp. Neurol. **7**, 112 (1947). — Zaffiro, A.: Forma singolare di mycosi cutanea da *Monosporium apiospermum* a sviluppo clinicamete setticemo: Considerazioni diagnostiche deduzioni mediocolegali. Giorn. ital. Med. Milit. **86**, 636 (1938).

Alternariosis

Ainsley, R., Smith, B.: Fungal flora of the conjunctival sac in healthy and diseased eyes. Brit. J. Ophthal. **49**, 505 (1965). — Borsook, M.E.: Skin infection due to *Alternaria tenuis* with the report of a case. Canad. med. Ass. J. **29**, 479 (1933). — Botticher, W.W.: *Alternaria* as a possible human pathogen. Sabouraudia **4**, 256 (1966). — Harris, L.H.:

Experimental reproduction of respiratory mold allergy. J. Allergy **12**, 279 (1941). — HENRICI, A.T.: *Molds, Yeasts, and Actinomycetes.* John Wiley and Sons, Inc., 2nd ed., 1947. — HOPKINS, J.G., BENHAM, R.W., KESTEN, B.M.: Asthma due to a fungus *Alternaria.* J. Amer. med. Ass. **94**, 6 (1930). — HYDE, H.A., WILLIAMS, D.A.: A daily census of *Alternaria* spores caught from the atmosphere at Cardiff in 1942 and 1943. Trans. Brit. Mycol. Soc. **29**, 78 (1946). — MILLER, W.S., TINDALL, J.P.: Presented at Fourteenth Annual Meeting of the Zola Cooper Memorial Clinico-Pathologic Seminar, Miami, Florida Nov. 13, 1967. — OHASHI, Y.: On a rare disease due to *Alternaria tenuis Nees* (Alternariasis). Tohoku J. exp. Med. **72**, 78 (1960). — PRATT, H.N., CROSSMAN, R.: The comparative atopic activity of *Alternaria* spores and mycelium. J. Allergy **13**, 227 (1942). — PRINCE, H.E., EPSTEIN, S., FIGLEY, K.D., WITTICH, F.W., HENRY, L.D., MORROW, M.B.: Mold fungi in the etiology of respiratory allergic disease. IX. Further studies with mold extracts. Ann. Allergy **7**, 301 (1949). — PRINCE, H.E., HALPIN, L.J., TALBOTT, G., ETTER, R.L., RAYMER, W.J., JACKSON, R.H., BARTLETT, L.L., MANSMANN, J.A., FRAYSER, L., EISENBERG, B.C., MORROW, M.B., MEYER, G.H.: Molds and bacteria in the etiology of respiratory allergic diseases. XXI. Studies with mold extracts produced from cultures grown in modified synthetic media. Ann. Allergy **19**, 259 (1961). — WALTON, C.H.A.: Respiratory allergy due to fungi. Canad. med. Ass. J. **60**, 272 (1949).

Arthrodermosis

EMMONS, C.W.: *Trichophyton mentagrophytes (Pinoyella simii)* isolated from dermatophytosis in the monkey. Mycopathologia (Den Haag) **2**, 317 (1940). — GUGNANI, H.C., MULAY, D.N., MURTY, D.K.: Fungus flora of dermatophytosis and *Trichophyton simii* infection in North India. Indian. J. Derm. Venerol. 33, 73 (1967a). — GUGNANI, H.C., SHRIVASTAV, J.B., GUPTA, N.P.: Occurrence of *Arthroderma simii* in soil and on hair of small mammals. Sabouraudia **6**, 77 (1967b). — KLOKKE, A.H., DURAIRAJ, P.: The causal agents of superficial mycoses in rural areas of South India. Sabouraudia **5**, 153 (1967). — PINOY, E.: Sur une teigne cutanée du singe. C. R. Soc. Biol. (Paris) **72**, 59 (1912a). ~ Epidermophyton du singe. Bull. Soc. Path. exot. **5**, 60 (1912b). — RIPPON, J.W., ENG, A., MALKINSON, F.D.: *Trichophyton simii* infection in the United States. Arch. Derm. **98**, 615 (1968). — SINGH, M.P.: Studies on the occurrence of mycotic infections in domestic animals and poultry. *M. V. Sc. Thesis, Agra University,* 1963 (cited by STOCKDALE et al., 1965). — STOCKDALE, P.M., MAC KENZIE, D.W.R., AUSTWICK, P.K.C.: *Arthroderma simii* sp. nov., the perfect stage of *Trichophyton simii* (Pinoy) Comb. Nov. Sabouraudia **4**, 112 (1965).

Beauveriosis

BASSI, A.: Del mal del segno calcinaccio o moscardino malattia che affligge: bachi da seta. Partel. Teorica Tip. Orcesi, Lodi, 1835. — ELLINGBOE, A.H., KERNKAMP, M.F., HAWS, B.A.: Sweetclover weevil parasitized by *Beauvaria bassiana* (Bals.) Vuill. in Minnesota. J. Econ. Entomol. **50**, 173 (1957). — FRÉOUR, P., LAFOURCADE, M., CHOMY, P.: Les champignons "*Beauveria*" en pathologic humaine: A propos d'un cas à localisation pulmonaire. Presse méd. **74**, 2317 (1966a). ~ Une mycose nuovelle: étude clinique et mycologique d'une localization pulmonaire de "*Beauveria*". Bull. Soc. méd. Hôp. Paris **177**, 197 (1966b). ~ Sur une mycose pulmonaire nouvelle due à "*Beauveria*". J. Méd. Bordeaux **143**, 823 (1966c). — GEORG, L.W., WILLIAMSON, W.M., TILDEN, E.B., GETTY, R.E.: Mycotic pulmonary disease of captive giant tortoises due to *Beaveria bassiana* and *Paecilomyces fumoso-roseus.* Sabouraudia **2**, 80 (1962). — LAHOURCADE, M.: *Bull. Soc. fran. Mycol. Méd.* **2**, 30 (1966). — LANGERON, M.: Mycose oculaire primitive due au *Beauveria brumpti.* Bull. Acad. Méd. (Paris) III, 133 (1934). — MADELIN, M.F.: *Disease caused by Hyphomycetous Fungi.* In: *Insect Pathology: An Advanced Treatise* (Edited by STEINHAUS E.A.). Vol. **2**, pp. 233—271. New York and London: Academic Press, 1963.

Cephalosporiosis

ALMEIDA, F. DE, SIMÕES-BARBOSA, F.A.: [*Cephalosporium recifei* isolated from case of Madura foot]. Arg. Inst. biol. S. Paulo **11**, 1 (1940a). ~ Contribução para o estudo general dos micetomas maduromicóticos observados no Brasil. Micetomas brasileires. An. Fac. Med, S. Paulo **16**, 235 (1940b). — ALMEIDA, F. DE, DA SILVA-LACAZ, C., RIBEIRO-OLIVEIRA, D.. CORDEIRO-DE AZEVEDO: Contribução para o estudo da micetomas ha Sao Paulo, Brazil, Madura foot a *Cephalosporium sp.* Rev. bras. Biol. **8**, 287 (1948). — ARÊA-LEÃO, A.E. DE. LÔBO, J.: Mycétome du pied a *Cephalosporium recifei* n. sp. Mycétome a grains blancs. C. R. Soc. Biol. (Paris) **117**, 203 (1934a). ~ Micetoma podal a *Cephalosporium recifei* n. sp. Micetoma de grãos biancos. C. R. Soc. Biol. Rio de S. **177**, 203 (1934b). — ARIEVITCH, A.M., TJUFILINA, O.V., TEPLITZ, V.V.: [Gummo-ulcerative cephalosporiosis of the legs]. Vestn. Derm. Vener. (Mosk.) **38**, 73 (1964). (Russian). — ARIEVITCH, A.M., STEPANISHCHEVA, Z.G., TJUFILINA, O.V., KUKOLEVA, Z.I., MALKINA, A.T., TEPLITZ, V.V.: Gummato-ulcerative cephalosporiosis of the skin. Mycopathologia (Den Haag) **28**, 113 (1966). — ASCHIERI, E.: Recerche sistematiche e fesiologiche su *Hyalopus* causa di onicomicosi. Att. lst Bot. Pavia **4**, 45 (1932). — Avram, A.: Étude clinique et mycologique concernant le premier cas européen de mycétome déterminé par *Cephalosporium sp.* Mycopathologia (Den Haag) **24**, 177 (1964). ~ Étude sur des mycétomas de Roumaniae. Mycopathologia (Den Haag) **28**, 1 (1966a). ~ Experimental induction of grains with *Cephalosporium falciforme.* Sabouraudia **5**, 89 (1966b). ~ Experimental infections with *Cephalosporium falciforme, Monosporium apiospermum, Madurella mycetomi* and *Nocardia asteroides.* Mycopathologia (Den Haag) **32**, 319 (1967). — Avram, A., Porojan, I., Cojocaru, I.: Trois nouveaux cas roumains de pied de Madura. Agents incriminés: *Nocardia asteroides, Cephalosporium falciforme* et *Madurella mycetomi.* Dermatologica (Basel) **136**, 311 (1968). — Ballagi, S.: Mykologische Beschreibung der Acremoniosis. Arch. Derm. Syph. (Berl.) **166**, 405 (1932). — Bastardo de Albornoz, M.C.: Micetoma (Pie de Madura) debido a *Cephalosporium recifei.* Derm. Venez **4**, 56 (1963/1964). — Batista, A., Lima, J.A., Oliveira, S.D., Artagaveytia-Allende, R.C.: Alguns *Hyalopus* assinalados sôbre suco gástrico e bile humana no Recife. Atas. Inst. Micol. Recife. **1**, 275 (1960). — Baylet, R., Camain, R., Bèzes, H.: Mycétomes à *Cephalosporium* au Sénégal. Bull. Soc. Path. exot. **54**, 802 (1961). — Bedell, A.J.: *Cephalosporium* keratitis. Trans. Amer. ophth. Soc. **44**, 80 (1946). ~ *Cephalosporium* keratitis. Amer. J. Ophthal. **30**, 997 (1947). — Benedek, T.: Über Cephalosporiose. Ein Beitrag zur Kenntnis der seltenen Mykosen, unter besonderer Berücksichtigung der Serumdiagnose. Arch. Derm. Syph. (Berl.) **154**, 96 (1927a). ~ Vergleichende Untersuchungen über einige Arten der Gattung „*Cephalosporium*" nebst Mitteilung einer neuen Art: „*Cephalospor. Niveolanosum*" nov. spec. Arch. Derm. Syph. (Berl.) **154**, 154 (1927b). — Bommer, S., Werner, W., Haufe, F.: Beitrag zur Pathologie der Cephalosporiose. Derm. Wschr. **144**, 952 (1961). — Borelli, D.: Esquemas de micologia medica: micosis y suo agentes principales en Venezuela. Derm. Venez **3**, 136 (1962). — Boucher, H.: Les mycoses gommeuses de la Côte d'Ivoire. Bull. Soc. Path. exot. **11**, 306 (1918). — Burda, C.D., Fisher, E., Jr.: Corneal destruction by extracts of *Cephalosporium* mycelium. Amer. J. Ophthal. **50**, 926 (1960). — Byers, J.L., Holland, M.G., Allen, J.H.: *Cephalosporium* keratitis. Amer. J. Ophthal. **49**, 267 (1960). — Cabrini, C., Redaelli, P.: Sopra un caso di cefalosporiosi tonsillare. (*Cephalosporium acremonium* Corda). Boll. Soc. med.-chir. Pavia **7**, 475 (1929). — Carrión, A.L.: Estudio micológico de un caso de micetoma por *Cephalosporium* en Puerto-Rico. Mycopathologia (Den Haag) **6**, 165 (1939). ~ *Cephalosporium falciforme* sp. nov. A new etiologic agent of maduromycosis. Mycologia **43**, 522 (1951). — Catanei, A.: Sur quelques champignons saprophytes isolés au cours de la recherche de mycoses superficielles. Arch. Inst. Pasteur Alger **22**, 119 (1944). — Ciferri, R.: *Cephalosporium pseudofermentum* n. sp., isolato della bocca dell'uomo. Arch. Protistenk. **78**, 227 (1932). — Coutelen, F., Cochet, G.: Étude biologique d'un *Cephalosporium*, agent pathogène d'une gomme cervico-maxillaire de l'homme. C. R. Soc. Biol. (Paris) **139**, 392 (1945). — Coutelen, F., Cochet, G., Biquet, J.: Les Céphalosporioses humaines. Révue critique à propos d'un cas. Ann. Parasit. hum. comp. **23**, 364 (1948). — Cowen, D.E.: *Cephalosporium* midline granuloma. Trans. Indiana Acad. Ophthal. Otolaryng. **48**, 78 (1965). — Cowen, D.E., Dines, D.E.: *Cephalosporium* midline granuloma. Ann. intern. Med. **60**, 731 (1964) (Abstract). — Cowen, D.E., Dines, D.E., Chessen, J., Proctor, H.H.: *Cephalosporium* midline granuloma. Ann. intern. Med. **62**, 791 (1965). — Debusmann, M.: Über das Vorkommen eines seltenen Pilzes (*Cephalosporium acremonium* Corda) im Blut bei tonsillogener Sepsis. Arch. Kinderheilk. **116**, 172 (1939). — Destombes, P., Segretain, G.: Les mycétomes fongigues: caractères histologiques et culturaux. Arch. Inst. Pasteur Tunis **39**, 273 (1962). — Douglass, R., Simpson, R.E.: *Cephalosporium* in pleural fluid. Amer. Rev. Tuberc. **48**, 237 (1943). — Drouhet, E., Martin, L., Segretain, G., Destombes, P.: Mycose méningocérébrale à "*Cephalosporium*". Presse méd. **31**, 1809 (1965). — Du Bois, C.: Onychomycoses non teigneuses. Dermatologica (Basel) **86**, 57 (1942). — Fine, B.S.: Intraocular mycotic infections. Lab. Invest. **11**, 1161 (1962). — Fine, B.S., Zimmerman, L.E.: Exogenous intraocular fungus infections: with particular reference to complications of intraocular surgery. Amer. J. Ophthal. **48**, 151 (1959). — Focosi, M.: Su di un caso di concrezione micotica del sacco lacrimale. Boll. Oculist. **5**, 554 (1932a). — Su di un caso di cheratomicosi da *Cephalosporium.* (Contributo clinico sperimentale). Boll. Oculist. **12**, 1250 (1932b). — Gaind, M.L., Padhye, A.A., Thirumalachar, M.J.: Madura foot in India caused by *Cephalosporium infestans* sp. nov. Sabouraudia **1**, 230 (1962). — Gamet, A., Brottes, H., Essomba, R.: Nouveaux cas de mycétomes dépistés au Cameroun. Bull. Soc. Path. exot. **57**, 1191 (1964). — Gingrich, W.D.: Keratomycosis. J. Amer. med. Ass. **179**, 602 (1962). — Gonzalez-Ochoa, A., Ruiloba, J.: Acción del propionato de socio, in vitro sobre *Actinomyces mexicanas* y *Cephalosporium sp.* Ensayo terapéutico en micetomas producidos por estos hongos. Rev. Inst. Salubr. Enferm. trop. (Méx.) **5**, 83 (1944). — Gougerot, H., Burnier, R., Duché, J.: Mycose végétante et ulcéreuse due au *Cephalosporium griseum.* Bull. Soc. franç. Derm. Syph. **40**, 417 (1933). — Green, W.R., Bennett, J.E., Goos, R.D.: Ocular penetration of amphotericin B: a report of laboratory studies and a case report post surgical *Cephalosporium* endophthalmitis. Arch. Ophthal. **73**,

769 (1965). — Grütz, O.: Beitrag zu den seltenen Mykosen: Über einen durch *Akremonium* verursachte Pilzerkrankung. Derm. Wschr. **80**, 765 (1925). — Haensch, R.: Cephalosporiose ein Beitrag zur Klinik der seltenen Mykosen. Haut- u. Geschlechtskrankh. **23**, 137 (1957). — Harada, S., Usui, Y.: [On a species of *Cephalosporium* isolated from the epidermophytosis between the toes]. Nagaoa **7**, 51 (1960) (Japanese). — Hartmann, E.: Über ein gemeinsames Vorkommen von *Cephalosporium* und *Trichophyton gypseum*. Derm. Wschr. **82**, 565 (1926). — Hoffmann, D.H., Naumann, G.: Ein Beitrag zur Pilzinfektion der Hornhaut. Klin. Mbl. Augenheilk. **142**, 286 (1963). — Janke, D.: Zur Klinik und Mykologie der Cephalosporiose. Ein Beitrag zur Kenntnis seltener Mykosen. Arch. Derm. Syph. (Berl.) **188**, 257 (1949). — Janke, D., Rohrschneider, W.: Beitrag zu den seltenen Mykosen: über eine Pilzerkrankung der Tränenröhren und der Oberhaut und Befund eines bisher unbekannten *Cephalosporiums*. Derm. Wschr. **123**, 49 (1951). — Kambayashi, T.: Botanische Untersuchungen über japanische Fadenpilze, die auf der Menschenhaut parasitieren. III. Mitteilung. Über *Cephalosporium nigrum* n. sp. isoliert von einer Dermatomycosis in Oethiopien. Botan. Mag. (Tokyo) **51**, 436 (1937). — Küper, J.: Zur Klinik postoperativer intraokutarer Mykosen. Klin. Mbl. Augenheilk. **140**, 827 (1962). — Kesteven, H.L.: The mycotic flora of "Surfer's foot" in Sydney. Med. J. Aust. **1**, 420 (1939). — Kukoleva, L.I., Malkuha, A.R.: [A case of extensive gumma-ulcerative cephalosporiosis]. Vestn. Derm. Vener. (Mosk.) **41**, 76 (1967). (Russian). — Lacaz, C. Da S., Fava Netto, C.: Contribution à l'étude des agents étiologiques des maduromycoses. Folia clín. biol. **21**, 331 (1954). — Latapí, F., Ortiz, Y.: Los micetomas en México. Memorias del primer congreso de dermatología. pp. 126—144 (1963). — Lehner, E. Von: Über einen Fall von *Acremoniosis*. Arch. Derm. Syph. (Berl.) **166**, 399 (1932). — Lôbo, J.: Micetomas podias em Pernambuco. An. Fac. Med. Recife **8—9**, 5 (1943). ~ Micetomas em Pernambuco. Rev. Fac. Med. Univ. Ceara **3**, 5 (1963). — MacKinnon, J.E.: Los agentes de maduromicosis de los géneros *Monosporium, Allescheria, Cephalosporium* y otros de dudosa identidad. An. Fac. Med. Montevideo **36**, 153 (1951). ~ A contribution to the study of the causal organisms of maduromycosis. Trans. roy. Soc. trop. Med. Hyg. **48**, 470 (1954). ~ Agentes de maduromicosis en la región neotropical. An. Fac. Med. Montevideo **48**, 453 (1963). — Maffei, O.L.: Nuova species di *Cephalosporium* causa di una cheratomicosi dell' chio. Atti. Ist. Bot. R. Univ. Pavia **1**, 183 (1929). — Male, O., Tappeiner, J.: Nagelveränderungen durch Schimmelpilze. Derm. Wschr. **151**, 212 (1965). — Mariat, F.: Sur la distribution geographique et la repartition des agents de mycetomes. Bull. Soc. Path. exot. **56**, 35 (1963). — Matruchot: Cited by Coutelen et al., 1948. — Miller, H.E., Morrow, H.: Cephalosporiosis: an unusual mycotic infection. Arch. Derm. Syph. (Chic.) **25**, 294 (1932). — Moore, M.: Onychomycosis caused by species of three separate genera: report of a case with a study of a species of *Hyalopus (Cephalosporium)*. J. invest. Derm. **24**, 489 (1955). — Mühlens, K.J.: Beobachtungen an drei Cephalosporienstämme die aus menschlichem Blut und Harn gezüchtet wurden. Zbl. Bakt., I. Abt. Ref. **142**, 160 (1938). — Murray, I.G., Holt, H.D.: Is *Cephalosporium acremonium* capable of producing maduromycosis? Mycopathologia (Den Haag) **22**, 335 (1964). — Nakamura, T.: Demonstration der Actinomyces Kulturen und ein Fall von Cephalosporiosis. *Jap. Dermato-Urol. Tochterges.*, Tokyo, Sitzung v. 13. VI. 1933 (Cited by Coutelen et al., 1948). — Naumann, G., Green, W.R., Zimmerman, L.E.: Mycotic keratitis a histopathologic study of 73 cases. Amer. J. Ophthal. **64**, 668 (1967). — Negroni, P.: Onicomicosis por *Cephalosporium spinosus* n. sp. Negroni. Rev. Soc. argent. Biol. **9**, 16 (1933a). ~ Onycomycose par *Cephalosporium spinosus* n. sp. Negroni. C. R. Soc. Biol. (Paris) **113**, 478 (1933b). — Padhye, A.A., Sukapure, R.S., Thirumalachar, M.J.: *Cephalosporium madurae* n. sp., Cause of Madura foot in India. Mycopathologia (Den Haag) **16**, 315 (1962). — Padhye, A.A., Gokhale, B.B., Thirumalachar, M.J.: Studies on some cases of deep mycoses with special reference to the in vitro activity of some new antifungal antibiotics. Hindustan Antibiot. Bull. **5**, 74 (1963). — Pisano, M.A.: Activities of the cephalosporia. Trans. N.Y. Acad. Sci. **25**, 716 (1963). — Pitotti, P.: Un caso di acremoniosi cutanea. Rif. med. **48**, 1567 (1932). — Ponnampalam, J.T.: The genus *Cephalosporium* as a cause of Madura foot in Malaya. Med. J. Malaya **18**, 229 (1964). — Segretain, G.: Some new or infrequent fungous pathogens. In *Fungi and Fungous diseases*. Springfield, Illinois: Charles C. Thomas 1962. — Schmidt, Van Beyma: (1933) (Cited by Coutelen et al., 1948). — Simôes-Barbosa, F.A.: Concerning a new hyphomycete parasite of man: *Cephalosporium cordoniformis* n. sp. Mycopathologia (Den Haag) **3**, 93 (1941). — Theodore, F.H., Littman, M.L., Almeda, E.: The diagnosis and management of fungus endophthalmitis following cataract extraction. Arch. Ophthal. **66**, 163 (1961). — Verghese, A., Klokke, A.H.: Histologic diagnosis of species of fungus causing mycetoma. Indian J. med. Res. **54**, 524 (1966). — Verkholomov, E.E.: [Methods for the acquisition of haemocultures of some hyphal fungi from the blood of cancer patients] Botan. Zh. SSSR **46**, 357 (1961) (Russian). — Walshe, M.M., English, M.P.: Fungi in nails. Brit. J. Derm. **78**, 198 (1966). — Ward, H.P., Martin, W.J., Ivins, J.C., Weed, L.A.: *Cephalosporium* arthritis. Proc. Mayo Clin. **36**, 337 (1961). — Weidman, F.D., Kligman, A.M.: A new species of *Cephalosporium* in Madura foot (*Cephalo-*

sporium granulomatis). J. Bact. **50**, 491 (1945). — ZAIAS, N.: Superficial white onychomycosis. Sabouraudia **5**, 99 (1966). — ZIMMERMANN, L. E.: Mycotic keratitis. Lab. Invest. **11**, 1151 (1962). ~ Keratomycosis. Surv. Ophthal. **8**, 1 (1963).

Cercosporamycosis

CHUPP, C.: The possible infection of the human body with *Cercospora apii*. Mycologia **49**, 773 (1957). — EMMONS, C. W., BINFORD, C. H., UTZ, J. P.: *Medical Mycology*, p. 346. Philadelphia: Lea and Febiger 1963. — EMMONS, C. W., LIE-KIAN-JOE, NJO-INJO TJOEI ENG, POHAN, A., KERTOPATI, S., VAN DER MEULEN, A.: *Basidiobolus* and *Cercospora* from human infections. Mycologia **49**, 1 (1957). — LIE-KIAN-JOE, NJO-INJO TJOEI ENG, KERTOPATI, S., EMMONS, C. W.: A new verrucous mycosis caused by *Cercospora apii*. Arch. Derm. **75**, 864 (1957).

Cryptostromosis

EMANUEL, D. A., LAWTON, B. R., WENZEL, F. J.: Maple-bark disease: pneumonitis due to *Coniosporium corticale*. New Engl. J. Med. **266**, 333 (1962). — EMANUEL, D. A., WENZEL, F. J., LAWTON, B. R.: Pneumonitis due to *Cryptostroma corticale* (Maple-bark disease). New Engl. J. Med. **274**, 1413 (1966). — GREGORY, P. H., WALLER, S.: *Cryptostroma corticale* and sooty bark disease of sycamore (*Acer pseudoplantanus*). Trans. Brit. Mycol. Soc. **34**, 579 (1951). — TOWEY, J. W., SWEANY, H. C., HURON, W. H.: Severe bronchial asthma apparently due to fungus spores found in maple bark. J. Amer. med. Ass. **99**, 453 (1932). — WENZEL, F. J., EMANUEL, D. A.: The epidemiology of maple bark disease. Arch. environm. Hlth. **14**, 385 (1967).

Curvulariosis

ANDERSON, B., ROBERTS, S. S., Jr., GONZALEZ, D., CHICK, E. W.: Mycotic ulcerative keratitis. Arch. Ophthal. **62**, 169 (1959). — BAYLET, J., CAMAIN, R., SEGRETAIN, G.: Identification des agents des maduromycoses du Sénégal et de la Mauritanie: description d'une espèce nouvelle. Bull. Soc. Path. exot. **52**, 447 (1959). — BRIDGES, C. H.: Maduromycotic mycetomas in animals. *Curvularia geniculata* as an etiologic agent. Amer. J. Path. **33**, 411 (1957). ~ Maduromycosis of bovine nasal mucosa (nasal granuloma of cattle). Cornell Vet. **50**, 469 (1960). — GEORG, L. K.: *Curvularia geniculata*, a cause of mycotic keratitis. J. med. Ass. Ala. **31**, 234 (1964). — NITYANANDA, K., SIVASUBRAMANIAM, P., AJELLO, L.: Mycotic keratitis caused by *Curvularia lunata:* case report. Sabouraudia **2**, 35 (1962). ~ A case of mycotic keratitis caused by *Curvularia geniculata*. Arch. Ophthal. **71**, 456 (1964). — PARMELEE, J. A.: The identification of the *Curvularia* parasite of Gladiolus. Mycologia **48**, 558 (1956). — ROBERTS, E. D., McDANIEL, H. A., CARBREY, E. A.: Maduromycosis of the bovine nasal mucosa. J. Amer. vet. med. Ass. **142**, 42 (1963). — SPRAGUE, R.: *Diseases of Cereals and Grasses in North America*. New York: The Ronald Press, 1950. — WARREN, C. M., Jr.: Dangers of steroids in ophthalmology with report of a case of mycotic perforating corneal ulcer. J. med. Ass. Ala. **33**, 229 (1964).

Dermatophilosis

AUSTWICK, P. K. C.: Cutaneous streptothricosis, mycotic dermatitis, and strawberry foot rot and the genus *Dermatophilus* Van Saceghem. Vet. Rev. Annot. **4**, 33 (1958). — DEAN, D. J., GORDON, M. A., SEVERINGHAUS, C. W., KROLL, E. T., REILLY, J. R.: Streptothricosis: a new zoonotic disease. N. Y. St. J. Med. **61**, 1283 (1961). — GORDON, M. A.: The genus *Dermatophilus*. J. Bact. **88**, 509 (1964). — GORDON, M. A., EDWARDS, M. R.: Micromorphology of *Dermatophilus congolensis*. J. Bact. **86**, 1101 (1963). — HART, C. B., TYSZKIEWICZ, K., ROGERS, B. A., KANE, G. J.: Mycotic dermatitis in sheep. II. *Dermatophilus congolensis* and its reaction to compounds *in vitro*. Vet. Rec. **81**, 623 (1967). — HUDSON, J. R.: Cutaneous streptothricosis. Proc. roy. Soc. Med. **30**, 1457 (1937). — KAPLAN, W.: Dermatophilosis: a recently recognized disease in the United States. Southwest. Vet. **20**, 14 (1966). — KAPLAN, W., JOHNSTON, W. J.: Equine dermatophilosis (cutaneous streptothricosis) in Georgia. J. Amer. vet. med. Ass. **149**, 1162 (1966). — LE RICHE, P. D.: The transmission of dermatophilosis (mycotic dermatitis) in sheep. Aust. vet. J. **44**, 64 (1968). — MACADAM, I.: The effects of ectoparasites and humidity on natural lesions of streptothricosis. Vet. Rec. **76**, 345 (1964). — PIER, A. C., NEAL, F. C., CYSEWSKI, S. J.: Cutaneous Streptothricosis in Iowa Cattle. J. Amer. vet. med. Ass. **142**, 995 (1963). — RICHARD, J. L., PIER, A. C.: Transmission of *Dermatophilus congolensis* by *Stomoxys calcitrans* and *Musca domestica*. Amer. J. vet. Res. **27**, 419 (1966). — ROBERTS, D. S.: Barriers to *Dermatophilus dermatonomus* infection on the skin of sheep. Aust. J. Agric. Res. **14**, 492 (1963). ~ The role of granulocytes in resistance to *Dermatophilus congolensis*. Brit. J. exp. Path. **46**, 643 (1965a). ~ The histopathology of epidermal infection with the actinomycete *Dermatophilus congolensis*. J. Path. Bact. **90**, 213 (1965b). ~ Cutaneous actinomycosis due to the single species *Dermatophilus congolensis*. Nature (Lond.) **206**, 1068 (1965c). ~ Chemotherapy of epidermal infection with *Dermatophilus congolensis*. J. comp. Path. **77**,

129 (1967). — SEARCY, G.P., HULLAND, T.J.: Dermatophilus dermatitis (streptotrichosis) in Ontario. I. Clinical observations. Canad. vet. *J.* **9**, 7 (1968a). ~ Dermatophilus dermatitis (streptotrichosis) in Ontario. II. Laboratory findings. Canad. vet. *J.* **9**, 16 (1968b). — SMITH, J.M.B., DANIEL, R.C.W., BRUERE, A.N.: *Dermatophilosis:* an emerging disease in New Zealand. N. Z. vet. J. **15**, 88 (1967). — VAN SACEGHEM, R.: Dermatose contagieuse (Impetigo contagieux). Bull. Soc. Path. exot. **8**, 354 (1915). — VANDEMAELE, F.P.: Investigation of cutaneous streptothricosis in Africa. Bull. epizoot. Dis. Afr. **9**, 251 (1961). — ZLOTNIK, I.: Cutaneous streptothricosis in cattle. Vet. Rec. **67**, 613 (1955).

Fusariosis

ANDERSON, B., CHICK, E.W.: Mycokeratitis: treatment of fungal corneal ulcers with amphotericin B and mechanical debridement. South med. J. (Birmingham, Ala.) **56**, 270 (1963). — ANDERSON, B., ROBERTS, S.S., GONZALEZ, C., CHICK, E.W.: Mycotic ulcerative keratitis. Arch. Ophthal. **62**, 169 (1959). — ARNSTEIN, H.R., COOK, A.H., LACEY, M.S.: Production of antibiotics by fungi. II. Production of *Fusarium javanicum* and other fusaria. Brit. J. exp. Path. **27**, 349 (1946). — BOISSEVAIN, C.H.: Growth inhibition of tubercle bacilli by *Fusarium sp.* Proc. Soc. exp. Biol. Med. (N.Y.) **63**, 555 (1946). — BOTTICHER, W.W.: *Alternaria* as a possible human pathogen. Sabouraudia **4**, 256 (1966). — CAJORI, F.A., OTANI, T.T., HAMILTON, M.A.: Isolation and some properties of antibiotic from *Fusarium bostrycoides.* J. biol. Chem. **208**, 107 (1954). — CHICK, E.W., CONANT, N.F.: Mycotic ulcerative keratitis: a review of 148 cases from the literature. Invest. Ophthal. **1**, 419 (1962). (Abstract). — COOK, A.H., COX, S.F., FARMER, T.H., LACEY, M.S.: Production of antibiotics by fusaria. Nature (Lond.) **160**, 31 (1947). — EMMONS, C.W.: Misuse of name "*Trichophyton Rosaceum*" for saprophytic fusarium. J. Bact. **47**, 107 (1944). — FORGACS, J.: Mycotoxicoses: neglected diseases. Feedstuffs **34**, 124 (1962a). ~ Mycotoxicoses in animal and human health. U. S. Livestock Sanitary A. 426 (1962b). — FORGACS, J., CARLL, W.T.: Mycotoxicoses. Advanc. vet. Sci. **7**, 273 (1962). — FREI, W.: Urethritis posterior chronica mycotica. Derm. Wschr. **80**, 411 (1925). — GAMET, A., BROTTES, H., ESSOMBA, R.: [New cases of mycetoma detected in the Cameroons] Bull. Soc. Path. exot. **57**, 1191 (1964). — GILLESPIE, F.D.: Fungus corneal ulcer: report of a case. Amer. J. Ophthal. **56**, 823 (1963). — GINGRICH, W.D.: Keratomycosis. J. Amer. med. Ass. **179**, 602 (1962). — HOLZEGEL, K., KEMPF, H.F.: [Fusarium mycosis of the skin of a burned patient.] Derm. Wschr. **150**, 651 (1964). — JOFFE, A.Z.: The mycoflora of overwintered cereals and its toxicity. Bull. Res. Coun. Israel 9D, 101 (1960). ~ Biological properties of some toxic fungi isolated from overwintered cereals. Mycopathologia (Den Haag) **16**, 201 (1962). ~ Toxicity of overwintered cereals. Plant Soil **18**, 31 (1963). ~ Toxin production by cereal fungi causing toxic alimentary aleukia in man. In: *Mycotoxins in Foodstuffs*, pp. 77—85. WOGEN, G.N., Ed. Cambridge, Massachusetts: M. I. T. Press 1965. — LACEY, M.S.: The antibiotic properties of 52 strains of *Fusarium.* J. gen. Microbiol. **4**, 122 (1950). — LAZARUS, J.A., SCHWARZ, L.H.: Infestation of urinary bladder with unusual fungus (*Fusarium*) Urol. cutan. Rev. **52**, 185 (1948). — LYNN, J.R.: *Fusarium* keratitis treated with cycloheximide. Amer. J. Ophthal. **58**, 637 (1964). — MAYER, C.F.: Endemic panmyelotoxicosis in Russian grain belt. I. Clinical aspects of alimentary toxic aleukia (ATA): comprehensive review. Milit. Surg. **113**, 173 (1953a). — MIKAMI, R., STEMMERMANN, G.N.: Keratomycosis: caused by *Fusarium oxysporum.* Amer. J. clin. Path. **29**, 257 (1958). — MING, Y.N., YU, T.F.: Identification of a pathogenic *Fusarium* isolated from foot ulcer of a male patient. Acta. Microbiol. Sinica **10**, 409 (1964). — PETERSON, J.E., BAKER, T.J.: An isolate of *Fusarium roseum* from human burns. Mycologia **51**, 453 (1959). — PERZ, M.: *Fusarium nivale* as a cause of corneal mycosis. Klin. oczna **36**, 609 (1966). — SARKISOV, A.C., KVASHNINA, E.S.: [Toxicological properties of *Fusarium sporotrichioides*, in "cereal crops wintered under snow."] Publ. Min. Agr. Moscow 89 (1948) (Russian). — SIGTENHORST, M.L., GINGRICH, W.D.: Bacteriologic studies of keratitis. South. med. J. (Birmingham, Ala.) **50**, 346 (1957). — STOKES, J.A.: Fusarium keratomycosis. report of a case successfully treated with griseofulvin 1959, cited by HILDICK-SMITH, G., BLANK, H., and SARKANY, I. *Fungus diseases and their management*, p. 355. Boston: Little Brown 1963. — WOLLENWEBER, H.W., SHERBAKOFF, O.A., REINKING, O.A., JOHANN, H., BAILEY, A.A.: Fundamentals for taxonomic studies of *Fusarium.* J. Agr. Res. **30**, 833 (1925). — WOLLENWEBER, H.W., REINKING, O.A.: *Die Fusarien.* Berlin: Paul Parey 1935. — ZAIAS, N.: Superficial white onychomycosis. Sabouraudia **5**, 99 (1966). — ZIMMERMAN, L.E.: Keratomycosis. Surv. Ophthal. **8**, 1 (1963).

Helminthosporiosis

DOLAN, C.T., WEED, L.A.: Opportunistic mycotic pulmonary infection by the family Dematiaceae. Amer. J. clin. Path. **50**, 600 (1968) (Abstract). — DOLAN, C.T., WEED, L.A., DINES, D.E.: Bronchopulmonary Helminthosporiosis. Amer. J. clin. Path. **53**, 235 (1970). — GEORG, L.K.: *Curvularia geniculata*, a cause of mycotic keratosis. J. med. Ass. Ala. **31**, 234 (1964). — ROBERTS, E.D., MCDANIEL, H.A., CARBREY, E.A.: Maduromycosis of the bovine nasal mucosa. J. Amer. vet. med. Ass. **142**, 42 (1963). — WALTON, C.H.A.: Respiratory allergy due to fungi. Canad. med. Ass. J. **60**, 272 (1949).

Keratomycosis

AGARWAL, L.P., MALIK, S.R.K., MOHAN, M., KHOSLA, P.K.: Mycotic corneal ulcer. Brit. J. Ophthal. **47**, 109 (1963). — AINLEY, R., SMITH, B.: Fungal flora of the conjunctival sac in healthy and diseased eyes. Brit. J. Ophthal. **49**, 505 (1965). — ANDERSON, B., CHICK, E.W.: Mycokeratitis: treatment of fungal corneal ulcers with amphotericin B and mechanical debridement. South. med. J. (Birmingham, Ala.) **56**, 270 (1963). — ANDERSON, B., JR., ROBERTS, S.S., GONZALEZ, C., CHICK, E.W.: Mycotic ulcerative keratitis. Arch. Ophthal. **62**, 169 (1959). — BEDELL, A.J.: *Cephalosporium* keratitis. Trans. Amer. ophthal. Soc. **44**, 80 (1946). ~ *Cephalosporium* keratitis. Amer. J. Ophthal. **30**, 997 (1947). — BURDA, C.D., FISHER, E., Jr.: Corneal destruction by extract of *Cephalosporium* mycelium. Amer. J. Ophthal. **50**, 926 (1960). — BYERS, J.L., HOLLAND, M.G., ALLEN, J.H.: *Cephalosporium* keratitis. Amer. J. Ophthal. **49**, 267 (1960). — CHICK, E.W., CONANT, N.F.: Mycotic ulcerative keratitis: a review of 148 cases from the literature. Invest. Ophthal. **1**, 419 (1962) (Abstract). — ERNEST, J.T., RIPPON, M.W.: Keratitis due to *Allescheria boydii (Monosporium apiospermum)*. Amer. J. Ophthal. **62**, 1202 (1966). — FAZAKAS, S. VON: Zusammenfassender Bericht über die sekundären Mykosen bei Erkrankungen des Augenlidrandes, der Bindehaut und der Hornhaut. Ophthal. **138**, 108 (1959). — FOCOSI, M.: Su di un caso di cheratomicosi dá *Cephalosporium* (Contributo clinico sporimentale). Boll. Oculist. **12**, 1250 (1932). — GEORG, L.K.: *Curvularia geniculata*, a cause of mycotic keratitis. J. med. Ass. Ala. **31**, 234 (1964). — GILLESPIE, F.D.: Fungus corneal ulcer: report of a case. Amer. J. Ophthal. **56**, 823 (1963). — GINGIRCH, W.D.: Keratomycosis. J. Amer. med. Ass. **179**, 602 (1962). — GORDON, M.A., VALLOTTON, W.W., CROFFEAD, G.S.: Corneal allescheriosis: a case of keratomycosis treated successfully with nystatin and amphotericin B. Arch. Ophthal. **62**, 758 (1959). — HAGGERTY, T.E., ZIMMERMAN, L.E.: Mycotic keratitis. South. med. J. (Birmingham, Ala.) **51**, 153 (1958). — HILDICK-SMITH, G., BLANK, H., SARKANY, I.: *Fungus Diseases and their Treatment*, Little, Brown and Company, Boston: 1st ed. pp. 347—357, 1964. — HOFFMANN, D.H., NAUMANN, G.: Ein Beitrag zur Pilzinfektion der Hornhaut. Klin. Mbl. Augenheilk. **42**, 286 (1963). — KINNAS, J.S.: Ophthalmic diseases caused by a mycete of the giant cane. Brit. J. Ophthal. **49**, 327 (1965). — LEBER, T.: Keratomycosis aspergillina als Ursache von Hypopyonkeratitis. Arch. Ophthal. [v. Graefes] **25**, 285 (1879). — LEY, A.P., SANDERS, T.E.: Fungus keratitis: report of three cases. A.M.A. Arch. Ophthal. (N.S.) **56**, 257 (1956). — LYNN, J.R.: *Fusarium* keratitis treated with cycloheximide. Amer. J. Ophthal. **58**, 637 (1964). — MAFFEI, O.L.: Nuova specie di *Cephalosporium* causa di una cheratomicosi dell' echio. Atti. Ist. Bot. R. Univ. Pavia **1**, 183 (1928). — MANCHESTER, P.T., JR., GEORG, L.K.: Corneal ulcer due to *Candida parapsilosis* (*C. parakrusei*). J. Amer. med. Ass. **171**, 1339 (1959). — MANGIARACINE, A.B., LIEBMAN, S.D.: Fungus keratitis (*Aspergillus fumigatus*): treatment with nystatin (mycostatin). Arch. Ophthal. (N.S.) **58**, 695 (1957). — MENDELBLATT, D.L.: Moniliasis: a review and a report of the first case demonstrating the *Candida albicans* in the cornea. Amer. J. Ophthal. **36**, 379 (1953). — MIKAMI, R., STEMMERMANN, G.N.: Keratomycosis caused by *Fusarium oxysporum*. Amer. J. clin. Path. **29**, 257 (1958). — MINODA, Y.: Effect of antibiotics on some fungi isolated from ocular lesions. Acta Soc. Ophthal. Japon. **61**, 562 (1957). — MITSUI, Y., HANABUSA, J.: Corneal infections after cortisone therapy. Brit. J. Ophthal. **39**, 244 (1955). — NAUMANN, G., GREEN, W.R., ZIMMERMAN, L.E.: Mycotic keratitis: a histopathologic study of 73 cases. Amer. J. Ophthal. **64**, 668 (1967). — NITYANANDA, K., SIVASUBRAMANIAM, P., AJELLO, L.: Mycotic keratitis caused by *Curvularia lunata:* case report. Sabouraudia **2**, 35 (1962). — PAUTLER, E.E., ROBERTS, R.W., BEAMER, P.R.: Mycotic infection of the eye: *Monosporium apiospermum* associated with corneal ulcer. Arch. Ophthal. **53**, 385 (1955). — PERZ, M.: [*Fusarium nivale* as a cause of corneal mycosis]. Klin. oczna **36**, 609 (1966). — SELIGMANN, E.: Virulence enhancing activities of Aureomycin on *Candida albicans*. PROC. Soc. exp. Biol. (N.Y.) **79**, 481 (1952). ~ Virulence enhancement of *Candida albicans* by antibiotics and cortisone. Proc. Soc. exp. Biol. (N.Y.) **83**, 778 (1953). — SUIE, T., HAVENER, W.H.: Mycology of the eye: review. Amer. J. Ophthal. **56**, 63 (1963). — ZIMMERMAN, L.E.: Mycotic keratitis. Lab. Invest. **11**, 1151 (1962). ~ Keratomycosis. Surv. Ophthal. **8**, 1 (1963).

Otomycosis

ANANTHANARAYAN, R.: Otomycosis in South Indians. Indian J. Surg. **13**, 344 (1951), — BATISTA, A.C., MAIA, H.S., CAVALCANTI, W.: Otomycosis caused by *Waldemaria pernambucensis* n. gen. n. sp. Atas. Inst. Micol. Recife 5—12 (1960). — BELDING, D.L., UMANZIO, C.B.: A new species of the genus *Monosporium* associated with chronic otomycois. Amer. J. Path. **11**, 856 (1935). — BLANK, F., STUART, E.A.: *Monosporium apiospermum* Sacc., 1911, associated with otomycosis. Canad. med. J. **72**, 601 (1955). — BRICEÑO-MAAZ, T., BRICEÑO-MAAZ, C.: Otomicois. Gac. méd. Caracas **72**, 509 (1964). — ENGLISH, M.P.: Otomycosis caused by a ringworm fungus. J. Laryng. Otol. **71**, 207 (1957). — GILL, W. D.: Otomycosis: Some comments concerning its incidence, symptomatology and treatment. South. med. J. (Birmingham, Ala.) **40**, 637 (1947). — GREGSON, A. E. W., LATOUCHE, C. J.: Otomycosis: a neglected disease. J. Laryng. Otol. **75**, 45 (1961). — HALEY, L. D.: Etiology of otomycosis. I. Mycologic flora of the ear. Arch. Otolaryng. **52**, 202 (1950). — KEOGH, C., RUSSELL, B.: The problem of otitis externa. Brit. med. J. **1**, 1068 (1956). — KINGERY, F. A.: The myth of otomycosis. J. Amer. med. Ass. **191**, 129 (1965). — KUNEL'SKAYA, V. Y., STEPANISHCHEVA, Z. G.: [Mycotic flora in otitis externa]. Vestn. Derm. Vener. **38**, 51 (1964) (Russian). — LAKSHMIPATHI, G., MURTI, B.: Otomycosis. J. Indian med. Ass. **34**, 439 (1960). — LEA, W. A., SCHUSTER, D. S., HARRELL, E. R.: Mycological flora of the healthy external auditory canal: a study of 120 human subjects. J. invest. Derm. **31**, 137 (1958). — MCBURNEY, R., GULLEDGE, M. A.: Investigation of certain fungicides and bactericides for use in otitis externa. Ann. Otol. Rhinol. Laryng. (St. Louis) **64**, 1009 (1955). — MOHAPATRA, L. M.: A study of the fungi isolated from cases of otomycosis. Indian J. Microbiol. **1**, 103 (1961). — SHARP, W. B., JOHN, M. B., ROBISON, J. M.: Etiology of otomycosis. Tex. St. J. Med. **42**, 380 (1946). — SMYTH, G. D. L.: A preliminary report on fungal infections of mastoid and fenestration cavities. J. Laryng. Otol. **75**, 703 (1961). — STUART, E. A., BLANK, F.: Aspergillosis of the ear: a report of twenty-nine cases. Canad. med. Ass. J. **72**, 334 (1955). — TOMIĆ-KAROVIĆ, V. K., POPADIC, V.: Mittelohrentzündungen. Verursacht durch *Candida albicans*. Schweiz. med. Wschr. **83**, 59 (1953). — WOLF, F. T.: Relation of various fungi to otomycosis. Arch. Otolaryng. **46**, 361 (1947).

Paecilomycosis

PODEDWORNY, W., SUIE, T.: Mycotic infection of the sclera. Amer. J. Ophthal. **58**, 494 (1964). — ŠAULOV, I.: Die Nagelpilzformen im Gebiet von Plovdiv/Südbulgarien. Derm. Wschr. **151**, 716 (1965). — UYS, C. J., DON, P. A., SCHRIRE, V., BARNARD, C. N.: Endocarditis following cardiac surgery due to the fungus *Paecilomyces*. S. Afr. med. J. **37**, 1276 (1963).

Penicilliosis

AIMÉ, P., CREUZÉ, P., KRESSER, H.: Mycose pulmonaire à "*Penicillium crustaceum*" avec signes clinique et aspect radiologique d'abcès du poumon. Presse méd. **41**, 761 (1933). — AINLEY, R., SMITH, B.: Fungal flora of the conjunctival sac in healthy and diseased eyes. Brit. J. Ophthal. **49**, 505 (1965). — ANDERSON, B., ROBERTS, S. S., GONZALEZ, C., CHICK, E. W.: Mycotic ulcerative keratitis. Arch. Ophthal. **62**, 169 (1959). — ANDREEV, V. G.: [Causal agents of dermatomycoses in the Kursk region.] Vestn. Derm. Vener. **38**, 85 (1964) (Russian). — ARIEVICH, A. M., STEPANISHCHEVA, Z. G.: [The first experience of using antibiotic grisemin for the treatment of patients with mold mycoses]. Antibiotiki (Moskva) **9**, 186 (1964) (Russian). — BUGYI, B.: Coniomicosi dei lavoratori del riso. Med. d. Lavoro **49**, 368 (1958). — CAPPONI, M., SUREAU, P., SEGRETAIN, G.: Pénicillose de *Rhizomys sinensis*. Bull. Soc. Path. exot. **49**, 418 (1956). — CASTELLANI, A.: The higher fungi in relation to human pathology. Lancet **1**, 895 (1920). — CHICK, E. W., CONANT, N. F.: Mycotic ulcerative keratitis: a review of 148 cases from the literature. Invest. Ophthal. **1**, 419 (1962) (Abstract). — CHUTE, A. L.: An infection of the bladder with *Penicillium glaucum*. Boston Surg. J. **164**, 420 (1911). — DA SILVA, LACAZ, D.: Considerações sobre um caso di peniciliose pulmonar. Hospital (Rio de J.) **14**, 327 (1939). — DELORE, P., COUDERT, J., LAMBERT, R., FAYOLLE, J.: Un cas de mycose bronchique avec localisations muscularies septicémiques. Presse méd. **63**, 1580 (1955). — DE MONTEMAYOR, L., RODRIQUEZ, R., SALAS, M.: Hongos alergenicos en los esputos (estudio sorbre 92 casos de asma y rinitis). Mycopathologia (Den Haag) **26**, 410 (1965). — ENOMOTO, M.: Histopathological studies on adenomatous nodules of liver induced in mice by *Penicillium islandicum* Sopp. Acta path. jap. **9**, 189 (1959). — FAZAKAS, S.: Zusammenfassender Bericht über die sekundären Mykosen bei Erkrankungen des Augenlidrandes, der Bindehaut und der Hornhaut. Ophthalmologica (Basel) **138**, 108 (1959). — GARRETÓN, U. I.: Un caso de bronquitis micotica à *Penicillium*. An. méd. Concepción **2**, 103 (1945). — GERNEZ-RIEUX, C., BIGUET, P., CAPRON, A., VOISIN, C., ANDRIEU, M.: Étude de la flore mycologique des bronches par examen des sécrétions bronchiques prélevées sous bronchoscopie. Rev. Tuberc. Pneum. (Paris) **28**, 439 (1964). — GILLIAM, J. S., Jr., Vest, S. A.: *Penicillium* infection of the urinary tract. J. Urol. (Baltimore) **65**, 484 (1951). — GINGRICH, W. D.: Keratomycosis. J. Amer. med. Ass. **179**, 602 (1962). — GIORDANO, M.: Un caso di micosi polmonare da "*Penicillium glaucum*." Ann. Med. nav. Roma **2**, 912 (1918). — HOŘEJŠÍ, M., SACH, J., TOMŠÍKOVÁ, A., MECL, A.: A syndrome resembling farmer's lung in workers inhaling spores of *Aspergillus* and penicillia moulds. Thorax **15**, 212 (1960). — HUANG, S. N., HARRIS, L. S.: Acute disseminated penicilliosis: report of a case and review of pertinent literature. Amer. J. clin. Path. **39**, 167 (1963). — ISHIKO, T.: Histological studies in the injuries of various organs of mice and albino rats except liver, fed with "Yellowed Rice" artificially polluted by *P. islandicum* Sopp. Acta. path. jap. **7**, 368 (1957). — JACKSON, D., YOW, E.: Pulmonary infiltration with eosinophilia: report of two cases of farmer's lung. New Engl. J.

Med. **264**, 1271 (1961). — KESTEN, B.M., ASHFORD, B.K., BENHAM, R.W., EMMONS, C.W., MOSS, M.C.: Fungus infections of the skin and its appendages occurring in Puerto Rico: a clinical and mycologic study. Arch. Derm. Syph. (Chic.) **24**, 1046 (1932). — KIRK, R.: Mycoses of Malaya and Singapore. J. trop. Med. Hyg. **67**, 10 (1959). — KOBAYASHI, Y., URAGUCHI, K., SAKAI, F., TATSUNO, T., TSUKIOKA, M., SAKAI, Y., SATO, T., MIYAKE, M., SAITO, M., ENOMOTO, M., SHIKATA, T., ISHIKO, T.: Toxicological studies on the yellowed rice by *P. isl.* Sopp. I. Experimental approach to liver injuries by long term feedings with the noxious fungus on mice and rats. Proc. Japan Acad. **34**, 139 (1958). ~ Toxicological studies on the yellowed rice by *P. isl.* Sopp. II. Isolation of the two toxic substances from the noxious fungus, and their chemical and biological properties. Proc. Japan Acad. **34**, 736 (1959a). — KOBAYASHI, Y., URAGUCHI, K., SAKAI, F., TATSUNO, T., TSUKIOKA, M., NOGUCHI, Y., MIYAKE, M., SAITO, M., ENOMOTO, M., SHIKATA, T., ISHIKO, T.: Toxicological studies on the yellowed rice by *P. islandicum* Sopp. III. Experimental verification of primary hepatic carcinoma of rats by long term feeding with the fungus-growing rice. Proc. Japan Acad. **35**, 501 (1959b). — KUNEL'SKAIA, V. YA: [Fungus disease of the external ear]. Vestn. Oto-rinolaring. **2**, 37 (1964) (Russian). — KUNEL'SKAIA, V. YA, STEPANISHCHEVA, Z.G.: [Fungus flora of external otitis]. Vestn. Derm. Vener. **38**, 51 (1964) (Russian). — LINEBACK, M.: The bioserologic approach to black hairy tongue with new concepts of its treatment: secondary mycosis linguae of pathologic significance. Laryngoscope (St. Louis) **69**, 1194 (1959). — MANTELLI, C., NEGRI, G.: Ricerche sperimentali sull' agente eziologico di' un micetoma a grani neri (*Penicillium mycetogenum* n. f.). Giorn. R. Accad. Med. Torino **21**, 161 (1915). — MCBURNEY, R., GULLEDGE, M.A.: Investigation of certain fungicides and bactericides for use in otitis externa. Ann. Otol. Rhinol. Laryng. (St. Louis) **64**, 1009 (1955). — MINODA, Y.: Effect of antibiotics on some fungi isolated from ocular lesions. Acta Soc. Ophthal. Japon **61**, 562 (1957). — MITSUI, Y., HANABUSA, J.: Corneal infection after cortisone therapy. Brit. J. Ophthal. **39**, 224 (1955). — MIYAKE, M., SAITO, M., ENOMOTO, M., URAGUCHI, K., TSUKIOKA, M., IKEDA, Y., OMORI, Y.: Histological studies on the liver injury due to the toxic substances of *Penicillium islandicum*, Sopp. Acta path. jap. **5**, 208 (1955). — MIYAKE, M., SAITO, M., ENOMOTO, M., SHIKATA, T., ISHIKO, T., URAGUCHI, K., SAKAI, F., TATSUNO, T., TSUKIOKA, M., NOGUCHI, Y.: Development of primary hepatic carcinoma in rats by long-term feeding with the yellowed rice by *Penicillium islandicum* Sopp — with study on influence of fungus-growing rice on DAB carcinogenesis in rats. Gann. **50**, 117 (1959). — MIYAKE, M., SAITO, M., ENOMOTO, M., SHIKATA, T., ISHIKO, T., URAGUCHI, K., SAKAI, F., TATSUNO, T., TSUKIOKA, M., SAKAI, Y.: Toxic liver injuries and liver cirrhosis induced in mice and rats through long term feeding with *Penicillium islandicum* Sopp — growing rice. Acta path. jap. **10**, 75 (1960). — MOHAPATRA, L.M.: A study of the fungi isolated from cases of otomycosis. Indian. J. Microbiol. **1**, 103 (1961). — NIÑO, F.L.: Broncomicosis penicilliar. Semana méd. (B. Aires) **2**, 1015 (1932). — NUSSBAUM, R., BENEDEK, T.: Pneumonomycosis penicillina, eine Gewerbekrankheit zum Kapitel der Lungengeschwülste. Beitr. Klin. Tuberk. **67**, 756 (1927). — OHMORI, Y., ISONO, C., UCHIDA, H.: [On toxicity of Thailand and yellowsis rice and Islandia Yellowsis rice.] Jap. J. Pharmacol. **50**, 246 (1954) (Japan.). — PEZZALI, G.: Contributo di casistica clinica: un caso di penicillosi del polmone secondaria. Clin. vet. (Milano) **44**, 201 (1921). — POLYANSKIY, L.N.: [A case of otogenic abscess of the temporal lobe of the brain produced by *Penicillium*]. Zh. ushn. nos. gorlov. Bolezn. **15**, 138 (1938) (Russian). — SAITO, M.: Liver cirrhosis induced by metabolites of *Penicillium islandicum* Sopp. Acta path. jap. **9**, 785 (1959). — SAKAI, F.: [Experimental studies on rice yellowsis caused by *P. citrinum* Thom and toxicity especially kidney-damaging effect of citrinin pigment produced by the fungus.] Jap. J. Pharmacol. **51**, 431 (1955) (Japan.). — SAKAI, F., URAGUCHI, K.: [Studies by long-term feeding experiments with rats on development of chronic poisoning by toxic substance from yellowsis rice. VII. Pharmacological studies on toxicity of yellowsis rice.] Nisshin Igaku **42**, 609 (1955) (Japan.). — SAKAKI, J.: [Toxicological studies on moldy rice: Report I.] J. Tokyo Med. Soc. **5**, 1097 (1891) (Japan.). — SALISBURY, J.H.: On the parasitic forms developed in parent epithelial cells of the urinary and genital organs, and their secretions. Amer. J. med. Sci. **55**, 371 (1868). — ŠAULOV, I.: Die Nagelpilzformen im Gebiet von Plovdiv/Südbulgarien. Derm. Wschr. **151**, 716 (1965). — SEGRETAIN, G.: *Penicillium marneffei* n. sp., agent d'une mycose du système réticulo-endothélial. Mycopathologia (Den Haag) **11**, 327 (1959). ~ Some new or infrequent fungous pathogens. In: *Fungi and Fungous Diseases.* Ed., DALLDORF, G. Springfield, Illinois: Charles C. Thomas 1962. — SHELLEY, W.B.: Erythema annulare centrifugum: a case due to hypersensitivity to blue cheese *Penicillium.* Arch. Derm. **90**, 54 (1964). — SMYTH, G.D.L.: A preliminary report on fungal infections of mastoid and fenestration cavities. J. Laryng. Otol. **75**, 703 (1961). ~ Fungal infection of the postoperative mastoid cavity. J. Laryng. Otol. **76**, 797 (1962). — SOOD, V.P., SINHA, A., MOHAPATRA, L.N.: Otomycosis: a clinical entity-clinical and experimental study. J. Laryng. Otol. **81**, 999 (1967). — SOUCHERAY, P.: Farmer's lung: form of bronchopulmonary moniliasis. Minn. Med. **37**, 251 (1954). — TALICE, R.V., MACKINNON, J.E.: *Penicillium bertai* (I), n. sp. agent d'une mycose broncho-pulmonaire de l'homme. Ann. parasitol. **7**, 97

(1929). — TSUNODA, H.: [Study on damage of stored rice, caused by microorganisms. III. On yellowsis rice from Thailand, Food Research Institute, Report 8, 77.] Trans. Jap. Phytopathol. Soc. **13**, 3 (1953) (Japan.). ~ [Microorganisms of spoiled rice in storage.] *Food — Its Science and Technology*, Suppl. p. 161, May 1963. (Japan.). — URAGUCHI, K., SAKAI, F., TSUKIOKA, M., NOGUCHI, Y., TATSUNO, T., SAITO, M., ENOMOTO, M., ISHIKO, T., SHIKATA, T., MIYAKE, M.: Acute and chronic toxicity in mice and rats of the fungus mat of *Penicillium islandicum* Sopp added to the diet. Jap. J. exp. Med. **31**, 435 (1961a). — URAGUCHI, K., TATSUNO, T., SAKAI, F., TSUKIOKA, M., SAKAI, Y., YONEMITSU, O., ITO, H., MIYAKE, M., SAITO, M., ENOMOTO, M., SHIKATA, T., ISHIKO, T.: Isolation of two toxic agents, luteoskyrin and chlorine-containing peptide, from the metabolites of *Penicillium islandicum* Sopp, with some properties thereof. Jap. J. exp. Med. **31**, 19 (1961b). — URAGUCHI, K., TATSUNO, T., TSUKIOKA, M., SAKAI, Y., KOBAYASHI, Y., SAITO, M., ENOMOTO, M., MIYAKE, M.: Toxicological approach to the metabolites of *Penicillium islandicum* Sopp growing on the yellowed rice. Jap. J. exp. Med. **31**, 1 (1961c). — WALTON, C.H.A.: Respiratory allergy due to fungi. Canad. med. Ass. J. **60**, 272 (1949).

Rhodotorulosis

AHEARN, D.G., JANNACH, J.R., ROTH, F.J., Jr.: Speciation and densities of yeast in human urine specimens. Sabouraudia **5**, 110 (1966). — BERGMANN, M., LIPSKY, H.: Über die Pilzbesiedlung des Urogenitaltraktes. Med. Klin. **59**, 732 (1964). — CARETTA, G.: [Fungi isolated from dermatophytoses in Puglia in 1959]. G. ital. Derm. **102**, 389 (1961) (Ital.). — CHAKRAVARTY, S.C.: Role of fungi in chronic bronchitis. Indian J. Chest Dis. **8**, 6 (1966). — CISALPINO, E.O., MAYRINK, W., CARDOSO, J.P.: Micose intestinal. Éstudo dé treze casos. Hospital (Rio de J.) **65**, 237 (1964). — GERNEZ-RIEUX, C., BIQUET, P., CAPRON, A., VOISIN, C., ANDRIEU, S.: Étude dé la flóre mycologique des bronches par examen des sécrétions bronchiques, prélevées sous bronchoscope. Rev. Tuberc. Pneum. (Paris) **28**, 439 (1964). — LEEBER, D.A., SCHER, I.: *Rhodotorula* fungemia presenting as "endotoxic" shock. Arch. intern. Med. **123**, 78 (1969). — LINEBACK, M.: The bioserologic approach to black hairy tongue, with new concepts of its treatment: secondary mycosis linguae of pathologic significance. Laryngoscope (St. Louis) **69**, 1194 (1959). — LOURIA, D.B., GREENBERG, S.M., MOLANDER, D.W.: Fungemia caused by certain nonpathogenic strains of the family Cryptococcaceae. New Engl. J. Med. **263**, 1281 (1960). — LOURIA, D.B., BLEVINS, A., ARMSTRONG, D., BURDICK, R., LIEBERMAN, P.: Fungemia caused by "nonpathogenic" yeasts. Arch. intern. Med. **119**, 247 (1967). — MIGUENS, M.P.: Hongos levaduniformes de la flora vaginal. Proc. IV. *Internat. Congr. of Clinical Pathology*, Madrid, pp. 405—415, 1960. — MÜLLER, W.A., HOLTORFF, J., BLASCHKE-HELLMESSEN, R.: [Investigation of the occurrence and frequency of various types of microorganisms in human vagina during vaginitis]. Arch. Hyg. Bact. **151**, 610 (1967). — RIOPEDRE, R.N., DE CESARE, I., MIATELLO, E., CARÍA, M.A., ZAPATER, R.C.: Aislamiento de *Rhodotorula mucilaginosa* del L.C.R., heces, orina, exudado foringeo y piel de un lactante de 3 meses. Rev. Asoc. méd. argent. **74**, 431 (1960). — SHELBOURNE, P.F., CAREY, R.J.: *Rhodotorula* fungemia complicating staphylococcal endocarditis. J. Amer. med. Ass. **180**, 38 (1962). — SILVEIRA, J.S.: [Clinical experimental studies with trichomycin in genital fungal infection in women]. Rev. Gynec. Obstet. (Rio de J.) **105**, 439 (1959) (Port.). — VIEIRA, J.R., BATISTA, A.C.: [Importance of laboratory research into skin lesions of children]. Publ. Inst. Micol. Univ. Recife. **258**, 28 (1962). — WILSON, J.W.: Paronychia and onycholysis, etiology and therapy. Arch. Derm. **92**, 726 (1965).

Streptomycosis

ABBOTT, P.: Mycetoma in the Sudan. Trans. roy. Soc. trop. Med. Hyg. **50**, 11 (1956). — ADAMI, J.G., KIRKPATRICK, R.C.: Surgery of the extremities: a case of Madura foot disease. Trans. Ass. Amer. Phycns **10**, 92 (1895). — ANDRÉ, M.: Aspects cliniques et chirurgie des mycétomes. Bull. Soc. Path. exot. **51**, 817 (1958). — BLANC, G., BRUN, G.: Nouveau cas de Mycetome à grains noirs observé en Tunisie. Bull. Soc. Path. exot. **12**, 741 (1919). — BRAULT, J.: Mycétome à forme néoplasique. Bull. Soc. Chir. 1906. (Cited by SEGRETAIN and DESTOMBES, 1963). — BRUMPT, V.: A propos de 4 mycétomes observés à Bobo-Dioulasso. Bull. Soc. Path. exot. **53**, 610 (1960). — CAMAIN, R., SEGRETAIN, G., NAZIMOFF, O.: Étude histopathologique des mycétomes du Sénégal et de la Mauritanie. Arch. Biol. Med. **33**, 923 (1957). — CARTER, H.V.: On a new and striking form of fungus disease, principally affecting the foot, and prevailing endemically in many parts of India. Trans. Med. Phys. Soc. Bombay **5**, 104 (1860). ~ On a new and striking form of fungus disease, principally affecting the foot, and prevailing endemically in many parts of India. Trans. Med. Phys. Soc. Bombay (N.S.) **6**, 104 (1861). — CATANEI, A.: Étude parasitologique de trois mycétomes du pied observés en Algérie en 1933. Arch. Inst. Pasteur Algér. **12**, 169 (1934). ~ Sur des changements de caractères culturaux de *Nocardia madurae:* étude morphologique et expérimentale. Arch. Inst. Pasteur Algér. **20**, 299 (1942). ~ Variation in culture characters of *Nocardia madurae:* morphologic and experimental study. Trop. Dis. Bull. **40**, 940 (1943). — CATANEI, A., GOINARD, P.: Un nouveau cas algérien de mycétome du pied. Bull Soc. Path. exot. **27**, 176 (1934). — CATANEI, A., GROSSDEMANGE, L., LEGROUX, C.: Sur un cas de mycetome du pied observé en Algérie. Bull. Soc. Path. exot. **20**, 11 (1927). — CATANEI, A., LEGROUX, C.: Un nouveau cas de mycétome observè en Algérie. Arch. Inst. Pasteur Algér. **9**, 378 (1931). — CHADLI, A., JUMINER, B., HELDT, N., GUETAT, M.S., LADJIMI, R.: Contribution à l'étude des mycétomes tunisiens et de leurs agents. Arch. Inst. Pasteur Tunis **40**, 279 (1963). — CHICK, E.W., CONANT, N.F.: Mycotic keratitis: a review of 148 cases from the literature. Invest. Ophthal. **1**, 419 (1962) (Abstract). — CLARKE, P.R.R.: Mycetoma of the testis. Lancet **2**, 1341 (1953). — CLARKE, P.R.R., WARNOCK, G.B.R., BLOWERS, R., WILKINSON, M.: Brain abscess due to *Streptomyces griseus*. J. Neurol. Neurosurg. Psychiat. **27**, 553 (1964). — CORNWALL, J.E., LA FRENAIS, H.M.: A new variety of Streptothrix cultivated from mycetoma of the leg. Indian J. med. Res. **10**, 239 (1922). — CONVIT, J., BORELLI, D., ALBORNOZ, R., RODRÍGUEZ, G., HÓMEZ, J.: Micetomas, cromomicosis, esporotricosis y enfermedad de Jorge Lobo. Mycopathologia (Den Haag) **15**, 394 (1961). — DELAHAYE, R.P., DESTOMBES, P., MOUTOUNET, J.: Les aspects radiologiques des mycétomes. Ann. Radiol. **5**, 817 (1962). — DESTOMBES, P., ANDRÉ, M.: Contribution à l'étude des mycétomes en Afrique française. Bull. Soc. Path. exot. **51**, 815 (1958). — DESTOMBES, P., RANNOU, M., NEEL, R.: Mycétome à *Streptomyces somaliensis* observé en Algérie au sud de l'atlas. Bull. Soc. Path. exot. **58**, 1017 (1965). — DESTOMBES, P., MARIAT, R., ROSATI, L., SEGRETAIN, G.: Les mycétomes en Republique de Somalie. C. R. Acad. Sci. (Paris) Ser. D **263**, 2062 (1966). — DEY, N.C.: Epidemiology and incidence of fungus diseases in India. Indian J. Derm. **8**, 21 (1962). — DIOUF, B., FUSTEC, R., POUYE, I., GOUDOTE, E., FOURNIER, J.P., BENIER, J., CAUE, L., SERAFINO, X.: Formes anatomo-cliniques des mycétomes à Dakar. Problems thérapeutique Travail utilisant 89 dossiers. Bull. Soc. méd. Afr. noire Langue franç. **10**, 564 (1965). — GAMET, A., BROTTES, H., ESSOMBA, R.: Nouveaux cas de mycétomes dépistés au Cameroun. Bull. Soc. Path. exot. **57**, 1191 (1964). — GEMY, VINCENT, H.: Sur une affection parasitaire du pied non encore décrite (variété de pied de Madura). Annals Dermat. **5**, 577 (1892). ~ Sur un nouveau cas ce "Pied de Madura". Annals Dermat. **7**, 1253 (1896). — GUICHARD, F., JAUSION, H.: Un cas de pied de Madura observé à Marrakech. Arch. Inst. Pasteur Algér. **1**, 641 (1923). — GONZALEZ-OCHOA, A., AHUMADA PIDALLA, M.: Tratmiento del micetoma actinomicosico por la inyeccion local de diaminodifenilsulfona. Rev. Inst. Salubr. Enferm. trop. (Méx.) **18**, 41 (1958). — HALDE, C., RINGROSE, E.J.: Mycetoma originating in Northern California: disease caused by a fungus, resembling *Nocardia madurae*. Arch. Derm. **74**, 80 (1958). — HOMEZ, J.: Primer caso de micetoma por *S.pellieri* en Venezuela. Cited by LATAPÍ, F.: Das Mycetom. In: *Die Pilzkrankheiten der Haut durch Hefen, Schimmel, Aktinomyceten und verwandte Erreger*. [MARCHIONINI, A., and GÖTZ, H. (Eds.)]. Berlin-Göttingen-Heidelberg: Springer 1963. — HYDE, J.N.: A contribution to the study of mycetoma of the foot as it occurs in America. Trans. Amer. Derm. Ass. **19**, 74 (1896). — KLÜKEN, N., CAMAIN, R., BAYLET, M., BASSET, A.: Zur Epidemiologie Klinik und Therapie der Mycetome in West-afrika. Hautarzt **1**, 1 (1965). — KOHN, P.M., TAGER, M., SIEGEL, M.L., ASHE, R.: Aerobic *Actinomyces* septicemia: report of a case. New Engl. J. Med. **245**, 640 (1951). — LATAPÍ, F.: Micetoma. Analisis de 100 casos estudiados en la Ciudad de México. Mem. Congr. ib.-latin.-amer. Dermat. (Méx.) **3**, 203 (1959). — LATAPÍ, F., ORTIZ, Y.: Los micetomas en Mexico. Memorias del primer congreso de dermatología, pp. 126—144 (1963). — LATAPÍ, F., MARIAT, F., LAVALLE, P., ORTIZ, Y.: Micetoma por *Streptomyces somaliensis* localizado a un dedo de la mano. Dermatologia (Méx.) **5**, 257 (1961). — LE DANTEC: Étude bactériologique sur le pied de madura du Senegal (variete truffoide). Arch. Méd. nav. **62**, 447 (1894). — LEGRAIN, E.: Sur quelques affections parasitaires observées en algérie. Arch. Parasit. **1**, 148 (1898). — LINCOLN, C., NORDSTROM, R.C., GOODMAN, S.: Actinomycotic mycetoma (*Streptomyces madurae*). Arch. Derm. **91**, 189 (1965). — LÔBO, J.: Micetomes em Pernambuco. Rev. Fac. Med. Univ. Ceara **3**, 5 (1963). — LYNCH, J.B.: Mycetoma in the Sudan. Ann. roy. Coll. Surg. Engl. **35**, 319 (1964). — LYNCH, J.B., MOGHRABY, I.: Mycetoma in the Sudan associated with *Streptomyces Madurae*. Trans. roy. Soc. trop. Med. Hyg. **55**, 446 (1961). — MACKINNON, J.E.: A contribution to the study of the causal organism of maduromycosis. Trans. roy. Soc. trop. Med. Hyg. **48**, 470 (1954). — MACKINNON, J.E., ARTAGAVEYTIA-ALLENDE, R.C.: The main species of pathogenic aerobic actinomycetes causing mycetomas. Trans. roy. Soc. trop. Med. Hyg. **50**, 31 (1956). — MAEGRAITH, B.: Unde venis. Lancet **1**, 401 (1963). — MARIAT, F.: Sur la distribution geographique et al repartition des agents de mycetomas. Bull. Soc. Path. exot. **56**, 35 (1963). — MONTPELLIER, J., LACROIX, A.: Encore un mycétome du pied, type, „Pied de Madure" observé en Algérie, et du au *Nocardia madurae*. Bull. Soc. Path. exot. **14**, 357 (1921). — NICOLAU, St. G., AVRAM, A., STOIAN, M.: Micetom cu grǎunt: albi al piciorului *Streptomyces madurae*. Derm. Vener (Buc.) **1**, 65 (1959). — ORIO, J., DESTOMBES, P., MARIAT, F., SEGRETAIN, G.: Les mycétomes en Côte Francaise des Somalis — revue de 50 cas. Bull. Soc. Path. exot. **56**, 161 (1963). — PORRITT, A.: An African surgical pilgrimage. Ann. roy. Coll. Surg. Engl. **31**, 330 (1962). — PROCTOR, A.G.: Pathogenic streptomycetes.

J. med. Lab. Technol. **23**, 109 1966. — Raynaud, M., Montpellier, J., Lacroix, A.: Un cas de mycétome du pied a *Nocardia madurae* chez un indigène algérien. Bull. Soc. Path. exot. **15**, 379 (1922). — Remlinger, P.: Un cas de pied de Madura observé au Maroc. Bull. Soc. Path. exot. **5**, 707 (1912) (Also in Maroc méd. **78**, 188, 1938). ~ Contribution à l'étude de "*Discomyces madurae*" Vincent. C. R. Soc. Biol. (Paris) **74**, 516 (1913). — Rey, M.: *Les Mycétomes dans L'Quest Africain.* Thesis, Paris: R. Foulon and Co. 1961. — Rippon, J.W., Lorincz, A.L.: Collagenase activity of *Streptomyces (Nocardia) madura.* J. invest. Derm. **43**, 483 (1964). — Rippon, J.W., Peck, G.L.: Experimental infection with *Streptomyces madurae* as a function of collagenase. J. invest. Derm. **49**, 371 (1967). — Saul, A.: Mycetoma caused by *S. madurae.* Dermatologia (Méx.) **5**, 206 (1961). — Segretain, G., Destombes, P.: Les mycétomes en Afrique du Nord. Maroc. méd. **42**, 445 (1963). — Segretain, G., Mariat, F.: Contribution à l'étude de la mycologie et de la bactériologie des mycétomes du Tchad et de la Côte des Somalis. Bull. Soc. Path. exot. **51**, 833 (1958). — Symmers, D., Sporer, A.: Maduromycosis of the hand, with special reference to heretofore undescribed foreign body granulomas formed around disintegrated chlamydospores. Arch. Path. **37**, 309 (1944). — Umezawa, H., Kondo, S., Maeda, K., Okami, Y., Okuda, T., Takeda, K.: *Index of Antibiotics from Actinomycetes.* University of Tokyo Press, Tokyo and University Park Press, State College, Pennsylvania, 1967. — Vanbreuseghem, R.: Epidémiologie et thérapeutique des pieds du madura au Congo belge. Bull. Soc. Path. exot. **51**, 793 (1958). — Verghese, A., Klokke, A.H.: Histologic diagnosis of species of fungus causing mycetoma. Indian J. med. Res. **54**, 524 (1966). — Victoria, R.V.: [Study of a case of mycetoma due to *Streptomyces somaliensis*]. Rev. lat.-amer. Microbiol. **9**, 61 (1967). — Vincent, M.H.: Étude sur le parasite du "Pied de Madura". Ann. Inst. Pasteur **8**, 129 (1894). — Waksman, S.A.: *The Actinomycetes. A Summary of Current Knowledge.* New York: The Ronald Press Company, 1967.

Torulopsiosis

Ahearn, D.G., Jannach, J.R., Roth, F.J., Jr.: Speciation and densities of yeasts in human urine specimens. Sabouraudia **5**, 110 (1966). — Artagaveytia-Allende, R.C., Silveira, J.S.: [*Torulopsis glabrata*, its isolation from various human sources]. Ciencia (Méx.) **21**, 59 (1961). — Batista, A.C., Pereira, V.: Infectacãe do liquido amniótico da mulher por fungos levedoriformes. Publ. Inst. Micol. Univ. Recife **243**, 39 (1960). — Batista, A.C., Passos, G.M., Charifker, M.: Um caso de parodontite micotica. Publ. Inst. Micol. Univ. Recife **303**, 16 (1961). — Batista, A.C., Silveira, G., Oliveira, D.: Fungos *Candida* e *Torulopsis* na bile em colecistites. Rev. Asoc. méd. bras. **4**, 360 (1958). — Bergmann, M., Lipsky, H.: Über die Pilzbesiedlung des Urogenitaltraktes. Med. Klin. **59**, 732 (1964). — Bernhardt, H.: Die Pilzarten des Sputums. Zbl. Bakt., I. Abt. Orig. **178**, 515 (1960). — Black, R.A., Fisher, C.B.: Cryptococcic bronchopneumonia. Amer. J. Dis. Child. **54**, 81 (1937). — Brunner, A.: Die Lungenmykosen in Chirurgischer Sicht. Chirurg. **35**, 149 (1964). — Caretta, G.: Funghi isolatida dermatofizienellie Puglie nel 1959. G. ital. Derm. Sif. **5**, 389 (1961). — Cernik, L., Doanhoag-Hoa, Vudinh-Hai: [The effect of mycotic flora on bronchial asthma in tropical conditions]. Čas. Lék. čes. **102**, 1313 (1963) (Russian). — Cohen, D., Nierman, M.M., Goldin, M.: Cutaneous moniliasis: topical treatment with amphotericin B lotion. Illinois med. J. **123**, 47 (1963). — Daglio, C.A.N., Martinez-Montez, E.A.: Estudio de los hongos blastoesporados aislados de vulvovaginitis. Rev. Asoc. méd. argent. **71**, 45 (1957). — Edebo, L., Jönsson, L.E., Hallén, A., Lindbom, G.: A fatal case of *Torulopsis glabrata* fungaemia. Acta. chir. scand. **131**, 473 (1966). — Edebo, L., Spetz, A.: Urinary tract infection with *Torulopsis glabrata* treated by alkalization of urine. Brit. med. J. **2**, 983 (1965). — Földvari, F., Florian, E.: Über unsere die tiefen Mykosen betreffenden 15 jährigen Erfahrungen. Derm. Wschr. **151**, 707 (1965). — Grimley, P.M., Wright, L.D., Jr., Jennings, A.E.: *Torulopsis glabrata* infection in man. Amer. J. clin. Path. **43**, 216 (1965). — Guze, L.B., Haley, L.D.: Fungus infections of urinary tract. Yale J. Biol. Med. **30**, 292 (1958). — Haley, L.D.: Yeasts of medical importance. Amer. J. clin. Path. **36**, 227 (1961). — Hasenclever, H.F., Mitchell, W.O.: Pathogenesis of *Torulopsis glabrata* in physiologically altered mice. Sabouraudia **2**, 87 (1962). — Hermanek, P.: Zur Mykologie der Galle bzw. Gallenblase. I. Mykologisch-bakteriologische Untersuchungen an intraoperativ gewonnenen Gallenflüssigkeiten Langenbeck. Arch. klin. Chir. **307**, 277 (1964). — Holmstrom, B., Wallersten, S., Frisk, A.: Presence of fungi in gastric and duodenal ulcers. Acta chir. scand. **117**, 215 (1959). — Hurley, R,. Morris, E.D.: The pathogenicity of *Candida* species in the human vagina. J. Obstet. Gynaec. **71**, 692 (1964). — Katz, D., Pickard, R.E.: Systemic *Torulopsis glabrata* infection causing shock, fever, and coma. Amer. J. Med. **43**, 151 (1967). — Kearns, P.R., Gray, J.E.: Mycotic vulvovaginitis: incidence and persistence of specific yeast species during infection. Obstet. and Gynec. **22**, 621 (1963). — López-Fernández, J.R.: Acción patógena experimental de la levadura *Torulopsis glabrata* (Anderson, 1917) Lodder y De Vries (1938) productora de lesiones sema j antes a la histoplasmosis. An. Fac. Med. Montevideo **37**,

470 (1952). — LOURIA, D.B., GREENBERG, S.M., MOLANDER, D.W.: Fungemia caused by certain non-pathogenic strains of the family cryptococcaceae. New Engl. J. Med. **263**, 1281 (1960). — LOURIA, D.B., BLEVINS, A., ARMSTRONG, D., BURDICK, R., LIEBERMAN, P.: Fungemia caused by "non-pathogenic" yeasts. Arch. intern. Med. **119**, 247 (1967). — MALICKE, H.: [What is the role of yeast infections in pregnant women and newborn infants?] Med. Welt (Stuttg.) **33**, 1725 (1964) (Ger.). — MARKS, M.I., LANGSTON, C., EICKHOFF, T.C.: *Torulopsis glabrata* — an opportunistic pathogen of man. New Engl. J. Med. **283**, 1131 (1970). — MIGUENS, M.P.: Hongos levaduniformes de la Flora vaginal. *Proc. IV. Internal. Congr. of Clinical Pathology*. Madrid, pp. 405—415 (1960). — MINKOWITZ, S., KOFFLER, D., ZAK, F.G.: *Torulopsis glabrata* septicemia. Amer. J. Med. **34**, 252 (1963). — MÜLLER, W.A., HOLTORFF, J., BLASCHKE-HELLMESSEN, R.: Untersuchungen über das Vorkommen und die Häufigkeit verschiedenartiger Mikro-organismen (anaerobe Bakterien, Trichomonaden, Mycoplasmen und Sproßpilze) in der menschlichen Vagina bei Scheidenentzündungen. Arch. Hyg. Bact. **151**, 610 (1967). — NEGRONI, R., DE OBRUTSKY, C.W., GONZALEZ, R.O.: Estudio de un caso de septicemia por *Torulopsis glabrata*. Sabouraudia **4**, 244 (1966). — OLDFIELD, F.S.J., KAPICA, L., PIROZYNSKI, W.J.: Pulmonary infection due to *Torulopsis glabrata*. Canad. med. Ass. J. **98**, 165 (1968). — PASSOS, G. DA M.: Processos cariόticos micόticos: técnicas de defesa. *Thesis Fac. Odont. Univ. Recife*, 187 pp. 1961. — PEREZ, J.S., GIL, R.A.: Estudios sobre la debariocidina. I. Acción in vitro sobre levaduras aisladas de enfermos tratados con antibioticos. Microbiol. esp. **13**, 323 (1960). — PERRY, J.E.: Opportunistic fungal infections of the urinary tract. Tex. St. J. Med. **60**, 146 (1964). — PLAUT, A.: Human infection with *Cryptococcus glabratus:* report of case involving uterus and fallopian tube. Amer. J. clin. Path. **20**, 377 (1950). — REIERSÖL, S.: *Torulopsis norwegica* nov. spec. Leeuwenhoek J. Microbiol. Serol. **24**, 111 (1958). — RIDDELL, R.W., CLAYTON, Y.M.: Pulmonary mycoses occurring in Britain. Brit. J. Tuberc. **52**, 34 (1958). — ROSE, H.D., HECKMAN, M.G.: Persistent fungemia caused by *Torulopsis glabrata*: Treatment with amphotericin B. Amer. J. clin. Path. **54**, 205 (1970). — ŠAULOV, I.: Die Nagelpilzformen im Gebiet von Plovdiv/Südbulgarien. Derm. Wschr. **151**, 716 (1965). — SILVA-HUTNER, M.: Yeasts. *Manual of Clinical Microbiology*. Ed. by J.E. BLAIR, E.H. LENNETTE, J.P. TRUANT, Bethesda, Am. Soc. for Microbiology, 1970, pp. 352—363. — SILVEIRA, J.S.: Ensaios clínico-experimentais com tricomycina (cabimicina) en ginecomicopatolgia. Rev. Ginec. Obstet. (Rio de J.) **105**, 439 (1959). — SMITH, A.G., TAUBERT, H.D., MARTIN, C.W.: The use of trichomycin in the treatment of vulvovaginal mycosis in pregnant women. Amer. J. Obstet. Gynec. **87**, 455 (1963). — VIEIRA, J.R., BATISTA, A.C.: Importância da pesquisa laboratorial nas lesões epidérmicas da crianca. Publ. Inst. Micol. Univ. Recife **258**, 28 (1962). — WICKERHAM, L.J.: Apparent increase in frequency of infections involving *Torulopsis glabrata*. J. Amer. med. Ass. **165**, 47 (1957).

Ustilagosis

ANDERSON, B., ROBERTS, S.S., GONZALEZ, C., CHICK, E.W.: Mycotic ulcerative keratitis. Arch. Ophthal. **62**, 169 (1959). — LE COULANT, M.P., LOPES, G.: Asthme endémique régional par sensibilisation à l'Ustilago: étude expérimentale. J. Méd. Bordeaux **134**, 1094 (1957). — MOORE, M., RUSSELL, W.O., SACHS, E.: Chronic leptomeningitis and ependymitis caused by Ustilago, probably *U.zeae* (corn smut): ustilagomycosis, the second reported instance of human infection. Amer. J. Path. **22**, 761 (1946). — PREININGER, T.: Durch Maisbrand (*Ustilago maydis*) bedingte Dermatomykose. Arch. Derm. Syph. (Berl.) **176**, 109 (1937).

Actinomycosis

D. J. Guidry, New Orleans/Louisiana, USA

With 18 Figures

Definition

Actinomycosis is a chronic suppurative infection which occurs in man and in domestic animals such as horses, cattle and swine. It is characterized by the formation of abscesses and multiple draining sinus tracts, by peripheral spread to contiguous tissues, and by the presence in pus of firm, yellowish tangled mycelial masses which are referred to as "sulfur granules". These represent colonies of the etiologic agent *Actinomyces israelii*; they may reach a diameter of 2 mm and are a prime diagnostic feature of actinomycotic infection (Figs. 1, 5).

Human infections are usually classified as cervico-facial, thoracic, or abdominal types, each may represent a primary infection, or, in the case of thoracic and abdominal lesions, be the result of contiguous spread from a primary site elsewhere in the body. Hematogenous dissemination may result in infection of the liver, kidneys, and brain. Osteomyelitis may occur in the mandible, vertebral bodies, and the neighboring parts of adjacent ribs; this may be accompanied by extensive destruction of bone.

In *cattle*, lesions occur most frequently in the bones of the face and jaw, although other tissues such as the testes may be affected. Actinomycotic infection is also known to occur in the intermandibular space in the horse, in lung, cheek bone, and mandible of dogs, and in the udder of swine. A case of actinomycotic encephalitis in a deer has also been recorded.

Actinomycosis is worldwide in distribution. It is endogenously acquired and man-to-man or animal-to-man transmission has not been reported.

Historical

The earliest description of typical actinomycosis were provided by Bollinger in 1877 from cases of "lumpy jaw" in cattle. He observed that pus from the lesions contained numerous yellowish, coarsely granular bodies which were shown to consist of a fungus. Harz, a botanist with whom he consulted, concluded that the organism was a true mold and named it *Actinomyces bovis* (ray fungus). Subsequently, Harz (1879) presented additional descriptions of *A. bovis* in material obtained from cases of "lumpy jaw". Israel (1878) described a similar mycotic disease in man based on his studies of human autopsy material. The first human case was described by Ponfick in 1880 and in 1885 Israel, reporting on 38 human cases, clearly defined the clinical manifestations of the disease (Fig. 1).

The etiologic agent was successfully isolated from "lumpy jaw" in cattle in 1890 by Mosselman and Lienaux and from the human disease in 1891 by Wolff and Israel. Because of the many similarities in the pathology of the two diseases as shown by Israel (1878) and Ponfick (1880b), including the essentially identical appearance of granules in tissue and pus, the etiologic agent was considered to be the same. This opinion was supported by subsequent cultural and morphological studies as well as observations of pathological materials (Silberschmidt, 1901; Wright, 1905; Breed and Conn, 1919).

The single species, *A. bovis*, was used to designate the etiologic agent in both human and animal infections as late as 1954 in spite of the fact that Erikson (1940) had shown that separation into two species was justified and had subsequently presented a detailed description and classification of the Actinomycetes (Erikson, 1949). However, separation of the two species *A. bovis* and *A. israelii*

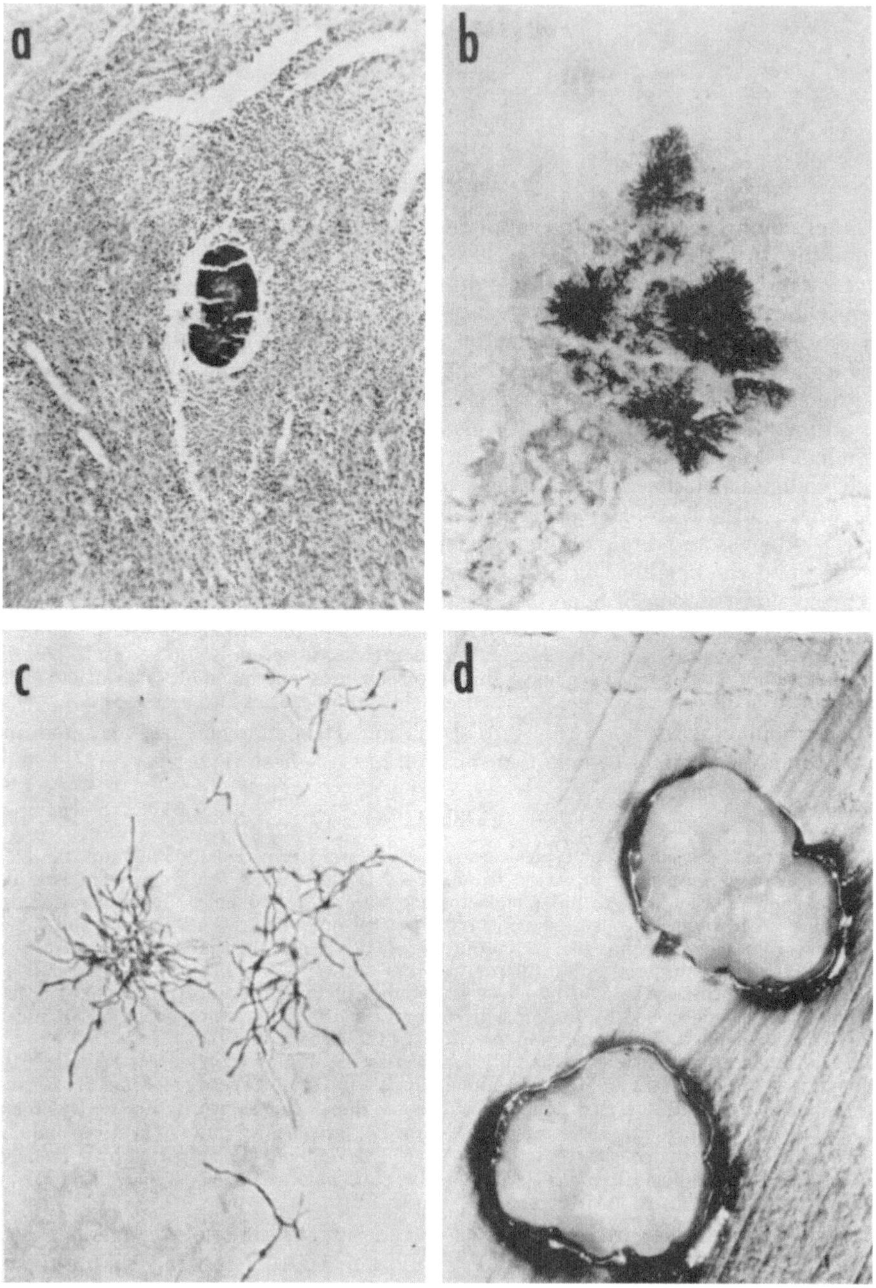

Fig. 1. *Actinomyces israelii*. a) "Sulfur granule" in pus. H & E, × 120. b) Granule showing the darkly-stained hyphae to be oriented radially towards the periphery and lying in a pale-gray matrix of inflammatory cells and necrotic debris. Note that the matrix extends well past the hyphal tips. GMS, × 450. c) 24-hour "spider colonies" on brain-heart infusion agar. × 380. d) 7-day colonies. The smooth, convex surface of the younger colonies has now become granular and lobulated. Penetration of hyphae into the medium has caused dimpling of the agar surface. (c and d courtesy of Dr. LUCILLE K. GEORG, Mycology Section, National Communicable Disease Center; National Audiovisual Center. Atlanta, Ga.)

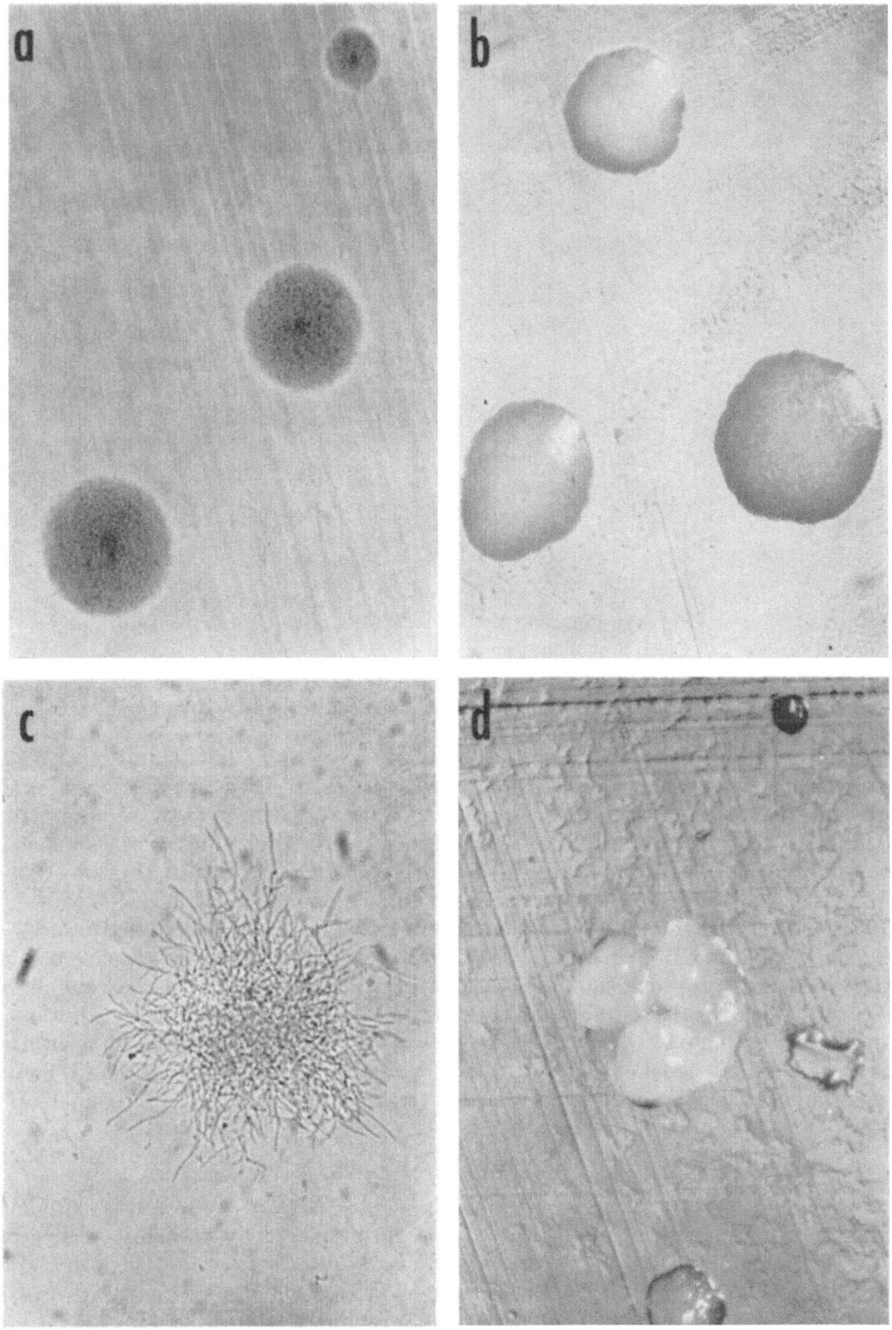

Fig. 2. *Actinomyces bovis*. a) 24-hour colonies ("S" form) on brain-heart infusion agar. Their appearance suggests dense hyphal formation, but microscopic preparations show only masses of diphtheroids. × 125. b) 7-day colonies of "S" form. The smooth, convex colonies yield only diphtheroidal forms. c) 24-hour colonies of "R" form. Note the similarity to microcolonies of *A. israelii*. × 475. d) 7-day colony of "R" form. Mature colonies of the R variant resemble those of *A. israelii*. However, microscopic preparations of the R variant also reveal masses of diphtheroids rather than branched hyphae. (Photographs courtesy of Dr. LUCILLE K. GEORG)

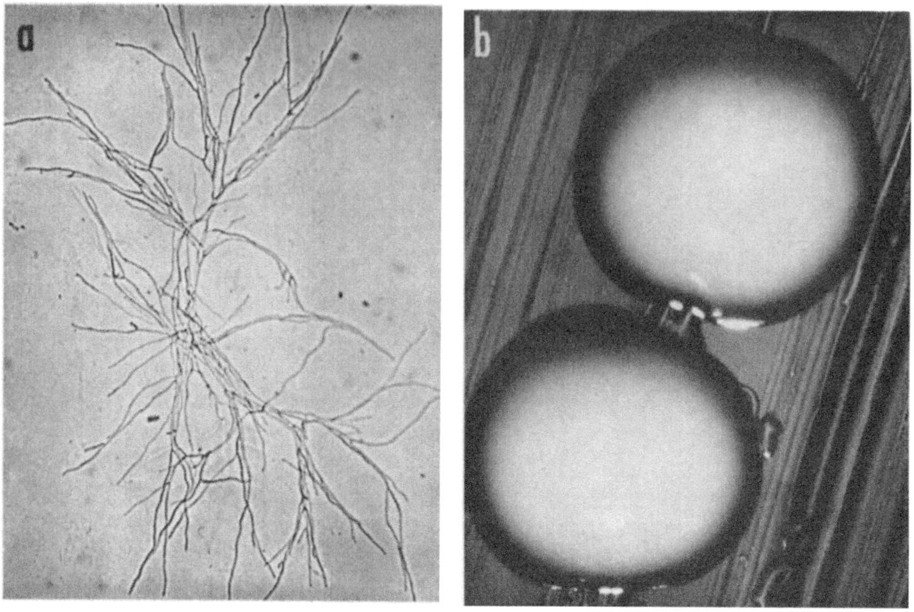

Fig. 3. *Actinomyces propionicus.* a) 24-hour colony on brain-heart infusion agar. The mycelial nature of this species is readily apparent. × 380. b) 7-day colonies. The smooth, convex colonies yield long, branching hyphae. (Photographs courtesy of Dr. LUCILLE K. GEORG)

was firmly established following detailed studies of their morphological and physiological characteristics (PINE et al., 1960a), their growth requirements (CHRISTIE and PORTEOUS, 1962a, 1962b, 1962c), and antigenic structure, chemical composition, and ultrastructure of their cell walls (CUMMINS and HARRIS, 1958; MACLENNAN, 1961; CUMMINS, 1962; OVERMAN and PINE, 1963; KWAPINSKI, 1964) (Fig. 2). In the meantime some oral strains of *Actinomyces* were described by THOMPSON and LOVESTEDT (1951) which grew in air on primary isolation and for which they proposed the name *Actinomyces naeslundii*; this species was more clearly defined by the studies of HOWELL et al. (1959) (Fig. 4). Two additional species have been described; *Actinomyces propionicus* was isolated from a human case of lacrimal canaliculitis (PINE and HARDIN, 1959; BUCHANAN and PINE, 1962) and *Actinomyces eriksonii* was isolated from a lung abscess (GEORG et al., 1965) (Fig. 3).

Geographic Pathology

Although actinomycosis is world-wide in distribution, it is not a common disease of man. In the United States, it is outnumbered by both histoplasmosis and coccidioidomycosis and possibly by blastomycosis as well (PEABODY and SEABURY, 1957). Some of the earliest reported cases in the United States were those cited by RUHRAH (1899); SANFORD's review (1923) cited approximately 700 cases. PUTMAN et al. (1950) reported on 122 cases diagnosed at the Mayo Clinic over a period of 35 years. PORTER (1953) cited 183 cases in Scotland over a 13 year period, while in England and Wales 151 cases were listed during the five years from 1957—1961 (WILSON and MILES, 1964). GLAHN (1954) reported on 90 cases of cervico-facial actinomycosis observed in Copenhagen over a five-year period.

Most surveys show that the disease is approximately twice as common in males as in females; in the 90 cases reported by Glahn (1945b) 78% were males. While some surveys show the disease to be more common among agricultural workers and residents of small towns (Porter, 1953; Wilson and Miles, 1964), other reviews have failed to demonstrate a clear occupational association. In the 122 cases reviewed by Putman et al. (1950) 40% of the patients were farmers, 25% were housewives, and 12% were students. The remaining cases were distributed among many trades and professions. Fifty-three patients lived in rural homes and 69 were urban dwellers.

Most cases of actinomycosis occur in patients over 20 years of *age*; in the series by Putman and his associates 93 of the cases were patients between 20 and 60 years of age.

Although the disease is not well-known in some countries, differential race susceptibilities are not established; in many instances the low incidence may well reflect lack of diagnostic and reporting facilities rather than a paucity of cases.

Cervico-facial actinomycosis is generally conceded to be the most common form of the disease. In the series by Cope (1938) the distribution of cases was cervico-facial 56.8%, thoracic 22.3%, and abdominal 15.0%; in the data tabulated by Wilson and Miles (1964) the distribution was 61.2—, 12.0—, and 18.7% respectively.

Actinomycosis is an endogenously-acquired infection which occurs only sporadically and the disease is not contagious.

Mycology

Etiology. Most workers are agreed that human cases of actinomycosis are usually caused by *Actinomyces israelii* whereas the disease in cattle is usually caused by *A.bovis* (Figs. 1, 2). As previously mentioned, there has been a great deal of reluctance in accepting separation into two species and a survey of the literature makes it quickly apparent why this is so. Not only are there legitimate exceptions to the usual etiology in both man and animals, but there are numerous reports which contain confusing and conflicting descriptions of the etiologic agent. The actinomycotic lesions almost invariably contain a mixed flora (Holm, 1951) and anaerobic diphtheroids and other bacteria frequently have been identified erroneously as *Actinomyces*. In addition, it is suspected that some reports of cultural and physiologic studies are based on work done using impure cultures. Add to this the fact that both the morphologic and physiologic characteristics of the *Actinomyces* are somewhat erratic, with occasional isolates showing characteristics of both species, and the difficulty of definitive identification can be appreciated. Nevertheless, as indicated in the historical review, a workable definition has been established not only for *A.bovis* and *A.israelii*, but also for the three additional species designated as *A.naeslundii*, *A.propionicus* and *A.eriksonii* (Figs. 3, 4).

A discussion of etiology is not complete without consideration of the possible role played by *associated microorganisms* in the pathogenesis of actinomycotic infection. Holm (1948, 1950, 1951) conducted a careful bacteriologic study of specimens from 650 patients with closed lesions and found that *A.israelii* was never present in pure culture. Lentze (1958) studied 608 cases and found associated microorganisms in every case. These microbial associates are usually anaerobic organisms and include *Actinobacillus actinomycetemcomitans*, *Fusiformis melaninogenicus*, other fusobacteria, anaerobic streptococci, and staphylococci (Lentze, 1958; Heinrich and Pulverer, 1959; Brede, 1959). This association of *A. israelii* with other microbes suggests the possibility of synergistic action. Nearly

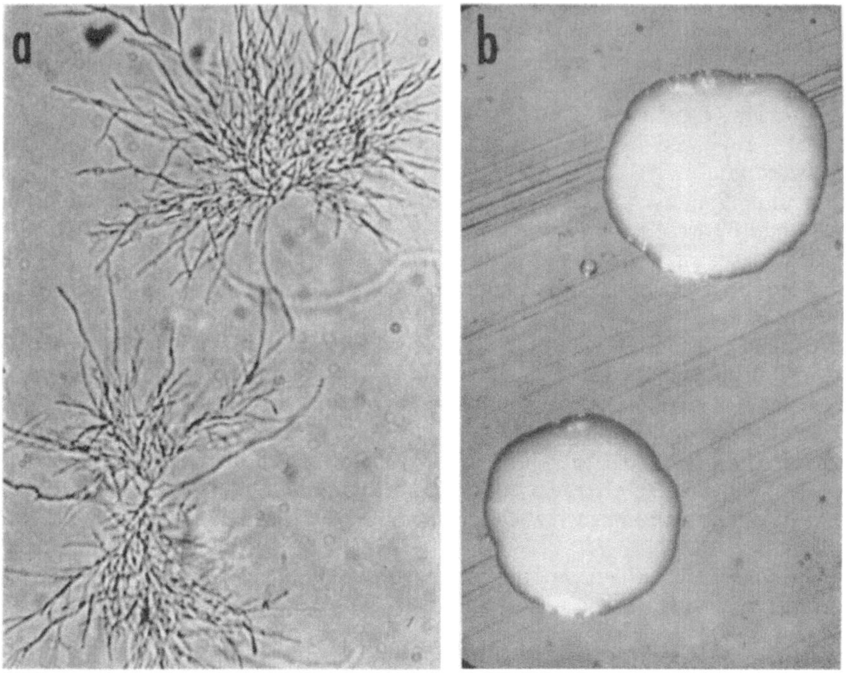

Fig. 4. *Actinomyces naeslundii*. a) 24-hour colonies on brain-heart infusion agar. Although *A. naeslundii* produces true mycelium, as shown here, an occasional isolate may yield only diphtheroidal forms. × 380. b) 7-day colonies. Older colonies may be rough, granular and pitted or they may appear as creamy and soft white with a ground glass to finely lobate surface. (Photographs courtesy Dr. LUCILLE K. GEORG)

all of the microbial associates studied by BREDE formed hyaluronidase and other depolymerizing enzymes, a property absent in *A. israelii*, and it appears that their contribution to the infectious process consists of lowering the redox potential and of facilitating the invasion of tissues. HOLM (1951) has shown that patients with actinomycosis may be treated adequately with penicillin, insofar as *A. israelii* is concerned, without being cured of the disease and that this is due to persistance of microbial associates in the lesions. *Actinobacillus actinomycetemcomitans* quite frequently appears to be responsible for continued disease in penicillin-treated patients, although other gram-negative anaerobes may produce the same syndrome.

Laboratory Diagnosis. Anaerobic bacteriology is usually the least developed area in most hospital laboratories. For this reason and because of inherent difficulties in working with the *Actinomyces*, isolation and identification of this group presents a severe test in laboratory technique. PINE (1963) reviewed the current status of knowledge concerning the anaerobic actinomycetes and presented some recommendations for the isolation and identification of etiologic agents from actinomycotic lesions.

The only completely definitive procedure for species identification of *Actinomyces* rests on cell wall analysis (CUMMINS and HARRIS, 1958; PINE and BOONE, 1967) and serologic methods (GEORG et al., 1965; LAMBERT et al., 1967). BLANK and GEORG (1968) have applied the fluorescent antibody technique to direct detection and identification of *Actinomyces* species in tissues and in exudates

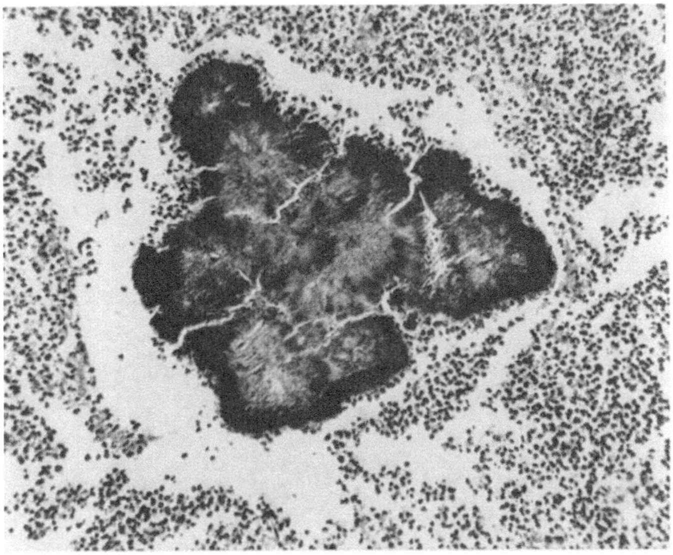

Fig. 5. Granule of *Actinomyces* in an abscess. The peripheral eosinophilic material of the granule stains intensely and includes some of the "clubs". The surrounding pus consists of polymorphonuclear neutrophils, many of which are fragmenting or losing their staining properties because of necrosis. H & E, × 970

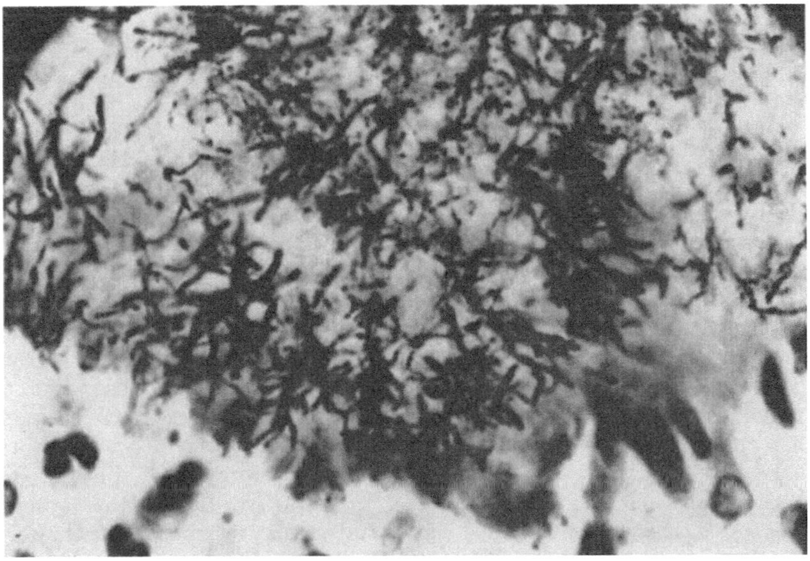

Fig. 6. Granule of *Actinomyces* stained by the method of Gram. Note the tangled filaments with the branching clearly visible. × 1500

(Fig. 7). However, neither of these is currently available to most hospital laboratories. A practical approach is to let positive identification rest upon microscopic morphology, morphology of micro- and mature colonies, physiological properties of the isolate in pure culture, and consideration of the clinical syndrome in the patient (Figs. 5, 6, 7).

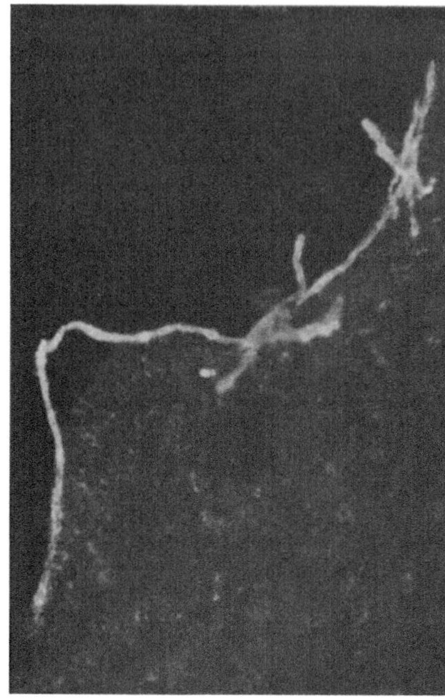

Fig. 7. Fluorescent antibody technique for detection of *Actinomyces*. Positive fluorescence reaction obtained in a direct smear prepared from tonsillar material. The FA test may be used to detect *Actinomyces* in clinical material, to monitor its presence in mixed cultures, and for making final and definitive identification of the organism. (From BLANK and GEORG: Journal of Laboratory and Clinical Medicine *71*, 283—293 (1968), The C.V. Mosby Company, St. Louis, Missouri)

Direct Examination. This consists primarily of looking for mycelial clumps or sulfur granules. These may be present in sputum or pus, but are rarely if ever found in the spinal fluid and the circulating blood. Pus which has been collected by aspiration may be spread out on a glass surface and exxamined with a hand lens for the presence of white to yellowish granules. Gauze dressings from draining sinus tracts should be rinsed with sterile water and the washings examined. Sputum may also be spread thin in a petri dish for examination; dilution with sterile water is recommended. The presence of granules in sputum is not necessarily of diagnostic significance, especially if they are associated with large numbers of contaminating bacteria. Such granules usually lack the typical eosinophilic "clubs" and represent saprophytic growths of *Actinomyces* which were expressed from the tonsillar crypts during coughing and expectoration of sputum. On the other hand, absence of granules does not necessarily rule out the presence of *Actinomyces* in the specimen, since the organism may be present in the form of small, poorly-organized mycelial clumps. Such specimens should be cultured even if the direct examination is negative.

If granules are found, samples should be removed, crushed between two microscope slides, and examined as a cover glass preparation. Additional slides should be prepared, heat-fixed and stained by the Gram and Ziehl-Neelsen technics. In fresh preparations, the granules will appear as lobulated bodies composed of delicate, branching and intertwined hyphae radiating towards the periphery. These are approximately 1 micron in diameter and the tips are frequently surrounded by a gelatinous matrix imparting a club-shaped appearance to the ends of the hyphae. The Gram stain will show gram-positive branching filaments, some of which may stain irregularly (Fig. 6). *Actinomyces is not acid-fast.* Frequently, the stained smears will reveal the presence of associated bacteria within and around the granule. In pathologic material in which granules cannot be demonstrated, stained smears should be prepared and examined for short, branching gram-positive filaments which are not acid-fast.

Direct Culture. Some effort should be devoted to removing as much contamination as possible from the granules prior to culture. EMMONS et al. (1963) recommend placing 10—15 drops of thioglycollate (containing 0.05% cysteine hydrochloride) in a sterile petri dish and washing granules by manipulating them with sterile needles and forceps in successive drops of the broth. This should be repeated until debris and pus as well as most of the microbial contaminants have been removed. Finally, the granules should be crushed and additional thioglycollate added to yield a dilute suspension of hyphal fragments.

A number of *suitable media* are available for isolation of *Actinomyces* and it is desirable to employ several of these in culturing a clinical specimen. Brain-heart infusion glucose broth, glucose thioglycollate broth, or brain-heart infusion glucose agar shake tubes may be used for culturing uncontaminated specimens. For contaminated material, Garrod's starch agar with 10% blood (EMMONS et al., 1963), brain-heart infusion agar with and without 10% rabbit blood, and the medium of PINE and WATSON (1959) are recommended. The latter, while it requires more work for its preparation, is purported to be superior on all counts (PINE, 1963). It is recommended that two series of isolations be made from each clinical specimen; one should be a series of streak plates (1.5—2.0% agar) and the other a series of dilution tubes containing media with 0.7% agar. While the streak plates can be incubated in a candle jar, an anaerobic chamber which can be evacuated three times with a mixture of 5% CO_2, 95% N_2 is greatly preferred. The agar shake cultures are sealed by cutting the top of the cotton plug and pushing the remainder of the plug into the tube, to a point just above the surface of the medium. An adsorbent cotton plug is then inserted on top of the first, 5 drops each of saturated pyrogallol solution and 10% Na_2CO_3 solution are added, and the tube sealed with a tight rubber stopper. Cultures should be incubated at 30—37°C and examination of both series should begin on the third to fourth day. The agar plates are best examined at the surface or from the bottom, first using a hand lens and then under a low power objective. For agar shake cultures, the sides of the tube should be warmed sufficiently to melt agar adjacent to the glass and a Pasteur pipette inserted along the wall to the bottom of the tube. Air is then blown through the pipette, expelling the entire agar plug into a sterile Petri dish. The plug is then cut into slices, cutting so as to separate individual colonies. These can then be placed on microscopic slides and examined with the low or high magnifications. It should be emphasized that only the young colonies of *Actinomyces* will be characteristic of the species; colonies become similar, if not identical, after 7—8 days. Several subcultures, working with discrete colonies, are desirable before any physiological studies are performed.

Associated Microbial Flora. Quite frequently these comprise the majority of colonies obtained in anaerobic cultures from actinomycotic lesions and one must be able to distinguish them from the colonies of *Actinomyces*. Specimens from cervico-facial lesions may contain *Actinobacillus actinomycetemcomitans*, fusiform bacilli, anaerobic streptococci, and various anaerobic gram-negative bacilli in association with *A. israelii*. Thoracic lesions usually have the first three named as associated while abdominal lesions most often yield *Escherichia coli* and various anaerobic gram-negative bacteria (EMMONS et al., 1963). These must be carefully excluded from *A. bovis* and *A. israelii* before any cultural and physiological studies can be performed.

Animal Inoculation. This is of no practical value in the primary isolation of *Actinomyces* from clinical material. However, hamsters have been used for differentiating anaerobic diphtheroids from *A. israelii* and *A. bovis* isolates (HAZEN and LITTLE, 1951). GEORG et al. (1965) used 4—6 week old white mice for comparing the pathogenicity of strains of *A. eriksonii* with that of *A. israelii*. Groups of mice were inoculated by the subcutaneous, intraperitoneal, and intravenous routes. None of the mice inoculated with strains of *A. eriksonii* died. Subcutaneous abscesses developed several days following subcutaneous inoculation; gram-positive branched filaments could be demonstrated by smear and retrocultures were positive up to 2 weeks. Small, well-encapsulated abscesses developed in the mesenteries and peritoneal surfaces, but there was no evidence of tissue invasion and smears from infected tissues revealed gram-negative disintegrating organisms. Results with 2 recently isolated strains of *A. israelii* were more impressive. Four out of 20 mice inoculated intraperitoneally died within 2 weeks. Five out of 10 mice inoculated intravenously died within one week. All of the mice inoculated intraperitoneally with *A. israelii* developed numerous massive abscesses in the peritoneum; microscopic granules with clubbed hyphae were demonstrated in every case.

BUCHANAN and PINE (1962) concluded that *A. propionicus* was slightly more pathogenic for mice than were *A. israelii*, *A. bovis*, and *A. naeslundii*. PEGRUM (1964) found that *A. israelii*, and *A. bovis* produced small, usually single nodules on the chorioallantoic membrane of embryonated eggs. Histologic sections showed invasion of the membrane with proliferation of the *Actinomyces* within the mesoderm. A proliferative tissue reaction with edema occurred; the infected tissue was later walled off by fibroblasts and became ulcerated. The pathogenicity of *A. israelii* for albino mice is considerably enhanced by the addition of hog gastric mucin (MEYER and VERGES, 1950).

Serology. At present, serologic methods are of little practical value. Although agglutinins and complement-fixing antibodies have been demonstrated and positive skin tests have been reported, results in general have been inconsistent or equivocal. Indeed, MATHIESON et al. (1935) found that normal individuals gave more frequent and more marked skin reactions to *A. bovis* antigen than did actinomycotic patients. Very likely the current unsatisfactory status of serologic diagnosis in actinomycosis, like the earlier experiences with mycoses such as histoplasmosis and sporotrichosis, is due to poorly standardized antigens and technics rather than lack of an immune response on the part of the patient. Certainly it is clear that antibodies can be produced experimentally in animals and species specificity demonstrated when careful attention to detail is exercised by a competent investigator (GEORG et al., 1965; LAMBERT et al., 1967). Admittedly, cross reactions between *Actinomyces* species have been reported along with demonstrable relationships with other genera. SLACK et al. (1955) employing a reciprocal agglutinin adsorption technique were able to separate 20 isolates of microaerophilic *Actinomyces* into two broad serologic groups. Serologic group A included isolates from cases of human, bovine, equine and porcine actinomycosis along with isolates from human non-actinomycotic tonsils and one case of human pyorrhea. Six of the 20 isolates comprised serologic group B and these were from human, bovine, and equine infections. KING and MEYER (1957) reported that anaerobic diphtheroids could be differentiated from *A. bovis* and *A. israelii* on the basis of metabolic and serologic tests and indicated that the anaerobic diphtheroids are more closely related to the genus *Corynebacterium* than to *Actinomyces*. CUMMINS (1962) reported that seven out of eight isolates of *Actinomyces* yielding a cell-wall analysis identical to that for *A. israelii* also proved to be homogeneous by serological tests. KWAPINSKI (1964) found that the cell walls of *Actinomyces* species were related serologically to those of *Corynebacterium* and to some strains of *Mycobacterium* and *Nocardia*. Cell wall antigens of *A. israelii* reacted with sera from *A. bovis*, *C. diphtheriae*, mycobacteria, nocardiae, and *Waksmania*. *A. bovis* cell wall antigens reacted in a similar pattern with the exception that it failed to react with sera prepared with *Nocardia asteroides*. Undoubtedly, the manner in which antigens and antisera are prepared has a graet deal to do with the reactive

Table 1. *Morphological and physiological characteristics of Actinomyces species*[1]

Character	Results obtained with strains of Actinomyces: israelii	bovis	propionicus	eriksonii	naeslundii
Oxygen requirement	Anaerobic to micro-aerophilic	Anaerobic to micro-aerophilic	Facultative	Obligate anaerobe	Facultative in presence of increased carbon dioxide
Maximum growth	3—7 days	2—3 days	2—3 days	3—4 days	1—2 days
Color of colony	white	white	dull orange	white-cream	white
Biochemical reactions:					
Catalase	0	0	0	0	0
Nitrate reduct.	+	0	+	0	+
Starch hydrol.	0	+	0	+	0
Gelatin liquef.	0	0	0	0	0
Litmus milk	reduced	reduced	reduced	reduced and firm curd	reduced
Propionic acid	0	0	+	0	0
Acid without gas:					
Glucose	+	+	+	+	+
Mannitol	+	0	±	+	0
Mannose	+	0 or ±	+	+	+
Raffinose	Variable	0	+	+	+
Xylose	+	Variable	0	+	0

1 Based on data from GEORG et al. (1965) and PINE (1963).

patterns obtained in serologic tests. One additional point should be made. GEORG et al. (1964) have shown that smooth-rough (S—R) variation can occur with *A.bovis* with the R variant producing microcolonies which resembled those of *A.israelii* whereas the S variant was more typical of *A.bovis*. Both S and R variants had a cell wall composition similar to that of *A.bovis*; however, antigens from the S variant reacted with homologous sera as well as sera prepared from *A.bovis* and the R variant, whereas antigens from the R variant reacted with homologous serum only.

Species Description. Morphological and physiological characteristics pertinent to identification of the *Actinomyces* are presented in Table 1. Their oxygen requirements, rates of growth, and microscopic and colonial morphology are features which can be determined and compared in most laboratories. In physiological studies, the catalase reaction is most useful in excluding anaerobic diphtheroids which are catalase positive. The remainder of the biochemical tests listed in Table 1 are best performed as a battery since an occasional isolate may yield variable results, especially in the case of carbohydrate fermentations. Data on cell wall composition, believed to be the most definitive criteria for differentiating *Actinomyces* species, are listed in table 2 along with ultrastructural features which show some morphologic differences between species.

Actinomyces israelii (HARZ) KRUSE (1896)

A.israelii can be isolated from the normal saprophytic flora of the oral cavity. HOWELL et al. (1962) recovered the organism in 48% of non-salivary oral samples and from 28.9% of salivary samples. In 40—50% of plaques taken from early or shallow carious lesions, *A.israelii* was the predominant organism. The organism has also been isolated from tonsillar crypts (EMMONS, 1935, 1936, 1938) where it occurs in association with other microorganisms.

The *granules* which occur in tissues and pus vary in color from white to yellowish and may be round or lobulated (Fig. 1A). The center of the granule is comprised of hyphae and cellular debris (Fig. 1B). Hyphae are oriented radially towards the periphery of the granule; some extend beyond the surface while others end short of the surface and are surrounded by a sheath of eosinophilic material. When granules are crushed and stained they show gram-positive hyphal fragments 0.5—0.1 μ in diameter. These vary in length from intact hyphal elements measuring several microns (rarely with branches) to short fragments resembling diphtheroids. Associated bacteria may be present in the stained preparations. *A. israelii*, like all of the other *Actinomyces*, is not acid-fast.

In *culture*, the importance of the highly diagnostic microcolonies cannot be overemphasized (PINE et al., 1960a; BUCHANAN and PINE, 1962; PINE, 1963). At 48 hours *A. israelii* appears as delicate, branching hyphae radiating towards the periphery to form so-called "spider colonies" (Fig. 1C). These gradually increase in size to about 2.0 mm, becoming white and opaque with a surface which is at first convex and smooth but later becomes granular and lobulated (Fig. 1D). Growth often penetrates the agar to cause dimpling of the surface and adherance of the colony to the agar. The colony remains intact when removed from the agar surface and has a cheesy consistency when crushed. *A. israelii* forms discrete colonies in liquid media; there is no turbidity since diffuse growth does not occur.

Actinomyces bovis (HARZ, 1877)

It is generally accepted that *A. bovis* occurs frequently as part of the normal oral flora of cattle. Very likely, its mode of saprophytic existence in the oral cavity of cattle is quite similar to that demonstrated for *A. israelii* in humans. Trauma often seems to play a part in initiating actinomycotic infection in cattle; lesions frequently contain barley spikes or awns of grass, suggesting their traumatic introduction into the tissues during chewing. These probably carry *A. bovis* and other microorganisms into the tissues with them and by eliciting a foreign body reaction create an environment favorable to growth of the anaerobes.

The *granules* formed by *A. bovis* in tissues and exudate are indistinguishable morphologically from those of *A. israelii*. Hyphae of *A. bovis* are more fragile than those of *A. israelii* and tend to break up into short bacillary elements. Experimental inoculations into hamsters (PINE et al., 1960a) showed that while *A. bovis* formed typical branching mycelial elements, its capacity for forming mycelium *in vivo* was much more limited than that of *A. israelii*.

Microcolonies of *A. bovis* are very characteristic on brain-heart infusion agar and are easily differentiated from those of *A. israelii* and *A. naeslundii*. Two colonial types occur, smooth and rough. The *smooth* colony at 24 hours is small and circular with an entire edge (Fig. 2A). The surface is smooth, slightly convex and moist. Examination by transmitted light shows dense hyphal formation in the center. The smaller smooth colonies are easily broken with a needle, but the larger ones maintain their structure during such manipulation (Fig. 2B).

The *rough* type of microcolony at 24—48 hours is more dense when viewed in transmitted light, more irregular in its margin, and has a granular appearance (Fig. 2C). As these become older, they are considerably more elevated than the *smooth* colonies and the surfaces may be folded and irregular (Fig. 2D). The *rough* colonies remain intact during manipulation. However, microscopic examination of both *rough* and *smooth* colonies show only masses of diphtheroids. Growth in liquid medium is usually diffuse and also yields only diphtheroidal forms of *A. bovis*.

Table 2. *Cell wall composition and ultrastructural features of Actinomyces species*

	Results obtained with strains of *Actinomyces*:			
	israelii[1]	*bovis*[1]	*propionicus*[1]	*eriksonii*[2]
Compound:				
Arabinose	0	0	0	0
Rhamnose	0	±	0	0
Galactose	±	0	±	±
Mannose	0	0	0	0
Fucose	0	±	±	0
6-L-desoxytalose	0	±	±	0
Glucose	0	0	±	0
Galactosamine	0	±	±	0
Aspartic acid	0	±	±	±
Alanine	±	±	±	±
Glutamic acid	±	±	±	±
Lysine	±	±	0	±
Diaminopimelic acid	0	0	±	0
Ultrastructure:[3]				
Cell wall thickness (A°)	290	100	114	NA[4]
Membrane coils	Simple coils	None	Complex coils	NA

1 Based on data from PINE (1963).
2 Based on strain CDC X407 from GEORG et al. (1965).
3 Based on data from OVERMAN and PINE (1963).
4 Data not available.

A. bovis hydrolizes starch rapidly and completely. It does not reduce nitrate or utilize mannitol and raffinose.

Actinomyces propionicus Sp. nov.

A. propionicus was first isolated from yellowish-white concretions removed from the left superior canaliculus of a patient (PINE et al., 1960b). In the original description of this isolate (PINE and HARDIN, 1959) it was designated as *A. israelii* strain 699, but the authors noted that it had the unique property among *Actinomyces* of producing propionic acid from glucose. Further studies (BUCHANAN and PINE, 1962; OVERMAN and PINE, 1963) revealed other distinctive properties which are presented in Tables 1 and 2.

The mycelial nature of the microcolony is readily apparent even at 24 hours (Fig. 3A). Examination of the growth under high power shows long hyphal elements which branch repeatedly. The mycelial structure is retained even when a cover glass is placed over the colonies. Older colonies are smooth and convex (Fig. 3B).

A. propionicus generally produces branched hyphae on standard media. However, under special conditions unusual morphologic forms have been obtained which ranged from long thread-like filaments and hyphae with bulbous ends to masses of spherical cells. Under these circumstances the morphological similarity of *A. propionicus* to *Propionibacterium pentosaceum* in particular as well as to certain other propionic bacteria was quite striking (BUCHANAN and PINE, 1962).

A. propionicus grows well aerobically, but smaller inocula are needed to initiate anaerobic growth. Carbon dioxide has no stimulatory effect on aerobic growth.

A. propionicus is pathogenic for mice. Intraperitoneal inoculation produces lesions in the stomach, kidney, intestinal tract and liver.

Actinomyces eriksonii (Georg, Robertstad, Brinkman, and Hicklin, sp. nov.)

Thus far, the habitat of this organism seems confined to human clinical material (Georg et al., 1964, 1965). It has been isolated from pleural fluid and from exudate obtained from both subcutaneous and lung abscesses. Typical sulfur granules are not formed in tissues, although microscopic colonies of the organism have been demonstrated in exudate. It also appears singly as a gram-positive, elongate, rodshaped organism which exhibits frequent branching.

In *culture A.eriksonii* may appear in diphtheroidal form or as filamentous and highly branched. Clubbed or bifurcated ends are common. The organism is non-motile and not acid-fast.

At 24—48 hours, microcolonies have a glistening, dewdrop appearance. They are circular, flat, and granular with a central core of denser growth and with finely serrated to occasionally fuzzy edges. After 7—10 days, the colonies are white to cream, convex to conical, with smooth to pebbly surfaces. Additional characteristics are listed in Tables 1 and 2.

Actinomyces naeslundii (Thompson and Lovestedt, 1951)

This inhibitant of the normal oral cavity was shown by Thompson and Lovestedt (1951) to differ significantly from *Actinomyces* isolated from actinomycotic lesions. Subsequently, Howell et al. (1959) studied approximately 200 oral strains and presented a detailed description of this species.

In the majority of cases, *A.naeslundii* can be isolated using blood agar plates which have been incubated aerobically. However, increased carbon dioxide tension is required for good growth in air. With a very small inoculum, most strains grow better anaerobically with carbon dioxide than aerobically.

In *culture A.naeslundii* may appear as a gram-positive filamentous organism or as a diphtheroidal form. The filaments are slender, of variable thickness, and possess either straight, clavate, or tapered ends. They stain solidly, show segmentations and may be beaded or stippled. In an occasional culture, only gram-positive bacillary forms will be present. These may be slender, long or short, curved, straight or bent.

At 18—24 hours microcolonies usually show a dense mass of diphtheroid cells or a dense clump of filaments surrounded by a rudimentary mycelium (Fig. 4A). Older colonies may be either rough or smooth, the latter reaching a diameter of 2—3 mm (Fig. 4B). The organism is reported to be nonpathogenic for experimental animals.

Pathogenesis

Since *A.israelii* is a normal inhabitant of the oral cavity, it would seem reasonable to assume that all cases, cervico-facial, thoracic, or abdominal, stem directly or indirectly from this source. The fact is that there are still widely divergent views concerning pathogenesis even of the cervico-facial type of infection.

Cervico-facial Actinomycosis. Israel (1878) expressed the view that human actinomycosis, especially the cervico-facial type, originated in the oral cavity. He was the first to notice the relationship between carious teeth and cervico-facial actinomycosis. Subsequently, many other workers were to point out the relationship between dental infections such as periapical osteitis (Harbitz and Backer-Grondahl, 1910; Axhausen, 1935, 1936; Bjerrum and Sv. Hansen, 1931; Nielsen, 1942), dental extractions (Brofeldt, 1926; Havens, 1933; MacGregor, 1951; Zitka, 1949), and jaw fractures (MacGregor, 1945). The trauma of extraction was suggested as predisposing to actinomycotic infection (Jacobson, 1930; Kolouch and Peltier, 1946; Naeslund, 1931). The oral cavity has also been implicated as the source of infection in cases of skin actinomycosis following bite wounds inflicted by human teeth (COLEBROOK, 1920; MORTON, 1940; ROBINSON, 1944; ZISKIN et al., 1943). GLAHN (1954a) has challanged the early literature on occurrence of pathogenic actinomycetes in the healthy human mouth (NAESLUND, 1925; EMMONS, 1935, 1936, 1938; LORD and TREVETT, 1936; BIBBY and KNIGHTON, 1941; SLACK, 1942; ROSEBURY et al., 1944), mainly on the grounds that pathogenicity of the isolates was not demonstrated and identity with strains of *Actinomyces* from actinomycotic lesions was not established. Only in the case of SULLIVAN and GOLDSWORTHY (1940) was a comparison made between oral strains and isolates from actinomycotic lesions, and experiments included morphology, oxygen tolerance, and fermentation reactions. GLAHN's objections hold some merit, but in the light of our present knowledge and ability to identify *A. israelii*, work such as that of HOWELL et al. (1962) leaves no doubt that *A. israelii* exists in the oral cavity and particularly around early or shallow carious lesions. It is really a question of *how* the organism invades the tissues to initiate cervico-facial infection.

GLAHN (1954a) suggested *two main routes of infection*, namely (a) via gates or ducts which are normally present and lead from the mucous membranes into deeper tissue and (b) gaps produced by pathological conditions and their sequelae. That *A. israelii* acts in synergism with a well-defined group of anaerobic microorganisms has already been established. Assuming that both the fungus and its microbial associates exist in the oral cavity, they would have to compete with other members of the oral flora in invasion of tissues. Conditions which promote anaerobiasis would favor development of the actinomycotic lesion.

The mucous membrane of the oral cavity, in its normal state, is highly resistant to microorganisms. Slight traumatic wounds usually heal quickly without becoming infected. Presumably microorganisms which do enter such temporary gaps are quickly destroyed. The case for entry and localization in tissues by *A. israelii* via the ducts of the salivary glands is not a convincing one. Indeed, attempts to demonstrate a relationship between sialolithiasis, sialoadenitis and actinomycotic infection have failed (GANNER, 1929; GRABNER, 1936; HUSTED, 1953). Likewise, the presence of *A. israelii* in tonsillar crypts most likely represents saprophytic growth on accumulations of necrotic epithelium and there is no evidence to support the view that the tonsils may represent an actual portal of entry.

Actinomycosis of Dental Origin. This leads to consideration of entry through gaps produced by pathological conditions. These are listed by GLAHN (1954a) to include: deep caries with periapical osteitis, dentigerous cysts, impacted teeth, open alveoli after tooth extraction, jaw fractures, marginal paradentitis and partial eruption of teeth, especially the wisdom teeth of the lower jaw.

Proving a connection between infectious foci of dental origin and cervico-facial actinomycosis by microbiologic studies is especially difficult. Problems with overgrowth from contaminants as well as the chance isolation of *A. israelii* from contaminated food particles and saliva can readily be appreciated. Nevertheless GLAHN (1954a) was able to trace a case of actinomycosis of the cheek to two impacted canines in the upper jaw. A sinus tract led from the cheek lesion into the cavity containing the impacted teeth. *A. israelii* was isolated from the cheek lesion, the granulation tissue at the base of the two teeth, and from the surface of one of the teeth. In examining 90 cases of cervico-facial actinomycosis, GLAHN (1954a) was able to implicate one of the routes listed above in 72 instances. The one exception was marginal paradentitis which is rather surprising since the gingival pockets would seem a likely place for anaerobes to grow.

Thoracic Actinomycosis. WAYL et al. (1958) suggest that there are three ways by which *A. israelii* can reach the lung from the oral cavity: (1) via the tracheobronchial tree by aspiration; (2) via the blood stream; (3) along the fascial planes of the neck and mediastinal structures.

It is generally agreed that the tracheobronchial tree represents the usual route (GARROD, 1952; PEABODY and SEABURY, 1957; EMMONS et al., 1963; HILDICK-SMITH et al., 1964) and that this results from aspiration of hyphal fragments from carious lesions or other infectious foci of dental origin, or of actinomycotic granules growing saprophytically in tonsillar crypts. A case resulting from aspiration of an entire tooth has been reported (WARWICK, 1923). Presumably, the conditions set up by bronchiolar occlusion secondary to simultaneously aspirated particulate matter are favorable to growth of *A. israelii* and its microbial associates.

Hematogenous dissemination from a septic phlebitis in a small vessel associated with a dental infection would provide a rapid means by which *A. israelii* could reach the lung; the condition of infarction resulting from occlusion of a blood vessel by the septic embolus would favor establishment of an actinomycotic lesion (GARROD, 1952; WAYL et al., 1958).

Actinomyces israelii exhibits a propensity for spreading along contiguous surfaces and the possibility of extension from an oral lesion or cervico-facial lesion to the lungs via the esophagus and mediastinal structures has been considered by PEABODY and SEABURY (1957) and by WAYL et al. (1958). The latter presented a case history of descending actinomycosis in which the order of progression was: sinus of the gum of the lower jaw; right lower lobe lung abscess with pleural effusion; sinus in left submammary region; sinuses in left lumbar region and over the left greater trochanter, left upper gluteal region, and left lower gluteal region. While lung involvement could have resulted from aspiration, it could also have resulted from direct spread from the mediastinum which would account for the sternal and sub-mammary sinuses. The mediastinal lesion could then have extended directly through the diaphragm to the retroperitoneal tissues of the abdomen, then through the posterior abdominal wall in the region of the lumbar triangle of Petit and then to the trochanteric and gluteal areas.

Pulmonary actinomycosis may develop in the hilar region or in basal parenchymatous areas. TURNER (1926) divided pulmonary actinomycosis into essentially two types; (1) a first or early form beginning with involvement of the hilum and extending into the lung along the bronchi and (2) a secondary parenchymatous involvement, with or without abscess formation, which may later extend to the pleural cavity, ribs, vertebrae, pericardium, and sternum (DECKER, 1946; KAY, 1947; PEABODY and SEABURY, 1957; WARTHIN and BUSHUEFF, 1958).

Abdominal Actinomycosis. There is evidence that *A. israelii* is harbored at all levels of the intestinal tract and can, in the absence of trauma, invade the cecum or appendix and occasionally the stomach, gallbladder, or liver (CAMPBELL and BRADFORD, 1948). However, actinomycotic infection of the abdominal viscera seldom occurs as a primary disease. It most often becomes evident several weeks or months after an acute, perforative gastrointestinal disease such as acute appendicitis, perforating colonic diverticulum, perforated peptic ulcer, acute ulcerative disease of the intestinal tract, or traumatic perforation of the bowel (PUTMAN et al., 1950; RAPER, 1950; PEABODY and SEABURY, 1957; ASHTON and SLANEY, 1963; EMMONS et al., 1963). Typical acute appendicitis, usually of the perforative type, preceded the onset of actinomycotic infection in 72% of the 122 cases reviewed by PUTMAN and his associates (1950). In only 7 of the 122 cases was there no history of some previous acute illness and 103 of their cases had undergone some type of emergency surgical procedure for acute gastrointestinal lesions prior to onset of actinomycotic infection.

Structures of the right side of the body are most frequently involved, especially the right iliac fossa (Fig. 9). The frequency of region and organ involvement has been calculated from the data of Putman et al. (1950) and is presented in Table 3. The authors also reported occasional involvement of the epigastrium, buttocks and peri-anal region, left psoas muscle, scrotal and supra-pubic regions, vertebral column, spleen and brain.

Table 3. *Regions of the body or organs involved in cases of primary abdominal actinomycosis*[1]

Region or Organ	% of cases showing involvement	Region or Organ		% of cases showing involvement
RLQ abdominal wall	82	Subdiaphragmatic	Abscesses	19
LLQ abdominal wall	25	Pelvic	Abscesses	10
RUQ abdominal wall	17	Right psoas	Abscesses	7
LUQ abdominal wall	4	Right perinephritic	Abscesses	8
Right lumbar	20	Liver		16
Left lumbar	6	Right kidney		7
Right hip	7	Tube and ovary		8
Left hip	3	Right lung		16
Right chest wall	21	Left lung		16
Left chest wall	7			

1 Adapted from the data of Putman et al. (1950).

Extension of the actinomycotic lesion occurs along contiguous surfaces and spread is usually directed away from the bowel and peritoneal cavity. Connective tissue and muscle are infiltrated with resultant necrosis and fibrosis, but bone is seldom attacked. The infection seldom spreads along lymphatic channels although lymph node involvement has been described (Brickner, 1925; Fairley, 1947).

Rarely, the primary site may be other than the three described above. *Cerebral involvement* (Jacobson and Cloward, 1948), primary *skin involvement* from bites and wounds (McWilliams, 1917; Robinson, 1944; Burrows, 1945; Andleigh, 1951; Cullen and Sharp, 1951; Banerjee, 1952; Finch, 1953; Montgomery and Welton, 1959; Brainey, 1965) and anorectal lesions (Gordon and DuBose, 1951) have all occurred as primary infections.

The role of immunity in the pathogenesis of actinomycosis remains essentially undefined (Wilson, 1957). In general, attempts at sensitizing experimental animals have been unsuccessful. On the other hand, repeated injections of *A. israelii* into guinea pigs and rabbits have, in some instances, resulted in increasing the severity and duration of the actinomycotic lesions (Mathieson et al., 1935; Emmons, 1938; Rosebury et al., 1944). What this means in terms of human infections is not clear. Mathieson and his associates suggested that human cases may represent individuals who have become allergic to *A. israelii* as a result of repeated absorption of products from that fungus and from other antigenically related saprophytes. However, the experimental evidence to support this concept is completely inadequate and clinical evidence is virtually non-existent. Definition of the role played by hypersensitivity in actinomycotic infections must await preparation of more specific skin-testing antigens.

No significant metabolites, including exotoxins and endotoxins, have been demonstrated for *A. israelii*. This would seem to emphasize the importance of the

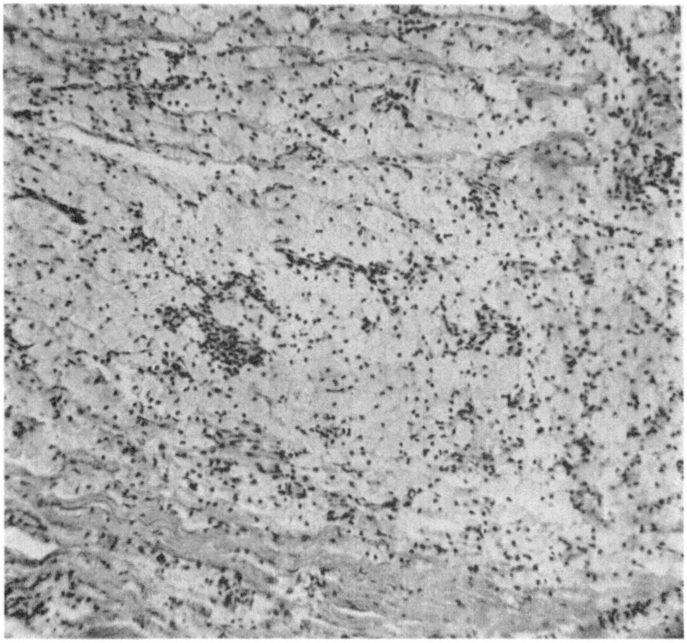

Fig. 8. The wall of an actinomycotic abscess. The pus-filled center of the abscess, not visible here, is oriented towards the top of this figure. Fat-laden macrophages, or lipophages, comprise the inner zone of the wall followed by a zone of scar tissue with collagen fibers. There is a sprinkling of lymphocytes in both zones. H & E, × 450

synergism observed in actinomycotic lesions, for most of the microorganisms found in association with *A. israelii* have been shown to produce hyaluronidase and other depolymerizing enzymes (Brede, 1959). Indeed, some of them have been responsible for persistence of actinomycotic lesions which had been rendered free of *A. israelii* with penicillin therapy (Holm, 1951). Both Holm (1950) and Glahn (1954b) have referred to actinomycosis as a *group of infectious diseases* caused by *Actinomyces* in synergism with certain other microbes, and Glahn has suggested that variations in clinical severity of the disease are to a certain extent dependent upon the combination of organisms involved.

Gross and Microscopic Appearances

Gross Features. Although the anatomic location may tend to influence the appearance of the actinomycotic lesion, it is generally characterized by suppuration, sinus formation and scarring. The external appearance may first be that of a firm, indurated mass, poorly circumscribed, and covered by a tight, brawny, dusky red skin. More advanced lesions tend to soften and develop cutaneous sinuses, some of which may be open and draining while others are fibrosed and closed. On cut sections, the lesions present a honeycombed appearance due to the multiple abscesses and dense fibrous tissue (Putman et al., 1950). With visceral lesions, suppuration, sinus formation, intercommunicating abscesses, extensive fibrous tissue reaction and a yellow color are the most constant gross characteristics (Weed and Baggenstoss, 1949).

Microscopic Features. The microscopic picture is that of chronic neutrophilic response similar to that seen in chronic infections due to cocci or to *Nocardia*

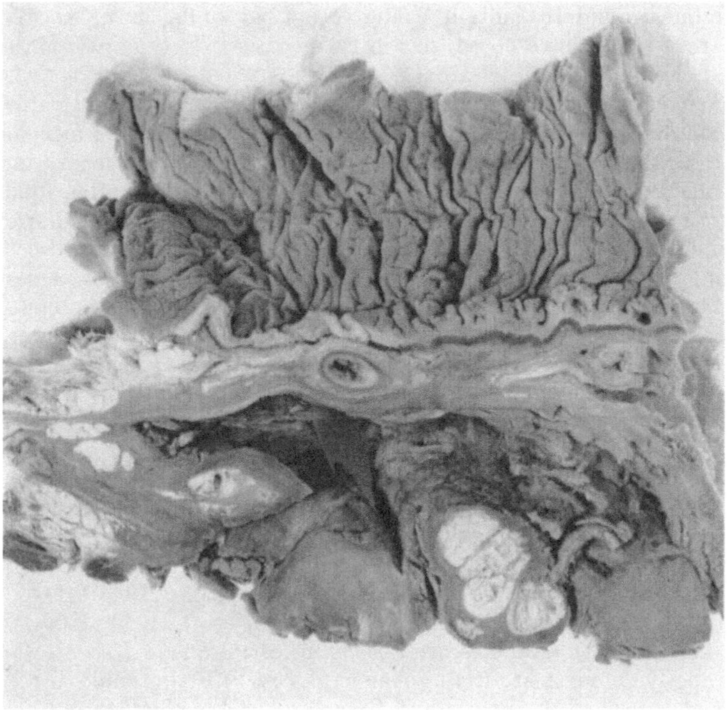

Fig. 9. Periappendiceal actinomycotic abscess. This is an autopsy specimen from a 16 year old colored boy, a resident of North Carolina, USA. His symptoms of actinomycosis began 4 months before he died. Note the appendix shown in 2 cross sections in the horizontal midplane between the abscess cavity below and the cecum above. The infection probably developed when an ordinary acute appendicitis perforated, spilling *Actinomyces israelii* into the abdominal cavity along with coliform bacilli and streptococci of the fecal material in the appendix. The appendix is now encased in dense scar tissue but the large abscess persists below. It has burrowed in the retroperitoneal muscle and other tissues. The rounded white and lobulated areas are adipose tissue. In the cut surface of the wall of the cecum, to the right in the photograph, two intramural actinomycotic abscesses are clearly shown. These represent burrowing pus rather than a cecal origin of the process

asteroides (Figs. 1A, 5). Histologic features include micro-abscesses, ramifying sinuses containing purulent exudate, and fibrosis (BINFORD, 1962). There is *liquefactive* rather than caseous necrosis. The inflammatory cells in the walls of the abscesses and sinuses include neutrophils, lymphocytes, plasma cells and histiocytes. Giant cells are rare, but an occasional one may be seen in the proximity of or in contact with a sulfur granule (BAKER, 1947). *Foamy, fat-laden, macrophages* (lipophages) occur in large numbers in the fibrosing walls around lesions, and are responsible for the yellow color observed during gross examination (Fig. 8). The abscesses vary considerably in size and may show both intercommunication and coalescence. Generally, there is extensive fibrosis in the walls of the abscesses. The connective tissue reaction is most extensive in lesions involving the liver, kidneys, ovaries, subcutaneous and muscular tissues. It is less extensive in the lungs and is minimal in the central nervous system (WEED and BAGGENSTOSS, 1949).

The extent to which the accompanying bacteria influence the histopathologic picture is not clear. However, abscesses well removed from the central infection are sometimes found to contain bacteria, but no sulfur granules. WEED and BAGGENSTOSS (1949) have noted that in many cases of actinomycosis there is an associated chronic active interstitial nephritis.

Actinomycosis can be diagnosed histopathologically by demonstrating microcolonies (sulfur granules) of *A. israelii*. These are easily detected in hematoxylin and eosin sections where they appear as round to oval bodies measuring several hundred microns in diameter (Figs. 1A, 5, 6). They are most often found in the center of a micro-abscess. The actinomycotic granule is surrounded first by a zone of neutrophils, then histiocytes, and lipophages, then fibroblasts and collagenous fibers, and finally dense fibrous tissue. In other words, the abscesses usually present fibrous encapsulation. The granule stains deeply with the hematoxylin, but at the periphery it usually exhibits radial hyphae with eosin-staining hyaline clubs (Fig. 5). The nature or significance of these clubs in not completely known. They very likely represent a host response to the fungus, for they are not seen in microcolonies grown on synthetic medium.

It must be emphasized that a final histopathologic diagnosis of actinomycosis should not be made solely from the appearance of the granules stained with hematoxylin and eosin. Similar granules with eosinophilic clubs at the periphery are produced in tissues in cases of "woody tongue" (*Actinobacillus lignieresi*), botryomycosis (*Staphylococcus*), mycetomas due to *Streptomyces*, and in infections due to *Nocardia brasiliensis* (TRIBEDI and MUKHERJEE, 1939; HAGAN, 1943; CALERO, 1946; MOORE, 1946; AUGER, 1948; WEED et al., 1949; LAVALLE, 1962; EMMONS et al., 1963). Granules are absent in lesions due to *Nocardia asteroides*.

To properly demonstrate the hyphae of *A. israelii* in a granule, use the Gram stain. This is a part of the Brown and Brenn and of the MacCallum-Goodpasture methods. With these stains an actinomycotic granule is shown to be composed of gram-positive, delicate hyphal filaments with occasional branching (Fig. 6). The filaments tend to stain irregularly. Associated bacteria, either cocci, bacilli, or fusiforms, may be found intermingled with the hyphae.

The hyphae can also be demonstrated clearly with the Gomori-methenamine-silver stain, but it has the disadvantage of not being able to distinguish gram-positive from gram-negative elements (Fig. 1B). The PAS and Gridley stains do not demonstrate the hyphae satisfactorily.

Single, isolated hyphal filaments of *A. israelii* are seldom found outside the granules. When they do occur, they are indistinguishable from filaments of *Nocardia asteroides*. However, the pyogenic infiltrate from lesions of nocardiosis, due to *N. asteroides*, do contain numerous filaments of the fungus but granules are not present.

The advantage of microbiologic confirmation of a histopathologic diagnosis is, of course, obvious. However, the pathologist is frequently confronted with the problem of making a diagnosis solely from a fixed specimen which has been obtained from an operation, a curettage, or a biopsy. It should be emphasized that granules tend to occur in the pus so that biopsies which include only the walls of sinuses and abscesses will frequently be negative. Furthermore, granules are often sparse in infected tissues and numerous sections may be required before they can be demonstrated. The gross specimen should be preserved until microscopic studies are complete as additional blocks from other sites may be needed (BAKER, 1957).

Cervico-facial Actinomycosis. The features of oral, facial, and maxillary actinomycosis have been discussed at length by HERTZ (1957). The actinomycotic inflammation produces a firm or hard, tumor-like mass. The numerous abscesses and sinuses within the mass contain thick and yellow or hemorrhagic pus with a

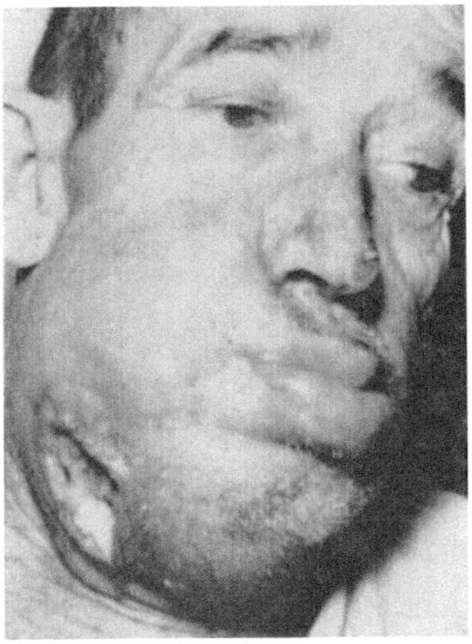

Fig. 10

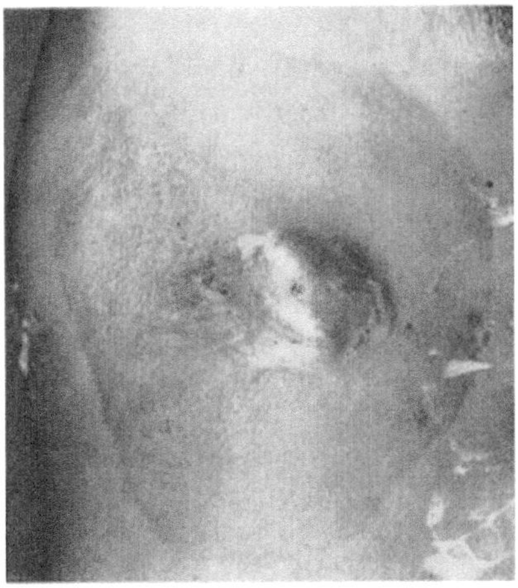

Fig. 11

Fig. 10. Cervico-facial actinomycosis. Extensive involvement of the mandible with coalescing sinus tracts resulting in a deep ulcerative lesion. (Courtesy Doctors ALFREDO A. NAVARRO and RAUL MENA)

Fig. 11. Extension of a sinus tract to the surface of the right thoracic wall. A 33-year-old alcoholic complained of back pain, weight loss, night sweats, and cough for two months prior to the development of a tumor on the right lateral thorax. An x-ray of her chest showed a right upper lobe pneumonia with basilar empyema. An abscess was present in the right lower lobe of the lung, and the sinus tract on the thorax communicated with a subdiaphragmatic abscess as well as the empyema. Cultures were positive for *Actinomyces israelii*. (Courtesy of Dr. JOHN H. SEABURY)

putrid odor. Sulfur granules are usually present. Inflammatory foci near the surface ulcerate to form sinuses, the openings of which come to be surrounded by elevated, soft, yellow, fungating granulation tissue. New foci develop in adjacent areas and the condition advances slowly. The fibrosis and the active purulent process produce a complex of abscesses and fistulae. The infiltration procedes without regard to anatomic boundaries. Burrowing carries the infection along relentlessly through various tissues. When the periosteum is involved, there follows destruction of bone cortex, and tumor-like masses of osseous tissue are formed around irregular cavities. Bone involvement usually occurs early, preceding the appearance of sinus tracts, and may progress to osteomyelitis.

The primary lesion is frequently a *periodontal abscess* (VILLA, 1957) and extension of the infection usually follows extraction of the affected tooth. The process then extends to involve subcutaneous tissues forming a hard mass and eventually reaches the skin producing numerous fistulae. In other instances the primary lesion develops as a result of trauma from a carious or broken tooth or from an accidental fracture of the jaw (Fig. 10, 11, 12).

By the time the infection is noticed, it has usually left the site of invasion, but healing and scarring has left a palpable string of connective tissue extending from the initial site of the active process. Generally, the lymph nodes are not affected.

Maxillary involvement, with rarefaction of the maxilla, is quite common (LUDWIG, 1955; MAIN and MACPHEE, 1964). The lesion may extend to the skin to form a tumor with draining sinuses in the upper cheek, parotid area and neck,

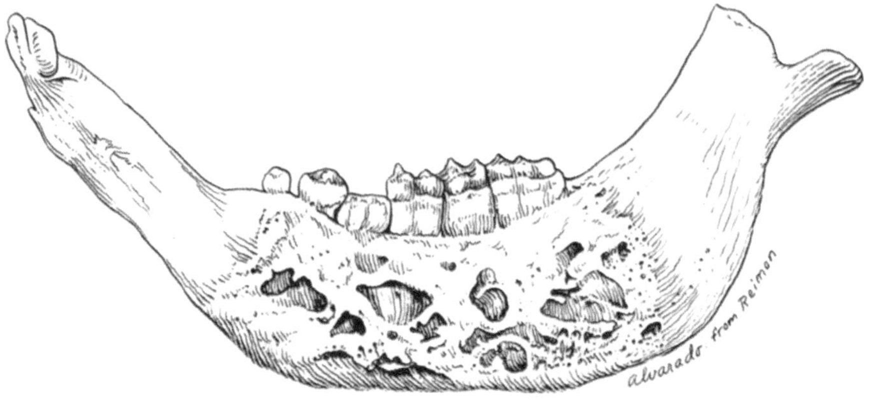

Fig. 12. Bovine actinomycosis. The lower jaw of a cow showing perforated tumor-like distension of the bone. In cattle, the most frequent site of infection is the mucosa of the mouth. The process commonly extends to the jaws to produce a tumor-like, central, destructive and ossifying osteomyelitis. (Redrawn from Reimann: Pathology for Students and Physicians, vol. 1, p. 558, 1929; P. Blakiston's Son & Co., Philadelphia)

or it may extend to the orbit, cranial bones, meninges, and brain (Zitka, 1952). Sazama (1965) found the parotid gland involved in 11% of his cases.

Stanton (1966) has reported primary involvement of the paranasal sinuses; in one case the patient had been edentulous for 17 years.

Lesions in the lower jaw may be extensions from periodontal or osseous lesions of the madible or from tonsillar or regional lymphatic tissue (Choukas, 1958; Hanratty and Naeve, 1964; Owen and Macansh, 1965) (Fig. 15).

The histopathologic picture is one of granulation tissue with many plasma cells, lipophages, mocrophages (epithelioid cells), and occasionally giant cells. As such, it carries no features specific for actinomycosis, but instead presents a picture which is typical of any inflammation accompanied by granulation tissue. Occasionally, sulfur granules cannot be demonstrated.

Osteomyelitis of the alveolar process may follow pulpal involvement. Gold and Doyne (1952) described a case involving the mandibular right first molar and the alveolar bed of bone. The bone was thin, porous, and white. The roots were surrounded by sparse bone trabeculae and canals contained remnants of *gutta-percha*. The marrow spaces were large and empty except for soft tissue fragments found in the interradicular and interdental areas. Peridontal fibers were completely absent. One of the root canals was incompletely filled with *gutta-percha* and contained dentine debris, filamentous organisms and inflammatory cells. The lacunae of the alveolar bone were devoid of osteocytes and appeared to be necrotic; the normal cellular and fibrous components of the marrow spaces were not present. The cementum of the tooth, like the bone, contained no viable cells. Classic sulfur granules were found within the soft tissue fragments in the marrow spaces and interstices of bone; they were also found to be protruding from the apex of the root canal and firmly attached to the surface of the cementum.

Osteomyelitis of the *humerus*, apparently resulting from hematogenous dissemination, has been reported (McCormack et al., 1954).

Paralaryngeal involvement may stem from the tonsils, via direct extension through the mucous membranes of the pharynx and hypopharynx, or via the inferior constrictor muscle. In the case reported by Davies (1952) the abscess was located in the lower half of the anterior triangle of the neck, lateral to the

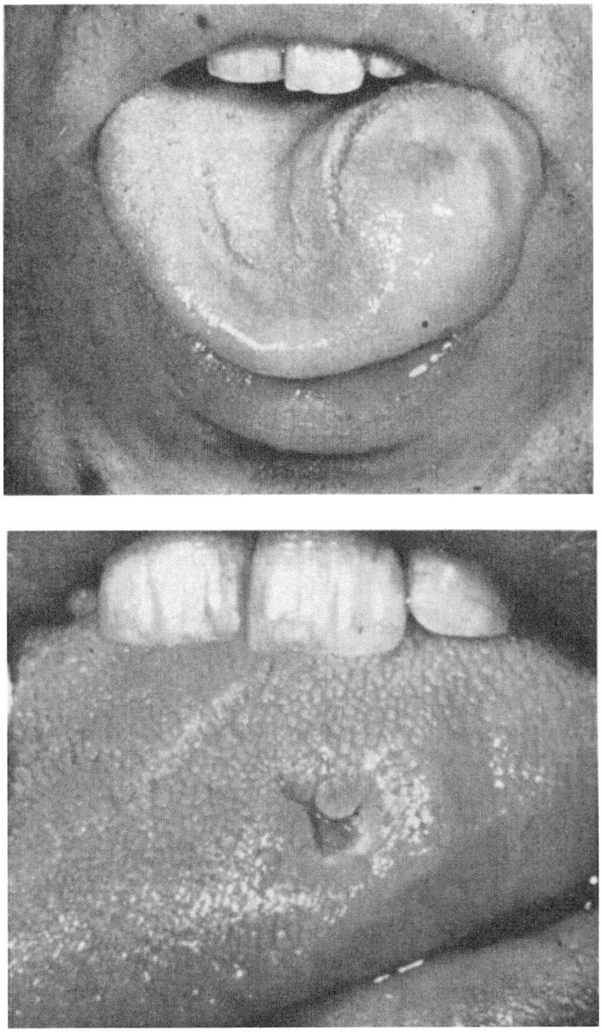

a

b

Fig. 13. Actinomycosis of the tongue. a) Nodule on the left lateral aspect of the dorsum linguae in a 27-year-old white male. Two days prior to admission he suddenly developed a painful swelling of the left side of the tongue. Incision of the nodule yielded yellow, thick, odorless pus. Direct examination showed *Actinomyces*-like gram positive organisms and cultures yielded an anaerobic actinomycete resembling *A. israelii* with regard to biochemical properties. b) Same lesion after eleven daily treatments with 800,000 units of penicillin intramuscularly. (From Dorph-Petersen and Pindborg: Oral Surgery, Oral Medicine, and Oral Pathology, *7*, 1178—1182 (1954), The C.V. Mosby Company, St. Louis, Missouri)

trachea and infrahyoid muscle. A fibrous band could be traced from the abscess upwards in the neck towards the angle of the jaw. The histopathology was remarkable in that a strikingly large number of eosinophils were present.

Actinomycosis of the Tongue. Involvement of the tongue usually occurs in the anterior third and near the tip on one side of the mid line (Dorph-Petersen and Pendborg, 1954). The process is at first a hard, nodular swelling which increases in size and becomes softer (Fig. 13). The overlying mucosa develops a bluish color.

Eventually the nodule ruptures and drains and increases in size to become a multilocular abscess. Histopathologic sections show an irregular epithelium covering the tongue. Below this is a zone of loose connective tissue with many thin blood vessels. At the border of the abscess, the epithelium is replaced by fibrin and polymorphonuclear neutrophils. The deeper muscle layers may be partially replaced by fibrous tissue, lymphocytes, and neutrophils.

FLETCHER (1956) reported an interesting case of actinomycotic infection which manifested itself initially as a *chronic otorrhea*. Although a number of biopsies were made from the middle ear, mastoid antrum, and nasopharynx — all of which revealed a granulomatous process — the etiologic diagnosis was not established until the patient came to autopsy. At that time a mass 3 cm by 1 cm with an ill-defined border surrounded the entire right petrous tip area and extended into the foramen magnum. It enveloped the cranial nerves of the right side, and had dissected downward near the mid-line to a point just above the nasopharynx on the right. Histopathologic sections were typical for actinomycosis. Ear involvement is rare, Fletcher's report being the 36th case to be added to the literature.

VERHOEFF (1926) described a case of *intraocular* actinomycosis in a patient who later died of an actinomycotic liver abscess. Eye involvement is virtually always extraocular and most frequently occurs as a *lacrimal canaliculitis* (SMITH, 1953; PINE et al., 1960; PINE and HARDIN, 1959).

Pulmonary Actinomycosis. Actinomycosis of the lungs begins in the bronchial mucosa, or in bronchopneumonic or peribronchial foci (TURNER, 1926; ARNDT, 1931; KAY, 1947, 1948; PRINSLEY, 1957; WARTHIN and BUSHUEFF, 1958). The infection usually starts with central hilar involvement and extends along the bronchi into the lung parenchyma. In the cases studied by KAY (1948) the initial lesion was lobular in distribution and pneumonic in type. The pneumonia, believed to be secondary to atelectasis, progresses to suppuration and abscess formation. The abscesses are burrowing in character and are surrounded by a highly vascularized granulation tissue. There is marked fibrosis and many alveoli are replaced by scar tissue. As the infection spreads, interlobar fissures are obliterated and adjacent lobes are involved. Pleural involvement may occur early with dense fibrous adhesions to the chest wall. In one case cited by WAYL et al. (1958) the apex of the lung was found to be infiltrated with hard masses and its separation from the thoracic wall was rendered difficult by dense adhesions, especially posteriorly and medially. These involved the bronchial plexus, subclavian artery, and superior vena cava. Empyema and chest wall sinuses may follow (Fig. 14).

Older lesions tend to be hard, consolidated and fibrotic. Within these firm cicatricial areas of the lungs, the granulation tissue masses are preserved as branching fistulous tracts and as abscesses filled with pus.

Usually the lower lobes are affected, but apical involvement also occurs (WINGO and WILLIAMS, 1957; WAYL et al., 1958; LEE, 1966). A tendency towards bilateral involvement has been cited (CONANT et al., 1954), but PEABODY and SEABURY (1957) feel that this feature is associated primarily with well-advanced cases. KAY (1948) reported involvement of the mid lung field in 11 cases, upper lung field in 8 cases, and lower lung field in 3 cases. Involvement was unilateral in 16 cases and bilateral in 6 cases. One of the cases reported by HOLLIS and HARGROVE (1947) presented as a posterior mediastinal abscess.

The microscopic findings are essentially those described under General Features. WINGO and WILLIAMS (1957) described a small actinomycotic nodule removed during thorocotomy as showing masses of young and mature granulation tissue mingled with alveoli containing macrophages. In many areas the alveoli were compressed and the septa were markedly thickened. There was diffuse infiltration with eosinophils and lymphocytes. Occasional basophilic masses of tangled filaments were lying free in the alveoli. These were gram-positive and

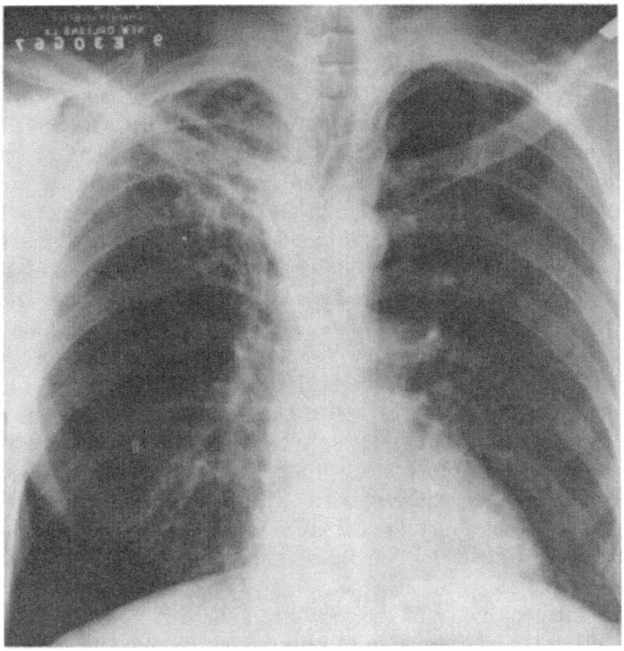

Fig. 14. Right upper lobe involvement associated with a perinephric abscess. A 39-year-old male gave a history of a 40 lb. weight loss during five months, right shoulder pain for two months, swelling of the shoulder and blood-tinged sputum production for two weeks. One week prior to admission he developed a rapidly enlarging left flank mass with chills and fever. A perinephric abscess arising from the lower pole of the kidney was drained with the removal of three liters of pus. Both his sputum and the pus were positive for *Actinomyces israelii*. Precipitating antibody was present in his serum. The right upper lobe infiltrate appearing in this figure is indistinguishable from tuberculosis. (Courtesy Dr. JOHN H. SEABURY)

non-acid fast. Some of these masses were surrounded by a single ring of polymorphonuclear leucocytes.

Extension through the *thoracic wall*, via the intercostal spaces, is quite characteristic. Extensive areas of marked fibrosis and scarring may be found in the musculature, subcutaneous tissues and in the skin; all are characterized by numerous fistulous tracts.

Pericardial Actinomycosis. Extension to the *pericardium* may occur from pulmonary, cervical, or mediastinal lesions. Typically, actinomycotic pericarditis appears as fibrous adhesions between the pericardial layers, pleura and mediastinal tissues. These fibrotic masses contain abscesses and fistulous tracts. KAUFMANN (1929) cites a case showing invasion of the mediastinum and pericardium from the lung, penetrating the cavity of the right ventricle and entering the large coronary vein.

CORNELL and SHOOKHOFF (1944) reviewed the world literature and were able to collect 68 cases of actinomycosis in which the *heart* was involved. The *pericardium* was found to be most commonly involved by direct spread, usually from the lungs. An occasional case of pericarditis followed active infection of the myocardium which in turn had resulted from hematogenous dissemination. In two-thirds of the cases of actinomycotic pericarditis the myocardium was also involved and in many instances this led to penetration of the endocardium.

Projection of the granulomatous mass into the cardiac chambers with perforation of the endocardium and subsequent pyemia occurred in six cases.

Clinical evidence of cardiac involvement was present in less than half of the cases reviewed by CORNELL and SHOOKHOFF. However, some did present as a *chronic constrictive pericarditis* as in the case reported by HARA and PIERCE (1957). In this instance, there was a left pleural empyema and a 3 cm abscess in the soft tissues anterior to the left fifth rib. The heart was encased in a thick fibrous capsule and was markedly restricted in its activity. The pericardium was purplish-tan and 3—10 mm thick. Histologic sections revealed markedly fibrotic tissue with many microabscesses, some of which contained granules of *A. israelii*. Very few cases have been added to those collected by CORNELL and SHOOKHOFF (SAVIDGE and DAVIES, 1953; ZOECKLER, 1951).

Spinal Actinomycosis. Pulmonary lesions may extend to involve tissues of the neck and then proceed as a prevertebral process to the spinal column where it may ascend or descend. It may ascend by continuity along vessels and nerves to to reach the cranial cavity, meninges, and brain; however, involvement of the central nervous system is more often due to hematogenous metastasis (BOLTON and ASHENHURST, 1964). Characteristic fibrotic processes with accompanying abscesses and sinus tracts may appear over the entire back and may involve the ribs and vertebrae. Usually, bone involvement remains essentially superficial, i.e., peri- and parosteal, and it is only rarely that the actinomycotic process extends deeply into the body of the vertebra (Fig. 18).

Finally, there may be extension of pulmonary lesions through the diaphragm into the abdominal cavity.

Abdominal Actinomycosis. PUTMAN and his associates (1950) as well as others (BROGDEN, 1922; COPE, 1938; CAMPBELL and BRADFORD, 1948; PHEILS et al., 1964) have presented strong evidence to support the contention that abdominal actinomycosis is almost invariably preceded by some disease process or by trauma (See Pathogenesis). Actinomycotic lesions usually arise in structures adjacent to the site of ulceration or perforation of a viscus, or in some pocket or recess in which gastrointestinal contents have collected (Fig. 9). In most cases involvement of the intestinal mucosa cannot be demonstrated. Most often, the initial actinomycotic lesion is found in the retroperitoneal tissues or tissues of the abdominal wall. BROCKMAN (1923) observed that the appendix and intestine are usually free of infection and that the initial lesion involves the muscles and connective tissues of the right iliac fossa (Fig. 15).

Periappendiceal Actinomycotic Abscess. Since the appendix is the commonest site of perforative gastrointestinal disease in man (PUTMAN et al., 1950), it is not surprising that the right iliac fossa (right lower quadrant of the abdominal wall) is the site most frequently affected (Table 3). The early lesion presents externally as a firm, indurated mass near the previous operative site. In time, this mass softens and the formation of abscesses and sinuses ensues (Fig. 9). Gross examination reveals destruction of subperitoneal and subcutaneous connective tissue as well as muscle. Varying quantities of pus, oftentimes very little, are to be found. The direction of spread of the lesion appears to be away from the bowel.

Actinomycotic Abscesses of the Liver. The liver may be involved by direct extension through the diaphragm from foci in the thoracic cavity (WAGENSTEEN et al., 1952), or from a primary lesion in the abdominal cavity. Usually, involvement is via the portal vein and secondary to an abdominal lesion (FORBUS, 1943). Actinomycotic pylephlebitis may develop. The process may continue from the liver through the diaphragm to involve the pleura and lungs (BONNEY, 1947; FOREMAN, 1963). Commonly, there is one or several foci; these appear on cut

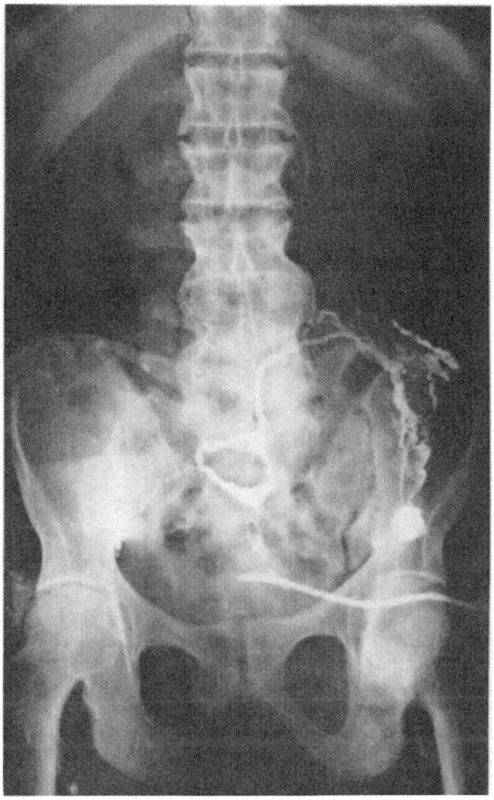

Fig. 15. Multiple sinus tracts in the pelvic region. At the time of diagnosis, this patient had a history of multiple draining sinuses in the lumbar, sacroiliac, and inguinal areas. Fistula-in-ano had been present intermittently. He had been treated on three occasions for the presumptive diagnosis of tuberculosis without any response to streptomycin. Direct examination of drainage and curettings from the sinus tract were positive for sulfur granules. Cultures were positive for *Actinomyces bovis*; treatment with penicillin was curative. The sinogram presented here shows marked involvement of the pelvis with extensions along the prevertebral fascial plane, the iliopsoas muscles and sigmoid colon. (Courtesy Dr. JOHN H. SEABURY)

surface as yellow, soft, fibrillar, meshy, or honeycombed masses saturated with pus (Fig. 16). Occasionally, numerous metastatic nodules may be present. An abscess may have a wall up to 1 cm in thickness. The creamy yellow pus can be expressed by slight pressure and contains numerous sulfur granules. Histologic sections show complete destruction of hepatic tissue with replacement by the typical abscesses. Generally, there are no inflammatory or degenerative changes in the liver parenchyma outside of the abscess although there may be some scarring and collapse of liver tissue along with proliferation of Kupffer cells (WAGENSTEEN et al., 1952).

Actinomycosis of the Spleen. Multiple actinomycotic abscesses of the spleen with spontaneous rupture of that organ has been observed (WEED et al., 1949; PUTMAN et al., 1950). In the case illustrated by PUTMAN et al. the gastrosplenic ligament was thickened and fibrotic and was found to contain several spicules of chicken bone embedded in the fibrous tissue. These had apparently perforated through the gastric wall, carrying *A. israelii* and other microorganisms with them.

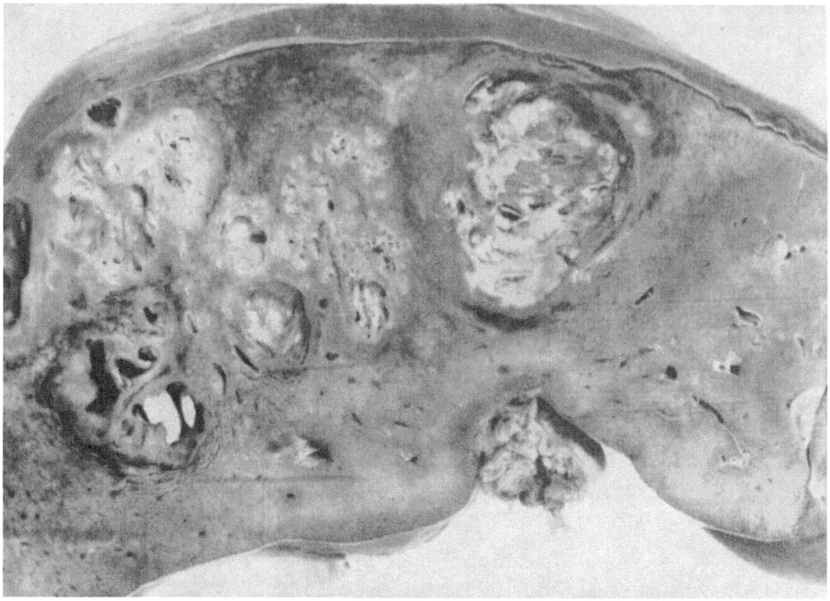

Fig. 16. Multiple actinomycotic abscesses of the liver. Note the pus still filling smaller abscesses and clinging to the wall of the larger ones. Although sulfur granules are not discernable at this magnification, they were readily demonstrable in the pus at the postmortem examination. The surrounding zone of lipophages is visible around smaller abscesses, and the dense fibrosis of scarring is well shown about abscesses to the left. This liver specimen, while taken from a different case, is strikingly similar to that seen in the 16 year old boy presented in Fig. 9. At the periappendiceal region the fungus enters venules and infects the liver via the portal vein. Sometimes the hepatic abscesses perforate the capsule of the liver and produce a subdiaphragmatic abscess or even a penetration of the diaphragm with empyema or abscess of the lung

Renal Actinomycosis. Involvement of the kidneys may be the result of contiguous spread from some primary focus in the abdomen or may represent hematogenous dissemination from some focus which can be either apparent or inapparent. A unique case was reported by LEBRUN and GILMOUR (1953) in which actinomycosis of the kidney followed a ureterocecostomy. The infection is presumed to have originated from the cecum and extended via the lumen of the ureter.

When renal actinomycosis occurs in the absence of any other demonstrable foci, it is considered to be a "primary infection". Primary renal involvement may occur as: 1) a chronic suppurative lesion resembling a renal carbuncle, 2) pyelonephritis, and 3) pyonephrosis.

Gross changes in the kidney may be classified as *localized* or *diffuse*. The localized lesion will appear on cut surface as a pyramidal area of granulation and scar tissue with the apex pointing towards the renal pelvis; within this area, yellowish streaks or granules and abscess cavities of various size occur. The process may extend through the capsule to invade the perinephric fat and produce a perirenal abscess while sparing the pelvis and calyces (WHISENAND and MOORE, 1951). In diffuse involvement, the entire kidney may be converted into a suppurating, granulomatous mass with sinus tracts discharging to the skin.

When involvement begins in the lower pole of the kidney, obstruction of the ureter occurs and hydronephrosis develops (DAMMGAARD-MORCH, 1956). Calculus formation and pyonephrosis are natural sequelae (WILSON-PEPPER, 1951). Extension from the infected kidney to the lower urinary tract is uncommon.

Tubo-ovarian Actinomycosis. In tubo-ovarian lesions the right adnexal structures are usually involved first; extension to the left occurs later. Multilocular abscesses are usually present and sinus tracts may extend in various directions to involve the other ovary and retroperitoneal tissues, or to re-enter the gastrointestinal tract as an anal fistula (HARTL, 1951; MURPHY, 1954; LOTH, 1956; PHEILS et al., 1964; SWEENEY and BLACKWELDER, 1965; KRIEG and STAIB, 1966). The review of INGALLS and MERENDINO (1952) showed bilateral involvement in 44.4% of cases; right side only in 37.8%; and left side only in 17.8%. Typically, histologic sections will show chronic salpingitis with fibrosis of the ovary and areas of necrosis with miliary abscesses. The abscesses are intercommunicating, frequently contain granules, and have the characteristic cellular constituents. Most workers agree that involvement of the female genitalia is the result of extension from the intestine (WAGNER, 1910; ROBINSON, 1919; FRANK, 1922; BRICKNER, 1925; STEVENSON, 1957). HAZELHORST (1928) reported cases which followed instrumentation for abortion. BARTH (1928) reported actinomycosis in women fitted with intrauterine pessaries. The incidence of pelvic organ involvement is not high; RASHBAUM and McINTOSH (1944) were able to collect only 85 cases from the literature, but later a review by MACCARTHY (1955) cited 157 recorded cases.

In the case cited by CAMPBELL and BRADFORD (1948) generalized actinomycotic involvement of the pelvic organs followed surgery for a ruptured ovarian cyst by 11 months. The patient died approximately 13 months after original surgery. Autopsy revealed completely necrotic pelvic organs; the uterus; oviducts, and ovaries were identified with great difficulty. The mass of necrotic pelvic structures was adherent to the anterior surface of the rectum and rectosigmoid, the posterior aspect of the bladder, the tip of the cecum and one loop of ileum. There was an obstruction of the left ureterovesical junction with accompanying hydroureter and severe pyelonephritis. One sinus tract communicated with and drained through the lower part of the abdomen.

Involvement of the *uterus* is exceedingly rare; only 14 of the 157 cases reviewed by MACCARTHY (1955) showed uterine involvement and in only one case was an *endometrial* lesion specifically mentioned. BAGATKO (1958) has reported involvement of the *cervix*.

In the case presented by MACCARTHY, the uterus was enlarged (147 gm) with a localized mass in the right postero-lateral region comprised of necrotic and friable material. The lesion had penetrated the myometrium and extended into the cavity of the body of the uterus. The muscle coat showed scattered yellow foci. Histologically, the outer two-thirds of the myometrium was replaced by multiple abscesses, each surrounded by a wall of granulation tissue. There were numerous sulfur granules. The inflammatory process extended from the parametrium to a junction between the outer two-thirds and inner one-third of uterine muscle. At one point, there was a fistula opening on the endometrial surface. The infection had obliterated the right ovary and the right fallopian tube was enlarged and contained pus.

Perforation into the *urinary bladder* is also uncommon. KUSUNOKI et al. (1958) found only 14 cases in the literature and were able to contribute an additional case.

Actinomycosis of the Breast. DAVIES (1951) found primary involvement of the breasts to be exceptionally rare, his case representing the 14th in the literature. Three degrees of involvement are recognized, viz., an early stage of recurrent acute abscess formation, a stage of multiple sinus formation with distortion of the breast, and a quiescent stage of chronic abscess formation. The portal of entry seems to be the nipple.

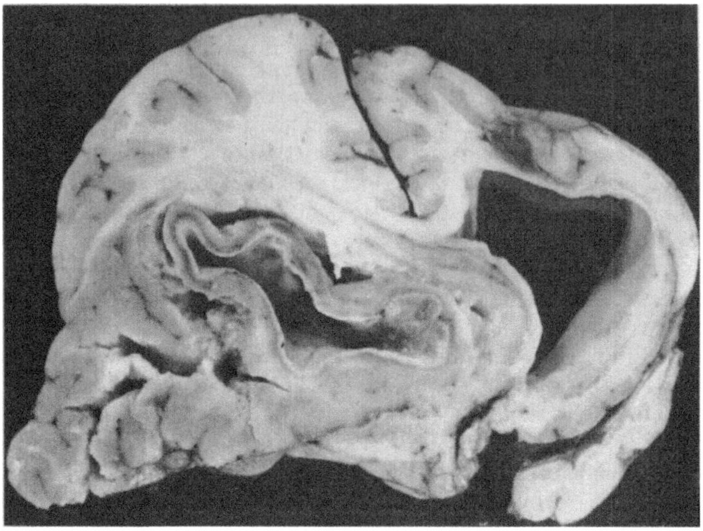

Fig. 17. Actinomycotic abscess of the brain. Coronal section of the brain showing a left frontal lobe abscess, focal granulomatous meningoencephalitis, and obstructive hydrocephalus. From a two-year-old white female who died four months after a right ventriculojugular shunt for a left parieto-occipital abscess. The initial abscess was associated with a urinary tract infection, but the exact organism is not known. (Courtesy Dr. PAUL A. MCGARRY)

Actinomycosis of the Testis. SCORER (1952) reported actinomycotic infection of the left testicle, his case representing the fifth to be added to the literature (SCHNEIDER, 1945; BAKER and REGINS, 1946). Involvement is unilateral, with swelling of the cord and general enlargement of the testicle, particularly of the globus minor. On cut surface multiple abscesses with typical gross and histologic features are seen. The *epididymis* is apparently spared by the disease process.

Subdiaphragmatic and Other Abdominal Abscesses. Numerous abscesses may develop in the abdominal cavity; those most commonly encountered include the subdiaphragmatic or subphrenic abscess, pelvic abscess, psoas abscess, and perinephric abscess (Table 3). While these occur most frequently on the right side or medially, involvement of the left side can occur (SILBER, 1953).

MACHAFFIE et al. (1957) described an actinomycotic abscess positioned in the mesentery between the junction of the descending and sigmoid colon and crest of the ileum. Grossly, it resembled a sarcoma, but on cut surface it proved to be a multiloculated cyst with numerous bands of white fibrous tissue and contained a large amount of purulent material. MILLER (1964) reported the occurence of an actinomycotic abscess at the site of anastomosis following resection of the left colon for adenocarcinoma.

Psoas and Retroperitoneal Abscesses. Sinus tracts extend to the lumbar, greater trochanter, and gluteal regions (ROSS and KNIGHT, 1954; WAYL et al., 1958; CUMMINGS et al., 1959) (Fig. 13).

Anorectal Lesions. These may arise either as a primary infection (GORDON and DUBOSE, 1951; ANSCOMBE and HOFMEYR, 1954; SWINTON and SCHATMAN, 1964; FRY et al., 1965) or secondarily to some focus in the pelvis (PHEILS et al., 1964). The case reported by GORDON and DUBOSE apparently had its origin at the site of an external hemorrhoid. The process had persisted for 30 years, initially producing an anal fistula which was excised, then multiple subcutaneous nodules and fistulae in the perianal region, posterior buttocks, intracrural region and posterior thighs.

Malignant Changes in Sinuses. Malignant degenerative changes, usually carcinomatous in nature, may occur in chronic discharging sinuses. GOODWIN (1955) reported an unusual case where the malignant change took the form of a granulation tissue sarcoma.

Actinomycosis of the Central Nervous System. The actual incidence of central nervous system involvement with *A. israelii* cannot be determined from the early literature since these investigators failed to distinguish actinomycotic lesions from those caused by *Nocardia*. Indications are that such involvement is an infrequent complication of actinomycosis in man. BOLTON and ASHENHURST (1964) were able to collect only 17 adequately documented cases from the world literature for the period 1937—1964 and were able to add a case of their own. Their survey revealed that there was evidence of hematogenous spread from teeth or lungs in 15 cases, direct extension of cervico-facial lesions in 2 cases, and no apparent primary source in one case.

Fig. 18. Actinomycosis of the vertebral column. The result of extension from the apices of the lungs and the deep cervical tissues. Usually, bone involvement is superficial and it is rare that the actinomycotic process extends deeply into the body of the vertebra. (Redrawn from REIMANN: Pathology for Students and Physicians, vol. 2, p. 1146, 1929; P. Blakiston's Son & Co., Philadelphia)

Actinomycotic Abscesses of the Brain. Commonly, there was a single abscess involving a cerebral hemisphere (Fig. 17). Multiple abscesses were present in 2 cases and in only one was meningitis the only manifestation of the disease. The abscess wall is usually thick and fibrotic. In the case reported by BOLTON and ASHENHURST there were three thick-walled abscess cavities, the largest measuring 1.5 cm and the two smaller ones 0.5 cm in diameter. The abscess wall contained an inner zone of granulation tissue which was infiltrated by lymphocytes, plasma cells, and an occasional polymorphonuclear leucocyte. Numerous nests of lipophages were present. The outer layer was composed of dense glial tissue and there was some perivascular chronic infiltrate associated with adjacent blood vessels. A few granules of *A. israelii* were found within the cavities.

Not included in the review by BOLTON and ASHENHURST was a case by PANTAZOPOULOS (1964) in which the actinomycotic abscess presented as a small, cyst-like tumor in the third ventricle. It was suspended from the inferior surface of the corpus callosum and obstructed the third ventricle causing hydrocephalus. The mass was surrounded by a capsule except for a small area at the base of its pedicle. It was soft and when incised a dense purulent yellow-grey fluid flowed out. It had the features of an abscess; its wall was thin and consisted mostly of vascular connective tissue and showed the structure of choroid plexus or choroid tela. There were numerous leucocytes, remnants of red blood cells, and necrotic neuroglial cells. Sulfur granules were demonstrated. No primary focus could be demonstrated at autopsy.

Spinal Actinomycosis. Involvement of the vertebral column is attended by both necrosis and hyperplasia (COPE, 1951; BRETT, 1951a, 1951b; WINSTON, 1951; BAYLIN and WEAR, 1953). It is usually secondary to involvement of contiguous tissues with several vertebrae being affected. Suppuration may spread under the anterior common ligament to involve larger segments of the vertebral column. The infection extends to adjacent pedicles and transverse processes and also to the heads of neighboring ribs. Characteristically, there is bone adsorption around the focus of infection with new bone being formed distally. This combination of rarefaction and sclerosis occurs in varying proportions (COPE, 1951); when the disease is progressing, rarefaction predominates, during regression sclerosis (Fig. 18). The infection may involve the entire vertebral body resulting in a network of suppurating chambers which are bounded by bone of increased density imparting a honeycombed effect which shows up on X-ray as a lattice work or "soap-bubble" effect (BAYLIN and WEAR, 1953).

Differential Diagnosis

The need to differentiate actinomycotic granules in tissue from those produced by other microorganisms has already been discussed.

Clinically, actinomycosis must be differentiated from tuberculosis, syphilis, neoplasm, glanders, tularemia, granuloma inguinale, osteomyelitis, botryomycosis, sarcoidosis, chronic appendicitis, amebiasis, typhoid fever, carcinoma of the intestine, intestinal tuberculosis, liver abscess, psoas abscess and sarcoma of the retroperitoneal tissue or iliac bones (CONANT et al., 1954). It must also be differentiated from other mycoses, especially nocardiosis, North American blastomycosis, paracoccidiodal granuloma, coccidioidomycosis, cryptococcosis, and sporotrichosis.

Clinical Features

Cervico-facial type. Commonly, the onset is insidious and first becomes apparent as a persistent swelling in the parotid or mandibular regions. Pain is minimal unless secondary infection develops. Where the actinomycotic involvement follows trauma or a surgical procedure, healing of the wound will be retarded or arrested with a concomitant increase in edema and induration. As the infection progresses, there appears the classical external manifestations of a dark purplish skin overlying a hard, tumorous process. At least in the early stages, fever is usually minimal or absent and the patient's general health remains unaffected so long as the process does not become too extensive (Fig. 15).

HERTZ (1957) has divided cervico-facial infections into the classical form, atypical form, and a tumor-like process which avoids bone and fails to form an abscess.

The *classical form* is characterized by a chronic course, accompanied by firm, extensive infiltrations which alternate with purulent foci, and by the formation of numerous fistulae. It often starts with a toothache, presenting as a hard tumor-like mass. Gradually, an abscess is formed and extends towards the surface of the skin. Fistulae appear which discharge a foul-smelling pus at the surface. Trismus develops where the muscles of mastication are involved. A febrile reponse is absent or minimal except in extensive involvement. Generally the lymph nodes are not involved, but the process may invade blood vessels with subsequent hematogenous dissemination. In this form, extension usually avoids bony tissue.

The *atypical form* may be acute or subacute and is associated with a marked febrile response. These cases yield a foul-smelling pus and an osseous focus can always be established. The fever is accompanied by painful swelling, then abscess formation and finally perforation with formation of a fistula. In many respects, the early stages of both the classical and atypical forms resemble an ordinary dental inflammation.

The tumor-like form of involvement is painless, grows continually without abscess formation and without involving the bone or skin, and fails to respond to the conventional doses of penicillin. In the case described by HERTZ (1957) sulfur granules were absent but anaerobic gram negative bacilli and anaerobic streptococci were isolated. This form of cervico-facial involvement is the hardest to diagnose clinically. The painless swelling and slow steady growth will suggest a tumor. Roentgenologic studies are helpful to the extent that they rule out bone involvement.

Cervico-facial infections must be differentiated from glanders, tularemia, tuberculosis, and bacterial osteomyelitis.

Thoracic type. As the pulmonary lesion extends to include consolidation, sinus tracts and pleural involvement, the patient will experience night sweats and high fever and sustain a substantial weight loss (Fig. 11). Febrile episodes of 101 F. to 103 F. and white blood count of 20,000 to 30,000 cells/cu mm are not uncommon.

Radiologic findings may be suggestive but are never diagnostic (PEABODY and SEABURY, 1957). Usually they are indistinguishable from those found in pulmonary tuberculosis (LEE, 1966; WINGO and WILLIAMS, 1957), although in some cases the area of lung involved may be of some help in the differential diagnosis (Fig. 12). In KAY's series (1948) the disease involved the mid lung field in 11 cases, upper lung field in 8 cases and base of the lungs in 3 cases; involvement was unilateral in 16 cases and bilateral in 6 cases. Commonly, X-ray findings will reveal a large area of infiltration or consolidative pneumonia with hazy margins along with an associated pleuritis. Frank cavitation is unusual, although multiple abscesses appearing as small round lucencies can be seen within the consolidated areas (PEABODY and SEABURY, 1957). Empyema is usually present; pleural effusion, though rare, has been reported (BONNEY, 1947). Parahilar lesions may present some difficulties in differentiating actinomycosis from bronchial neoplasm (WARTHIN and BUSHUEFF, 1958). Vertebral involvement along with the characteristic "soap bubble" appearance on roentgenograms has already been discussed.

Pulmonary lesions may extend to involve the pericardium and heart with consequent cardiac constriction. This can manifest itself as chronic constrictive pericarditis (CORNELL and SHOOKHOFF, 1944; ZOECKLER, 1951; HARA and PIERCE, 1957).

Pulmonary actinomycosis must also be differentiated from other mycotic infections of the lung, particularly nocardiosis.

Abdominal type. Abdominal actinomycosis is usually preceded by some type of acute, perforative gastrointestinal disease; most frequently this occurs in the form of an acute appendicitis (See Pathogenesis). Carious teeth and diseased tonsils may also serve as predisposing factors (BROGDEN 1922). Symptoms most frequently associated with abdominal involvement are weight loss, cachexia, fever, and a palpable mass. There may be localized pain and tenderness, muscular rigidity, induration, vomiting and jaundice. A purplish discoloration of the skin may appear over the mass and around the sinus tract openings. Kidney involvement may occur as an abscess, pyelonephritis, or pyonephrosis; the signs are obscure and suggest a malignancy. Abdominal actinomycosis must be differentiaed from tuberculosis, malignancy and amebiasis.

Prognosis

The prognosis for cervico-facial involvement was relatively good even before the advent of antibiotics. HERTZ (1957) quotes from BROFELDT'S statistics of 1926 which showed that 67% of 151 cervico-facial cases were cured. Currently, recovery rates are approximately 90% for cervico-facial cases, 80% for abdominal cases, and 40% for thoracic cases.

Therapy

Treatment of actinomycotic lesions has included simple needle aspiration (MACCONNELL, 1944), surgical drainage (COLEBROOK, 1921), and complete excision (WAGENSTEEN, 1936). Irradiation was recommended by LAMB et al. (1947) and NEUBER (1940) reported excellent results from the use of freshly prepared, polyvalent autogenous vaccines. Some of the drugs used prior to penicillin included potassium iodide, copper sulfate, methylene blue, and arsenicals (HUNTER and WESTRICK, 1957).

The use of *penicillin* has had the most profound influence on the prognosis of actinomycotic infections, particularly the thoracic and abdominal types (MACGREGOR, 1945; DOBSON and CUTTING, 1945; NICHOLS and HERRELL, 1948). Actinomycosis has also been treated successfully with aureomycin (WRIGHT and LOWEN, 1950; MCVAY et al., 1951; MCVAY and SPRUNT, 1953a; SVANE, 1966), Chloramphenicol (LITTMAN et al., 1950, 1952; KELLY, 1951), oxytetracycline (PULASKI et al., 1952; LANE et al., 1953), tetracycline (MARTIN et al., 1956), achromycin (HINDS and DEGNAN, 1955), erythromycin (HERRELL et al., 1955), streptomycin (TORRENS and WOOD, 1949), isoniazid (MCVAY and SPRUNT, 1953b; GREEN and BLACK, 1955), and stilbamidine (MILLER et al., 1952).

While an occasional strain of *A. israelii* has been found to be resistant to penicillin therapy (PEABODY and SEABURY, 1960), most strains are inhibited by 0.01—0.6 units of the drug. GARROD (1952) tested the *in vitro* sensitivity of *A. israelii* to various antibiotics and found effective inhibitory concentration to be 0.01—0.6 unit for penicillin, 2.2 μg for terramycin, 2.8 μg for chloramphenicol, 4.2 μg for aureomycin, and 23.7 μg for streptomycin.

Penicillin remains the drug of choice, but it must be administered in adequate amounts if treatment failures are to be avoided (RAPER, 1950; HUNTER and WESTRICK, 1957; PEABODY and SEABURY, 1957; HERTZ, 1957; ASHTON and SLANEY, 1963; SWEENEY and BLACKWELDER, 1965; STANTON, 1966).

For cervico-facial cases, HERTZ (1957) recommends 1.5—5 million units of penicillin given intramuscularly daily for 14 days. Local injection of penicillin (300,000 units) into the actinomycotic foci is sometimes desirable.

For systemic infections, SEABURY and DASCOMB (1964) recommend a minimum of 1—2 million units of penicillin daily for approximately one month. Dosages of 1—20 million units daily for 6—8 weeks have also been suggested (PEABODY and SEABURY, 1957; UTZ, 1967). The practice of administering sulfadiazine or a broad-spectrum antibiotic along with the penicillin is not justified. Most of the treatment failures in the past have been due to inadequate dosage and not to lack of sensitivity of *A. israelii* to penicillin.

The role of surgical procedures in the treatment of actinomycosis continues to be of major importance (PUTMAN et al., 1950). These include incision and drainage of abscesses, curettage of sinuses and fistulae, and excision of sinus tracts.

References

ANDLEIGH, H.S.: Two rare cases of fungus infection of skin in Rajasthan: actinomycosis and rhinosporidiosis. Indian med. Gaz. **86**, 100 (1951). — ANSCOMBE, A.R., HOFMEYR, J.: Perianal actinomycosis complicating pilonidal sinus. Brit. J. Surg. **41**, 666 (1954). — ARNDT, H.J.: Die aktinomykotischen Veränderungen der Lunge und des Brustfells und das Verhalten der Lunge und des Brustfells bei Aktinomykose. In: Handbuch der Speziellen Pathologischen Anatomie und Histologie, Vol. **3** (3), pp. 397—458. F. HENKE and O. LUBARSCH, Editors. Berlin: Springer 1931. — ASHTON, F., SLANEY, G.: Post-gastrectomy duodenal fistula due to actinomycosis. Brit. J. Surg. **50**, 884 (1963). — AUBERT: Observation d'un cas d'actinomycose maxillaire chez le chien. Rev. Path. comp. **51**, 276 (1951). — AUGER, C.: Human actinobacillary and staphylococcic actinophytosis. Amer. J. clin. Path. **18**, 645 (1948). — AXHAUSEN, G.: Die Pathogenese und Klinik der Kieferaktinomykose. Dtsch. Zahn-, Mund- u. Kieferheilk. **2**, 197 (1935). ~ Das Frühbild der Kieferaktinomykose. Dtsch. med. Wschr. **62**, 1449 (1936). — BAKER, R.D.: Tissue changes in fungous disease. Arch. Path. **44**, 459 (1947). ~ The diagnosis of fungus diseases by biopsy. J. chron. Dis. **5**, 552 (1957). — BAKER, W.J., RAGINS, A.B.: Actinomycosis of the testicle. J. Urol. **56**, 547 (1946). — BANERJEE, B.N.: Actinomycosis of the skin. Indian med. Gaz. **87**, 253 (1952). — BARTH, H.: Über Parametritis Actinomycotica und ihre Entstehung. Arch. Gynäk. **134**, 310 (1928). — BAYLIN, G.J., WEAR, J.M.: Blastomycosis and actinomycosis of the spine. Amer. J. Roentgenol. **69**, 395 (1953). — BERTI, P.: Sulla patologia granulomatosa della parotide tuberculosi, sarcoidosi, actinomicosi. Arch. De Vecchi Anat. pat. **35**, 889 (1961). — BIBBY, B.G., KNIGHTON, H.T.: The actinomyces of the human mouth. J. infect. Dis. **69**, 148 (1941). — BIGGS, J.S.G.: Actinomycosis of the liver: a report of two cases. Med. J. Aust. **47**, II, 939 (1960). — BINFORD, C.H.: Tissue reactions elicited by fungi. In: Fungi and Fungous Diseases, pp. 220—238. G. DALLDORF, editor. Springfield, Ill.: Charles C. Thomas 1962. — BJERRUM, O., HANSEN, SV.: Undersogelser over forekomsten af actinomyceten i mundhulen og deres betydning for den acute form af actinomycose. Ugeskr. Laeg. **94**, 1075 (1931). — BLANK, C.H., GEORG, L.K.: The use of fluorescent antibody methods for the detection and identification of *Actinomyces* species in clinical material. J. Lab. clin. Med. **71**, 283 (1968). — BOGATKO, F.L.H.: Actinomycosis of cervix. J. Amer. med. Wom. Ass. **13**, 268 (1958). — BOLLINGER, O.: Über eine neue Pilzkrankheit beim Rinde. Zbl. med. Wiss. **15**, 481 (1877). — BOLTON, C.F., ASHENHURST, E.M.: Actinomycosis of the brain. Case report and review of the literature. Canad. med. Ass. J. **90**, 922 (1964). — BONNEY, G.L.W.: Actinomycosis of liver; report of an unusual case. Brit. J. Surg. **34**, 316 (1947). — BRAINEY, R.R.: Primary cutaneous actinomycosis. J. Amer. med. Ass. **194**, 679 (1965). — BREDE, H.D.: Zur Ätiologie und Mikrobiologie der Aktinomykose; *in-vitro* Versuche zur Frage der fermentativen Unterstützung des *Actinomyces israeli* durch Begleitbakterien. Zbl. Bakt. **174**, 110 (1959). — BREED, R.S., CONN, H.J.: The nomenclature of the Actinomycetaceae. J. Bact. **4**, 585 (1919). — BRETT, M.S.: Advanced actinomycosis of the spine treated with penicillin and streptomycin; report of a case. J. Bone Jt Surg. **33-B**, 215 (1951a). ~ Advanced actinomycosis of the spine treated with penicillin and streptomycin. J. Bone Jt Surg. **33-B**, 482 (1951b). — BRICKNER, W.M.: Pelvic actinomycosis; a study of five consecutive cases successfully treated by operation. Ann. Surg. **81**, 343 (1925). — BROCKMAN, R.: Actinomycosis of the right iliac fossa. Brit. J. Surg. **10**, 456 (1923). — BROFELDT, S.A.: Über Actinomycose in Finnland. Duodecim (Helsinki) **7**, (1926). — BRODGEN, J.C.: Actinomycosis of the gastrointestinal tract; a study of fourteen cases. J. Lab. clin. Med. **8**, 180 (1922). — BUCHANAN, B.B., PINE, L.: Characterization of a propionic acid producing actinomyces, *Actinomyces propionicus* sp. nov. J. gen. Microbiol. **28**, 305 (1962). — BURROWS, H.J.: Actinomycosis from punch injuries: with report of a case affecting a metacarpal bone. Brit. J. Surg. **32**, 506 (1945). — CALERO, C.: Pulmonary actinomycosis; report of the first case observed in the Isthmus of Panama. Dis. Chest **12**, 402 (1946). — CAMPBELL, D.A., BRADFORD, B., Jr.: Actinomycosis of thorax and abdomen. A.M.A. Arch. Surg. **57**, 202 (1948). — CHOUKAS, N.C.: Actinomycosis of the mandible. Oral Surg. **11**. (1), 14 (1958). — CHRISTIE, A.O., PORTEOUS, J.W.: Growth of several strains of *Actinomyces israelii* in chemically defined media. Nature (Lond.) **195**, 408 (1962a). ~ The cultivation of a single strain of *Actinomyces israelii* in a simplified and chemically defined medium. J. gen. Microbiol. **28**, 443 (1962b). ~ The growth factor requirements of the Willis strain of *Actinomyces israelii* growing in a chemically defined medium. J. gen. Microbiol. **28**, 455 (1962c). — CINTI, G.: Le malattie granulomatose (tubercolosi, actinomicosi e granuloma maligno di Sternberg) della lingua. Arch. De Vecchi Anat. pat. **22**, 693 (1954). — COLEBROOK, L.: The mycelial and other micro-organisms associated with human actinomycosis. Brit.J. Path. **1**, 197 (1920). ~ A report upon 25 cases of actinomycosis, with especial reference to vaccine therapy. Lancet **1**, 893 (1921). — CONANT, N.F., SMITH, D.T., BAKER, R.D., CALLOWAY, J.L., MARTIN, D.S.: Actinomycosis. In: Manual of Clinical Mycology, pp. 1—25. Philadelphia: W.B. Saunders Co. 1954. — COPE, V.Z.: Actinomycosis. London-New York-Toronto: Oxford University

Press 1938. ~ Actinomycosis involving the colon and rectum. J. int. Coll. Surg. **12**, 401 (1949). ~ Actinomycosis of bone with special reference to infection of the vertebral column. J. Bone Jt Surg. **33-B**, 205 (1951). — CORNELL, A., SHOOKHOFF, H.B.: Actinomycosis of the heart simulating rheumatic fever. Arch. intern. Med. **74**, 11 (1944). — CULLEN, C.H., SHARP, M.E.: Infection of wounds with *Actinomyces*. J. Bone Jt Surg. **33-B**, 221 (1951). — CUMMINGS, J.R., BEEKLEY, M.E., EARLEY, N.: Abdominal actinomycosis; a case report. Ohio St. med. J. **55** (3), 350 (1959). — CUMMINS, C.S., HARRIS, H.: Studies on the cell wall composition and taxonomy of Actinomycetales and related groups. J. gen. Microbiol. **18**, 173 (1958). — CUMMINS, C.S.: Chemical composition and antigenic structure of cell walls of *Corynebacterium, Mycobacterium, Nocardia, Actinomyces*, and *Arthrobacter*. J. gen. Microbiol. **28**, 35 (1962). — DAMMGAARD-MORCH, P.: Abdominal actinomycosis; three cases of atypical localization. Acta chir. scand. **110**, 458 (1956). — DAVIES, D.F.: Paralaryngeal actinomycosis in a child. Brit. med. J. **1**, 363 (1952). — DAVIES, J.A.L.: Primary actinomycosis of the breast. Brit. J. Surg. **38**, 378 (1951). — DECKER, H.R.: Treatment of thoracic actinomycosis by penicillin and sulfonamides. J. thorac. Surg. **15**, 430 (1946). — DOBSON, L., CUTTING, W.C.: Penicillin and sulfonamides in the therapy of actinomycosis; report of 16 additional cases and *in vitro* tests of susceptibility of *Actinomyces* to penicillin and sulfadiazine. J. Amer. med. Ass. **128**, 856 (1945). — DORPH-PETERSEN, L., PINDBORG, J.J.: Actinomycosis of the tongue; report of a case. Oral Surg. **7**, 1178 (1954). — ELSAESSER, K.H.: Über die Aktinomykose und ihre Lokalisation im Zentralnervensystem. Klinik, Bakteriologie, pathologische Anatomie. Dtsch. Z. Nervenheilk. **164**, 123 (1950). — EMMONS, C.W.: *Actinomyces* and actinomycosis. Puerto Rico J. publ. Hlth **11**, 63 (1935). ~ Strains of *Actinomyces bovis* isolated from tonsils. Puerto Rico J. publ. Hlth **11**, 720 (1936). ~ The isolation of *Actinomyces bovis* from tonsillar granules. Publ. Hlth Rep. (Wash.) **53**, 1967 (1938). — EMMONS, C.W., BINFORD, C.H., UTZ, J.P.: Actinomycosis. In: Medical Mycology, pp. 55—71. Philadelphia: Lea & Febiger 1963. — ERIKSON, D.: Pathogenic anaerobic microorganisms of the actinomyces group. Med. Res. Council Spec. Rept. Ser. **240**, 1 (1940). ~ The morphology, cytology, and taxonomy of the actinomycetes. Ann. Rev. Microbiol. **3**, 23 (1949). — FAIRLEY, K.D.: Actinomycosis of the paraaortic lymph glands. Med. J. Aust. **1**, 799 (1947). — FEGELER, F.: Die Aktinomykose. S. 527—588. — FINCH, P.G.: Actinomycosis of forearm. Canad. med. Ass. J. **68**, 595 (1953). — FLETCHER, R.: A rare case of chronic otorrhea with intracranial complications. Laryngoscope (St. Louis) **66**, 702 (1956). — FOREMAN, J.M.: Actinomycosis of the liver. N. Z. med. J. **62**, 536 (1963). — FRANK, R.T.: Gynaecological and Obstetrical Pathology. D. Appleton and Co. 1922. — FRY, G.A., MARTIN, W.J., DEARING, W.H., CULP, C.E.: Primary actinomycosis of the rectum with multiple perianal and perineal fistulae. Mayo Clin. Proc. **40**, 296 (1965). — GANNER, H.: Ein Fall von Aktinomykose der Unterkieferspeicheldrüse, zugleich ein Beitrag zur Frage der Ätiologie der sogenannten Küttnerschen Speicheldrüsentumoren. Arch. klin. Chir. **155**, 495 (1929). — GARROD, L.P.: Actinomycosis of the lung: aetiology, diagnosis, and chemotherapy. Tubercle (Lond.) **33**, 258 (1952). — GENESI, M.: Pneumomicosi actinomicotica con ascesso cerebrale. Minerva pediat. (Torino) **4**, 39 (1952). — GEORG, L.K., ROBERTSTADT, G.W., BRINKMAN, S.A.: Identification of species of *Actinomyces*. J. Bact. **88**, 477 (1964). — GEORG, L.K., ROBERTSTADT, G.W., BRINKMAN, S.A., HICKLIN, M.D.: A new pathogenic anaerobic *Actinomyces* species. J. infect. Dis. **115**, 88 (1965). — GINSBERG, A., LITTLE, A.C.W.: Actinomycosis in dogs. J. Path. Bact. **60**, 563 (1948). — GLAESMER-ZAFF, M.: Ein Schulfall einer Ovarialaktinomykose. Zbl. Gynäk. **74**, 1927 (1952). — GLAHN, M.: The pathogenesis of cervico-facial actinomycosis. Acta chir. scand. **108**, 193 (1954a). ~ Cervico-facial actinomycosis — etiology and diagnosis. Acta chir. scand. **108**, 183 (1954b). — GOLD, L., DOYNE, E.E.: Actinomycosis with osteomyelitis of the alveolar process. Oral Surg. **5**, 1056 (1952). — GOODWIN, M.A.: Sarcoma arising in a chronic actinomycotic sinus. Brit. J. Surg. **44**, (187) 489 (1957). — GORDON, M.A., DUBOSE, H.M.: Anorectal actinomycosis. Amer. J. clin. Path. **21**, 460 (1951). — GRABNER, A.: Aktinomykose in Speichelsteinen. Z. Stomat. **34**, 862 (1936). — GREEN, R., BOLTON, T.C., WOOLSEY, C.I.: Mycetoma, madura foot; a case of mycetoma pedis in Chicago. Ann. Surg. **128**, 1015 (1948). — GREENE, L.W., BLACK, W.C.: Treatment of cervico-facial actinomycosis with isoniazid. Rocky Mtn med. J. **52**, 43 (1955). HAGAN, W.A.: The pathogenic actinomycetes. In: The Infectious Diseases of Domestic Animals, with Special Reference to Etiology, Diagnosis and Biologic Therapy, pp. 321—338. Ithaca, N.Y.: Comstock Publishing Co 1943. — HANRATTY, W.J., NAEVE, H.F.: Actinomycosis with pathologic fracture of mandible; report of a case. Oral Surg. **18**, 303 (1964). — HARA, M., PIERCE, J.A.: Chronic constrictive pericarditis due to *Actinomyces bovis*; report of a case treated by pericardectomy. J. thorac. Surg. **33**, 730 (1957). — HARBITZ, F., GRONDAHL, N.B.: Actinomykosen i Norge. Oslo: Kristiania 1910. — HARTL, H.: Zur Kenntnis der Genitalaktinomykose der Frau. Arch. Gynäk. **179**, 677 (1951). — HARZ, C.O.: *Actinomyces bovis*, ein neuer Schimmel in den Geweben des Rindes. Jahresb. d. k. Centralbl. Tierarznei, Schule in München **5**, 125 (1879). — HAVENS, F.Z.: Actinomycosis of the head and neck. J. Amer. dent. Ass. **20**, 478 (1933). — HAZELHORST, G.: Aktinomykose der weiblichen Genital-organe als Abtreibungsfolge. Arch. Gynäk, **134**. 561 (1928). — Hazen, E.L., Little, G.N.: *Actinomyces bovis* and 'anaerobic diphtheroids'. Pathogenicity for hamsters and some other differentiating characteristics. J. Lab. clin. Med. **51**, 968 (1951). — Heinrich, S., Pulverer, G.: Zur Ätiologie und Mikrobiologie der Aktinomykose: III. Die Pathogenbedeutung des *Actinobacillus Actinomycetem-comitans* den „Begleitbakterien" des *Actinomyces israeli*. Zbl. Bakt. **176**, 91 (1959). — Heite, H.J.: Krankheiten durch Aktinomyceten und verwandte Erreger. Wechselwirkung zwischen pathogenen Pilzen und Wirtsorganismus. (4. Wiss. Tag. d. Deutschsprach. Mykolog. Gesellsch. in Freiburg/Br., 30. und 31. 10. 1964). Berlin-Heidelberg-New York: Springer 1967. — Hepburn, R.H.: Actinomycosis of the testicle. J. Urol. (Baltimore) **63**, 113 (1950). — Herrell, W.E., Belows, A., Dailey, J.S.: Erythromycin in the treatment of actinomycosis. Antibiot. Med. **1**, 507 (1955). — Herrmann, H.: Zur Pathogenese und Therapie der Aktinomykose. Dtsch. zahnärztl. Z. **7**, 725 (1952). — Hertz, J.: Actinomycosis: oral, facial and maxillary manifestations. J. int. Coll. Surg. **28**, 539 (1957). — Hildick-Smith, G., Blank, H., Sarkany, I.: Actinomycosis. In: Fungus Diseases and Their Treatment, pp. 293—302. Boston: Little Brown and Co. 1964. — Hinds, E.C., Degnan, E.J.: The use of achromycin and neomycin in the treatment of actinomycosis. Oral Surg. **8**, 1034 (1955). — Hollis, W.J., Hargrove, M.D.: Actinomycosis; a report of 12 cases with special reference to a mediastinal case. New Orleans med. surg. J. **99**, 499 (1947). — Holm, P.: Some investigations into the penicillin sensitivity of human pathogenic actinomycetes. Acta path. microbiol. scand. **25**, 376 (1948). ~ Studies on the aetiology of human actinomycosis. I. The "other microbes" of actinomycosis and their importance. Acta path. microbiol. scand. **27**, 736 (1950). ~ Studies on the aetiology of human actinomycosis. II. Do the "other microbes" of actinomycosis posses virulence? Acta path. microbiol. scand. **28**, (4) 391 (1951). — Howell, A., Jr., Murphy III, W.C., Paul, F., Stephan, R.M.: Oral strains of *Actinomyces*. J. Bacteriol. **78**, 82 (1959). — Howell, A., Jr., Stephan, R.M., Paul, F.: Prevalence of *Actinomyces israelii*, *A.naeslundii*, *Bacterionema matruchotii*, and *Candida albicans* in selected areas of the oral cavity and saliva. J. dent. Res. **41**, 1050 (1962). — Hunter, G.C., Jr., Westrick, C.M.: Cervicofacial abscess by *Actinomyces*; report of a case. Oral Surg. **10**, 793 (1957). — Husted, E.: Sialolithiasis. Acta chir. scand. **105**, 161 (1953). — Ingalls, E.G., Merendino, K.A.: Actinomycosis of the female genitalia; case report and review of the literature. West. J. Surg. **60**, 476 (1952). — Israel, J.: Neue Beobachtungen auf dem Gebiete der Mykosen des Menschen. Virchows Arch. path. Anat. 15 (1878). ~ Klinische Beiträge zur Kenntnis der Aktinomykose des Menschen. Berlin: A. Hirschwald 1885. — Jacobsen, H.P.: Actinomycosis; a clinical, pathological and bacteriological study. Med. J. Rec. **132**, 342 ,379 (1930). — Jacobson, J.R., Cloward, R.B.: Actinomycosis of the central nervous system. J. Amer. med. Ass. **137**, 769 (1948). — Jantschew, W.G.: Über die primäre Aktinomykose des Mastdarms. Kasuistischer Beitrag. Z. ges. inn. Med. **16**, 476 (1961). — Jutzler, G.A., Leppla, W., Brunck, H.J.: Primäre doppelseitige Nierenaktinomykose mit Papillennekrose als Ursache eines Nierenversagens. Ärztl. Forsch. **15**, I, 340 (1961). — Kaufmann, E.: Pathology for Students of Medicine. Translation by S.P. Reimann. Philadelphia: P. Blakiston's Son and Co. 1929. — Kay, E.B.: Bronchopulmonary actinomycosis. Ann. intern. Med. **26**, 581 (1947). ~ *Actinomyces* in chronic bronchopulmonary infections. Amer. Rev. Tuberc. **57**, 322 (1948). — Kelly, H.H.D.: Intestinal actinomycosis treated with chloramphenicol and aureomycin. Brit. med. J. No. 4734, 779 (1951). — King, S., Meyer, E.: Metabolic and serologic differentiation of *Actinomyces* bovis and anaerobic diphtheroids. J. Bact. **74**, 234 (1957). — Knake, H.-J., Zeiss, K.-H.: Primäre Abdominalaktinomykose mit sekundärer Ausbreitung auf beide Tuben und Netz. Geburtsh. u. Frauenheilk. **15**, 816 (1955). — Köhlmeier, W., Niel, K.: Aktinomykose-Endokarditis. Wien. klin. Wschr. **1953**, 25. — Kolouch, F., Peltier, L.F.: Actinomycosis. Surgery **20**, 401 (1946). — Krieg, H., Staib, F.: Zur Abdominalaktinomykose der Frau. Gynäcologia (Basel) **162**, 156 (1966). — Kusunoki, T., Hayashi, I., Kashiwai, K.: Actinomycosis of the ovary perforating into the urinary bladder; report of a case. Urol. int. (Basel) **6**, 251 (1958). — Kwapinski, J.B.: Antigenic structure of the Actinomycetales. VII. Chemical and serological similarities of cell walls from 100 actinomycetales strains. J. Bact. **88**, 1211 (1964). — Lamb, J.H., Lain, E.S., Jones, P.E.: Actinomycosis of the face and neck. J. Amer. med. Ass. **134**, 351 (1947). — Lambert, F.W., Jr., Brown, J.M., Georg, L.K.: Identification of *Actinomyces israelii* and *Actinomyces naeslundii* by fluorescent antibody and agar-gel diffusion techniques. J. Bact. **94**, 1287 (1967). — Lane, S.L., Kutscher, A.H., Chaves, R.: Oxytetracycline in the treatment of oro-cervical-facial actinomycosis; report of seven cases. J. Amer. med. Ass. **151**, 986 (1953). — Lavalle, P.: Agents of mycetoma. In: Fungi and Fungous Diseases, pp. 50—68. G. Dalldorf, Editor. Springfield, Ill.: Charles C. Thomas 1962. — LeBrun, H., Gilmour, I.E.: Hypokalaemia and renal actinomycosis following ureterocaecostomy; report of a case. Brit. J. Urol. **25**, 132 (1953). — Lee, B.Y.: Actinomycosis of the lung coexisting with pulmonary tuberculosis; report of a case. Dis. Chest **50**, 211 (1966). — Lentze, F.: 1958. Quoted by Wilson and Miles (1964). — Ley, E., Peraita, P., Ley, E.

hijo: Granuloma actinomicótico cerebral Rev. clín. esp. **41**, 234 (1951). — Littman, M.L., Phillips, G.E., Fusillo, M.H.: *In vitro* susceptibility of human pathogenic actinomycetes to chloramphenicol (chloromycetin). Amer. J. clin. Path. **20**, 1076 (1950). — Littman, M.L., Paul, J.S., Fusillo, M.H.: Treatment of pulmonary actinomycosis with chloramphenicol; report of a case. J. Amer. med. Ass. **148**, 608 (1952). — Lord, F.T., Trevett, L.D.: The pathogenesis of actinomycosis. Recovery of actinomyces-like organisms from the human mouth. J. infect. Dis. **58**, 115 (1936). — Lorenz, O.: Neue Erkenntnisse zur Pathogenese und Klinik der cervico-Facialen Aktinomykose. Med. Klin. **54**, 9 (1959). — Loth, M.F.: Actinomycosis of the fallopian tube. Amer. J. Obstet. **72**, (4) 919 (1956). — Ludwig, T.G.: Actinomycosis originating in the maxillary region. Oral Surg. **8**, 877 (1955). — MacCarthy, J.: Actinomycosis of the female pelvic organs with involvement of the endometrium. J. Path. Bact. (Lond.) **69**, 175 (1955). — McConnell, O.H.: Actinomycosis; report of cases. J. oral Surg. **2**, 173 (1944). — McCormack, L.J., Dickson, J.A., Reich, A.R.: Actinomycosis of the humerus. J. Bone Jt Surg. **36-A**, 1255 (1954). — MacGregor, A.B.: Cervico-facial actinomycosis. Proc. roy. Soc. Med. **38**, 639 (1945). ~ Cervico-facial actinomycosis. Proc. roy. Soc. Med. **44**, 55 (1951). — MacHaffie, R.A., Zasyer, R.L., Salchek, H., Sciortino, A.L.: An unusual case of actinomycosis manifested as an abdominal wall abscess. Gastroenterology **33**, (5) 830 (1957). — MacLennan, A.P.: Composition of the cell wall of *Actinomyces bovis*: the isolation of 6-deoxy-L-talose. Biochem. Biophys. Acta **48**, 600 (1961). — McVay, L.V., Jr., Guthrie, F., Sprunt, D.H.: Aureomycin in the treatment of actinomycosis. New Engl. J. Med. **245**, 91 (1951). — McVay, L.V., Jr., Sprunt, D.H.: A long term evaluation of aureomycin in the treatment of actinomycosis. Ann. intern. Med. **38**, 955 (1953a). ~ Treatment of actinomycosis with isoniazid. J. Amer. med. Ass. **153**, 95 (1953b). — McWilliams, C.A.: Actinomycosis of phalanx of finger. Ann. Surg. **66**, 117 (1917). — Main, J.H.P., MacPhee, I.T.: Actinomycosis of the maxilla in relation to a peridontal abscess. Oral Surg. **17**, 299 (1964). — Martin, W.J., Nichols, D.R., Wellman, W.E., Weed, L.A.: Disseminated actinomycosis treated with tetracycline. A.M.A. Arch. Int. Med. **97**, 252 (1956). — Mathieson, D.R., Harrison, R., Hammond, C., Henrici, A.T.: Allergic reactions to actinomycetes. Amer. J. Hyg. **21**, 405 (1935). — Miller, A.G.: Actinomycosis of the colon; case report. Dis. Colon Rect. **7**, 207 (1964). — Miller, J.M., Long, P.H., Schoenbach, E.B.: Successful treatment of actinomycosis with "stilbamidine". J. Amer. med. Ass. **150**, 35 (1952). — Meyer, E., Verges, P.: Mouse pathogenicity as a diagnostic aid in the identification of *Actinomyces bovis*. J. Lab. clin. Med. **36**, 667 (1950). — Montgomery, R.M., Welton, W.A.: Primary actinomycosis of the upper extremity. A.M.A. Arch. Derm. **79**, 578 (1959). — Moore, M.: Radiate formation on pathogenic fungi in human tissue. Arch. Path. **42**, 115 (1946). — Morton, H.S.: Actinomycosis. Canad. med. Ass. J. **42**, 231 (1940). — Mosselman, G., Lienaux, E.: L'actinomycose et son agent infectieux. Ann. Méd. vét. **39**, 409 (1890). — Murphy, L.C.: A case of actinomycosis of ovary. J. Kans. med. Soc. **55**, 254 (1954). — Naeslund, C.: Studies of *Actinomyces* from the oral cavity. Acta path. microbiol. scand. **2**, 110 (1925). ~ Experimentelle Studien über die Ätiologie und Pathogenese der Aktinomykose. Acta path. microbiol. scand. **8**, (Suppl. 6) 1 (1931). — Neuber, E.: Spezifische Diagnostik und Therapie der Aktinomykose. Klin. Wschr. **19**, 736 (1940). — Nichols, D.R., Herrell, W.E.: Penicillin in the treatment of actinomycosis. J. Lab. clin. Med. **33**, 521 (1948). — Nielsen, J.: Dosierungsfragen bei der Röntgenbehandlung von zervico-fazialer Aktinomykose. Acta Radiol. **23**, 303 (1942). — Overman, J.R., Pine, L.: Electron microscopy of cytoplasmic structures in facultative and anaerobic *Actinomyces*. J. Bact. **86**, 656 (1963). — Owen, M.D., Macansh, J.: Actinomycosis of the mandible. Med. J. Aust. **2**, 962 (1965). — Pantazopoulos, P.E.: Actinomycosis of brain manifested by vestibular symptoms. Arch. Otolaryng. **80**, 309 (1964). — Peabody, J.W., Jr., Seabury, J.H.: Actinomycosis and Nocardiosis. J. chron. Dis. **5**, 374 (1957). ~ Actinomycosis and nocardiosis; a review of basic differences in therapy. Amer. J. Med. **28**, 99 (1960). — Pegrum, G.D.: Actinomycotic lesions in the chorioallantoic membrane of the chick embryo. J. Path. Bact. **88**, 323 (1964). — Pheils, M.T., Reid, D.J., Ross, C.F.: Abdominal actinomycosis. Brit. J. Surg. **51**, 345 (1964). — Pine, L., Hardin, H.: *Actinomyces israelii*, a cause of lacrimal canaliculitis in man. J. Bact. **78**, 164 (1959). — Pine, L., Watson, S.J.: Evaluation of an isolation and maintenance medium for *Actinomyces* species and related organisms. J. Lab. clin. Med. **54**, 107 (1959). — Pine, L., Howell, A., Jr., Watson, S.J.: Studies of the morphological, physiological, and biochemical characteristics of *Actinomyces bovis*. J. gen. Microbiol. **23**, 403 (1960a). —Pine, L., Hardin, H., Turner, L.: Actinomycotic lacrimal canaliculitis. A report of two cases with a review of the characteristics which identify the causal organism, *Actinomyces israelii*. Amer. J. Ophthal. **49**, 1278 (1960b). — Pine, L.: Recent developments on the nature of the anaerobic actinomycetes. Ann. Soc. belge Méd. trop. **3**, 247 (1963). — Pine, L., Boone, C.J.: Comparative cell wall analyses of morphological forms within the genus *Actinomyces*. J. Bact. **94**, 875 (1967). — v. Plothe, O.: Farbstoffbildung bei Actinomyceten. Naturwissenschaften **34**, 190 (1947). — Ponfick, E.: Ein Vortrag über Actinomykose des Menschen.

Ärztl. Z. (Breslau) **2**, 151 (1880a). ~ Über Actinomykose. Berl. klin. Wschr. **17**, 660 (1880b). — PORTER, I.A.: Actinomycosis in Scotland. Brit. med. J. **2**, 1084 (1953). — PRINSLEY, D.M.: Pulmonary actinomycosis; recovery in a mongol. Brit. J. Tuberc. **51**, (1) 40 (1957). — PULASKI, E.J., ARTZ, C.P., REISS, E.: Terramycin and aureomycin in surgical infections; report of 200 cases. J. Amer. med. Ass. **149**, 35 (1952). — PUTMAN, H.C., Jr., DOCKERTY, M.C., WAUGH, J.M.: Abdominal actinomycosis: an analysis of 122 cases. Surgery **28**, 781 (1950). — RAPER, F.P.: Abdominal actinomycosis following a perforated duodenal ulcer. Brit. J. Surg. **38**, 240 (1950). — RASHBAUM, M., MCINTOSH, H.C.: Pelvic actinomycosis treated by surgery and roentgen ray, with recovery. Amer. J. Obstet. Gynec. **47**, 849 (1944). — ROBINSON, M.R.: Actinomycosis of both ovaries and fallopian tubes. Surg. Gynec. Obstet. **29**, 569 (1919). — ROBINSON, R.A.: Actinomycosis of the subcutaneous tissue of the forearm secondary to a human bite. J. Amer. med. Ass. **124**, 1049 (1944). — ROSEBURY, T., EPPS, L.J., CLARK, A.R.: A study of the isolation, cultivation, and pathogenicity of *Actinomyces israelii* recovered from the human mouth and from actinomycosis in man. J. infect. Dis. **74**, 131 (1944). — ROSS, J.A., KNIGHT, I.C.: Actinomycosis of the subphrenic space. Edinb. med. J. **61**, 170 (1954). — RUHRÄH, J.: Actinomycosis in man, with special reference to the cases which have been observed in America. Ann. Surg. **30**, 417, 605, 722 (1899). — SANFORD, A.H.: Distribution of actinomycosis in the United States. J. Amer. med. Ass. **81**, 655 (1923). — SAZAMA, L.: Actinomycosis of the parotid gland: report of five cases. Oral Med. **19**, 197 (1965). — SAVIDGE, R.S., DAVIES, D.M.: Generalized actinomycosis with possible cardiac involvement. Brit. med. J. **2**, 136 (1953). — SCHNEIDER, D.H.: Actinomycosis of the testicle; case report. J. Urol. (Baltimore) **54**, 296 (1945). — SCORER, C.G.: Actinomycosis of the testis. Brit. J. Surg. **40**, 244 (1952). — SEABURY, J.H., DASCOMB, H.E.: Results of the treatment of systemic mycoses. J. Amer. med. Ass. **188**, 509 (1964). — SILBER, W.: Left perinephric actinomycosis. S. Afr. med. J. **27**, 264 (1953). — SILBERSCHMIDT, W.: Über Actinomykose. Z. Hyg. Infekt.-Kr. **37**, 345 (1901). — SLACK, J.: The source of infection in actinomycosis. J. Bact. **43**, 193 (1942). — SLACK, J.M.: Studies with microaerophilic actinomycetes. II. Serological groups as determined by the reciprocal agglutinin adsorption technique. J. Bact. **70**, 400 (1955). — SMITH, C.H.: Ocular actinomycosis. Proc. roy. Soc. Med. (Lond.) **46**, 209 (1953). — STANTON, M.B.: Actinomycosis of the maxillary sinus. J. Laryng. **80**, 168 (1966). — STEVENSON, A.E.M.: Actinomycosis of ovaries and fallopian tubes; report of a case. J. Obstet. Gynaec. Brit. Emp. **64**, 365 (1957). — SULLIVAN, H.R., GOLDSWORTHY, N.E.: A comparative study of anaerobic strains of *Actinomyces* from clinically normal mouths and from actinomycotic lesions. J. Path. Bact. **51**, 253 (1940). — SVANE, S.: Visceral actinomycosis, report on 6 cases with special reference to aureomycin treatment. Acta chir. scand. **131**, 160 (1966). — SWEENEY, D.F., BLACKWELDER, T.F.: Pelvic actinomycosis, report of a case. Obstet. and Gynec. **25**, 690 (1965). — SWINTON, N.W., SCHATMAN, B.H.: Actinomycosis, a rare cause of fistula-in-ano. Dis. Colon Rect. **7**, 315 (1964). — THOMPSON, L., LOVESTEDT, B.A.: An actinomyces-like organism obtained from the human mouth. Proc. Mayo Clin. **26**, 169 (1951). — TORRENS, J.A., WOOD, M.W.W.: Streptomycin in the treatment of actinomycosis; report of three cases. Lancet **1**, 1091 (1949). — TRIBEDI, B.P., MUKHERJEE, B.N.: Actinomycotic and mycotic lesions, with special reference to madura foot. Brit. J. Surg. **27**, 256 (1939). — TRIVEDI, B.P., SARKAR, S.K.: Pulmonary actinomycosis. Indian J. med. Sci. **6**, 595 (1952). — TURNER, G.: Actinomycosis of the lungs. Radiology **7**, 39 (1926). — UTZ, J.P.: Actinomycosis. In: Cecil-Loeb Textbook of Medicine, pp. 308—309. BEESON, P.B., MCDERMOTT, W., Editors. 12th Ed. Philadelphia: W.B. Saunders Co. 1967. — VERHOEFF, F.H.: A case of metastatic intraocular mycosis. Arch. Ophthal. **55**, 225 (1926). — VERME, G., CONTU, L.: L'actinomicosi del cuore. Osservazioni su un caso clinico e rassegna della letteratura. Arch. De Vecchi Anat. pat. **37**, 13 (1962). — VILLA, V.G.: Pulp abscess associated with actinomycosis. Oral Surg. **10**, (2) 207 (1957). — WAGENSTEEN, O.H.: The role of surgery in the treatment of actinomycosis. Ann. Surg. **104**, 752 (1936). — WAGENSTEEN, O.H., SNAPPER, I., POPPER, H.: Clinical pathological conference, actinomycosis of the lung with extension to the liver. Rev. Gastroent. **19**, 950 (1952). — WAGNER, C.: Actinomycosis of the uterine appendages. Surg. Gynec. Obstet. **10**, 148 (1910). — WARTHIN, T.A., BUSHUEFF, B.: Pulmonary actinomycosis. A.M.A. Arch. Int. Med. **101**, 239 (1958). — WARWICK, W.T.: A clinical contribution to the etiology of actinomycosis. Lancet **2**, 497 (1923). — WAYL, P., RAKOWER, J., HOCHMAN, A.: Pulmonary ray fungus disease; clinical aspects and pathogenesis. Dis. Chest **34**, 506 (1958). — WEED, L.A., BAGGENSTOSS, A.H.: Actinomycosis; a pathologic and bacteriologic study of 21 fatal cases. Amer. J. clin. Path. **19**, 201 (1949). — WEED, L.A., BAGGENSTOSS, A.H., BAUGHER, L.: Some problems in the diagnosis of actinomycosis. Proc. Mayo Clin. **24**, 463 (1949). — WHISENAND, J.M., MOORE, E.V.: Renal actinomycosis; with a report of a primary case. Calif. Med. **74**, 133 (1951). — WILLE-BAUMKAUFF, H.: Aktinomykose der Niere. Z. Urol. **43**, 240 (1950). — WILSON, G.S., MILES, A.A.: Actinomycosis. In: Topley and Wilson's Principles of Bacteriology and Immunity, pp. 1563—1578. 5th ed. Baltimore: Williams and Wilkins Co. 1964. — WILSON, J.W.: Actinomycosis. In:

Clinical and Immunologic Aspects of Fungous Diseases, pp. 190—204. Springfield, Ill.: C.C. Thomas 1957. — WILSON-PEPPER, J.K.: Report of renal actinomycosis. Brit. J. Urol. **23**, 160 (1951). — WINGO, C.F., WILLIAMS, R.O.: Pulmonary actinomycosis diagnosed by lung biopsy. Amer. Rev. Tuberc. **76**, 660 (1957). — WINSTON, M.E.: Actinomycosis of the spine. Lancet **1**, 945 (1951). — WOLFF, M., ISRAEL, J.: Über Reincultur des *Actinomyces* und seine Übertragbarkeit auf Thiere. Virchows Arch. path. Anat. **126**, 11 (1891). — WRIGHT, J.H.: The biology of the microorganism of actinomycosis. J. med. Res. **13**, 349 (1905). — WRIGHT, L.T., LOWEN, H.J.: Aureomycin hydrochloride in actinomycosis. J. Amer. med. Ass. **144**, 21 (1950). — ZANDER, E., BARONTINI, F.: Über die Aktinomykose des Nervensystems. Schweiz. med. Wschr. **1956**, 1409. — ZETTERGREN, L.: On the pathogenesis of appendix-actinomycosis. Acta path. scand. (Kobenh.) **25**, 543 (1948). — ZISKIN, D.E., SHORAM, B.S., HANFORD, J.M.: Actinomycosis, a report of 26 cases. Amer. J. Orthodont. **29**, 193 (1943). — ZITKA, E.: Vakzinebehandlung der Kieferaktinomykose. Z. Stomat. **46**, 202 (1949). ~ Tödlich verlaufende Fälle von zervikofazialer Aktinomykose. Wien. med. Wschr. **102**, 939 (1952). — ZOECKLER, S.J.: Cardiac actinomycosis; a case report and survey of the literature. Circulation **3**, 854 (1951).

Nocardiosis

PHILIP PIZZOLATO — New Orleans/Louisiana, USA

With 15 Figures

Definition

Nocardiosis is an infrequent, subacute or chronic, suppurative, deep actinomycetic infection of man and animals usually caused by *Nocardia asteroides*. Suppurative pneumonia, cerebral abscess, or disseminated abscesses (pyemia) are anatomical forms of the disease. Granules are usually not present in the pus. Occasionally nocardiosis is an opportunistic infection, developing within a host with a lowered resistance. Nocardiosis includes nocardial mycetoma, in which granules are present. (See Chapter by Dr. WINSLOW, page 589.)

Mycology

The genus *Nocardia* is composed of free-living or parasitic aerobic mycelium-forming microorganisms belonging to the family Actinomycetaceae and is separated from the other single genus *Actinomyces* which is anaerobic or microaerophilic and parasitic. These organisms are members of the order Actinomycetales which includes Mycobacteriaceae, Streptomycetaceae and Actinoplanaceae, and which may cause confusion with the Actinomycetaceae because of similar mycelial structures.

Recognition of the genus *Nocardia* has been relatively recent (BERGEY, 1957). NOCARD in 1888 isolated branched filaments from caseating lesions of lymph nodes from cattle and horses in farcy, a condition simulating tuberculosis and glanders. He identified this fungus as *Streptothrix farcinica* and it was renamed *Nocardia farcinica* by TREVISAN in 1889. In 1890, EPPINGER isolated an aerobic, Gram-positive, acid-fast fungus from a fatal case of multiple brain abscesses and meningitis in a 52-year-old glass blower. Because of the starlike shape of its colonies, he called the pathogen *Cladothrix asteroides*; however, BLANCHARD in 1895 recommended that the organism be classified as *Nocardia*.

In 1909 LINDENBERG reported a new species isolated from the leg of an Italian patient in Brazil; this fungus produced grains similar to those found in Madura foot and in actinomycosis and he named it *Discomyces brasiliensis*.

Confusion arose as to the characteristics of these organisms and they were placed in various genera such as *Streptothrix*, *Oospora*, *Actinomyces*, *Proactinomyces*, *Discomyces*, *Cladothrix* and *Mycobacterium*. BREED and CONN in 1919 decided that these terms should be eliminated and only the genus *Actinomyces* be used.

WAKSMAN and HENRICI in 1943 categorized the aerobic organism as *Nocardia* to distinguish it from the anaerobic *Actinomyces*. Although many strains of Nocardia have been encountered, WAKSMAN (1961) has accepted 59 species, of which 26 have been encountered in human or mammalian diseased organs. Among these, *N.asteroides* and *N.brasiliensis* are the most common (Fig. 1). Because of certain characteristics with respect to the grains, aerial mycelium, and spore formation, GONZALEZ OCHOA and SANDOVAL have transferred *N.madurae* to *Streptomyces madurae*. More recently, GONZALEZ OCHOA (1962) has classified *N.pelletieri* and *N.paraguayensis* as species of the genus *Streptomyces*.

Nocardia is Gram-positive (Figs. 2, 15), partially acid-fast or nonacid-fast. It branches, and has nonseptate, filamentous microorganisms with a diameter

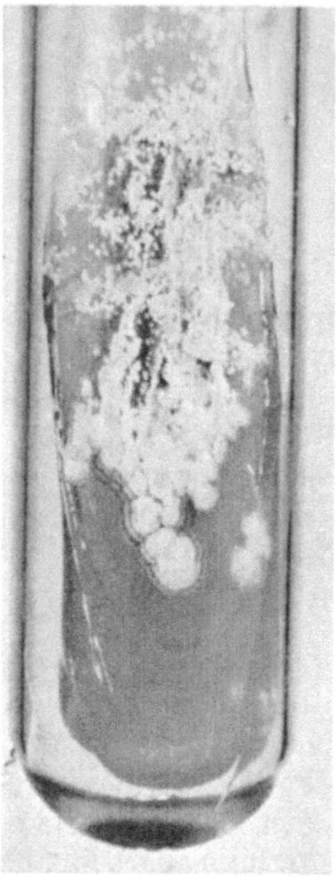

Fig. 1

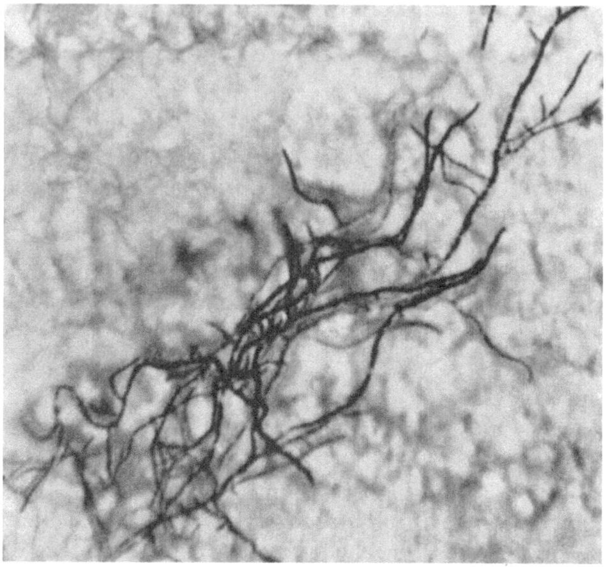

Fig. 2

Fig. 1. Culture of *N.asteroides* on Sabouraud medium, 4 weeks duration, 37°C. (From previously unpublished case of the author, PHILIP PIZZOLATO, New Orleans Veterans Administration Hospital No. A-67-120)

Fig. 2. Smear of *N.asteroides* from colony of Fig. 1, Gram stain, showing branching and segmented character of filaments, × 1500

of 1 micron or less. Some strains fragment into bacillary or coccoid forms when cultured. They are aerobic organisms, growing readily at 24° or 37°C. GEORG et al. (1961) found most strains of *N.asteroides* grew at 46°C whereas no strains of *N.brasiliensis* grew at this temperature. They grow on a variety of media such as beef infusion glucose agar, Sabouraud's glucose agar, blood agar, Czapek agar. The *colonies* are slow-growing (Fig. 1). It requires 2—4 weeks to produce conspicuous colonies. These may be pale yellow, orange, pink, red, or chalky white and wrinkled or granular. Occasional strains produce an aerial mycelium giving the colonies a white, chalky or powdery appearance. GORDON et al. and GEORG et al. have described features distinguishing Nocardia from *Streptomyces* and *Mycobacterium*. *N.asteroides* was found to show acid-fastness, branched mycelia, pellicle formation in thioglycollate broth, paraffin utilization (MCCLUNG) but no casein hydrolysis or tyrosine utilization. *N.brasiliensis* had similar features but these organisms revealed casein hydrolysis and tyrosine utilization and good growth in 0.4% gelatin (BOJALIL and CERBON, and CONANT).

SCHNEIDAU and SHAFFER made extensive cultural and biochemical studies on 51 strains of *Nocardia* species, 10 strains of *Mycobacterium* species and 4 strains of *Streptomyces* species. They divided the genus *Nocardia* into two types: "True" *Nocardia* and *Streptomyces*-like *Nocardia* as the former showed mycelium which was able to fragment into bacillary and coccoid forms and nearly all forms were acidfast. They further divided the "true" *Nocardia* into four groups: 1. *Nocardia asteroides* 2. *N.brasiliensis* 3. *N.corallima* 4. *N.opaca*. They observed that *N.intracellularis* failed to produce branching mycelium and it behaved much like a *Mycobacterium*.

McClung (1960) examined 102 soil samples from the southern area of the United States by means of the paraffin baiting technique and isolated 48 strains of *Nocardia asteroides* and 16 other species of *Nocardia*.

They were all acid fast, displayed characteristic colonial morphology and color, produced nitrite from nitrate, were nonproteolytic on gelatin and casein and non diastasic on starch. The remaining 16 isolates were not identified as to species but were representatives of "soft *Nocardias*" belonging to the morphological Group II described by McClung (1949).

Gonzales-Ochoa (1962) isolated *N.brasiliensis* from the soil by the simple procedure of seeding soil samples in Botcher-Conn agar. From these 21 samples six strains of *N.brasiliensis* were isolated and they showed typical fragmentation, acid fastness and enzymatic activities. The pathogenicity of these soil strains for mice was similar to those isolated from patients. In a study of 134 samples of the soil obtained in the region of New Dehli, India, Kumar and Mohapatra, using the "paraffin baiting" technique, isolated 52 strains of *N.asteroides*, 8 strains of *N.brasiliensis* and 4 strains of *N.caviae*. Of these, 42 strains were pathogenic for mice.

Incidence

Because of the unreliability of case reporting due to variations in morphology, culture, and clinical features, such as disseminated or localized infection, as well as terminology used, the incidence of nocardiosis is difficult to evaluate. Musgrave et al. in 1908 and Claypole in 1914 in their reviews of the literature included all cases belonging to the genera *Streptothrix*, *Cladothrix*, *Actinomyces*, *Oospora*, *Nocardia*, and *Sphaerotilus*. Henrici and Gardner in a review of the world literature up to 1921 considered that only 26 cases fulfilled the criteria of *Nocardia asteroides*. Ballenger and Goldring in a review of the literature prior to 1957, found 95 cases, and Murray et al. in 1960 found 179 cases. Gilligan et al. in 1961 encountered 33 verified cases of nocardiosis of the central nervous system in the literature. We have isolated *Nocardia* species from 12 patients over a 20-year period at the New Orleans Veterans Administration Hospital. Four of these among 6,000 autopsies during this period showed systemic pathologic lesions as well as central nervous system disease. Carlile et al. in 1963 reported the seventeenth fatal case of nocardiosis in a 7-year-old boy. This is also the fifth case of nocardiosis of the central nervous system in children.

As far as geographic distribution is concerned, nearly two-thirds of the reported cases of infection with *N.asteroides* were from North America and most of these were from the region of the Great Lakes, Mississippi Valley, and Eastern United States. About two-thirds of the patients were between 20 and 60 years of age; Ballenger and Goldring found 11 cases in the pediatric age group and reported a case in a 9-month-old child. The sex incidence showed a predominance of males to females of 2.5 to 1.

Cutaneous nocardiosis may be *mycetoma (Madura foot)* or a cutaneous manifestation of a systemic disease. According to Abbott, the term "mycetoma" was first used in 1860 by Van Dyke Carter to denote a fungus tumor of the foot or Madura foot as the disease was commonly called owing to its prevalence in that province of India. In 1916 Chalmers and Archibald carefully defined the term as covering "all growths and granulations producing enlargement, deformity and destruction in any part of the body of man brought about by the invasion of the affected area by certain species of fungi which give rise to variously shaped bodies called 'grains' which are found embedded in the pathological tissue or in the discharge from the diseased area". Mycetomas which contain granules are caused by *Nocardia* and by various Eumycetes, or true fungi. Refer to Chapter 12.

Much confusion has arisen as to the etiology as well as to the *classification* of the micro-organisms of nocardial mycetoma. However, the consensus is that among the various genera isolated, two species of *Nocardia* are noted: *N.brasiliensis* and *N.asteroides*. Some investigators have doubted the ability of *N.*

asteroides to produce localized mycetoma and they believe that the identification had been poorly determined and confused with *N. brasiliensis*.

LACAZ in 1945 reviewed 25 cases from Brazil in which *N. brasiliensis* had been identified. Nine were cases of mycetoma of the leg or foot, four were generalized cases, three involved the knee and the remainder came from miscellaneous sites. GONZALEZ-OCHOA (1962) studied 103 cases of mycetomas from the Institute of Health and Tropical Diseases in Mexico City and noted that 94% were due to *N. brasiliensis*, 1% to *N. asteroides* and the remainder to other genera. The lower extremities were involved in 65% of the patients, lesions of the trunk were seen in 25% and other sites were implicated in the remainder.

WILSON at the University of Ibadan, Nigeria, obtained biopsies and cultures from nine patients. From six patients, *N. asteroides* was identified and from one, *N. brasiliensis*. The identification was confirmed by other authorities.

Clinical Manifestations

Fever is the most constant sign of the disease. Frequently present are night sweats, chills, headaches and anorexia. In pulmonary nocardiosis, cough is a frequent symptom and is associated with the production of variable amounts of white mucoid to yellow or green purulent sputum, which may be streaked with blood or may actually be blood. Chest pain, with and without pleural involvement, and dyspnea may occur.

Lesions in the skin and subcutaneous tissues are encountered as a primary manifestation with or without the characteristic features of mycetoma (Madura foot) or secondary to pulmonary involvement. As a complication frequently associated with pulmonary disease, the central nervous system may become involved. Headaches develop or get progressively worse, followed by lethargy, mental confusion, hallucinations, convulsions, paralyses, or nuchal rigidity. Nausea and vomiting may be symptoms of nocardiosis of the central nervous system.

Laboratory Findings

Sputum examination is the most valuable laboratory procedure in the diagnosis of pulmonary nocardiosis. Gram stain of the sputum frequently shows Gram-positive branching mycelia which are acid fast when decolorized by the acid-alcohol solution for a short time, or with a weak (1% sulfuric acid) decolorizing agent. Cultures of the sputum, with and without prior digestion of the mucus and saprophytic microorganisms, frequently reveal *N. asteroides*. (AJELLO, 1951). BALLINGER and GOLDRING were able to isolate acid fast *Nocardia* from gastric washings from a 9-month-old child. Direct examination of exudates from sinus tracts of nocardial mycetoma frequently reveals white, yellow, red or brown "grains" which are heavily loaded with branching mycelia.

If the meninges are not involved, the spinal fluid is usually clear and not diagnostic. If there is meningitis or there is an abscess near the leptomeninges, the fluid may be yellow, cloudy or turbid, with an increase of protein and cells with a predominence of neutrophils. The sugar of the spinal fluid is frequently reduced (GILLIGAN et al.).

The peripheral blood reveals a mild leukocytosis of 10,000—20,000 per cmm often with a neutrophilia. A patient of FREESE et al. had a white count of 50,000 per cmm. A mild anemia may exist. The sedimentation rate is reported to be rapid.

Blood cultures may be positive for *N. asteroides*.

The blood and urine chemistry determinations are not diagnostic.

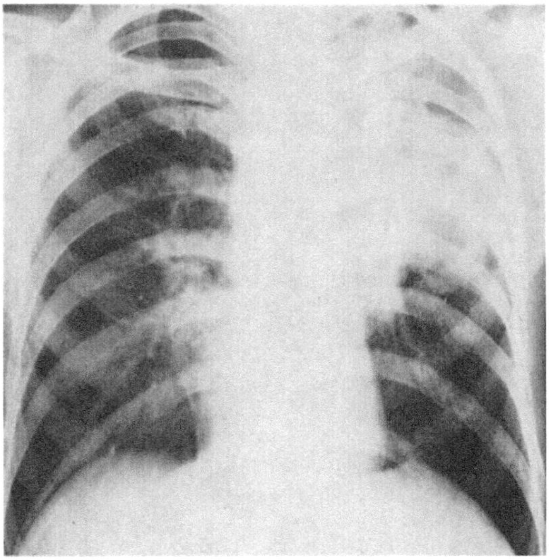

Fig. 3. Chest X-ray of a 49-year-old man showing a dense opacity in the right chest (New Orleans Veterans Administration Hospital Case No. A-46-29)

Late in the course of the disease, KIRBY and MCNAUGHT isolated *Nocardia* from sputum, the subcutaneous abscesses that developed, and the blood, from a 63-year-old man who had been admitted to the hospital for pneumonia. A report of three positive blood cultures for *Nocardia* associated with pneumonia was made by KOHN et al. The fungus was not pathogenic for mice, guinea pigs, and one rabbit. The patient recovered with sodium iodide, sulfadiazine, streptomycin and penicillin. HIDDLESTONE observed a patient who developed a loud, harsh systolic murmur audible over the mitral and tricuspid area and was found to have a positive blood culture.

Roentgen Findings

Roentgenograms of the chest are of great value. Lesions vary greatly, depending upon the stage of the disease, but, unfortunately, they are not diagnostic of pulmonary nocardiosis. One or both lungs may be involved. The hilum on both sides may be considerably enlarged. The lungs may show scattered, innumerable spot-like shadows, the size of a pinhead or larger. The spots may become confluent to form large non-homogeneous shadows or large areas of consolidation (Fig. 3). A single cavity may occur or multiple small cavities may become confluent to produce a multilocular large cavity. Pleural effusions and atelectasis may be found. An abscess may rupture into the mediastinum or pleura to form pleural empyema.

SALTZMAN et al. reported a 36-year-old white man with disseminated lupus erythematosus for which he received corticosteroid therapy. He developed a small coin lesion in the upper lobe of the right lung. One month later, the lesion had undergone cavitation and the sputum revealed *N. asteroides*. He was given sulfisoxazole (Gantrisin) and the cavity disappeared radiologically. A patient of WILKITE and COLE had a lung abscess with cavity formation in which a moveable mass resembling a "fungus ball" was found.

With a brain abscess there is shifting of the lateral ventricles or pineal gland as observed by pneumo-encephalogram or angiogram.

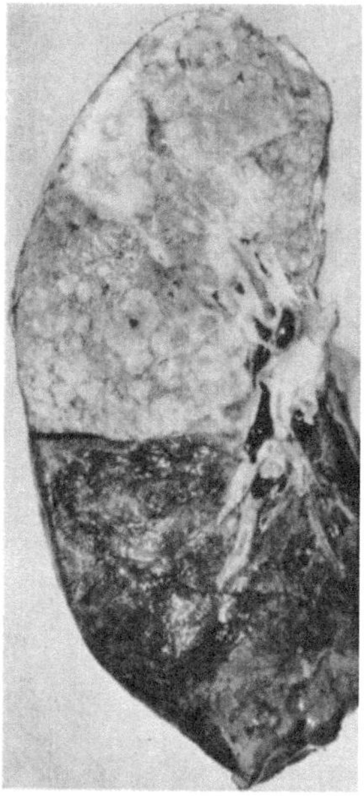

Fig. 4

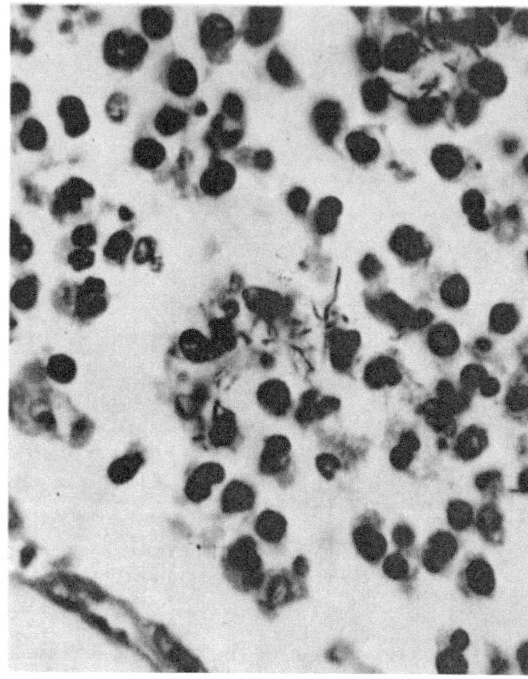

Fig. 5

Fig. 4. Pulmonary nocardiosis. Coronal section of the lung of a 49-year-old Negro man showing purulent consolidation of the right upper lobe. (Case A-46-29 of PIZZOLATO et al.)

Fig. 5. Pulmonary nocardiosis. Microscopic section of lung showing branching *Nocardia asteroides* in purulent exudate. The exudate is predominantly polymorphonuclear, but many mononuclear cells and lymphocytes are present. Gram stain, × 800

A 15-year-old boy was presented to DOLAN et al. with swelling of the knee of nine months' duration. There was radiologic evidence of a lytic lesion of the distal femur, with periosteal elevation and adjacent soft tissue calcification. An incision and drainage of the mass over the knee revealed necrosis of subcutaneous fat, muscle and fascia, with destruction of the periosteum. Culture of this material yielded *N. asteroides*.

Gross Lesions

Pulmonary Nocardiosis: The pulmonary lesions are characterized by a rather rapid progressive consolidation simulating caseous pneumonia or miliary tuberculosis. The lungs may be very heavy, the left weighing 1430 grams (normally 375) and the right weighing 1500 (normally 450) as in the case of KIRBY and MCNAUGHT. The cut surface of an entire lung or lobe may be homogeneous, firm, pale, yellow or gray, and airless or there may be numerous firm, white, or pinkish gray nodules, 1—3 cm in diameter, as in the case of PIZZOLATO et al. (Fig. 4). The nodules may become confluent producing larger nodules or there may be a diffuse consolidation (Figs. 4, 5). The solid areas may undergo liquefaction necrosis yielding poorly defined 1—5 cm single or multiloculated abscesses filled with brown, green or yellow pus. There may be communications with the bronchi and when the contents have been discharged, the lining of the abscess cavity is rough, shaggy, gray to brown. There is consolidation of the adjacent alveolar spaces. Also the adjacent lung may be edematous and hyperemic. WILHITE and COLE

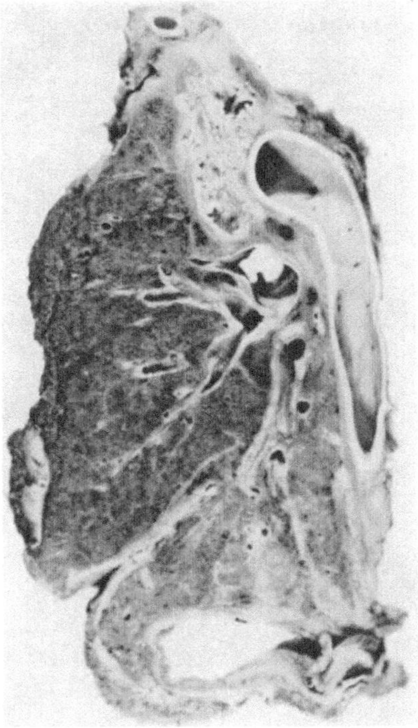

Fig. 6. Pleural and mediastinal nocardiosis. Coronal section of lung of a 52-year-old Negro man showing extensive pleuritis and mediastinitis extending from apex to diaphragm, and complete atelectasis. (From previously unpublished case of the author, PHILIP PIZZOLATO. Autopsy No. A-67-120.) Culture shown in Figs. 1 and 2 is from this case

reported five cases of *Nocardia* invading a pre-formed lung abscess and producing "fungus balls". The patients had thoracotomies with the removal of the affected area in which a large cavity was found in two patients.

Pleuritis may develop locally, forming a ragged or thickening pleura with or without the formation of an effusion. It may remain localized or spread to involve the entire pleural cavity forming the classical *empyema* (HAGER et al.). Fibrinous and fibrous adhesions may develop between the visceral and parietal pleurae, and the infection may spread through the intercostal spaces to produce subcutaneous abscesses or it may extend into the mediastinum (Fig. 6).

In a previously unpublished case PIZZOLATO (1967) observed a 52-year-old man with a five day history of cough, fever and back pain (New Orleans Veterans Administration Hospital Autopsy No. A-67-120). He had been a heavy beer drinker with a past history of visual and auditory hallucinations. Examination revealed a temperature of 101°F., crepitant and subcrepitant rales, dullness, diminished breath sounds and clubbing of the fingers. A chest X-ray showed some interstitial fibrotic changes bilaterally. He was given penicillin and his temperature varied from normal to 100° or 101°F. A repeat chest X-ray revealed consolidation of the left chest wall and its configuration was more in keeping with loculated pleural effusion. The fluid was aspirated and found to be sterile on culture. To insure better drainage, a tube was inserted into the left chest cavity, and pus began to drain. A few days later, a standard thoracotomy incision was made and the seventh rib was resected. Pleural exudate on culture revealed *N. asteroides* (Figs. 1, 2). Three blood cultures were negative. Sulfadiazine was given. About two months after admission he became stuporous, the pupils measured 2 mm in diameter and reacted sluggishly to light; the eye movements deviated to the right. He became progressively worse and expired. At *autopsy*, the left pleural cavity appeared as an empyema sac filled with dense yellow and red masses of necrotic tissue and liquid pus compressed the

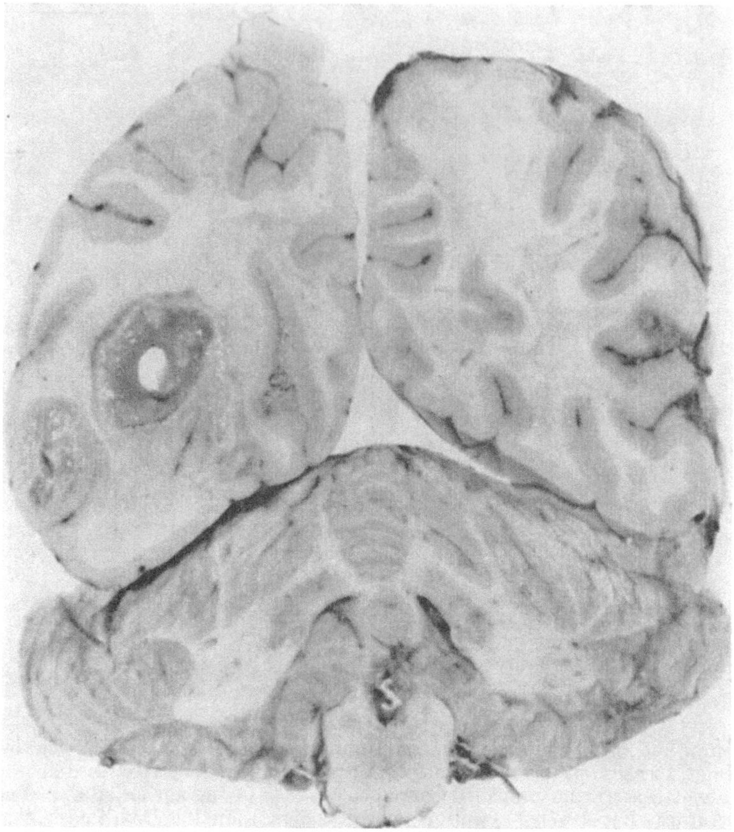

Fig. 7. Nocardiosis of brain. Coronal section of brain of a 61-year-old white man showing two abscesses of the left cerebral hemisphere. (Case of PIZZOLATO et al.)

left lung against the mediastinum (Fig. 6). The left lung was dark bluish-violet and atelectactic. No lung abscesses were found. The brain showed symmetrical hemispheres. The cut sections of the cerebrum and cerebellum revealed hundreds of tiny abscesses, some measuring up to 5 mm in diameter. They were pale tan and were surrounded by a rim of gray ragged tissue and a zone of hyperemia and swelling of the surrounding brain tissue.

A unique case of lung abscess due to *N. brasiliensis* was reported by BOBBITT et al. This 61-year-old man was admitted for fever, malaise and cough. Chest X-ray revealed a density in the left apex. Recovery was uneventful following thoracotomy, penicillin and sulfadiazine.

Nocardiosis of the Brain: The fungus may enter the blood stream to be disseminated to other organs. The brain is the extra-pulmonary organ most frequently infected (BERNSTEIN et al., BINFORD and LANE, CUPP et al., ECKHARDT and PILCHER, ERCHUL and KOCH, JACOBSON and CLOWARD, KIRBY and MCNAUGHT, KRUEGER et al., LIST et al., MUNSLOW, PIZZOLATO et al. (Figs. 7, 8), TUCKER and HIRSCH, WEED et al., WELSH et al., WICHELHAUSEN et al.).

Solitary and multiple abscesses of varying sizes and locations are encountered in the cerebrum and cerebellum (Fig. 7). Abscesses 3 cm in diameter are not uncommon; and BINFORD and LANE described a multiloculated abscess 6 cm in diameter extending from the cortex to the lateral ventricle. The abscesses are poorly encapsulated and filled with viscid or gelatinous, green or gray pus, and the lining of the abscess cavities is rough and irregular.

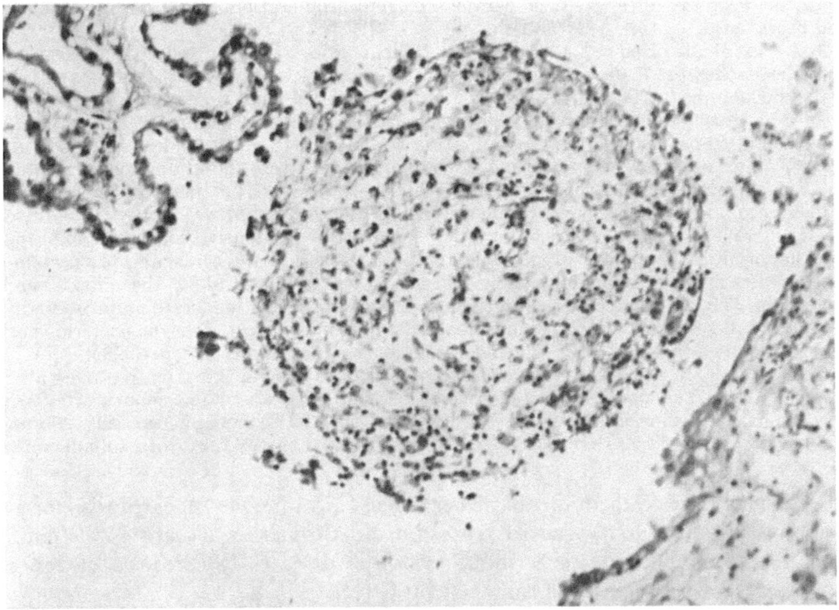

Fig. 8. Nocardiosis of brain. Cellular infiltrate, polymorphonuclear neutrophils, lymphocytes, and macrophages, in the choroid plexus. H & E, × 370. (Case A-58-212, same as in Fig. 7, Case 2 of PIZZOLATO et al.)

A 49-year-old female patient of KIRBY and MCNAUGHT was admitted to the hospital for headaches, nausea and vomiting for one month. Bilateral choked disks, central scotomas and enlarged blindspots, engorged fundic vessels and poor hearing were noted. The spinal fluid had no white cells and 25 mg% protein. She died suddenly and autopsy revealed that the cerebellum had distinct markings where it had been forced into the foramen magnum. From the right cerebellum a thick green exudate was expressed. The exudate was semitranslucent and contained no granules. A section of the cerebellum revealed an irregular 2.5 cm abscess with a soft wall extending to within 1 cm of the posterior medial border of the right lobe and not communicating with the fourth ventricle. Culture of the exudate disclosed *N. asteroides.*

The abscesses may invade the ventricles and choroid plexus (Fig. 8). They may break through the cortical surface to produce a meningitis.

A 47-year-old man reported by CUPP et al., developed anorexia, restlessness and lethargy following transfusion hepatitis and ulcerative colitis. Spinal fluid revealed 15,000 white cells per cmm with 100% neutrophils, 21 mg % sugar, and a negative culture. A repeat spinal fluid showed faintly acid fast branching filaments. The patient subsequently died. At autopsy, multiple abscesses were seen at the insulotemporal area. One abscess had ruptured into the lateral ventricle, resulting in meningitis.

A 02-year-old female patient of KEPES and SCHOOLMAN, after an injury in an automobile accident, disclosed quadriparesis and died after 26 days. Autopsy revealed a roughly triangular area of necrosis and punctate hemorrhages in the brain. The center of the lesion was yellow-green and paste-like. Microscopic examination showed purulent inflamation with the formation of an abscess. The periphery of the abscess revealed a wide area of fibroblastic proliferation, gliosis, petechiae, and perivascular inflammatory cells. Gram stain demonstrated Gram-positive branching filaments which were slender, slightly beaded and morphologically consistent with *Nocardia.*

WELSH et al. encountered a patient who was admitted for a pulmonary infiltrate. She subsequently developed an epigastric subcutaneous nodule, progressive numbness and weakness of the lower extremities, and later complete paralysis. The epigastric mass was found to be an abscess which on culture revealed *Nocardia.* In spite of streptomycin and sulfadiazine, she died and was found to have abscesses in the fourth and fifth lumbar vertebrae. The middle thoracic spinal cord was destroyed. Microscopically, the remaining cord showed liquefactive necrosis, gitter cells, moderate astrocyte proliferation. Gram positive fungi were found in the cord.

GILLIGAN et al. observed a 32-year-old patient who was admitted to the hospital for intense headaches for 5 days, associated with fever, neck stiffness and Kernig's sign. The spinal fluid showed 4,000 neutrophils per cmm. No organisms were seen on smears. The patient was given antibiotics and sulfadrugs. He had periods without headaches and no Kernig's sign. Repeat spinal fluid examination revealed 1345 neutrophiles and 185 lymphocytes per cmm and proteins of 105 mg %. No organisms were found on smears or culture. After 5 days, intense headaches and neck stiffness returned. The spinal fluid exhibited 63 cells per cmm, mostly mononuclear, and protein of 260 mg/%. *Nocardia* grew on culture. In spite of vigorous treatment, he died. Autopsy demonstrated extensive collections of granular, yellow material in the leptomeninges over the basal aspect of the cerebrum, brain stem and cerebellar hemispheres. Granular yellow material was also found along the middle cerebral artery in the Sylvian fissures. Serial sections of the brain showed moderate uniform dilatation of the ventricular system and 2—3 mm areas of yellow material in the right posterior portion of the mid-pons adjacent to the thickened meninges. The fungus was pathogenic for mice.

A successfully treated case of nocardial meningitis was reported by KING et al. in a 50-year-old man. The patient developed signs of meningitis about one month after a head injury. Spinal fluid showed 570 white cells per cmm with 99% mononuclear cells, 260 mg/% protein and glucose of less than 25 mg/%. He was treated successfully with sulfadiazine and urea.

Cutaneous and Subcutaneous Nocardiosis: Cutaneous nocardiosis may be mycetoma (Madura foot) due to *Nocardia brasiliensis* or *Nocardia asteroides* or a skin involvement with or without systemic disease. Nocardiosis of the skin resembling sporotrichosis was reported by Guy.

His patient was a 50-year-old man who developed a "simple pimple" of the thumb. Other pimples appeared and they formed ulcers which had ragged edges and the base was covered with an indolent appearing granulation tissue. Inflammatory nodules with soft centers ascended towards the elbow. RAICH et al. reported a 38-year-old farmer who had a red draining granuloma of the index finger with subcutaneous nodules along the forearm. Exudate from the finger showed numerous intracellular acid fast organisms in bundles or "globi" which on culture yielded *Nocardia*.

A progressively enlarging breast mass following a two-year history of an abrasion with a rosebush thorn in a 36-year-old woman was reported by LARSEN et al. The mass was firm and tender and the nipple retracted. Aspiration showed beaded Gram positive branching structures which on culuture grew *N. asteroides*. The lesion did not respond to tetracycline and developed an indurated sinus tract with a recurrent abscess which necessitated wide local excision. HICKEY and BERGLUND were able to isolate *N. asteroides* from three patients with abscesses associated with a branchiogenic cyst, a pilonidal cyst and ischio-rectal abscess, respectively, without systemic lesions.

Frequently, skin and subcutaneous lesions are noticed in the chest in association with pulmonary nocardiosis. MURRAY et al. observed a 36-year-old man complaining of left anterior chest pain, night sweats, fever of a week's duration, and cough productive of a small amount of mucoid sputum. The chest was asymmetrical with a tender prominence of the left pectoral area. A pleural friction rub and moist rales were heard over the left chest wall. Chest X-ray showed an infiltration of the midportion of the left lung field. Culture of pus aspirated from the left chest abscess yielded *N. asteroides*. The patient was treated with antibiotics and sulfadiazine without improvement. Two weeks later, under local anesthesia, the left chest lesion was explored and a small amount of pus was found under the pectoral muscle. Later, under general anesthesia, he was explored, and a wide exposure of the subpectoral space demonstrated a large amount of thick pus and a sinus tract that extended beneath the ribs into the pleural cavity. The diseased ribs were resected. He received more sulfadiazine and improved.

FREESE et al. reported that among 11 cases of nocardiosis, 6 patients had cutaneous lesions as "cold abscesses", tender subcutaneous nodules, lymphadenopathy and draining sinus tracts. The fluctuant swellings were related in position to underlying empyema or abscesses of the forehead or epigastrium.

A patient with ischiorectal abscess was observed by BINFORD and LANE. BOBBITT et al. encountered a patient with draining sinuses of the hip and buttocks; autopsy revealed a large psoas abscess and multiple abscesses in both lungs with fibrinous pleuritis. A patient of WICHELHAUSEN et al. had a peri-anal abscess. Later, he developed a chest mass from which

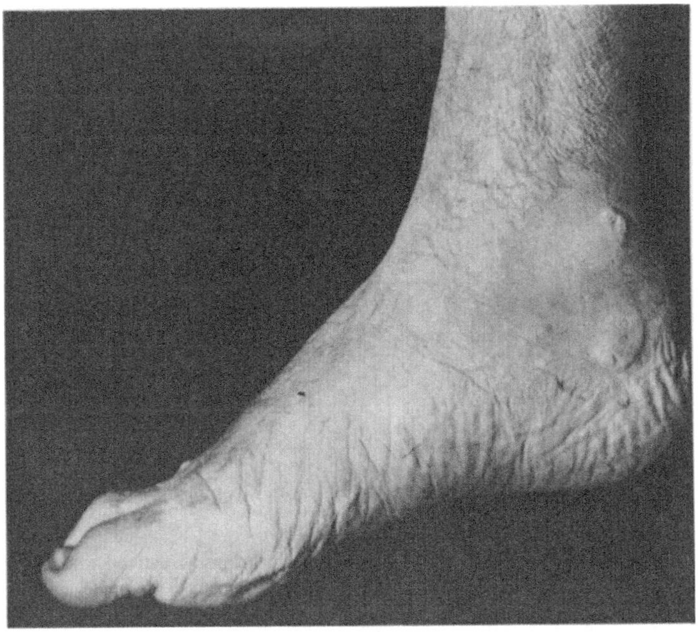

Fig. 9. Nocardial mycetoma of foot. Note the sinus openings on the elevated mounds of granulation and scar tissue. Pus from these sinuses contains granules or grains. These should be examined in fresh and in gram-stained preparations for the presence of the slender, branching, segmented filaments of *Nocardia*. Case from Durham, North Carolina, USA

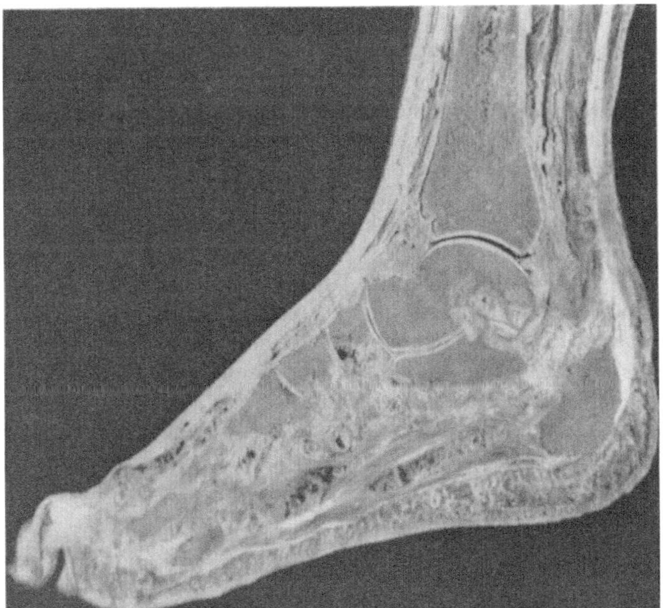

Fig. 10. Nocardial mycetoma. Sagittal section of foot shown in Figure 9. Note the abscesses of bone and the ramifying sinuses which discharge pus and granules to the skin surface

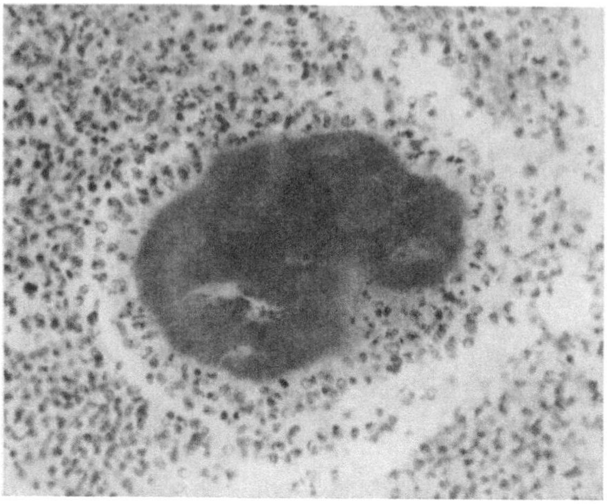

Fig. 11. Nocardial mycetoma. Granule in pus H & E, × 100

N. asteroides was isolated. Central nervous system symptoms occurred and the patient died. Autopsy demonstrated abscesses only in the chest wall and brain.

A portal of entry of the fungus was probably in the hand of the patient of BOBBITT et al. This 55-year-old woman received an injury to her right hand in which a firm violaceous raised 3 cm abscess occurred. From a central crater a freely draining purulent material was cultured to reveal *N. asteroides*. She developed a brain abscess which on aspiration showed *N. asteroides*.

Nocardial Mycetoma: Mycetoma pedis (Madura foot) occurs frequently in the lower extremities (Figs. 9, 10). Involvement of the upper extremities and back is fairly common. GONZALEZ-OCHOA (1962) has observed lesions on the back of Mexican laborers who carry soil-contaminated fiber sacks on their backs. The lesion may start as a papule or vesicle and later develops into a cutaneous or subcutaneous firm mass. Ulceration occurs in the center of the mass from which a thin, watery, serous or purulent material oozes. One or more sinus tracts form; they may heal leaving a depressed center. Other nodules and sinuses develop in the adjacent areas and the intervening tissues become indurated with the characteristics of brawny edema. Lesions between the toes and heel result in great distortions, deformities and contractures. Simultaneously, sinus tracts penetrate the deep soft tissues demonstrating no resistance of the tissue other than the formation of large amounts of scar tissue with intervening zones of necrosis. Infection extends to the adjacent bone to form a chronic osteomyelities.

In the exudate from the sinuses are found "grains" or clusters of mycelial threads (Figs. 11, 12). They are usually yellow when caused by *Nocardia*, in contrast to black, white or red when the disease is caused by other fungi.

The disease remains localized for months and even years, spreading by continuity. According to GONZALEZ-OCHOA (1962), visceral invasion by contiguity is not rare; there is pulmonary involvement in 25% of the cases of thoracic mycetoma caused by *N. brasiliensis*.

The incubation period is unknown; MACKINNON et al. observed two cases of *N. brasiliensis* 15 and 20 days after minor injury to the hand. The wound was granulomatous with lymphangitis of the hand and forearm extending to the upper arm in one case and with nodules along the lymphatics. One patient had an abscess with pus formation. Filaments of an actinomycete were present but no grains. Culture yielded colonies of *N. brasiliensis* which on intraperitoneal injec-

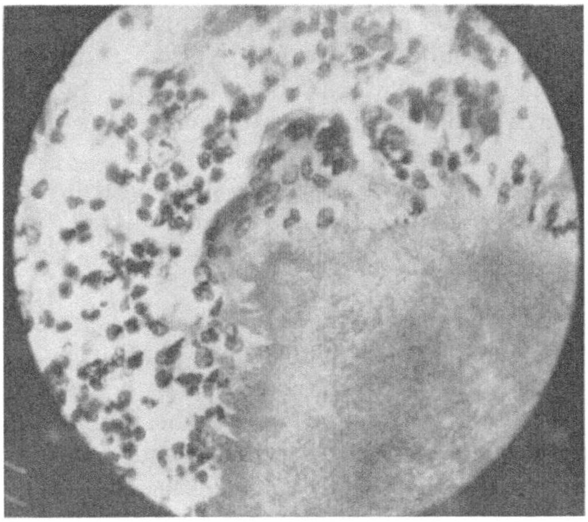

Fig. 12. Nocardial mycetoma. Granule in pus. Giant cell applied to the periphery of the granule. Club formation at the periphery. H & E, × 300. Gram stain would color the filaments of the granule as in Figs. 2 and 5

tions into mice produced typical grains of *N. brasiliensis*. The patients responded to sulfonamide therapy.

Miscellaneous Nocardial Lesions: Other single organs may be involved as a primary site. HENDERSON et al. reported a case of eye involvement in a 49-year-old woman who was struck in the eye by the tail of a cow which she was milking. The eye became red and a mass appeared. She was treated with chloramphenicol, staphylococcus toxoid and typhoid vaccine but the mass continued to enlarge. The mass was later excised; microscopically it showed "inflammatory granulation tissue" and culture revealed *N. asteroides*. Later a small daughter granuloma appeared. After sulfadiazine therapy, this second granuloma became localized and spontaneous resolution occurred. A 23-year-old female patient of BENEDICT and IVERSON developed episodes of ocular pain, photophobia, dimness of vision and conjunctival congestion after an attack of scarlet fever. The inner surfaces of the eyelids presented either large patches of granulation tissue or patches of scar tissue. The right cornea showed interstitial keratitis. Eight cultures over a period of four years revealed nocardial organisms which were not pathogenic for guinea pigs, mice or rabbits.

Skin lesions caused by *N. brasiliensis* have been observed in the United States. A case resembling clinical sporotrichosis was reported by RAPAPORT in a 56-year-old man who pricked his index finger with a rose thorn two or three weeks before examination. A chancriform ulceration with advancing lymphangitis, mild axillary lymph-adenopathy and multiple erythematous nodules along the course of the lymphatics on the exterior surface of the forearm were noted. He received erythromycin and potassium iodide and after one week, the lesions were generally improved, but a fluctuant small bubo developed on the forearm. Pus from the chancre and abscess revealed *N. brasiliensis*. Skin lesions due to *N. brasiliensis* were encountered by M. A. GORDON, MOORE et al., and MOORE and CONRAD.

A patient observed by MAHVI was admitted for epigastric pain and tenderness associated with X-ray changes in the bones showing a "punched-out" appearance suggestive of multiple myeloma. He was treated with urethane, cortisone, chloramphenicol and deep X-ray therapy. He developed a perirenal abscess and multiple ulcerating cutaneous abscesses from which *N. brasiliensis* was isolated. He was then treated with sulfonamides but expired. Autopsy revealed abscesses in the skin, kidney and lungs with an acute suppurative inflammatory reaction containing Gram-positive branching masses, some of which were coccoid and others bacillary.

BAUMGARTEN and GUY reported cases of skin nocardiosis resembling clinical sporotrichosis in which *N. asteroides* had been isolated. LONG and CAMPANA isolated a *Nocardia* species in a granuloma with draining sinuses following a dog bite.

The lacrimal gland was surgically removed by CHRISTOPHERSON and ARCHIBALD because of swelling of the eye for $3^1/_2$ years. The eyelids could not be everted, but when raised, a thick yellow discharge poured out. The whole conjunctival surface of the upper lid appeared to be rough and granular. Smears showed "bacilliform hyphae" which were not pathogenic for the gray monkey.

The eye may be a part of a systemic nocardiosis, as observed in Hiddlestone's case. An acute chalazion of the left upper lid developed and culture of the exudate exhibited *N. asteroides*.

A patient of CRUZ and CLANCY was admitted to the hospital for pain in the right thigh for six weeks accompanied by weight loss. She also had fever, night sweats, and chilly sensations but no cough. She had a blowing systolic murmur at the cardiac apex transmitted over the entire precordium. Examination of the thigh revealed acute tenderness, swelling, and induration over the anterolateral area. X-ray of the thigh revealed osteomyelitis of several weeks duration. X-ray of the chest was normal. Surgical incision was made into the osteomyelitic cavity; smears showed filaments with beaded Gram-positive granules. *N. asteroides* was isolated on culture.

LARSEN et al. observed a patient who had Hodgkin's disease and was treated with nitrogen mustard and cortisone. She developed a purulent discharge from the ear from which *Nocardia* was isolated.

BIANCO et al. observed a 26-year-old man who developed soreness of both calves of the leg after exercise Redness and swelling of the right calf later developed. Draining revealed a "quart of bloody pus" from which *N. asteroides* was cultured. Osteomyelitis of the proximal part of the fibula was noted.

A patient of UTZ had osteomyelitis and a severe burrowing cellulitis with abscess and sinus tract formation in the thigh and leg. The infection responded to surgical drainage and prolonged sulfadiazine and penicillin therapy.

MURRAY et al. observed a 43-year-old man with glomerulonephritis which was treated with cortisone. He developed a purulent pleural effusion from which *N. asteroides* was isolated. In the course of treatment with sulfadimethoxine, symptoms of peritonitis were noted. A peritoneal tap yielded purulent fluid which on smear showed Gram-positive filaments suggestive of *Nocardia* but no cultures were made. He was continued on steroid therapy, sulfadimethoxine, antibiotics, and drainage of the empyema cavity and discharged improved.

Nearly every organ has been observed at autopsy to be involved in nocardiosis. The heart of the patient reported by CRUZ and CLANCY showed extensive necrosis of a papillary muscle of the mitral valve. Sections throughout the papillary muscle revealed 1—3 mm abscesses containing greenish-yellow, thick, odorless pus. LARSEN et al. also observed in a patient with Hodgkin's disease, after treatment by nitrogen mustard and cortisone, degenerative changes with acute purulent inflammation in the cardiac papillary muscles. Abscesses due to *Nocardia* were encountered in the liver, pancreas, kidney, brain and lung. A patient of HIDDLESTONE revealed in the mitral valve firm pale vegetations from which

Nocardia was cultured. The spleen had a large abscess filled with a thick, glairy, greenish-yellow pus.

Nocardia As an "Opportunistic" Fungus: At the International Symposium held at Duke University in June, 1962, *Nocardia* was included among the opportunistic fungi. These were defined as "ubiquitous saprophytes and occasional pathogens that invade the tissues of man or animals with (1) predisposing diseases such as diabetes, leukemia, lymphoma, cancer, and aplastic anemia; or (2) predisposing conditions such as agammaglobulinemia, neutropenia, splenectomy, and roentgen therapy, and the use of steroids and antileukemic and antibiotic drugs" (BAKER, 1962).

Among 202 cases of fungus infections encountered in cancer patients at the Memorial Hospital for Cancer and Allied Diseases and the James Ewing Hospital in New York, HUTTER and COLLINS found 28 cases of infections due to actinomycetes. From these cases *Nocardia* species was cultured in eight.

From the Armed Forces Institute of Pathology at Washington, D.C., CROSS and BINFORD studied 44 cases of nocardiosis. Of these, 17 cases revealed previous diseases or conditions that could have predisposed to fungus infection. Eight patients received prolonged steroid therapy. Leukemia or lymphoma was the primary disease in six cases.

Similar observations have been made by CASAZZA et al., HATHAWAY and MASON, and SALTZMAN et al. Nocardiosis has been encountered with other pathologic states such as lupus erythematosus (PILLAY et al., and SANTEN and WRIGHT), sarcoidosis (STEINBERG), intestinal lymphangiectasia (HARGROVE et al.) and other blood dyscrasias (UTZ, WHITMORE et al., GYDELL et al.) and Cushing's syndrome (DANOWSKI et al.).

There has been a progressively more frequent reporting of cases of pulmonary alveolar proteinosis associated with a focus of nocardiosis. No explanation had been advanced except that the fungus may favor or stimulate the formation or deposition of the periodic acid – Schiff positive material in the alveolar sacs of the lung. Cases have been reported by ANDRIOLE et al., BURBANK et al., CARLSEN et al., MARTINEZ-MALDONADO and RAMIREZ DE ARELLANO, ROSEN et al., and ANDERSON et al.

Microscopic Lesions

Histologic sections of material from the draining sinuses of nocardial mycetomas stained with hematoxylin and eosin often reveal oval or irregular masses of pink or yellow staining mycelial filaments — so called "sulfur granules" (Figs. 11, 12). Tissue sections of aspirated or biopsy material or the wall of an abscess from other sites do not show the granules. There is a loss of tissue structure and cellular detail and an extensive infiltration of neutrophils, lymphocytes, plasma cells, and macrophages in various stages of degeneration. Beyond this zone of suppuration the inflammatory cells are more viable and discernible and are enmeshed in a network of fibrin. Epithelioid and multinucleated giant cells are seldom and sporadically found. There is a poorly defined wall of fibroblastic activity, with some deposition of collagen, proliferation of capillaries or a pseudocapsule from compression of the surrounding stroma. The brain shows numerous gitter cells and a variable amount of gliosis. No micro-organisms are seen with hematoxylin and eosin stains. Gram stain of an adjacent section reveals well outlined Gram positive branching, occasional beaded, non-septate, slender, 0.5—1 μ thick filaments. (Fig. 5). Gomori methenamine silver stain also shows well-outlined fungi. Bacillary forms of varying lengths and coccoid bodies are occasionally found. Phagocytized organisms may be found in macrophages. Acid fast stains reveal some strains that are positive especially if a weak decolorizing agent such as 1—5% aqueous sulfuric acid is used. These filaments are often beaded (RIDELL). The periodic acid - Schiff stain is not valuable as the filaments stain like the surrounding material (BALLINGER and GOLDRING, BAKER, 1957; BINFORD and LANE, CALERO, KADE and KAPLAN).

A unique case of nocardiosis was reported by CUTTINO and MCCABE in a 34-month-old girl who was admitted to the hospital for anorexia, nausea, vomiting and weight loss. A lymph node showed complete alteration of its architecture with proliferation of macrophages, the cytoplasm of which appeared foamy and granular resembling Gaucher's cells and containing massive numbers of acid fast bacilli. She did not respond to treatment and expired. Autopsy revealed that the abdominal lymph nodes were matted together, and yellow on section with centers exhibiting irregular zones of necrosis. Microscopic examination of the spleen showed great distortion of the architecture by an abundant proliferation of large, pale, foamy-appearing macrophages to the extent that lymphoid remnants appeared only as focal collections of cells. The malpighian corpuscles were replaced by epithelioid cells, concentrically arranged about the central artery. There were groups of multinucleated giant cells, usually in clusters of a dozen or more. The nuclei were in a peripheral position and were small and hyperchromatic. In the central zone of the cytoplasm there was a clear spherical homogeneous area. Acid fast bacilli were found in the epithelioid cells. The liver showed scattered tubercle-like nodules of epithelioid cells with many acid fast organisms. Many Kupffer cells showed phagocytosis of the organisms but there was no apparent damage to the liver cells. Culture revealed a new species: *N. intracellularis* which was not pathogenic for guinea pigs, rats, mice or rabbits.

Animal Pathogenicity

All of the 69 strains of *N. asteroides* studied by GEORG et al. were pathogenic for guinea pigs. Other workers have had variable results with guinea pigs, as well as with rabbits, mice and monkeys. When the fungus is pathogenic for guinea pigs, the animals succumb after intraperitoneal infection, in 10—30 days. STRAUS and KLIGMAN found that the addition of 5% gastric mucin to the inoculating medium enhanced the infectivity for mice. They noted many small nodules scattered throughout the peritoneal cavity, and adhesions around the liver, stomach and spleen. Similar lesions are seen in guinea pigs and from these lesions *Nocardia* was isolated on smears and cultures (DRAKE and HENRICI).

GONZALEZ-OCHOA (1967) found that the intramuscular, subcutaneous or intradermal injection of a light inoculum of *N. brasiliensis* into mice produced only transient lesions with a tendency to spontaneous cure. When the inoculum was given intravenously or intraperitoneally, the animals died. A single injection of the organisms into the foot pad of the mouse resulted in a chronic fistulous swelling containing actinomycotic grains resembling clinical mycetoma. Some mice showed invasion of the inguinal and intra-abdominal lymphnodes and lesions in the lungs and liver.

MARIAT and MACOTELA-RUIZ were able to produce true mycetoma in hamsters by the intradermal injection of *N. asteroides* and *N. brasiliensis*.

BRUECK and BUDDINGH were able to cultivate *Nocardia* by the use of the chick embryo. They noted that *N. asteroides* grew very readily in the chorioallantois as well as in the yolk sac. In both of these sites *N. asteroides* proliferated so rapidly the embryo died within 3—4 days. The fungus apparently could be maintained indefinitely in serial passage in the membrane or within the yolk sac. *N. intra-*

cellularis thrived poorly in the chorioallantois but grew readily in the yolk sac. Within 48—72 h following inoculation numerous organisms were demonstrated with Gram and acid fast stained smears from the yolk.

Nocardia species have been found in dogs (BOHL et al., THORDAL-CHRISTENSEN and CLIFFORD), in a cat (AJELLO et al., 1961), cows (NOCARD) and marsupials (TUCKER and MILLAR). Bovine farcy produced by a species of *Nocardia* resembling *N. farcinica* which was originally described by Nocard had been observed in regions south of African Sudan by Mostafa.

KINCH reported a fatal infection in a dog by *N. caviae*. This 18-month-old dog was treated with corticosteroid for severe muscle stiffness in the hind quarters following excessive exercise. Temperature of 106° F. developed and intramuscular chloramphenicol was given. The dog became rapidly worse and died eight days after the onset of illness. At autopsy, adjacent to the mandible was a 2.5 cm circumscribed abscess filled with a foul-smelling greyish-white pus. The liver was dark reddish purple and contained numerous grey nodules up to 4 mm in diameter. The kidneys were reddish brown and contained a few small greyish-white nodules up to 5 mm in diameter. Microscopic examination of the liver showed small granulomatous abscesses. The smallest abscesses were composed of cores of lymphocytes and macrophages surrounded by a few epithelioid cells and fibroblasts. Small amounts of collagen and reticulum were around the periphery of the lesions. In larger lesions the cores were composed of neutrophils and free red cells and these were surrounded by a typical granulomatous type of reaction. The kidney showed abscesses composed of neutrophils and a small number of macrophages and epithelioid cells. There was no encapsulation by fibrous tissue. Direct smears from the mandibular abscess, liver, lung and kidney revealed large numbers of branching filamentous Gram-positive organisms. The organisms in stained section of the liver, lung and kidney were Gram-negative, periodic acid-Schiff negative and non-acid fast. Cultures revealed characteristics of *N. caviae* according to the criteria of GORDON and MIHM (1962).

Diagnosis

A positive diagnosis of nocardiosis depends on the isolation of *Nocardia* and the study of its cultural and biochemical characteristics as well as its pathogenicity for laboratory animals. A tentative diagnosis of nocardial mycetoma may be made by the recognition of pink or yellow "sulfur granules" in the exudate from the draining sinus. Since the pulmonary variety is rarely associated with sinus formation, aspiration or needle biopsy is the method of choice (PIZZOLATO et al.). Any material aspirated should be cultured and smears made. The Gram method reveals Gram-positive, branching, non septate, slender, filaments or hyphae among the numerous leukocytes in various stages of degeneration (Fig. 14). Centrifuged sediments of pleural fluid and spinal fluid should be cultured before smears are made since therapy is vastly different in bacterial infections from what it is in nocardiosis. A definite diagnosis must be made (MOGABGAB; WEBSTER, 1962; HUNTER et al.; HOSTY et al.; GLOVER et al.; HALL and COOLEY).

Treatment

Sulfonamides are the drugs of choice and if the dose is sufficiently high, the prognosis is good (SEABURY and DASCOMB). HOEPRICH et al. recommend surgical drainage and combined antimicrobic therapy for nocardial cerebral abscess and use cycloserine and sulfonamides. Utz's patient with osteomyelites and sinus tract formation in the thigh and leg responded to surgical drainage and prolonged sulfadiazine and penicillin therapy. GONZALEZ-OCHOA (1960) did not get encouraging results in the treatment of nocardiosis with griseofulvin.

Mixed Infections: Baker reported a fatal case of combined pulmonary nocardiosis and cryptococcosis (Figs. 13, 15). The gross appearance was that of pulmonary nocardiosis but microscopically there were areas rich in *Cryptococcus neoformans*. Fig. 15 shows only the necrotizing nocardial pneumonia.

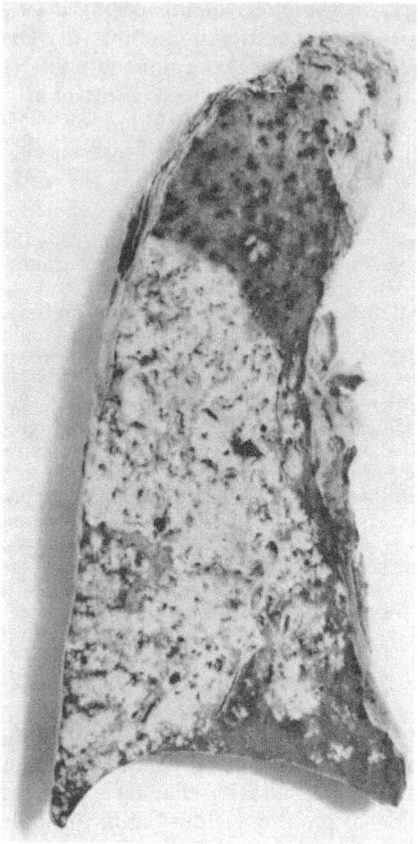

Fig. 13. Mixed nocardiosis and cryptococcosis of lower lobe of lung. The predominant appearance is that of pulmonary nocardiosis, but there are areas of cryptococcal organisms visible microscopically. This is an example of an opportunistic infection as the patient's resistence had been lowered by multiple myeloma and cytotoxic drug therapy

The case was that of a 52-year-old Negro woman who suffered from multiple myeloma for three years and received 200 mg of Fluorouracil three times a week during the last month of life. Three weeks before death she became febrile and had a slight cough productive of pink sputum. Hemoglobin terminally, 5.4 gm.; white blood cell count, 3,715; 60% neutrophils, 36% small lymphocytes; 4% monocytes. There was 2 plus albuminuria.

At necropsy there were white and red gelatinous nodules of multiple myeloma throughout the marrow of the ribs, sternum and spine, with collapse of a vertebral body. Caseation and abscesses involved the entire left lobe and a portion of the left upper lobe of the lung (Fig. 13); fibrin covered the pleural surfaces on this side. The right lung was similarly involved but to a lesser extent. Microscopically the regions of nocardiosis showed purulent necrotizing pneumonia in H & E stain (Fig. 15), and branching filaments of *Nocardia* in Gram stain. The regions of cryptococcosis, not illustrated here, were composed of foci of *Cryptococcus neoformans* with much gelatinous capsular material and minimal inflammatory response. The mucicarmine stain, specific for *Cryptococcus*, stained the organisms nicely.

This case is a good example of an opportunistic nocardial infection in a patient with lowered resistence due to the multiple myeloma and the cytotoxic drug.

HALL and COOLEY reported a case of pulmonary nocardiosis complicated by tuberculosis, another mixed infection.

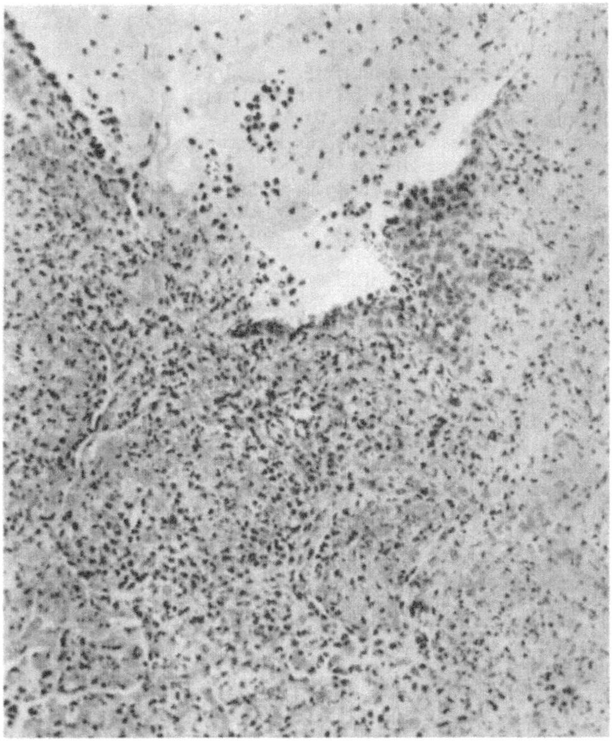

Fig. 14. *Nocardia asteroides* in smear of pus. Gram stain

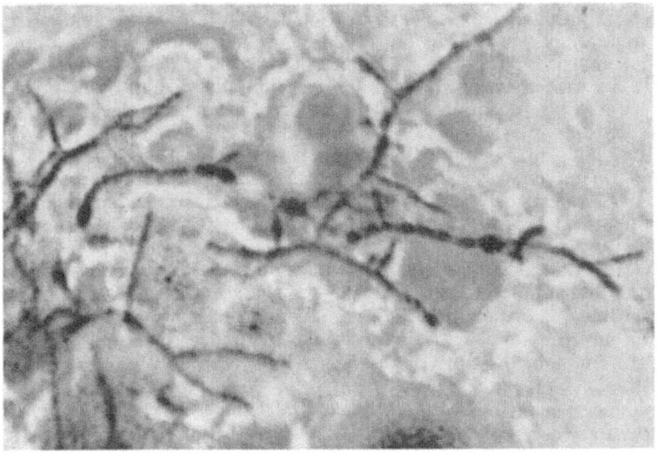

Fig. 15. Nocardiosis of the lung. This is a field from the lung of Figure 13 with necrotizing pneumonia and bronchiolitis due to *Nocardia*. There is no cryptococcosis in this particular field. H & E, × 60

References

Abbott, P.: Mycetoma in the Sudan. Trans. roy. Soc. trop. Med. Hyg. **50**, 11 (1956). — Ajello, L., Grant, V.Q., Gutske, M.A.: The effects of tubercle bacillus concentration procedures on fungi causing pulmonary mycosis. J. Lab. clin. Med. **38**, 486 (1951). — Ajello, L., Walker, W.W., Dungworth, D.L., Brumfield, G.: Isolation of *Nocardia brasiliensis* from a cat, with review of its prevalence and geographic distribution. J. Amer. vet. med. Ass. **188**, 370 (1961). — Anderson, B.R., Echlund, R.E., Kellow, W.E.: Pulmonary alveolar proteinosis with systemic nocardiosis. A case report. J. Amer. med. Ass. **174**, 28 (1960). — ANDRIOLE, V.T., BALLAS, M., WILSON, G.L.: The association of nocardiosis and pulmonary alveolar proteinosis. A case study. Ann. intern. Med. **50**, 266 (1964). — ARAGONA, F., DE PASQUALE, N.: Sulla nocardiasi pulmonare. G. Mal. infett. **5**, 310 (1953). — BAKER, R.D.: The diagnosis of fungus diseases by biopsy. J. chron. Dis. **5**, 552 (1957). ~ Foreword. International Symposium on Opportunistic Fungus Infections. Lab. Invest. **11**, 1017 (1962). ~ In Seminar on Virus and Fungus Diseases of the A.S.C.P. September 7, 1962. Published by the American Society of Clinical Pathology, Chicago 1963. — BALLENGER, C.N., Jr., GOLDRING, D.: Nocardiosis in childhood. J. Pediat. **50**, 145 (1957). — BAUMGARTEN, A.: Nocardiosis: Report of a case resembling sporotrichosis. Med. J. Aust. **2**, 321 (1961). — BENEDICT, W.L., IVERSON, H.A.: Chronic keratoconjunctivitis associated with *Nocardia*. Arch. Ophthal. **32**, 89 (1944). — BERGEY'S Manual of Determinative Bacteriology. Ed. by R.S. BREED, E.G.D. MURRAY and N.R. SMITH, p. 713. Baltimore, Maryland: Williams and Wilkins Co. 1957. — BERNSTEIN, I.L., COOK, J.E., PLOTNICH, H., TENCZAR, F.J.: Nocardiosis. Three case reports. Ann. intern. Med. **36**, 852 (1952). — BIANCO, A.J., Jr., JOHNSON, E.W., Jr., MARTIN, W.J., NICHOLS, D.R.: Nocardiosis without involvement of pulmonary or central nervous system. Proc. Mayo Clin. **32**, 119 (1957). — BINFORD, C.H., LANE, J.D.: Actinomycosis due to *Nocardia asteroides*. Report of a case. Amer. J. clin. Path. **15**, 17 (1945). — BLANCHARD, R.: Parasites vegetaux a l'exclusion des bacteries. Traite de pathologie generale. pp. 811—932, Tome II. Ed. by CH. BOUCHARD, G. MASSON, Paris. Quoted by Gorden and Mihm 1957. — BOBBITT, O.B., FRIEDMAN, I.H., LUPTON, C.: Nocardiosis. Report of three cases. New Engl. J. Med. **252**, 893 (1955). — BOHL, E.H., JONES, D.O., FARRELL, R.L., CHAMBERLAIN, D.M., COLE, C.R., FERGUSON, L.C.: Nocardiosis in the dog. A case report. J. Amer. vet. med. Ass. **122**, 81 (1953). — BOJALIL, L.F., CERBON, J.: Schema for the differentation of *Nocardia asteroides*, and *Nocardia brasiliensis*. J. Bact. **78**, 852 (1959). — BREED, R.S., CONN, H.J.: The nomenclature of the Actinomycetaceae. J. Bact. **4**, 585 (1919). — BRUECK, J.W., BUDDINGH, G.J.: Propagation of pathogenic fungi in yolk sac of embryonated eggs. Proc. Soc. exp. Biol. (N.Y.). **76**, 258 (1951). — BURBANK, B., MORRIONE, T.G., CULLEN, S.S.: Pulmonary alveolar proteinosis and nocardiosis. Amer. J. Med. **28**, 1002 (1960). — CALERO, C.: Pulmonary actinomycosis (Report of the first case observed in the Isthmus of Panama). Dis. Chest **12**, 402 (1946). — CARLILE, W.K., HOLLEY, K.E., LOGAN, G.B.: Fatal acute disseminated nocardiosis in a child. J. Amer. med. Ass. **184**, 477 (1963). — CARLSEN, E.T., HILL, R.B., ROWLANDS, D.T., Jr.: Nocardiosis and pulmonary alveolar proteinosis. Ann. intern. Med. **60**, 275 (1964). — CASAZZA, A.R., DUVALL, C.P., CARBONE, P.P.: Infection in lymphoma. Histology, treatment and duration in relation to incidence and survival. J. Amer. med. Ass. **197**, 710 (1966). — CHALMERS, A.J., ARCHIBALD, R.G.: A sudanese maduromycosis. Ann. trop. Med. Parasit. **10**, 169 (1916). — CHRISTOPHERSON, J.B., ARCHIBALD, R.G.: Primary nocardiosis of the lacrymal gland. Lancet **1918** II, 847. — CLAYPOLE, E.J.: Human streptothricosis and its differentiation from tuberculosis. Arch. intern. Med. **14**, 104 (1914). — CONANT, N.F.: Bacterial and Mycotic Infections of Man. Ed. by R.J. DUBOS. Pp. 584. Third edition. Philadelphia: Lippincott 1958. — COTTON, R.E., LLOYD, H.E.D.: Lipid pneumonia and infection with *Nocardia asteroides* complicating achalasia of the cardia. J. Path. Bact. **79**, 251 (1960). — CROSS, R.M., BINFORD, C.H.: Is *Nocardia asteroides* an opportunist? Lab. Invest. **11**, 1103 (1962). — CRUZ, P.T., CLANCY, C.F.: Nocardiosis. Nocardial osteomyelitis and septicemia. Amer. J. Path. **28**, 607 (1952). — CUPP, E.M., EDWARDS, W.M., CLEVE, E.A.: Nocardiosis of the central nervous system. Report of two fatal cases. Ann. intern. Med. **52**, 223 (1960). — CUTTINO, J.T., MCCABE, A.M.: Pure granulomatous nocardiosis. A new fungus disease distinguished by intracellular parasitism. A description of a new disease in man due to a hitherto undescribed organism, *Nocardia intracellularis*, N. sp., including a study of the biologic and pathogenic properties of this species. Amer. J. Path. **25**, 1 (1949). — DANOWSKI, T.S., COOPER, W.M., BRAUDE, A.: Cushing's syndrome in connection with *Nocardia asteroides*. Metabolism **3**, 265 (1962). — DOLAN, T.F., Jr., MCCULLOUGH, N.B., GIBSON, L.E.: Nocardiosis. Report of two cases in children. J. Dis. Child. **99**, 234 (1960). — DRAKE, C.H., HENRICI, A.T.: *Nocardia asteroides*. Its pathogenicity and allergic properties. Amer. Rev. Tuberc. **48**, 184 (1943). — DROPMANN, K.: Nokardiose unter Berücksichtigung des Diabetes mellitus. Verh. dtsch. Ges. Path. **1959**, 154—155. — ECKHARDT, K., PILCHER, J.: Brain abscess due to *Nocardia asteroides*. Report of case. Tex. St. J. Med. **46**, 915 (1950). — EPPINGER, H.: Über eine neue pathogene Cladothrix und eine durch sie hervorgerufene Pseudo-Tuberculosis. Wien. klin. Wschr. **3**, 321 (1890, 1891). — ERCHUL, J.W., KOCH, M.L.: Cerebral nocardiosis with coexistant tuberculosis. Report of a fatal case. Amer. J. clin. Path. **25**, 775 (1955). — FREESE, J.W., YOUNG, W.G., Jr., SEALY, W.C., CONANT, N.F.: Pulmonary infection by *Nocardia asteroides*. Findings in eleven cases. J. thorac. cardiovasc. Surg. **46**, 537 (1963). — GEORG, L.K., AJELLO, A., MCDURMONT, C., HOSTY, T.S.: The identification of *Nocardia asteroides* and *Nocardia brasiliensis*. Amer. Rev. resp. Dis. **84**, 337 (1961). — GILLIGAN, B.S., WILLIAMS, I., PERCEVAL,

A.K.: Nocardial meningitis: report of a case with bacteriological studies. Med. J. Aust. **2**, 747 (1962). — GLOVER, R.P., WALLACE, E.H., HERRELL, W.E., HEILMAN, F.R., PFUETZE, K.H.: Nocardiosis: *Nocardia asteroides* simulating pulmonary tuberculosis. J. Amer. med. Ass. **136**, 172 (1948). — GOLDSWORTHY, N.E.: Pulmonary actinomycosis caused by acid-fast species of actinomycosis. J. Path. Bact. **45**, 17 (1937). — GONZALEZ-OCHOA, A.: Mycetoma caused by *Nocardia brasiliensis*: with a note on the isolation of the causative organism from soil. Lab. Invest. **11**, 1118 (1962). ~ In Systemic Mycosis, Ciba Foundation Symposium. Ed. by G.E.W. WALSTENHOLME, and R. PORTER p. 86. Boston: Little Brown and Co 1967. — GONZALEZ-OCHOA, A., SANDOVAL, M.A.: Revisión determinación de algunas especies de actinomicetes patógenes descritas como diferentes. Rev. Inst. Salubr. Enferm. trop. (Méx.) **16**, 17, 1956 (quoted by WAKSMAN). — GORDON, R.E., HAGEN, W.A.: A study of some acid fast actinomycetes from soil with special references to pathogenicity for animals, J. infect. Dis. **59**, 200 (1936). — GORDON, R.E., MIHM, J.M.: A comparative study of some strains received as *Nocardiae*. J. Bact. **73**, 15 (1957). ~ A comparison of *Nocardia asteroides* and *Nocardia brasiliensis* J. gen. Microbiol. **20**, 129 (1959). ~ The type species of the genus *Nocardia*. J. gen. Microbiol. **27**, 1 (1962). — GORDON, R.E., SMITH, E.: Proposed group of characters for the separation of *Streptomyces* and *Nocardia* J. Bact. **69**, 147 (1955). — GUY, W.H.: Nocardiosis cutis resembling sporotrichosis. Report of a case. Arch. Derm. Syph. (Chic.) **2**, 137 (1920). — GYDELL, K., JOHLIN, I., LJUNGGREN, H., NORDEN, J.G., FORS, B.: Nocardiosis during steroid treatment of auto-immune haemolytic anaemia. Report of fatal case. Acta med. scand. **178**, 221 (1965). — HAGER, H.F., MIGLIACCIO, A.C., YOUNG, R.M.: Nocardiosis: Pneumonia and empyema due to *N.astoides*. New Engl. J. Med. **241**, 226 (1949). — HALL, E.R., Jr., COOLEY, D.A.: Pulmonary nocardiosis: report of case complicated by tuberculosis. Dis. Chest. **31**, 453 (1957). — HARGROVE, M.D., MATHEWS, W.R., MCINTYRE, P.A.: Intestinal lymphangiectasia with response to corticosteroids. Arch. intern. Med. **119**, 206 (1967). — HATHAWAY, B.H., MASON, K.N.: Nocardiosis. Study of fourteen cases. Amer. J. Med. **32**, 903 (1962). — HENDERSON, J.W., WELLMAN, W.E., WEED, L.A.: Nocardiosis of the eye: Report of case. Proc. Mayo Clin. **35**, 614 (1960). — HENRICI, A.T., GARDNER, E.L.: The acid-fast actinomycetes with a report of a case from which a new species was isolated. J. infect. Dis. **28**, 232 (1921). — HICKEY, R.G., BERGLUND, E.M.: Nocardiosis. Aerobic actinomycosis with emphasis on alimentary tract as portal of entry. Arch. Surg. **67**, 381 (1953). — HIDDLESTONE, H.J.H.: Nocardiosis in New Zealand; Report of a case. N.Z. med. J. **56**, 399 (1957). — HOEPRICH, P.D., BRANDT, D., PARKER, R.H.: Nocardial brain abscess cured with cycloserine and sulfonamides. Amer. J. med. Sci. **255**, 208 (1968). — HOSTY, T.S., MCDURMONT, C., AJELLO, L., GEORG, L., BLUMFIELD, G.L., CALIX, A.A.: Prevalence of *Nocardia asteroides* in sputa examined by a tuberculosis diagnostic laboratory. J. Lab. clin. Med. **58**, 107 (1961). — HUNTER, R.A., WILLCOX, D.R.C., WOOLF, A.L.: Aerobic actinomycosis with a report of a case resembling miliary tuberculosis. Guy's Hosp. Rep. **103**, 196 (1954). — HUTTER, R.V.P., COLLINS, H.S.: The occurrence of opportunistic fungus infections in a cancer hospital. Lab. Invest. **11**, 1035 (1962). — JACOBSON, J.R., CLOWARD, R.B.: Actinomycosis of the central nervous system. A case of meningitis with recovery. J. Amer. med. Ass. **137**, 769 (1948). — KADE, H., KAPLAN, L.: Evaluation of staining techniques in the histologic diagnosis of fungi. Arch. Path. **59**, 571 (1955). — KEPES, J.J., SCHOOLMAN, A.: Post-traumatic abscess of the medulla oblongata containing *Nocardia asteroides*. J. Neurosurg. **22**, 511 (1965). — KINCH, D.A.: A rapidly fatal infection caused by *Nocardia caviae* in a dog. J. Path. Bact. **95**, 540 (1968). — KING, R.B., STOOPS, W.L., FITZGIBBONS, J., BUNN, P.: *Nocardia asteroides* meningitis. A case successfully treated with large doses of sulfadiazine and urea. J. Neurosurg. **24**, 749 (1966). — KIRBY, W.M.M., MCNAUGHT, J.B.: Actinomycosis due to *Nocardia asteroides*. Arch. intern. Med. **78**, 578 (1946). — KOHN, P.M., TAGER, M., SIEGEL, M.L., ASKE, R.: Aerobic actinomyces septicemia. Report of a case. New Engl. J. Med. **245**, 640 (1951). — KRUEGER, E.G., NORSA, L., KENNEY, M., PRICE, A.: Nocardiosis of the central nervous system. J. Neurosurg. **11**, 226 (1954). — KUMAR, R., MOHAPATRA, L.N.: Studies on aerobic actinomycetes isolated from soil. I. Isolation and identification of strains. Sabouraudia **6**, 140 (1968). — LACAZ, C.S.: Contribuicas para o estudo dos actinomicetos productores de micetomas. Thesis, Faculdade de Medicina de Sao Paulo, Brazil, 1945. Quoted by MOORE, LANE and GAUL. — LARSEN, M.C., DIAMOND, H.D., COLLINS, H.S.: *Nocardia asteroides* infection. A report of seven cases. Arch. intern. Med. **103**, 712 (1959). — LINDENBERG, A.: Un nouveau mycetome. Arch. Parasit. **13**, 265 (1909). — LIST, C.F., WILLIAMS, J.R., BEEMAN, C.B., PAYNE, C.A.: Nocardiosis with multilocular cerebellar abscess. J. Neurosurg. **11**, 394 (1954). — LONG, P.I., CAMPANA, H.A.: An unusual mycetoma. Arch. Derm. **93**, 341 (1966). — MACKINNON, J.: Systemic Mycosis, Ciba Foundation Symposium, p. 89. Ed. G.E.W. WOLSTENHOLME and R. PORTER Boston: Little, Brown and Co. 1967. — MAHVI, T.A.: Disseminated nocardiosis caused by *Nocardia brasiliensis*. Arch. Derm. **89**, 426 (1964). — MARIAT, F., MACOTELA-RUIZ, E.: In Systemic Mycosis, Ciba Foundation Symposium, p. 90. Ed. G.E.W. WOLSTENHOLME and R. PORTER Boston: Little, Brown and

Co. 1967. — MARTINEZ-MALDONADO, M., RAMIREZ DE ARELLANO, G.: Pulmonary alveolar proteinosis, nocardiosis and granulocytic leukemia. Sth. med. J. (Bgham, Ala.) **59**, 901 (1966). — McCLUNG, N.M.: The utilization of carbon compounds by Nocardia species. J. Bact. **68**, 231 (1954). ~ Morphological studies in the genus *Nocardia*. I. Developmental studies. Lloydia **12**, 137 (1949). ~ Isolation of *Nocardia asteroides* from soils. Mycologia **52**, 154 (1960). — MOGABGAB, W.J., FLOYD, J.L.: Acid-fast Nocardia causing erroneous diagnosis of tracheobronchial tuberculosis. New Orleans med. surg. J. **104**, 28 (1951). — MOORE, M., CONRAD, A.H.: Sporotrichoid nocardiosis caused by *Nocardia brasiliensis*. Arch. Derm. **95**, 390 (1967). — MOORE, M., LANE, C.W., GAUL, L.E.: Nocardiosis of the knee caused by *Nocardia brasiliensis*. Report of first case in a native of the United States. Arch. Derm. Syph. (Chic.) **70**, 302 (1954). — MOSTAFA, I.E.: In: Vet. Bull. (Weybridge) **36**, 189 (1966) quoted in Systemic Mycosis, Ciba Foundation Symposium, p. 90. Ed. by G.E.W. WOLSTENHOLME and R. PORTER. Boston: Little Brown and Co. 1967. — MUNSLOW, R.A.: Actinomycotic (*Nocardia asteroides*) brain abscess with recovery. J. Neurosurg. **11**, 399 (1954). — MURRAY, J.F., FINEGOLD, S.M., FROMAN, S., WILL, D.W.: The changing spectrum of nocardiosis. A review and presentation of nine cases. Amer. Rev. resp. Dis. **83**, 315 (1961). — MUSGRAVE, W.E., CLEGG, M.T., POLK, M.: Streptothricosis with special reference to the etiology and classification of mycetoma. Philipp. J. Sci. B., Med. Sci. **3**, 477 (1908). — NOCARD, E.: Note sur la maladie des boeufs de la Guadeloupe connue sans le nom de farcin. Ann. Inst. Pasteur **2**, 293 (1888). — PILLAY, V.K., WILSON, D.M., ING, T.S., KARK, R.M.: Fungus infection in steroid treated systemic lupus erythematosus. J. Amer. med. Ass. **205**, 261 (1968). — PIZZOLATO, P., ZISKIND, J., DERMAN, H., BUFF, E.E.: Nocardiosis of the brain. Report of three cases. Amer. J. clin. Path. **36**, 151 (1961). — RAICH, R.A., CASEY, R., HALL, W.H.: Pulmonary and cutaneous nocardiosis. The significance of the laboratory isolation of *Nocardia*. Amer. Rev. resp. Dis. **83**, 505 (1961). — RAPAPORT, J.: Primary chancriform syndrome caused by *Nocardia brasiliensis*. Arch. Derm. **98**, 62 (1966). — RIDDELL, R.W.: Permanent stained mycological preparation obtained by slide culture. Mycologica **42**, 265 (1950). — ROSEN, S.H., CASTLEMAN, B., LIEBOW, A.A.: Pulmonary alveolar proteinosis. New Engl. J. Med. **258**, 1124 (1958). — RUNYON, E.H.: *Nocardia asteroides*: Studies of its pathogenicity and drug sensitivities. J. Lab. clin. Med. **37**, 713 (1951). — SALTZMAN, H.A., CHICK, E.W., CONANT, N.F.: Nocardiosis as a complication of other disease. Lab. Invest. **11**, 1110 (1962). — SEABURY, J.H., DASCOMB, H.E.: Results of treatment of systemic mycosis. J. Amer. med. Ass. **188**, 509 (1964). — SANTEN, R.J., WRIGHT, I.S.: Systemic lupus erythematosis associated with pulmonary nocardiosis. Arch. intern. Med. **119**, 202 (1967). — SCHNEIDAU, J.D., SCHAFFER, M.F.: Studies on *Nocardia* and other actinomycetales: I. Cultural studies. Amer. Rev. Tuberc. **76**, 770 (1957). — STAFOVÀ, J., MOTLIKOVÀ, M., KRUMPL, J., DVORSKÝ, K., VIKLICHÝ, J., VACEK, V.: Generalisierte Nokardiose. Eine zentrale Nokardia-Sepsis bei einer Frau mit tödlichem Ausgang. Zbl. allg. Path. path. Anat. **110**, 351 (1967). — STEINBERG, I.: Fatal fungus infection in sarcoidosis: Report of two cases treated with antibiotics and cortisone. Ann. intern. Med. **48**, 1359 (1958). — STRAUSS, R.E., KLIGMAN, A.M.: The use of gastric mucin to lower resistance of laboratory animals to systemic fungus infections. J. infect. Dis. **88**, 151 (1951). — THORDAL-CHRISTENSEN, A., CLIFFORD, D.H.: Actinomycosis (Nocardiosis) in dog with brief review of this disease. Amer. J. vet. Res. **14**, 298 (1953). — TREVISAN, V.I.: Generi e le specie delle Batteriacee. Milan 1889, quoted by S.A. WAKSMAN. — TUCKER, F.C., HIRSCH, E.F.: Nocardiosis with report of three cases of actinomycosis due to *Nocardia asteroides*. J. infect. Dis. **85**, 72 (1949). — TUCKER, R., MILLAR, R.: Outbreak of nocardiosis in marsupials in Brisbane. Botanical Gardens J. Comp. Path. Therap. **63**, 143 (1953). — UTZ, J.P.: The spectrum of opportunistic fungus infections. Lab. Invest. **11**, 1018 (1962). — WAKSMAN, S.A.: The Actinomycetes. II. Classification, identification and descriptions of general and species. Baltimore: Williams and Wilkins 1961. — WAKSMAN, S., HENRICI, A.T.: The nomenclature and classification of the actinomycetes. J. Bact. **46**, 337 (1943). — WEBSTER, B.H.: Pulmonary nocardiosis. A review of a report of seven cases. Amer. Rev. Tuberc. **73**, 485 (1956). ~ Pulmonary nocardiosis simulating pulmonary tuberculosis in the aged. J. Amer. Geriat. Soc. **10**, 192 (1962). — WEED, L.A., ANDERSON, H.A., GOOD, C.A., BAGGENSTOSS, A.H.: Nocardiosis. Clinical, bacteriologic and pathological aspects. New Engl. J. Med. **253**, 1137 (1955). — WELSH, J.D., RHOADES, E.R., JAQUES, W.: Disseminated nocardiosis involving spinal cord. Arch. intern. Med. **108**, 73 (1961). — WICHELHAUSEN, R.H., ROBINSON, L.B., MAZZARA, J.R., EVERDING, C.J.: Nocardiosis. Report of a fatal case. Amer. J. Med. **16**, 295 (1954). — WHITMORE, D.N., GRESHAM, G.A., GRAYSON, M.J.: Nocardiosis in anaemic patients given steroids. J. clin. Path. **14**, 259 (1961). — WILHITE, J.E., Cole, F.H.: Invasion of pulmonary cavities by *Nocardia asteroides*. Report of 5 cases. Amer. Surg. **32**, 107 (1966). — WILSON, A.M.M.: E. Afr. med. J. **42**, 182 (1965) quoted in Systemic Mycosis, Ciba Foundation Symposium, p. 108. Ed. by G.E.W., WOLSTENHOLME and R. PORTER. Boston: Little, Brown and Co. 1967.

Protothecosis – Algal Infection

Bernard F. Fetter, Gordon K. Klintworth, and **Harry S. Nielsen, Jr.**

Durham/North Carolina, USA

With 15 Figures

Numerous microscopic agents, including bacteria, fungi, protozoa, spirochetes, and viruses, are well recognized as pathogens. Algae, however, have seldom been associated with disease, and are rarely mentioned in the standard texts in clinical microbiology, pathology, and other branches of medicine. In the present report we have attempted to review the evidence concerning the pathogenicity of algae, with particular reference to *Prototheca*, which is thus far the only genus known to invade the tissues of man. Mention has not been made of the role algae play in gastroenteritis and death following ingestion, and in cutaneous and upper respiratory allergic reactions (Gorham, 1964; Schwimmer and Schwimmer, 1964).

Organism

The genus *Prototheca* was described in 1894 by Krüger to designate a group of non-pigmented unicellular organisms isolated from the mucous flux of trees. Based on a yeast-like appearance in culture, early investigators, including Krüger (1894 a, b), considered the organism to be a fungus. This view was generally accepted until West (1916) directed attention to its alga-like mode of reproduction. Unlike most yeasts, *Prototheca* does not propagate by budding, but by internally produced spores which are morphologically identical to the parent cell. This method of sporulation is indistinguishable from that observed in the green alga *Chlorella*. Based on this observation West (1916) classified the organism in the Chlorophyaceae. Printz (1927) later restudied the genus and recommended that it be placed in the Protothecaceae because of its achlorotic nature. Ciferri, Montemartini and Ciferri (1957) endorsed this taxon with a minor alteration in spelling, changing Protothecaeae to Prototheceae as recommended by the standards of botanical nomenclature.

It is generally believed that *Prototheca* originated from the Chlorophyceae through a sequence of events involving a loss of chlorophyll and adaption to heterotropic habit (Cooke, 1968a). This conclusion is based on the observation that colorless strains of *Chlorella* occur in nature and in the laboratory (Beijerinck, 1904); that x-irradiated cells of *Chlorella* produce achloric mutants (Granick, 1948); and that non-pigmented variants of *Chlorella* continue to propagate colorless cells in subsequent generations (Butler, 1954). Other investigators have questioned this phylogenetic derivation of the genus since species of *Prototheca* are better adapted to the heterotrophic life than the colorless strains of *Chlorella*, and, unlike the latter, are unable to grow in the absence of exogenously supplied thiamine (Ciferri, 1956; Ciferri, Montemartini and Ciferri, 1957).

Isolation: *Prototheca* is readily cultivated at room temperature on conventional media such as Sabouraud's glucose agar, brain heart infusion agar, blood agar, and beef infusion broth. The organism is fast growing, producing macroscopic colonies within 24—36 hours. Routine mycological procedures are satisfactory for the laboratory study and maintenance of the organism. It is important to note, however, that the antibiotic cyclohexamide which is frequently employed in diagnostic mycology as one of the selective agents against saprophytic fungi may inhibit the growth of some strains of the alga.

The organism can be retained in culture for extended periods under sterile mineral oil. Hartsell (1956) has shown that one strain of *P. zopfii* survived preservation for 5 years on a malt-yeast extract medium overlayed with oil.

Cultural Morphology: On solid media, isolates of *Prototheca* are similar to many yeasts or yeast-like fungi. Cultures vary from white to cream colored and may be smooth, wrinkled, or pasty depending upon the strain (Fig. 1).

In the diagnostic laboratory, the organism must be distinguished from species of *Candida* and *Cryptococcus*, and to a lesser extent from the yeast forms of *Histoplasma* and *Blastomyces*.

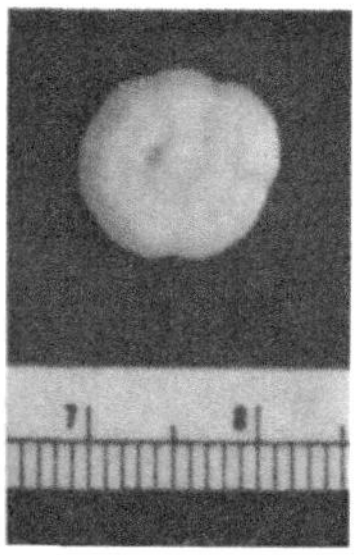

Fig. 1. Colony of *P. wickerhamii* on Sabouraud's glucose agar after growth for 14 days at room temperature (22° C)

Microscopic Morphology: Aqueous or lactophenol cotton blue preparations of the alga reveal round to oval cells which appear hyaline with granular protoplasts and which have thin, but highly refractile cell walls. The organism is non-encapsulated and lacks chloroplastids which characterize other algae. This point has been confirmed for *P. ciferri*, *P. segbwema*, and *P. wickerhamii* by electron microscopy (Brown, 1967; Menke and Fricke, 1962; and unpublished personal observations) (Fig. 2).

Life Cycle: Reproduction occurs asexually by the formation of autospores within cells (Figs. 3 and 4); sexual reproduction is unknown. The organism becomes sporogenous through a series of internal changes beginning with nuclear division, cytoplasmic cleavage, and cell wall synthesis. In contrast to the fungi, the process is highly coordinated in that cytokinesis follows immediately after daughter nuclei have separated. Daughter cells are liberated after rupture of the parent structure and increase in size before reinitiating the life cycle. Using the light microscope Tubaki and Soneda (1959) estimated that there were one to several autospores in *P. wickerhamii*; however, calculations based on the dimensions of the autospores and parent cells of this species as seen by electron microscopy indicate that the sporangium contains approximately 50 autospores (Klintworth, Fetter and Nielsen, 1968b). The data indicate that previous estimates of the maximum number of autospores in various species of *Prototheca* could be invalid.

Under conditions, yet undefined, some vegetative cells undergo encystment. These cells develop a thick, smooth wall around the entire protoplast or around

2

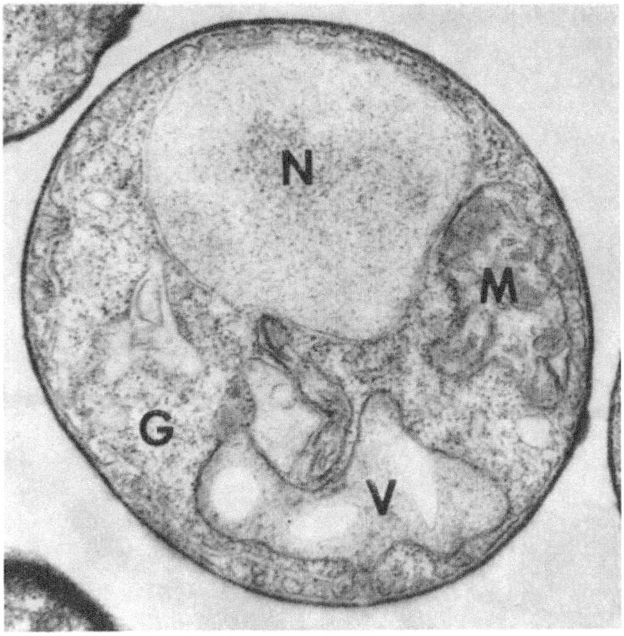

3

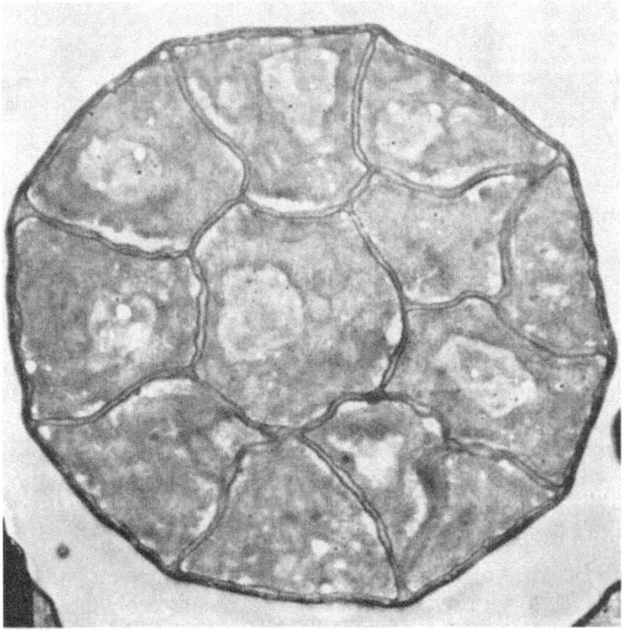

Fig. 2. Electron micrograph showing an immature form of *Prototheca wickerhamii*. Nucleus (N), mitochondrion (M), intracytoplasmic vesicle (V), granular material (G). × 36,000. (From Klintworth, G. K., Fetter, B. F. and Nielsen, H. S.: J. Medical Microbiology: **1**, 211—216, 1968)

Fig. 3. Electron micrograph showing a mature sporangium of *Prototheca wickerhamii* containing eleven autospores in a single plane of section. × 11,000. (From Klintworth, G. K., Fetter, B. F. and Nielsen, H. S.: J. Medical Microbiology: **1**, 211—216, 1968)

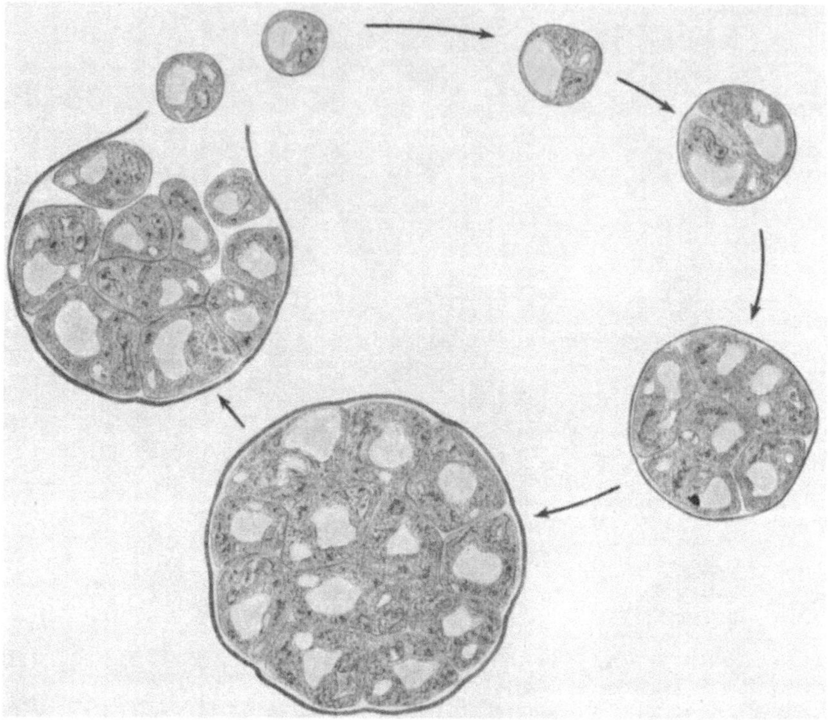

Fig. 4. Schematic reconstruction of *Prototheca*, as exemplified by *P. wickerhamii*. (From KLINTWORTH, G.K., FETTER, B.F. and NIELSEN, H.S.: J. Medical Microbiology: **1**, 211—216, 1968)

several daughter cells. The resulting structure (hypnospore) is easily differentiated from an autospore by its larger size and by the thickness of its wall (COOKE, 1968b).

Habitat: Species of *Prototheca* have been found in such geographically separated areas as Argentina, Germany, Puerto Rico, Sierra Leone, and the USA. The alga has been identified in varied environments, including the slime flux and frass on trees (KRÜGER, 1894a, b; TUBAKI and SONEDA, 1959; PHAFF, YONEYAMA and DO CARMO SOUSA, 1964), potato skin (NEGRONI and BLAISTEN, 1940); from the intestinal content of the tadpoles of *R. pipiens* (RICHARDS, 1958), fresh and marine water (AHEARN, 1957), and in water and sewage treatment systems (COOKE, 1968b). It has also been isolated from cutaneous scrapings, nails, sputum, and fecal specimens of man (AHEARN, 1967; ASHFORD, CIFERRI and DALMAU, 1930; GORDON, 1966; ADLER, 1967).

Speciation: Because of the inordinate variation between the strains of single species, classification within the genus is a difficult problem. CIFERRI, MONTEMARTINI and CIFERRI (1957) recognized four species and one variety to include *P. moriformis*, *P. portoricensis*, *P. trispora*, *P. zopfii*, and *P. portoricensis* var. ciferri. TUBAKI and SONEDA (1959) subsequently reduced *P. portoricensis* to the synonymy of *P. zopfii*, but elevated *P. portoricensis* var. ciferri to species level (*P. ciferri*) and described a new alga which was designated as *P. wickerhamii*. A sixth species, *P. segbwema*, was reported by DAVIS, SPENCER and WAKELIN in 1964, and COOKE (1968b) recently reported another, which he called *P. stagnora*.

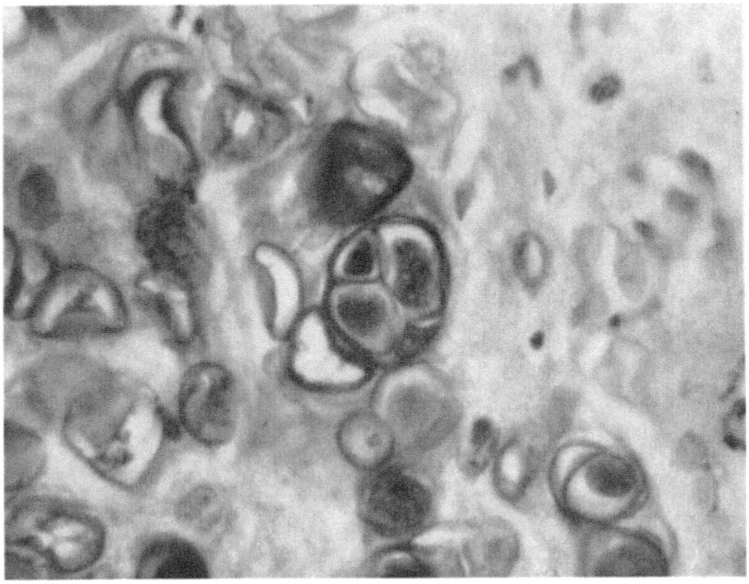

Fig. 5. Organism from lesion of case reported by DAVIES et al. (1964) and DAVIES and WILKINSON (1967) and designated by culture as *P. segbwema*. Periodic acid Schiff. × 1000

On the basis of existing reports in which cell size and shape are considered, it is believed that the genus contains the following seven different species:

Prototheca ciferii (NEGRONI and BLAISTEN, 1940): Cells ovoid or ellipsoid, 13—16 × 19 to 20 μ. Autospores ellipsoid or rarely cylindrical, 8—13 × 7.5—10 μ.

Prototheca moriformis (KRÜGER, 1894): Cells spheroid or ovoid, 8—12 μ in diameter, occasionally 16 μ in diameter. Autospores usually spheroid, 4—5 μ in diameter. In contrast to all other species with cells of these dimensions, *P. moriformis* will not survive at 30°C.

Prototheca segbwema (DAVIES, SPENCER and WAKELIN, 1964): Cells ovoid, 17—24 × 22—30 μ.

Prototheca stagnora (COOKE, 1968b): Cells ellipsoid and of two overlapping sizes 10.8 to 18.0 × 14.4—21.6 μ or 14.4—21.6 × 18.0—27.0 μ; autospores ellipsoid and usually 3.6 to 7.2 × 7.2—14.4 μ.

Prototheca trispora (ASHFORD, CIFERRI and DALMAU, 1930): Cells ovoid, ellipsoid, often cylindrical, 10—20 × 20—40 μ. Autospores 4—5 × 10—12.5 μ.

Prototheca wickerhamii (TUBAKI and SONEDA, 1959): Cells spherical or spheroid, 8—13.5 μ, usually 11.5—13 μ in diameter. Autospores spherical or spheroid, 4—5 μ in diameter.

Prototheca zopfii (KRÜGER, 1894): Cells spherical or ovoid, 10—18 μ in diameter. Autospores usually spherical, 9—11 μ in diameter.

Identification of Prototheca in Tissue: *Prototheca* is readily identifiable in tissue sections if the mature forms are observed (Figs. 5—7). These show characteristic septations resulting from the formation of autospores. The walls of the latter do not usually react with hematoxylin and eosin, but stain well with Grocott's methenamine silver, Mowry's modification of Hale's colloidal iron, and Kligman's modification of the periodic acid-Schiff techniques. Secondary fluorescence is demonstrated after acridine orange staining. Differentiation of species can be made on the size of the organisms and the number of autospores. In histologic sections individual organisms of the same species often manifest variability in staining.

At first glance *Prototheca* may be confused with *Blastomyces dermatitidis*, *Paracoccidioides braziliensis*, and *Histoplasma duboisii*. The latter organisms are readily differentiated because they reproduce in tissue by budding whereas

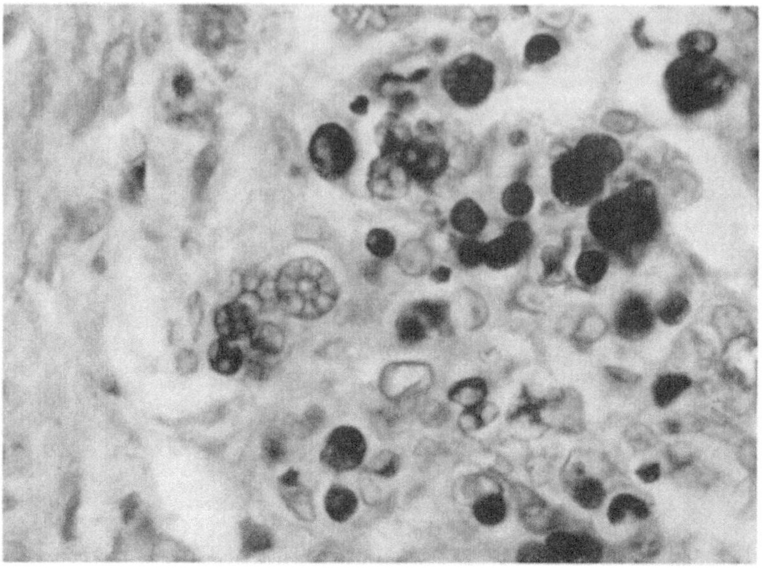

Fig. 6. *P. wickerhamii*, established by culture, is shown in tissue. Periodic acid Schiff. × 1000

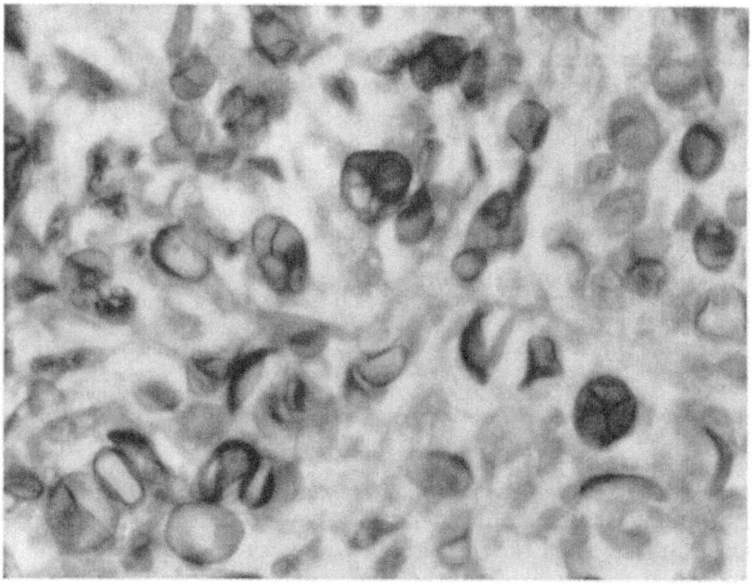

Fig. 7. Organisms with this morphology have been seen in tissue of two dogs. They are intermediate in size between *P. segbwema* and *P. wickerhamii* and presumably represent a different species. Periodic acid Schiff. × 1000

Prototheca does not. *Coccidioides immitis* would also have to be considered in the differential diagnosis because of its reproduction by endospore formation. Distinction between these two organisms is made on the basis of the fact that the endospores of *C. immitis* are smaller (1—2 μ), more numerous, and generally spherical. The cell walls of the autospores in *Prototheca* are more prominent.

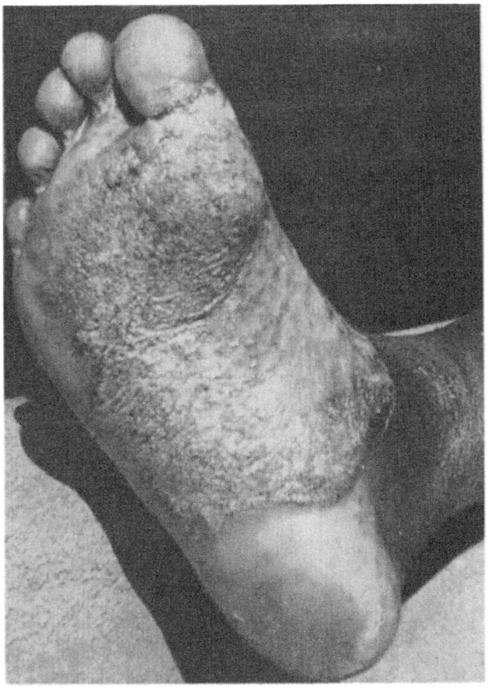

Fig. 8. Fungating lesion due to *P.segbwema*. (Courtesy of Dr. DAVIES, R.R.)

Prototheca and Disease

In 1930, ASHFORD, CIFERRI and DALMAU drew attention to the association of *Prototheca* with disease, when they isolated *P.portoricensis* (currently designated *P.ciferri*) from two patients with sprue. This relationship was not confirmed and is generally regarded as coincidental.

The first proven infection by *Prototheca* was reported by DAVIES, SPENCER and WAKELIN (1964), following the isolation of *Prototheca segbwema* from a cutaneous lesion of an African rice farmer. These investigators clearly demonstrated the alga within the tissue and provided ample proof of its association with a lesion of the foot that began as a small itching, weeping papule. The eruption gradually increased in size, causing the patient to seek medical advice five years later. By this time, the process had extended over most of the foot (Fig. 8). Histologic examination of the skin revealed hyperkeratosis and pseudoepitheliomatous hyperplasia. *Prototheca* was identified in the papillary and reticular dermis as well as in the epidermis (Fig. 9). Associated with the organisms were numerous plasma cells and macrophages. A previously undescribed species of *Prototheca* was isolated from the cutaneous lesion and designated *P.segbwema*. The cutaneous lesions progressed (Fig. 10), and the alga subsequently spread to the regional lymph nodes and could be identified by biopsy and culture procedures ten years after the infection began (DAVIES and WILKINSON, 1967).

The pathogenic potential of *Prototheca* for man was underscored a few years later when *P.wickerhamii* was cultured from, and identified in, ulcerating cutaneous papulopustular lesions on the legs of a 45-year-old woman from North Carolina, USA (KLINTWORTH, FETTER and NIELSEN, 1968a, 1968b). The patient had a past history of carcinoma of the breast with widespread metastases, and diabetes

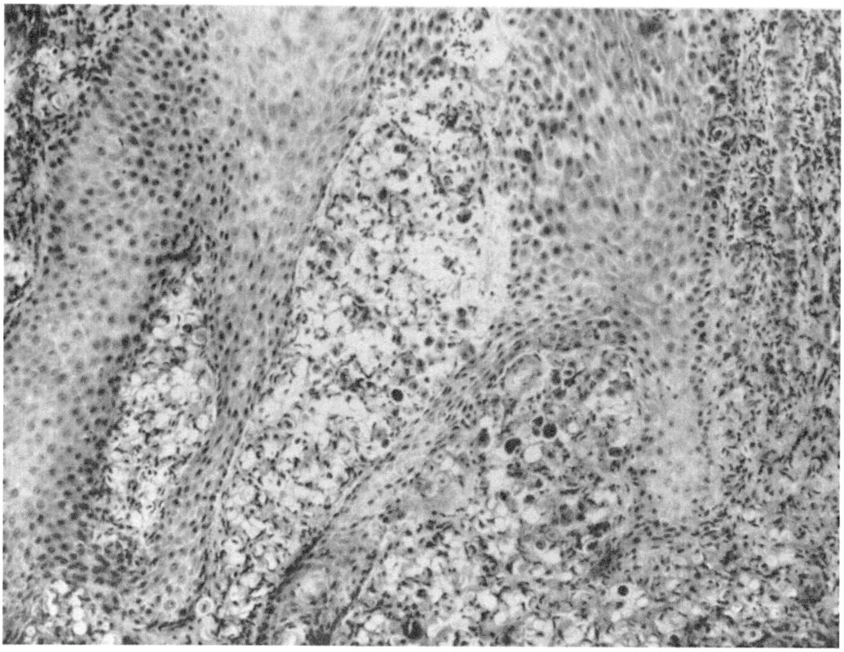

Fig. 9. *P. segbwema* appears here as dark dots and clear circles. There is an extensive epithelial hyperplasia. Hematoxylin and eosin × 110

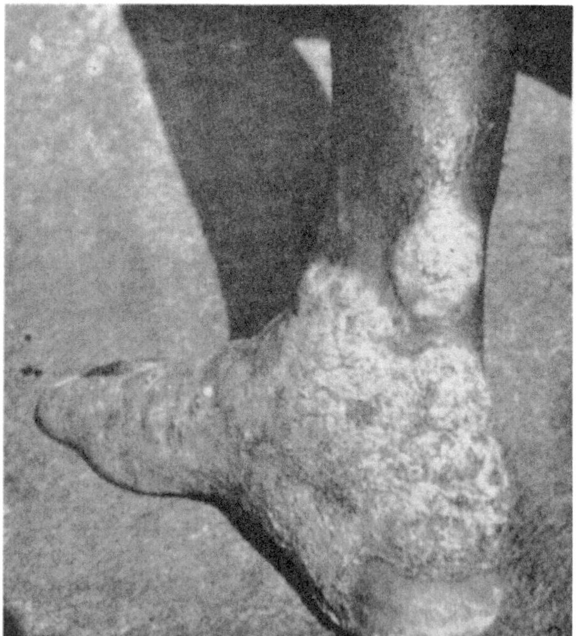

Fig. 10. Photograph of same leg as Fig. 8, infected by *P. segbwema*, shown 2 years later. (Courtesy of Drs. DAVIES, R.R. and WILKINSON, I.L.: Annals of Tropical Medicine and Parasitology **61**, 112—115, 1967)

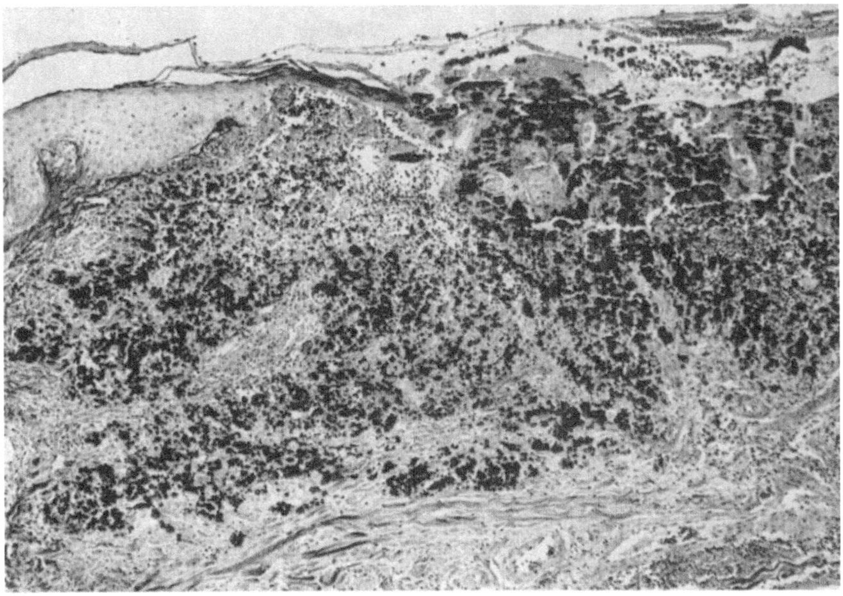

Fig. 11. A cutaneous ulcer due to *P. wickerhamii* is shown. The black dots in the photograph represent organisms. Grocott's methenamine silver. × 75. (From Klintworth, G.K., Fetter, B.F. and Nielsen, H.S.: J. Medical Microbiology: **1**, 211—216, 1968)

mellitus. Prior to the onset of the cutaneous disorder, the patient's treatment had included corticosteroids, cytotoxic agents, antibacterials, and antibiotics. The lesions on the leg were small cutaneous papulopustules 1—5 mm in diameter. Some were elevated, some were crusted, and some were umbilicated and covered by a small amount of purulent exudate. Biopsies of the lesions were performed on separate occasions and showed numerous organisms in varying stages of development. The tissue reaction consisted of a small focus of neutrophils and macrophages underlying a thin necrotic epidermis (Fig. 11). No neoplastic cells were identified in the lesions. *Prototheca wickerhamii* was obtained from three separate lesions and represented the sole agent cultured or identified in the tissue by light and electron microscopy. The lesions continued to progress despite treatment which included x-irradiation (Figs. 12 and 13). The patient died withcarcinomatosis at home, 15 months after the onset of her cutaneous lesions. An autopsy was not performed.

In addition to these examples of human infection, *Prototheca* has been isolated from ulcerated nodular masses in a two-year-old deer, which was shot by a hunter in Germany (Frese and Gedek, 1968). The lesions were situated on the lower part of all limbs and about the nose and mouth (Fig. 14). *Prototheca* was identified at post mortem examination in the skin, subcutaneous tissue, regional lymph nodes, and bone of the lower legs, as well as in the skin and mucous membranes of the nose and mouth. The alga cultured was *Prototheca zopfii*.

Despite the fact that the aforementioned cases are the only ones in which *Prototheca* has been cultured from and observed in tissues, there has been suggestive evidence that this disease may be more widespread. The alga has been cultured from man in independent laboratories on several occasions, although histologic confirmation was lacking in each instance (Ashford, Ciferri and Dalmau, 1930; Gordon, 1966; Ahearn, 1967; Adler, 1967). In addition, organisms believed to be *Prototheca* have been identified in, but not isolated from,

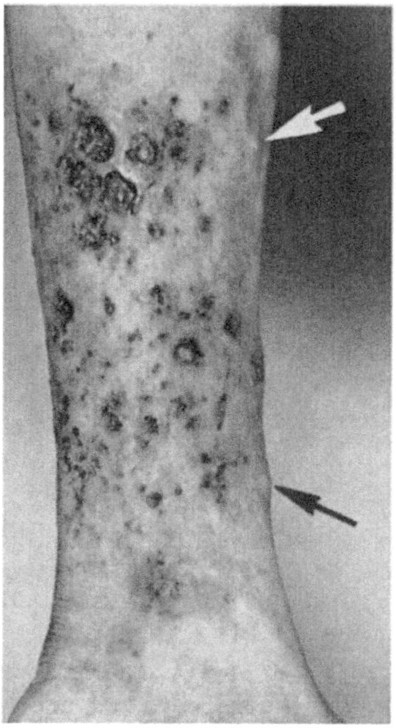

12

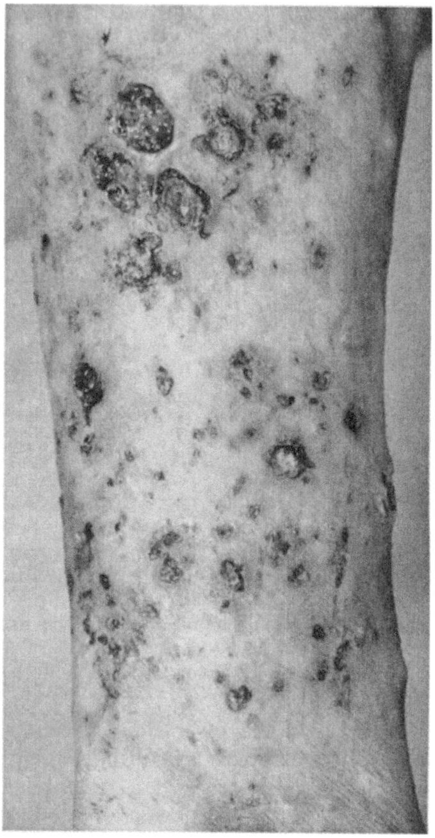

13

Figs. 12/13. Ulcerations and papules (arrows) in skin due to *P. wickerhamii*

lesions in man and other animals on a few occasions. For example, a bursa which had been aspirated several times in the past, was excised from the elbow of a male patient in Florida, USA (MILLARD, 1968). The excised tissue was necrotic and contained numerous organisms identified as *Prototheca*. The bursal wall was formed by fibrous tissue containing foci of lymphocytes, macrophages, and multinucleated giant cells. Small areas of necrosis were present in part of the wall. In the surrounding tissue, especially the dermis, was a perivascular lymphocytic infiltrate. Organisms were in an intracellular and extracellular position in the cavity as well as in the wall. Three years later there was no evidence of further disease at the site (BLANK, 1968).

Organisms morphologically resembling *Prototheca* have been observed in lesions of dogs (GARNER, 1968; VAN KRUININGEN and SCHIEFER, 1969). One animal, a 9-year-old female boxer, had polyuria, polydipsia, diarrhea, and bilateral iritis, which ultimately progressed to blindness. Post mortem examination following euthanasia revealed numerous organisms in the brain, eye, heart, kidney, liver, and periadrenal tissue, sometimes in association with non-caseating tubercle formation (Fig. 15). In most areas containing the alga, the reaction was minimal and consisted of a few lymphocytes and/or plasma cells. The organism was larger than *P. wickerhamii*, but smaller than *P. segbwema*, and may be *Prototheca zopfii*. The second dog was an 8-year-old keeshund from California, USA with ocular protothecosis. The retina and most of the vitreous was replaced by numerous

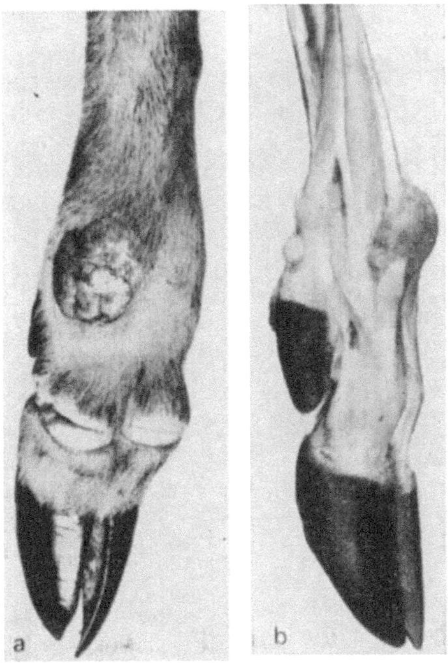

Fig. 14. The foot of a deer infected with *P. zopfii*. The nodular mass involves the skin extending into the subcutaneous tissues and adjacent bone. (From FRESE, K. and GEDEK, B.: Berl. Münch. tierärztl. Wschr. 5, 174—178, 1968)

organisms. The cellular reaction to the *Prototheca* consisted of scant neutrophils and macrophages with the organisms being more numerous than the reacting cells. The algae resembled that in the previously mentioned dog, in appearance and in size, suggesting that the same species was involved in both canine cases.

Prototheca has also been identified in milk from cows with mastitis (LERCHE, 1952; AINSWORTH and AUSTWICK, 1955) and in gastrointestinal lesions of the cow (BINFORD, 1968).

It is clear that at least three species of *Prototheca* have been related to spontaneously occuring lesions in which they were identified viz.; *P. segbwema*, *P. wickerhamii*, and *P. zopfii*. Although the organisms in the canine cases of protothecosis were not cultured, they are intermediate in size between *P. wickerhamii* and *P. segbwema* and presumably represent a different pathogenic species.

Of these species, most have evoked disease in the laboratory animals following inoculation. However, none are consistently pathogenic, and lesions have not yet been produced with *Prototheca segbwema* (DAVIES, SPENCER and WAKELIN, 1964). GORDON (1966) injected a species of *Prototheca*, isolated from a cutaneous lesion (later identified as *P. wickerhamii*) intraperitoneally into mice, and caused subcutaneous edema, splenomegaly, adrenal congestion, and small abscesses in the lower peritoneum. He was unable to identify the alga in lesions, but cultured *Prototheca* from the livers and spleen at autopsy. However, KLINTWORTH, FETTER and NIELSEN (1968a, b) were not able to infect mice, rats, or rabbits by intraperitoneal or subcutaneous injection with the strain of *P. wickerhamii* isolated from their case. Nor were they able to produce disease by the instillation of the alga into the anterior chamber of the rabbit eye. Granulomatous lesions have been elicited in the guinea pig with *P. portoricensis* (currently designated *P. zopfii*) by subcutaneous and intraperitoneal injections (REDAELLI and CIFERRI,

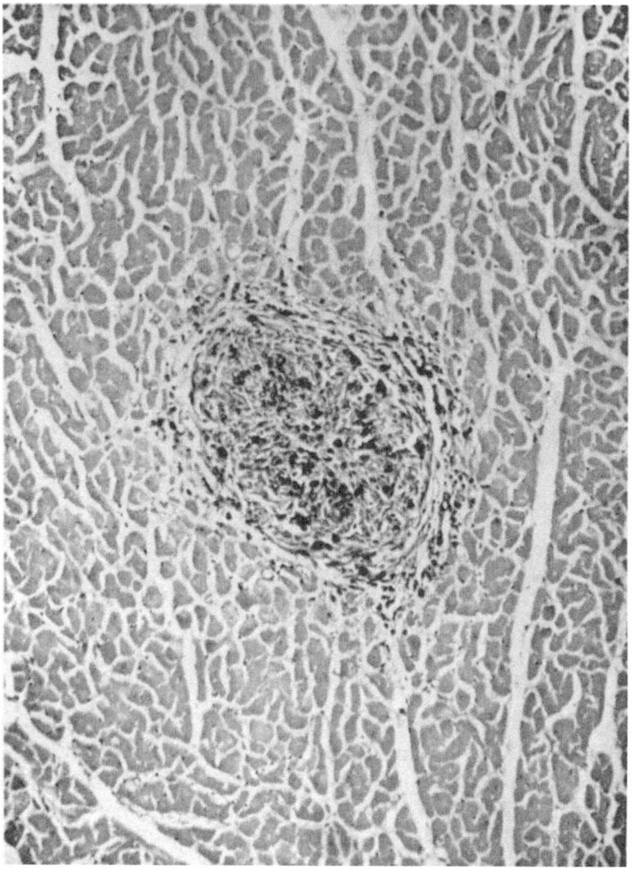

Fig. 15. A non-caseating tubercle in the cardiac muscle of a dog. The dark spots represent organisms, presumed to be *P. zopfii*. Periodic acid Schiff. × 110

1935; MARIANI, 1942). SCHIEFER and GEDEK (1968) produced mastitis in a cow with *P. moriformis*, but were unable to do so with *P. zopfii*. These authors also injected *P. moriformis* and *P. zopfii* subcutaneously, intramuscularly, and intraperitoneally into mice and reported infections at the sites of inoculation, and also metastatic lesions in the brain and kidney with both species.

One of the species which has not yet been associated with spontaneous disease is *P. ciferri*. Nevertheless, microabscesses, from which the alga was recovered, have been documented in the guinea pig and rabbit following intratesticular inoculation of the organism (NEGRONI and BLAISTEN, 1941).

It is somewhat difficult to reconcile the paucity of spontaneous disease with the pathogenicity of laboratory animals. Certainly the organism is widespread. The lack of reported cases may be due, at least in part, to a failure of recognition. It seems highly likely that *Prototheca* is relatively non-pathogenic and requires a distinct alteration in host resistance in order to cause disease. This premise is supported by the difficulty that most investigators have experienced in producing lesions in normal laboratory animals with different species of the alga. Also, the patient reported by KLINTWORTH, FETTER and NIELSEN (1968a, b) clearly had a lowered resistance to infection contributed to by underlying diseases and treatment.

Acknowledgements: We would like to thank all those who assisted in the accummulation of data for this review and particularly Drs.: D.L. ADLER, D.G. AHEARN, H. BLANK, R.M. BROWN, W.B. COOKE, R.R. DAVIES, C.W. EMMONS, K. FRESE, F.M. GARNER, M.A. GORDON. M. MILLARD, B. SCHIEFER and J.P. TINDALL, who supplied us with illustrative material, histologic slides, cultures as well as invaluable information. We wish to thank the editors of *Berliner und Münchener Tierärztliche Wochenschrift*, *Annals of Tropical Medicine and Parasitology*, and *Journal of Medical Microbiology* for permission to reproduce previously published illustrations.

References

ADLER, D.L.: Personal communication, 1967. — AHEARN, D.G.: Personal communication, 1967. — AINSWORTH, G.C., AUSTWICK, P.K.C.: A survey of animal mycoses in Britain; mycological aspects. Trans. Brit. Mycol. Soc. **38**, 369 (1955). — ASHFORD, B.K., CIFERRI, R., DALMAU, L.M.: A new species of *Prototheca* and a variety of the same isolated from human intestine. Arch. Protistenk. **70**, 619 (1930). — BEIJERINCK, M.W.: Chlorella variegata, ein bunter Mikrobe. Rec. Trav. Bot. Neerl. **1**, 14 (1904). — BINFORD, C.H.: Personal communication, 1968. — BLANK, H.: Personal communication, 1968. — BROWN, R.M.: Personal communication, 1967. — BUTLER, E.E.: Radiation-induced chlorophyll-less mutants of Chlorella. Science **120**, 274 (1954). — CIFERRI, O.: Thiamine deficiency of *Prototheca*, a yeast-like achloric alga. Nature (Lond.) **178**, 1475 (1956). — CIFERRI, R., MONTEMARTINI, A., CIFERRI, O.: Caratteristiche morfologiche e assimilative e speciologia delle Protothecae. Nuovi Ann. Ig. **8**, 554 (1957). — COOKE, W.B.: Studies in the genus *Prototheca*. Literature review. J. Elisha Mitchell Sci. Soc. **84**, 213 (1968a). ~ Studies in the genus *Prototheca* II. Taxonomy. J. Elisha Mitchell Sci. Soc. **84**, 217 (1968b). — DAVIES, R.R., SPENCER, H., WAKELIN, P.O.: A case of human protothecosis. Trans. roy. Soc. trop. Med. Hyg. **58**, 448 (1964). — DAVIES, R.R., WILKINSON, J.L.: Human Protothecosis: Supplementary studies. Ann. trop. Med. Parasit. **61**, 112 (1967). — FRESE, K., GEDEK, B.: Ein Fall von Protothecosis beim Reh. Berl. Münch. tierärztl. Wschr. 81. Jahrgang, Heft 9, S. 174—178 (Pagination 1—12), 1968. — GARNER, F.M.: Personal communication, 1968. — GORDON, M.A.: Protothecosis. In: Annual Report of Division of Laboratories and Research, New York State Department of Health, 119 (1966). — GORHAM, P.R.: Toxic Algae. In: Algae and *Man*, pp. 307—326. Ed. by JACKSON, D.F. New York: Plenum Press 1964. — GRANICK, S.: Protoporphyrin 9 as a precursor of Chlorophyll. J. biol. Chem. **172**, 717 (1948). — HARTSELL, S.E.: Microbiological process report. Maintenance of cultures under paraffin oil. Appl. Microbiol. **4**, 350 (1956). — KLINTWORTH, G.K., FETTER, B.F., NIELSEN, H.S., Jr.: Protothecosis: An algal infestation. Lab. Invest. **18**, 11 (1968a) (Abstract). ~ Protothecosis, an algal infection: report of a case in man. J. Medical Microbiology **1**, 211 (1968b). — KRÜGER, W.: Kurze Characteristik einiger niederen Organismen in Saftflüsse der Laubbäume. 1. Über einen neuen Pilztypus, repräsentiert durch die Gattung *Prototheca (P.moriformis (sic) et P.zopfii)* 2. Über zwei aus Saftflüssen rein gezüchtete Algen. Hedwigia **33**, 241 (1894a). ~ Beiträge zur Kenntnis der Organismen des Saftflusses (sog. Schleimflüsse) der Laubbäume. 1. Über einen neuen Pilztypus, repräsentiert durch die Gattung *Prototheca* 2. Über zwei aus Saftflüssen rein gezüchtete Algen. Zopf's Beitr. A. Physiol. u. Morph. **4**, 69 (1894b). — LERCHE, M.: Eine durch Algen (Prototheca) hervorgerufene Mastitis der Kuh. Berl. Münch. tierärztliche Wochenschrift **65**, 64 (1952). — MARIANI, P.L.: Ricerche sperimentali intorno ad alcune alghe parassite dell' uonomo. Boll. Soc. ital. microbiol. **14**, 113 (1942). — MENKE, W., FRICKE, B.: Einige Beobachtungen an *Prototheca ciferri*. Portugaliae Acta Biologica Serie A **6**, 243 (1962). — MILLARD, M.: Personal communication, 1968. — NEGRONI, P., BLAISTEN, R.: Estudio morfologio y fisiologico de una nueva especie de *Prototheca*: *Prototheca ciferri* N. sp., aislada de epidermis de papa. Mycopath. **3**, 94 (1940). — PHAFF, H.J., YONEYAMA, M., DO CARMO SOUSA, L.: A one year quantitative study of the yeast flora in a single slime flux of *Ulmus carpinifolia*. Gled. Riv. Patol. Veg. Ser. III **4**, 485 (1964). — PRINTZ, A.: In: Engler and Prantl. Die natürlichen Pflanzenfamilien. 2nd Ed. Chlorophyceae **3**, 131 (1927). — REDAELLI, P., CIFERRI, R.: La patogenicita per gli animali di alge achloriche coprofite del genere *Prototheca*. Boll. Soc. ital. Biol. sper. **10**, 809. Also in French: Pouvoir pathogene pour les animaux des algues coprophytes achloriques du genre *Prototheca*. Observations sur les Protothecaceae. Boll. sez. ital. soc. internaz. microbiol. fasc. **8—9**, 1 (1935). — RICHARDS, C.M.: The inhibition of growth in crowded *Rana pipiens* tadpoles. Physiol. Zool. **31**, 138 (1958). — SCHIEFER, B., GEDEK, B.: Zum Verhalten von *Prototheca — Species* im Gewebe von Säugetieren. Berl. Münch. tierärztl. Wschr. 81. Jahrg., **24**, 485 (1968). — SCHWIMMER, D., SCHWIMMER, M.: Algae and Medicine. In: *Algae and Man*, pp. 368—412. Ed. by JACKSON, D.F. New York: Plenum Press 1964. — TUBAKI, K., SONEDA, M.: Cultural and taxonomical studies on *Prototheca*. Nagaoa **6**, 25 (1959). — VAN KRUININGEN, H., SCHIEFER, B.: Protothecosis in the dog. Path. vet. **6**, 348 (1969). — WEST, G.S.: *Algae*. Cambridge: Cambridge University Press 1916. Vol. 1 ×, 475 pp.

Authors Index

Costa, S. 629, *665*

Subject Index

GPSR Compliance

The European Union's (EU) General Product Safety Regulation (GPSR) is a set of rules that requires consumer products to be safe and our obligations to ensure this.

If you have any concerns about our products, you can contact us on ProductSafety@springernature.com

In case Publisher is established outside the EU, the EU authorized representative is:

Springer Nature Customer Service Center GmbH
Europaplatz 3
69115 Heidelberg, Germany

Batch number: 10838062

Printed by Printforce, the Netherlands